STUDENT'S SOLUTIONS MANUAL

EDGAR REYES
Southeastern Louisiana University

COLLEGE ALGEBRA AND TRIGONOMETRY: A UNIT CIRCLE APPROACH

SIXTH EDITION

Mark Dugopolski
Southeastern Louisiana University

PEARSON

Boston Columbus Indianapolis New York San Francisco Upper Saddle River
Amsterdam Cape Town Dubai London Madrid Milan Munich Paris Montreal Toronto
Delhi Mexico City São Paulo Sydney Hong Kong Seoul Singapore Taipei Tokyo

Reproduced by Pearson from electronic files supplied by the author.

Publishing as Pearson, 75 Arlington Street, Boston, MA 02116.

ISBN-13: 978-0-321-91653-2
ISBN-10: 0-321-91653-0

1 2 3 4 5 6 OPM 17 16 15 14

www.pearsonhighered.com

PEARSON

Table of Contents

For Thought

1. False, 0 is not an irrational number.

2. True

3. True

4. False, 0 does not have a multiplicative inverse.

5. True

6. False, the case $a = w = z$ satisfies $a \le w$ and $w \le z$ but it does not satisfy $a < z$.

7. False, since $a - (b - c) = a - b + c$.

8. False, the distance is $|a - b|$.

9. False, π is an irrational number.

10. False, the opposite is $-(a+b)$, which is $-a-b$.

P.1 Exercises

1. rational numbers

3. variable

5. additive inverse

7. difference

9. absolute value

11. e, true

13. h, true

15. g, true

17. c, true

19. All

21. $\{-\sqrt{2}, \sqrt{3}, \pi, 5.090090009...\}$

23. $\{0, 1\}$ **25.** $x + 7$ **27.** $5x + 15$

29. $5(x + 1)$ **31.** $(-13 + 4) + x$

33. $\dfrac{1}{0.125}$ or 8

35. $\sqrt{3}$ **37.** $y^2 - x^2$ **39.** 7.2

41. $\sqrt{5}$ **43.** $d(13, 8) = |13 - 8| = 5$

45. $d(-5, 17) = |-5 - 17| = |-22| = 22$

47. $d(-6, -18) = |-6 - (-18)| = |12| = 12$

49. $d\left(-\frac{1}{2}, \frac{1}{4}\right) = \left|-\frac{1}{2} - \frac{1}{4}\right| = \left|-\frac{3}{4}\right| = \frac{3}{4}$

51. 8 **53.** -49

55. $(-4)(-4) = 16$

57. $-\dfrac{1}{64}$

59. $2 \cdot 5 - 3 \cdot 6 = 10 - 18 = -8$

61. $|3 - 4 \cdot 5| - 5 = |3 - 20| - 5 = 12$

63. $|-4 \cdot 3| - |-3 \cdot 5| = 12 - 15 = -3$

65. $\dfrac{4}{4} = 1$

67. $4 - 5 \cdot 3^2 = 4 - 5 \cdot 9 = 4 - 45 = -41$

69. -7

71. $3 \cdot 6 + 2 \cdot 4 = 18 + 8 = 26$

73. $\dfrac{26}{5} \div \dfrac{1}{2} \cdot 5 = \dfrac{52}{5} \cdot 5 = 52$

75. $(12 - 1)(1 + 8) = 11 \cdot 9 = 99$

77. $2 - 3|3 - 24| = 2 - 3 \cdot 21 = -61$

79. $49 - 36 = 13$

81. $3 - 4(-1)^2 = 3 - 4(1) = -1$

83. $\dfrac{2(9)}{25 - 16} = \dfrac{18}{9} = 2$

85. $b^2 - 4ac = 3^2 - 4(-2)(4) = 9 - (-32) = 41$

87. $\dfrac{a - c}{b - c} = \dfrac{-2 - 4}{3 - 4} = \dfrac{-6}{-1} = 6$

89. $a^2 - b^2 = (-2)^2 - (3)^2 = 4 - 9 = -5$

91. $(a - b)(a + b) = (-2 - 3)(-2 + 3) = (-5)(1) = -5$

93. $(a - b)(a^2 + ab + b^2) = (-2 - 3)((-2)^2 + (-2)(3) + (3)^2) = (-5)(4 - 6 + 9) = (-5)(7) = -35$

95. $a^b + c^b = (-2)^3 + (4)^3 = -8 + 64 = 56$

97. $-2x$ **99.** $0.85x$

101. $-6xy$

103. $3 - 2x$

105. $\frac{6x}{2} - \frac{2y}{2} = 3x - y$

107. $-3x - 6$

109. $-8 + 2x - 9 + 9x = 11x - 17$

111. a)

$$\begin{aligned} 0.60(220 - a - r) + r &= \\ 132 - 0.60a - 0.60r + r &= \\ 0.40r - 0.60a + 132 & \end{aligned}$$

b) If $a = 20$ and $r = 70$, then the target heart rate is

$$\begin{aligned} 0.40r - 0.60a + 132 &= \\ 0.40(70) - 0.60(20) + 132 &= \\ 148 \text{ beats/min} & \end{aligned}$$

c)

$$\begin{aligned} 0.60\left(205 - \frac{a}{2} - r\right) + r &= \\ 123 - 0.30a - 0.60r + r &= \\ 0.40r - 0.30a + 123 & \end{aligned}$$

113. Since $\frac{1}{3} = \frac{4}{12}$, we find

$$-\frac{1}{2} < -\frac{5}{12} < -\frac{1}{3} < 0 < \frac{1}{3} < \frac{5}{12} < \frac{1}{2}.$$

$-\frac{1}{2}$ $-\frac{5}{12}$ $-\frac{1}{3}$ 0 $\frac{1}{3}$ $\frac{5}{12}$ $\frac{1}{2}$

115. Place 16 red cubes on opposite faces but not in the center of a face. Place the 17th red cube anywhere except on the center of a face. The percentage of the surface area that is red is

$$\frac{42}{54} \times 100\% = 77\frac{7}{9}\%.$$

For Thought

1. True, since $\frac{1}{2} + \frac{1}{2} = 1$.

2. True, since $2^{100} = (2^2)^{50} = 4^{50}$.

3. True, since $9^8\dot{9}^8 = (9 \cdot 9)^8 = 81^8$.

4. True, since $(0.25)^{-1} = (1/4)^{-1} = 4$.

5. False, since $\frac{5^{10}}{5^{-12}} = 5^{10-(-12)} = 5^{22}$.

6. False, since $2^3 \cdot 2^{-1} = 2^2 = 4$.

7. True, since $-\frac{1}{3^3} = -\frac{1}{27}$.

8. True, since $a^{-2} = (1/a)^2$ if $a \neq 0$.

9. False, since $10^{-4} = 0.0001$.

10. False, since $98.6 \times 10^8 = 9.86 \times 10^9$.

P.2 Exercises

1. product rule

3. power of a power rule

5. $3^{-1} = \frac{1}{3}$

7. $-4^{-2} = -\frac{1}{16}$

9. $\frac{1}{2^{-3}} = (2)^3 = 8$

11. $\left(\frac{2}{3}\right)^3 = \frac{8}{27}$

13. $\left(-\frac{2}{1}\right)^2 = (-2)^2 = 4$

15. $\frac{1}{2} \cdot 16 \cdot \frac{1}{10} = \frac{8}{10} = \frac{4}{5}$

17. $\frac{6^3}{3^2} = 24$

19. $1 + \frac{1}{2} = \frac{3}{2}$

21. $-\frac{2}{1000} = -\frac{1}{500}$

23. $-6x^{11}y^{11}$

25. $x^6 + x^6 = 2x^6$

27. $-6b^5 - 15b^5 = 21b^5$

29. $-2a^{11} + 4a^{11} = 2a^{11}$

31. $m^3 + m^2 + 3m^3 = 4m^3 + m^2$

33. $25 \cdot 9 = 225$

35. $\frac{1}{6}x^0y^{-3} = \frac{1}{6y^3}$

37. $3x^4$

39. $\frac{-3}{-6} \cdot \frac{m^{-1}}{m^{-1}} \cdot \frac{n}{n^{-1}} = \frac{n^2}{2}$

41. $8a^6 + 9a^6 = 17a^6$

43. $-4x^6$

45. $\frac{(-2)^3(x^2)^3}{27} = \frac{-8x^6}{27}$

47. $\frac{25}{y^4}$

49. $\left(\frac{3x^5}{4y}\right)^{-3} = \frac{64y^3}{27x^{15}}$

51. $(x^{3b-3})(x^{8-2b}) = x^{b+5}$

53. $-125a^{6t}b^{-9t} = -\frac{125a^{6t}}{b^{9t}}$

55. $\frac{-3y^{6v}}{2x^{5w}}$

57. $\left(a^{-s+5}\right)^4 = a^{-4s+20}$

59. 43,000 **61.** 0.0000356 **63.** 5×10^6

65. 6.72×10^{-5}

67. 0.000000007 **69.** 2×10^{10}

71. $20 \times 10^{15} = 2 \times 10^{16}$

73. $2 \times 10^{-6+3} = 2 \times 10^{-3}$

75. $40 \times 10^{-30} = 4 \times 10^{-29}$

77. $\frac{(8 \times 10^{18})(5 \times 10^{-6})}{(4 \times 10^{-10})} = \frac{40 \times 10^{12}}{4 \times 10^{-10}} = 1 \times 10^{23}$

79. 9.936×10^{-5}

81. Using $\pi \approx 3.14$, we find

$$\frac{(178.45 \times 10^{-18})(0.00666 \times 10^{-28})}{3.14(79.21 \times 10^{-8})} \approx$$

$$\frac{1.188 \times 10^{-46}}{248.72 \times 10^{-8}} \approx 4.78 \times 10^{-41}$$

83. BMI$= \frac{703(335)}{79^2} \approx 37.7.$

85. Since 1 ton = 2000 pounds, the ratio is

$$D = \frac{2200(2000)175^{-3}10^6}{2240} \approx 366.5$$

87. Each person owes $\frac{15.885 \times 10^{12}}{3.131 \times 10^8} = \$50,735$

89. The number of seconds in 10 billion years is $60^2 \cdot 24 \cdot 365 \cdot 10^{10} = 3.1536 \times 10^{17}$. So, the amount of mass transformed by the sun in 10 billion years is 5 million $\times\ 3.1536 \times 10^{17} \approx 1.577 \times 10^{24}$ tons

91. The earth's radius is $\frac{6.9599 \times 10^5}{109.2} \approx 6379$ km.

93. $\frac{1.989 \times 10^{30}}{5.976 \times 10^{24}} \approx 3.3 \times 10^5.$

95. **a)** 5365.4

b) 9702.38

c) Let a, b, c, d, e, f be whole numbers. Then

$$a \cdot 10^3 + b \cdot 10^2 + c \cdot 10^1 + d \cdot 10^0 + e \cdot 10^{-1} + f \cdot 10^{-2}$$

is the number $abcdef$ obtained by juxtaposition.

d) $9 \cdot 10^3 + 6 \cdot 10^1 + 3 \cdot 10^0 + 2 \cdot 10^{-1} + 4 \cdot 10^{-2} + 1 \cdot 10^{-3}$

Not unique for $9063.241 = 90 \cdot 10^2 + 63 \cdot 10^1 + 241 \cdot 10^{-3}$.

e) $4 \cdot 10^4 + 3 \cdot 10^3 + 2 \cdot 10^0 + 1 \cdot 10^{-1} + 9 \cdot 10^{-2}$

97. $-3, 0, \frac{9}{2}$

99. associative

101. $1 - 3(102) = 1 - 306 = -305$

103. Rewriting the equation, we find

$$\begin{aligned}
\frac{1}{3^3}\cdot 3^{100}\cdot\frac{1}{3^4}\cdot 3^{2x} &= \frac{1}{3}\cdot 3^x\\
\frac{1}{3^3}\cdot 3^{100}\cdot\frac{1}{3^4}\cdot 3^{2x} &= 3^{x-1}\\
3^{2x+93} &= 3^{x-1}\\
2x+93 &= x-1\\
x &= -94.
\end{aligned}$$

For Thought

1. False, since $8^{-\frac{1}{3}} = \frac{1}{2}$.

2. True, since $16^{\frac{1}{4}} = 2 = 4^{\frac{1}{2}}$.

3. False, since $\sqrt{\frac{4}{6}} = \frac{2}{\sqrt{6}}$. **4.** True

5. False, since $(-1)^{\frac{1}{2}}$ is not a real number.

6. False, since $\sqrt[3]{7^2} = 7^{\frac{2}{3}}$. **7.** False, since $9^{\frac{1}{2}} = 3$.

8. True, since $\frac{1}{\sqrt{3}}\cdot\frac{\sqrt{3}}{\sqrt{3}} = \frac{\sqrt{3}}{3}$.

9. True, since $\frac{2^{\frac{1}{2}}}{2^{\frac{1}{3}}} = 2^{\frac{1}{2}-\frac{1}{3}} = 2^{\frac{1}{6}} = \sqrt[6]{2}$.

10. True, since $\sqrt[3]{7^5} = \sqrt[3]{7^3\cdot 7^2} = 7\sqrt[3]{49}$.

P.3 Exercises

1. cube root

3. radicand, index

5. -3 **7.** 8

9. -4 **11.** 81

13. $(8^{1/3})^{-4} = \frac{1}{16}$

15. $\left(\frac{1}{4}\right)^{1/2} = \frac{1}{2}$

17. $\left(\frac{4}{9}\right)^{3/2} = \left(\left(\frac{4}{9}\right)^{1/2}\right)^3 = \left(\frac{2}{3}\right)^3 = \frac{8}{27}$

19. $|x|$ **21.** a^3 **23.** a^2

25. $x^{3\cdot(1/3)}y^{6\cdot(1/3)} = xy^2$

27. $y^{2/3+7/3} = y^{9/3} = y^3$

29. $x^2y^{1/2}$

31. $6a^{1/2+1} = 6a^{3/2}$

33. $3a^{1/2-1/3} = 3a^{1/6}$

35. $a^{2+1/3}b^{1/2-1/2} = a^{7/3}$

37. $\frac{x^2y}{z^3}$

39. 30 **41.** -2

43. -2 **44.** 2

45. $\frac{2}{3}$ **47.** 0.1

49. $\frac{\sqrt[3]{-8}}{\sqrt[3]{1000}} = \frac{-2}{10} = -\frac{1}{5}$

51. $\sqrt[4]{(2^4)^3} = \sqrt[4]{2^{12}} = 2^{12\cdot(1/4)} = 2^3 = 8$

53. $10^{2/3} = \sqrt[3]{10^2}$

55. $\frac{3}{y^{3/5}} = \frac{3}{\sqrt[5]{y^3}}$ **57.** $x^{-1/2}$

59. $(x^3)^{1/5} = x^{3/5}$ **61.** $4x$ **63.** $2y^3$

65. $\frac{\sqrt{xy}}{10}$

67. $\frac{-2a}{b^{15\cdot(1/3)}} = -\frac{2a}{b^5}$

69. $\sqrt{4(7)} = 2\sqrt{7}$

71. $\frac{1}{\sqrt{5}}\cdot\frac{\sqrt{5}}{\sqrt{5}} = \frac{\sqrt{5}}{5}$

73. $\frac{\sqrt{x}}{\sqrt{8}}\cdot\frac{\sqrt{2}}{\sqrt{2}} = \frac{\sqrt{2x}}{\sqrt{16}} = \frac{\sqrt{2x}}{4}$

75. $\sqrt[3]{8\cdot 5} = 2\sqrt[3]{5}$

77. $\sqrt[3]{-250x^4} = \sqrt[3]{(-125)(2)x^3x} = -5x\sqrt[3]{2x}$

79. $\sqrt[3]{\frac{1}{2}}\cdot\sqrt[3]{\frac{4}{4}} = \sqrt[3]{\frac{4}{8}} = \frac{\sqrt[3]{4}}{2}$

81. $2\sqrt{2} + 2\sqrt{5} - 2\sqrt{3}$

83. $-10\sqrt{18} = -30\sqrt{2}$

85. $12(5a) = 60a$

87. $(-5)^2(3) = 75$

89. $\sqrt{\dfrac{9}{a^3}} = \dfrac{3}{a\sqrt{a}} \cdot \dfrac{\sqrt{a}}{\sqrt{a}} = \dfrac{3\sqrt{a}}{a^2}$

91. $\dfrac{5}{\sqrt{x}} \cdot \dfrac{\sqrt{x}}{\sqrt{x}} = \dfrac{5\sqrt{x}}{x}$

93. $2x\sqrt{5x} + 3x\sqrt{5x} = 5x\sqrt{5x}$

95. $\sqrt[6]{3^2}\sqrt[6]{2^3} = \sqrt[6]{72}$

97. $\sqrt[12]{3^4}\sqrt[12]{x^3} = \sqrt[12]{81x^3}$

99. $\sqrt[6]{(xy)^3}\sqrt[6]{(2xy)^2} = \sqrt[6]{x^3y^3}\sqrt[6]{4x^2y^2} = \sqrt[6]{4x^5y^5}$

101. $(7^{1/2})^{1/3} = \sqrt[6]{7}$

103. Since $E = \sqrt{\dfrac{2(25)(6000)}{140}} \approx 46.3$, the most economic order quantity is $E = 46$.

105. $S = \dfrac{16(42,700)}{[(2200)(2000)]^{2/3}} \approx 25.4$

107. **a)** 50%, 30%

b) If $n = 5$, the depreciation rate is

$$r = 1 - \left(\frac{200}{5000}\right)^{1/5} \approx 0.47469 \text{ or } 47.5\%.$$

If $n = 10$, the depreciation rate is

$$r = 1 - \left(\frac{200}{5000}\right)^{1/10} \approx 0.2752 \text{ or } 27.5\%.$$

109. $\text{D} = \sqrt{4^2 + 6^2 + 12^2} = 14$ in.

111. $A = \sqrt{12(6)(5)(1)} = \sqrt{360} \approx 19.0 \text{ ft}^2$

113. Each expression is equivalent to $t^{4/5}$ except for $\sqrt[4]{t^5}$ which is $t^{5/4}$, the expression in (b).

115. No, since $\sqrt{9+16} = 5 \neq 7 = \sqrt{9} + \sqrt{16}$.

117. distributive

119. $25 - 4(-8) = 25 + 32 = 57$

121. $-6x^4y^4 - x^4y^4 = -7x^4y^4$

123. Note the prime factorization below

$$349,272,000 = 2^6 \cdot 3^4 \cdot 5^3 \cdot 7^2 \cdot 11.$$

The perfect squares dividing 2^6 are 1, 2^2, 2^4, and 2^6; That is, there are four distinct perfect squares that divide 2^6.

Similarly, there are three distinct perfect squares that divide 3^4; two perfect squares that divide 5^3; two perfect squares that divide 7^2; one perfect square divides 11;

Then the product of the number of perfect squares that divide the 2^6, 3^4, 5^3, 7^2, and 11 is

$$4 \cdot 3 \cdot 2 \cdot 2 \cdot 1 = 48$$

Thus, 48 perfect squares divide 349,272,000.

For Thought

1. False, polynomials do not have negative exponents.

2. False, since the degree is 4.

3. False, since $(8)^2 \neq 9 + 25$.

4. True, since $50^2 - 1^2 = 2499$.

5. False, since $(x+3)^2 = x^2 + 6x + 9$.

6. False, since the left-hand side is $-x - 6$.

7. False, since $(a+b)^3 = a^3 + 3a^2b + 3ab^2 + b^3$.

8. False, since divisor times quotient plus the remainder is equal to the dividend.

9. False, since the equation is not defined at $x = -2$.

10. True, since $P(17) + Q(17) = 2310 = S(17)$.

P.4 Exercises

1. constant

3. binomial

5. cubic

7. linear

9. Degree 3, leading coefficient 1, trinomial

11. Degree 2, leading coefficient -3, binomial

13. Degree 0, leading coefficient 79, monomial

15. $P(-2) = 4 + 6 + 2 = 12$

17. $M(-3) = 27 + 45 + 3 + 2 = 77$

19. $8x^2 + 3x - 1$

21. $4x^2 - 3x - 9x^2 + 4x - 3 = -5x^2 + x - 3$

23. $4ax^3 - a^2x - 5a^2x^3 + 3a^2x - 3 =$
$(-5a^2 + 4a)x^3 + 2a^2x - 3$

25. $2x - 1$

27. $3x^2 - 3x - 6$

29. $-18a^5 + 15a^4 - 6a^3$

31. $(3b^2 - 5b + 2)b - (3b^2 - 5b + 2)3 =$
$3b^3 - 5b^2 + 2b - 9b^2 + 15b - 6 =$
$3b^3 - 14b^2 + 17b - 6$

33. $2x(4x^2 + 2x + 1) - (4x^2 + 2x + 1) =$
$8x^3 + 4x^2 + 2x - 4x^2 - 2x - 1 =$
$8x^3 - 1$

35. $(x - 4)z + (x - 4)3 = xz - 4z + 3x - 12$

37. $a(a^2 + ab + b^2) - b(a^2 + ab + b^2) =$
$a^3 + a^2b + ab^2 - a^2b - ab^2 - b^3 = a^3 - b^3$

39. $a^2 - 2a + 9a - 18 = a^2 + 7a - 18$

41. $2y^2 + 18y - 3y - 27 = 2y^2 + 15y - 27$

43. $4x^2 + 18x - 18x - 81 = 4x^2 - 81$

45. $4x^2 + 10x + 10x + 25 = 4x^2 + 20x + 25$

47. $6x^4 + 10x^2 + 12x^2 + 20 = 6x^4 + 22x^2 + 20$

49. $(1+\sqrt{2})(3+\sqrt{2}) = 3+\sqrt{2}+3\sqrt{2}+2 = 5+4\sqrt{2}$

51. $20 - 15\sqrt{2} + 4\sqrt{2} - 3\sqrt{2}\sqrt{2} = 14 - 11\sqrt{2}$

53. $6\sqrt{2}\sqrt{2} - 3\sqrt{6} + 2\sqrt{6} - \sqrt{3}\sqrt{3} = 9 - \sqrt{6}$

55. $\sqrt{5}\sqrt{5} + \sqrt{15} + \sqrt{15} + \sqrt{3}\sqrt{3} = 8 + 2\sqrt{15}$

57. $(3x)^2 + 2(3x)(5) + (5)^2 = 9x^2 + 30x + 25$

59. $(x^n)^2 - 3^2 = x^{2n} - 9$

61. $(\sqrt{2})^2 - 5^2 = -23$

63. $(3\sqrt{6})^2 - 2(3\sqrt{6})(1) + 1 = 55 - 6\sqrt{6}$

65. $(3x^3)^2 - 2(3x^3)(4) + 4^2 = 9x^6 - 24x^3 + 16$

67. $\dfrac{5}{1-\sqrt{7}} \cdot \dfrac{1+\sqrt{7}}{1+\sqrt{7}} = \dfrac{5+5\sqrt{7}}{1-7} = \dfrac{-5-5\sqrt{7}}{6}$

69. $\dfrac{\sqrt{10}}{\sqrt{5}-2} \cdot \dfrac{\sqrt{5}+2}{\sqrt{5}+2} = \dfrac{\sqrt{50}+2\sqrt{10}}{5-4} = 5\sqrt{2}+2\sqrt{10}$

71. $\dfrac{\sqrt{6}}{6+\sqrt{3}} \cdot \dfrac{6-\sqrt{3}}{6-\sqrt{3}} = \dfrac{6\sqrt{6}-3\sqrt{2}}{33} = \dfrac{2\sqrt{6}-\sqrt{2}}{11}$

73.
$\dfrac{1+\sqrt{2}}{2-\sqrt{3}} \cdot \dfrac{2+\sqrt{3}}{2+\sqrt{3}} = \dfrac{2+2\sqrt{2}+\sqrt{3}+\sqrt{6}}{4-3} =$
$2+2\sqrt{2}+\sqrt{3}+\sqrt{6}$

75.
$\dfrac{\sqrt{2}+\sqrt{3}}{\sqrt{6}} \cdot \dfrac{\sqrt{6}}{\sqrt{6}} = \dfrac{\sqrt{12}+\sqrt{18}}{6} =$
$\dfrac{2\sqrt{3}+3\sqrt{2}}{6}$

77. $\dfrac{36x^6}{-4x^3} = -9x^3$

79. $\dfrac{3x^2}{-3x} + \dfrac{6x}{3x} = -x + 2$

81. Factor & simplify: $\dfrac{(x+3)^2}{x+3} = x + 3$

83. Factor & simplify: $\dfrac{(a-1)(a^2+a+1)}{a-1} =$
$a^2 + a + 1$

85. Quotient $x + 5$, remainder 13 since

$$\begin{array}{r|l} & x+5 \\ x-2 & x^2+3x+3 \\ & \underline{x^2-2x} \\ & \quad 5x+3 \\ & \quad \underline{5x-10} \\ & \qquad 13 \end{array}$$

87. Quotient $2x - 6$, remainder 13 since

$$\begin{array}{r|l} & 2x-6 \\ x+3 & 2x^2+0x-5 \\ & \underline{2x^2+6x} \\ & \quad -6x-5 \\ & \quad \underline{-6x-18} \\ & \qquad 13 \end{array}$$

89. Quotient $x^2 + x + 1$, remainder 0 since

$$\begin{array}{r|l} & x^2 + x + 1 \\ \hline x - 3 & x^3 - 2x^2 - 2x - 3 \\ & \underline{x^3 - 3x^2} \\ & \quad x^2 - 2x \\ & \quad \underline{x^2 - 3x} \\ & \qquad x - 3 \\ & \qquad \underline{x - 3} \\ & \qquad\quad 0 \end{array}$$

91. $x + 1 + \dfrac{-1}{x - 1}$ since

$$\begin{array}{r|l} & x + 1 \\ \hline x - 1 & x^2 + 0x - 2 \\ & \underline{x^2 - x} \\ & \quad x - 2 \\ & \quad \underline{x - 1} \\ & \qquad -1 \end{array}$$

93. $\dfrac{2x^2}{x} - \dfrac{3x}{x} + \dfrac{1}{x} = 2x - 3 + \dfrac{1}{x}$

95. $\dfrac{x^2}{x} + \dfrac{x}{x} + \dfrac{1}{x} = x + 1 + \dfrac{1}{x}$

97. $x + \dfrac{1}{x - 2}$ since

$$\begin{array}{r|l} & x \\ \hline x - 2 & x^2 - 2x + 1 \\ & \underline{x^2 - 2x} \\ & \qquad\quad 1 \end{array}$$

99. $x^2 + 2x - 24$

101. $2a^{10} - 3a^5 - 27$

103. $-y - 9$ **105.** $6a - 6$

107. $w^2 + 8w + 16$

109. $16x^2 - 81$

111. $3y^5 - 9xy^2$

113. $2b - 1$

115. $x^6 - 64$

117. $9w^4 - 12w^2n + 4n^2$

119. 1

121. The area is $(x + 3)(2x - 1) = 2x^2 + 5x - 3$.

123. Divide the volume by the height.

$$\begin{array}{r|l} & x^2 + 5x + 6 \\ \hline x + 4 & x^3 + 9x^2 + 26x + 24 \\ & \underline{x^3 + 4x^2} \\ & \quad 5x^2 + 26x \\ & \quad \underline{5x^2 + 20x} \\ & \qquad 6x + 24 \\ & \qquad \underline{6x + 24} \\ & \qquad\quad 0 \end{array}$$

Then the area of the bottom is $x^2 + 5x + 6$ cm^2.

125. The dimensions of the rectangular habitat are $6 - 2x$ and $4 - 2x$. The area is the product of the dimensions which is $4x^2 - 20x + 24$ square miles.

127. **a)** Approximately \$0.40

b) $A(.1442) - A(.0284) \approx \$0.41$

129. The fine is $F(25) = 500 + 200(25) = \$5,500$.

131. Let x be the length of a side of the square. The areas of the square and rectangle are x^2 and $(x + 10)(x - 10) = x^2 - 100$, respectively. Wilson had 100 ft^2 less area than he thought.

133. $\dfrac{100 - 8}{4} = \dfrac{92}{4} = 23$

135. $-\dfrac{1}{4}(1296) = -324$

137. $\left((-125)^{1/3}\right)^{-2} = (-5)^{-2} = \dfrac{1}{25}$

139. Notice, if $a > b > 0$ and $n > 0$, then

$$\frac{a}{n} + \frac{b}{n+1} > \frac{b}{n} + \frac{a}{n+1}.$$

Thus, the arrangement with the largest sum is

$$\frac{2015}{1} + \frac{2014}{2} + \cdots + \frac{2}{2014} + \frac{1}{2015}.$$

For Thought

1. False, since it factors as $(x + 8)(x - 2)$.

2. True **3.** True

4. False, since $a^2 - 1 = (a - 1)(a + 1)$.

5. False, since $a^3 - 1 = (a - 1)(a^2 + a + 1)$.

6. False, since $x^3 - y^3 = (x - y)(x^2 + xy + y^2)$.

7. False, it factors as $(2a + 3b)(4a^2 - 6ab + 9b^2)$.

8. False, since $(x + 2)(x - 6) = x^2 - 4x - 12$.

9. True **10.** True

P.5 Exercises

1. factoring

3. perfect square trinomial

5. $6x^2(x - 2)$, $-6x^2(-x + 2)$

7. $4a(1 - 2b)$, $-4a(2b - 1)$

9. $ax(-x^2 + 5x - 5), -ax(x^2 - 5x + 5)$

11. $1(m - n), -1(n - m)$

13. $x^2(x + 2) + 5(x + 2) = (x^2 + 5)(x + 2)$

15. $y^2(y - 1) - 3(y - 1) = (y^2 - 3)(y - 1)$

17. $ady + d - awy - w = d(ay + 1) - w(ay + 1) =$
$(d - w)(ay + 1)$

19. $x^2y^2 - ay^2 - (bx^2 - ab) = y^2(x^2 - a) - b(x^2 - a) =$
$(y^2 - b)(x^2 - a)$

21. $(x + 2)(x + 8)$

23. $(x - 6)(x + 2)$

25. $(m - 2)(m - 10)$

27. $(t - 7)(t + 12)$

29. $(2x + 1)(x - 4)$

31. $(4x + 1)(2x - 3)$

33. $(3y + 5)(2y - 1)$

35. $(t - u)(t + u)$

37. $(t+1)^2$ **39.** $(2w-1)^2$ **41.** $(y^{2t}-5)(y^{2t}+5)$

43. $(3zx + 4)^2$ **45.** $(t - u)(t^2 + tu + u^2)$

47. $(a - 2)(a^2 + 2a + 4)$

49. $(3y + 2)(9y^2 - 6y + 4)$

51. $(3xy^2)^3 - (2z^3)^3 =$
$(3xy^2 - 2z^3)(9x^2y^4 + 6xy^2z^3 + 4z^6)$

53. $(x^n)^3 - 2^3 = (x^n - 2)(x^{2n} + 2x^n + 4)$

55. Replace y^3 by w and y^6 by w^2.
Then $w^2 + 10w + 25 = (w + 5)^2 = (y^3 + 5)^2$.

57. Replace $2a^2b^4$ by w and $4a^4b^8$ by w^2.
So, $w^2 - 4w - 5 = (w + 1)(w - 5) =$
$(2a^2b^4 + 1)(2a^2b^4 - 5)$.

59. Replace $(2a + 1)$ by w and $(2a + 1)^2$ by w^2.
Then $w^2 + 2w - 24 = (w + 6)(w - 4) =$
$((2a + 1) + 6)((2a + 1) - 4) = (2a + 7)(2a - 3)$.

61. Replace $(b^2 + 2)$ by w and $(b^2 + 2)^2$ by w^2.
We find $w^2 - 5w + 4 = (w - 4)(w - 1) =$
$((b^2 + 2) - 4)((b^2 + 2) - 1) = (b^2 - 2)(b^2 + 1)$.

63. $-3x^3 + 27x = -3x(x^2 - 9) = -3x(x - 3)(x + 3)$

65. $2t(8t^3 + 27w^3) = 2t(2t + 3w)(4t^2 - 6tw + 9w^2)$

67. $a^3 + a^2 - 4a - 4 = a^2(a + 1) - 4(a + 1) =$
$(a^2 - 4)(a + 1) = (a - 2)(a + 2)(a + 1)$

69. $x^4 - 2x^3 - 8x + 16 = x^3(x - 2) - 8(x - 2) =$
$(x - 2)(x^3 - 8) = (x - 2)^2(x^2 + 2x + 4)$

71. $-2x(18x^2 - 9x - 2) = -2x(6x + 1)(3x - 2)$

73. $a^7 - a^6 - 64a + 64 = a^6(a - 1) - 64(a - 1) =$
$(a^6 - 64)(a - 1) = (a^3 - 8)(a^3 + 8)(a - 1) =$
$(a - 2)(a^2 + 2a + 4)(a + 2)(a^2 - 2a + 4)(a - 1)$

75. $-(3x + 5)(2x - 3)$

77. Replace $(a^2 + 2)$ by w and $(a^2 + 2)^2$ by w^2.
$w^2 - 4w + 3 = (w - 3)(w - 1) =$
$((a^2 + 2) - 3)((a^2 + 2) - 1) = (a^2 - 1)(a^2 + 1) =$
$(a - 1)(a + 1)(a^2 + 1)$

79. Yes, since

$$\begin{array}{r@{}l}
 & x^2 + x + 1 \\
x + 3 & \overline{)x^3 + 4x^2 + 4x + 3} \\
 & \underline{x^3 + 3x^2} \\
 & \qquad x^2 + 4x \\
 & \qquad \underline{x^2 + 3x} \\
 & \qquad\qquad x + 3 \\
 & \qquad\qquad \underline{x + 3} \\
 & \qquad\qquad\quad 0
\end{array}$$

and $x^3 + 4x^2 + 4x + 3 = (x + 3)(x^2 + x + 1)$.

81. No, since

$$
\begin{array}{rl}
 & 3x^2+8x-4 \\
x-1 & \overline{)3x^3+5x^2-12x-9} \\
 & \underline{3x^3-3x^2} \\
 & 8x^2-12x \\
 & \underline{8x^2-8x} \\
 & -4x-9 \\
 & \underline{-4x+4} \\
 & -13
\end{array}
$$

and the remainder is not 0.

83.

$$
\begin{array}{rl}
 & x^2+5x+6 \\
x-1 & \overline{)x^3+4x^2+x-6} \\
 & \underline{x^3-x^2} \\
 & 5x^2+x \\
 & \underline{5x^2-5x} \\
 & 6x-6 \\
 & \underline{6x-6} \\
 & 0
\end{array}
$$

Then $x^3+4x^2+x-6=(x-1)(x^2+5x+6)=(x-1)(x+2)(x+3)$.

85.

$$
\begin{array}{rl}
 & x^2+2x+2 \\
x-3 & \overline{)x^3-x^2-4x-6} \\
 & \underline{x^3-3x^2} \\
 & 2x^2-4x \\
 & \underline{2x^2-6x} \\
 & 2x-6 \\
 & \underline{2x-6} \\
 & 0
\end{array}
$$

So $x^3-x^2-4x-6=(x-3)(x^2+2x+2)$.

87.

$$
\begin{array}{rl}
 & x^3+3x^2-x-3 \\
x+2 & \overline{)x^4+5x^3+5x^2-5x-6} \\
 & \underline{x^4+2x^3} \\
 & 3x^3+5x^2 \\
 & \underline{3x^3+6x^2} \\
 & -x^2-5x \\
 & \underline{-x^2-2x} \\
 & -3x-6 \\
 & \underline{-3x-6} \\
 & 0
\end{array}
$$

Then we obtain

$$
\begin{aligned}
x^4+5x^3+5x^2-5x-6 &= \\
(x+2)(x^3+3x^2-x-3) &= \\
(x+2)(x^2(x+3)-(x+3)) &= \\
(x+2)(x+3)(x^2-1) &= \\
(x+2)(x+3)(x-1)(x+1).
\end{aligned}
$$

89. The area of the bottom is the volume divided by the height.

$$
\begin{array}{rl}
 & x^2+x+1 \\
x-1 & \overline{)x^3+0x^2+0x-1} \\
 & \underline{x^3-x^2} \\
 & x^2+0x \\
 & \underline{x^2-x} \\
 & x-1 \\
 & \underline{x-1} \\
 & 0
\end{array}
$$

Then x^2+x+1 ft^2 is the area of the bottom.

91. The volume is $V=x(6-2x)(7-2x)$. From the tabulated values below

x	0.5	1	2
Volume	15 in.^{3}	20 in.^{3}	12 in.^{3}

we see that $x=1$ produces the largest volume.

93. **b** is not a perfect square trinomial since $(\sqrt{1000}a-b)^2\neq 1000a^2-200b+b^2$. The others are perfect squares.

95. No, since $1^3+1^3=2\neq n^3$ for any integer n.

97. $3+\sqrt{3}$

99. $3-5|-4|=3-20=-17$

101. $\dfrac{5}{2\sqrt{3}}=\dfrac{5\sqrt{3}}{6}$

103. No, since the product of the numbers from 1 through 9 is

$$1\cdot 2\cdot 3\cdots 9=362,880$$

and $71^3=357,911$.

For Thought

1. False, since $\dfrac{2x+5}{2y}=\dfrac{x+\frac{5}{2}}{y}$. **2.** True

3. False, since the first expression is not defined at $x=0$.

4. False, since $\dfrac{(a-b)(a+b)}{a-b} = a+b$.

5. False, the LCD is $x(x+1)$. **6.** True

7. True **8.** True

9. True, since $\dfrac{2(500)+1}{500-3} = \dfrac{1001}{497} \approx 2.014$

10. True, since $\dfrac{5x+1}{x} \approx \dfrac{5x}{x} = 5$ when $|x|$ is large.

P.6 Exercises

1. rational expression

3. least common denominator

5. reducing

7. $\{x \mid x \neq -2\}$

9. $\{x \mid x \neq 4, -2\}$

11. $\{x \mid x \neq \pm 3\}$

13. All real numbers since $x^2+3 \neq 0$ for any real number x.

15. $\dfrac{3(x-3)}{(x-3)(x+2)} = \dfrac{3}{x+2}$

17. $\dfrac{10a-8b}{12b-15a} = \dfrac{2(5a-4b)}{-3(5a-4b)} = -\dfrac{2}{3}$

19. $\dfrac{a^3b^6}{a^2b^3-a^4b^2} = \dfrac{a^3b^6}{a^2b^2(b-a^2)} = \dfrac{ab^4}{b-a^2}$

21. $\dfrac{y^2z}{x^3}$

23. $\dfrac{a^3-b^3}{a^2-b^2} = \dfrac{(a-b)(a^2+ab+b^2)}{(a-b)(a+b)} = \dfrac{a^2+ab+b^2}{a+b}$

25. $\dfrac{(3y-1)^2}{(1-3y)(1+3y)} = \dfrac{-(3y-1)}{1+3y}$

27. $\dfrac{2a}{3b^2} \cdot \dfrac{9b}{14a^2} = \dfrac{1}{b} \cdot \dfrac{3}{7a} = \dfrac{3}{7ab}$

29. $\dfrac{12a}{7} \cdot \dfrac{49}{2a^3} = \dfrac{42}{a^2}$

31. $\dfrac{(a-3)(a+3)}{3(a-2)} \cdot \dfrac{(a-2)(a+2)}{(a-3)(a+2)} = \dfrac{a+3}{3}$

33. $\dfrac{(x-y)(x+y)}{9} \cdot \dfrac{9(2)}{(x+y)^2} = \dfrac{2x-2y}{x+y}$

35. $\dfrac{(x-y)(x+y)}{-3xy} \cdot \dfrac{3xy(2xy^2)}{-2(x-y)} = x^2y^2+xy^3$

37. $\dfrac{(b-3)(b+2)}{(3-b)(3+b)} \cdot \dfrac{2(b+3)}{(b+4)(b+2)} = \dfrac{-2}{b+4}$

39. $\dfrac{16a}{12a^2}$

41. $\dfrac{x-5}{x+3} \cdot \dfrac{x-3}{x-3} = \dfrac{x^2-8x+15}{x^2-9}$

43. $\dfrac{x}{x+5} \cdot \dfrac{x+1}{x+1} = \dfrac{x^2+x}{x^2+6x+5}$

45. $12a^2b^3$

47. Since $3a+3b = 3(a+b)$ and $2a+2b = 2(a+b)$, the LCD is $6(a+b)$.

49. Since $x^2+5x+6 = (x+3)(x+2)$ and $x^2-x-6 = (x-3)(x+2)$, the LCD is $(x+2)(x+3)(x-3)$.

51. $\frac{9}{3x-6} = \frac{3}{x-2}$

53. $\dfrac{3(3)}{2x(3)} + \dfrac{x}{6x} = \dfrac{9+x}{6x}$

55. $\dfrac{(x+3)(x+1)}{(x-1)(x+1)} - \dfrac{(x+4)(x-1)}{(x-1)(x+1)} = \dfrac{x^2+4x+3}{(x-1)(x+1)} - \dfrac{x^2+3x-4}{(x-1)(x+1)} = \dfrac{x+7}{(x-1)(x+1)}$

57. $\dfrac{3a}{a} + \dfrac{1}{a} = \dfrac{3a+1}{a}$

59. $\dfrac{(t-1)(t+1)}{t+1} - \dfrac{1}{t+1} = \dfrac{t^2-2}{t+1}$

61.
$$\frac{x}{(x+2)(x+1)}+\frac{x-1}{(x+3)(x+2)}=$$
$$\frac{x(x+3)}{(x+2)(x+1)(x+3)}+\frac{(x-1)(x+1)}{(x+3)(x+2)(x+1)}=$$
$$\frac{2x^2+3x-1}{(x+1)(x+2)(x+3)}$$

63.
$$\frac{1}{x-3}-\frac{5}{-2(x-3)}=\frac{2}{2(x-3)}-\frac{-5}{2(x-3)}=$$
$$\frac{7}{2x-6}$$

65.
$$\frac{y^2}{(x-y)(x^2+xy+y^2)}+\frac{(x+y)(x-y)}{(x^2+xy+y^2)(x-y)}=$$
$$\frac{y^2}{(x-y)(x^2+xy+y^2)}+\frac{x^2-y^2}{(x^2+xy+y^2)(x-y)}=$$
$$\frac{x^2}{x^3-y^3}$$

67.
$$\frac{(x+1)(x-1)}{x(x+1)(x-1)}+\frac{x(x+1)}{(x-1)(x)(x+1)}-\frac{x(x-1)}{(x+1)(x)(x-1)}=$$
$$\frac{x^2-1}{x(x+1)(x-1)}+\frac{x^2+x}{(x-1)(x)(x+1)}-\frac{x^2-x}{(x+1)(x)(x-1)}=$$
$$\frac{x^2+2x-1}{x(x^2-1)}$$

69.
$$\frac{5(x+1)(x-3)+2x(x-3)-6x(x+1)}{x(x+1)(x-3)}=$$
$$\frac{5(x^2-2x-3)+2x^2-6x-6x^2-6x}{x(x+1)(x-3)}=$$
$$\frac{x^2-22x-15}{x(x+1)(x-3)}$$

71. $$\frac{25}{36a}\cdot\frac{27}{10a}=\frac{5^2\cdot 3\cdot 9}{4\cdot 9\cdot 5\cdot 2a^2}=\frac{15}{8a^2}$$

73.
$$\frac{\left(\frac{4}{a}-\frac{3}{b}\right)(ab^2)}{\left(\frac{1}{ab}+\frac{2}{b^2}\right)(ab^2)}=\frac{4b^2-3ab}{b+2a}$$

75.
$$\frac{\frac{1}{b^2}-\frac{1}{ab^2}}{\frac{3}{a^2}+\frac{1}{a^2b}}=\frac{\left(\frac{1}{b^2}-\frac{1}{ab^2}\right)(a^2b^2)}{\left(\frac{3}{a^2}+\frac{1}{a^2b}\right)(a^2b^2)}=\frac{a^2-a}{3b^2+b}$$

77.
$$\frac{\left(a+\frac{4}{a+4}\right)(a+4)}{\left(a-\frac{4a+4}{a+4}\right)(a+4)}=$$
$$\frac{a^2+4a+4}{a^2+4a-(4a+4)}=$$
$$\frac{(a+2)^2}{(a+2)(a-2)}=\frac{a+2}{a-2}$$

79.
$$\frac{\left(\frac{t+2}{t-1}-\frac{t-3}{t}\right)((t-1)t)}{\left(\frac{t+4}{t}+\frac{t-2}{t-1}\right)((t-1)t)}=$$
$$\frac{(t^2+2t)-(t-3)(t-1)}{(t+4)(t-1)+(t^2-2t)}=$$
$$\frac{(t^2+2t)-(t^2-4t+3)}{(t^2+3t-4)+(t^2-2t)}=\frac{6t-3}{2t^2+t-4}$$

81.
$$\frac{(x^{-1}+1)x}{(x^{-1}-1)x}=\frac{1+x}{1-x}$$

83. $$\frac{(a^2+a^{-1}b^{-3})\,ab^3}{ab^3}=\frac{a^3b^3+1}{ab^3}$$

85.
$$\frac{(x^2-y^2)xy}{(x^{-1}-y^{-1})xy}=\frac{(x-y)(x+y)xy}{y-x}=$$
$$-(x+y)xy=-x^2y-xy^2$$

87.
$$\left(\frac{1}{m}-\frac{1}{n}\right)^{-2}=\left(\frac{n-m}{mn}\right)^{-2}=$$
$$\left(\frac{mn}{n-m}\right)^2=\frac{m^2n^2}{n^2-2mn+m^2}$$

89. $\frac{3}{7}$ **91.** $\frac{3}{506}$

93. $S(2) = \dfrac{4-5}{4-9} = \dfrac{1}{5}$

95. $\dfrac{1200-5}{1200-9} = \dfrac{1195}{1191}$

97. $\dfrac{9(16)-1}{3(16)-2} \approx 3.1087$

99. $T(-400) = \dfrac{9(160,000)-1}{3(160,000)-2} \approx 3.00001$

101. **(a)** Average cost decreases as capacity increases

(b) The average costs are

n	7	12	22
$A(n)$	\$27.14	\$24.17	\$22.27

103. **a)** The costs are tabulated below.

p	$C(p)$
50%	\$6 million
75%	\$18 million
99%	\$594 million

b) The cost of cleaning up goes up without bound. A 100% clean up is impossible.

c) domain is $\{p \mid 0 \le p < 100\}$

105. The portion of the invoices Gina and Bert can file in one hour are $\frac{1}{4}$ and $\frac{1}{6}$, respectively. The part they can file together in one hour is $\frac{1}{4} + \frac{1}{6} = \frac{5}{12}$.

107. Let d be the distance between the restaurant and his home. The number of hours dashing home and returning to the restaurant are $\frac{d}{250}$ and $\frac{d}{300}$, respectively. His average speed is

$$\frac{2d}{\frac{d}{250}+\frac{d}{300}} = \frac{2}{\frac{1}{250}+\frac{1}{300}} \approx 272.7 \text{ mph.}$$

109. If a rational expression simulates an application, then the domain describes the extent to which the simulation applies. This is one reason.

111. **(a)** Since $1+\frac{1}{2} = \frac{3}{2}$, $1+\frac{1}{3/2} = \frac{5}{3}$, and so on..., the exact answer is $\frac{8}{13}$.

(b) Since $1-\frac{1}{3} = \frac{2}{3}$, $1-\frac{1}{2/3} = -\frac{1}{2}$, and so on..., the exact answer is $\frac{3}{2}$.

113.

a) $9x^2(1-x^2) = 9x^2(1-x)(1+x)$

b) Difference of two cubes: $(w-1)(w^2+w+1)$

115.

a) $3x^3 + x^2 - 3$

b) $x^3 - x^2 + 6x + 7$

117. 9.8×10^4

119. The domain of $N(x) = \sqrt{5x+6} + \sqrt{7x+8}$ is $[-8/7, \infty)$. We note that

$$N(s) < N(t) \text{ for all } -\frac{8}{7} \le s < t.$$

Thus, the minimum value of N is

$$N(-8/7) = \sqrt{5\left(-\frac{8}{7}\right)+6} = \sqrt{\frac{2}{7}} = \frac{\sqrt{14}}{7}.$$

For Thought

1. True, since $i \cdot (-i) = 1$.

2. True, since $\overline{0+i} = 0 - i = -i$.

3. False, the set of real numbers is a subset of the complex numbers.

4. True, $(\sqrt{3} - i\sqrt{2})(\sqrt{3} + i\sqrt{2}) = 3 + 2 = 5$.

5. False, since $(2+5i)(2+5i) = 4 + 20i + 25i^2 = 4 + 20i - 25 = -21 + 20i$.

6. False, $5 - \sqrt{-9} = 5 - 3i$.

7. True, since $(3i)^2 + 9 = (-9) + 9 = 0$.

8. True, since $(-3i)^2 + 9 = (-9) + 9 = 0$.

9. True, since $i^4 = i^2 \cdot i^2 = (-1)(-1) = 1$.

10. False, $i^{18} = (i^4)^4 i^2 = (1)^4(-1) = -1$.

P.7 Exercises

1. complex numbers

3. imaginary number

5. $0 + 6i$, imaginary

7. $\frac{1}{3} + \frac{1}{3}i$, imaginary

9. $\sqrt{7} + 0i$, real

11. $\frac{\pi}{2} + 0i$, real

13. $7 + 2i$

15. $1 - i - 3 - 2i = -2 - 3i$

17. $1 + 3 - i\sqrt{2} + 2i\sqrt{2} = 4 + i\sqrt{2}$

19. $5 - \frac{1}{2} + \frac{1}{3}i + \frac{1}{2}i = \frac{9}{2} + \frac{5}{6}i$

21. $-18i + 12i^2 = -12 - 18i$

23. $8 + 12i - 12i - 18i^2 = 26 + 0i$

25. $(4 - 5i)(6 + 2i) = 24 + 8i - 30i + 10 = 34 - 22i$

27. $(5 - 2i)(5 + 2i) = 25 - 4i^2 = 25 - 4(-1) = 29$

29. $(\sqrt{3} - i)(\sqrt{3} + i) = 3 - i^2 = 3 - (-1) = 4$

31. $9 + 24i + 16i^2 = -7 + 24i$

33. $5 - 4i\sqrt{5} + 4i^2 = 1 - 4i\sqrt{5}$

35. $(i^4)^4 \cdot i = (1)^4 \cdot i = i$

37. $(i^4)^{24} i^2 = 1^{24}(-1) = -1$

39. Since $i^4 = 1$, we get $i^{-1} = i^{-1}i^4 = i^3 = -i$.

41. Since $i^4 = 1$, we get $i^{-3} = i^{-3}i^4 = i^1 = i$.

43. Since $i^{16} = 1$, we get $i^{-13} = i^{-13}i^{16} = i^3 = -i$.

45. Since $i^{-4} = 1$, we get $i^{-38} = i^2 i^{-40} = i^2(i^{-4})^{10} = i^2(1) = -1$.

47. $(3 - 9i)(3 + 9i) = 9 - 81i^2 = 90$

49. $\left(\frac{1}{2} + 2i\right)\left(\frac{1}{2} - 2i\right) = \frac{1}{4} - 4i^2 = \frac{1}{4} + 4 = \frac{17}{4}$

51. $i(-i) = -i^2 = 1$

53. $(3 - i\sqrt{3})(3 + i\sqrt{3}) = 9 - 3i^2 = 9 - 3(-1) = 12$

55. $\frac{1}{2-i} \cdot \frac{2+i}{2+i} = \frac{2+i}{5} = \frac{2}{5} + \frac{1}{5}i$

57. $\frac{-3i}{1-i} \cdot \frac{1+i}{1+i} = \frac{-3i+3}{2} = \frac{3}{2} - \frac{3}{2}i$

59.
$$\frac{-3+3i}{i} \cdot \frac{-i}{-i} = \frac{3i - 3i^2}{1} = 3i - 3(-1) = 3 + 3i$$

61.
$$\frac{1-i}{3+2i} \cdot \frac{3-2i}{3-2i} = \frac{3 - 5i - 2}{13} = \frac{1}{13} - \frac{5}{13}i$$

63.
$$\frac{2-i}{3+5i} \cdot \frac{3-5i}{3-5i} = \frac{6 - 10i - 3i - 5}{34} = \frac{1}{34} - \frac{13}{34}i$$

65. $2i - 3i = -i$

67. $-4 + 2i$

69. $\left(i\sqrt{6}\right)^2 = -6$

71. $(i\sqrt{2})(i\sqrt{50}) = i^2\sqrt{2}\cdot 5\sqrt{2} = (-1)(2)(5) = -10$

73.
$$\frac{-2}{2} + \frac{i\sqrt{20}}{2} = -1 + i\frac{2\sqrt{5}}{2} = -1 + i\sqrt{5}$$

75. $-3 + \sqrt{9 - 20} = -3 + i\sqrt{11}$

77. $2i\sqrt{2}\left(i\sqrt{2} + 2\sqrt{2}\right) = 4i^2 + 8i = -4 + 8i$

79. $\frac{-2 + \sqrt{-16}}{2} = \frac{-2 + 4i}{2} = -1 + 2i$

81. $\frac{-4 + \sqrt{16 - 24}}{4} = \frac{-4 + 2\sqrt{2}i}{4} = \frac{-2 + i\sqrt{2}}{2}$

83. $\frac{-6 - \sqrt{-32}}{2} = \frac{-6 - 4i\sqrt{2}}{2} = -3 - 2i\sqrt{2}$

85. $\frac{-6 - \sqrt{36 + 48}}{-4} = \frac{-6 - 2\sqrt{21}}{-4} = \frac{3 + \sqrt{21}}{2}$

87. $(3 - 5i)(3 + 5i) = 3^2 + 5^2 = 34$

89. $(3 - 5i) + (3 + 5i) = 6$

91. $\dfrac{3-5i}{3+5i}\cdot\dfrac{3-5i}{3-5i}=\dfrac{9-15i-15i-25}{34}=$
$-\dfrac{16}{34}-\dfrac{30}{34}i=-\dfrac{8}{17}-\dfrac{15}{17}i$

93. $(6-2i)-(7-3i)=6-7-2i+3i=-1+i$

95. $i^5(i^2-3i)=i(-1-3i)=-i+3=3-i$

97. If r is the remainder when n is divided by 4, then $i^n=i^r$. The possible values of r are $0,1,2,3$ and for i^r they are $1,i,-1,-i$, respectively.

99. Note, $w+\overline{w}=(a+bi)+(a-bi)=2a$ is a real number and $w-\overline{w}=(a+bi)-(a-bi)=2bi$ is an imaginary number.

When a complex number is added to its complex conjugate the sum is twice the real part of the complex number. When the complex conjugate of a complex number is subtracted from the complex number, the difference is an imaginary number.

101. The reciprocal is $\dfrac{1}{a+bi}=\dfrac{a-bi}{(a+bi)(a-bi)}=$
$\dfrac{a-bi}{a^2+b^2}=\dfrac{a}{a^2+b^2}-\dfrac{b}{a^2+b^2}i$

103. Since $x^2-1\neq 0$, the domain is $(-\infty,-1)\cup(-1,1)\cup(1,\infty)$

105. $\dfrac{\frac{3}{6x}+\frac{4}{6x}}{\frac{6}{6x}+\frac{5x}{6x}}=\dfrac{\frac{7}{6x}}{\frac{6+5x}{6x}}=\dfrac{7}{6+5x}$

107. $\dfrac{1}{1.2}=\dfrac{1}{\frac{12}{10}}=\dfrac{10}{12}=\dfrac{5}{6}$

109. 2178

Review Exercises

1. False, since $\sqrt{2}$ is an irrational number.

3. False, since -1 is a negative number.

5. False, since terminating decimal numbers are rational numbers.

7. False, since $\{1,2,3,...\}$ is the set of natural numbers.

9. False, since $\dfrac{1}{3}=0.3333...$

11. False, since the additive inverse of 0.5 is -0.5

13. $-3x-12+20x=17x-12$

15. $\dfrac{2x}{10}+\dfrac{x}{10}=\dfrac{3x}{10}$

17. $\dfrac{3(x-2)}{3\cdot 3}=\dfrac{x-2}{3}$

19. $\dfrac{-6}{8}=-\dfrac{3}{4}$

21. $3-5=-2$ **23.** $|-4|=4$

25. $8-18\div 3+5=8-6+5=7$

27. $3\cdot 3\div 6+3^3=9\div 6+27=1.5+27=28.5$

29. 625 **31.** $4+40=44$

33. $\dfrac{1}{2}+1=\dfrac{3}{2}$

35. $\dfrac{-2^3\cdot 2^1}{3^1}=-\dfrac{16}{3}$

37. $(8^2)^{-1/3}=64^{-1/3}=\dfrac{1}{4}$

39. $5x^2$

41. 11

43. $\sqrt{4\cdot 7s^2\cdot s}=2s\sqrt{7s}$

45. $\sqrt[3]{-1000(2)}=-10\sqrt[3]{2}$

47. $\sqrt{\dfrac{5}{2a}}\cdot\sqrt{\dfrac{2a}{2a}}=\dfrac{\sqrt{10a}}{2a}$

49. $\sqrt[3]{\dfrac{2}{5}}\sqrt[3]{\dfrac{25}{25}}=\dfrac{\sqrt[3]{50}}{5}$

51. $3n\sqrt{2n}+5n\sqrt{2n}=8n\sqrt{2n}$

53. $\dfrac{2\sqrt{3}}{\sqrt{3}-1}\cdot\dfrac{\sqrt{3}+1}{\sqrt{3}+1}=\dfrac{6+2\sqrt{3}}{3-1}=3+\sqrt{3}$

55. $\dfrac{\sqrt{6}}{2\sqrt{2}+3\sqrt{2}}=\dfrac{\sqrt{6}}{5\sqrt{2}}=\dfrac{\sqrt{12}}{10}=\dfrac{\sqrt{3}}{5}$

57. 320,000,000

59. 0.000185 **61.** 5.6×10^{-5} **63.** 2.34×10^{6}

65. $125\cdot 10^{18}=1.25\times 10^{20}$

67. $\dfrac{(8\times 10^2)^2(10^{-5})^{-3}}{(2\times 10^6)^3(2\times 10^{-5})} =$
$\dfrac{(64\times 10^4)(10^{15})}{(8\times 10^{18})(2\times 10^{-5})} =$
$\dfrac{64}{8(2)}\times 10^6 = 4\times 10^6$

69. $3x^2 - x^2 - x + 2x - 2 - 5 = 2x^2 + x - 7$

71. $-4x^4 - 3x^3 + x - x^4 + 6x^3 + 2x =$
$-5x^4 + 3x^3 + 3x$

73. $(3a^2 - 2a + 5)a - (3a^2 - 2a + 5)\cdot 2 =$
$3a^3 - 2a^2 + 5a - 6a^2 + 4a - 10 =$
$3a^3 - 8a^2 + 9a - 10$

75. $b^2 - 6by + 9y^2$

77. $3t^2 - 7t - 6$

79. $-\dfrac{35y^5}{7y^2} = -5y^3$

81. $9 - 2 = 7$

83. $1 + 2\sqrt{3} + \sqrt{3}^2 = 1 + 2\sqrt{3} + 3 = 4 + 2\sqrt{3}$

85. $2^2\sqrt{5}^2 + 4\sqrt{15} + \sqrt{3}^2 = 20 + 4\sqrt{15} + 3 =$
$23 + 4\sqrt{15}$

87. Quotient $x^2 + 4x - 1$, remainder 1, since

$$\begin{array}{r|l} & x^2 + 4x - 1 \\ x - 2 & x^3 + 2x^2 - 9x + 3 \\ & x^3 - 2x^2 \\ & \quad 4x^2 - 9x \\ & \quad 4x^2 - 8x \\ & \qquad -x + 3 \\ & \qquad -x + 2 \\ & \qquad\qquad 1 \end{array}$$

89. Quotient $3x + 2$, remainder 4, since

$$\begin{array}{r|l} & 3x + 2 \\ 2x - 1 & 6x^2 + x + 2 \\ & 6x^2 - 3x \\ & \quad 4x + 2 \\ & \quad 4x - 2 \\ & \qquad 4 \end{array}$$

91. $x - 2 + \dfrac{1}{x+2}$ since

$$\begin{array}{r|l} & x - 2 \\ x + 2 & x^2 + 0x - 3 \\ & x^2 + 2x \\ & \quad -2x - 3 \\ & \quad -2x - 4 \\ & \qquad 1 \end{array}$$

93. $2 + \dfrac{13}{x-5}$ since

$$\begin{array}{r|l} & 2 \\ x - 5 & 2x + 3 \\ & 2x - 10 \\ & \quad 13 \end{array}$$

95. $6x(x^2 - 1) = 6x(x-1)(x+1)$ **97.** $(3h + 4t)^2$

99. $(t + y)(t^2 - ty + y^2)$

101. $x^2(x+3) - 9(x+3) = (x^2 - 9)(x+3) =$
$(x-3)(x+3)(x+3) = (x-3)(x+3)^2$

103. $t^6 - 1 = (t^3 - 1)(t^3 + 1) =$
$(t-1)(t^2 + t + 1)(t+1)(t^2 - t + 1)$

105. $(6x + 5)(3x - 4)$

107. $ab(a^2 + 3a - 18) = ab(a+6)(a-3)$

109. $x^2(2x + y) - (2x + y) = (x^2 - 1)(2xy + 1) =$
$(x-1)(x+1)(2x+y)$

111. $\dfrac{2x+6}{x+3} = \dfrac{2(x+3)}{x+3} = 2$

113.
$\dfrac{(x-1)(x+4)}{(x-2)(x+4)} - \dfrac{(x+3)(x-2)}{(x+4)(x-2)} =$
$\dfrac{x^2 + 3x - 4 - (x^2 + x - 6)}{(x-2)(x+4)} = \dfrac{2x+2}{(x-2)(x+4)}$

115. $\dfrac{(x-3)(x+3)}{x+3}\cdot\dfrac{1}{-2(x-3)} = -\dfrac{1}{2}$

117. $\dfrac{c^7}{a^6b^2}\cdot\dfrac{a^2b^6c^{10}}{a^4b^3} = \dfrac{bc^{17}}{a^8}$

119. $\dfrac{1}{(x-2)(x+2)} + \dfrac{3(x+2)}{(x-2)(x+2)} = \dfrac{3x+7}{x^2-4}$

121. $\dfrac{5x}{30x^2} - \dfrac{21}{30x^2} = \dfrac{5x-21}{30x^2}$

123. $\dfrac{(a-5)(a+5)}{(a-5)(a+1)}\cdot\dfrac{(a-1)(a+1)}{2(a+5)} = \dfrac{a-1}{2}$

125. $\dfrac{(x-4)(x+4)}{(x+4)(x+1)} \cdot \dfrac{(x+1)(x^2-x+1)}{-2(x-4)} =$
$\dfrac{-x^2+x-1}{2}$

127. $\dfrac{a-2}{(a+5)(a+1)} + \dfrac{2a+1}{(a-1)(a+1)} =$
$\dfrac{(a-2)(a-1)}{(a+5)(a+1)(a-1)} +$
$\dfrac{(2a+1)(a+5)}{(a-1)(a+1)(a+5)} =$
$\dfrac{(a^2-3a+2)+(2a^2+11a+5)}{(a+5)(a+1)(a-1)} =$
$\dfrac{3a^2+8a+7}{(a+1)(a-1)(a+5)}$

129. $\dfrac{\left(\frac{5}{2x}-\frac{3}{4x}\right)(4x)}{\left(\frac{1}{2}-\frac{2}{x}\right)(4x)} = \dfrac{10-3}{2x-8} = \dfrac{7}{2x-8}$

131. $\dfrac{\left(\frac{1}{y^2-2}-3\right)(y^2-2)}{\left(\frac{5}{y^2-2}+4\right)(y^2-2)} = \dfrac{1-3(y^2-2)}{5+4(y^2-2)} =$
$\dfrac{-3y^2+7}{4y^2-3}$

133. $\dfrac{(a^{-2}-b^{-3})\,a^2b^3}{(a^{-1}b^{-1})\,a^2b^3} = \dfrac{b^3-a^2}{ab^2}$

135. $(p^{-1}+pq^{-3})\,\dfrac{pq^3}{pq^3} = \dfrac{q^3+p^2}{pq^3}$

137. $P(2) = (2)^3 - 3(2)^2 + 2 - 9 = 8 - 3(4) - 7 = -11$
139. $P(0) = -9$

141. $R(-1) = \dfrac{3(-1)-1}{2(-1)-9} = \dfrac{-4}{-11} = \dfrac{4}{11}$

143. $R(50) = \dfrac{3(50)-1}{2(50)-9} = \dfrac{149}{91}$

145. $3 - 4 - 7i + 6i = -1 - i$

147. $16 - 40i - 25 = -9 - 40i$

149. $2 + 6i - 6i + 18 = 20$

151. $\dfrac{2-3i}{i} \cdot \dfrac{-i}{-i} = \dfrac{-2i-3}{1} = -3-2i$

153. $\dfrac{1-i}{2+i} \cdot \dfrac{2-i}{2-i} = \dfrac{1-3i}{5} = \dfrac{1}{5} - \dfrac{3}{5}i$

155. $\dfrac{1+i}{2-3i} \cdot \dfrac{2+3i}{2+3i} = \dfrac{-1+5i}{13} = -\dfrac{1}{13} + \dfrac{5}{13}i$

157. $\dfrac{6+2i\sqrt{2}}{2} = 3 + i\sqrt{2}$

159. $\dfrac{-6+\sqrt{-20}}{-8} = \dfrac{-6+2i\sqrt{5}}{-8} = \dfrac{3}{4} - \dfrac{\sqrt{5}}{4}i$

161. $i^{32}i^2 + i^{16}i^3 = (1)(-1) + (1)(-i) = -1 - i$

163. **a)** 300 ft

b) $\dfrac{130^2}{32} - \dfrac{125^2}{32} \approx 39.8$ feet

165. The number of hydrogen atoms in one kilogram of hydrogen is

$$\frac{1000}{1.7 \times 10^{-24}} \approx 5.9 \times 10^{26}.$$

167. $|-2.35 - 8.77| = |-11.12| = 11.12$

169. The parts of the lawn Howard and Will can mow in 2 hours are $\frac{2}{6}$ and $\frac{2}{4}$, respectively. In 2 hours, the portion they will mow together is $\frac{2}{6} + \frac{2}{4} = \frac{5}{6}$.

171. Since there will be no 1's left after the 162nd house, the first house that cannot be numbered correctly is 163.

Chapter P Test

1. All **2.** $\{-1.22, -1, 0, 2, 10/3\}$

3. $\{-\pi, -\sqrt{3}, \sqrt{5}, 6.020020002...\}$ **4.** $\{0, 2\}$

5. $|6-25| - 6 = |-19| - 6 = 13$ **6.** $8^{1/3} = 2$

7. $-\dfrac{1}{(27^{1/3})^2} = -\dfrac{1}{3^2} = -\dfrac{1}{9}$

8. $\dfrac{9+12+9}{(-5)(6)} = \dfrac{30}{-30} = -1$

9. $6x^5y^7$ **10.** $2x^2 - 2x^2 = 0$

11. $\dfrac{(ab(b+a))^2}{a^2b^2} = \dfrac{a^2b^2(b+a)^2}{a^2b^2} = a^2+2ab+b^2$

12. $\dfrac{-8a^{-3}b^{18}}{2^{-2}a^{-6}b^8} = -32a^3b^{10}$

13. $3\sqrt{3}-2\sqrt{2}+4\sqrt{2} = 3\sqrt{3}+2\sqrt{2}$

14.
$$\frac{2\sqrt{2}}{\sqrt{6}-\sqrt{2}}\cdot\frac{\sqrt{6}+\sqrt{2}}{\sqrt{6}+\sqrt{2}} = \frac{2\sqrt{12}+4}{4} =$$
$$\frac{2(2\sqrt{3})+4}{4} = \sqrt{3}+1$$

15.
$$\frac{1}{x\sqrt[3]{4x}} = \frac{\sqrt[3]{2x^2}}{x\sqrt[3]{4x}\sqrt[3]{2x^2}} = \frac{\sqrt[3]{2x^2}}{2x^2}$$

16. $\sqrt{4(3)x^2xy^8y\cdot 1} = 2xy^4\sqrt{3xy}$

17. $16-24i-9 = 7-24i$

18. $\dfrac{2-i}{3+i}\cdot\dfrac{3-i}{3-i} = \dfrac{5-5i}{10} = \dfrac{1}{2}-\dfrac{1}{2}i$

19. $i^4i^2 - i^{32}i^3 = (1)(-1)-(1)(-i) = -1+i$

20. $2i\sqrt{2}(i\sqrt{2}+\sqrt{6}) = -4+2i\sqrt{12} = -4+4i\sqrt{3}$

21. $3x^3+3x^2-12x$

22. $-x^2+3x-4-4x^2+6x-9 = -5x^2+9x-13$

23. $x(x^2-2x-1)+3(x^2-2x-1) =$
$x^3-2x^2-x+3x^2-6x-3 = x^3+x^2-7x-3$

24. $\dfrac{(2h-1)(4h^2+2h+1)}{2h-1} = 4h^2+2h+1$

25. $x^2-6xy-27y^2$

26.
$$\begin{array}{r|l} & x^2+2x+2 \\ \hline x-2 & x^3+0x^2-2x-4 \\ & \underline{x^3-2x^2} \\ & \quad 2x^2-2x \\ & \quad \underline{2x^2-4x} \\ & \qquad 2x-4 \\ & \qquad \underline{2x-4} \\ & \qquad\quad 0 \end{array}$$

Then $\dfrac{x^3-2x-4}{x-2} = x^2+2x+2.$

27. $9x^2-48x+64$ **28.** $4t^8-1$

29.
$$\frac{x(x^2-5x+6)}{2x(x-3)}\cdot\frac{4(x^3+8)}{2x(x^2-4)} =$$
$$\frac{x(x-3)(x-2)}{2x(x-3)}\cdot\frac{4(x+2)(x^2-2x+4)}{2x(x-2)(x+2)} =$$
$$\frac{x^2-2x+4}{x}$$

30.
$$\frac{(x+5)(x+4)}{(x-3)(x-1)(x+4)}+$$
$$\frac{(x-1)^2}{(x+4)(x-3)(x-1)} =$$
$$\frac{(x^2+9x+20)+(x^2-2x+1)}{(x+4)(x-3)(x-1)} =$$
$$\frac{2x^2+7x+21}{(x+4)(x-3)(x-1)}$$

31.
$$\frac{a-1}{(2a-3)(2a+3)}+\frac{a-2}{2a-3} =$$
$$\frac{a-1}{(2a-3)(2a+3)}+\frac{(a-2)(2a+3)}{(2a-3)(2a+3)} =$$
$$\frac{(a-1)+(2a^2-a-6)}{(2a-3)(2a+3)} = \frac{2a^2-7}{4a^2-9}$$

32.
$$\frac{\dfrac{1}{2a^2b}-2a}{\dfrac{1}{4ab^3}+\dfrac{1}{3b}}\cdot\frac{12a^2b^3}{12a^2b^3} = \frac{6b^2-24a^3b^3}{3a+4a^2b^2}$$

33. $a(x^2-11x+18) = a(x-9)(x-2)$

34. $m(m^4-1) = m(m^2-1)(m^2+1) =$
$m(m-1)(m+1)(m^2+1)$

35. $(3x-1)(x+5)$

36. $bx(x-3)+w(x-3) = (bx+w)(x-3)$

37. The U.S. per capita income is
$$\frac{1.304\times 10^{13}}{3.131\times 10^8} \approx \$41,648$$

38. The annual appreciation rate is
$$\left(\frac{2.5}{1.3}\right)^{1/3} - 1 \approx 0.24355 \text{ or } 24.4\%.$$

39. The altitude is $A(2) = -64+240 = 176$ feet.

For Thought

1. True, since $5(1) = 6 - 1$.

2. True, since $x = 3$ is the solution to both equations.

3. False, -2 is not a solution of the first equation since $\sqrt{-2}$ is not a real number.

4. True, since $x - x = 0$.

5. False, $x = 0$ is the solution. **6.** True

7. False, since $|x| = -8$ has no solution.

8. False, $\dfrac{x}{x-5}$ is undefined at $x = 5$.

9. False, since we should multiply by $-\dfrac{3}{2}$.

10. False, $0 \cdot x + 1 = 0$ has no solution.

1.1 Exercises

1. equation

3. equivalent

5. identity

7. conditional equation

9. No, since $2(3) - 4 = 2 \neq 9$.

11. Yes, since $(-4)^2 = 16$.

13. Since $3x = 5$, the solution set is $\left\{\dfrac{5}{3}\right\}$.

15. Since $-3x = 6$, the solution set is $\{-2\}$.

17. Since $14x = 7$, the solution set is $\left\{\dfrac{1}{2}\right\}$.

19. Since $7 + 3x = 4x - 4$, the solution set is $\{11\}$.

21. Since $x = -\dfrac{4}{3} \cdot 18$, the solution set is $\{-24\}$.

23. Multiplying by 6 we get

$$\begin{aligned} 3x - 30 &= -72 - 4x \\ 7x &= -42. \end{aligned}$$

The solution set is $\{-6\}$.

25. Multiply both sides of the equation by 12.

$$\begin{aligned} 18x + 4 &= 3x - 2 \\ 15x &= -6 \\ x &= -\frac{2}{5}. \end{aligned}$$

The solution set is $\left\{-\dfrac{2}{5}\right\}$.

27. Note, $3(x - 6) = 3x - 18$ is true by the distributive law. It is an identity and the solution set is R.

29. Note, $5x = 4x$ is equivalent to $x = 0$. The latter equation is conditional whose solution set is $\{0\}$.

31. Equivalently, we get $2x + 6 = 3x - 3$ or $9 = x$. The latter equation is conditional whose solution set is $\{9\}$.

33. Using the distributive property, we find

$$\begin{aligned} 3x - 18 &= 3x + 18 \\ -18 &= 18. \end{aligned}$$

The equation is inconsistent and the solution set is $\emptyset$.

35. An identity and the solution set is $\{x | x \neq 0\}$.

37. Multiplying by $2(w - 1)$, we get

$$\begin{aligned} \frac{1}{w-1} - \frac{1}{2w-2} &= \frac{1}{2w-2} \\ 2 - 1 &= 1. \end{aligned}$$

An identity and the solution set is $\{w | w \neq 1\}$

39. Multiply by $6x$.

$$\begin{aligned} 6 - 2 &= 3 + 1 \\ 4 &= 4 \end{aligned}$$

An identity with solution set $\{x | x \neq 0\}$.

41. Multiply by $3(z - 3)$.

$$\begin{aligned} 3(z + 2) &= -5(z - 3) \\ 3z + 6 &= -5z + 15 \\ 8z &= 9 \end{aligned}$$

A conditional equation with solution set $\left\{\dfrac{9}{8}\right\}$.

43. Multiplying by $(x-3)(x+3)$.

$$\begin{aligned}(x+3)-(x-3) &= 6\\ 6 &= 6\end{aligned}$$

An identity with solution set $\{x|x \neq 3, x \neq -3\}$.

45. Multiply by $(y-3)$.

$$\begin{aligned}4(y-3)+6 &= 2y\\ 4y-6 &= 2y\\ y &= 3\end{aligned}$$

Since division by zero is not allowed, $y = 3$ does not satisfy the original equation. We have an inconsistent equation and so the solution set is $\emptyset$.

47. Multiply by $t+3$.

$$\begin{aligned}t+4t+12 &= 2\\ 5t &= -10\end{aligned}$$

A conditional equation with solution set $\{-2\}$.

49. Since $-4.19 = 0.21x$ and $\dfrac{-4.19}{0.21} \approx -19.952$, the solution set is approximately $\{-19.952\}$.

51. Divide by 0.06.

$$\begin{aligned}x-3.78 &= \frac{1.95}{0.06}\\ x &= 32.5+3.78\\ x &= 36.28\end{aligned}$$

The solution set is $\{36.28\}$.

53.

$$\begin{aligned}2a &= -1-\sqrt{17}\\ a &= \frac{-1-\sqrt{17}}{2}\\ a &\approx \frac{-1-4.1231}{2}\\ a &\approx -2.562\end{aligned}$$

The solution set is approximately $\{-2.562\}$.

55.

$$\begin{aligned}0.001 &= 3(y-0.333)\\ 0.001 &= 3y-0.999\\ 1 &= 3y\\ \frac{1}{3} &= y\end{aligned}$$

The solution set is $\left\{\dfrac{1}{3}\right\}$.

57. Factoring x, we get

$$\begin{aligned}x\left(\frac{1}{0.376}+\frac{1}{0.135}\right) &= 2\\ x(2.6596+7.4074) &\approx 2\\ 10.067x &\approx 2\\ x &\approx 0.199\end{aligned}$$

The solution set is approximately $\{0.199\}$.

59.

$$\begin{aligned}x^2+6.5x+3.25^2 &= x^2-8.2x+4.1^2\\ 14.7x &= 4.1^2-3.25^2\\ 14.7x &= 16.81-10.5625\\ 14.7x &= 6.2475\\ x &= 0.425\end{aligned}$$

The solution set is $\{0.425\}$.

61.

$$\begin{aligned}(2.3\times10^6)x &= 1.63\times10^4-8.9\times10^5\\ x &= \frac{1.63\times10^4-8.9\times10^5}{2.3\times10^6}\\ x &\approx -0.380\end{aligned}$$

The solution set is approximately $\{-0.380\}$.

63. Solution set is $\{\pm 8\}$.

65. Since $x-4 = \pm 8$, we get $x = 4 \pm 8$. The solution set is $\{-4, 12\}$.

67. Since $x-6 = 0$, we get $x = 6$. The solution set is $\{6\}$.

69. Since the absolute value of a real number is not a negative number, the equation $|x+8| = -3$ has no solution. The solution set is $\emptyset$.

71. Since $2x - 3 = 7$ or $2x - 3 = -7$, we get $2x = 10$ or $2x = -4$. The solution set is $\{-2, 5\}$.

73. Multiplying $\frac{1}{2}|x - 9| = 16$ by 2 we obtain $|x - 9| = 32$. Then $x - 9 = 32$ or $x - 9 = -32$. The solution set is $\{-23, 41\}$.

75. Since $2|x + 5| = 10$, we find $|x + 5| = 5$. Then $x + 5 = \pm 5$ or $x = \pm 5 - 5$.

The solution set is $\{-10, 0\}$.

77. Dividing $8|3x - 2| = 0$ by 8, we obtain $|3x - 2| = 0$. Then $3x - 2 = 0$ and the solution set is $\{2/3\}$.

79. Subtracting 7, we find $2|x| = -1$ and $|x| = -\frac{1}{2}$. Since an absolute value is not equal to a negative number, the solution set is $\emptyset$.

81. Since $0.95x = 190$, the solution set is $\{200\}$.

83.

$$\begin{aligned} 0.1x - 0.05x + 1 &= 1.2 \\ 0.05x &= 0.2 \end{aligned}$$

The solution set is $\{4\}$.

85. Simplifying $x^2 + 4x + 4 = x^2 + 4$, we obtain $4x = 0$. The solution set is $\{0\}$.

87. Since $|2x - 3| = |2x + 5|$, we get $2x - 3 = 2x + 5$ or $2x - 3 = -2x - 5$. Solving for x, we find $-3 = 5$ (an inconsistent equation) or $4x = -2$. The solution set is $\{-1/2\}$.

89. Multiply by 4.

$$\begin{aligned} 2x + 4 &= x - 6 \\ x &= -10 \end{aligned}$$

The solution set is $\{-10\}$.

91. Multiply by 30.

$$\begin{aligned} 15(y - 3) + 6y &= 90 - 5(y + 1) \\ 15y - 45 + 6y &= 90 - 5y - 5 \\ 26y &= 130 \end{aligned}$$

The solution set is $\{5\}$.

93. Since $7|x + 6| = 14$, $|x + 6| = 2$. Then $x + 6 = 2$ or $x + 6 = -2$. The solution set is $\{-4, -8\}$.

95. Since $-4|2x-3| = 0$, we get $|2x-3| = 0$. Then $2x - 3 = 0$ and the solution set is $\{3/2\}$.

97. Since $-5|5x + 1| = 4$, we find $|5x + 1| = -4/5$. Since the absolute value is not a negative number, the solution set is $\emptyset$.

99. Multiply by $(x - 2)(x + 2)$.

$$\begin{aligned} 3(x + 2) + 4(x - 2) &= 7x - 2 \\ 3x + 6 + 4x - 8 &= 7x - 2 \\ 7x - 2 &= 7x - 2 \end{aligned}$$

An identity with solution set $\{x \mid x \neq 2, x \neq -2\}$.

101. Multiply $(x + 3)(x - 2)$ to both sides of

$$\frac{4}{x + 3} + \frac{3}{x - 2} = \frac{7x + 1}{(x + 3)(x - 2)}.$$

Then we find

$$\begin{aligned} 4(x - 2) + 3(x + 3) &= 7x + 1 \\ 4x - 8 + 3x + 9 &= 7x + 1 \\ 7x + 1 &= 7x + 1. \end{aligned}$$

An identity and the solution set is $\{x \mid x \neq 2 \text{ and } x \neq -3\}$.

103. Multiply by $(x - 3)(x - 4)$.

$$\begin{aligned} (x - 4)(x - 2) &= (x - 3)^2 \\ x^2 - 6x + 8 &= x^2 - 6x + 9 \\ 8 &= 9 \end{aligned}$$

An inconsistent equation and so the solution set is $\emptyset$.

105. **a)** About 1995

b) Increasing

c) Let $y = 0.95$. Solving for x, we find

$$\begin{aligned} 0.95 &= 0.0102x + 0.644 \\ \frac{0.95 - 0.644}{0.0102} &= x \\ 30 &= x. \end{aligned}$$

In the year 2020 ($= 1990 + 30$), 95% of mothers will be in the labor force.

107. Since $B = 21,000 - 0.15B$, we obtain $1.15B = 21,000$ and the bonus is

$$B = \frac{21,000}{1.15} = \$18,260.87.$$

109. Rewrite the left-hand side as a sum.

$$\begin{aligned} 10,000 + \frac{500,000,000}{x} &= 12,000 \\ \frac{500,000,000}{x} &= 2,000 \\ 500,000,000 &= 2000x \\ 250,000 &= x \end{aligned}$$

Thus, 250,000 vehicles must be sold.

111. The third side of the triangle is $\sqrt{3}$ by the Pythagorean Theorem. Then draw radial lines from the center of the circle to each of the three sides. Consider the square with side r that is formed with the $90°$ angle of the triangle. Then the side of length $\sqrt{3}$ is divided into two segments of length r and $\sqrt{3} - r$. Similarly, the side of length 1 is divided into segments of length r and $1 - r$.

Note, the center of the circle lies on the bisectors of the angles of the triangle. Using congruent triangles, the hypotenuse consists of line segments of length $\sqrt{3} - r$ and $1 - r$. Since the hypotenuse is 2, we have

$$\begin{aligned} (\sqrt{3} - r) + (1 - r) &= 2 \\ \sqrt{3} - 1 &= 2r. \end{aligned}$$

Thus, the radius is

$$r = \frac{\sqrt{3} - 1}{2}.$$

115. $-3, 0, 7$

117. $-5x + 20 - 24 + 12x = 7x - 4$

119. $\dfrac{\sqrt{5}}{\sqrt{3}} \cdot \dfrac{\sqrt{3}}{\sqrt{3}} = \dfrac{\sqrt{15}}{3}$

121. \$9, \$99, \$999, \$9,999, \$99,999, \$999,999, \$9,999,999 \$99,999,999, \$999,999,999

For Thought

1. False, $P(1 + rt) = S$ implies $P = \dfrac{S}{1 + rt}$.

2. False, since the perimeter is twice the sum of the length and width. **3.** False, since $n + 1$ and $n + 3$ are even integers if n is odd.

4. True **5.** True, since $x + (-3 - x) = -3$.

6. False, since $P = 2S$

7. False, for if the house sells for x dollars then

$$\begin{aligned} x - 0.09x &= 100,000 \\ 0.91x &= 100,000 \\ x &= \$109,890.11. \end{aligned}$$

8. True

9. False, a correct equation is $4(x - 2) = 3x - 5$.

10. False, since 9 and $x + 9$ differ by x.

1.2 Exercises

1. formula

3. uniform

5. $r = \dfrac{I}{Pt}$

7. Since $F - 32 = \dfrac{9}{5}C$, $C = \dfrac{5}{9}(F - 32)$.

9. Since $2A = bh$, we get $b = \dfrac{2A}{h}$.

11. Since $By = C - Ax$, we obtain $y = \dfrac{C - Ax}{B}$.

13. Multiplying by $RR_1R_2R_3$, we find

$$R_1R_2R_3 = RR_2R_3 + RR_1R_3 + RR_1R_2$$

$$\begin{aligned} R_1R_2R_3 - RR_1R_3 - RR_1R_2 &= RR_2R_3 \\ R_1(R_2R_3 - RR_3 - RR_2) &= RR_2R_3. \end{aligned}$$

Then $R_1 = \dfrac{RR_2R_3}{R_2R_3 - RR_3 - RR_2}$.

15. Since $a_n - a_1 = (n-1)d$, we obtain

$$\begin{aligned} n-1 &= \frac{a_n - a_1}{d} \\ n &= \frac{a_n - a_1}{d} + 1 \\ n &= \frac{a_n - a_1 + d}{d}. \end{aligned}$$

17. Since $S = \dfrac{a_1(1-r^n)}{1-r}$, we obtain

$$\begin{aligned} a_1(1-r^n) &= S(1-r) \\ a_1 &= \frac{S(1-r)}{1-r^n}. \end{aligned}$$

19. Multiplying by 2.37, one finds

$$\begin{aligned} 2.4(2.37) &= L + 2D - F\sqrt{S} \\ 5.688 - L + F\sqrt{S} &= 2D \end{aligned}$$

and $D = \dfrac{5.688 - L + F\sqrt{S}}{2}$.

21. $R = D/T$

23. Since $LW = A$, we have $W = A/L$.

25. $r = d/2$

27. By using the formula $I = Prt$, one gets

$$\begin{aligned} 51.30 &= 950r \cdot 1 \\ 0.054 &= r. \end{aligned}$$

The simple interest rate is 5.4%.

29. Since $D = RT$, we find

$$\begin{aligned} 5570 &= 2228 \cdot T \\ 2.5 &= T. \end{aligned}$$

The surveillance takes 2.5 hours.

31. Note, $C = \dfrac{5}{9}(F - 32)$. If $F = 23^oF$, then

$C = \dfrac{5}{9}(23 - 32) = -5^o\text{C}.$

33. If x is the cost of the car before taxes, then

$$\begin{aligned} 1.08x &= 40,230 \\ x &= \$37,250. \end{aligned}$$

35. Let S be the saddle height and let L be the inside measurement.

$$\begin{aligned} S &= 1.09L \\ 37 &= 1.09L \\ \frac{37}{1.09} &= L \\ 33.9 &\approx L \end{aligned}$$

The inside leg measurement is 33.9 inches.

37. Let x be the sale price.

$$\begin{aligned} 1.1x &= 50,600 \\ x &= \frac{50,600}{1.1} \\ x &= \$46,000 \end{aligned}$$

39. Let x be the amount of her game-show winnings.

$$\begin{aligned} 0.14\frac{x}{3} + 0.12\frac{x}{6} &= 4000 \\ 6\left(0.14\frac{x}{3} + 0.12\frac{x}{6}\right) &= 24,000 \\ 0.28x + 0.12x &= 24,000 \\ 0.40x &= 24,000 \\ x &= \$60,000 \end{aligned}$$

Her winnings is $60,000.

41. If x is the length of the shorter piece in feet, then the length of the longer side is $2x + 2$. Then we obtain

$$\begin{aligned} x + x + (2x + 2) &= 30 \\ 4x &= 28 \\ x &= 7. \end{aligned}$$

The length of each shorter piece is 7 ft and the longer piece is $(2 \cdot 7 + 2)$ or 16 ft.

43. If x is the length of the side of the larger square lot then $2x$ is the amount of fencing needed to divide the square lot into four smaller lots. The solution to $4x + 2x = 480$ is $x = 80$. The side of the larger square lot is 80 feet and its area is 6400 ft^2.

45. Let d denote the distance from Fairbanks to Coldfoot, which is the same distance from Coldfoot to Deadhorse.

$$\begin{aligned} \frac{d}{50} + \frac{d}{40} &= 11.25 \text{ hr} \\ 90d &= 11.25(2000) \\ d &= 250 \text{ miles} \end{aligned}$$

47. Note, Bobby will complete the remaining 8 laps in $\frac{8}{90}$ of an hour. If Ricky is to finish at the same time as Bobby, then Ricky's average speed s over 10 laps must satisfy $\frac{10}{s} = \frac{8}{90}$. $\left(\text{Note: } time = \frac{distance}{speed}\right)$. The equation is equivalent to $900 = 8s$. Thus, Ricky's average speed must be 112.5 mph.

49. Let d be the halfway distance between San Antonio and El Paso, and let s be the speed in the last half of the trip. Junior took $\frac{d}{80}$ hours to get to the halfway point and the last half took $\frac{d}{s}$ hours to drive. Since the total distance is $2d$ and $distance = rate \times time$,

$$\begin{aligned} 2d &= 60\left(\frac{d}{80} + \frac{d}{s}\right) \\ 160sd &= 60\,(sd + 80d) \\ 160sd &= 60sd + 4800d \\ 100sd &= 4800d \\ 100d(s - 48) &= 0. \end{aligned}$$

Since $d \neq 0$, the speed for the last half of the trip was $s = 48$ mph.

51. If x is the part of the start-up capital invested at 5% and $x + 10,000$ is the part invested at 6%, then

$$\begin{aligned} 0.05x + 0.06(x + 10,000) &= 5880 \\ 0.11x + 600 &= 5880 \\ 0.11x &= 5280 \\ x &= 48,000. \end{aligned}$$

Norma invested \$48,000 at 5% and \$58,000 at 6% for a total start-up capital of \$106,000.

53. Let x and $1500 - x$ be the number of employees from the Northside and Southside, respectively. Then

$$\begin{aligned} (0.05)x + 0.80(1500 - x) &= 0.5(1500) \\ 0.05x + 1200 - 0.80x &= 750 \\ 450 &= 0.75x \\ 600 &= x. \end{aligned}$$

There were 600 and 900 employees at the Northside and Southside, respectively.

55. Let x be the number of hours it takes both combines working together to harvest an entire wheat crop.

	rate
old	1/72
new	1/48
combined	$1/x$

Then $\frac{1}{72} + \frac{1}{48} = \frac{1}{x}$. Multiply both sides by $144x$ and get $2x + 3x = 144$. The solution is $x = 28.8$ hr, which is the time it takes both combines to harvest the entire wheat crop.

57. Let t be the number of hours since 8:00 a.m.

	rate	time	work completed
Batman	1/8	$t-2$	$(t-2)/8$
Robin	1/12	t	$t/12$

$$\begin{aligned} \frac{t-2}{8} + \frac{t}{12} &= 1 \\ 24\left(\frac{t-2}{8} + \frac{t}{12}\right) &= 24 \\ 3(t-2) + 2t &= 24 \\ 5t - 6 &= 24 \\ t &= 6 \end{aligned}$$

At 2 p.m., all the crimes have been cleaned up.

59. Since there are 5280 feet to a mile and the circumference of a circle is $C = 2\pi r$, the radius r of the race track is $r = \dfrac{5280}{2\pi}$.

Since the length of a side of the square plot is twice the radius, the area of the plot is

$$\left(2 \cdot \frac{5280}{2\pi}\right)^2 \approx 2,824,672.5 \text{ ft}^2.$$

Dividing this number by 43,560 results to 64.85 acres which is the acreage of the square lot.

61. The area of a trapezoid is $A = \dfrac{1}{2}h(b_1 + b_2)$.

$$\begin{aligned} 90,000 &= \frac{1}{2}h(500+300) \\ 90,000 &= 400h \\ 225 &= h \end{aligned}$$

Thus, the streets are 225 ft apart.

63. Since the volume of a circular cylinder is $V = \pi r^2 h$, we have $\dfrac{22,000}{7.5} = \pi 15^2 \cdot h$.

Solving for h, we get $h = 4.15$ ft, the depth of water in the pool.

65. Let r be the radius of the semicircular turns. Since the circumference of a circle is given by $C = 2\pi r$, we have $514 = 2\pi r + 200$. Solving for r, we get

$$r = \frac{157}{\pi} \approx 49.9747 \text{ m}.$$

Note, the width of the rectangular lot is $2r$. Then the dimension of the rectangular lot is 99.9494 m by 199.9494 m; its area is 19,984.82 m^2, which is equivalent to 1.998 hectares.

67. Let x be Lorinda's taxable income.

$$\begin{aligned} 17,442 + 0.28(x - 85,650) &= 22,549 \\ 0.28x - 23,982 &= 5,107 \\ 0.28x &= 29,089 \\ x &\approx 103,889. \end{aligned}$$

Lorinda's taxable income is \$103,889.

69. Let x be the amount of water to be added. The volume of the resulting solution is $4 + x$ liters and the amount of pure baneberry in it is $0.05(4)$ liters. Since the resulting solution is a 3% extract, we have

$$\begin{aligned} 0.03(4+x) &= 0.05(4) \\ 0.12 + 0.03x &= 0.20 \\ x &= \frac{0.08}{0.03} \\ x &= \frac{8}{3}. \end{aligned}$$

The amount of water to be added is $\dfrac{8}{3}$ liters.

71. Let x be the number of gallons of gasoline with 12% ethanol that needs to be added. The volume of the resulting solution is $500 + x$ gallons and the amount of ethanol is $(500 + x)0.1$ gallons.

$$\begin{aligned} 500(0.05) + 0.12x &= (500 + x)0.1 \\ 0.02x &= 25 \\ x &= 1250. \end{aligned}$$

The amount of gasoline with 12% ethanol to be added is 1250 gallons.

73. The costs of x pounds of dried apples is $(1.20)4x$ and the cost of $(20 - x)$ pounds of dried apricots is $4(1.80)(20 - x)$. Since the 20 lb-mixture costs \$1.68 per quarter-pound, we obtain

$$\begin{aligned} 4(1.68)(20) &= (1.20)4x + 4(1.80)(20 - x) \\ 134.4 &= 4.80x + 144 - 7.20x \\ 2.40x &= 9.6 \\ x &= 4. \end{aligned}$$

The mix needs 4 lb of dried apples and 16 lb of dried apricots.

75. Let x and $8 - x$ be the number of dimes and nickels, respectively. Since the candy bar costs 55 cents, we have $55 = 10x + 5(8 - x)$. Solving for x, we find $x = 3$. Thus, Dana has 3 dimes and 5 nickels.

77. Let x be the amount of water needed. The volume of the resulting solution is $200 + x$ ml and the amount of active ingredient in it is $0.4(200)$ or 80 ml. Since the resulting solution is a 25% extract, we have

$$\begin{aligned} 0.25(200 + x) &= 80 \\ 200 + x &= 320 \\ x &= 120. \end{aligned}$$

The amount of water needed is 120 ml.

79. Let x be the number of gallons of the stronger solution. The amount of salt in the new solution is $5(0.2) + x(0.5)$ or $(1 + 0.5x)$ lb, and the volume of new solution is $(5 + x)$ gallons. Since the new solution contains 0.3 lb of salt per gallon, we obtain

$$\begin{aligned} 0.30(5 + x) &= 1 + 0.5x \\ 1.5 + 0.3x &= 1 + 0.5x \\ 0.5 &= 0.2x. \end{aligned}$$

Then $x = 2.5$ gallons, the required amount of the stronger solution.

81. Let x be the number of hours it takes both pumps to drain the pool simultaneously.

	The part drained in 1 hr
Together	$1/x$
Large pump	$1/5$
Small pump	$1/8$

It follows that $\frac{1}{5} + \frac{1}{8} = \frac{1}{x}$. Multiplying both sides by $40x$, we find $8x + 5x = 40$. Then $x = 40/13$ hr, or about 3 hr and 5 min.

83. Let x and $20 - x$ be the number of gallons of the needed 15% and 10% alcohol solutions, respectively. Since the resulting mixture is 12% alcohol, we find

$$\begin{aligned} 0.15x + 0.10(20 - x) &= 20(0.12) \\ 0.05x &= 0.4 \\ x &= 8. \end{aligned}$$

Then 8 gallons of the 15% alcohol solution and 12 gallons of the 10% alcohol solution are needed.

85. The amount of salt is

$$10 \text{ kg} \times 1\% = 0.1 \text{ kg}.$$

Let x be the amount of salt water after the evaporation. Since 2% salt is the new concentration, we obtain $\frac{0.1}{x} = 0.02$. Then $x = 5$ kg salt water after the evaporation.

87. a) Decreasing

b) If $M = 40$, then $40 = -1.64n + 74.48$. Solving for n, we find $n \approx 21$. In the year 2021 $(= 2000 + n)$, the mortality rate is 40 per 1000 live births.

89. If h is the number of hours it will take two hikers to pick a gallon of wild berries, then

$$\begin{aligned} \frac{1}{2} + \frac{1}{2} &= \frac{1}{h} \\ 1 &= \frac{1}{h} \\ 1 &= h. \end{aligned}$$

Two hikers can pick a gallon of wild berries in 1 hr.

If m is the number of minutes it will take two mechanics to change the oil of a Saturn, then

$$\begin{aligned} \frac{1}{6} + \frac{1}{6} &= \frac{1}{m} \\ \frac{1}{3} &= \frac{1}{m} \\ m &= 3. \end{aligned}$$

Two mechanics can change the oil in 3 minutes.

If w is the number of minutes it will take 60 mechanics to change the oil, then

$$\begin{aligned} 60 \cdot \frac{1}{6} &= \frac{1}{w} \\ w &= \frac{1}{10} \text{ min} \\ w &= 6 \text{ sec}. \end{aligned}$$

So, 60 mechanics working together can change the oil in 6 sec (an unreasonable situation and answer).

91.

$$\begin{aligned} \frac{x}{2} - \frac{x}{9} &= \frac{1}{6} - \frac{1}{3} \\ \frac{7x}{18} &= -\frac{3}{18} \\ 7x &= -3 \\ x &= -\frac{3}{7} \end{aligned}$$

The solution set is $\{-3/7\}$.

93.

$$\begin{aligned} 0.999x &= 9990 \\ x &= 10,000 \end{aligned}$$

The solution set is $\{10,000\}$.

95. $\left((-27)^{1/3}\right)^{-5} = (-3)^{-5} = \dfrac{-1}{243}$

97. Let x be the rate of the current in the river, and let $2x$ be the rate that Milo and Bernard can paddle. Since the rate going upstream is x and it takes 21 hours going upstream, the distance going upstream is $21x$.

Note, the rate going downstream is $3x$. Then the time going downstream is

$$\text{time} = \frac{\text{distance}}{\text{rate}} = \frac{21x}{3x} = 7 \text{ hours.}$$

Thus, in order to meet Vince at 5pm, Milo and Bernard must start their return trip at 10am.

For Thought

1. False, the point $(2, -3)$ is in Quadrant IV.

2. False, the point $(4, 0)$ does not belong to any quadrant.

3. False, since the distance is $\sqrt{(a-c)^2 + (b-d)^2}$.

4. False, since $Ax + By = C$ is a linear equation.

5. True, since the x-intercept can be obtained by replacing y by 0.

6. False, since $\sqrt{7^2 + 9^2} = \sqrt{130} \approx 11.4$

7. True

8. True

9. True

10. False, it is a circle of radius $\sqrt{5}$.

1.3 Exercises

1. ordered

3. Cartesian

5. circle

7. linear equation

9. $(4, 1)$, Quadrant I

11. $(1, 0)$, x-axis

13. $(5, -1)$, Quadrant IV

15. $(-4, -2)$, Quadrant III

17. $(-2, 4)$, Quadrant II

19. Distance is $\sqrt{(4-1)^2 + (7-3)^2} = \sqrt{9+16} = \sqrt{25} = 5$, midpoint is $(2.5, 5)$

21. Distance is $\sqrt{(-1-1)^2 + (-2-0)^2} = \sqrt{4+4} = 2\sqrt{2}$, midpoint is $(0, -1)$

23. Distance is $\sqrt{(12-5)^2 + (-11-13)^2} = \sqrt{49+576} = \sqrt{625} = 25$, and the midpoint is $\left(\dfrac{12+5}{2}, \dfrac{-11+13}{2}\right) = \left(\dfrac{17}{2}, 1\right)$

25. Distance is $\sqrt{(-1+3\sqrt{3}-(-1))^2 + (4-1)^2} = \sqrt{27+9} = 6$, midpoint is $\left(\dfrac{-2+3\sqrt{3}}{2}, \dfrac{5}{2}\right)$

27. Distance is $\sqrt{(1.2+3.8)^2 + (4.8+2.2)^2} = \sqrt{25+49} = \sqrt{74}$, midpoint is $(-1.3, 1.3)$

29. Distance is $\sqrt{(a-b)^2 + 0} = |a-b|$, midpoint is $\left(\dfrac{a+b}{2}, 0\right)$

31. Distance is $\dfrac{\sqrt{\pi^2+4}}{2}$, midpoint is $\left(\dfrac{3\pi}{4}, \dfrac{1}{2}\right)$

33. Center$(0,0)$, radius 4

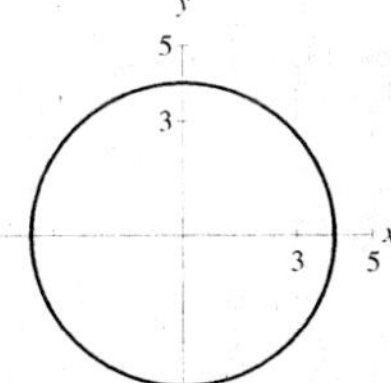

35. Center $(-6,0)$, radius 6

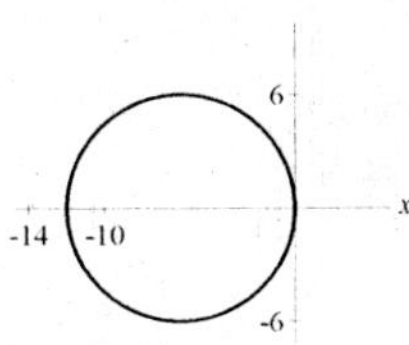

37. Center $(-1,0)$, radius 5

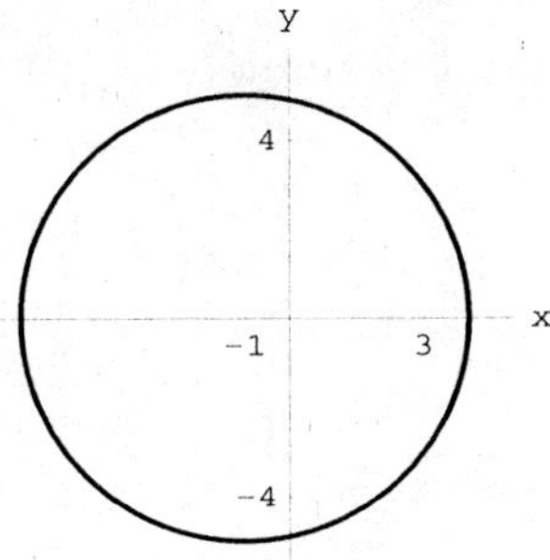

39. Center $(2,-2)$, radius $2\sqrt{2}$

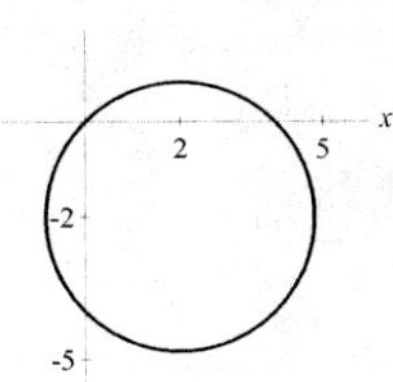

41. $x^2 + y^2 = 49$

43. $(x+2)^2 + (y-5)^2 = 1/4$

45. The distance between $(3,5)$ and the origin is $\sqrt{34}$ which is the radius. The standard equation is $(x-3)^2 + (y-5)^2 = 34$.

47. The distance between $(5,-1)$ and $(1,3)$ is $\sqrt{32}$ which is the radius. The standard equation is $(x-5)^2 + (y+1)^2 = 32$.

49. Center $(0,0)$, radius 3

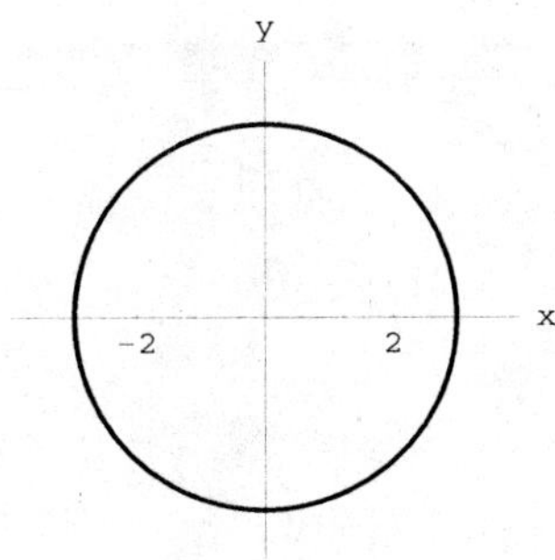

51. Completing the square, we have

$$\begin{aligned} x^2 + (y^2 + 6y + 9) &= 0 + 9 \\ x^2 + (y+3)^2 &= 9. \end{aligned}$$

The center is $(0,-3)$ and the radius is 3.

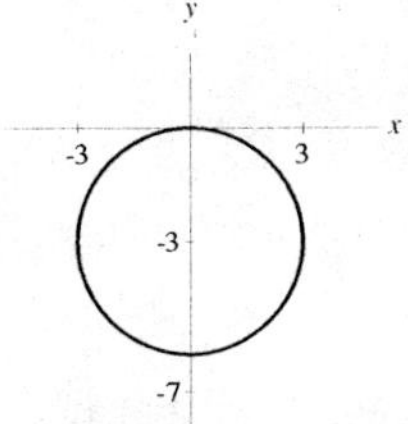

53. Completing the square, we obtain

$$\begin{aligned} (x^2 + 6x + 9) + (y^2 + 8y + 16) &= 9 + 16 \\ (x+3)^2 + (y+4)^2 &= 25. \end{aligned}$$

The center is $(-3,-4)$ and the radius is 5.

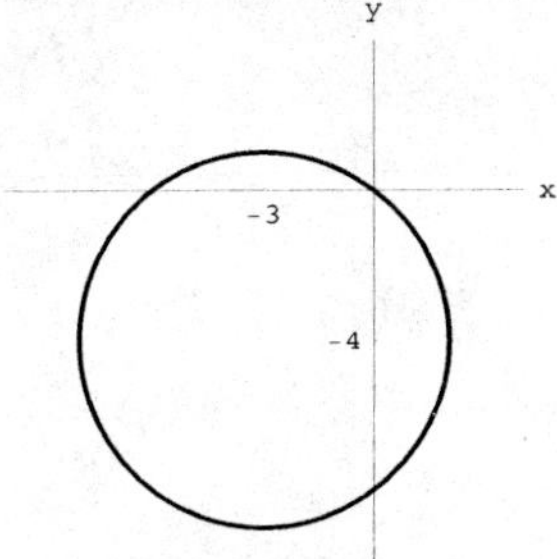

55. Completing the square, we find

$$\begin{aligned} \left(x^2 - 3x + \frac{9}{4}\right) + (y^2 + 2y + 1) &= \frac{3}{4} + \frac{9}{4} + 1 \\ \left(x - \frac{3}{2}\right)^2 + (y+1)^2 &= 4. \end{aligned}$$

The center is $\left(\frac{3}{2}, -1\right)$ and the radius is 2.

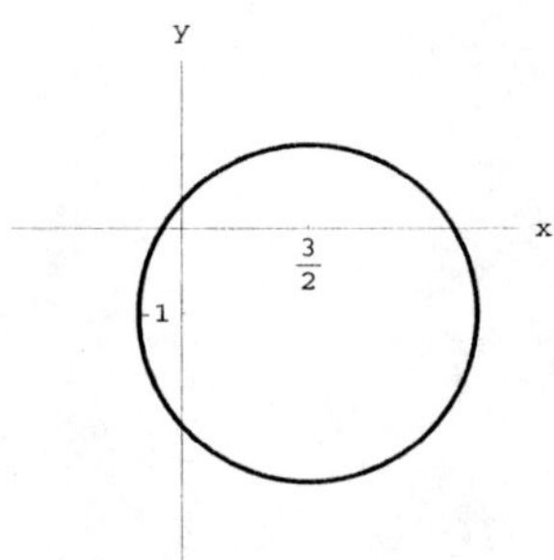

57. Completing the square, we obtain

$$\begin{aligned}(x^2-6x+9)+(y^2-8y+16) &= 9+16\\ (x-3)^2+(y-4)^2 &= 25.\end{aligned}$$

The center is $(3,4)$ and the radius is 5.

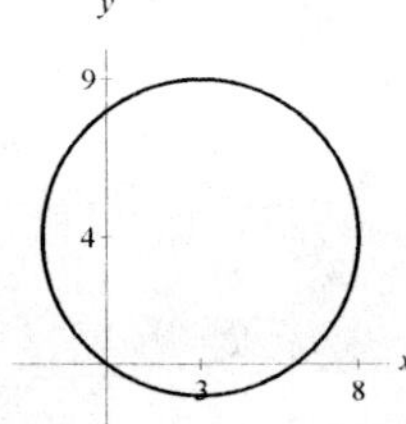

59. Completing the square, we obtain

$$\begin{aligned}(x^2-4x+4)+\left(y^2-3y+\frac{9}{4}\right) &= 4+\frac{9}{4}\\ (x-2)^2+\left(y-\frac{3}{2}\right)^2 &= \frac{25}{4}.\end{aligned}$$

The center is $\left(2,\frac{3}{2}\right)$ and the radius is $\frac{5}{2}$.

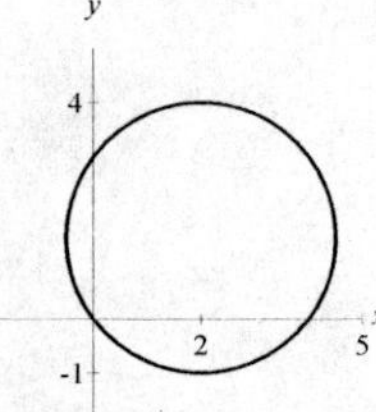

61. Completing the square, we obtain

$$\begin{aligned}\left(x^2-\frac{1}{2}x+\frac{1}{16}\right)+\left(y^2+\frac{1}{3}y+\frac{1}{36}\right) &= \frac{1}{36}\\ \left(x-\frac{1}{4}\right)^2+\left(y+\frac{1}{6}\right)^2 &= \frac{1}{36}.\end{aligned}$$

The center is $\left(\frac{1}{4},-\frac{1}{6}\right)$ and the radius is $\frac{1}{6}$.

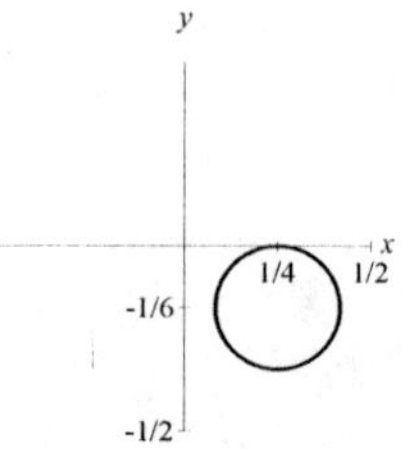

63. a. Since the center is $(0,0)$ and the radius is 7, the standard equation is $x^2+y^2=49$.

b. The radius, which is the distance between $(1,0)$ and $(3,4)$, is given by

$$\sqrt{(3-1)^2+(4-0)^2}=\sqrt{20}.$$

Together with the center $(1,0)$, it follows that the standard equation is

$$(x-1)^2+y^2=20.$$

c. Using the midpoint formula, the center is

$$\left(\frac{3-1}{2},\frac{5-1}{2}\right)=(1,2).$$

The diameter is

$$\sqrt{(3-(-1))^2+(5-(-1))^2}=\sqrt{52}.$$

Since the square of the radius is

$$\left(\frac{1}{2}\sqrt{52}\right)^2=13,$$

the standard equation is

$$(x-1)^2+(y-2)^2=13.$$

65. a. Since the center is $(2,-3)$ and the radius is 2, the standard equation is

$$(x-2)^2+(y+3)^2=4.$$

b. The center is $(-2,1)$, the radius is 1, and the standard equation is

$$(x+2)^2+(y-1)^2=1.$$

c. The center is $(3,-1)$, the radius is 3, and the standard equation is

$$(x-3)^2+(y+1)^2=9.$$

d. The center is $(0, 0)$, the radius is 1, and the standard equation is

$$x^2 + y^2 = 1.$$

67. $y = 3x - 4$ goes through $(0, -4)$, $\left(\frac{4}{3}, 0\right)$.

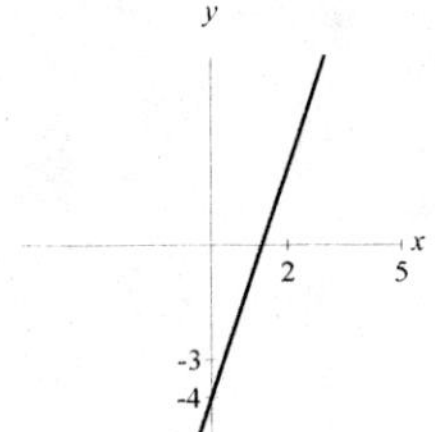

69. $3x - y = 6$ goes through $(0, -6)$, $(2, 0)$.

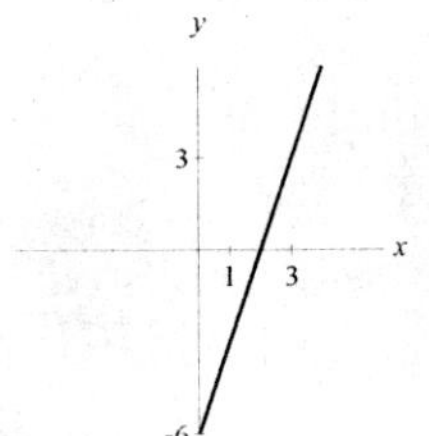

71. $x = 3y - 90$ goes through $(0, 30), (-90, 0)$.

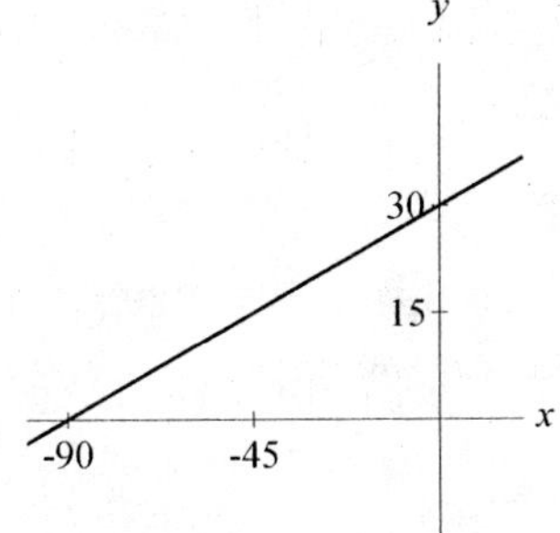

73. $\frac{2}{3}y - \frac{1}{2}x = 400$ goes through $(0, 600), (-800, 0)$.

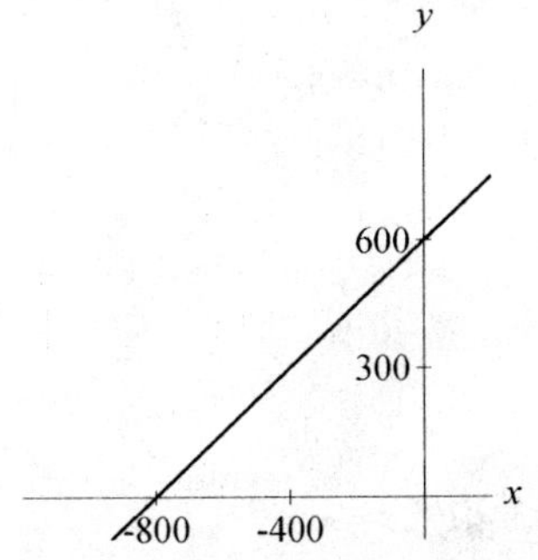

75. The intercepts are $(0, 0.0025), (0.005, 0)$.

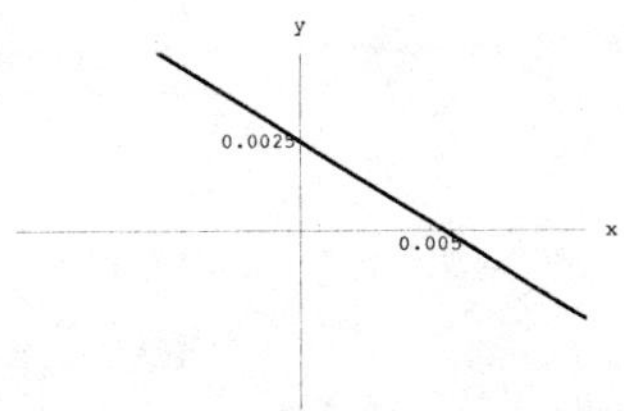

77. The intercepts are $(0, 2500), (5000, 0)$.

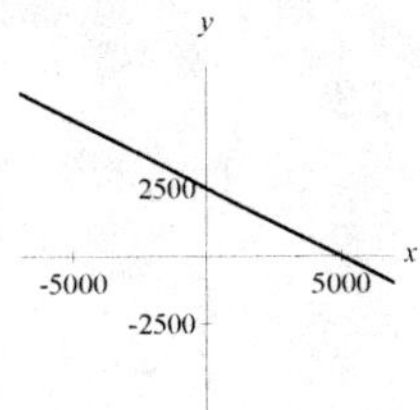

79. $x = 5$

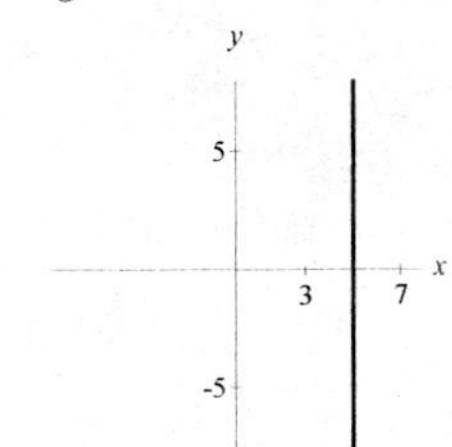

81. $y = 4$

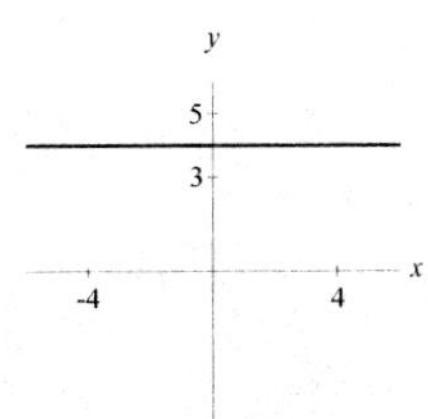

83. $x = -4$

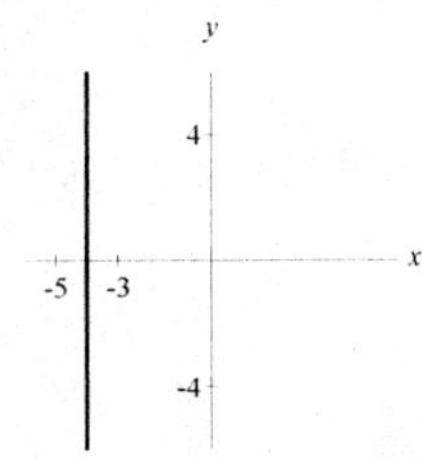

85. Solving for y, we have $y = 1$.

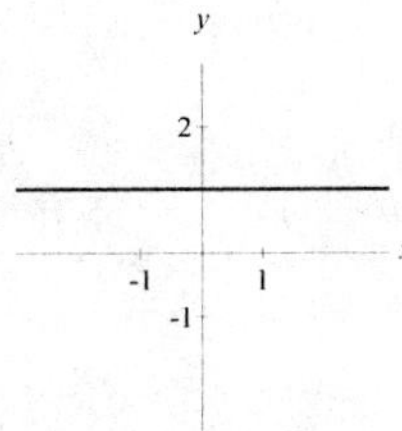

87. Since the x-intercept of $y = 2.4x - 8.64$ is $(3.6, 0)$, the solution set of $2.4x - 8.64 = 0$ is $\{3.6\}$.

89. Since the x-intercept of $y = -\frac{3}{7}x + 6$ is $(14, 0)$, the solution set of $-\frac{3}{7}x + 6 = 0$ is $\{14\}$.

91. The solution is $x = -\frac{3.4}{12} \approx -2.83$.

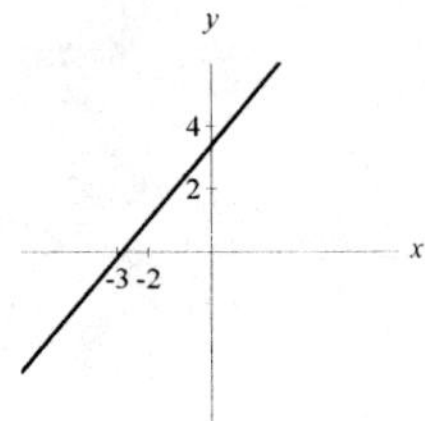

93. The solution is $\frac{687}{1.23} \approx 558.54$

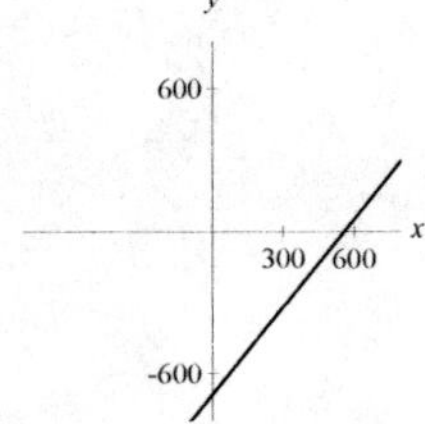

95. The solution is $\frac{3497}{0.03} \approx 116,566.67$

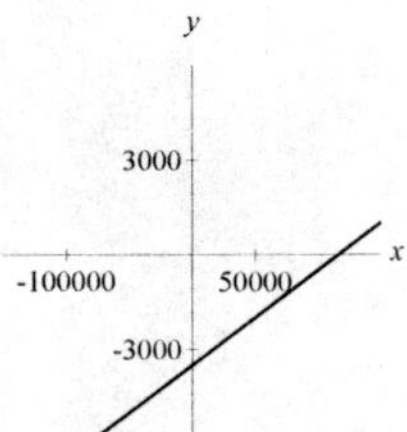

97. Note,

$$\begin{aligned} 4.3 - 3.1(2.3x) + 3.1(9.9) &= 0 \\ 4.3 - 7.13x + 30.69 &= 0 \\ 34.99 - 7.13x &= 0 \\ x &= \frac{34.99}{7.13} \\ x &\approx 4.91. \end{aligned}$$

The solution set is $\{4.91\}$.

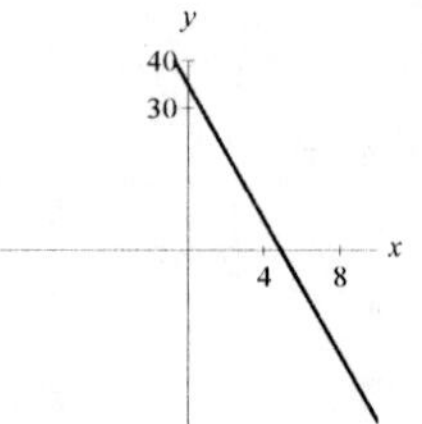

99. a) Let $0 < r < 2$ be the radius of the smallest circle centered at $(2 - r, 0)$. Apply the Pythagorean theorem to the right triangle with vertices at $(2 - r, 0)$, $(0, 0)$, and $(0, -1)$. Then

$$\begin{aligned} 1 + (2 - r)^2 &= (r + 1)^2 \\ 1 + (4 - 4r + r^2) &= r^2 + 2r + 1 \\ 4 &= 6r. \end{aligned}$$

Then $r = 2/3$, and the diameter of each smallest circle is $4/3$ cm.

b) Since $r = 2/3$, the centers of the smallest circles are at $(2 - r, 0)$ and $(-2 + r, 0)$. Equivalently, the centers are at $(\pm 4/3, 0)$. Thus, the equations of the smallest circles are

$$(x - 4/3)^2 + y^2 = 4/9$$

and

$$(x + 4/3)^2 + y^2 = 4/9.$$

101. a) Substitute $h = 0$ into $h = 0.229n + 2.913$. Solving for n, we get

$$n = -\frac{2.913}{0.229} \approx -12.72.$$

Then the n-intercept is near $(-12.72, 0)$. There were no unmarried-couple households in 1977 (i.e., 13 years before 1990). The answer does not make sense.

b) Let $n = 0$. Solving for h, one finds $h = 0.229(0) + 2.913 = 2.913$.
The h-intercept is $(0, 2.913)$.
In 1990 (i.e., $n = 0$), there were 2,913,000 unmarried-couple households.

c) If $n = 25$, then

$$h = 0.229(25) + 2.913 \approx 8.6.$$

In 2015 (i.e., $n = 25$), it is predicted that there will be 8.6 million unmarried-couple households.

103. Given $D = 22,800$ lbs, the graph of
$C = \frac{4B}{\sqrt[3]{22,800}}$ is given below.

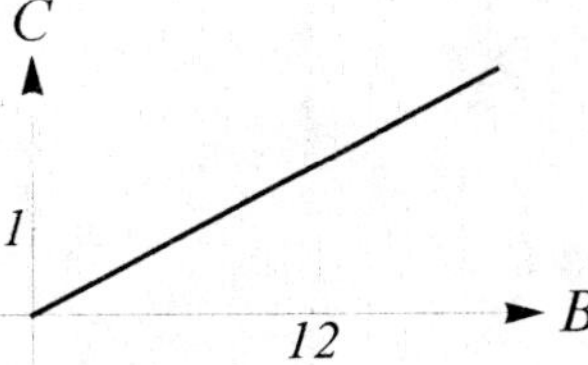

For Island Packet 40, $C = \frac{4(12 + \frac{11}{12})}{\sqrt[3]{22,800}} \approx 1.8.$

105. By the distance formula, we find

$$AB = \sqrt{(1+4)^2 + (1+5)^2} = \sqrt{25+36} = \sqrt{61}.$$

Similarly, we obtain

$$BC = \sqrt{61}$$

and

$$AC = \sqrt{244} = 2\sqrt{61}.$$

Since

$$AB + BC = AC$$

we conclude that A, B, and C are collinear.

107. The distance between $(10, 0)$ and $(0, 0)$ is 10. The distance between $(1, 3)$ and the origin is $\sqrt{10}$.

If two points have integer coordinates, then the distance between them is of the form $\sqrt{s^2 + t^2}$ where s^2, t^2 lies in the set

$$\{0, 1, 2^2, 3^2, 4^2, ...\} = \{0, 1, 4, 9, 16, ...\}.$$

Note, there exists no pair s^2 and t^2 in $\{0, 1, 4, 9, 16, ...\}$ satisfying $s^2 + t^2 = 19$. Thus, one cannot find two points with integer coordinates whose distance between them is $\sqrt{19}$.

111.

a) Conditional equation, solution is $x = \frac{1}{2}$.

b) Identity, both equivalent to $2x + 4$.

c) Inconsistent equation, i.e., no solution.

113. Cross-multiply to obtain

$$\begin{aligned} (x-2)(x+9) &= (x+3)(x+4) \\ x^2 + 7x - 18 &= x^2 + 7x + 12 \\ -18 &= 12 \end{aligned}$$

Inconsistent, the solution set is $\emptyset$.

115.

$$\begin{aligned} ax + b &= cx + d \\ ax - cx &= d - b \\ x(a - c) &= d - b \\ x &= \frac{d - b}{a - c} \end{aligned}$$

117. Let x be the uniform width of the swath. When Eugene is half done, we find

$$(300 - 2x)(400 - 2x) = \frac{300(400)}{2}.$$

We could rewrite the equation as

$$\begin{aligned} x^2 - 350x + 15,000 &= 0 \\ (x - 50)(x - 300) &= 0. \end{aligned}$$

Then $x = 50$ or $x = 300$. Since $x = 300$ is not possible, the width of the swath is $x = 50$ feet.

For Thought

1. False, the slope is $\frac{3-2}{3-2} = 1$.

2. False, the slope is $\frac{5-1}{-3-(-3)} = \frac{4}{0}$ which is undefined.

3. False, slopes of vertical lines are undefined.

4. False, it is a vertical line. **5.** True

6. False, $x = 1$ cannot be written in the slope-intercept form.

7. False, the slope is -2.

8. True **9.** False **10.** True

1.4 Exercises

1. rise

3. slope

5. slope-intercept

7. perpendicular

9. $\frac{5-3}{4+2} = \frac{1}{3}$

11. $\frac{3+5}{1-3} = -4$

13. $\frac{2-2}{5+3} = 0$

15. $\frac{1/2-1/4}{1/4-1/8} = \frac{1/4}{1/8} = 2$

17. $\frac{3-(-1)}{5-5} = \frac{4}{0}$, no slope

19. The slope is $m = \frac{4-(-1)}{3-(-1)} = \frac{5}{4}$. Since $y+1 = \frac{5}{4}(x+1)$, we get $y = \frac{5}{4}x + \frac{5}{4} - 1$ or $y = \frac{5}{4}x + \frac{1}{4}$.

21. The slope is $m = \frac{-1-6}{4-(-2)} = -\frac{7}{6}$. Since $y+1 = -\frac{7}{6}(x-4)$, we obtain $y = -\frac{7}{6}x + \frac{14}{3} - 1$ or $y = -\frac{7}{6}x + \frac{11}{3}$.

23. The slope is $m = \frac{5-5}{-3-3} = 0$. Since $y - 5 = 0(x-3)$, we get $y = 5$.

25. Since $m = \frac{12-(-3)}{4-4} = \frac{15}{0}$ is undefined, the equation of the vertical line is $x = 4$.

27. The slope of the line through $(0,-1)$ and $(3,1)$ is $m = \frac{2}{3}$. Since the y-intercept is $(0,-1)$, the line is given by $y = \frac{2}{3}x - 1$.

29. The slope of the line through $(1,4)$ and $(-1,-1)$ is $m = \frac{5}{2}$. Solving for y in $y+1 = \frac{5}{2}(x+1)$, we get $y = \frac{5}{2}x + \frac{3}{2}$.

31. The slope of the line through $(0,4)$ and $(2,0)$ is $m = -2$. Since the y-intercept is $(0,4)$, the line is given by $y = -2x + 4$.

33. The slope of the line through $(1,4)$ and $(-3,-2)$ is $m = \frac{3}{2}$. Solving for y in $y - 4 = \frac{3}{2}(x-1)$, we get $y = \frac{3}{2}x + \frac{5}{2}$.

35. $y = \frac{3}{5}x - 2$, slope is $\frac{3}{5}$, y-intercept is $(0,-2)$

37. Since $y - 3 = 2x - 8$, $y = 2x - 5$. The slope is 2 and y-intercept is $(0,-5)$.

39. Since $y + 1 = \frac{1}{2}x + \frac{3}{2}$, $y = \frac{1}{2}x + \frac{1}{2}$. The slope is $\frac{1}{2}$ and y-intercept is $\left(0, \frac{1}{2}\right)$.

41. Since $y = 4$, the slope is $m = 0$ and the y-intercept is $(0,4)$.

43.

$$\begin{aligned} y - 5 &= \frac{1}{4}(x+8) \\ y - 5 &= \frac{1}{4}x + 2 \\ y &= \frac{1}{4}x + 7 \end{aligned}$$

45.

$$\begin{aligned} y + 2 &= -\frac{1}{2}(x+3) \\ y + 2 &= -\frac{1}{2}x - \frac{3}{2} \\ y &= -\frac{1}{2}x - \frac{7}{2} \end{aligned}$$

47. $y = \frac{1}{2}x - 2$ goes through the points $(0,-2)$, $(2,-1)$, and $(4,0)$.

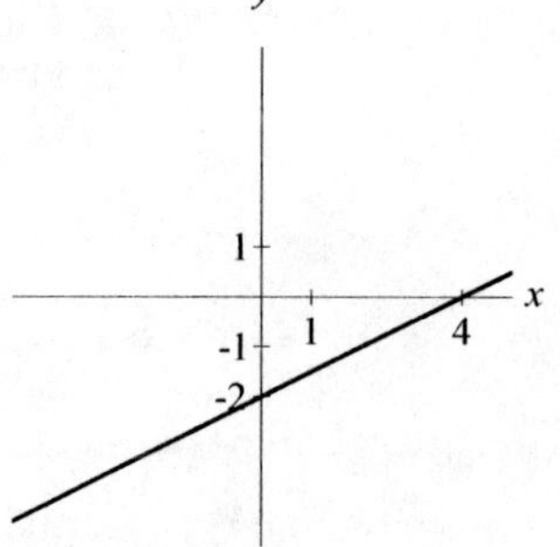

49. $y = -3x + 1$ goes through $(0, 1), (1, -2)$

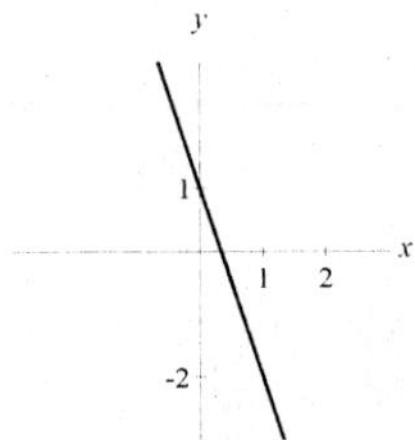

51. $y = -\frac{3}{4}x - 1$ goes through $(0, -1), (-4/3, 0)$

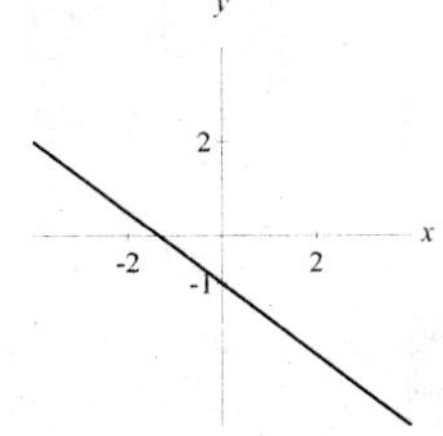

53. $x - y = 3$ goes through $(0, -3), (3, 0)$

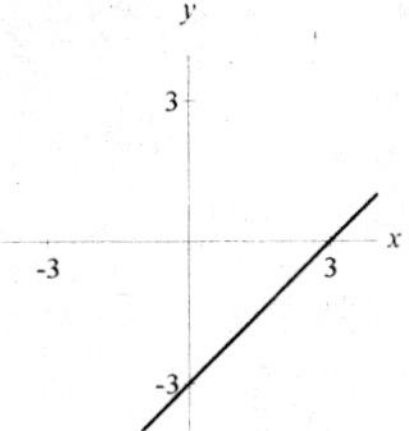

55. $y = 5$ is a horizontal line

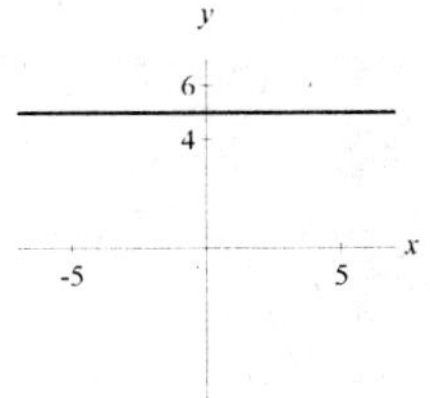

57. Since $m = \frac{4}{3}$ and $y - 0 = \frac{4}{3}(x - 3)$, we have $4x - 3y = 12$.

59. Since $m = \frac{4}{5}$ and $y - 3 = \frac{4}{5}(x - 2)$, we obtain $5y - 15 = 4x - 8$ and $4x - 5y = -7$.

61. $x = -4$ is a vertical line.

63. Note, the slope is

$$m = \frac{\frac{2}{3} + 2}{2 + \frac{1}{2}} = \frac{8/3}{5/2} = \frac{16}{15}.$$

Using the point-slope form, we obtain a standard equation of the line using only integers.

$$\begin{aligned} y - \frac{2}{3} &= \frac{16}{15}(x - 2) \\ 15y - 10 &= 16(x - 2) \\ 15y - 10 &= 16x - 32 \\ -16x + 15y &= -22 \\ 16x - 15y &= 22 \end{aligned}$$

65. The slope is

$$m = \frac{\frac{1}{4} - \frac{1}{5}}{\frac{1}{2} + \frac{1}{3}} = \frac{1/20}{5/6} = \frac{3}{50}.$$

Using the point-slope form, we get a standard equation of the line using only integers.

$$\begin{aligned} y - \frac{1}{4} &= \frac{3}{50}\left(x - \frac{1}{2}\right) \\ 100y - 25 &= 6\left(x - \frac{1}{2}\right) \\ 100y - 25 &= 6x - 3 \\ -22 &= 6x - 100y \\ 3x - 50y &= -11 \end{aligned}$$

67. 0.5 **69.** -1 **71.** 0

73. Since $y + 2 = 2(x - 1)$, $2x - y = 4$

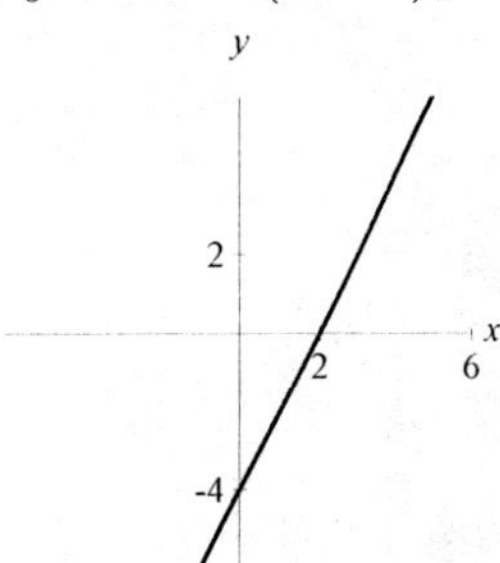

75. Since slope of $y = -3x$ is -3 and $y - 4 = -3(x - 1)$, we obtain $3x + y = 7$.

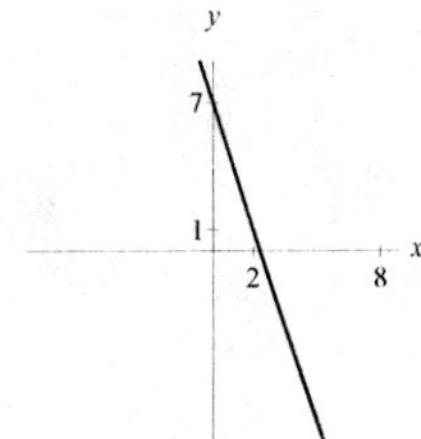

77. Since the slope of $5x - 7y = 35$ is $\frac{5}{7}$, we obtain $y - 1 = \frac{5}{7}(x - 6)$. Multiplying by 7, we get $7y - 7 = 5x - 30$ or equivalently $5x - 7y = 23$.

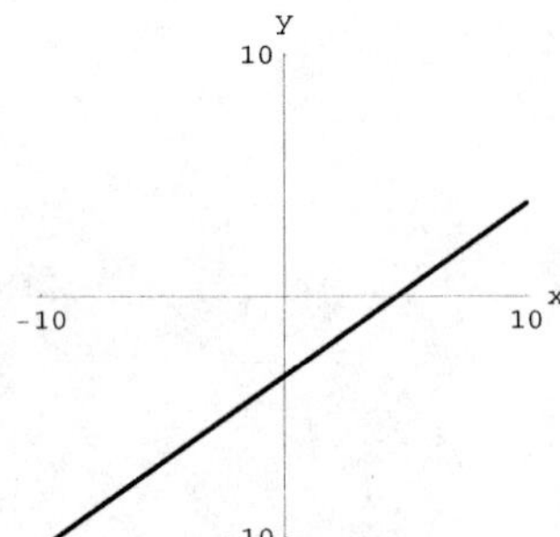

79. Since the slope of $y = \frac{2}{3}x + 5$ is $\frac{2}{3}$, we obtain $y + 3 = -\frac{3}{2}(x - 2)$. Multiplying by 2, we find $2y + 6 = -3x + 6$ or equivalently $3x + 2y = 0$.

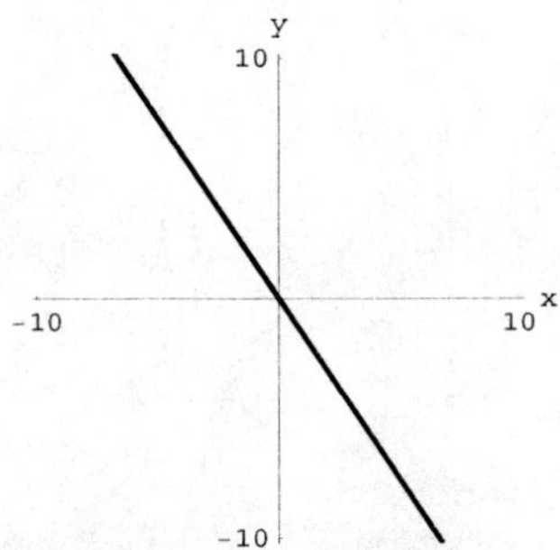

81. Since slope of $y = \frac{1}{2}x - \frac{3}{2}$ is $\frac{1}{2}$ and $y - 1 = -2(x + 3)$, we find $2x + y = -5$.

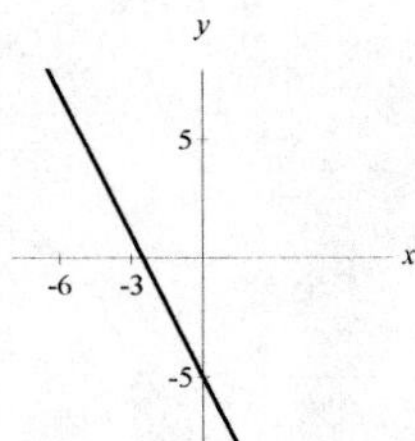

83. Since $x = 4$ is a vertical line, the horizontal line through $(2, 5)$ is $y = 5$.

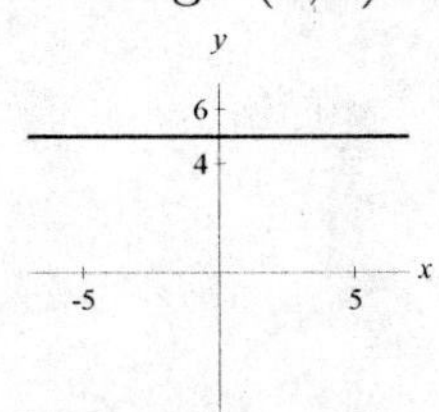

85. Since $\frac{5-3}{8+2} = \frac{1}{5} = -\frac{1}{a}$, we find $a = -5$.

87. Since $\frac{a-3}{-2-a} = -\frac{1}{2}$, we obtain $2a - 6 = 2 + a$ and $a = 8$.

89. Plot the points $A(-1, 2)$, $B(2, -1)$, $C(3, 3)$, and $D(-2, -2)$, respectively. The slopes of the opposite sides are $m_{AC} = m_{BD} = 1/4$ and $m_{AD} = m_{BC} = 4$. Since the opposite sides are parallel, it is a parallelogram.

91. Plot the points $A(-5, -1)$, $B(-3, -4)$, $C(3, 0)$, and $D(1, 3)$, respectively. The slopes of the opposite sides are

$$m_{AB} = m_{CD} = -3/2$$

and

$$m_{AD} = m_{BC} = 2/3.$$

Since the adjacent sides are perpendicular, it is a rectangle.

93. Plot the points $A(-5, 1)$, $B(-2, -3)$, and $C(4, 2)$, respectively. The slopes of the sides are $m_{AB} = -4/3$, $m_{BC} = 5/6$ and $m_{AC} = 1/9$. It is not a right triangle since no two sides are perpendicular.

95. Yes, they appear to be parallel. However, they are not parallel since their slopes are not equal, i.e., $\frac{1}{3} \neq 0.33$.

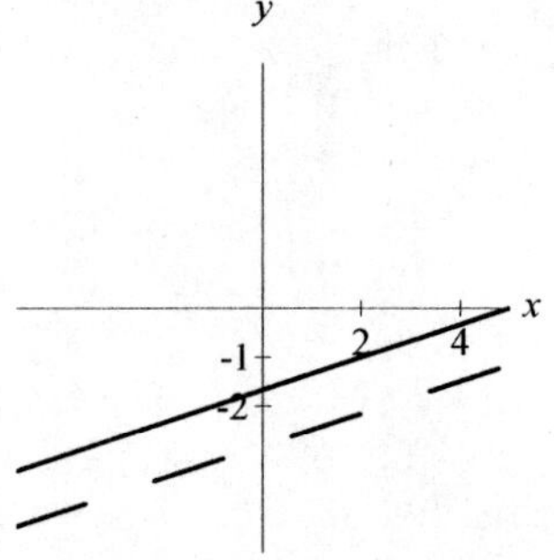

97. Since $x^3 - 8 = (x - 2)(x^2 + 2x + 4)$, we obtain $\frac{x^3 - 8}{x^2 + 2x + 4} = x - 2$. A linear function for the graph is $y = x - 2$.

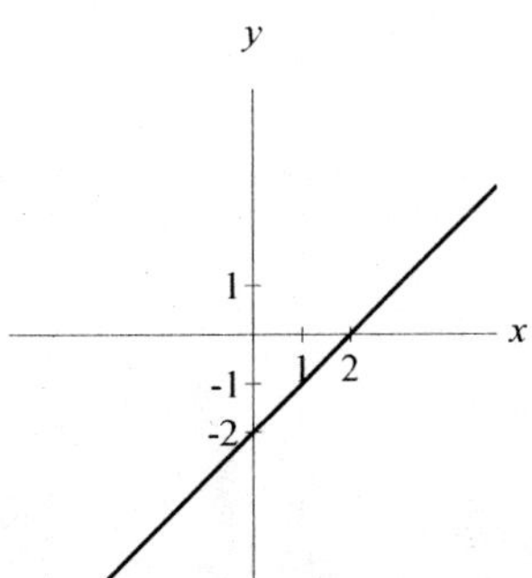

99. The slope is $\frac{212-32}{100-0} = \frac{9}{5}$.

Since $F - 32 = \frac{9}{5}(C - 0)$, $F = \frac{9}{5}C + 32$.

When $C = 150$, $F = \frac{9}{5}(150) + 32 = 302^o$F.

101. In 10 years, the new 18-wheeler depreciated by \$110,000. Then the slope of the linear function is $-\$11,000$/year. Thus, the linear value V is

$$V = 130,000 - 11,000n$$

If $n = 7$ years old, the value of truck is

$$130,000 - 11,000(7) = \$53,000.$$

103. The linear function through $(1, 49)$ and $(2, 48)$ is $c = 50 - n$. With $n = 40$ people in a tour, she charges \$10 each and make \$400.

105. The slope is $\frac{75-95}{4000} = -0.005$. Since $S - 95 = -0.005(D-0)$, we obtain $S = -0.005D + 95$.

107. Let c and p be the number of computers and printers, respectively. Since $60,000 = 2000c + 1500p$, we have

$$\begin{aligned} 2000c &= -1500p + 60,000 \\ c &= -\frac{3}{4}p + 30. \end{aligned}$$

The slope is $-\frac{3}{4}$, i.e., if 4 more printers are purchased then 3 fewer computers must be purchased.

109. Using the equation of the line given by $y = -\frac{3x}{5} + \frac{43}{5}$, the y-values are integers exactly for $x = -4, 1, 6, 11, 16$ in $[-9, 21]$. The points with integral coordinates are $(-4, 11)$, $(1, 8)$, $(6, 5)$, $(11, 2)$, and $(16, -1)$.

111.

$$\begin{aligned} d &= \frac{|5(3) - 12(-6) - 2|}{\sqrt{5^2 + (-12)^2}} \\ d &= \frac{85}{\sqrt{169}} \\ d &= \frac{85}{13} \end{aligned}$$

113.

$$\begin{aligned} d &= \frac{|(-5)(1) + (1)(3) + 4|}{\sqrt{(-5)^2 + 1^2}} \\ d &= \frac{2}{\sqrt{26}} \\ d &= \frac{2\sqrt{26}}{26} \\ d &= \frac{\sqrt{26}}{13} \end{aligned}$$

115. Let $b_1 \neq b_2$. If $y = mx + b_1$ and $y = mx + b_2$ have a point (s, t) in common, then $ms + b_1 = ms + b_2$. After subtracting ms from both sides, we get $b_1 = b_2$; a contradiction. Thus, $y = mx + b_1$ and $y = mx + b_2$ have no points in common if $b_1 \neq b_2$.

117.

$$\begin{aligned} 3 &= 5|x-4| \\ \frac{3}{5} &= |x-4| \\ \pm\frac{3}{5} &= x - 4 \\ 4 \pm \frac{3}{5} &= x \end{aligned}$$

The solution set is $\{17/5, 23/5\}$.

119. The midpoint or center is $((1+3)/2, (3+9)/2)$ or $(2, 6)$. The radius is $\sqrt{(2-1)^2 + (6-3)^2} = \sqrt{10}$. The circle is given by

$$(x-2)^2 + (x-6)^2 = 10.$$

121.

$$\begin{aligned} (x^2 + wx) - (2w + 2x) &= \\ x(x + w) - 2(w + x) &= \\ (x + w)(x - 2) \end{aligned}$$

123. Let x be the number of ants. Then

$$x = 10a + 6 = 7b + 2 = 11c + 2 = 13d + 2$$

for some positive integers a, b, c, d.

The smallest positive x satisfying the system of equations is $x = 4006$ ants.

For Thought

1. False, since $x = 1$ is a solution of the first equation and not of the second equation.

2. False, since $x^2 + 1 = 0$ cannot be factored with real coefficients.

3. False, $\left(x + \frac{2}{3}\right)^2 = x^2 + \frac{4}{3}x + \frac{9}{4}$.

4. False, the solutions to $(x - 3)(2x + 5) = 0$ are $x = 3$ and $x = -\frac{5}{2}$.

5. False, $x^2 = 0$ has only $x = 0$ as its solution.

6. True, since $a = 1, b = -3$, and $c = 1$, then by the quadratic formula we obtain

$$x = \frac{3 \pm \sqrt{9-4}}{2} = \frac{3 \pm \sqrt{5}}{2}.$$

7. False, the quadratic formula can be used to solve any quadratic equation.

8. False, $x^2 + 1 = 0$ has only imaginary zeros.

9. True, for $b^2 - 4ac = 12^2 - 4(4)(9) = 0$.

10. True, $x^2 - 6x + 9 = (x - 3)^2 = 0$ has only one real solution, namely, $x = 3$.

1.5 Exercises

1. quadratic

3. discriminant

5. Since $(x - 5)(x + 4) = 0$, the solution set is $\{5, -4\}$.

7. Since $a^2 + 3a + 2 = (a + 2)(a + 1) = 0$, the solution set is $\{-2, -1\}$.

9. Since $(2x + 1)(x - 3) = 0$, the solution set is $\left\{-\frac{1}{2}, 3\right\}$.

11. Since $(2x - 1)(3x - 2) = 0$, the solution set is $\left\{\frac{1}{2}, \frac{2}{3}\right\}$.

13. Note, $y^2 + y - 12 = 30$. Subtracting 30 from both sides, one obtains $y^2 + y - 42 = 0$ or $(y+7)(y-6) = 0$. The solution set is $\{-7, 6\}$.

15. Since $x^2 = 5$, the solution set is $\{\pm\sqrt{5}\}$.

17. Since $x^2 = -\frac{2}{3}$, we find $x = \pm i\frac{\sqrt{2}}{\sqrt{3}}$.

The solution set is $\left\{\pm i\frac{\sqrt{6}}{3}\right\}$.

19. Since $x - 3 = \pm 3$, we get $x = 3 \pm 3$.

The solution set is $\{0, 6\}$.

21. By the square root property, we get $3x - 1 = \pm 0 = 0$. Solving for x, we obtain $x = \frac{1}{3}$. The solution set is $\left\{\frac{1}{3}\right\}$.

23. Since $x - \frac{1}{2} = \pm\frac{5}{2}$, it follows that $x = \frac{1}{2} \pm \frac{5}{2}$.

The solution set is $\{-2, 3\}$.

25. Since $x + 2 = \pm 2i$, the solution set is $\{-2 \pm 2i\}$.

27. Since $x - \frac{2}{3} = \pm\frac{2}{3}$, we get

$$x = \frac{2}{3} \pm \frac{2}{3} = \frac{4}{3}, 0.$$

The solution set is $\left\{0, \frac{4}{3}\right\}$.

29. $x^2 - 12x + \left(\frac{12}{2}\right)^2 = x^2 - 12x + 6^2 = x^2 - 12x + 36$

31. $r^2 + 3r + \left(\frac{3}{2}\right)^2 = r^2 + 3r + \frac{9}{4}$

33. $w^2 + \frac{1}{2}w + \left(\frac{1}{4}\right)^2 = w^2 + \frac{1}{2}w + \frac{1}{16}$

35. By completing the square, we derive

$$\begin{aligned} x^2+6x &= -1 \\ x^2+6x+9 &= -1+9 \\ (x+3)^2 &= 8 \\ x+3 &= \pm 2\sqrt{2}. \end{aligned}$$

The solution set is $\{-3 \pm 2\sqrt{2}\}$.

37. By completing the square, we find

$$\begin{aligned} n^2-2n &= 1 \\ n^2-2n+1 &= 1+1 \\ (n-1)^2 &= 2 \\ n-1 &= \pm\sqrt{2}. \end{aligned}$$

The solution set is $\{1 \pm \sqrt{2}\}$.

39.

$$\begin{aligned} h^2+3h &= 1 \\ h^2+3h+\frac{9}{4} &= 1+\frac{9}{4} \\ (h+\frac{3}{2})^2 &= \frac{13}{4} \\ h+\frac{3}{2} &= \pm\frac{\sqrt{13}}{2} \end{aligned}$$

The solution set is $\left\{\dfrac{-3\pm\sqrt{13}}{2}\right\}$.

41.

$$\begin{aligned} x^2+\frac{5}{2}x &= 6 \\ x^2+\frac{5}{2}x+\frac{25}{16} &= 6+\frac{25}{16} \\ \left(x+\frac{5}{4}\right)^2 &= \frac{121}{16} \\ x &= -\frac{5}{4}\pm\frac{11}{4} \end{aligned}$$

The solution set is $\left\{-4, \dfrac{3}{2}\right\}$.

43.

$$\begin{aligned} x^2+\frac{2}{3}x &= -\frac{1}{3} \\ x^2+\frac{2}{3}x+\frac{1}{9} &= -\frac{3}{9}+\frac{1}{9} \\ \left(x+\frac{1}{3}\right)^2 &= -\frac{2}{9} \\ x &= -\frac{1}{3}\pm i\frac{\sqrt{2}}{3} \end{aligned}$$

The solution set is $\left\{\dfrac{-1\pm i\sqrt{2}}{3}\right\}$.

45. Since $a=1, b=3, c=-4$ and

$x = \dfrac{-3\pm\sqrt{3^2-4(1)(-4)}}{2(1)} = \dfrac{-3\pm\sqrt{25}}{2} =$

$\dfrac{-3\pm 5}{2}$, the solution set is $\{-4, 1\}$.

47. Since $a=2, b=-5, c=-3$ and

$x = \dfrac{5\pm\sqrt{(-5)^2-4(2)(-3)}}{2(2)} = \dfrac{5\pm\sqrt{49}}{4} =$

$\dfrac{5\pm 7}{4}$, the solution set is $\left\{-\dfrac{1}{2}, 3\right\}$.

49. Since $a=9, b=6, c=1$ and

$x = \dfrac{-6\pm\sqrt{6^2-4(9)(1)}}{2(9)} = \dfrac{-6\pm 0}{18}$,

the solution set is $\left\{-\dfrac{1}{3}\right\}$.

51. Since $a=2, b=0, c=-3$ and

$x = \dfrac{0\pm\sqrt{0^2-4(2)(-3)}}{2(2)} = \dfrac{\pm\sqrt{24}}{4} =$

$\dfrac{\pm 2\sqrt{6}}{4}$, the solution set is $\left\{\pm\dfrac{\sqrt{6}}{2}\right\}$.

53. In $x^2-4x+5=0$, $a=1, b=-4, c=5$.

Then $x = \dfrac{4\pm\sqrt{(-4)^2-4(1)(5)}}{2(1)} =$

$\dfrac{4\pm\sqrt{-4}}{2} = \dfrac{4\pm 2i}{2}$.

The solution set is $\{2\pm i\}$.

55. Note, $a=1, b=-2$, and $c=4$.

Then $x = \dfrac{2\pm\sqrt{4-16}}{2} = \dfrac{2\pm 2i\sqrt{3}}{2}$.

The solution set is $\{1\pm i\sqrt{3}\}$.

57. Since $2x^2-2x+5=0$, we find $a=2, b=-2$,

and $c=5$. Then $x = \dfrac{2\pm\sqrt{4-40}}{4} = \dfrac{2\pm 6i}{4}$.

The solution set is $\left\{\dfrac{1}{2}\pm\dfrac{3}{2}i\right\}$.

59. Since $a = 4, b = -8, c = 7$ and
$x = \dfrac{8 \pm \sqrt{64 - 112}}{8} = \dfrac{8 \pm \sqrt{-48}}{8} =$
$\dfrac{8 \pm 4i\sqrt{3}}{8}$, the solution set is $\left\{1 \pm \dfrac{\sqrt{3}}{2}i\right\}$.

61. Since $a = 3.2, b = 7.6$, and $c = -9$,
$x = \dfrac{-7.6 \pm \sqrt{(7.6)^2 - 4(3.2)(-9)}}{2(3.2)} \approx$
$\dfrac{-7.6 \pm \sqrt{172.96}}{6.4} \approx \dfrac{-7.6 \pm 13.151}{6.4}$.
The solution set is $\{-3.24, 0.87\}$.

63. Note, $a = 3.25, b = -4.6$, and $c = -22$.
Then $x = \dfrac{4.6 \pm \sqrt{(-4.6)^2 - 4(3.25)(-22)}}{2(3.25)}$
$= \dfrac{4.6 \pm \sqrt{307.16}}{6.5}$. The solution set is $\{-1.99, 3.40\}$.

65. The discriminant is

$$(-30)^2 - 4(9)(25) = 900 - 900 = 0.$$

Only one solution and it is real.

67. The discriminant is

$$(-6)^2 - 4(5)(2) = 36 - 40 = -4.$$

There are no real solutions.

69. The discriminant is

$$12^2 - 4(7)(-1) = 144 + 28 = 172.$$

There are two distinct real solutions.

71. Note, x-intercepts are $\left(-\dfrac{2}{3}, 0\right)$ and $\left(\dfrac{1}{2}, 0\right)$.

The solution set is $\left\{-\dfrac{2}{3}, \dfrac{1}{2}\right\}$.

73. Since the x-intercepts are $(-3, 0)$ and $(5, 0)$, the solution set is $\{-3, 5\}$.

75. Note, the graph of $y = 1.44x^2 - 8.4x + 12.25$ has exactly one x-intercept.

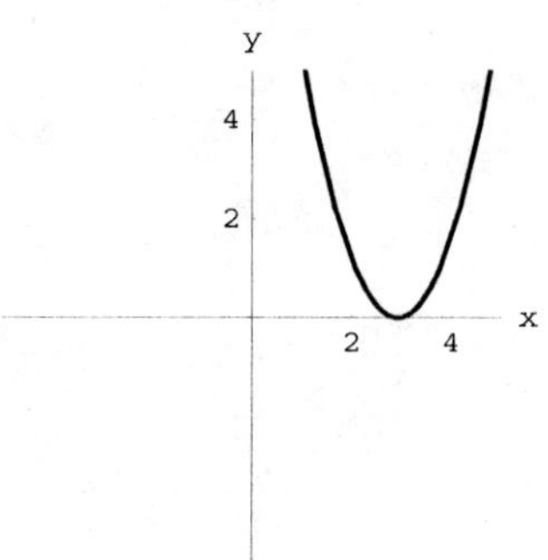

77. Note, the graph of $y = x^2 + 3x + 15$ has no x-intercept.

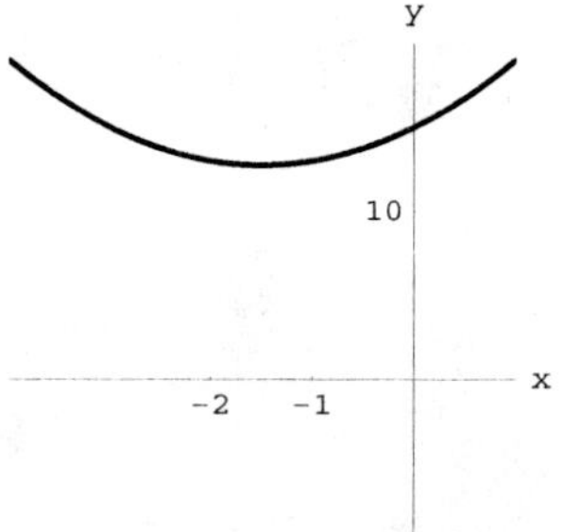

Thus, $x^2 + 3x + 15 = 0$ has no real solution.

79. The graph of $y = x^2 + 3x - 160$ has two x-intercepts.

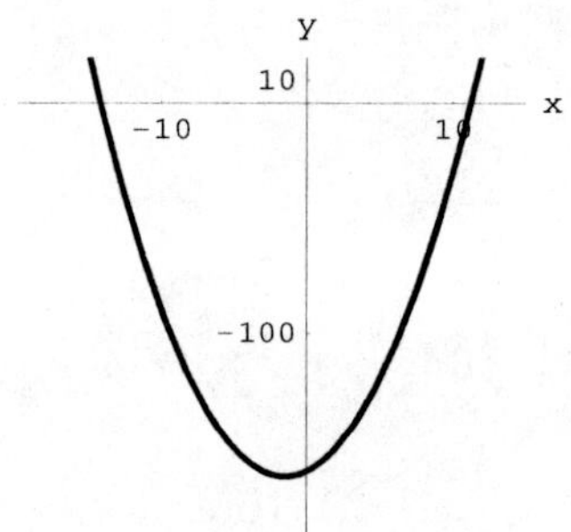

Then $x^2 + 3x - 160 = 0$ has two real solutions.

81. Set the right-hand side to 0.

$$\begin{aligned} x^2 - \frac{4}{3}x - \frac{5}{9} &= 0 \\ 9x^2 - 12x - 5 &= 0 \\ (3x + 1)(3x - 5) &= 0 \end{aligned}$$

The solution set is $\left\{-\dfrac{1}{3}, \dfrac{5}{3}\right\}$.

83. Since $x^2 = \sqrt{2}$, $x = \pm\sqrt{\sqrt{2}} = \pm\sqrt[4]{2}$.
The solution set is $\{\pm\sqrt[4]{2}\}$.

85. By the quadratic formula,

$$x = \frac{-\sqrt{6} \pm \sqrt{(-\sqrt{6})^2 - 4(12)(-1)}}{2(12)} =$$

$$\frac{-\sqrt{6} \pm \sqrt{54}}{24} = \frac{-\sqrt{6} \pm 3\sqrt{6}}{24} = \frac{2\sqrt{6}}{24}, \frac{-4\sqrt{6}}{24}.$$

The solution set is $\left\{-\frac{\sqrt{6}}{6}, \frac{\sqrt{6}}{12}\right\}$.

87. Since $x^2 + 6x - 72 = (x + 12)(x - 6) = 0$, the solution set is $\{-12, 6\}$.

89. Multiply by x to get $x^2 = x + 1$. So $x^2 - x - 1 = 0$ and by the quadratic formula,

$$x = \frac{1 \pm \sqrt{(-1)^2 - 4(1)(-1)}}{2} = \frac{1 \pm \sqrt{5}}{2}.$$

The solution set is $\left\{\frac{1 \pm \sqrt{5}}{2}\right\}$.

91. Multiply by x^2 to get $28x - 7 = 7x^2$. Applying the quadratic formula to $7x^2 - 28x + 7 = 0$, we obtain

$$x = \frac{28 \pm \sqrt{588}}{14} = \frac{28 \pm 14\sqrt{3}}{14}.$$

The solution set is $\{2 \pm \sqrt{3}\}$.

93. Multiplying by $(3 - x)(x + 7)$, we get

$$\begin{aligned} (x - 12)(x + 7) &= (x + 4)(3 - x) \\ x^2 - 5x - 84 &= -x^2 - x + 12 \\ 2x^2 - 4x - 96 &= 0 \\ 2(x - 8)(x + 6) &= 0. \end{aligned}$$

The solution set is $\{8, -6\}$.

95. Multiplying by $(x + 2)(x + 3)$, we find

$$\begin{aligned} (x - 8)(x + 3) &= (x + 2)(2x - 1) \\ x^2 - 5x - 24 &= 2x^2 + 3x - 2 \\ 0 &= x^2 + 8x + 22 \\ -22 + 16 &= x^2 + 8x + 16 \\ -6 &= (x + 4)^2. \end{aligned}$$

Since the right side is not a negative number, the solution set is the empty set $\emptyset$.

97. Multiplying by $(2x + 1)(2x + 3)$, we find

$$\begin{aligned} (2x + 3)^2 &= 8(2x + 1) \\ 4x^2 + 12x + 9 &= 16x + 8 \\ 4x^2 - 4x + 1 &= 0 \\ (2x - 1)^2 &= 0. \end{aligned}$$

Thus, the solution set is $\left\{\frac{1}{2}\right\}$.

99. Since $r^2 = \frac{A}{\pi}$, $r = \pm\sqrt{\frac{A}{\pi}}$.

101. We use the quadratic formula to solve

$$x^2 + (2k)x + 3 = 0.$$

Since $a = 1, b = 2k$, and $c = 3$, we obtain

$$\begin{aligned} x &= \frac{-2k \pm \sqrt{(2k)^2 - 4(1)(3)}}{2(1)} \\ x &= \frac{-2k \pm \sqrt{4k^2 - 12}}{2} \\ x &= \frac{-2k \pm \sqrt{4(k^2 - 3)}}{2} \\ x &= \frac{-2k \pm 2\sqrt{k^2 - 3}}{2} \\ x &= -k \pm \sqrt{k^2 - 3}. \end{aligned}$$

103. We use the quadratic formula to solve

$$2y^2 + (4x)y - x^2 = 0.$$

Since $a = 2, b = 4x$, and $c = -x^2$, we obtain

$$y = \frac{-4x \pm \sqrt{(4x)^2 - 4(2)(-x^2)}}{2(2)} =$$

$$\frac{-4x \pm \sqrt{16x^2 + 8x^2}}{4} = \frac{-4x \pm \sqrt{24x^2}}{4}$$

$$\frac{-4x \pm 2|x|\sqrt{6}}{4} = x\left(-1 \pm \frac{\sqrt{6}}{2}\right).$$

Note that we used $|x| = \sqrt{x^2}$.

105. From the revenue function,

$$\begin{aligned} x(40 - 0.001x) &= 175,000 \\ 40x - 0.001x^2 &= 175,000 \end{aligned}$$

By applying the quadratic formula to $0.001x^2 - 40x + 175,000 = 0$, we get

$$\begin{aligned} x &= \frac{40 \pm \sqrt{(-40)^2 - 4(0.001)(175,000)}}{0.002} \\ x &= \frac{40 \pm \sqrt{900}}{0.002} = \frac{40 \pm 30}{0.002} \\ x &= 5000 \text{ or } 35,000. \end{aligned}$$

Then 5000 units or 35,000 units must be produced weekly.

107. The height S (in feet) of the ball from the ground t seconds after it was tossed is given by $S = -16t^2 + 40t + 4$. When the height is 4 feet,

$$\begin{aligned} -16t^2 + 40t + 4 &= 4 \\ -16t^2 + 40t &= 0 \\ -8t(2t - 5) &= 0 \\ t &= 0, \frac{5}{2}. \end{aligned}$$

The ball returns to a height of 4 ft in 2.5 sec.

109. Let d be the diagonal distance across the field from one goal to the other. By the Pythagorean Theorem, we obtain $d = \sqrt{300^2 + 160^2} = 340$ ft.

111. Let w and $2w + 2$ be the width and length. From the given area of the court, we obtain

$$\begin{aligned} (2w + 2)w &= 312 \\ 2w^2 + 2w - 312 &= 0 \\ w^2 + w - 156 &= 0 \\ (w - 12)(w + 13) &= 0. \end{aligned}$$

Then $w = 12$ yd and the length is 26 yd. The distance between two opposite corners (by the Pythagorean Theorem) is $\sqrt{12^2 + 26^2} = 2\sqrt{205} \approx 28.6$ yd.

113. Substituting the values of S and A, we find that the displacement is

$$\begin{aligned} \frac{1}{2^{12}} d^2 (18.8)^3 - 822^3 &= 0 \\ \frac{1}{2^{12}} d^2 (18.8)^3 &= 822^3 \\ d &= \sqrt{\frac{822^3(2^{12})}{18.8^3}} \\ d &\approx 18,503.4 \text{ lbs.} \end{aligned}$$

115. By choosing an appropriate coordinate system, we can assume the circle is given by $(x+r)^2 + (y-r)^2 = r^2$ where $r > 0$ is the radius of the circle and $(-5, 1)$ is the common point between the block and the circle. Note, the radius is less than 5 feet. Substitute $x = -5$ and $y = 1$. Then we obtain

$$\begin{aligned} (-5 + r)^2 + (1 - r)^2 &= r^2 \\ r^2 - 10r + 25 + 1 - 2r + r^2 &= r^2 \\ r^2 - 12r + 26 &= 0. \end{aligned}$$

The solutions of the last quadratic equation are $r = 6 \pm \sqrt{10}$. Since $r < 5$, the radius of the circle is $r = 6 - \sqrt{10}$ ft.

117. Let x be the normal speed of the tortoise in ft/hr.

	distance	rate	time
hwy	24	$x + 2$	$24/(x + 2)$
off hwy	24	x	$24/x$

Since 24 minutes is 2/5 of an hour, we get

$$\begin{aligned} \frac{2}{5} + \frac{24}{x + 2} &= \frac{24}{x} \\ 2x(x + 2) + 24(5)x &= 24(5)(x + 2) \\ 2x^2 + 4x + 120x &= 120x + 240 \\ x^2 + 2x - 120 &= 0 \\ (x + 12)(x - 10) &= 0 \\ x &= -12, 10. \end{aligned}$$

The normal speed of the tortoise is 10 ft/hr.

119. Using $v_1^2 = v_0^2 + 2gS$ with $S = 1.07$ and $v_1 = 0$, we find that

$$\begin{aligned} v_0^2 + 2(-9.8)(1.07) &= 0 \\ v_0^2 - 20.972 &= 0 \\ v_0 = \pm\sqrt{20.972} &\approx \pm 4.58. \end{aligned}$$

His initial upward velocity is 4.58 m/sec.

Using $S = \frac{1}{2}gt^2 + v_0 t$ with $S = 0$ and $v_o = 4.58$, we find that his time t in the air satisfies

$$\frac{1}{2}(-9.8)t^2 + 4.58t = 0$$

$$\begin{aligned} t(4.58 - 4.9t) &= 0 \\ t &\approx 0, 0.93. \end{aligned}$$

Carter is in the air for 0.93 seconds.

121. Let x and $x - 2$ be the number of days it takes to design a direct-mail package using traditional methods and a computer, respectively.

	rate
together	$2/7$
computer	$1/(x-2)$
traditional	$1/x$

$$\begin{aligned} \frac{1}{x-2} + \frac{1}{x} &= \frac{2}{7} \\ 7x + (7x - 14) &= 2(x^2 - 2x) \\ 0 &= 2x^2 - 18x + 14 \\ 0 &= x^2 - 9x + 7 \\ x &= \frac{9 \pm \sqrt{81 - 28}}{2} \\ x &= \frac{9 \pm \sqrt{53}}{2} \\ x &\approx 8.14, 0.86 \end{aligned}$$

Curt, using traditional methods, can do the job in 8.14 days.

Note, $x \approx 0.86$ days has to be excluded since $x - 2$ is negative when $x \approx 0.86$.

123. Let x and $x - 10$ be the number of pounds of white meat chicken in a Party Size bucket and a Big Family Size bucket, respectively. From the ratios, we obtain

$$\begin{aligned} \frac{8}{x} &= \frac{3}{x-10} + 0.10 \\ 8(x - 10) &= 3x + 0.10x(x - 10) \\ 0 &= 0.10x^2 - 6x + 80 \\ 0 &= x^2 - 60x + 800 \\ 0 &= (x - 40)(x - 20) \\ x &= 40, 20 \end{aligned}$$

A Party Size bucket weighs 20 or 40 lbs.

125. a) Let x be the number of years since 1980. With the aid of a graphing calculator, the quadratic regression curve is approximately

$$y = -0.054x^2 + 0.996x + 51.78.$$

b) Using the regression curve in part a) and the quadratic formula, we find that the positive solution to

$$0 = ax^2 + bx + c$$

is

$$x = \frac{-b - \sqrt{b^2 - 4ac}}{2a} \approx 42.$$

In the year 2022 (= 1980 + 42), the extrapolated birth rate will be zero.

127. Since the lines are parallel, we find C such that the point $(-2, 6)$ satisfies

$$4x - 5y = C$$

Then $-8 - 30 = C$ or $C = -38$. The standard form is $4x - 5y = -38$.

129. Let x be the amount she invested in a CD.

$$\begin{aligned} 0.05x + (x + 4000)0.06 &= 1230 \\ 0.11x + 240 &= 1230 \\ 0.11x &= 990 \\ x &= \$9000 \end{aligned}$$

131. $-\sqrt{2}$

133. a) Let $m = l + 1$ and $n = w + 1$ where l and w are relatively prime. The number of streets that the crow passes as it flies from $(1, 1)$ to (m, n) is l. When the crow crosses an avenue, it will cross over two blocks bounded by the same streets (and not over an intersection) since l and w are relatively prime. Since the crow will cross over $w - 1$ avenues between the first block and the last block, the crow flies over $w - 1$ additional blocks. Thus, the total number of blocks that the crow flies over is

$$l + w - 1 = m + n - 3.$$

b) Let $m = l + 1$ and $n = w + 1$ where d is the greatest common divisor of l and w. Write $l = dl_1$ and $w = dw_1$ where l_1 and w_1 are relatively prime. Then we can break the flight of the crow into d segments described as follows:

(a) $(1, 1)$ to $(l_1 + 1, w_1 + 1)$

(b) $(l_1 + 1, w_1 + 1)$ to $(2l_1 + 1, 2w_1 + 1)$, and so on, and finally from

(c) $((d - 1)l_1 + 1, (d - 1)w_1 + 1)$ to $(dl_1 + 1, dw_1 + 1)$.

By part a), the number of blocks that the crow flies over in each segment is

$$l_1 + w_1 - 1.$$

Thus, the total number of blocks that the crow flies over is

$$d(l_1 + w_1 - 1) = l + w - d = m + n - 2 - d.$$

For Thought

1. False, since
$$(\sqrt{x-1} + \sqrt{x})^2 = (x - 1) + 2\sqrt{x(x-1)} + x.$$

2. False, since -1 is a solution of the first and not of the second equation.

3. False, since -27 is a solution of the first equation but not of the second.

4. False, rather let $u = x^{1/4}$ and $u^2 = x^{1/2}$.

5. True, since $x - 1 = \pm 4^{-3/2}$.

6. False, $\left(-\dfrac{1}{32}\right)^{-2/5} = (-32)^{2/5} = (-2)^2 = 4$.

7. False, $x = -2$ is not a solution.

8. True

9. True

10. False, since $(x^3)^2 = x^6$.

1.6 Exercises

1. Factor: $x^2(x + 3) - 4(x + 3) = 0$
$(x^2 - 4)(x + 3) = (x - 2)(x + 2)(x + 3) = 0$
The solution set is $\{\pm 2, -3\}$.

3. Factor: $2x^2(x + 500) - (x + 500) = 0$
$(2x^2 - 1)(x + 500) = 0$
The solution set is $\left\{\pm\dfrac{\sqrt{2}}{2}, -500\right\}$.

5. Set the right-hand side to 0 and factor.
$$\begin{aligned} a(a^2 - 15a + 5) &= 0 \\ a = \frac{15 \pm \sqrt{(-15)^2 - 4(1)(5)}}{2} &\quad \text{or} \quad a = 0 \\ a = \frac{15 \pm \sqrt{205}}{2} &\quad \text{or} \quad a = 0 \end{aligned}$$
The solution set is $\left\{\dfrac{15 \pm \sqrt{205}}{2}, 0\right\}$.

7. Factor: $3y^2(y^2 - 4) = 3y^2(y - 2)(y + 2) = 0$
The solution set is $\{0, \pm 2\}$.

9. Factor: $(a^2 - 4)(a^2 + 4) =$
$(a - 2)(a + 2)(a - 2i)(a + 2i) = 0$.
The solution set is $\{\pm 2, \pm 2i\}$.

11. Squaring each side, we get
$$\begin{aligned} x + 1 &= x^2 - 10x + 25 \\ 0 &= x^2 - 11x + 24 = (x - 8)(x - 3). \end{aligned}$$
Checking $x = 3$, we get $2 \neq -2$. Then $x = 3$ is an extraneous root. The solution set is $\{8\}$.

13. Isolate the radical and then square each side.
$$\begin{aligned} x &= x^2 - 40x + 400 \\ 0 &= x^2 - 41x + 400 = (x - 25)(x - 16) \end{aligned}$$
Checking $x = 16$, we get $2 \neq -6$. Then $x = 16$ is an extraneous root. The solution set is $\{25\}$.

15. Isolate the radical and then square each side.
$$\begin{aligned} 2w &= \sqrt{1 - 3w} \\ 4w^2 &= 1 - 3w \\ 4w^2 + 3w - 1 &= (4w - 1)(w + 1) = 0 \\ w &= \frac{1}{4}, -1 \end{aligned}$$

Checking $w = -1$, we get $-1 \neq 1$.
Then $w = -1$ is an extraneous root.

The solution set is $\left\{\frac{1}{4}\right\}$.

17. Multiply both sides by $z\sqrt{4z+1}$ and square each side.

$$\begin{aligned} \sqrt{4z+1} &= 3z \\ 4z+1 &= 9z^2 \\ 0 &= 9z^2 - 4z - 1 \end{aligned}$$

By the quadratic formula,

$$z = \frac{4 \pm \sqrt{16 - 4(9)(-1)}}{18}$$

$$z = \frac{4 \pm \sqrt{52}}{18} = \frac{4 \pm 2\sqrt{13}}{18} = \frac{2 \pm \sqrt{13}}{9}.$$

Since $z = \frac{2 - \sqrt{13}}{9} < 0$ and the right-hand side of the original equation is nonnegative, $z = \frac{2 - \sqrt{13}}{9}$ is an extraneous root.

The solution set is $\left\{\frac{2 + \sqrt{13}}{9}\right\}$.

19. Squaring each side, one obtains

$$\begin{aligned} x^2 - 2x - 15 &= 9 \\ x^2 - 2x - 24 &= (x-6)(x+4) = 0. \end{aligned}$$

The solution set is $\{-4, 6\}$.

21. Isolate a radical and square each side.

$$\begin{aligned} \sqrt{x+40} &= \sqrt{x} + 4 \\ x + 40 &= x + 8\sqrt{x} + 16 \\ 24 &= 8\sqrt{x} \\ 3 &= \sqrt{x} \\ 9 &= x \end{aligned}$$

The solution set is $\{9\}$.

23. Isolate a radical and square each side.

$$\begin{aligned} \sqrt{n+4} &= 5 - \sqrt{n-1} \\ n+4 &= 25 - 10\sqrt{n-1} + (n-1) \\ -20 &= -10\sqrt{n-1} \\ 2 &= \sqrt{n-1} \\ 4 &= n-1 \end{aligned}$$

The solution set is $\{5\}$.

25. Isolate a radical and square each side.

$$\begin{aligned} \sqrt{2x+5} &= 9 - \sqrt{x+6} \\ 2x+5 &= 81 - 18\sqrt{x+6} + (x+6) \\ x - 82 &= -18\sqrt{x+6} \\ x^2 - 164x + 6724 &= 324(x+6) \\ x^2 - 488x + 4780 &= 0 \\ (x-10)(x-478) &= 0 \end{aligned}$$

Checking $x = 478$ we get $53 \neq 9$ and $x = 478$ is an extraneous root. The solution set is $\{10\}$.

27. Raise each side to the power $3/2$.
Then $x = \pm 2^{3/2} = \pm 8^{1/2} = \pm 2\sqrt{2}$.
The solution set is $\{\pm 2\sqrt{2}\}$.

29. Raise each side to the power $-3/4$.

Thus, $w = \pm(16)^{-3/4} = \pm(2)^{-3} = \pm\frac{1}{8}$.

The solution set is $\left\{\pm\frac{1}{8}\right\}$.

31. Raise each side to the power -2.
So, $t = (7)^{-2} = \frac{1}{49}$. The solution set is $\left\{\frac{1}{49}\right\}$.

33. Raise each side to the power -2.

Then $s - 1 = (2)^{-2} = \frac{1}{4}$ and $s = 1 + \frac{1}{4}$.

The solution set is $\left\{\frac{5}{4}\right\}$.

35. Since $(x^2 - 9)(x^2 - 3) = 0$, the solution set is $\{\pm 3, \pm\sqrt{3}\}$.

37. Since $(x^2 + 7)(x^2 - 1) = 0$, the solution set is $\{\pm 1, \pm i\sqrt{7}\}$.

39. Since $(x^2 + 9)(x^2 - 9) = 0$, the solution set is $\{\pm 3, \pm 3i\}$.

41. Let $u = \frac{2c-3}{5}$ and $u^2 = \left(\frac{2c-3}{5}\right)^2$. Then

$$\begin{aligned} u^2 + 2u - 8 &= 0 \\ (u+4)(u-2) &= 0 \\ u &= -4, 2 \end{aligned}$$

$$\frac{2c-3}{5} = -4 \quad \text{or} \quad \frac{2c-3}{5} = 2$$

$$2c - 3 = -20 \quad \text{or} \quad 2c - 3 = 10$$

$$c = -\frac{17}{2} \quad \text{or} \quad c = \frac{13}{2}.$$

The solution set is $\left\{-\frac{17}{2}, \frac{13}{2}\right\}$.

43. Let $u = \frac{1}{5x-1}$ and $u^2 = \left(\frac{1}{5x-1}\right)^2$. Then

$$\begin{aligned} u^2 + u - 12 &= (u+4)(u-3) = 0 \\ u &= -4, 3 \end{aligned}$$

$$\begin{array}{ccc} \frac{1}{5x-1} = -4 & \text{or} & \frac{1}{5x-1} = 3 \\ 1 = -20x + 4 & \text{or} & 1 = 15x - 3 \\ x = \frac{3}{20} & \text{or} & \frac{4}{15} = x. \end{array}$$

The solution set is $\left\{\frac{3}{20}, \frac{4}{15}\right\}$.

45. Let $u = v^2 - 4v$ and $u^2 = (v^2 - 4v)^2$. Then

$$\begin{aligned} u^2 - 17u + 60 &= (u-5)(u-12) = 0 \\ u &= 5, 12 \end{aligned}$$

$$\begin{array}{ccc} v^2 - 4v = 5 & \text{or} & v^2 - 4v = 12 \\ v^2 - 4v - 5 = 0 & \text{or} & v^2 - 4v - 12 = 0 \\ (v-5)(v+1) = 0 & \text{or} & (v-6)(v+2) = 0. \end{array}$$

The solution set is $\{-2, -1, 5, 6\}$.

47. Factor the left-hand side.

$$\begin{aligned} \left(\sqrt{x} - 3\right)\left(\sqrt{x} - 1\right) &= 0 \\ \sqrt{x} &= 3, 1 \\ x &= 9, 1 \end{aligned}$$

The solution set is $\{1, 9\}$.

49. Factor the left-hand side as $\left(\sqrt{q} - 4\right)\left(\sqrt{q} - 3\right) = 0$. Then $\sqrt{q} = 3, 4$ and the solution set is $\{9, 16\}$.

51. Set the right-hand side to 0 and factor.

$$\begin{aligned} x^{2/3} - 7x^{1/3} + 10 &= 0 \\ \left(x^{1/3} - 5\right)\left(x^{1/3} - 2\right) &= 0 \\ x^{1/3} &= 5, 2 \end{aligned}$$

The solution set is $\{8, 125\}$.

53. An equivalent statement is

$$\begin{array}{ccc} w^2 - 4 = 3 & \text{or} & w^2 - 4 = -3 \\ w^2 = 7 & \text{or} & w^2 = 1. \end{array}$$

The solution set is $\{\pm\sqrt{7}, \pm 1\}$.

55. An equivalent statement assuming $5v \geq 0$ is

$$\begin{array}{ccc} v^2 - 3v = 5v & \text{or} & v^2 - 3v = -5v \\ v^2 - 8v = 0 & \text{or} & v^2 + 2v = 0 \\ v(v-8) = 0 & \text{or} & v(v+2) = 0 \\ & v = 0, 8, -2. & \end{array}$$

Since $5v \geq 0$, $v = -2$ is an extraneous root and the solution set is $\{0, 8\}$.

57. An equivalent statement is

$$\begin{array}{ccc} x^2 - x - 6 = 6 & \text{or} & x^2 - x - 6 = -6 \\ x^2 - x - 12 = 0 & \text{or} & x^2 - x = 0 \\ (x-4)(x+3) = 0 & \text{or} & x(x-1) = 0. \end{array}$$

The solution set is $\{-3, 0, 1, 4\}$.

59. An equivalent statement is

$$\begin{array}{ccc} x + 5 = 2x + 1 & \text{or} & x + 5 = -(2x+1) \\ 4 = x & \text{or} & x = -2. \end{array}$$

The solution set is $\{-2, 4\}$.

61. An equivalent statement is

$$\begin{array}{ccc} x - 2 = 5x - 1 & \text{or} & x - 2 = -5x + 1 \\ -1 = 4x & \text{or} & 6x = 3 \\ -\frac{1}{4} = x & \text{or} & x = \frac{1}{2} \end{array}$$

Note, $x = -\frac{1}{4}$ is an extraneous root.

The solution set is $\{1/2\}$.

63. An equivalent statement is

$$\begin{array}{ccc} x - 4 = x - 2 & \text{or} & x - 4 = -x + 2 \\ -4 = -2 & \text{or} & 2x = 6 \\ \textit{inconsistent} & \text{or} & x = 3 \end{array}$$

The solution set is $\{3\}$.

65. Isolate a radical and square both sides.

$$\begin{aligned} \sqrt{16x+1} &= \sqrt{6x+13} - 1 \\ 16x + 1 &= (6x+13) - 2\sqrt{6x+13} + 1 \\ 10x - 13 &= -2\sqrt{6x+13} \end{aligned}$$

$$\begin{aligned} 100x^2 - 260x + 169 &= 4(6x + 13) \\ 100x^2 - 284x + 117 &= 0 \end{aligned}$$

$$\begin{aligned} x &= \frac{284 \pm \sqrt{284^2 - 4(100)(117)}}{200} \\ x &= \frac{284 \pm 184}{200} \\ x &= \frac{1}{2}, \frac{117}{50} \end{aligned}$$

Checking $x = \dfrac{117}{50}$ we get $\sqrt{\dfrac{1922}{50}} - \sqrt{\dfrac{1352}{50}} > 0$

and so $x = \dfrac{117}{50}$ is an extraneous root.

The solution set is $\left\{\dfrac{1}{2}\right\}$.

67. Factor as a difference of two squares and then as a sum and difference of two cubes.

$$\begin{aligned} (v^3 - 8)(v^3 + 8) &= 0 \\ (v - 2)(v^2 + 2v + 4)(v + 2)(v^2 - 2v + 4) &= 0 \end{aligned}$$

Then $v = \pm 2$ or

$$v = \frac{-2 \pm \sqrt{2^2 - 16}}{2} \quad \text{or } v = \frac{2 \pm \sqrt{2^2 - 16}}{2}$$

$$v = \frac{-2 \pm 2i\sqrt{3}}{2} \quad \text{or } v = \frac{2 \pm 2i\sqrt{3}}{2}$$

The solution set is $\{\pm 2, -1 \pm i\sqrt{3}, 1 \pm i\sqrt{3}\}$.

69. Raise both sides to the power 4. Then

$$\begin{aligned} 7x^2 - 12 &= x^4 \\ 0 &= x^4 - 7x^2 + 12 = (x^2 - 4)(x^2 - 3) \\ x &= \pm 2, \pm\sqrt{3} \end{aligned}$$

Since the left-hand side of the given equation is nonnegative, $x = -2, -\sqrt{3}$ are extraneous roots. The solution set is $\{\sqrt{3}, 2\}$.

71. Raise both sides to the power 3.

$$\begin{aligned} 2 + x - 2x^2 &= x^3 \\ x^3 + 2x^2 - x - 2 &= 0 \\ x^2(x + 2) - (x + 2) &= (x^2 - 1)(x + 2) = 0 \\ x &= \pm 1, -2 \end{aligned}$$

The solution set is $\{\pm 1, -2\}$.

73. Let $t = \dfrac{x-2}{3}$ and $t^2 = \left(\dfrac{x-2}{3}\right)^2$. Then

$$\begin{aligned} t^2 - 2t + 10 &= 0 \\ t^2 - 2t + 1 &= -10 + 1 \\ (t - 1)^2 &= -9 \\ t &= 1 \pm 3i \\ \frac{x-2}{3} &= 1 \pm 3i \\ x - 2 &= 3 \pm 9i \\ x &= 5 \pm 9i. \end{aligned}$$

The solution set is $\{5 \pm 9i\}$.

75. Raise both sides to the power 5/2. Then

$$\begin{aligned} 3u - 1 &= \pm 2^{5/2} \\ 3u - 1 &= \pm 32^{1/2} \\ 3u &= 1 \pm 4\sqrt{2} \end{aligned}$$

The solution set is $\left\{\dfrac{1 \pm 4\sqrt{2}}{3}\right\}$.

77. Factor this quadratic type expression.

$$\begin{aligned} (x^2 + 1) - 11\sqrt{x^2 + 1} + 30 &= 0 \\ \left(\sqrt{x^2 + 1} - 5\right)\left(\sqrt{x^2 + 1} - 6\right) &= 0 \end{aligned}$$

$$\sqrt{x^2 + 1} = 5 \text{ or } \sqrt{x^2 + 1} = 6$$

$$x^2 = 24 \text{ or } x^2 = 35$$

$$x = \pm 2\sqrt{6}, \pm\sqrt{35}$$

The solution set is $\{\pm\sqrt{35}, \pm 2\sqrt{6}\}$.

79. An equivalent statement is

$$\begin{aligned} x^2 - 2x = 3x - 6 \quad &\text{or} \quad x^2 - 2x = -3x + 6 \\ x^2 - 5x + 6 = 0 \quad &\text{or} \quad x^2 + x - 6 = 0 \\ (x - 3)(x - 2) = 0 \quad &\text{or} \quad (x + 3)(x - 2) = 0 \end{aligned}$$

$$x = 2, \pm 3$$

The solution set is $\{2, \pm 3\}$.

81. Raise both sides to the power $-5/3$. Then

$$\begin{aligned} 3m + 1 &= \left(-\frac{1}{8}\right)^{-5/3} \\ 3m + 1 &= \left(-\frac{1}{2}\right)^{-5} \\ 3m + 1 &= -32. \end{aligned}$$

The solution set is $\{-11\}$.

83. An equivalent statement assuming $x-2 \geq 0$ is

$$\begin{aligned} x^2-4=x-2 &\quad \text{or} \quad x^2-4=-x+2 \\ x^2-x-2=0 &\quad \text{or} \quad x^2+x-6=0 \\ (x-2)(x+1)=0 &\quad \text{or} \quad (x+3)(x-2)=0 \\ x &= 2,-1,-3. \end{aligned}$$

Since $x-2 \geq 0$, $x=-1,-3$ are extraneous roots and the solution set is $\{2\}$.

85. Solve for S.

$$\begin{aligned} 21.24+1.25S^{1/2}-9.8(18.34)^{1/3} &= 16.296 \\ 1.25S^{1/2}-25.84397 &\approx -4.944 \\ S^{1/2} &\approx 16.72 \\ S &\approx 279.56 \end{aligned}$$

The maximum sailing area is 279.56 m^2.

87. Solve for x with $C=83.50$.

$$\begin{aligned} 0.5x+\sqrt{8x+5000} &= 83.50 \\ \sqrt{8x+5000} &= 83.50-0.5x \\ 8x+5000 &= 6972.25-83.50x+0.25x^2 \\ 0 &= 0.25x^2-91.50x+1972.25 \end{aligned}$$

$$\begin{aligned} x &= \frac{91.50 \pm \sqrt{(-91.50)^2-4(0.25)(1972.25)}}{0.5} \\ x &= \frac{91.50 \pm 80}{0.5} \\ x &= 23, 343 \end{aligned}$$

Checking $x=343$, the value of the left-hand side of the first equation exceeds 83.50 and so $x=343$ is an extraneous root.
Thus, 23 loaves cost $83.50.

89. Let x and $x+6$ be two numbers. Then

$$\begin{aligned} \sqrt{x+6}-\sqrt{x} &= 1 \\ \sqrt{x+6} &= \sqrt{x}+1 \\ x+6 &= x+2\sqrt{x}+1 \\ 5 &= 2\sqrt{x} \\ 25 &= 4x \\ x &= \frac{25}{4} \end{aligned}$$

Since $\frac{25}{4}+6=\frac{49}{4}$, the numbers are $\frac{25}{4}$ and $\frac{49}{4}$.

91. Let x be the length of the short leg. Since $x+7$ is the other leg, by the Pythagorean Theorem the hypotenuse is $\sqrt{x^2+(x+7)^2}$. Then

$$x+(x+7)+\sqrt{x^2+(x+7)^2} = 30$$

$$\begin{aligned} 2x-23 &= -\sqrt{x^2+(x+7)^2} \\ 4x^2-92x+529 &= 2x^2+14x+49 \\ 2x^2-106x+480 &= 0 \\ x^2-53x+240 &= 0 \\ (x-5)(x-48) &= 0 \\ x &= 5, 48 \end{aligned}$$

Since the perimeter is 30 in., $x=48$ is an extraneous root. The short leg is $x=5$ in.

93. Let x be the length of one side of the original square foundation. From the 2100 ft^2 we have

$$\begin{aligned} (x-10)(x+30) &= 2100 \\ x^2+20x-2400 &= 0 \\ (x+60)(x-40) &= 0. \end{aligned}$$

Since x is nonnegative, $x=40$ and the area of the square foundation is $x^2=1600$ ft^2.

95. Solving for d, we find

$$\begin{aligned} 598.9\left(\frac{d}{64}\right)^{-2/3} &= 14.26 \\ \left(\frac{d}{64}\right)^{-2/3} &= \frac{14.26}{598.9} \\ \frac{d}{64} &= \left(\frac{14.26}{598.9}\right)^{-3/2} \\ d &= 64\left(\frac{14.26}{598.9}\right)^{-3/2} \\ d &\approx 17,419.4 \text{ lb.} \end{aligned}$$

97. Let $x^2(x+2)$ be the volume of shrimp to be shipped. Here, x is the length of one side of the square base and $x+2$ is the height. The volume of the styrofoam box is $(x+2)^2(x+4)$. Since the amount of shrimp is one-half of the volume of the styrofoam box, we have

$$\begin{aligned} 2x^2(x+2) &= (x+2)^2(x+4) \\ 2x^3+4x^2 &= (x^2+4x+4)(x+4) \\ x^3-4x^2-20x-16 &= 0. \end{aligned}$$

By using the Rational Zero Theorem and synthetic division, we obtain

$$x^3 - 4x^2 - 20x - 16 = (x+2)(x^2 - 6x - 8).$$

Note, we have to exclude the zero of $x+2$ which is -2 since a dimension is a positive number. Then by using the method of completing the square, we get

$$\begin{aligned} x^2 - 6x &= 8 \\ x^2 - 6x + 9 &= 17 \\ (x-3)^2 &= 17 \\ x &= 3 \pm \sqrt{17} \\ x &\approx -1.123, 7.123 \end{aligned}$$

Since $x > 0$, choose $x = 3 + \sqrt{17}$.
Thus, the volume of shrimp to be shipped is $(3+\sqrt{17})^2(5+\sqrt{17}) \approx 462.89$ in.^{3}

99. Let x be the number of hours after 10:00 a.m. so that the distance between Nancy and Edgar is 14 miles greater than the distance between Nancy and William.

By the Pythagorean Theorem, $\sqrt{(5x)^2 + (12x)^2}$ and $\sqrt{(5x)^2 + [4(x+2)]^2}$ are the distances between Nancy and Edgar and Nancy and William, respectively. Then

$$\begin{aligned} \sqrt{(5x)^2 + (12x)^2} - 14 &= \sqrt{(5x)^2 + [4(x+2)]^2} \\ \sqrt{169x^2} - 14 &= \sqrt{41x^2 + 64x + 64} \\ 13x - 14 &= \sqrt{41x^2 + 64x + 64} \\ 169x^2 - 364x + 196 &= 41x^2 + 64x + 64 \\ 128x^2 - 428x + 132 &= 0 \\ 32x^2 - 107x + 33 &= 0 \end{aligned}$$

$$x = \frac{107 \pm \sqrt{(-107)^2 - 4(32)(33)}}{64} = 3, \frac{11}{32}.$$

Since the left-hand side of the first equation is negative when $x = \frac{11}{32}$, we find that $x = \frac{11}{32}$ is an extraneous root. Thus, $x = 3$ hours and the time is 1:00 p.m.

101. **(a)** $\sqrt[3]{(28.6)(28.3)(46.3)} \approx \$33.5$ billion

(b) Let p be the net income for the fourth quarter. Then

$$\begin{aligned} \sqrt[4]{(28.6)(28.3)(46.3)p} &= 40 \\ p &= \frac{40^4}{(28.6)(28.3)(46.3)} \\ p &\approx \$68.3 \text{ billion.} \end{aligned}$$

103. The weight of the molasses in a cylindrical tank is its volume times its density. Then

$$\begin{aligned} \pi\left(\frac{d}{2}\right)^2 \cdot d \cdot 1600 &= 25,850,060 \\ 400\pi d^3 &= 25,850,060 \\ d &= \sqrt[3]{\frac{25,850,060}{400\pi}} \approx 27.4. \end{aligned}$$

The height of the tank is about 27.4 meters.

105. By choosing an appropriate coordinate system, we can assume the circle is given by

$$(x+r)^2 + (y-r)^2 = r^2$$

where $r > 0$ is the radius of the circle and $(-5, 1)$ is the common point between the block and the circle. Note, the radius is less than 5 feet. Substitute $x = -5$ and $y = 1$. Then we obtain

$$\begin{aligned} (-5+r)^2 + (1-r)^2 &= r^2 \\ r^2 - 10r + 25 + 1 - 2r + r^2 &= r^2 \\ r^2 - 12r + 26 &= 0. \end{aligned}$$

The solutions of the last quadratic equation are $r = 6 \pm \sqrt{10}$. Since $r < 5$, the radius of the circle is

$$r = 6 - \sqrt{10} \text{ ft.}$$

107. Note, $x - 5 = x + 7$ has no solution.

$$\begin{aligned} x - 5 &= \pm(x+7) \\ x - 5 &= -(x+7) \\ 2x &= -2 \\ x &= -1 \end{aligned}$$

The solution set is $\{-1\}$.

109. Completing the square:

$$(x+4)^2 + (y-5)^2 = 16+25$$
$$(x+4)^2 + (y-5)^2 = 41$$

The center is $(-4, 5)$ with radius $R = \sqrt{41}$.

111. Since $y = \frac{3}{5}x + \frac{11}{5}$, the slope is $\frac{3}{5}$.

113. Using the arrangement below, the maximum area that can be covered is 102 ft^2.

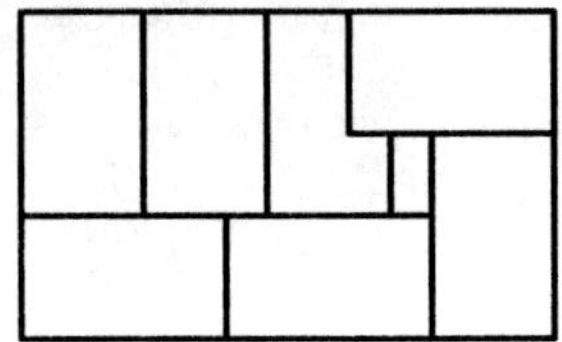

For Thought

1. True

2. False, since $-2x < -6$ is equivalent to

$$\frac{-2x}{-2} > \frac{-6}{-2}.$$

3. False, since there is a number between any two distinct real numbers.

4. True, since $|-6-6| = |-12| = 12 > -1$.

5. False, $(-\infty, -3) \cap (-\infty, -2) = (-\infty, -3)$.

6. False, $(5, \infty) \cap (-\infty, -3) = \emptyset$.

7. False, no real number satisfies $|x-2| < 0$.

8. False, it is equivalent to $|x| > 3$.

9. False, $|x| + 2 < 5$ is equivalent to $-3 < x < 3$.

10. True

1.7 Exercises

1. interval

3. closed

5. compound

7. $x < 12$

9. $x \geq -7$

11. $[-8, \infty)$

13. $(-\infty, \pi/2)$

15. Since $3x > 15$ implies $x > 5$, the solution set is $(5, \infty)$ and the graph is ← (→ with endpoint 5

17. Since $10 \leq 5x$ implies $2 \leq x$, the solution set is $[2, \infty)$ and the graph is ← [→ with endpoint 2

19. Multiply 6 to both sides of the inequality.

$$3x - 24 < 2x + 30$$
$$x < 54$$

The solution is the interval $(-\infty, 54)$ and the graph is ←) → with endpoint 54

21. Multiplying the inequality by 2, we find

$$7 - 3x \geq -6$$
$$13 \geq 3x$$
$$13/3 \geq x.$$

The solution is the interval $(-\infty, 13/3]$ and the graph is ←] → with endpoint 13/3

23. Multiply the inequality by -5 and reverse the direction of the inequality.

$$2x - 3 \leq 0$$
$$2x \leq 3$$
$$x \leq \frac{3}{2}$$

The solution is the interval $(-\infty, 3/2]$ and the graph is ←] → with endpoint 3/2

25. Multiply the left-hand side.

$$-6x + 4 \geq 4 - x$$
$$0 \geq 5x$$
$$0 \geq x$$

The solution is the interval $(-\infty, 0]$ and the graph is ←] → with endpoint 0

27. Using the portion of the graph below the x-axis, the solution set is $(-\infty, -3.5)$.

29. Using the part of the graph on or above the x-axis, the solution set is $(-\infty, 1.4]$.

31. By taking the part of the line $y = 2x - 3$ above the horizontal line $y = 5$ and by using $(4, 5)$, the solution set is $(4, \infty)$.

33. Note, the graph of $y = -3x - 7$ is above or on the graph of $y = x + 1$ for $x \leq -2$. Thus, the solution set is $(-\infty, -2]$.

35. $(-3, \infty)$

37. $(-3, \infty)$

39. $(-5, -2)$

41. $\emptyset$

43. $(-\infty, 5]$

45. Solve each simple inequality and find the intersection of their solution sets.

$$\begin{array}{lcl} x > 3 & \text{and} & 0.5x < 3 \\ x > 3 & \text{and} & x < 6 \end{array}$$

The intersection of these values of x is the interval $(3, 6)$ and the graph is

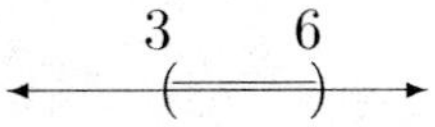

3 6

47. Solve each simple inequality and find the intersection of their solution sets.

$$\begin{array}{rcl} 2x - 5 > -4 & \text{and} & 2x + 1 > 0 \\ x > \frac{1}{2} & \text{and} & x > -\frac{1}{2} \end{array}$$

The intersection of these values of x is the interval $(1/2, \infty)$ and the graph is

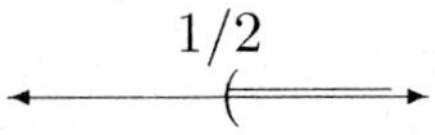

1/2

49. Solve each simple inequality and find the union of their solution sets.

$$\begin{array}{rcl} -6 < 2x & \text{or} & 3x > -3 \\ -3 < x & \text{or} & x > -1 \end{array}$$

The union of these values of x is $(-3, \infty)$ and the graph is

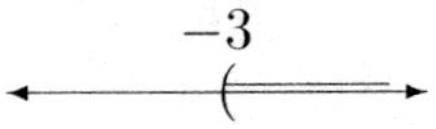

−3

51. Solve each simple inequality and find the union of their solution sets.

$$\begin{array}{rcl} x + 1 > 6 & \text{or} & x < 7 \\ x > 5 & \text{or} & x < 7 \end{array}$$

The union of these values of x is $(-\infty, \infty)$ and the graph is

53. Solve each simple inequality and find the intersection of their solution sets.

$$\begin{array}{rcl} 2 - 3x < 8 & \text{and} & x - 8 \leq -12 \\ -6 < 3x & \text{and} & x \leq -4 \\ -2 < x & \text{and} & x \leq -4 \end{array}$$

The intersection is empty.

The solution set is $\emptyset$.

55.

$$\begin{array}{ccccc} 6 & < & 3x & < & 12 \\ 2 & < & x & < & 4 \end{array}$$

The solution set is the interval $(2, 4)$ and the graph is

2 4

57.

$$\begin{array}{ccccc} -6 & \leq & -6x & < & 18 \\ 1 & \geq & x & > & -3 \end{array}$$

The solution set is the interval $(-3, 1]$ and the graph is

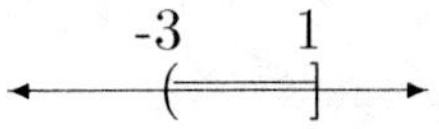

-3 1

59. Solve an equivalent compound inequality.

$$\begin{array}{rcl} -2 < & 3x - 1 & < 2 \\ -1 < & 3x & < 3 \\ -\frac{1}{3} < & x & < 1 \end{array}$$

The solution set is the interval $(-1/3, 1)$ and the graph is

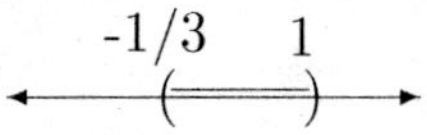

-1/3 1

61. Solve an equivalent compound inequality.

$$\begin{array}{rcl} -1 \leq & 5 - 4x & \leq 1 \\ -6 \leq & -4x & \leq -4 \\ \frac{3}{2} \geq & x & \geq 1 \end{array}$$

The solution set is the interval $[1, 3/2]$ and the graph is $\longleftarrow [1 \;\; 3/2] \longrightarrow$

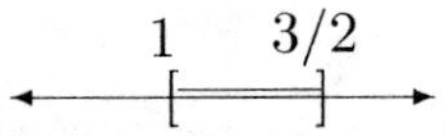

63. Solve an equivalent compound inequality.

$$\begin{array}{ccc} x-1\geq 1 & \text{or} & x-1\leq -1 \\ x\geq 2 & \text{or} & x\leq 0 \end{array}$$

The solution set is $(-\infty, 0] \cup [2, \infty)$ and the graph is $\longleftarrow 0] \;\; [2 \longrightarrow$

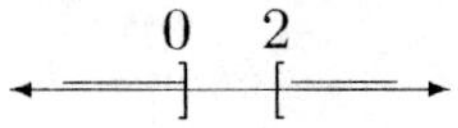

65. Solve an equivalent compound inequality.

$$\begin{array}{ccc} 5-x>3 & \text{or} & 5-x<-3 \\ 2>x & \text{or} & 8<x \end{array}$$

The solution set is $(-\infty, 2) \cup (8, \infty)$ and the graph is $\longleftarrow 2) \;\; (8 \longrightarrow$

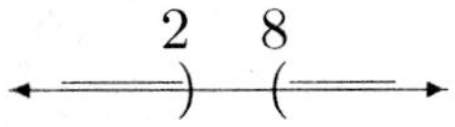

67. Solve an equivalent compound inequality.

$$\begin{array}{rcl} -5\leq & 4-x & \leq 5 \\ -9\leq & -x & \leq 1 \\ 9\geq & x & \geq -1 \end{array}$$

The solution set is the interval $[-1, 9]$ and the graph is $\longleftarrow [-1 \;\; 9] \longrightarrow$

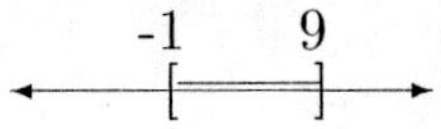

69. No solution since an absolute value is never negative. The solution set is $\emptyset$.

71. No solution since an absolute value is never negative. The solution set is $\emptyset$.

73. Note, $3|x-2| > 3$ or $|x-2| > 1$.
We solve an equivalent compound inequality.

$$\begin{array}{ccc} x-2>1 & \text{or} & x-2<-1 \\ x>3 & \text{or} & x<1 \end{array}$$

The solution set is $(-\infty, 1) \cup (3, \infty)$ and the graph is $\longleftarrow 1) \;\; (3 \longrightarrow$

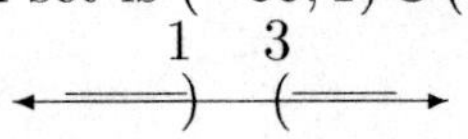

75. Solve an equivalent compound inequality.

$$\begin{array}{ccc} \dfrac{x-3}{2}>1 & \text{or} & \dfrac{x-3}{2}<-1 \\ x-3>2 & \text{or} & x-3<-2 \\ x>5 & \text{or} & x<1 \end{array}$$

The solution set is the interval $(-\infty, 1) \cup (5, \infty)$ and the graph is $\longleftarrow 1) \;\; (5 \longrightarrow$

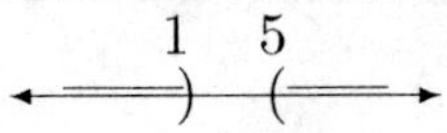

77. $|x| < 5$

79. $|x| > 3$

81. Since 6 is the midpoint of 4 and 8, the inequality is $|x-6| < 2$.

83. Since 4 is the midpoint of 3 and 5, the inequality is $|x-4| > 1$.

85. $|x| \geq 9$

87. Since 7 is the midpoint, the inequality is $|x-7| \leq 4$.

89. Since 5 is the midpoint, the inequality is $|x-5| > 2$.

91. Since $x-2 \geq 0$, the solution set is $[2, \infty)$.

93. Since $2-x > 0$ is equivalent to $2 > x$, the solution set is $(-\infty, 2)$.

95. Since $|x| \geq 3$ is equivalent to $x \geq 3$ or $x \leq -3$, the solution set is $(-\infty, -3] \cup [3, \infty)$.

97. If x is the price of a car excluding sales tax then it must satisfy $0 \leq 1.1x + 300 \leq 8000$.

This is equivalent to $0 \leq x \leq \dfrac{7700}{1.1} = 7000$.

The price range of Yolanda's car is the interval [\$0, \$7000].

99. Let x be Lucky's score on the final exam.

$$\begin{array}{rcl} 79 & \leq & \dfrac{65+x}{2} \leq 90 \\ 158 & \leq & 65+x \leq 180 \\ 93 & \leq & x \leq 115 \end{array}$$

Since $x \leq 100$, the final exam score must lie in $[93, 100]$.

101. Let x be Ingrid's final exam score. Since $\dfrac{2x+65}{3}$ is her weighted average, we obtain

$$\begin{array}{rcl} 79 & < & \dfrac{2x+65}{3} < 90 \\ 237 & < & 2x+65 < 270 \\ 172 & < & 2x < 205 \\ 86 & < & x < 102.5 \end{array}$$

Since $x \leq 100$, Ingrid's final exam score must lie in $(86, 100]$.

103. If h is the height of the box, then

$$\begin{aligned} 40 + 2(30) + 2h &\leq 130 \\ 100 + 2h &\leq 130 \\ 2h &\leq 30. \end{aligned}$$

The range of the height is (0 in., 15 in.].

105. By substituting $N = 50$ and $w = 27$ into $r = \dfrac{Nw}{n}$ we find $r = \dfrac{1350}{n}$. Moreover if $n = 14$, then $r = \dfrac{1350}{14} = 96.4 \approx 96$.
Similarly, the other gear ratios are the following.

n	14	17	20	24	29
r	96	79	68	56	47

Yes, the bicycle has a gear ratio for each of the four types.

107. Let x be the price of a CL 600.

a) $|x - 130,645| > 10,000$

b) The above inequality is equivalent to

$$\begin{aligned} x - 130,645 > 10,000 \quad &\text{or} \quad x - 130,645 < -10,000 \\ x > 140,645 \quad &\text{or} \quad x < 120,645. \end{aligned}$$

Thus, the price of a CL 600 is less than \$120,645 or more than \$140,645.

109. If x is the actual temperature, then

$$\begin{aligned} \left|\frac{x-35}{35}\right| &< 0.01 \\ -0.35 < x - 35 &< 0.35 \\ 34.65 < x &< 35.35. \end{aligned}$$

The actual temperature must lie in the interval $(34.65°, 35.35°)$.

111. If c is the actual circumference, then $c = \pi d$ and

$$\begin{aligned} |\pi d - 7.2| &\leq 0.1 \\ -0.1 \leq \pi d - 7.2 &\leq 0.1 \\ 7.1 \leq \pi d &\leq 7.3 \\ 2.26 \leq d &\leq 2.32. \end{aligned}$$

The actual diameter must lie in the interval [2.26 cm, 2.32 cm].

113. a) The inequality $|a - 40,584| < 3000$ is equivalent to

$$\begin{aligned} -3000 < \;& a - 40,584 < 3000 \\ 37,584 < \;& a < 43,584. \end{aligned}$$

The states within this range are Colorado, Iowa, and Vermont.

b) The inequality $|a - 40,584| > 5000$ is equivalent to

$$\begin{aligned} a - 40,584 > 5000 \quad &\text{or} \quad a - 40,584 < -5000 \\ a > 45,584 \quad &\text{or} \quad a < 35,584 \end{aligned}$$

The states satisfying the inequality are Alabama, Georgia, Maryland, New Jersey, and South Carolina.

115.

$$\begin{aligned} x(x+2) &= 0 \\ x &= 0, -2 \end{aligned}$$

The solution set is $\{-2, 0\}$.

117. Since the slope of $2x - y = 1$ is 2, the slope of a perpendicular line is $-\frac{1}{2}$. Then

$$\begin{aligned} y + 4 &= -\frac{1}{2}(x-3) \\ -2y - 8 &> x - 3 \\ -5 &> x + 2y. \end{aligned}$$

The standard form is $x + 2y = -5$.

119. Solving for y, we find

$$\begin{aligned} 3y - ay &= w + 9 \\ y(3-a) &= w + 9 \\ y &= \frac{w+9}{3-a} \end{aligned}$$

121. Consider the list of 6 digit numbers from 000,000 through 999,999. There are 1 million 6 digit numbers in this list for a total of 6 million digits. Each of the ten digits 0 through 9 occurs with the same frequency in this list. So there are 600,000 of each in this list. In particular there are 600,000 ones in the list. You need one more to write 1,000,000. So there are 600,001 ones used in writing the numbers 1 through 1 million.

Chapter 1 Review Exercises

1. Since $3x = 2$, the solution set is $\{2/3\}$.

3. Multiply by 60 to get $30y - 20 = 15y + 12$, or $15y = 32$. The solution set is $\{32/15\}$.

5. Multiply by $x(x-1)$ to get $2x - 2 = 3x$. The solution set is $\{-2\}$.

7. Multiply by $(x-1)(x-3)$ and get $-2x - 3 = x - 2$. Then $-1 = 3x$. The solution set is $\{-1/3\}$.

9. The distance is $\sqrt{(-3-2)^2 + (5-(-6))^2} = \sqrt{(-5)^2 + 11^2} = \sqrt{25+121} = \sqrt{146}$.

The midpoint is $\left(\dfrac{-3+2}{2}, \dfrac{5-6}{2}\right) = \left(-\dfrac{1}{2}, -\dfrac{1}{2}\right)$.

11. Distance is $\sqrt{\left(\dfrac{1}{2}-\dfrac{1}{4}\right)^2 + \left(\dfrac{1}{3}-1\right)^2} = \sqrt{\left(\dfrac{1}{4}\right)^2 + \left(-\dfrac{2}{3}\right)^2} = \sqrt{\dfrac{1}{16}+\dfrac{4}{9}} = \sqrt{\dfrac{73}{144}} = \dfrac{\sqrt{73}}{12}$. Midpoint is $\left(\dfrac{1/2+1/4}{2}, \dfrac{1/3+1}{2}\right) = \left(\dfrac{3/4}{2}, \dfrac{4/3}{2}\right) = \left(\dfrac{3}{8}, \dfrac{2}{3}\right)$.

13. Circle with radius 5 and center at the origin.

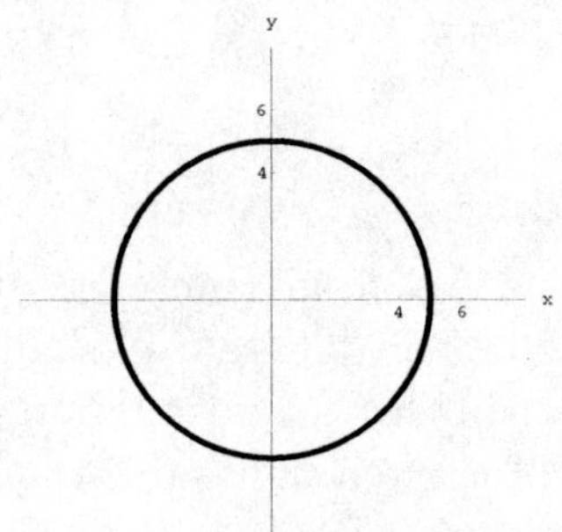

15. Equivalently, by using the method of completing the square, the circle is given by $(x+2)^2 + y^2 = 4$. It has radius 2 and center $(-2, 0)$.

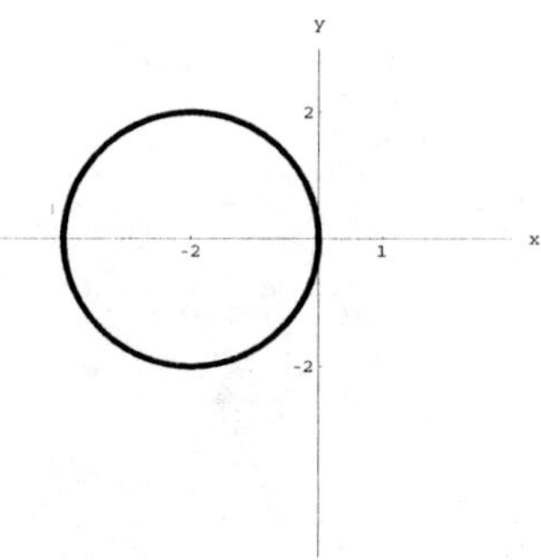

17. The line $y = -x + 25$ has intercepts $(0, 25)$, $(25, 0)$.

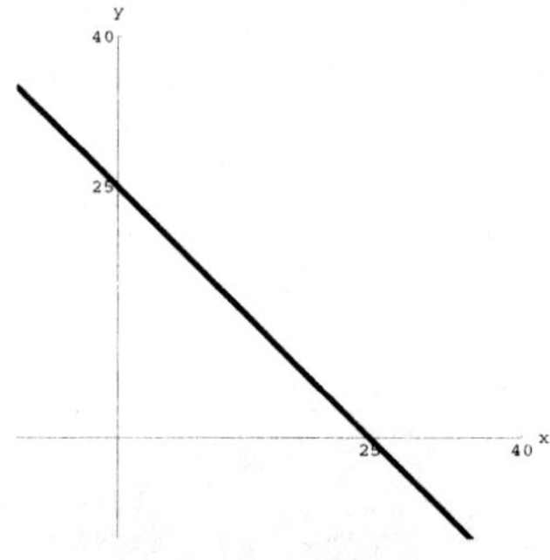

19. The line $y = 3x - 4$ has intercepts $(0, -4)$, $(4/3, 0)$.

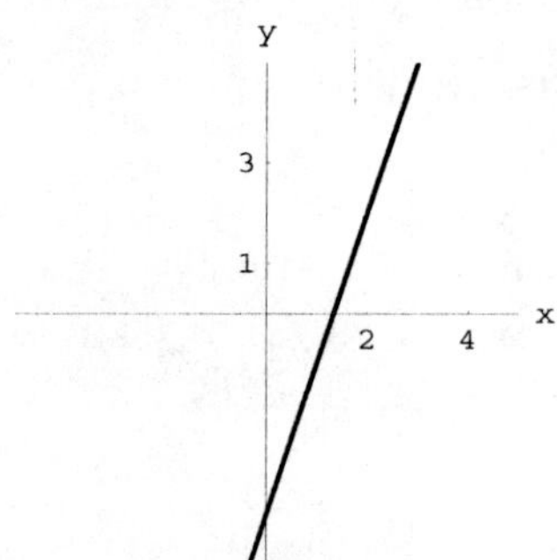

21. Vertical line $x = 5$ has intercept $(5, 0)$.

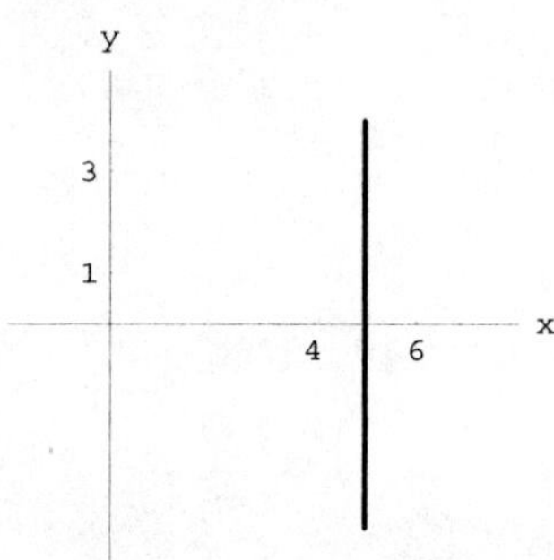

23. Simplify $(x-(-3))^2 + (y-5)^2 = \left(\sqrt{3}\right)^2$. The standard equation is $(x+3)^2 + (y-5)^2 = 3$.

25. Substitute $y = 0$ in $3x - 4y = 12$. Then $3x = 12$ or $x = 4$. The x-intercept is $(4, 0)$.

Substitute $x = 0$ in $3x - 4y = 12$ to get $-4y = 12$ or $y = -3$. The y-intercept is $(0, -3)$.

27. $\dfrac{2-(-6)}{-1-3} = \dfrac{8}{-4} = -2$

29. Note, $m = \dfrac{-1-3}{5-(-2)} = -\dfrac{4}{7}$. Solving for y in $y - 3 = -\dfrac{4}{7}(x+2)$, we obtain $y = -\dfrac{4}{7}x + \dfrac{13}{7}$.

31. Note, the slope of $3x + y = -5$ is -3. The standard form for the line through $(2, -4)$ with slope $\dfrac{1}{3}$ is derived below.

$$\begin{aligned} y + 4 &= \frac{1}{3}(x-2) \\ 3y + 12 &= x - 2 \\ -x + 3y &= -14 \\ x - 3y &= 14 \end{aligned}$$

33. Since $2x - 6 = 3y$, $y = \dfrac{2}{3}x - 2$.

35. Note, $y(x-3) = 1$. Then $y = \dfrac{1}{x-3}$.

37. Note, $by = -ax + c$. Then $y = -\dfrac{a}{b}x + \dfrac{c}{b}$ provided $b \neq 0$.

39. The discriminant of $x^2 - 4x + 2$ is $(-4)^2 - 4(2) = 8$. There are two distinct real solutions.

41. The discriminant is $(-20)^2 - 4(4)(25) = 0$. Only one real solution.

43. Since $x^2 = 5$, the solution set is $\{\pm\sqrt{5}\}$.

45. Since $x^2 = -8$, the solution set is $\{\pm 2i\sqrt{2}\}$.

47. Since $x^2 = -\dfrac{2}{4}$, the solution set is $\left\{\pm i\dfrac{\sqrt{2}}{2}\right\}$.

49. Since $x - 2 = \pm\sqrt{17}$, we get $x = 2 \pm \sqrt{17}$. The solution set is $\{2 \pm \sqrt{17}\}$.

51. Since $(x+3)(x-4) = 0$, the solution set is $\{-3, 4\}$.

53. We apply the method of completing the square.

$$\begin{aligned} b^2 - 6b + 10 &= 0 \\ b^2 - 6b + 9 &= -10 + 9 \\ (b-3)^2 &= -1 \\ b - 3 &= \pm i \end{aligned}$$

The solution set is $\{3 \pm i\}$.

55. We apply the method of completing the square.

$$\begin{aligned} s^2 - 4s &= -1 \\ s^2 - 4s + 4 &= -1 + 4 \\ (s-2)^2 &= 3 \\ s - 2 &= \pm\sqrt{3} \end{aligned}$$

The solution set is $\{2 \pm \sqrt{3}\}$.

57. Use the quadratic formula to solve $4x^2 - 4x - 5 = 0$.

$$\begin{aligned} x &= \frac{4 \pm \sqrt{(-4)^2 - 4(4)(-5)}}{2(4)} \\ &= \frac{4 \pm \sqrt{96}}{8} \\ &= \frac{4 \pm 4\sqrt{6}}{8} \\ &= \frac{1 \pm \sqrt{6}}{2} \end{aligned}$$

The solution set is $\left\{\dfrac{1 \pm \sqrt{6}}{2}\right\}$.

59. Subtracting 1 from both sides, we find

$$\begin{aligned} x^2 - 2x + 1 &= -1 \\ (x-1)^2 &= -1 \\ x - 1 &= \pm i. \end{aligned}$$

The solution set is $\{1 \pm i\}$.

61. Multiplying by $2x(x-1)$, we obtain

$$\begin{aligned} 2(x-1) + 2x &= 3x(x-1) \\ 0 &= 3x^2 - 7x + 2 \\ 0 &= (x-2)(3x-1). \end{aligned}$$

The solution set is $\left\{\dfrac{1}{3}, 2\right\}$.

63. Solve an equivalent statement

$$3q-4=2 \quad \text{or} \quad 3q-4=-2$$
$$3q=6 \quad \text{or} \quad 3q=2.$$

The solution set is $\{2/3, 2\}$.

65. We obtain

$$\begin{aligned} |2h-3| &= 0 \\ 2h-3 &= 0 \\ h = \frac{3}{2}. \end{aligned}$$

The solution set is $\left\{\frac{3}{2}\right\}$.

67. No solution since absolute values are nonnegative. The solution set is $\emptyset$.

69. Solve an equivalent statement assuming $3v \geq 0$.

$$2v-1=3v \quad \text{or} \quad 2v-1=-3v$$
$$-1=v \quad \text{or} \quad 5v=1$$

Since $3v \geq 0$, $v=-1$ is an extraneous root.

The solution set is $\{1/5\}$.

71. Let $w=x^2$ and $w^2=x^4$.

$$\begin{aligned} w^2+7w &= 18 \\ (w+9)(w-2) &= 0 \\ w &= -9, 2 \\ x^2=-9 \quad &\text{or} \quad x^2=2. \end{aligned}$$

Since $x^2=-9$ has no real solution, the solution set is $\{\pm\sqrt{2}\}$.

73. Isolate a radical and square both sides.

$$\begin{aligned} \sqrt{x+6} &= \sqrt{x-5}+1 \\ x+6 &= x-5+2\sqrt{x-5}+1 \\ 5 &= \sqrt{x-5} \\ 25 &= x-5 \end{aligned}$$

The solution set is $\{30\}$.

75. Let $w=\sqrt[4]{y}$ and $w^2=\sqrt{y}$.

$$\begin{aligned} w^2+w-6 &= 0 \\ (w+3)(w-2) &= 0 \\ w=-3 \quad &\text{or} \quad w=2 \\ y^{1/4}=-3 \quad &\text{or} \quad y^{1/4}=2 \end{aligned}$$

Since $y^{1/4}=-3$ has no real solution, the solution set is $\{16\}$.

77. Let $w=x^2$ and $w^2=x^4$.

$$\begin{aligned} w^2-3w-4 &= 0 \\ (w+1)(w-4) &= 0 \\ x^2=-1 \quad &\text{or} \quad x^2=4 \end{aligned}$$

Since $x^2=-1$ has no real solution, the solution set is $\{\pm 2\}$.

79. Raise to the power 3/2 and get $x-1=\pm(4^{1/2})^3$. So, $x=1\pm 8$. The solution set is $\{-7, 9\}$.

81. No solution since $(x+3)^{-3/4}$ is nonnegative.

83. Since $3x-7=4-x$, we obtain $4x=11$. The solution set is $\{11/4\}$.

85. The solution set of $x>3$ is the interval $(3,\infty)$ and the graph is

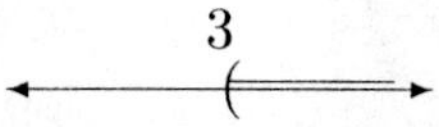

3

87. The solution set of $8>2x$ is the interval $(-\infty, 4)$ and the graph is

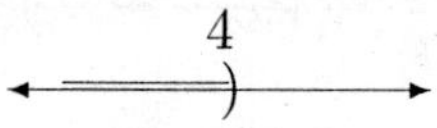

4

89. Since $-\frac{7}{3} > \frac{1}{2}x$, the solution set is $(-\infty, -14/3)$ and the graph is

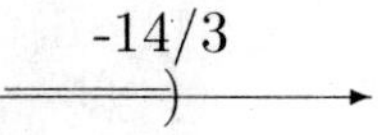

-14/3

91. After multiplying the inequality by 2 we have

$$\begin{aligned} -4<x-3 &\leq 10 \\ -1<x &\leq 13. \end{aligned}$$

The solution set is the interval $(-1, 13]$ and the graph is

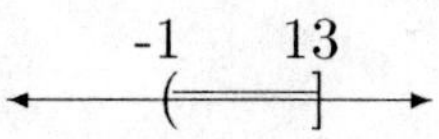

-1 13

93. The solution set of $\frac{1}{2}<x$ and $x<1$ is the interval $(1/2, 1)$ and the graph is

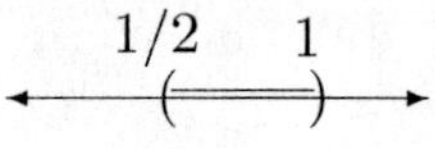

1/2 1

95. The solution set of $x > -4$ or $x > -1$ is the interval $(-4, \infty)$ and the graph is

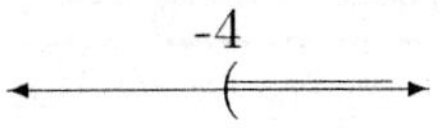

97. Solving an equivalent statement, we get

$$\begin{array}{ccc} x-3>2 & \text{or} & x-3<-2 \\ x>5 & \text{or} & x<1. \end{array}$$

The solution set is $(-\infty, 1) \cup (5, \infty)$ and the graph is

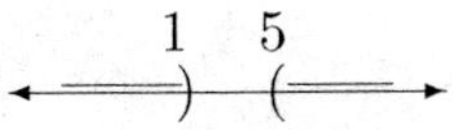

99. Since an absolute value is nonnegative, $2x - 7 = 0$. The solution set is $\{7/2\}$ and the graph is

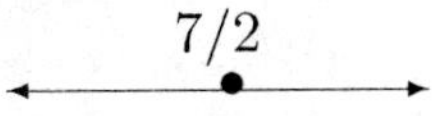

101. Since absolute values are nonnegative, the solution set is $(-\infty, \infty)$ and

the graph is

103. The solution set is $\{10\}$ since the x-intercept is $(10, 0)$.

105. Since the x-intercept is $(8, 0)$ and the y-values are negative in quadrants 3 and 4, the solution set is $(-\infty, 8)$.

107. Let x be the length of one side of the square. Since dimensions of the base are $8 - 2x$ and $11 - 2x$, we obtain

$$\begin{aligned} (11-2x)(8-2x) &= 50 \\ 4x^2 - 38x + 38 &= 0 \\ 2x^2 - 19x + 19 &= 0 \\ x &= \frac{19 \pm \sqrt{209}}{4}. \end{aligned}$$

But $x = \dfrac{19 + \sqrt{209}}{4} \approx 8.36$ is too big.

Then $x = \dfrac{19 - \sqrt{209}}{4} \approx 1.14$ in.

109. Let x be the number of hours it takes Lisa or Taro to drive to the restaurant. Since the sum of the driving distances is 300, we obtain

$300 = 50x + 60x$. Thus, $x = \dfrac{300}{110} \approx 2.7272$

and Lisa drove $50(2.7272) \approx 136.4$ miles.

111. Let x and $8000 - x$ be the number of fish in Homer Lake and Mirror Lake, respectively. Then

$$\begin{aligned} 0.2x + 0.3(8000 - x) &= 0.28(8000) \\ -0.1x + 2400 &= 2240 \\ 1600 &= x. \end{aligned}$$

There were originally 1600 fish in Homer Lake.

113. Let x be the distance she hiked in the northern direction. At 4 mph and 8 hr, she hiked 32 miles. Then she hiked $32 - x$ miles in the eastern direction. By the Pythagorean Theorem, we obtain

$$\begin{aligned} x^2 + (32 - x)^2 &= (4\sqrt{34})^2 \\ 2x^2 - 64x + 480 &= 0 \\ 2(x - 20)(x - 12) &= 0 \\ x &= 20, 12. \end{aligned}$$

Since the eastern direction was the shorter leg of the journey, the northern direction was 20 miles.

115. Let x and $x + 50$ be the cost of a haircut at Joe's and Renee's, respectively. Since 5 haircuts at Joe's is less than one haircut at Renee's, we have

$$5x < x + 50.$$

Thus, the price range of a haircut at Joe's is $x < \$12.50$ or $(0, \$12.50)$.

117. Let x and $x + 2$ be the width and length of a picture frame in inches, respectively. Since there are between 32 and 50 inches of molding, we get

$$\begin{aligned} 32 &< 2x + 2(x + 2) < 50 \\ 32 &< 4x + 4 < 50 \\ 28 &< 4x < 46 \\ 7 \text{ in.} &< x < 11.5 \text{ in.} \end{aligned}$$

The set of possible widths is (7 in, 11.5 in).

119. If the average gas mileage is increased from 29.5 mpg to 31.5 mpg, then the amount of gas saved is

$$\frac{10^{12}}{29.5} - \frac{10^{12}}{31.5} \approx 2.15 \times 10^9 \text{ gallons.}$$

Suppose the mileage is increased to x from 29.5 mpg. Then x must satisfy

$$\begin{aligned} \frac{10^{12}}{29.5} - \frac{10^{12}}{x} &= \frac{10^{12}}{27.5} - \frac{10^{12}}{29.5} \\ \frac{1}{29.5} - \frac{1}{x} &= \frac{1}{27.5} - \frac{1}{29.5} \\ -\frac{1}{x} &\approx -0.031433 \\ x &\approx 31.8. \end{aligned}$$

The mileage must be increased to 31.8 mpg.

121. a) The line is given by

$$y \approx 17.294x - 34,468$$

where x is the year and y is the number of millions of cell users. We use three decimal places.

b) If $x = 2016$, the number of millions of cell users is

$$y \approx 17.294(2016) - 34,468 \approx 397.$$

There will be 397 million cell users in 2016.

123. Let a be the age in years and p be the percentage. The equation of the line passing through $(20, 0.23)$ and $(50, 0.47)$ is

$$p = 0.008a + 0.07.$$

If $a = 65$, then $p = 0.008(65) + 0.07 \approx 0.59$. Thus, the percentage of body fat in a 65-year old woman is 59%.

125. a) Using a calculator, the regression line is given by

$$y \approx 3.52x + 48.03$$

where $x = 0$ corresponds to 2000.

b) If $x = 17$, then the average price of a prescription in 2017 is

$$y \approx 3.52(17) + 48.03 = \$107.87.$$

127. Circle A is given by

$$(x-1)^2 + (y-1)^2 = 1.$$

Draw a right triangle with sides 1 and x, and with hypotenuse 3 such that the hypotenuse has as endpoints the centers of circles A and B. Here, x is the horizontal distance between the centers of A and B. Since

$$1 + x^2 = 9$$

we obtain $x = 2\sqrt{2}$. Then the center of B is $(1 + 2\sqrt{2}, 2)$. Thus, circle B is given by

$$\left(x - 1 - 2\sqrt{2}\right)^2 + (y-2)^2 = 4.$$

Let r and (a, r) be the radius and center of circle C. Draw a right triangle with sides $a - 1$ and $1 - r$, and with hypotenuse $1 + r$ such that the hypotenuse has as endpoints the centers of circles A and C. Then

$$(1+r)^2 = (1-r)^2 + (a-1)^2.$$

Next, draw a right triangle with sides $1 + 2\sqrt{2} - a$ and $2 - r$, and with hypotenuse $2 + r$ such that the hypotenuse has as endpoints the centers of circles B and C. Then

$$(2+r)^2 = (2-r)^2 + (1 + 2\sqrt{2} - a)^2.$$

The solution of the two previous equations are

$$a = 5 - 2\sqrt{2}, \quad r = 6 - 4\sqrt{2}.$$

Hence, circle C is given by

$$(x - 5 + 2\sqrt{2})^2 + (y - 6 + 4\sqrt{2})^2 = (6 - 4\sqrt{2})^2$$

Chapter 1 Test

1. Since $2x - x = -6 - 1$, the solution set is $\{-7\}$.

2. Multiplying the original equation by 6, we get $3x - 2x = 1$. The solution set is $\{1\}$.

3. Since $x^2 = \dfrac{2}{3}$, one obtains $x = \pm\dfrac{\sqrt{2}}{\sqrt{3}} = \pm\dfrac{\sqrt{6}}{3}$.

The solution set is $\left\{\pm\dfrac{\sqrt{6}}{3}\right\}$.

4. By completing the square, we obtain

$$\begin{aligned} x^2 - 6x &= -1 \\ x^2 - 6x + 9 &= -1 + 9 \\ (x-3)^2 &= 8 \\ x - 3 &= \pm\sqrt{8}. \end{aligned}$$

The solution set is $\{3 \pm 2\sqrt{2}\}$.

5. Since $x^2 - 9x + 14 = (x-2)(x-7) = 0$, the solution set is $\{2, 7\}$.

6. After cross-multiplying, we get

$$\begin{aligned} (x-1)(x-6) &= (x+3)(x+2) \\ x^2 - 7x + 6 &= x^2 + 5x + 6 \\ -7x &= 5x \\ 0 &= 12x \end{aligned}$$

The solution set is $\{0\}$.

7. We use the method of completing the square.

$$\begin{aligned} x^2 - 2x &= -5 \\ x^2 - 2x + 1 &= -5 + 1 \\ (x-1)^2 &= -4 \\ x - 1 &= \pm 2i \end{aligned}$$

The solution set is $\{1 \pm 2i\}$.

8. Since $x^2 = -1$, the solution set is $\{\pm i\}$.

9. The line $3x - 4y = 120$ passes through $(0, -30)$ and $(40, 0)$.

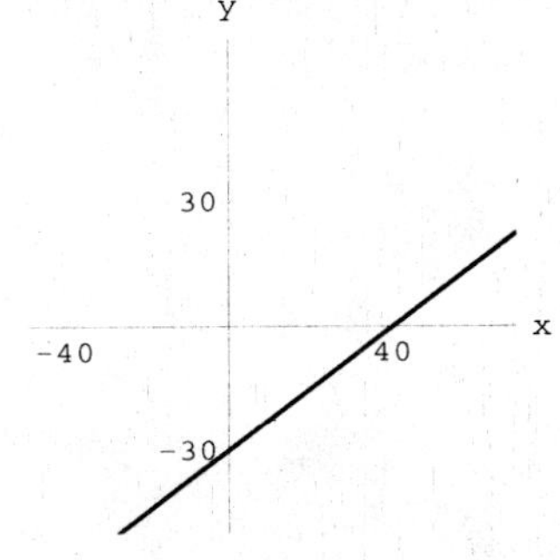

10. Circle with center $(0, 0)$ and radius 20.

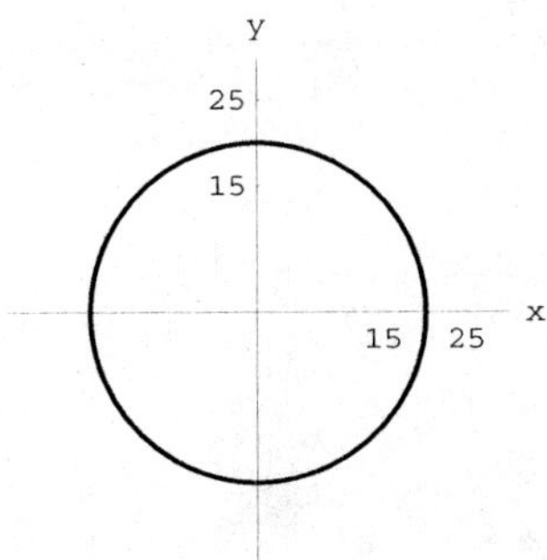

11. By using the method of completing the square, we obtain $x^2+(y+2)^2 = 4$. A circle with center $(0, -2)$ and radius 2.

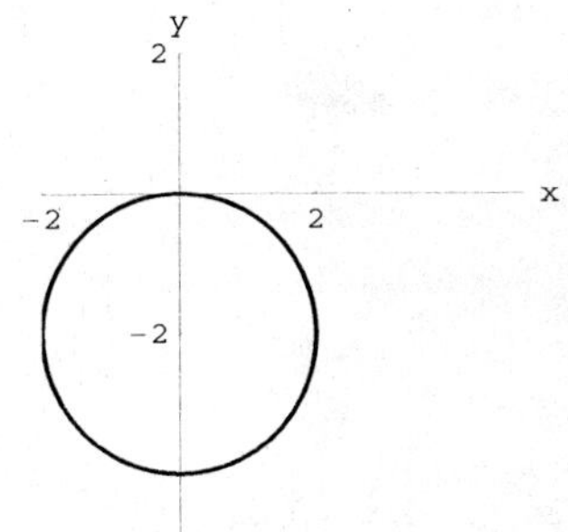

12. The line $y = -\frac{2}{3}x + 4$ passes through $(0, 4)$ and $(6, 0)$.

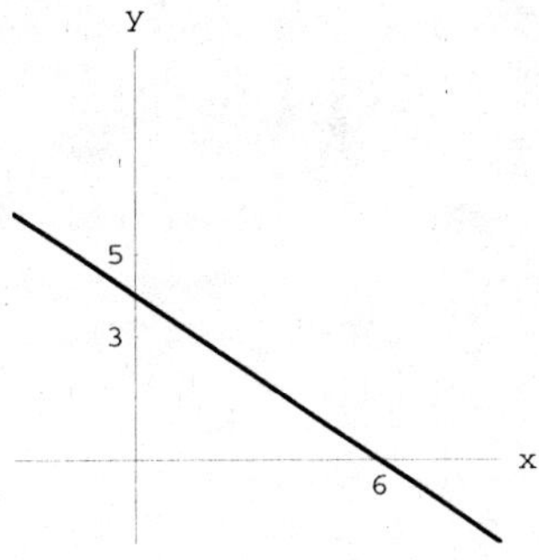

13. The horizontal line $y = 4$.

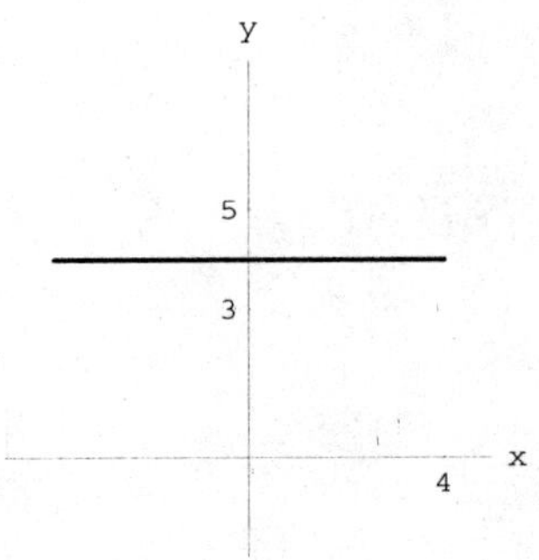

14. The vertical line $x = -2$.

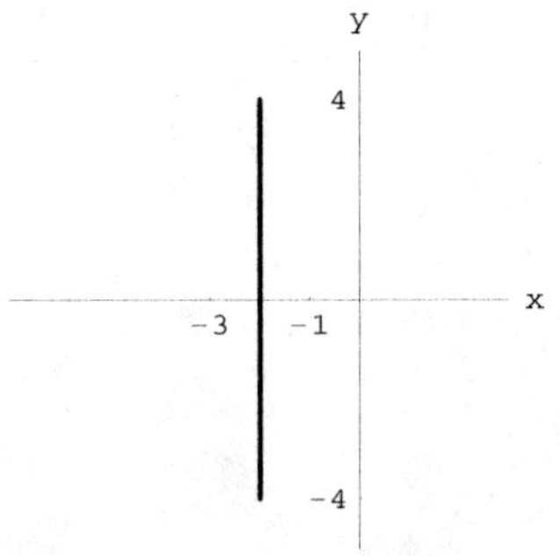

15. Since $y = \frac{3}{5}x - \frac{8}{5}$, the slope is $\frac{3}{5}$.

16. $\frac{-4-6}{5-(-3)} = \frac{-10}{8} = -\frac{5}{4}$

17. We rewrite $2x - 3y = 6$ as $y = \frac{2}{3}x - 2$. Note, the slope of $y = \frac{2}{3}x - 2$ is $\frac{2}{3}$. Then we use $m = -\frac{3}{2}$ and the point $(1, -2)$.

$$\begin{aligned} y+2 &= -\frac{3}{2}(x-1) \\ y &= -\frac{3}{2}x + \frac{3}{2} - 2. \end{aligned}$$

The perpendicular line is $y = -\frac{3}{2}x - \frac{1}{2}$.

18. The slope is $\frac{-1-2}{3-0} = -1$. We use the point $(3, -4)$.

$$\begin{aligned} y+4 &= -(x-3) \\ y &= -x + 3 - 4. \end{aligned}$$

The parallel line is $y = -x - 1$.

19. $\sqrt{(-3-2)^2 + (1-4)^2} = \sqrt{25+9} = \sqrt{34}$

20. $\left(\frac{-1+1}{2}, \frac{1+0}{2}\right) = \left(0, \frac{1}{2}\right)$

21. Since the discriminant is negative, namely, $(-5)^2 - 4(1)(9) = -11$, then there are no real solutions.

22. We solve for y.

$$\begin{aligned} 5-4 &= 3xy + 2y \\ 1 &= y(3x+2) \\ y &= \frac{1}{3x+2} \end{aligned}$$

23. If we raise each side of $(x-3)^{-2/3} = \frac{1}{3}$ to the power -3, then we obtain $(x-3)^2 = 27$. Thus, $x = 3 \pm \sqrt{27}$ and the solution set is $\{3 \pm 3\sqrt{3}\}$.

24. Isolate a radical and square each side.

$$\begin{aligned} \sqrt{x} - 1 &= \sqrt{x-7} \\ x - 2\sqrt{x} + 1 &= x - 7 \\ \sqrt{x} &= 4 \end{aligned}$$

The solution set is $\{16\}$.

25. Since $-4 > 2x$, the solution set is $(-\infty, -2)$ and the graph is 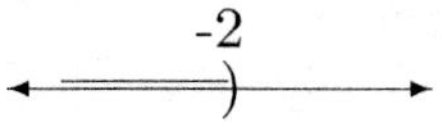

26. The solution set to $x > 6$ and $x > 5$ is the interval $(6, \infty)$ and the graph is 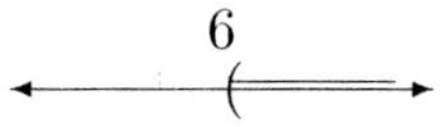

27. Solving an equivalent statement, we obtain

$$\begin{aligned} -3 &\le 2x - 1 \le 3 \\ -2 &\le 2x \le 4 \\ -1 &\le x \le 2. \end{aligned}$$

The solution set is the interval $[-1, 2]$ and the graph is 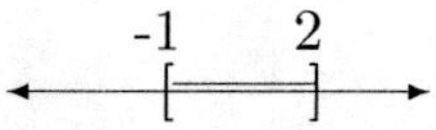

28. We rewrite $|x-3| > 2$ without any absolute values. Then

$$\begin{aligned} x - 3 > 2 &\quad \text{or} \quad x - 3 < -2 \\ x > 5 &\quad \text{or} \quad x < 1. \end{aligned}$$

The solution set is $(-\infty, 1) \cup (5, \infty)$ and the graph is 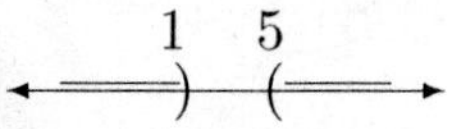

29. If x is the original length of one side of the square, then

$$\begin{aligned} (x+20)(x+10) &= 999 \\ x^2 + 30x + 200 &= 999 \\ x^2 + 30x - 799 &= 0 \\ \frac{-30 \pm \sqrt{900 + 4(799)}}{2} &= x \\ \frac{-30 \pm 64}{2} &= x \\ 17, -47 &= x. \end{aligned}$$

Thus, $x = 17$ and the original area is $17^2 = 289$ ft^2.

30. Let x be the number of gallons of the 20% solution. From the concentrations,

$$\begin{aligned} 0.3(10+x) &= 0.5(10)+0.2x \\ 3+0.3x &= 5+0.2x \\ 0.1x &= 2 \\ x &= 20. \end{aligned}$$

Then 20 gallons of the 20% solution are needed.

31. a) Using a calculator, the regression line is given by

$$y \approx 18.4x + 311$$

where $x = 0$ corresponds to 1997 and y is the median price of a home in thousands of dollars.

b) If $x = 18$, then

$$y \approx 18.4(18) + 311 \approx 642.$$

The predicted median price in 2015 is $642,000.

32. If $D = 10$, then $T = 0.07(10)^{3/2} \approx 2.2$ hr.

If $T = 4$, then $D = \left(\frac{4}{0.07}\right)^{2/3} \approx 14.8$ mi.

Tying It All Together

1. $7x$ **2.** $30x^2$ **3.** $\dfrac{2}{2x} + \dfrac{1}{2x} = \dfrac{3}{2x}$

4. $x^2 + 6x + 9$ **5.** $6x^2 + x - 2$

6. $\dfrac{x^2 + 2xh + h^2 - x^2}{h} = \dfrac{2xh + h^2}{h} = 2x + h$

7. $\dfrac{x+1}{(x-1)(x+1)} + \dfrac{x-1}{(x+1)(x-1)} = \dfrac{2x}{x^2-1}$

8. $x^2 + 3x + \dfrac{9}{4}$

9. An identity, the solution set is $(-\infty, \infty)$ or $\mathbb{R}$.

10. Since $30x^2 - 11x = x(30x - 11) = 0$, the solution set is $\left\{0, \dfrac{11}{30}\right\}$.

11. Since $\dfrac{3}{2x} = \dfrac{3}{2x}$, the solution set is $(-\infty, 0) \cup (0, \infty)$.

12. Subtract $x^2 + 9$ from $x^2 + 6x + 9 = x^2 + 9$ and get $6x = 0$. The solution set is $\{0\}$.

13. Since $(2x - 1)(3x + 2) = 0$, the solution set is $\left\{-\dfrac{2}{3}, \dfrac{1}{2}\right\}$.

14. Multiply the equation by $8(x + 1)(x - 1)$.

$$\begin{aligned} 8x+8+8x-8 &= 5(x^2-1) \\ 0 &= 5x^2 - 16x - 5 \\ x &= \frac{16 \pm \sqrt{356}}{10} \\ x &= \frac{16 \pm 2\sqrt{89}}{10} \end{aligned}$$

The solution set is $\left\{\dfrac{8 \pm \sqrt{89}}{5}\right\}$.

15. Since $7x - 7x^2 = 7x(1 - x) = 0$, the solution set is $\{0, 1\}$.

16. Since $7x - 7 = 7(x - 1) = 0$, the solution set is $\{1\}$.

17. 0 **18.** $-1 - 3 + 2 = -2$

19. $-\dfrac{1}{8} - 3\left(\dfrac{2}{8}\right) + \dfrac{16}{8} = \dfrac{9}{8}$

20. $-\dfrac{1}{27} - 3\left(\dfrac{3}{27}\right) + \dfrac{54}{27} = \dfrac{44}{27}$

21. $-1 - 3 - 4 = -8$

22. $-4 + 6 - 4 = -2$ **23.** -4

24. $-0.25 + 1.5 - 4 = -2.75$

25. conditional

26. identity

27. inconsistent

28. distributive

29. associative

30. irrational

31. additive, multiplicative

32. difference

33. quotient

34. additive inverse, opposite

For Thought

1. False, since $\{(1,2),(1,3)\}$ is not a function.

2. False, since $f(5)$ is not defined. **3.** True

4. False, since a student's exam grade is a function of the student's preparation. If two classmates had the same IQ and only one prepared, then the one who prepared will most likely achieve a higher grade.

5. False, since $(x+h)^2 = x^2 + 2xh + h^2$

6. False, since the domain is all real numbers.

7. True **8.** True **9.** True

10. False, since $\left(\dfrac{3}{8}, 8\right)$ and $\left(\dfrac{3}{8}, 5\right)$ are two ordered pairs with the same first coordinate and different second coordinates.

2.1 Exercises

1. function

3. relation

5. independent, dependent

7. difference quotient

9. Note, $b = 2\pi a$ is equivalent to $a = \dfrac{b}{2\pi}$.

Then a is a function of b, and b is a function of a.

11. a is a function of b since a given denomination has a unique length. Since a dollar bill and a five-dollar bill have the same length, then b is not a function of a.

13. Since an item has only one price, b is a function of a. Since two items may have the same price, a is not a function of b.

15. a is not a function of b since it is possible that two different students can obtain the same final exam score but the times spent on studying are different.

b is not a function of a since it is possible that two different students can spend the same time studying but obtain different final exam scores.

17. Since 1 in $\approx$ 2.54 cm, a is a function of b and b is a function of a.

19. No **21.** Yes **23.** Yes **26.** Yes

27. Not a function since 25 has two different second coordinates.

29. Not a function since 3 has two different second coordinates.

31. Yes

33. Since the ordered pairs in the graph of $y = 3x - 8$ are $(x, 3x - 8)$, there are no two ordered pairs with the same first coordinate and different second coordinates. We have a function.

35. Since $y = (x + 9)/3$, the ordered pairs are $(x, (x+9)/3)$. Thus, there are no two ordered pairs with the same first coordinate and different second coordinates. We have a function.

37. Since $y = \pm x$, the ordered pairs are $(x, \pm x)$. Thus, there are two ordered pairs with the same first coordinate and different second coordinates. We do not have a function.

39. Since $y = x^2$, the ordered pairs are (x, x^2). Thus, there are no two ordered pairs with the same first coordinate and different second coordinates. We have a function.

41. Since $y = |x| - 2$, the ordered pairs are $(x, |x| - 2)$. Thus, there are no two ordered pairs with the same first coordinate and different second coordinates. We have a function.

43. Since $(2,1)$ and $(2,-1)$ are two ordered pairs with the same first coordinate and different second coordinates, the equation does not define a function.

45. Domain $\{-3,4,5\}$, range $\{1,2,6\}$

47. Domain $(-\infty,\infty)$, range $\{4\}$

49. Domain $(-\infty,\infty)$;
since $|x| \geq 0$, the range of $y = |x| + 5$ is $[5,\infty)$.

51. Since $x = |y| - 3 \geq -3$, the domain of $x = |y| - 3$ is $[-3,\infty)$; range $(-\infty,\infty)$

53. Since $\sqrt{x-4}$ is a real number whenever $x \geq 4$, the domain of $y = \sqrt{x-4}$ is $[4, \infty)$.
Since $y = \sqrt{x-4} \geq 0$ for $x \geq 4$, the range is $[0, \infty)$.

55. Since $x = -y^2 \leq 0$, the domain of $x = -y^2$ is $(-\infty, 0]$; range is $(-\infty, \infty)$.

57. 6

59. $g(2) = 3(2) + 5 = 11$

61. Since $(3, 8)$ is the ordered pair, one obtains $f(3) = 8$. The answer is $x = 3$.

63. Solving $3x + 5 = 26$, we find $x = 7$.

65. $f(4) + g(4) = 5 + 17 = 22$

67. $3a^2 - a$

69. $4(a+2) - 2 = 4a + 6$

71. $3(x^2 + 2x + 1) - (x + 1) = 3x^2 + 5x + 2$

73. $4(x + h) - 2 = 4x + 4h - 2$

75. $3(x^2 + 2x + 1) - (x + 1) - 3x^2 + x = 6x + 2$

77. $3(x^2 + 2xh + h^2) - (x + h) - 3x^2 + x =$
$6xh + 3h^2 - h$

79. The average rate of change is

$$\frac{8,000 - 20,000}{5} = -\$2400 \text{ per year.}$$

81. The average rate of change on $[0, 2]$ is
$\dfrac{h(2) - h(0)}{2 - 0} = \dfrac{0 - 64}{2 - 0} = -32$ ft/sec.
The average rate of change on $[1, 2]$ is
$\dfrac{h(2) - h(1)}{2 - 1} = \dfrac{0 - 48}{2 - 1} = -48$ ft/sec.
The average rate of change on $[1.9, 2]$ is
$\dfrac{h(2) - h(1.9)}{2 - 1.9} = \dfrac{0 - 6.24}{0.1} = -62.4$ ft/sec.
The average rate of change on $[1.99, 2]$ is
$\dfrac{h(2) - h(1.99)}{2 - 1.99} = \dfrac{0 - 0.6384}{0.01} = -63.84$ ft/sec.
The average rate of change on $[1.999, 2]$ is
$\dfrac{h(2) - h(1.999)}{2 - 1.999} = \dfrac{0 - 0.063984}{0.001} = -63.984$
ft/sec.

83. The average rate of change is $\dfrac{673 - 1970}{24} \approx$
-54.0 million hectares per year.

85.

$$\begin{aligned} \frac{f(x+h) - f(x)}{h} &= \frac{4(x+h) - 4x}{h} \\ &= \frac{4h}{h} \\ &= 4 \end{aligned}$$

87.

$$\begin{aligned} \frac{f(x+h) - f(x)}{h} &= \frac{3(x+h) + 5 - 3x - 5}{h} \\ &= \frac{3h}{h} \\ &= 3 \end{aligned}$$

89. Let $g(x) = x^2 + x$. Then we obtain

$$\begin{aligned} \frac{g(x+h) - g(x)}{h} &= \\ \frac{(x+h)^2 + (x+h) - x^2 - x}{h} &= \\ \frac{2xh + h^2 + h}{h} &= \\ 2x + h + 1. \end{aligned}$$

91. Difference quotient is

$$\begin{aligned} &= \frac{-(x+h)^2 + (x+h) - 2 + x^2 - x + 2}{h} \\ &= \frac{-2xh - h^2 + h}{h} \\ &= -2x - h + 1 \end{aligned}$$

93. Difference quotient is

$$\begin{aligned} &= \frac{3\sqrt{x+h} - 3\sqrt{x}}{h} \cdot \frac{3\sqrt{x+h} + 3\sqrt{x}}{3\sqrt{x+h} + 3\sqrt{x}} \\ &= \frac{9(x+h) - 9x}{h(3\sqrt{x+h} + 3\sqrt{x})} \\ &= \frac{9h}{h(3\sqrt{x+h} + 3\sqrt{x})} \\ &= \frac{3}{\sqrt{x+h} + \sqrt{x}} \end{aligned}$$

95. Difference quotient is

$$= \frac{\sqrt{x+h+2}-\sqrt{x+2}}{h}\cdot\frac{\sqrt{x+h+2}+\sqrt{x+2}}{\sqrt{x+h+2}+\sqrt{x+2}}$$
$$= \frac{(x+h+2)-(x+2)}{h(\sqrt{x+h+2}+\sqrt{x+2})}$$
$$= \frac{h}{h(\sqrt{x+h+2}+\sqrt{x+2})}$$
$$= \frac{1}{\sqrt{x+h+2}+\sqrt{x+2}}$$

97. Difference quotient is

$$= \frac{\dfrac{1}{x+h}-\dfrac{1}{x}}{h}\cdot\frac{x(x+h)}{x(x+h)}$$
$$= \frac{x-(x+h)}{xh(x+h)}$$
$$= \frac{-h}{xh(x+h)}$$
$$= \frac{-1}{x(x+h)}$$

99. Difference quotient is

$$= \frac{\dfrac{3}{x+h+2}-\dfrac{3}{x+2}}{h}\cdot\frac{(x+h+2)(x+2)}{(x+h+2)(x+2)}$$
$$= \frac{3(x+2)-3(x+h+2)}{h(x+h+2)(x+2)}$$
$$= \frac{-3h}{h(x+h+2)(x+2)}$$
$$= \frac{-3}{(x+h+2)(x+2)}$$

101. **a)** $A = s^2$ **b)** $s = \sqrt{A}$ **c)** $s = \dfrac{d\sqrt{2}}{2}$

d) $d = s\sqrt{2}$ **e)** $P = 4s$ **f)** $s = P/4$

g) $A = P^2/16$ **h)** $d = \sqrt{2A}$

103. $C = 500 + 100n$

105.

a) The quantity $C(4) = (0.95)(4) + 5.8 = \$9.6$ billion represents the amount spent on computers in year 2004.

b) By solving $0.95n + 5.8 = 20$, we obtain

$$n = \frac{14.2}{0.95} \approx 14.9.$$

Thus, spending for computers will be \$20 billion in year 2015.

107. Let a be the radius of each circle. Note, triangle $\triangle ABC$ is an equilateral triangle with side $2a$ and height $\sqrt{3}a$.

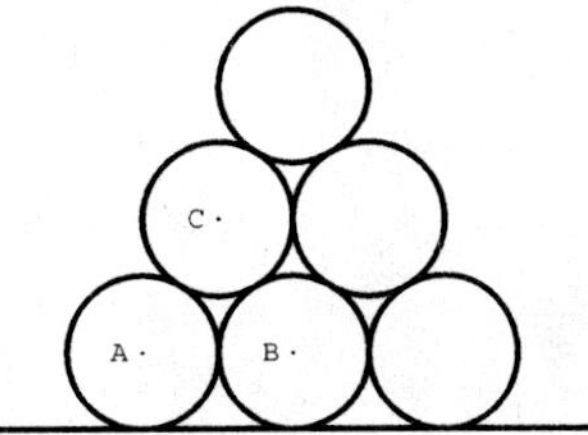

Thus, the height of the circle centered at C from the horizontal line is $\sqrt{3}a + 2a$. Hence, by using a similar reasoning, we obtain that height of the highest circle from the line is

$$2\sqrt{3}a + 2a$$

or equivalently $(2\sqrt{3}+2)a$.

109. When $x = 18$ and $h = 0.1$, we have

$$\frac{R(18.1) - R(18)}{0.1} = 1950.$$

The revenue from the concert will increase by approximately \$1,950 if the price of a ticket is raised from \$18 to \$19.

If $x = 22$ and $h = 0.1$, then

$$\frac{R(22.1) - R(22)}{0.1} = -2050.$$

The revenue from the concert will decrease by approximately \$2050 if the price of a ticket is raised from \$22 to \$23.

113.

$$\frac{3}{2}x - \frac{5}{9}x = \frac{1}{3} - \frac{5}{6}$$
$$\frac{17}{18}x = -\frac{1}{2}$$

$$x = -\frac{1}{2} \cdot \frac{18}{17}$$

$$x = -\frac{9}{17}$$

115. $\sqrt{(-4+6)^2 + (-3-3)^2} = \sqrt{4+36} = \sqrt{40} = 2\sqrt{10}$

117.

$$x^2 - x - 6 = 36$$
$$x^2 - x - 42 = 0$$
$$(x-7)(x+6) = 0$$

The solution set is $\{-6, 7\}$.

119. $(30+25)^2 = 3025$

For Thought

1. True, since the graph is a parabola opening down with vertex at the origin.

2. False, the graph is decreasing.

3. True

4. True, since $f(-4.5) = [-1.5] = -2$.

5. False, since the range is $\{\pm 1\}$.

6. True **7.** True **8.** True

9. False, since the range is the interval $[0, 4]$.

10. True

2.2 Exercises

1. square root

3. increasing

5. parabola

7. Function $y = 2x$ includes the points $(0,0), (1,2)$, domain and range are both $(-\infty, \infty)$

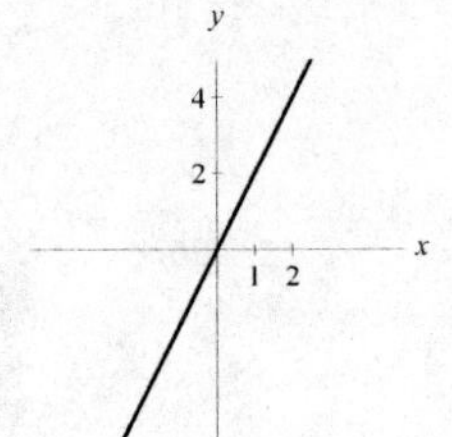

9. Function $x - y = 0$ includes the points $(-1,-1)$, $(0,0), (1,1)$, domain and range are both $(-\infty, \infty)$

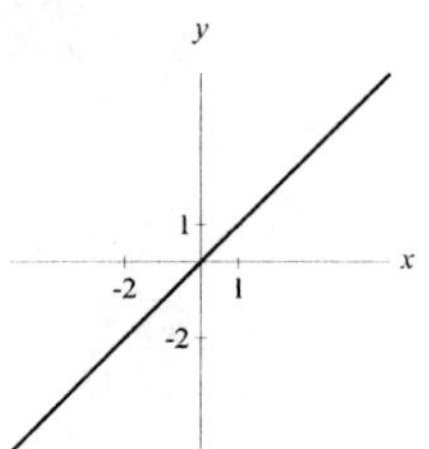

11. Function $y = 5$ includes the points $(0,5)$, $(\pm 2, 5)$, domain is $(-\infty, \infty)$, range is $\{5\}$

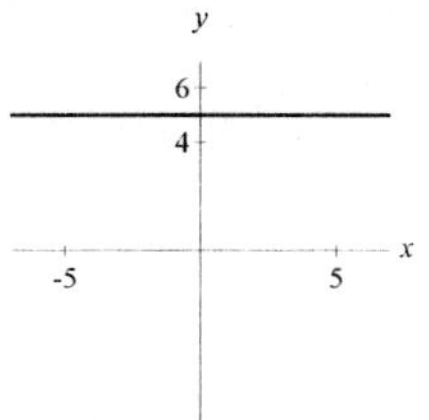

13. Function $y = 2x^2$ includes the points $(0,0)$, $(\pm 1, 2)$, domain is $(-\infty, \infty)$, range is $[0, \infty)$

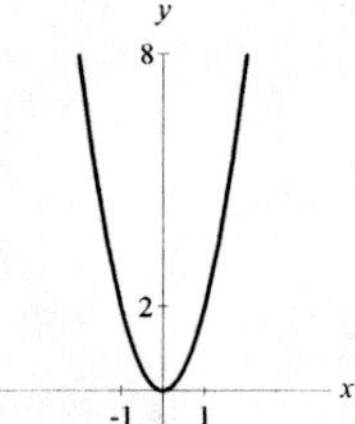

15. Function $y = 1 - x^2$ includes the points $(0,1)$, $(\pm 1, 0)$, domain is $(-\infty, \infty)$, range is $(-\infty, 1]$

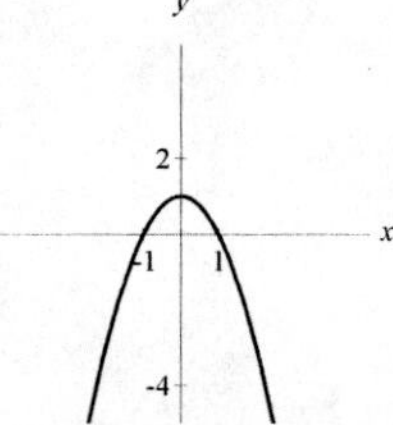

17. Function $y = 1 + \sqrt{x}$ includes the points $(0,1)$, $(1,2), (4,3)$, domain is $[0, \infty)$, range is $[1, \infty)$

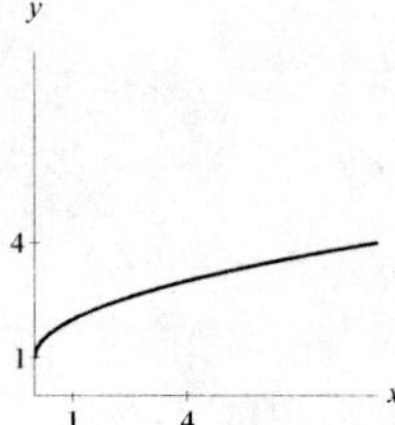

19. $x = y^2 + 1$ is not a function and includes the points $(1, 0), (2, \pm 1)$, domain is $[1, \infty)$, range is $(-\infty, \infty)$

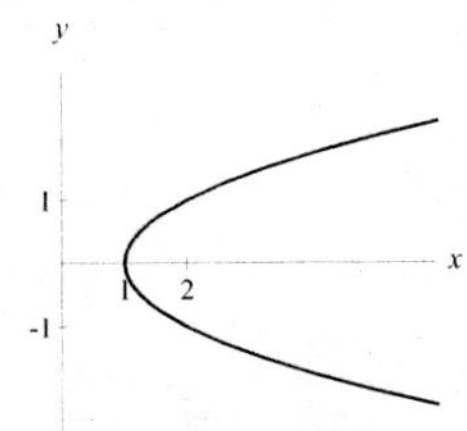

21. Function $x = \sqrt{y}$ goes through $(0, 0), (2, 4), (3, 9)$, domain and range is $[0, \infty)$

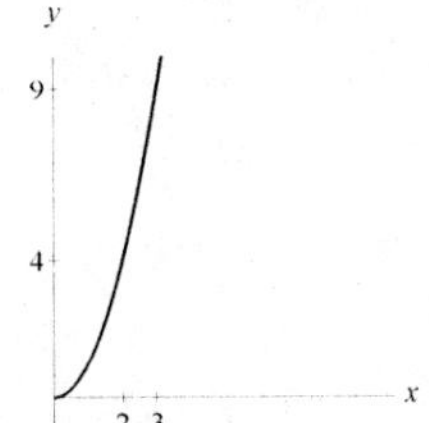

23. Function $y = \sqrt[3]{x} + 1$ goes through $(-1, 0), (1, 2), (8, 3)$, domain $(-\infty, \infty)$, and range $(-\infty, \infty)$

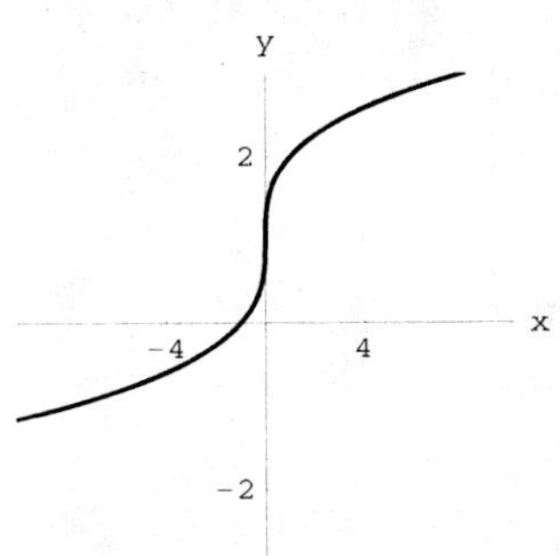

25. Function, $x = \sqrt[3]{y}$ goes through $(0, 0), (1, 1), (2, 8)$, domain $(-\infty, \infty)$, and range $(-\infty, \infty)$

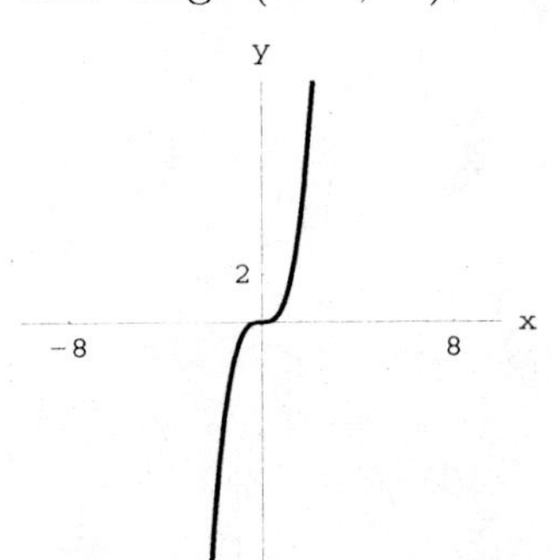

27. Not a function, $y^2 = 1 - x^2$ goes through $(1, 0), (0, 1), (-1, 0)$, domain $[-1, 1]$, and range $[-1, 1]$

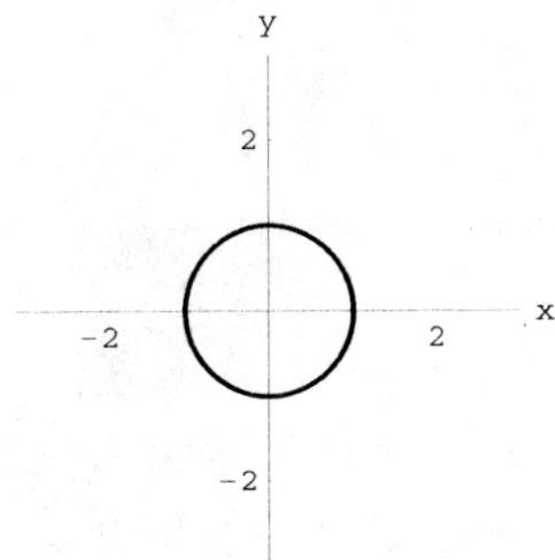

29. Function, $y = \sqrt{1 - x^2}$ goes through $(\pm 1, 0), (0, 1)$, domain $[-1, 1]$, and range $[0, 1]$

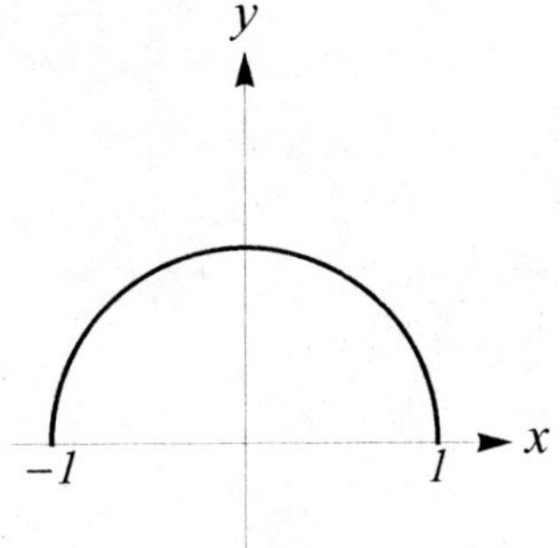

31. Function $y = x^3$ includes the points $(0, 0)$, $(1, 1), (2, 8)$, domain and range are both $(-\infty, \infty)$

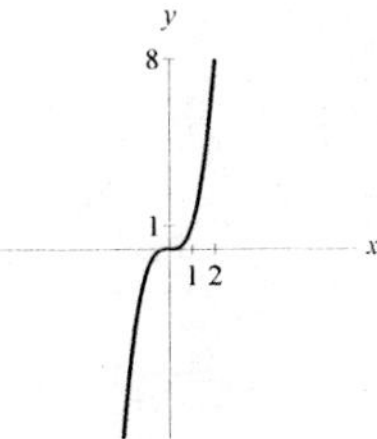

33. Function $y = 2|x|$ includes the points $(0, 0)$, $(\pm 1, 2)$, domain is $(-\infty, \infty)$, range is $[0, \infty)$

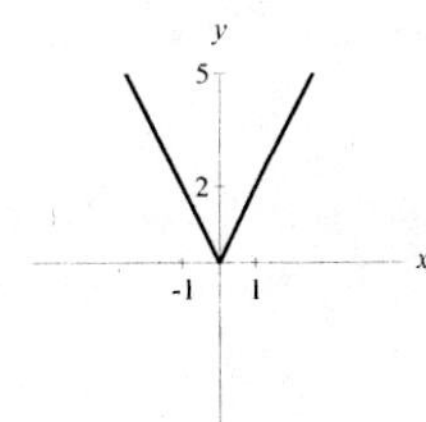

35. Function $y = -|x|$ includes the points $(0, 0)$, $(\pm 1, -1)$, domain is $(-\infty, \infty)$, range is $(-\infty, 0]$

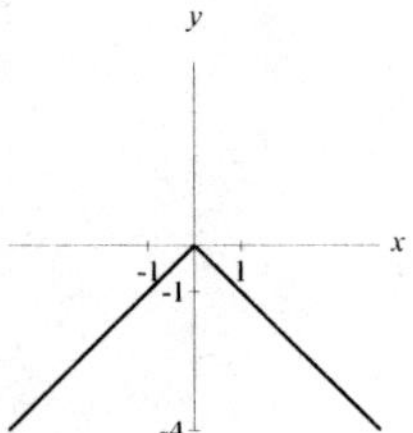

37. Not a function, graph of $x = |y|$ includes the points $(0, 0), (2, 2), (2, -2)$, domain is $[0, \infty)$, range is $(-\infty, \infty)$

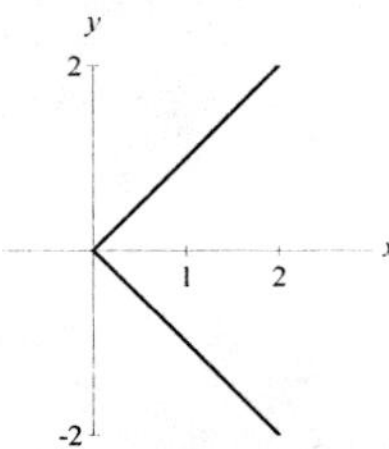

39. Domain is $(-\infty, \infty)$, range is $\{\pm 2\}$, some points are $(-3, -2)$, $(1, -2)$

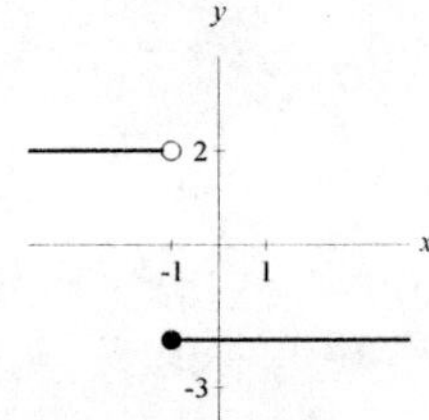

41. Domain is $(-\infty, \infty)$, range is $(-\infty, -2] \cup (2, \infty)$, some points are $(2, 3)$, $(1, -2)$

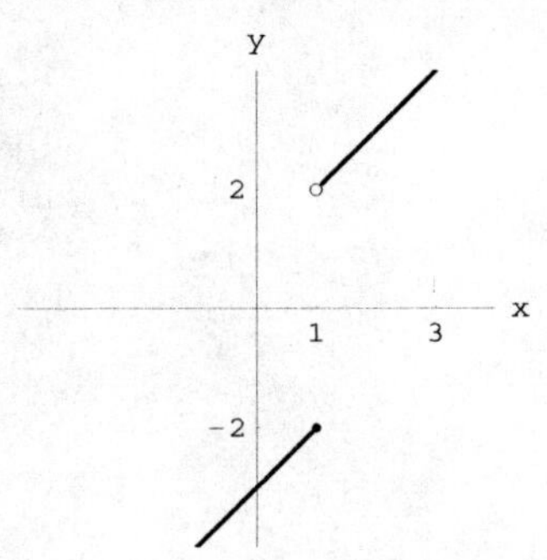

43. Domain is $[-2, \infty)$, range is $(-\infty, 2]$, some points are $(2, 2)$, $(-2, 0)$, $(3, 1)$

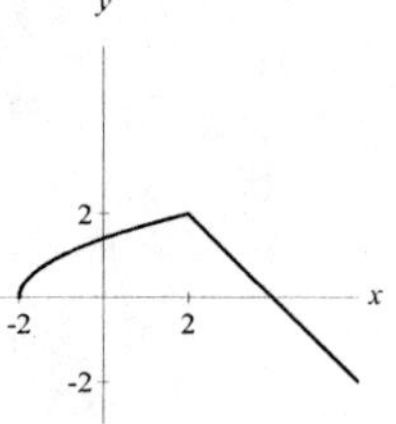

45. Domain is $(-\infty, \infty)$, range is $[0, \infty)$, some points are $(-1, 1)$, $(-4, 2)$, $(4, 2)$

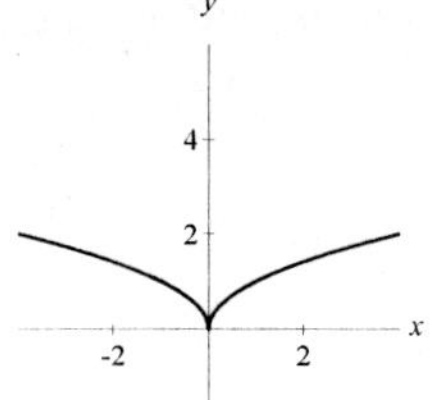

47. Domain is $(-\infty, \infty)$, range is $(-\infty, \infty)$, some points are $(-2, 4)$, $(1, -1)$

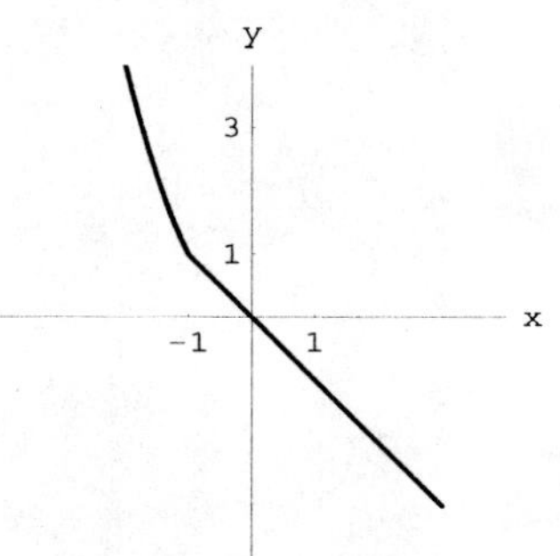

49. Domain is $(-\infty, \infty)$, range is the set of integers, some points are $(0, 1)$, $(1, 2)$, $(1.5, 2)$

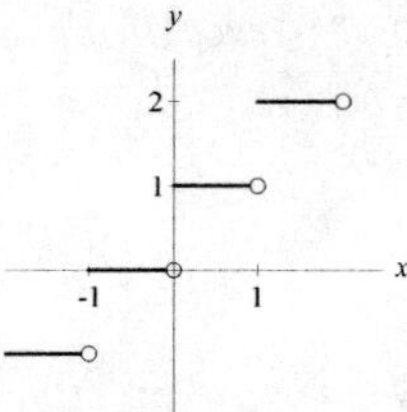

51. Domain $[0, 4)$, range is $\{2, 3, 4, 5\}$, some points are $(0, 2)$, $(1, 3)$, $(1.5, 3)$

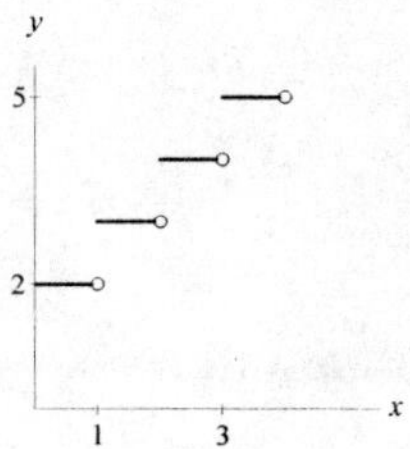

53. **a.** Domain and range are both $(-\infty, \infty)$, decreasing on $(-\infty, \infty)$

b. Domain is $(-\infty, \infty)$, range is $(-\infty, 4]$ increasing on $(-\infty, 0]$, decreasing on $[0, \infty)$

55. **a.** Domain is $[-2, 6]$, range is $[3, 7]$ increasing on $[-2, 2]$, decreasing on $[2, 6]$

b. Domain $(-\infty, 2]$, range $(-\infty, 3]$, increasing on $(-\infty, -2]$, constant on $[-2, 2]$

57. **a.** Domain is $(-\infty, \infty)$, range is $[0, \infty)$ increasing on $[0, \infty)$, decreasing on $(-\infty, 0]$

b. Domain and range are both $(-\infty, \infty)$ increasing on $[-2, -2/3]$, decreasing on $(-\infty, -2]$ and $[-2/3, \infty)$

59. **a.** Domain and range are both $(-\infty, \infty)$, increasing on $(-\infty, \infty)$

b. Domain is $[-2, 5]$, range is $[1, 4]$ increasing on $[1, 2]$, decreasing on $[-2, 1]$, constant on $[2, 5]$

61. Domain and range are both $(-\infty, \infty)$ increasing on $(-\infty, \infty)$, some points are $(0, 1)$, $(1, 3)$

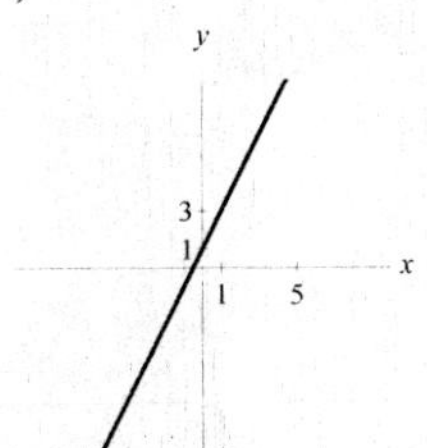

63. Domain is $(-\infty, \infty)$, range is $[0, \infty)$, increasing on $[1, \infty)$, decreasing on $(-\infty, 1]$, some points are $(0, 1)$, $(1, 0)$

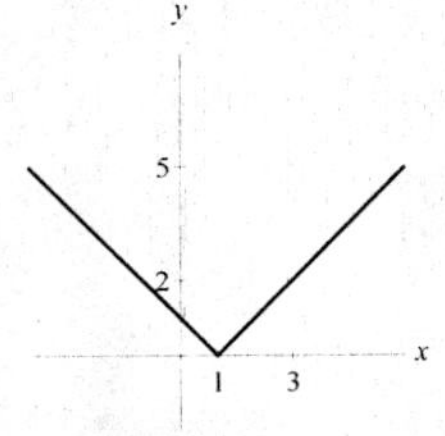

65. Domain is $(-\infty, 0) \cup (0, \infty)$, range is $\{\pm 1\}$, constant on $(-\infty, 0)$ and $(0, \infty)$, some points are $(1, 1)$, $(-1, -1)$

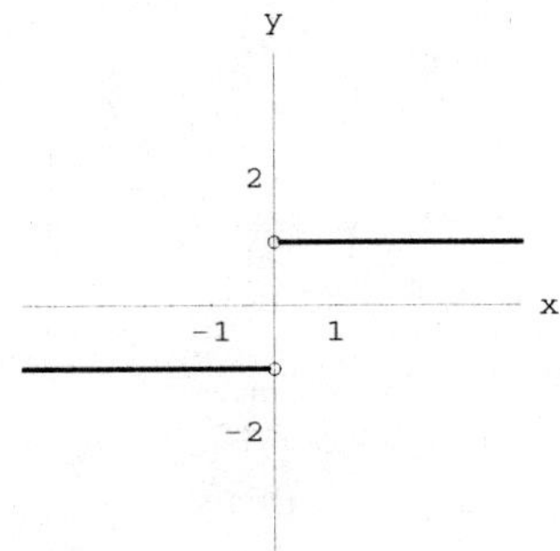

67. Domain is $[-3, 3]$, range is $[0, 3]$, increasing on $[-3, 0]$, decreasing on $[0, 3]$, some points are $(\pm 3, 0)$, $(0, 3)$

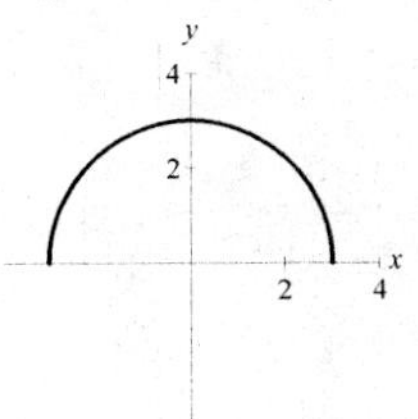

69. Domain and range are both $(-\infty, \infty)$, increasing on $(-\infty, 3)$ and $[3, \infty)$, some points are $(4, 5)$, $(0, 2)$

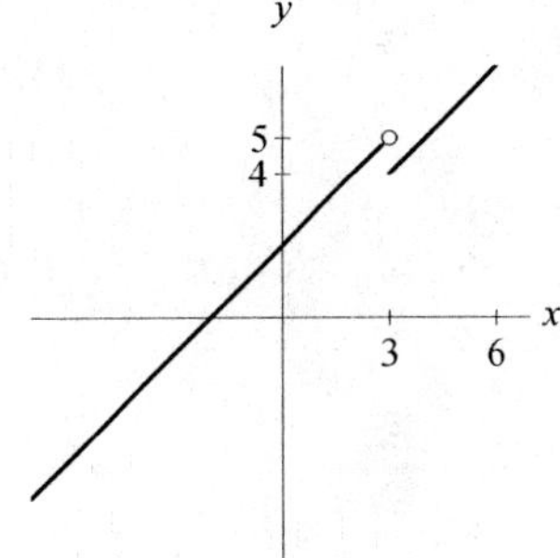

71. Domain is $(-\infty, \infty)$, range is $(-\infty, 2]$, increasing on $(-\infty, -2]$ and $(-2, 0)$, decreasing on $[0, 2)$ and $[2, \infty)$, some points are $(-3, 0)$, $(0, 2)$, $(4, -1)$

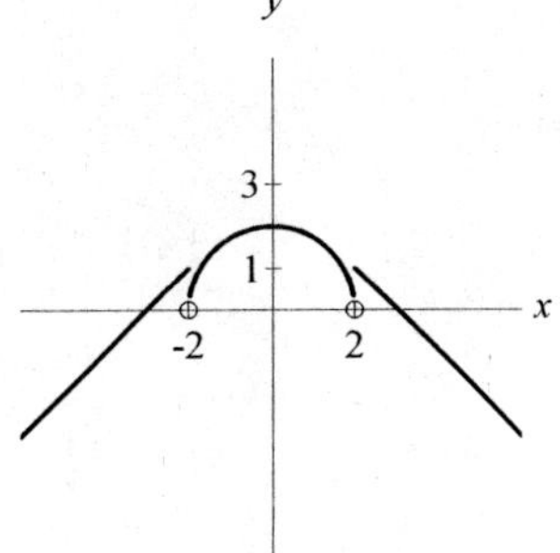

73. $f(x) = \begin{cases} 2 & \text{for} \quad x > -1 \\ -1 & \text{for} \quad x \leq -1 \end{cases}$

75. The line joining $(-1, 1)$ and $(-3, 3)$ is $y = -x$, and the line joining $(-1, -2)$ and $(3, 2)$ is $y = x - 1$. The piecewise function is

$$f(x) = \begin{cases} x - 1 & \text{for} \quad x \geq -1 \\ -x & \text{for} \quad x < -1. \end{cases}$$

77. The line joining $(0, -2)$ and $(2, 2)$ is $y = 2x - 2$, and the line joining $(0, -2)$ and $(-3, 1)$ is $y = -x - 2$. The piecewise function is

$$f(x) = \begin{cases} 2x - 2 & \text{for} \quad x \geq 0 \\ -x - 2 & \text{for} \quad x < 0. \end{cases}$$

79. increasing on the interval $[0.83, \infty)$, decreasing on $(-\infty, 0.83]$

81. increasing on $(-\infty, -1]$ and $[1, \infty)$, decreasing on $[-1, 1]$

83. increasing on $[-1.73, 0]$ and $[1.73, \infty)$, decreasing on $(-\infty, -1.73]$ and $[0, 1.73]$

85. increasing on $[30, 50]$, and $[70, \infty)$, decreasing on $(-\infty, 30]$ and $[50, 70]$

87. c, graph was increasing at first, then suddenly dropped and became constant, then increased slightly

89. d, graph was decreasing at first, then fluctuated between increases and decreases, then the market increased

91. The independent variable is time t where t is the number of minutes after 7:45 and the dependent variable is distance D from the holodeck.

D is increasing on the intervals $[0, 3]$ and $[6, 15]$, decreasing on $[3, 6]$ and $[30, 39]$, and constant on $[15, 30]$.

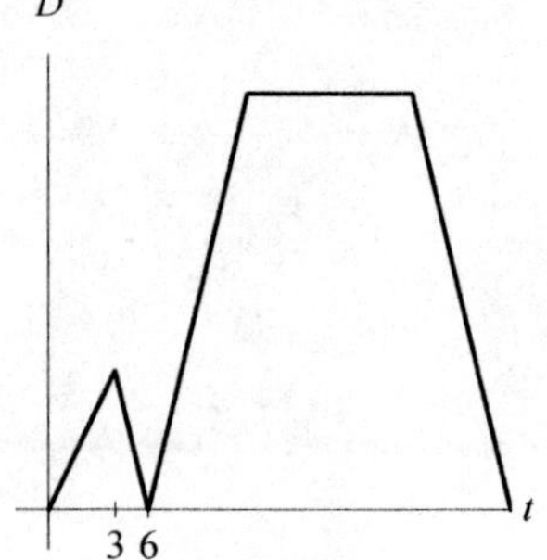

93. Independent variable is time t in years, dependent variable is savings s in dollars

s is increasing on the interval $[0, 2]$; s is constant on $[2, 2.5]$; s is decreasing on $[2.5, 4.5]$.

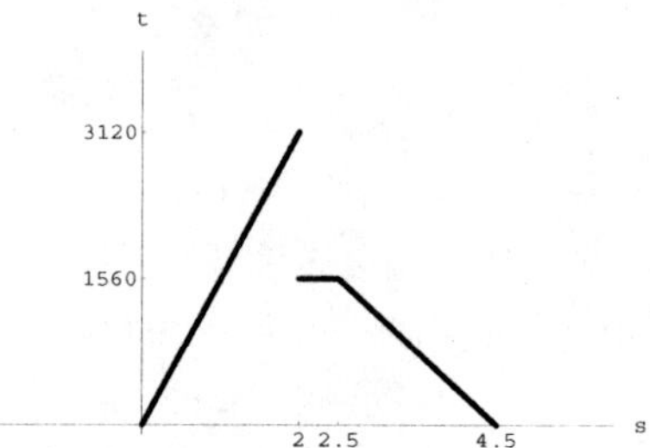

95. In 1988, there were $M(18) = 565$ million cars. In 2020, it is projected that there will be $M(50) = 900$ million cars.
The average rate of change from 1984 to 1994 is $\dfrac{M(24) - M(14)}{10} = 14.5$ million cars/year.

97. Constant on $[0, 10^4]$, increasing on $[10^4, \infty)$

99. The cost is over \$235 for t in $[5, \infty)$.

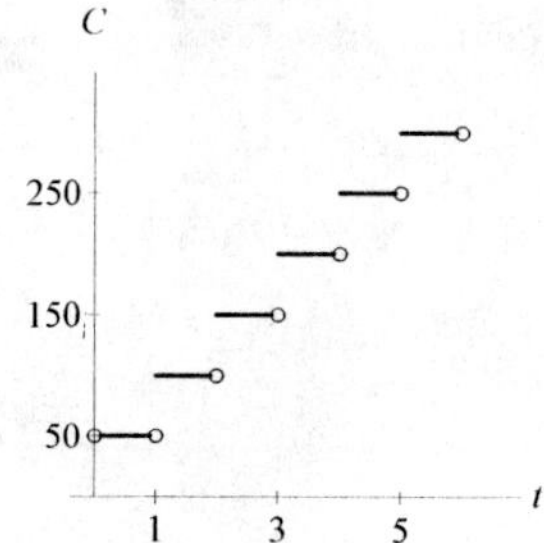

101. $f(x) = \begin{cases} 150 & \text{if} \quad 0 < x < 3 \\ 50x & \text{if} \quad 3 \leq x \leq 10 \end{cases}$

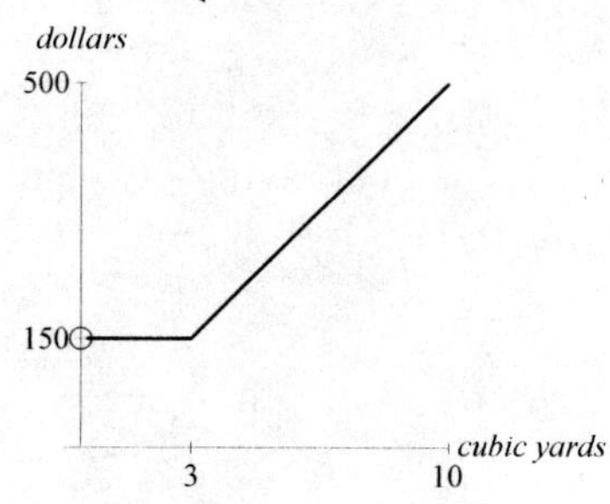

103. A mechanic's fee is \$20 for each half-hour of work with any fraction of a half-hour charged as a half-hour. If y is the fee in dollars and x is the number of hours, then $y = -20[-2x]$.

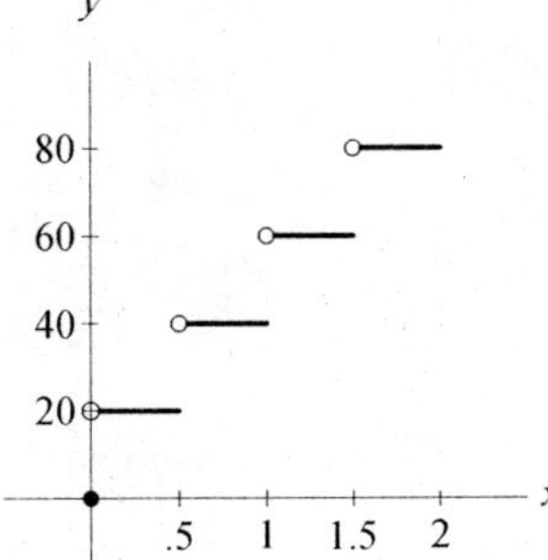

105. Since $x - 2 \geq 0$, the domain is $[2, \infty)$.

Since $\sqrt{x-2}$ is nonnegative, we find $\sqrt{x-2}+3$ is at least three. Then the range is $[3, \infty)$.

107. $6 + 4(24 - 20)^2 = 6 + 4(16) = 70$

109. **a)** $(w^2 - 1)(w^2 + 1) = (w-1)(w+1)(w^2+1)$

b) $a(10a^2 + 3a - 18) = a(5a - 6)(2a + 3)$

111. a) Consider the circle $x^2 + (y - r)^2 = r^2$ where $r > 0$. Suppose the circle intersects the parabola $y = x^2$ only at the origin. Substituting $y = x^2$, we obtain

$$\begin{aligned} y + (y-r)^2 &= r^2 \\ y^2 + y(1-2r) &= 0 \end{aligned}$$

Thus, $1 - 2r = 0$ since the circle and the parabola has exactly one point of intersection. The radius of the circle is $1/2$.

b) Consider the circle $x^2 + (y - r)^2 = r^2$ where $r > 0$. If the circle intersects the parabola $y = ax^2$ only at the origin, then the equation below must have exactly one solution, namely, $y = 0$.

$$\begin{aligned} \frac{1}{a}y + (y-r)^2 &= r^2 \\ y^2 + y\left(\frac{1}{a} - 2r\right) &= 0 \end{aligned}$$

Necessarily, we have $\frac{1}{a} - 2r = 0$ or $a = \frac{1}{2r}$. Thus, if $r = 3$ then $a = 1/6$.

For Thought

1. False, it is a reflection in the y-axis.

2. True **3.** False, rather it is a left translation.

4. True **5.** True

6. False, the down shift should come after the reflection. **7.** True

8. False, since their domains are different.

9. True **10.** True

2.3 Exercises

1. rigid

3. parabola

5. reflection

7. linear

9. odd

11. $f(x) = |x|, g(x) = |x| - 4$

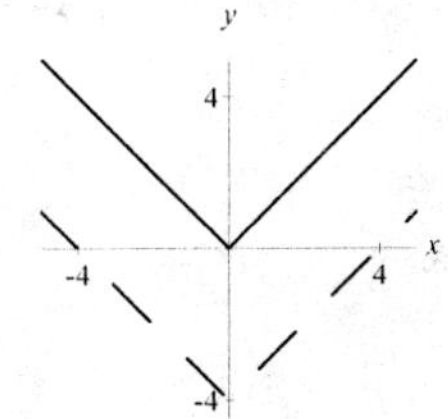

13. $f(x) = x, g(x) = x + 3$

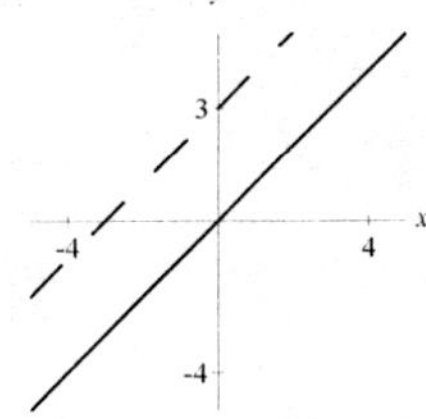

15. $y = x^2, y = (x-3)^2$

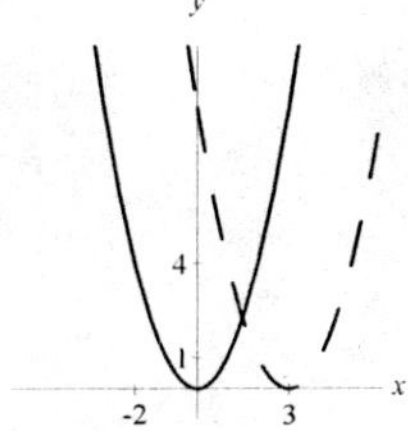

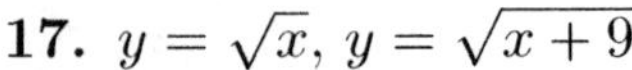

17. $y = \sqrt{x}$, $y = \sqrt{x+9}$

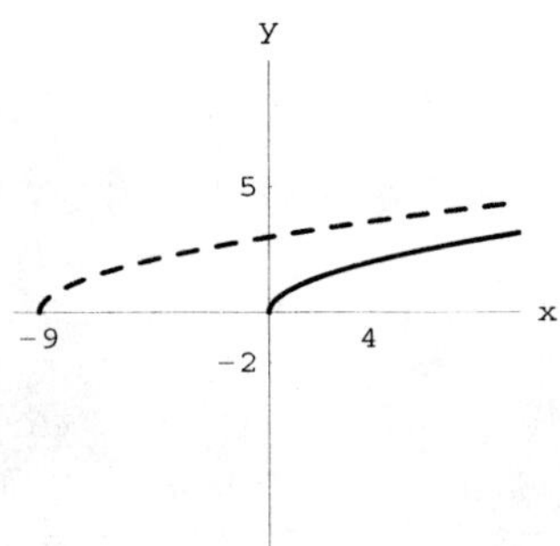

19. $f(x) = \sqrt{x}$, $g(x) = -\sqrt{x}$

21. $y = \sqrt{x}, y = 3\sqrt{x}$

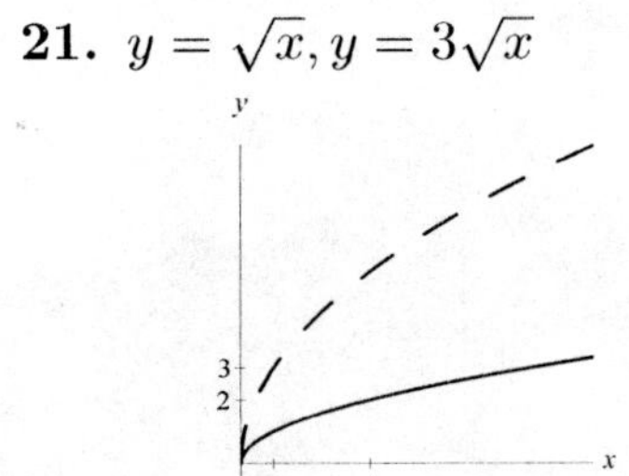

23. $y = x^2, y = \dfrac{1}{4}x^2$

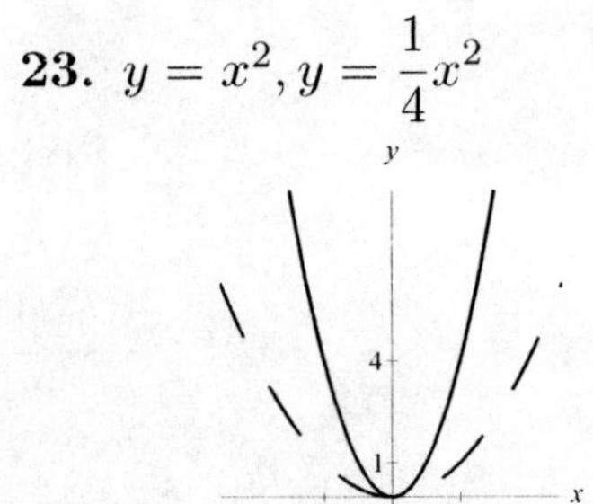

25. $y = \sqrt{4-x^2}, y = -\sqrt{4-x^2}$

27. g h **29.** b **31.** c **33.** f

35. $y = \sqrt{x} + 2$

37. $y = (x-5)^2$

39. $y = (x-10)^2 + 4$

41. $y = -(3\sqrt{x} + 5)$ or $y = -3\sqrt{x} - 5$

43. $y = -3|x-7| + 9$

45. $y = (x-1)^2 + 2$; right 1 unit, up 2 units, domain $(-\infty, \infty)$, range $[2, \infty)$

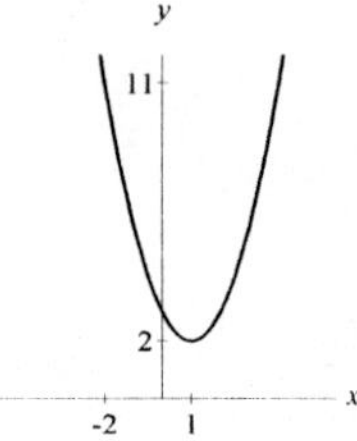

47. $y = |x-1| + 3$; right 1 unit, up 3 units domain $(-\infty, \infty)$, range $[3, \infty)$

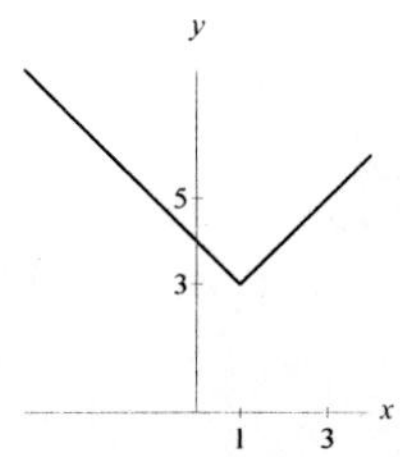

49. $y = 3x - 40$,
domain and range are both $(-\infty, \infty)$

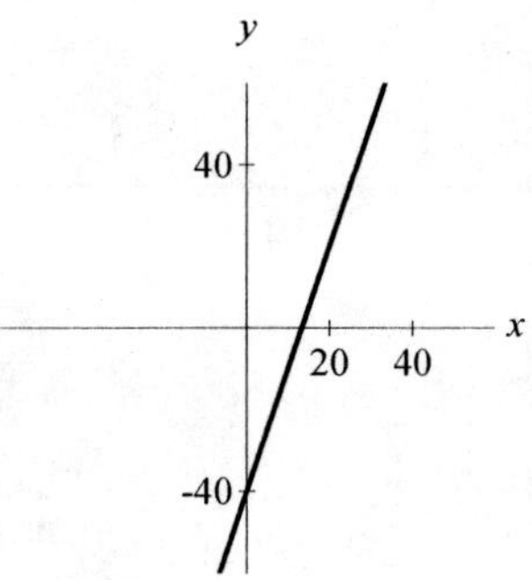

51. $y = \dfrac{1}{2}x - 20$,

domain and range are both $(-\infty, \infty)$

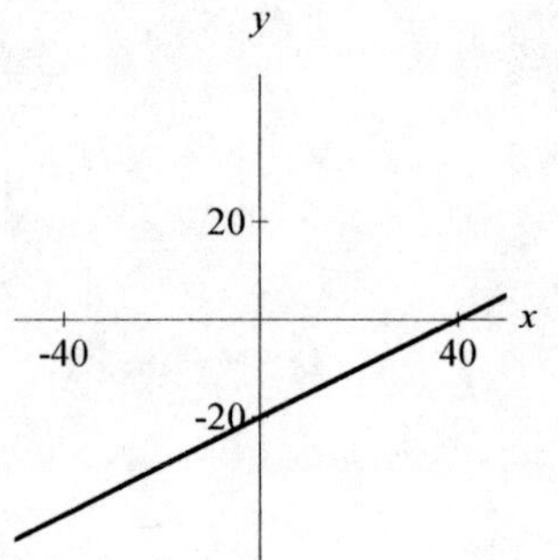

53. $y = -\dfrac{1}{2}|x| + 40$, shrink by 1/2, reflect about x-axis, up by 40, domain $(-\infty, \infty)$, range $(-\infty, 40]$

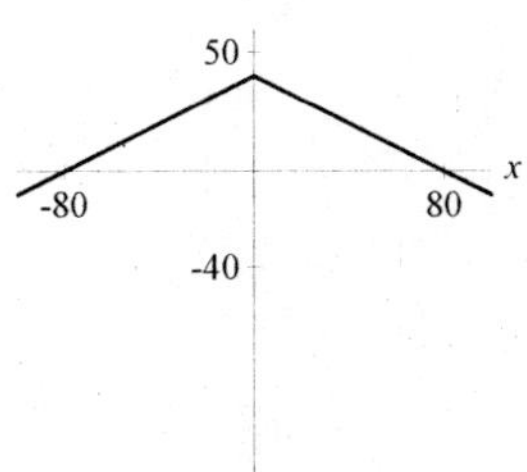

55. $y = -\dfrac{1}{2}|x+4|$, left by 4, reflect about x-axis, shrink by 1/2, domain $(-\infty, \infty)$, range $(-\infty, 0]$

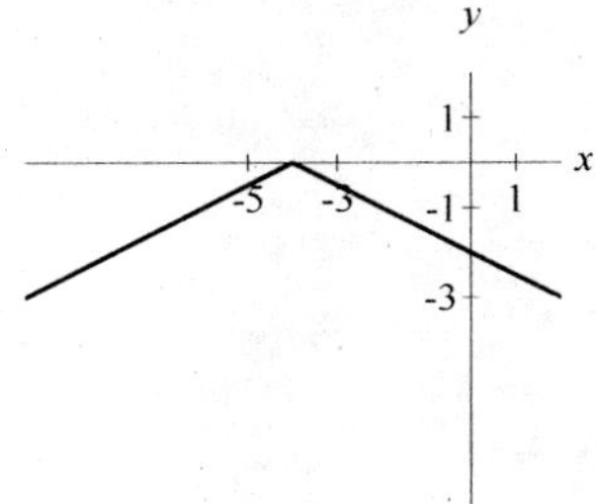

57. $y = -\sqrt{x-3} + 1$, right 3 units, reflect about x-axis, up 1 unit, domain $[3, \infty)$, range $(-\infty, 1]$

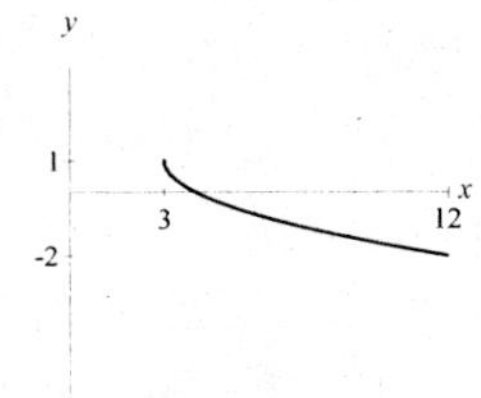

59. $y = -2\sqrt{x+3} + 2$, left 3 units, stretch by 2, reflect about x-axis, up 2 units, domain $[-3, \infty)$, range $(-\infty, 2]$

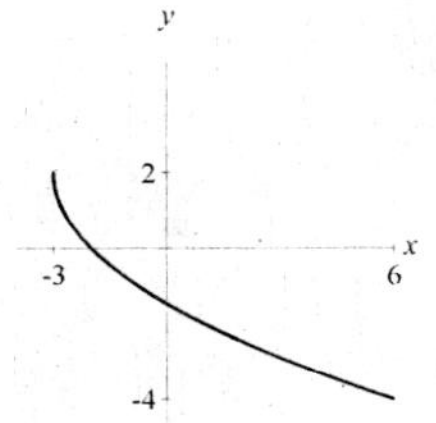

61. Symmetric about y-axis, even function since $f(-x) = f(x)$

63. No symmetry, neither even nor odd since $f(-x) \neq f(x)$ and $f(-x) \neq -f(x)$

65. Symmetric about $x = -3$, neither even nor odd since $f(-x) \neq f(x)$ and $f(-x) \neq -f(x)$

67. Symmetry about $x = 2$, neither an even or odd function since $f(-x) \neq f(x)$ and $f(-x) \neq -f(x)$

69. Symmetric about the origin, odd function since $f(-x) = -f(x)$

71. No symmetry, neither an even or odd function since $f(-x) \neq f(x)$ and $f(-x) \neq -f(x)$

73. No symmetry, neither an even or odd function since $f(-x) \neq f(x)$ and $f(-x) \neq -f(x)$

75. Symmetric about the y-axis, even function since $f(-x) = f(x)$

77. No symmetry, not an even or odd function since $f(-x) = -f(x)$ and $f(-x) \neq -f(x)$

79. Symmetric about the y-axis, even function since $f(-x) = f(x)$

81. e **83.** g

85. b **87.** c

89. $(-\infty, -1] \cup [1, \infty)$

91. $(-\infty, -1) \cup (5, \infty)$

93. Using the graph of $y = (x-1)^2 - 9$, we find that the solution is $(-2, 4)$.

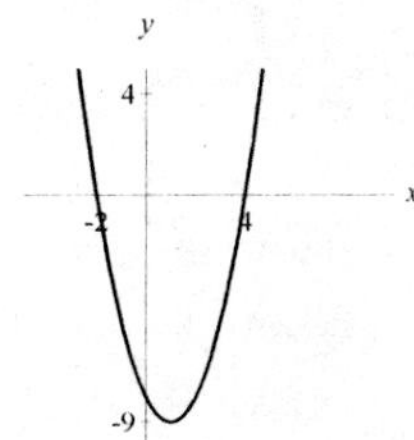

95. From the graph of $y = 5 - \sqrt{x}$, we find that the solution is $[0, 25]$.

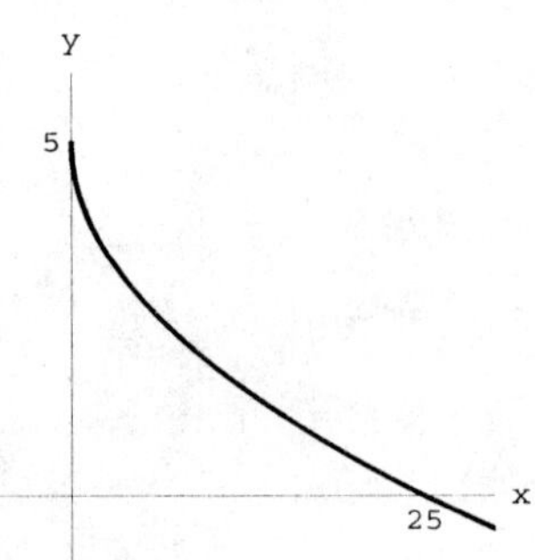

97. Note, the points of intersection of $y = 3$ and $y = (x-2)^2$ are $(2 \pm \sqrt{3}, 3)$. The solution set of $(x-2)^2 > 3$ is $\left(-\infty, 2-\sqrt{3}\right) \cup \left(2+\sqrt{3}, \infty\right)$.

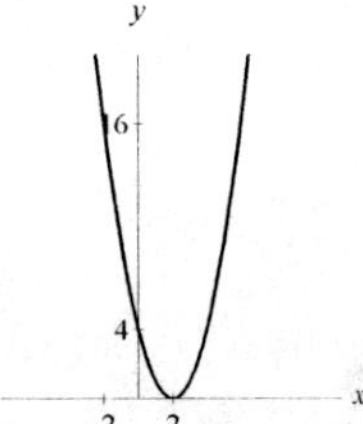

99. From the graph of $y = \sqrt{25 - x^2}$, we conclude that the solution is $(-5, 5)$.

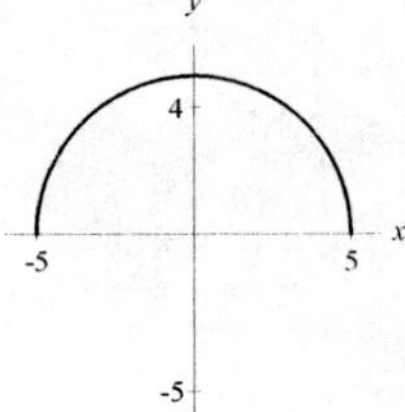

101. From the graph of $y = \sqrt{3}x^2 + \pi x - 9$,

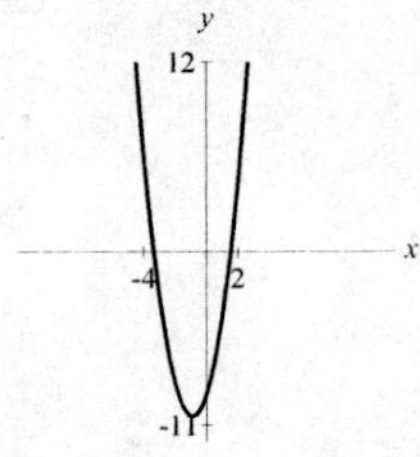

we observe that the solution set of $\sqrt{3}x^2 + \pi x - 9 < 0$ is $(-3.36, 1.55)$.

103. a. Stretch the graph of f by a factor of 2.

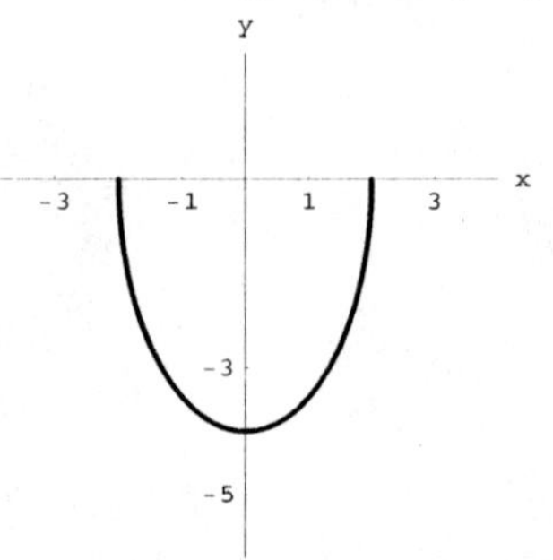

b. Reflect the graph of f about the x-axis.

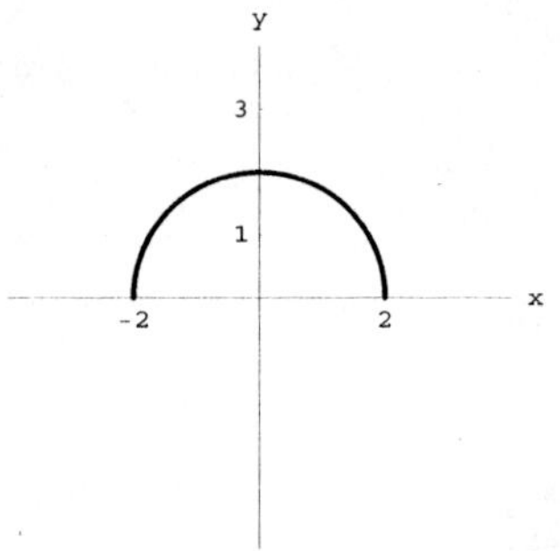

c. Shift the graph of f to the left by 1-unit.

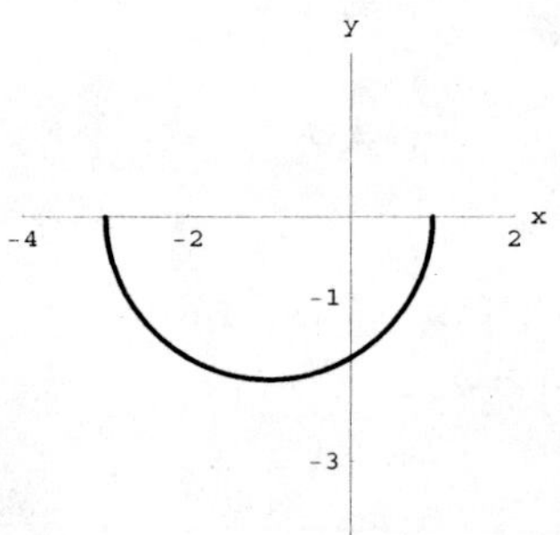

d. Shift the graph of f to the right by 3-units.

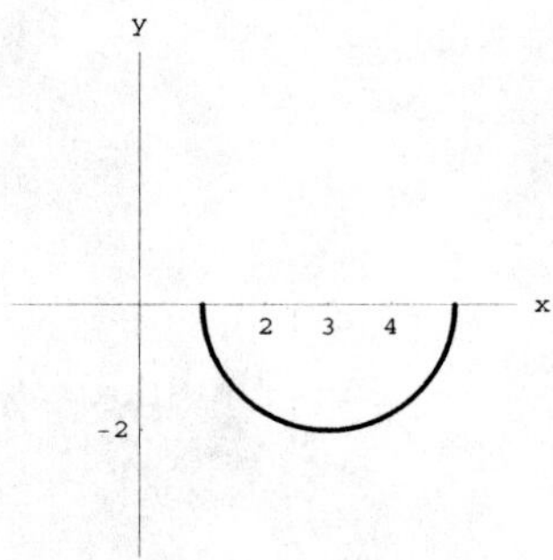

e. Stretch the graph of f by a factor of 3 and reflect about the x-axis.

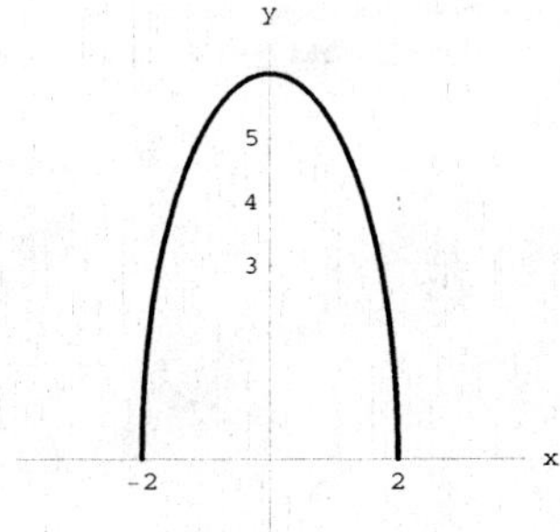

f. Translate the graph of f to the left by 2-units and down by 1-unit.

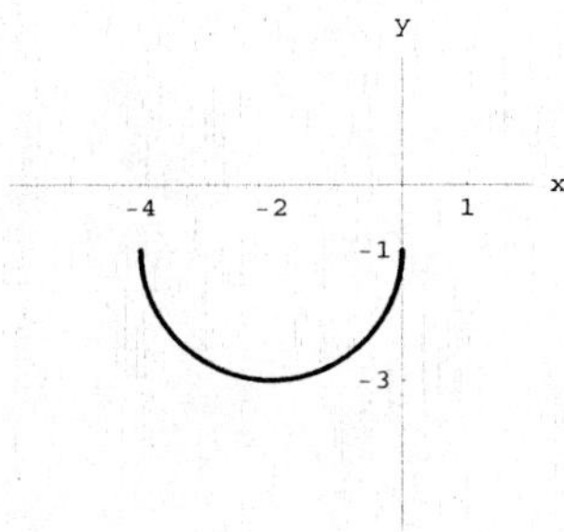

g. Translate the graph of f to the right by 1-unit and up by 3-units.

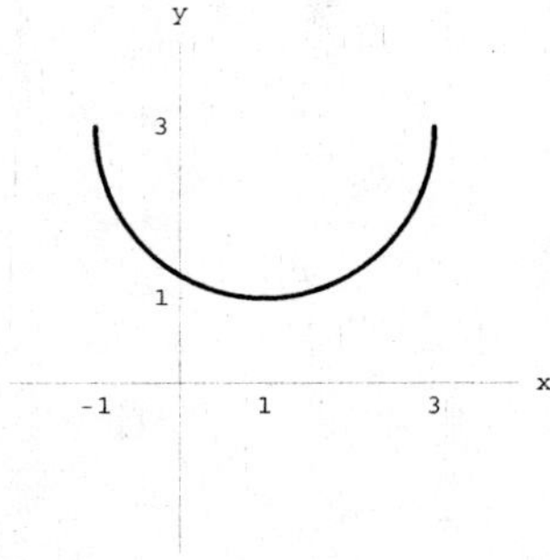

h. Translate the graph of f to the right by 2-units, stretch by a factor of 3, and up by 1-unit.

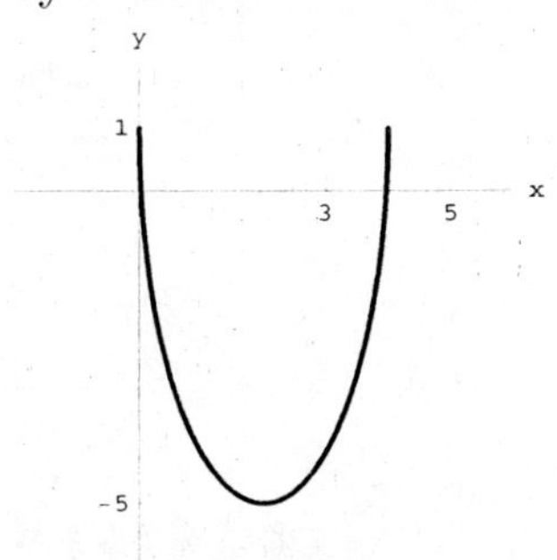

105. $N(x) = x + 2000$

107. If inflation rate is less than 50%, then $1 - \sqrt{x} < \frac{1}{2}$. This simplifies to $\frac{1}{2} < \sqrt{x}$. After squaring we have $\frac{1}{4} < x$ and so $x > 25\%$.

109.

(a) Both functions are even functions and the graphs are identical

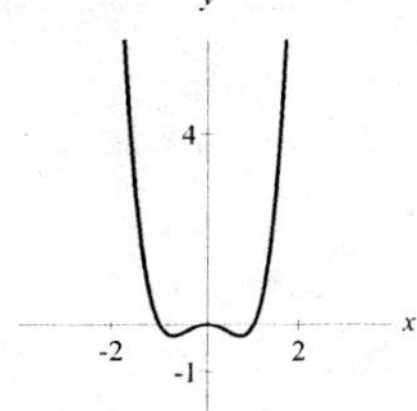

(b) One graph is a reflection of the other about the y-axis. Both functions are odd functions.

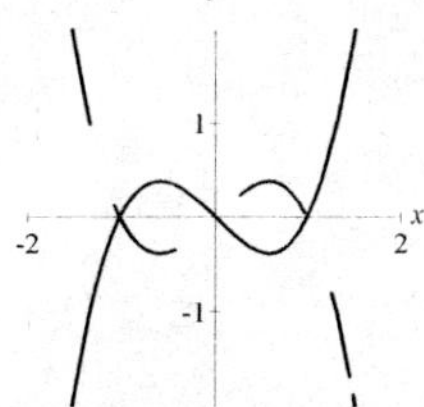

(c) The second graph is obtained by shifting the first one to the left by 1 unit.

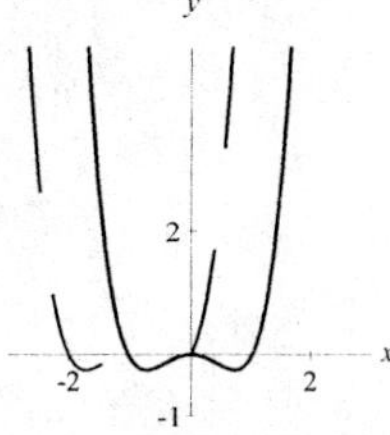

(d) The second graph is obtained by translating the first one to the right by 2 units and 3 units up.

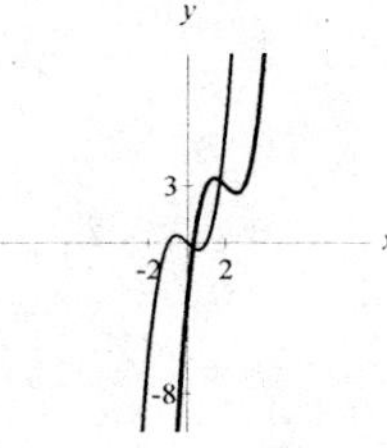

111. If $f(x) = x^2 + 1$, then

$$\begin{aligned} f(a+2) - f(a) &= \\ (a+2)^2 + 1 - a^2 - 1 &= \\ a^2 + 4a + 4 - a^2 &= \\ 4a + 4 & \end{aligned}$$

113. Note $|x| \geq 2/3$ is equivalent to $x \geq 2/3$ or $x \leq -2/3$. Then the solution set is

$$(-\infty, -2/3] \cup [2/3, \infty).$$

115.

$$\begin{aligned} ay + by - cy &= -2 \\ y &= \frac{-2}{a+b-c} \end{aligned}$$

117. Note, if $(x, x+h)$ is such an ordered pair then the average is $x + h/2$. Since the average is not a whole number, then $h = 1$. Then the ordered pairs are $(4,5)$, $(49,50)$, $(499,500)$, and $(4999,5000)$.

For Thought

1. False, since $f+g$ has an empty domain.

2. True **3.** True **4.** True

5. True, since $A = P^2/16$. **6.** True

7. False, since $(f \circ g)(x) = \sqrt{x-2}$ **8.** True

9. False, since $(h \circ g)(x) = x^2 - 9$.

10. True, since x belongs to the domain if $\sqrt{x-2}$ is a real number, i.e., if $x \geq 2$.

2.4 Exercises

1. sum

3. difference

5. intersection

7. $-1 + 2 = 1$

9. $-5 - 6 = -11$

11. $(-4) \cdot 2 = -8$

13. $1/12$

15. $(a-3) + (a^2 - a) = a^2 - 3$

17. $(a-3)(a^2-a) = a^3 - 4a^2 + 3a$

19. $f + g = \{(-3, 1+2), (2, 0+6)\} = \{(-3,3), (2,6)\}$, domain $\{-3, 2\}$

21. $f - g = \{(-3, 1-2), (2, 0-6)\} = \{(-3,-1), (2,-6)\}$, domain $\{-3, 2\}$

23. $f \cdot g = \{(-3, 1 \cdot 2), (2, 0 \cdot 6) = \{(-3,2), (2,0)\}$, domain $\{-3, 2\}$

25. $g/f = \{(-3, 2/1)\} = \{(-3,2)\}$, domain $\{-3\}$

27. $(f+g)(x) = \sqrt{x} + x - 4$, domain is $[0, \infty)$

29. $(f-h)(x) = \sqrt{x} - \dfrac{1}{x-2}$, domain is $[0,2) \cup (2, \infty)$

31. $(g \cdot h)(x) = \dfrac{x-4}{x-2}$, domain is $(-\infty, 2) \cup (2, \infty)$

33. $\left(\dfrac{g}{f}\right)(x) = \dfrac{x-4}{\sqrt{x}}$, domain is $(0, \infty)$

35. $\{(-3,0), (1,0), (4,4)\}$

37. $\{(1,4)\}$

39. $\{(-3,4), (1,4)\}$

41. $f(2) = 5$

43. $f(2) = 5$

45. $f(20.2721) = 59.8163$

47. $(g \circ h \circ f)(2) = (g \circ h)(5) = g(2) = 5$

49. $(f \circ g \circ h)(2) = (f \circ g)(1) = f(2) = 5$

51. $(f \circ h)(a) = f\left(\dfrac{a+1}{3}\right) = 3\left(\dfrac{a+1}{3}\right) - 1 = (a+1) - 1 = a$

53. $(f \circ g)(t) = f(t^2+1) = 3(t^2+1) - 1 = 3t^2 + 2$

55. $(f \circ g)(x) = \sqrt{x} - 2$, domain $[0, \infty)$

57. $(f \circ h)(x) = \dfrac{1}{x} - 2$, domain $(-\infty, 0) \cup (0, \infty)$

59. $(h \circ g)(x) = \dfrac{1}{\sqrt{x}}$, domain $(0, \infty)$

61. $(f \circ f)(x) = (x-2) - 2 = x - 4$, domain $(-\infty, \infty)$

63. $(h \circ g \circ f)(x) = h(\sqrt{x-2}) = \dfrac{1}{\sqrt{x-2}}$, domain $(2, \infty)$

65. $(h \circ f \circ g)(x) = h\left(\sqrt{x}-2\right) = \dfrac{1}{\sqrt{x}-2}$, domain $(0,4) \cup (4, \infty)$

67. $F = g \circ h$

69. $H = h \circ g$

71. $N = h \circ g \circ f$

73. $P = g \circ f \circ g$

75. $S = g \circ g$

77. If $g(x) = x^3$ and $h(x) = x - 2$, then

$$(h \circ g)(x) = g(x) - 2 = x^3 - 2 = f(x).$$

79. If $g(x) = x + 5$ and $h(x) = \sqrt{x}$, then

$$(h \circ g)(x) = \sqrt{g(x)} = \sqrt{x+5} = f(x).$$

81. If $g(x) = 3x - 1$ and $h(x) = \sqrt{x}$, then

$$(h \circ g)(x) = \sqrt{g(x)} = \sqrt{3x-1} = f(x).$$

83. If $g(x) = |x|$ and $h(x) = 4x + 5$, then

$$(h \circ g)(x) = 4g(x) + 5 = 4|x| + 5 = f(x).$$

85. $y = 2(3x+1) - 3 = 6x - 1$

87. $y = (x^2 + 6x + 9) - 2 = x^2 + 6x + 7$

89. $y = 3 \cdot \dfrac{x+1}{3} - 1 = x + 1 - 1 = x$

91. Since $m = n - 4$ and $y = m^2$, $y = (n-4)^2$.

93. Since $w = x + 16$, $z = \sqrt{w}$, and $y = \dfrac{z}{8}$, we obtain $y = \dfrac{\sqrt{x+16}}{8}$.

95. After multiplying y by $\dfrac{x+1}{x+1}$ we have

$$y = \frac{\dfrac{x-1}{x+1} + 1}{\dfrac{x-1}{x+1} - 1} = \frac{(x-1)+(x+1)}{(x-1)-(x+1)} = -x$$

The domain of the original function is $(-\infty, -1) \cup (-1, \infty)$ while the domain of the simplified function is $(-\infty, \infty)$. The two functions are not the same.

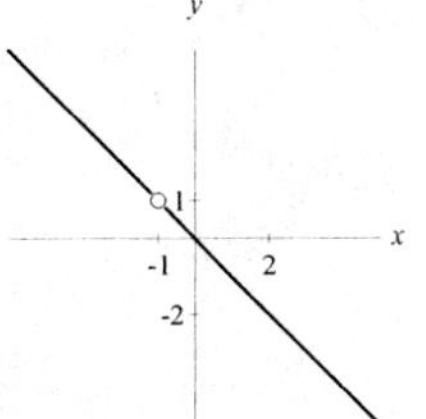

97. Domain $[-1, \infty)$, range $[-7, \infty)$

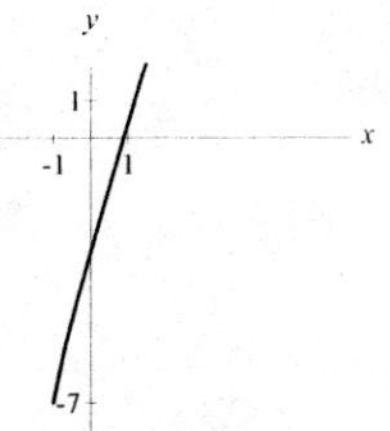

99. Domain $[1, \infty)$, range $[0, \infty)$

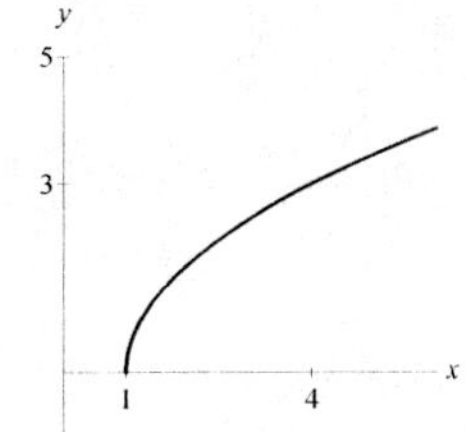

101. Domain $[0, \infty)$, range $[4, \infty)$

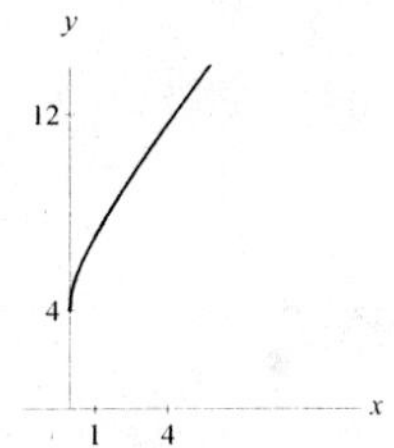

103. $P(x) = 68x - (40x + 200) = 28x - 200$.

Since $200/28 \approx 7.1$, the profit is positive when the number of trimmers satisfies $x \geq 8$.

105. $A = d^2/2$

107. $(f \circ f)(x) = 0.899x$ and $(f \circ f \circ f)(x) = 0.852x$ are the amounts of forest land at the start of 2014 and 2015, respectively.

109. Total cost is $(T \circ C)(x) = 1.05(1.20x) = 1.26x$.

111. Note, $D = \frac{d/2240}{x} = \frac{d/2240}{L^3/100^3} = \frac{100^3 d}{2240L^3}$
$= \frac{100^3(26,000)}{2240L^3} = \frac{100^4(26)}{224L^3} = \frac{100^4(13)}{112L^3}$.
Expressing D as a function of L, we write $D = \frac{(13)100^4}{112L^3}$ or $D = \frac{1.16 \times 10^7}{L^3}$.

113. The area of a semicircle with radius $s/2$ is $(1/2)\pi(s/2)^2 = \pi s^2/8$. The area of the square is s^2. The area of the window is

$$W = s^2 + \frac{\pi s^2}{8} = \frac{(8+\pi)s^2}{8}.$$

115. Form a right triangle with two sides of length s and a hypotenuse of length d. By the Pythagorean Theorem, we obtain

$$d^2 = s^2 + s^2.$$

Solving for s, we have $s = \frac{d\sqrt{2}}{2}$.

117. If a coat is on sale at 25% off and there is an additional 10% off, then the coat will cost $0.90(.75x) = 0.675x$ where x is the regular price. Thus, the discount sale is 32.5% off and not 35% off.

119. The difference quotient is

$$\frac{1 + \frac{3}{x+h} - 1 - \frac{3}{x}}{h} = \frac{\frac{3}{x+h} - \frac{3}{x}}{h} = \frac{3x - 3x - 3h}{xh(x+h)} = \frac{-3}{x(x+h)}$$

121. Since $x - 3 \geq 0$, the domain is $[3, \infty)$. Since $-5\sqrt{x-3}$ has range $(-\infty, 0]$, the range of $f(x) = -5\sqrt{x-3} + 2$ is $(-\infty, 2]$.

123. Since $5x > -1$, the solution set is $(-1/5, \infty)$.

125. $1 - (0.7)(0.7)^2(0.7)^3(0.7)^4 \approx 0.97175$ or 97.2%

For Thought

1. False, since the inverse function is $\{(3,2),(5,5)\}$.

2. False, since it is not one-to-one.

3. False, $g^{-1}(x)$ does not exist since g is not one-to-one.

4. True

5. False, a function that fails the horizontal line test has no inverse.

6. False, since it fails the horizontal line test.

7. False, since $f^{-1}(x) = \left(\frac{x}{3}\right)^2 + 2$ where $x \geq 0$.

8. False, $f^{-1}(x)$ does not exist since f is not one-to-one.

9. False, since $y = |x|$ is V-shaped and the horizontal line test fails.

10. True

2.5 Exercises

1. one-to-one

3. inverse

5. horizontal

7. Yes, since all second coordinates are distinct.

9. No, since there are repeated second coordinates such as $(-1,1)$ and $(1,1)$.

11. No, since there are repeated second coordinates such as $(1,99)$ and $(5,99)$.

13. Not one-to-one

15. One-to-one

17. Not one-to-one

19. One-to-one; since the graph of $y = 2x-3$ shows $y = 2x - 3$ is an increasing function, the Horizontal Line Test implies $y = 2x - 3$ is one-to-one.

21. One-to-one; for if $q(x_1) = q(x_2)$ then

$$\begin{aligned} \frac{1-x_1}{x_1-5} &= \frac{1-x_2}{x_2-5} \\ (1-x_1)(x_2-5) &= (1-x_2)(x_1-5) \\ x_2 - 5 - x_1x_2 + 5x_1 &= x_1 - 5 - x_2x_1 + 5x_2 \\ x_2 + 5x_1 &= x_1 + 5x_2 \\ 4(x_1 - x_2) &= 0 \\ x_1 - x_2 &= 0. \end{aligned}$$

Thus, if $q(x_1) = q(x_2)$ then $x_1 = x_2$. Hence, q is one-to-one.

23. Not one-to-one for $p(-2) = p(0) = 1$.

25. Not one-to-one for $w(1) = w(-1) = 4$.

27. One-to-one; for if $k(x_1) = k(x_2)$ then

$$\begin{aligned} \sqrt[3]{x_1+9} &= \sqrt[3]{x_2+9} \\ (\sqrt[3]{x_1+9})^3 &= (\sqrt[3]{x_2+9})^3 \\ x_1 + 9 &= x_2 + 9 \\ x_1 &= x_2. \end{aligned}$$

Thus, if $k(x_1) = k(x_2)$ then $x_1 = x_2$. Hence, k is one-to-one.

29. Invertible, $\{(3,9),(2,2)\}$

31. Not invertible

33. Invertible, $\{(3,3),(2,2),(4,4),(7,7)\}$

35. Not invertible

37. Not invertible, there can be two different items with the same price.

39. Invertible, since the playing time is a function of the length of a movie.

41. Invertible, assuming that cost is simply a multiple of the number of days. If cost includes extra charges, then the function may not be invertible.

43. $f^{-1} = \{(1,2),(5,3)\}$, $f^{-1}(5) = 3$, $(f^{-1} \circ f)(2) = 2$

45. $f^{-1} = \{(-3,-3),(5,0),(-7,2)\}$, $f^{-1}(5) = 0$, $(f^{-1} \circ f)(2) = 2$

47. Not invertible since it fails the Horizontal Line Test.

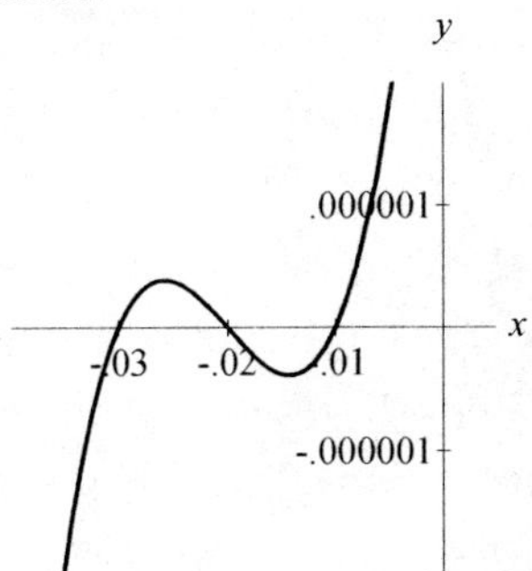

49. Not invertible since it fails the Horizontal Line Test.

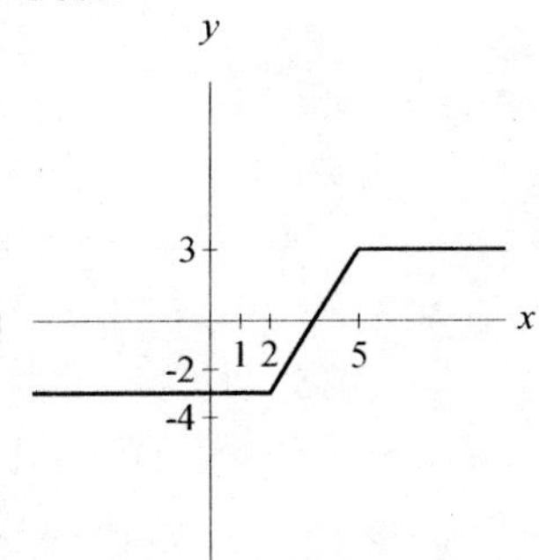

51.

a) $f(x)$ is the composition of multiplying x by 5, then subtracting 1. Reversing the operations, the inverse is $f^{-1}(x) = \dfrac{x-1}{5}$

b) $f(x)$ is the composition of multiplying x by 3, then subtracting 88. Reversing the operations, the inverse is $f^{-1}(x) = \dfrac{x+88}{3}$

c) $f^{-1}(x) = (x+7)/3$

d) $f^{-1}(x) = \dfrac{x-4}{-3}$

e) $f^{-1}(x) = 2(x+9) = 2x+18$

f) $f^{-1}(x) = -x$

g) $f(x)$ is the composition of taking the cube root of x, then subtracting 9. Reversing the operations, the inverse is $f^{-1}(x) = (x+9)^3$

h) $f(x)$ is the composition of cubing x, multiplying the result by 3, then subtracting 7.

Reversing the operations, the inverse is $f^{-1}(x) = \sqrt[3]{\dfrac{x+7}{3}}$

i) $f(x)$ is the composition of subtracting 1 from x, taking the cube root of the result, then adding 5.

Reversing the operations, the inverse is $f^{-1}(x) = (x-5)^3 + 1$

j) $f(x)$ is the composition of subtracting 7 from x, taking the cube root of the result, then multiplying by 2.

Reversing the operations, the inverse is $f^{-1}(x) = \left(\dfrac{x}{2}\right)^3 + 7$

53. No, since they fail the Horizontal Line Test.

55. Yes, since the graphs are symmetric about the line $y = x$.

57. Graph of f^{-1}

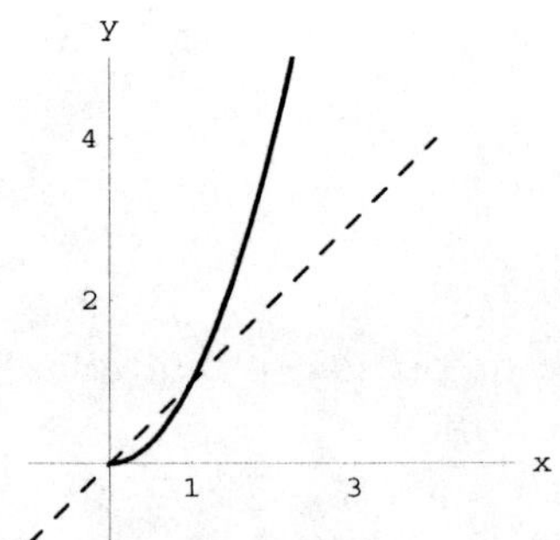

59. Graph of f^{-1}

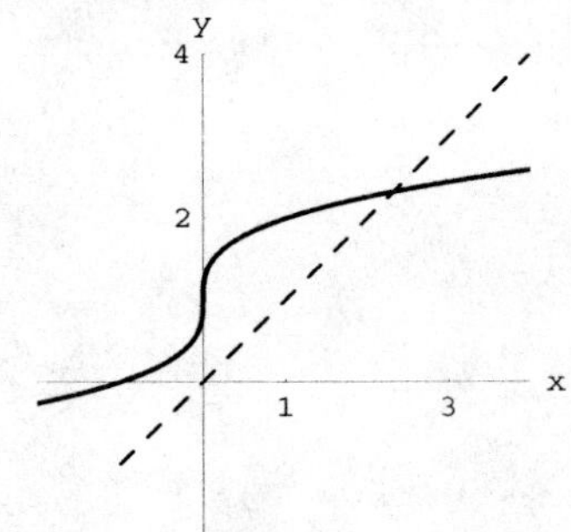

61. $f^{-1}(x) = \dfrac{x-2}{3}$

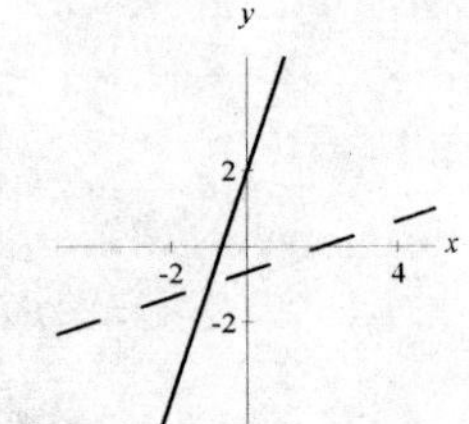

63. $f^{-1}(x) = \sqrt{x+4}$

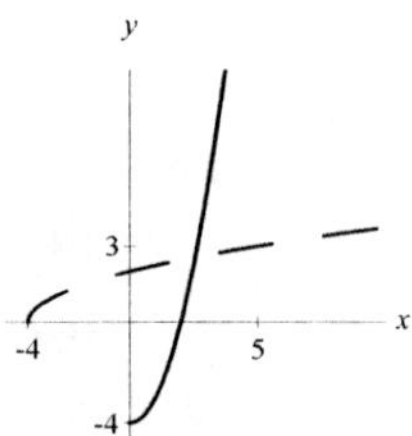

65. $f^{-1}(x) = \sqrt[3]{x}$

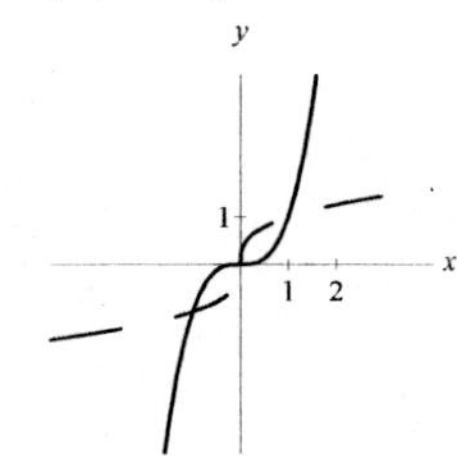

67. $f^{-1}(x) = (x+3)^2$ for $x \geq -3$

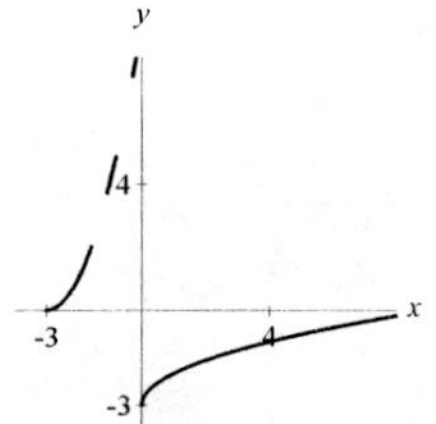

69. Interchange x and y then solve for y.

$$\begin{aligned} x &= 3y - 7 \\ \frac{x+7}{3} &= y \\ \frac{x+7}{3} &= f^{-1}(x) \end{aligned}$$

71. Interchange x and y then solve for y.

$$\begin{aligned} x &= 2 + \sqrt{y-3} \quad \text{for } x \geq 2 \\ (x-2)^2 &= y - 3 \quad \text{for } x \geq 2 \\ f^{-1}(x) &= (x-2)^2 + 3 \quad \text{for } x \geq 2 \end{aligned}$$

73. Interchange x and y then solve for y.

$$\begin{aligned} x &= -y - 9 \\ y &= -x - 9 \\ f^{-1}(x) &= -x - 9 \end{aligned}$$

75. Interchange x and y then solve for y.

$$x = \frac{y+3}{y-5}$$

$$
\begin{aligned}
xy - 5x &= y + 3 \\
xy - y &= 5x + 3 \\
y(x - 1) &= 5x + 3 \\
f^{-1}(x) &= \frac{5x + 3}{x - 1}
\end{aligned}
$$

77. Interchange x and y then solve for y.

$$
\begin{aligned}
x &= -\frac{1}{y} \\
xy &= -1 \\
f^{-1}(x) &= -\frac{1}{x}
\end{aligned}
$$

79. Interchange x and y then solve for y.

$$
\begin{aligned}
x &= \sqrt[3]{y - 9} + 5 \\
x - 5 &= \sqrt[3]{y - 9} \\
(x - 5)^3 &= y - 9 \\
f^{-1}(x) &= (x - 5)^3 + 9
\end{aligned}
$$

81. Interchange x and y then solve for y.

$$
\begin{aligned}
x &= (y - 2)^2 \quad x \geq 0 \\
\sqrt{x} &= y - 2 \\
f^{-1}(x) &= \sqrt{x} + 2
\end{aligned}
$$

83. Note, $(g \circ f)(x) = 0.25(4x + 4) - 1 = x$ and $(f \circ g)(x) = 4(0.25x - 1) + 4 = x$.

Yes, g and f are inverse functions of each other.

85. Since $(f \circ g)(x) = \left(\sqrt{x - 1}\right)^2 + 1 = x$ and and $(g \circ f)(x) = \sqrt{x^2 + 1 - 1} = \sqrt{x^2} = |x|$, g and f are not inverse functions of each other.

87. We find

$$
\begin{aligned}
(f \circ g)(x) &= \frac{1}{1/(x - 3)} + 3 \\
&= x - 3 + 3 \\
(f \circ g)(x) &= x
\end{aligned}
$$

and

$$
\begin{aligned}
(g \circ f)(x) &= \frac{1}{\left(\frac{1}{x} + 3\right) - 3} \\
&= \frac{1}{1/x} \\
(g \circ f)(x) &= x.
\end{aligned}
$$

Then g and f are inverse functions of each other.

89. We obtain

$$
\begin{aligned}
(f \circ g)(x) &= \sqrt[3]{\frac{5x^3 + 2 - 2}{5}} \\
&= \sqrt[3]{\frac{5x^3}{5}} \\
&= \sqrt[3]{x^3} \\
(f \circ g)(x) &= x
\end{aligned}
$$

and

$$
\begin{aligned}
(g \circ f)(x) &= 5\left(\sqrt[3]{\frac{x - 2}{5}}\right)^3 + 2 \\
&= 5\left(\frac{x - 2}{5}\right) + 2 \\
&= (x - 2) + 2 \\
(g \circ f)(x) &= x.
\end{aligned}
$$

Thus, g and f are inverse functions of each other.

91. y_1 and y_2 are inverse functions of each other and $y_3 = y_2 \circ y_1$.

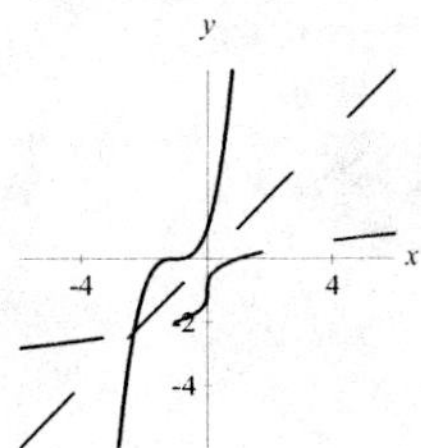

93. $C = 1.08P$ expresses the total cost as a function of the purchase price; and $P = C/1.08$ is the purchase price as a function of the total cost.

95. The graph of t as a function of r satisfies the Horizontal Line Test and is invertible. Solving for r we find,

$$
\begin{aligned}
t - 7.89 &= -0.39r \\
r &= \frac{t - 7.89}{-0.39}
\end{aligned}
$$

and the inverse function is $r = \dfrac{7.89 - t}{0.39}$.

If $t = 5.55$ min., then $r = \dfrac{7.89 - 5.55}{0.39} = 6$ rowers.

97. Solving for w, we obtain

$$\begin{aligned} 1.496w &= V^2 \\ w &= \frac{V^2}{1.496} \end{aligned}$$

and the inverse function is $w = \dfrac{V^2}{1.496}$. If $V = 115$ ft./sec., then $w = \dfrac{115^2}{1.496} \approx 8840$ lb.

99. **a)** Let $V = \$28,000$. The depreciation rate is

$$r = 1 - \left(\frac{28,000}{50,000}\right)^{1/5} \approx 0.109$$

or $r \approx 10.9\%$.

b) Writing V as a function of r we find

$$\begin{aligned} 1 - r &= \left(\frac{V}{50,000}\right)^{1/5} \\ (1-r)^5 &= \frac{V}{50,000} \end{aligned}$$

and $V = 50,000(1-r)^5$.

101. Since $g^{-1}(x) = \dfrac{x+5}{3}$ and $f^{-1}(x) = \dfrac{x-1}{2}$, we have

$$g^{-1} \circ f^{-1}(x) = \frac{\dfrac{x-1}{2} + 5}{3} = \frac{x+9}{6}.$$

Likewise, since $(f \circ g)(x) = 6x - 9$, we get

$$(f \circ g)^{-1}(x) = \frac{x+9}{6}.$$

Hence, $(f \circ g)^{-1} = g^{-1} \circ f^{-1}$.

103. One can easily see that the slope of the line joining (a, b) to (b, a) is -1, and that their midpoint is $\left(\dfrac{a+b}{2}, \dfrac{a+b}{2}\right)$. This midpoint lies on the line $y = x$ whose slope is 1. Then $y = x$ is the perpendicular bisector of the line segment joining the points (a, b) and (b, a)

105. Dividing we get $\dfrac{x-3}{x+2} = 1 - \dfrac{5}{x+2}$

107. $(f \circ g)(2) = f(g(2)) = f(1) = \dfrac{2+3}{5} = 1$

$(f \cdot g)(2) = f(2)g(2) = \dfrac{7}{5} \cdot 1 = \dfrac{7}{5}$

109. Observe, the graph of $f(x) = -\sqrt{9-x^2}$ is a lower semicircle of radius 3 centered at $(0, 0)$.

Then domain is $[-3, 3]$, the range is $[-3, 0]$, and increasing on $[0, 3]$

111.

$$\begin{aligned} 0.75 + 0.80 &= 0.225x - 0.125x \\ 1.55 &= 0.1x \\ 15.5 &= x \end{aligned}$$

The solution set is $\{15.5\}$.

113. Since $640,000 = 2^{10} \cdot 5^4$, we have either $x = 2^{10}$ and $y = 5^4$, or $x = 5^4$ and $y = 2^{10}$. In either case, $|x - y| = 399$.

For Thought

1. False, there is a fixed cost plus per minute rate.

2. False, since cost varies directly with the number of pounds purchased.

3. True **4.** True

5. True, since the area of a circle varies directly with the square of its radius.

6. False, since $y = k/x$ is undefined when $x = 0$.

7. True **8.** True **9.** True

10. False, the surface area is not equal to

$$(\text{Surface Area }) = k \cdot \text{length} \cdot \text{width} \cdot \text{height}$$

for some constant k.

2.6 Exercises

1. varies directly

3. varies inversely

5. $G = kn$ **7.** $V = k/P$ **9.** $C = khr$

11. $Y = \dfrac{kx}{\sqrt{z}}$

13. A varies directly as the square of r.

15. y varies inversely as x.

17. Not a variation expression.

19. a varies jointly as z and w.

21. H varies directly as the square root of t and inversely as s.

23. D varies jointly as L and J and inversely as W.

25. Since $y = kx$ and $5 = k \cdot 9$, $k = 5/9$. Then $y = 5x/9$.

27. Since $T = k/y$ and $-30 = k/5$, $k = -150$. Thus, $T = -150/y$.

29. Since $m = kt^2$ and $54 = k \cdot 18$, $k = 3$. Thus, $m = 3t^2$.

31. Since $y = kx/\sqrt{z}$ and $2.192 = k(2.4)/\sqrt{2.25}$, we obtain $k = 1.37$. Hence, $y = 1.37x/\sqrt{z}$.

33. Since $y = kx$ and $9 = k(2)$, we obtain

$$y = \frac{9}{2} \cdot (-3) = -27/2.$$

35. Since $P = k/w$ and $2/3 = \dfrac{k}{1/4}$, we find

$$k = \frac{2}{3} \cdot \frac{1}{4} = \frac{1}{6}. \text{ Thus, } P = \frac{1/6}{1/6} = 1.$$

37. Since $A = kLW$ and $30 = k(3)(5\sqrt{2})$,

we obtain $A = \sqrt{2}(2\sqrt{3})\dfrac{1}{2} = \sqrt{6}$.

39. Since $y = ku/v^2$ and $7 = k \cdot 9/36$,

we find $y = 28 \cdot 4/64 = 7/4$.

41. Let L_i and L_f be the length in inches and feet, respectively. Then $L_i = 12L_f$ is a direct variation.

43. Let P and n be the cost per person and the number of persons, respectively. Then $P = 20/n$ is an inverse variation.

45. Let S_m and S_k be the speeds of the car in mph and kph, respectively. Then $S_m \approx S_k/1.6 \approx 0.6S_k$ is a direct variation.

47. Not a variation

49. Let A and W be the area and width, respectively. Then $A = 30W$ is a direct variation.

51. Let n and p be the number of gallons and price per gallon, respectively. Since $np = 50$, we obtain that $n = \dfrac{50}{p}$ is an inverse variation.

53. If p is the pressure at depth d, then $p = kd$. Since $4.34 = k(10)$, $k = 0.434$. At $d = 6000$ ft, the pressure is $p = 0.434(6000) = 2604$ lb per square inch.

55. If h is the number of hours, p is the number of pounds, and w is the number of workers then $h = kp/w$. Since $8 = k(3000)/6$, $k = 0.016$. Five workers can process 4000 pounds in $h = (0.016)(4000)/5 = 12.8$ hours.

57. Since $I = kPt$ and $20.80 = k(4000)(16)$, we find $k = 0.000325$. The interest from a deposit of \$6500 for 24 days is

$$I = (0.000325)(6500)(24) \approx \$50.70.$$

59. Since $C = kDL$ and $18.60 = k(6)(20)$, we obtain $k = 0.155$. The cost of a 16 ft pipe with a diameter of 8 inches is

$$C = 0.155(8)(16) = \$19.84.$$

61. Since $w = khd^2$ and $14.5 = k(4)(6^2)$, we find $k = \dfrac{14.5}{144}$. Then a 5-inch high can with a diameter of 6 inches has weight

$$w = \frac{14.5}{144}(5)(6^2) = 18.125 \text{ oz.}$$

63. Since $V = kh/l$ and $10 = k(50)/(200)$, we get $k = 40$. The velocity, if the head is 60 ft and the length is 300 ft, is $V = (40)(60)/(300) = 8$ ft/year.

65. No, it is not directly proportional otherwise the following ratios $\dfrac{42,506}{1.34} \approx 31,720$, $\dfrac{59,085}{0.295} \approx 200,288$, and $\dfrac{738,781}{0.958} \approx 771,170$ would be the same but they are not.

67. Since $g = ks/p$ and $76 = k(12)/(10)$, $k = \dfrac{190}{3}$. If Calvin studies for 9 hours and plays for 15 hours, then his score is

$$g = \frac{190}{3} \cdot \frac{s}{p} = \frac{190}{3} \cdot \frac{9}{15} = 38.$$

69. Since $h = kv^2$ and $16 = k(32)^2$, we get $k = \dfrac{1}{64}$. To reach a height of $20'2.5''$, the velocity v must satisfy

$$20 + \frac{2.5}{12} = \frac{1}{64}v^2.$$

Solving for v, we find $v \approx 35.96$ ft/sec.

73. Interchange x and y then solve for y.

$$\begin{aligned} x &= \sqrt[3]{y-9}+1 \\ (x-1)^3 &= y-9 \\ (x-1)^3+9 &= y \\ f^{-1}(x) &= (x-1)^3+9 \end{aligned}$$

75. Let x be the average speed in the rain.Then

$$3x + 5(x+5) = 425$$

Solving for x, we find $x = 50$ mph.

77. The slope of the line is 1/2. Using $y = mx + b$ and the point $(-4, 2)$, we find

$$\begin{aligned} 2 &= \frac{1}{2}(-4)+b \\ 2 &= -2+b \\ 4 &= b \end{aligned}$$

The line is given by $y = \frac{1}{2}x + 4$, or $2y = x + 8$. A standard form is $x - 2y = -8$.

79. Let d be the distance Sharon walks. Since 2 minutes is the difference in the times of arrival at 4 mph and 5 mph, we obtain

$$\frac{d}{4} - \frac{d}{5} = \frac{2}{60}.$$

The solution of the above equation is $d = 2/3$ mile. Since $d/4 = (2/3)/4 = 10/60$, Sharon must walk to school in 9 minutes to arrive on time. Thus, Sharon's speed in order to arrive on time is

$$r = \frac{d}{t} = \frac{2/3}{9/60} = \frac{40}{9} \text{ mph}.$$

Review Exercises

1. Function, domain and range are both $\{-2, 0, 1\}$

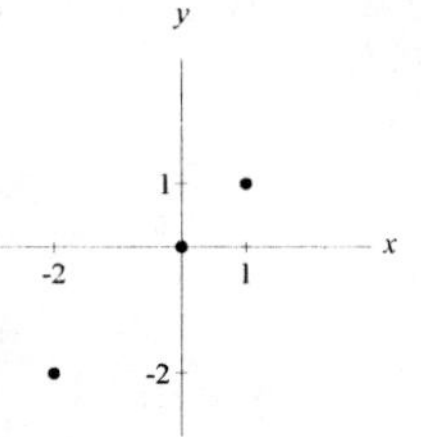

3. $y = 3 - x$ is a function, domain and range are both $(-\infty, \infty)$

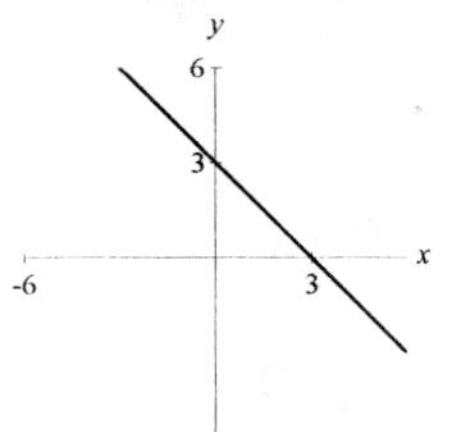

5. Not a function, domain is $\{2\}$, range is $(-\infty, \infty)$

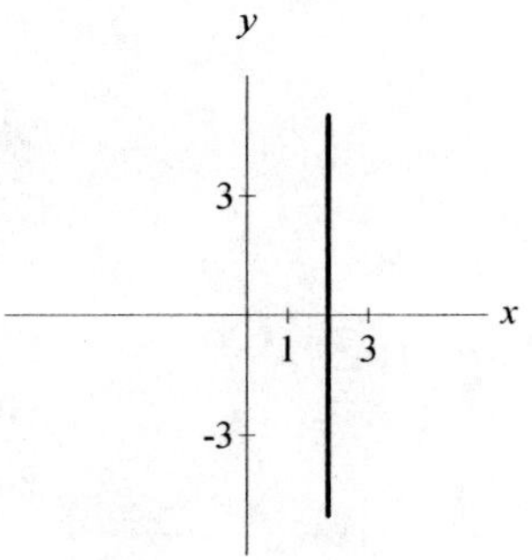

7. $x^2 + y^2 = 0.01$ is not a function, domain and range are both $[-0.1, 0.1]$

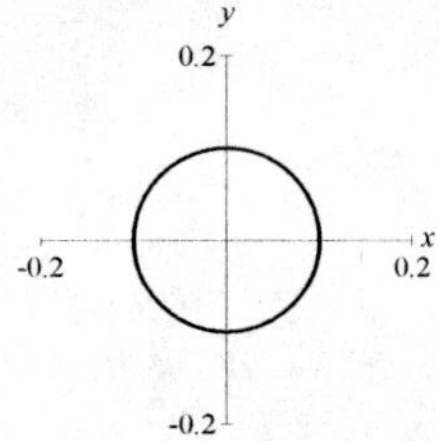

9. $x = y^2 + 1$ is not a function, domain is $[1, \infty)$, range is $(-\infty, \infty)$

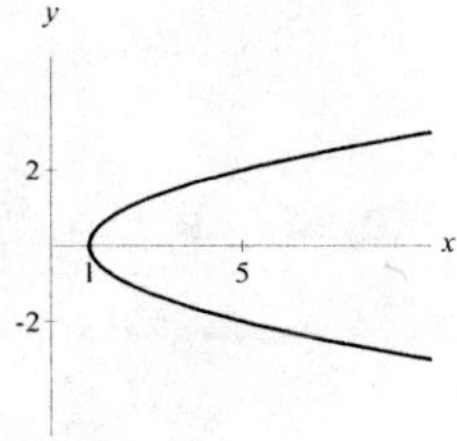

11. $y = \sqrt{x} - 3$ is a function, domain is $[0, \infty)$, range is $[-3, \infty)$

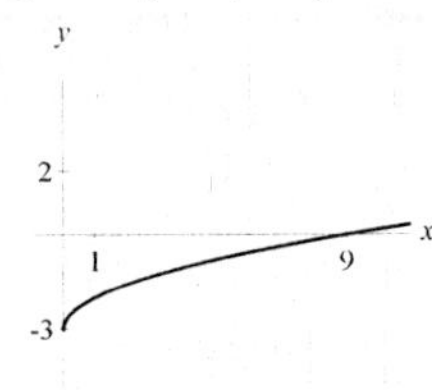

13. $9 + 3 = 12$

15. $24 - 7 = 17$

17. If $x^2 + 3 = 19$, then $x^2 = 16$ or $x = \pm 4$.

19. $g(12) = 17$

21. $7 + (-3) = 4$

23. $(4)(-9) = -36$

25. $f(-3) = 12$

27.

$$\begin{aligned} f(g(x)) &= f(2x-7) \\ &= (2x-7)^2 + 3 \\ &= 4x^2 - 28x + 52 \end{aligned}$$

29. $(x^2 + 3)^2 + 3 = x^4 + 6x^2 + 12$

31. $(a+1)^2 + 3 = a^2 + 2a + 4$

33.

$$\begin{aligned} \frac{f(3+h) - f(3)}{h} &= \frac{(9 + 6h + h^2) + 3 - 12}{h} \\ &= \frac{6h + h^2}{h} \\ &= 6 + h \end{aligned}$$

35.

$$\begin{aligned} \frac{f(x+h) - f(x)}{h} &= \\ \frac{(x^2 + 2xh + h^2) + 3 - x^2 - 3}{h} &= \\ \frac{2xh + h^2}{h} &= \\ 2x + h &= \end{aligned}$$

37. $g\left(\dfrac{x+7}{2}\right) = (x+7) - 7 = x$

39. $g^{-1}(x) = \dfrac{x+7}{2}$

41. $f(x) = \sqrt{x}, g(x) = 2\sqrt{x+3}$; left 3 units, stretch by 2

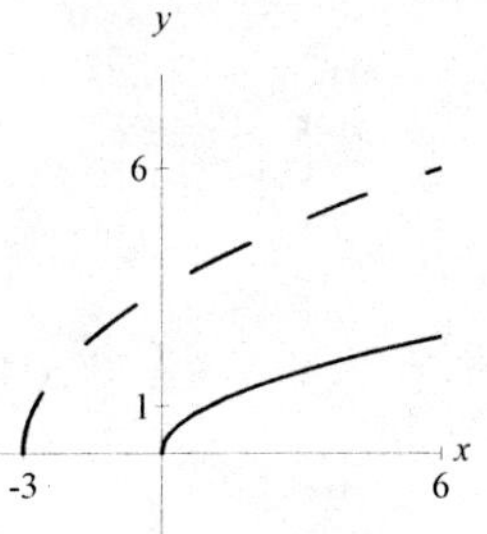

43. $f(x) = |x|, g(x) = -2|x+2| + 4$; left 2 units, stretch by 2, reflect about x-axis, up 4 units

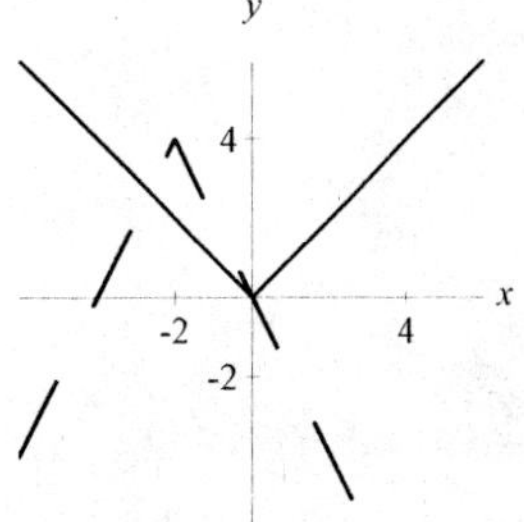

45. $f(x) = x^2, g(x) = \dfrac{1}{2}(x-2)^2 + 1$; right 2 units, stretch by $\dfrac{1}{2}$, up 1 unit

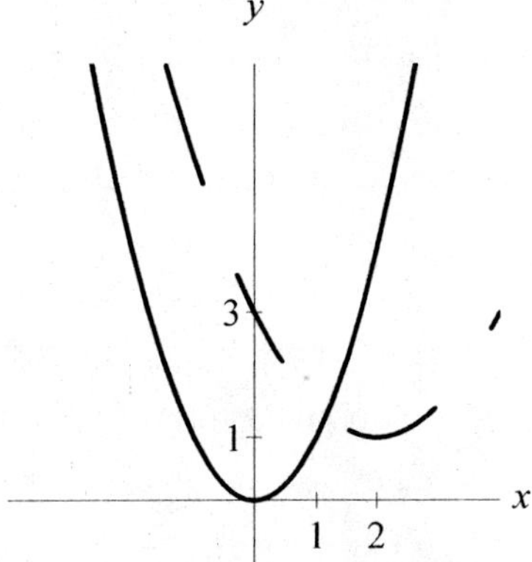

47. $(f \circ g)(x) = \sqrt{x} - 4$, domain $[0, \infty)$

49. $(f \circ h)(x) = x^2 - 4$, domain $(-\infty, \infty)$

51. $(g \circ f \circ h)(x) = g(x^2 - 4) = \sqrt{x^2 - 4}$.

To find the domain, solve $x^2 - 4 \geq 0$. Then the domain is $(-\infty, -2] \cup [2, \infty)$.

53. Translate the graph of f to the right 2 units, stretch by a factor of 2, shift up 1 unit.

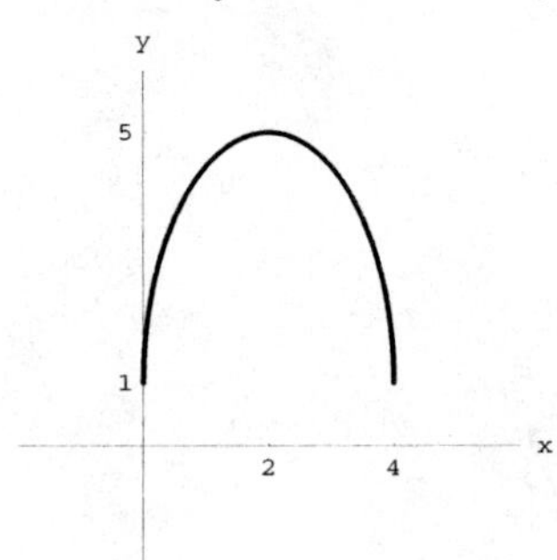

55. Translate the graph of f to the left 1 unit, reflect about the x-axis, shift down 3 units.

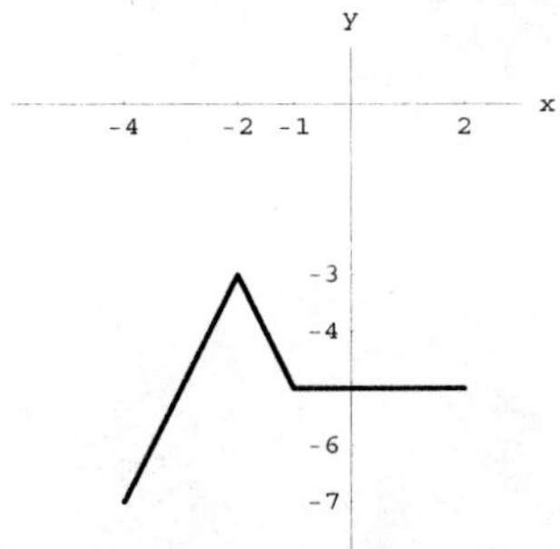

57. Translate the graph of f to the left 2 units, stretch by a factor of 2, reflect about the x-axis.

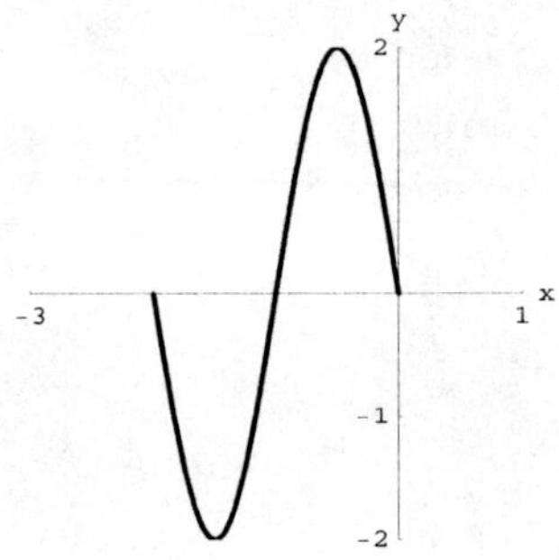

59. Stretch the graph of f by a factor of 2, reflect about the x-axis, shift up 3 units.

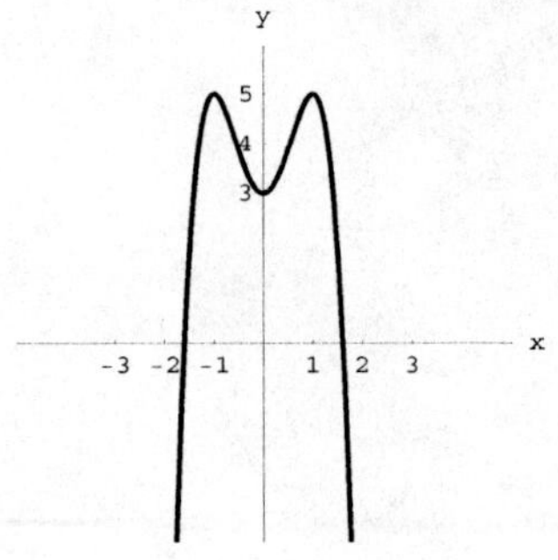

61. $F = f \circ g$

63. $H = f \circ h \circ g \circ j$

65. $N = h \circ f \circ j$ or $N = h \circ j \circ f$

67. $R = g \circ h \circ j$

69.

$$\begin{aligned}\frac{f(x+h)-f(x)}{h} &= \frac{-5(x+h)+9+5x-9}{h}\\ &= \frac{-5h}{h}\\ &= -5\end{aligned}$$

71.

$$\begin{aligned}\frac{f(x+h)-f(x)}{h} &= \\ &= \frac{\frac{1}{2x+2h}-\frac{1}{2x}}{h}\cdot\frac{(2x+2h)(2x)}{(2x+2h)(2x)}\\ &= \frac{(2x)-(2x+2h)}{h(2x+2h)(2x)}\\ &= \frac{-2}{(2x+2h)(2x)}\\ &= \frac{-1}{(x+h)(2x)}\end{aligned}$$

73. Domain is $[-10, 10]$, range is $[0, 10]$, increasing on $[-10, 0]$, decreasing on $[0, 10]$

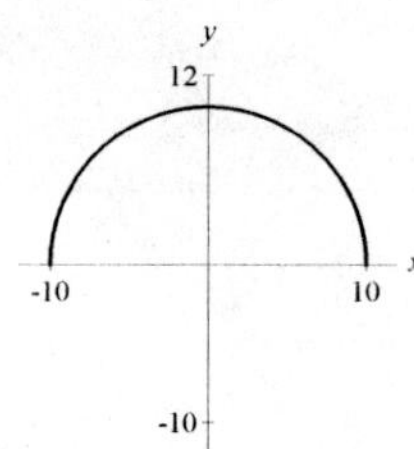

75. Domain and range are both $(-\infty, \infty)$, increasing on $(-\infty, \infty)$

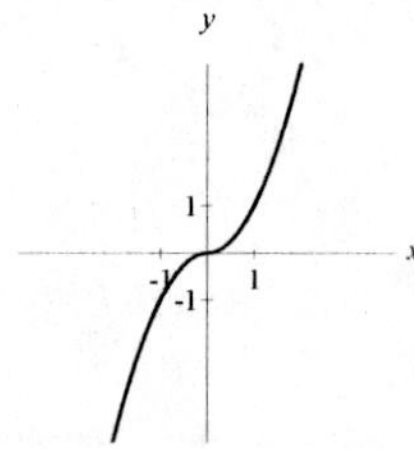

77. Domain is $(-\infty, \infty)$, range is $[-2, \infty)$, increasing on $[-2, 0]$ and $[2, \infty)$, decreasing on $(-\infty, -2]$ and $[0, 2]$

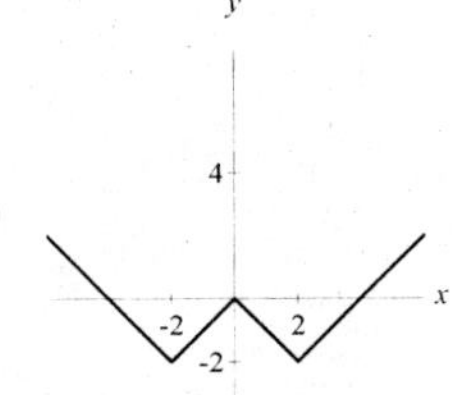

79. $y = |x| - 3$, domain is $(-\infty, \infty)$, range is $[-3, \infty)$

81. $y = -2|x| + 4$, domain is $(-\infty, \infty)$, range is $(-\infty, 4]$

83. $y = |x + 2| + 1$, domain is $(-\infty, \infty)$, range is $[1, \infty)$

85. Symmetry: y-axis

87. Symmetric about the origin

89. Neither symmetry

91. Symmetric about the y-axis

93. From the graph of $y = |x - 3| - 1$, the solution set is $(-\infty, 2] \cup [4, \infty)$

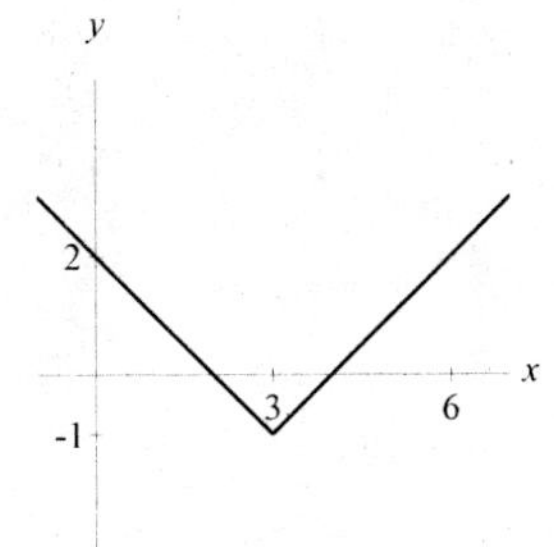

95. From the graph of $y = -2x^2 + 4$, the solution set is $(-\sqrt{2}, \sqrt{2})$

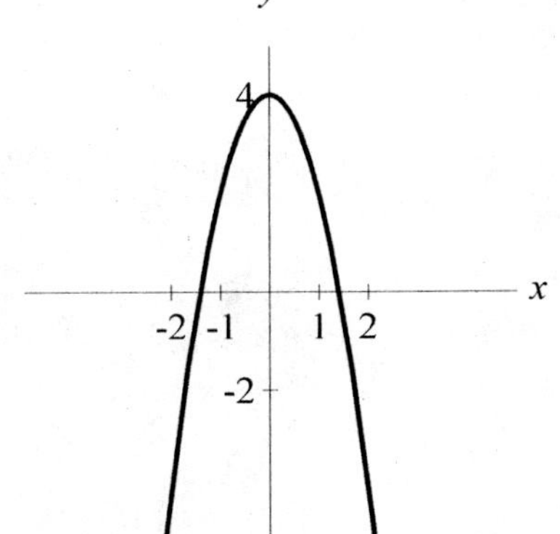

97. No solution since $-\sqrt{x+1} - 2 \leq -2$

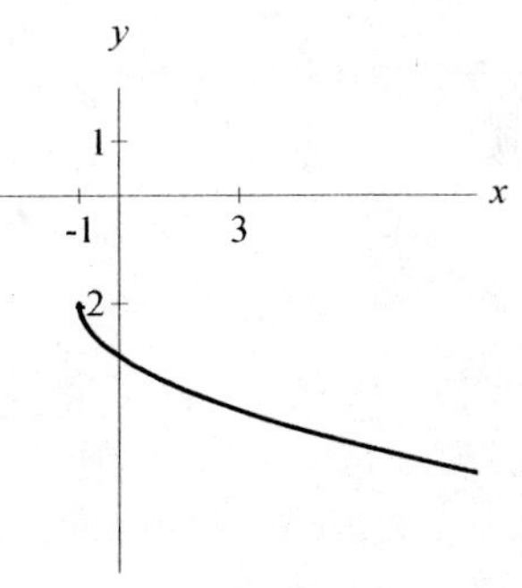

99. Inverse functions, $f(x) = \sqrt{x+3}, g(x) = x^2 - 3$ for $x \geq 0$

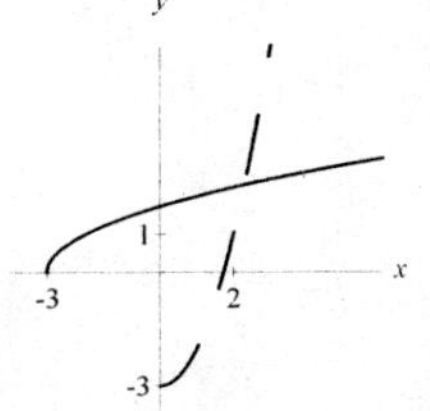

101. Inverse functions,

$f(x) = 2x - 4, g(x) = \frac{1}{2}x + 2$

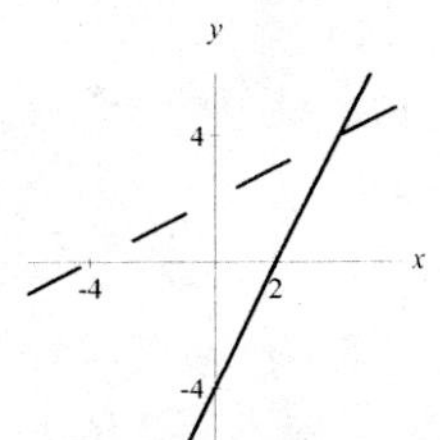

103. Not invertible, since there are two second components that are the same.

105. Inverse is $f^{-1}(x) = \frac{x + 21}{3}$ with domain and range both $(-\infty, \infty)$

107. Not invertible

109. Inverse is $f^{-1}(x) = x^2 + 9$ for $x \geq 0$ with domain $[0, \infty)$ and range $[9, \infty)$

111. Inverse is $f^{-1}(x) = \frac{5x + 7}{1 - x}$ with domain $(-\infty, 1) \cup (1, \infty)$, and range $(-\infty, -5) \cup (-5, \infty)$

113. Inverse is $f^{-1}(x) = -\sqrt{x - 1}$ with domain $[1, \infty)$ and range $(-\infty, 0]$

115. Let x be the number of roses. The cost function is $C(x) = 1.20x + 40$, the revenue function is $R(x) = 2x$, and the profit function is $P(x) = R(x) - C(x)$ or $P(x) = 0.80x - 40$.

Since $P(50) = 0$, to make a profit she must sell at least 51 roses.

117. Since $h(0) = 64$ and $h(2) = 0$, the range of $h = -16t^2 + 64$ is the interval $[0, 64]$.
Then the domain of the inverse function is $[0, 64]$. Solving for t, we obtain

$$\begin{aligned} 16t^2 &= 64 - h \\ t^2 &= \frac{64-h}{16} \\ t &= \frac{\sqrt{64-h}}{4}. \end{aligned}$$

The inverse function is $t = \dfrac{\sqrt{64-h}}{4}$.

119. Since $A = \pi \left(\dfrac{d}{2}\right)^2$, $d = 2\sqrt{\dfrac{A}{\pi}}$.

121. The average rate of change is
$\dfrac{8-6}{9-5} = 0.5$ inch/lb.

123. Since $D = kW$ and $9 = k \cdot 25$, we obtain
$D = \dfrac{9}{25}100 = 36$.

125. Since $V = k\sqrt{h}$ and $45 = k\sqrt{1.5}$, the velocity of a Triceratops is $V = \dfrac{45}{\sqrt{1.5}} \cdot \sqrt{2.8} \approx 61$ kph.

127. Since $C = kd^2$ and $4.32 = k \cdot 36$, a 16-inch diameter globe costs $C = \dfrac{4.32}{36} \cdot 16^2 = \$30.72$.

131. The quadrilateral has vertices $A(0,0)$, $B(0,-9/2)$, $C(82/17, -15/17)$, and $D(7/2, 0)$. The area of triangle ΔABC is

$$\frac{1}{2} \cdot \frac{9}{2} \cdot \frac{82}{17} = \frac{369}{34}$$

and the area of triangle ΔACD is

$$\frac{1}{2} \cdot \frac{7}{2} \cdot \frac{15}{17} = \frac{105}{68}.$$

The sum of areas of the two triangles is the area of the quadrilateral, i.e.,

$$\text{Area} = \frac{369}{34} + \frac{105}{68} = \frac{843}{68} \quad \text{square units.}$$

Chapter 2 Test

1. No, since $(0, 5)$ and $(0, -5)$ are two ordered pairs with the same first coordinate and different second coordinates.

2. Yes, since to each x-coordinate there is exactly one y-coordinate, namely, $y = \dfrac{3x-20}{5}$.

3. No, since $(1, -1)$ and $(1, -3)$ are two ordered pairs with the same first coordinate and different second coordinates.

4. Yes, since to each x-coordinate there is exactly one y-coordinate, namely, $y = x^3 - 3x^2 + 2x - 1$.

5. Domain is $\{2, 5\}$, range is $\{-3, -4, 7\}$

6. Domain is $[9, \infty)$, range is $[0, \infty)$

7. Domain is $[0, \infty)$, range is $(-\infty, \infty)$

8. Graph of $3x - 4y = 12$ includes the points $(4, 0), (0, -3)$

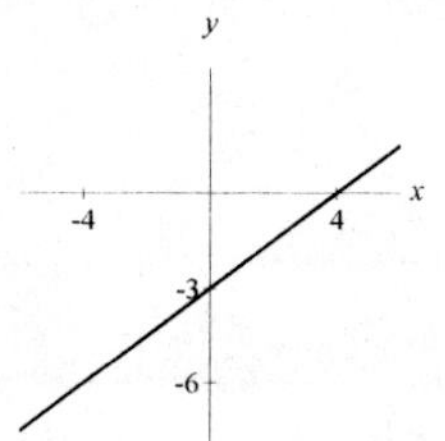

9. Graph of $y = 2x - 3$ includes the points $(3/2, 0), (0, -3)$

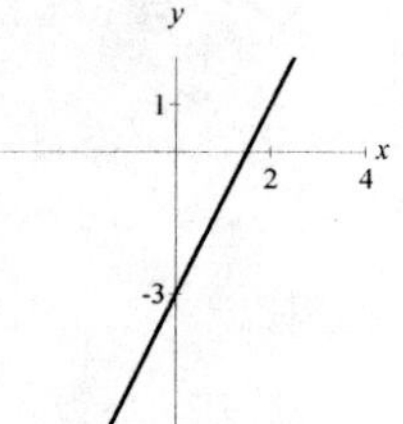

10. $y = \sqrt{25 - x^2}$ is a semicircle with radius 5

11. $y = -(x-2)^2 + 5$ is a parabola with vertex $(2,5)$

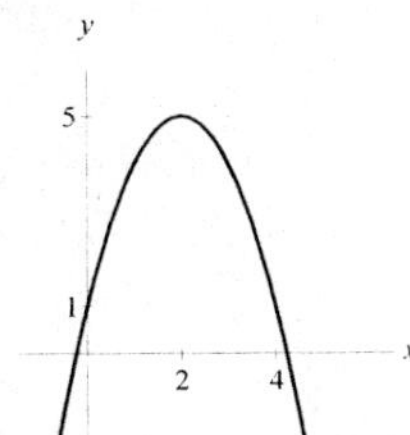

12. $y = 2|x| - 4$ includes the points $(0,-4)$, $(\pm 3, 2)$

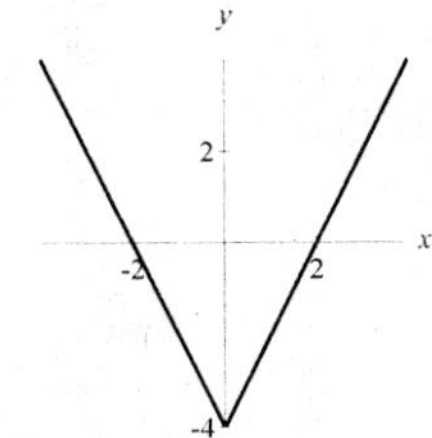

13. $y = \sqrt{x+3} - 5$ includes the points $(1,-3)$, $(6,-2)$

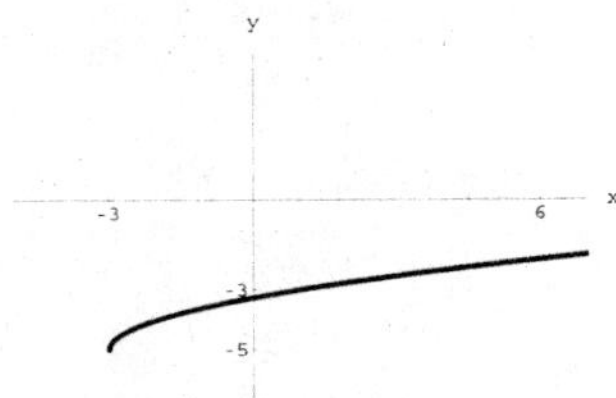

14. Graph includes the points $(-2,-2)$, $(0,2)$, $(3,2)$

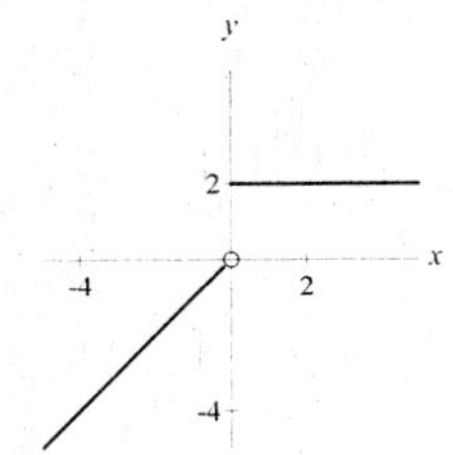

15. $\sqrt{9} = 3$ **16.** $f(5) = \sqrt{7}$

17. $f(3x-1) = \sqrt{(3x-1)+2} = \sqrt{3x+1}$.

18. $g^{-1}(x) = \dfrac{x+1}{3}$ **19.** $\sqrt{16} + 41 = 45$

20.

$$\begin{aligned} \frac{g(x+h)-g(x)}{h} &= \frac{3(x+h)-1-3x+1}{h} \\ &= \frac{3h}{h} \\ &= 3 \end{aligned}$$

21. Increasing on $[3,\infty)$, decreasing on $(-\infty,3]$

22. Symmetric about the y-axis

23. Add 1 to $-3 < x-1 < 3$ to obtain $-2 < x < 4$. Thus, the solution set $(-2,4)$.

24. $g(x)$ is the composition of subtracting 2 from x, taking the cube root of the result, then adding 3. Reversing the operations, the inverse is $g^{-1}(x) = (x-3)^3 + 2$.

25. The range of $f(x) = \sqrt{x-5}$ is $[0,\infty)$. Then the domain of $f^{-1}(x)$ is $x \geq 0$ Note, $f(x)$ is the composition of subtracting 5 from x, then taking the square root of the result. Reversing the operations, the inverse is $f^{-1}(x) = x^2 + 5$ for $x \geq 0$.

26. $\dfrac{60-35}{400-200} = \$0.125$ per envelope

27. Since $I = k/d^2$ and $300 = k/4$, we get $k = 1200$. If $d = 10$, then $I = 1200/100 = 12$ candlepower.

28. Let s be the length of one side of the cube. By the Pythagorean Theorem we have $s^2 + s^2 = d^2$. Then $s = \dfrac{d}{\sqrt{2}}$ and the volume is $V = \left(\dfrac{d}{\sqrt{2}}\right)^3 = \dfrac{\sqrt{2}d^3}{4}$.

Tying It All Together

1. Add 3 to both sides of $2x - 3 = 0$ to obtain $2x = 3$. Then the solution set is $\left\{\dfrac{3}{2}\right\}$.

2. Add $2x$ to both sides of $-2x + 6 = 0$ to get $6 = 2x$. Then the solution set is $\{3\}$.

3. Note, $|x| = 100$ is equivalent to $x = \pm 100$. The solution set is $\{\pm 100\}$.

4. The equation is equivalent to $\dfrac{1}{2} = |x + 90|$.

Then $x + 90 = \pm\dfrac{1}{2}$ and $x = \pm\dfrac{1}{2} - 90$.

The solution set is $\{-90.5, -89.5\}$.

5. Note, $3 = 2\sqrt{x+30}$. If we divide by 2 and square both sides, we get $x + 30 = \frac{9}{4}$. Since $x = \frac{9}{4} - 30 = -27.75$, the solution set is $\{-27.75\}$.

6. Note, $\sqrt{x-3}$ is not a real number if $x < 3$. Since $\sqrt{x-3}$ is nonnegative for $x \geq 3$, it follows that $\sqrt{x-3} + 15$ is at least 15. In particular, $\sqrt{x-3} + 15 = 0$ has no real solution.

7. Rewriting, we obtain $(x-2)^2 = \frac{1}{2}$. By the square root property, $x - 2 = \pm\frac{\sqrt{2}}{2}$. Thus, $x = 2 \pm \frac{\sqrt{2}}{2}$. The solution set is $\left\{\frac{4 \pm \sqrt{2}}{2}\right\}$.

8. Rewriting, we get $(x+2)^2 = \frac{1}{4}$. By the square root property, $x + 2 = \pm\frac{1}{2}$. Thus, $x = -2 \pm \frac{1}{2}$. The solution set is $\left\{-\frac{3}{2}, -\frac{5}{2}\right\}$.

9. Squaring both sides of $\sqrt{9 - x^2} = 2$, we get $9 - x^2 = 4$. Then $5 = x^2$. The solution set is $\left\{\pm\sqrt{5}\right\}$.

10. Note, $\sqrt{49 - x^2}$ is not a real number if $49 - x^2 < 0$. Since $\sqrt{49 - x^2}$ is nonnegative for $49 - x^2 \geq 0$, we get that $\sqrt{49 - x^2} + 3$ is at least 3. In particular, $\sqrt{49 - x^2} + 3 = 0$ has no real solution.

11. Domain is $(-\infty, \infty)$, range is $(-\infty, \infty)$, x-intercept $(3/2, 0)$

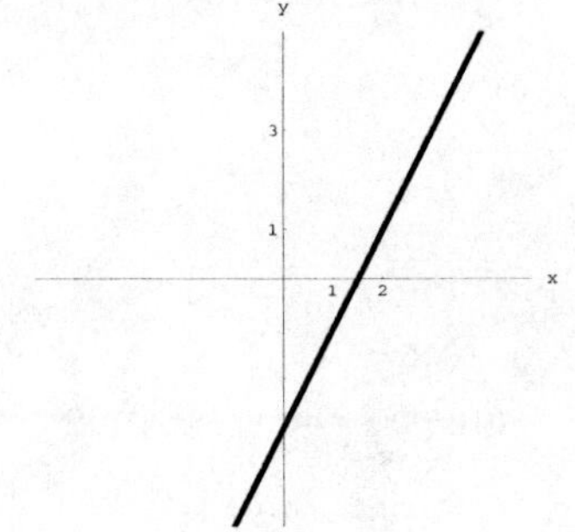

12. Domain is $(-\infty, \infty)$, range is $(-\infty, \infty)$, x-intercept $(3, 0)$

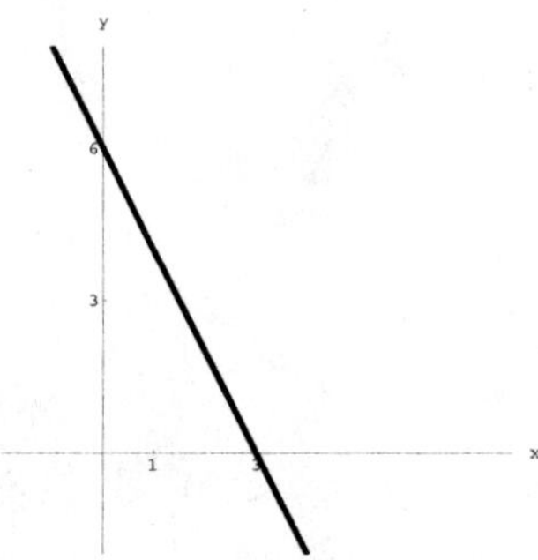

13. Domain is $(-\infty, \infty)$, range is $[-100, \infty)$, x-intercepts $(\pm 100, 0)$

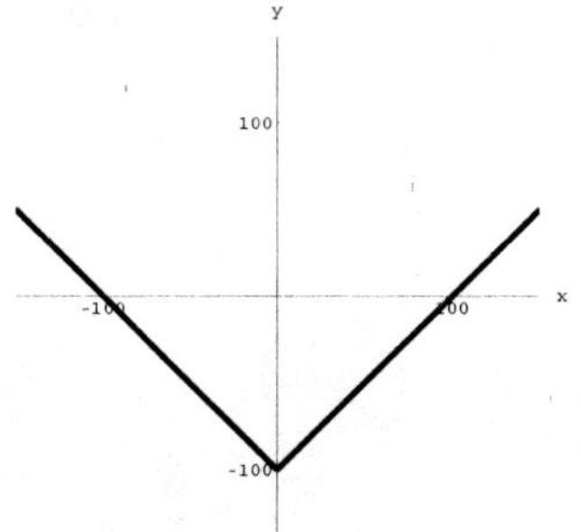

14. Domain is $(-\infty, \infty)$, range is $(-\infty, 1]$, x-intercepts $(-90.5, 0)$ and $(-89.5, 0)$.

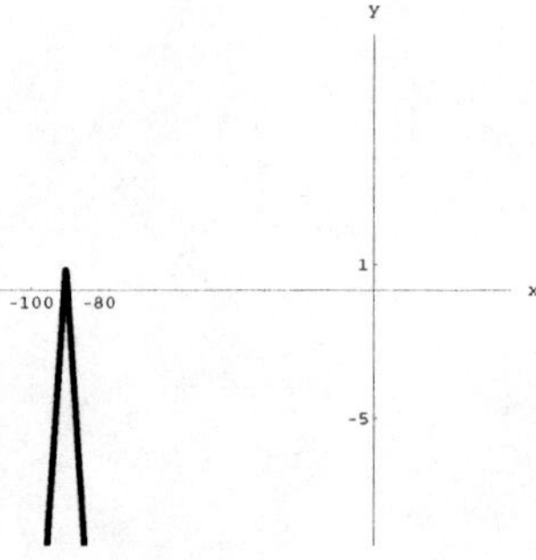

15. Domain is $[-30, \infty)$ since we need to require $x + 30 \geq 0$, range is $(-\infty, 3]$, x-intercept is $(-27.75, 0)$ since the solution to $3 - 2\sqrt{x+30} = 0$ is $x = -27.75$

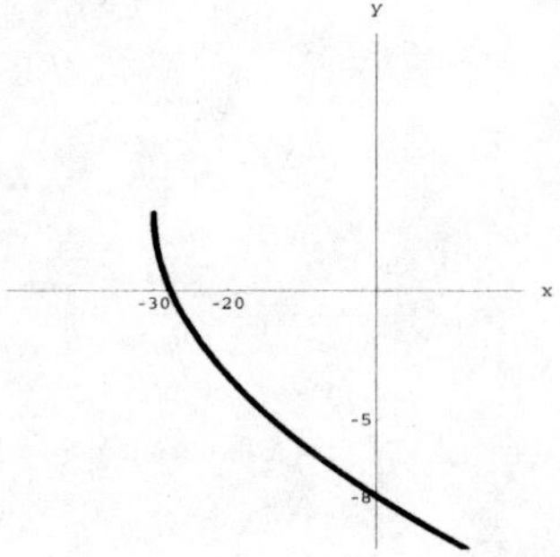

16. Domain is $[3, \infty)$ since we need to require $x - 3 \geq 0$, range is $[15, \infty)$, no x-intercept

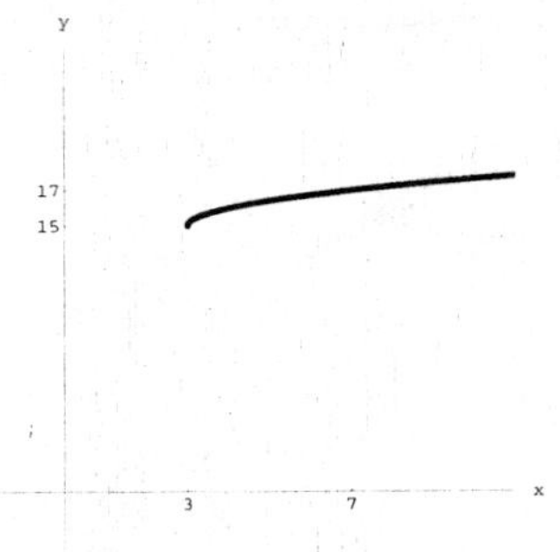

17. Domain is $(-\infty, \infty)$, range is $(-\infty, 1]$ since the vertex is $(2, 1)$, and the x-intercepts are $\left(\frac{4 \pm \sqrt{2}}{2}, 0\right)$ since the solutions of

$0 = -2(x-2)^2 + 1$ are $x = \dfrac{4 \pm \sqrt{2}}{2}$.

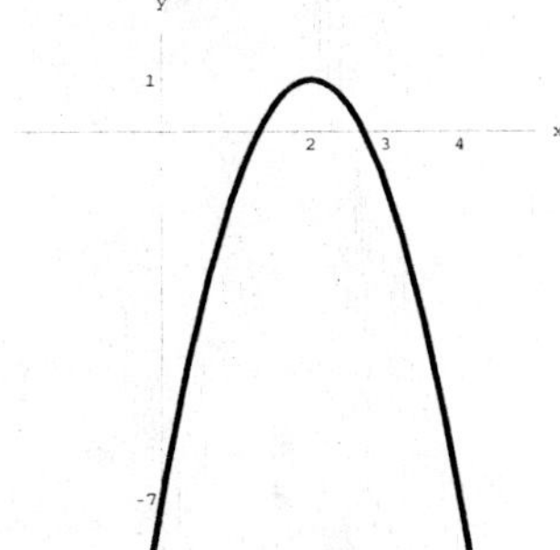

18. Domain is $(-\infty, \infty)$, range is $[-1, \infty)$ since the vertex is $(-2, -1)$, and the x-intercepts are $(-5/2, 0)$ and $(-3/2, 0)$ for the solutions to $0 = 4(x+2)^2 - 1$ are $x = -5/2, -3/2$.

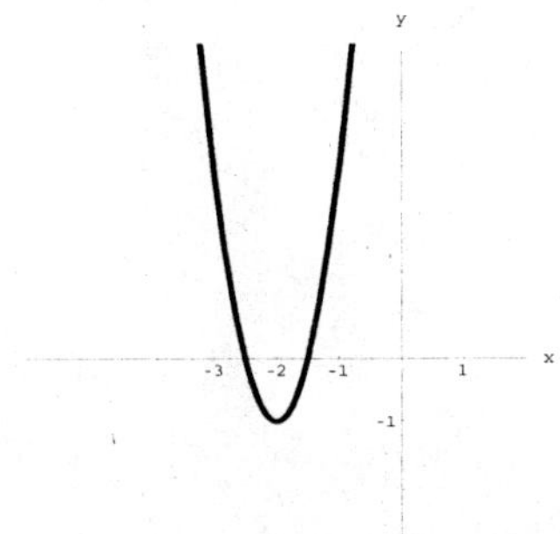

19. Domain is $[-3, 3]$, range is $[-2, 1]$ since the graph is a semi-circle with center $(0, -2)$ and radius 3. The x-intercepts are $(\pm\sqrt{5}, 0)$ for the solutions to $0 = \sqrt{9 - x^2} - 2$ are $x = \pm\sqrt{5}$. The graph is shown in the next column.

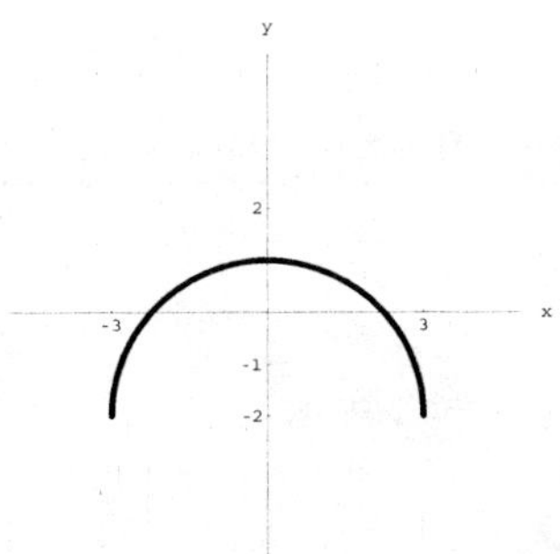

20. Domain is $[-7, 7]$, range is $[3, 10]$ since the graph is a semi-circle with center $(0, 3)$ and radius 7, there are no x-intercepts as seen from the graph.

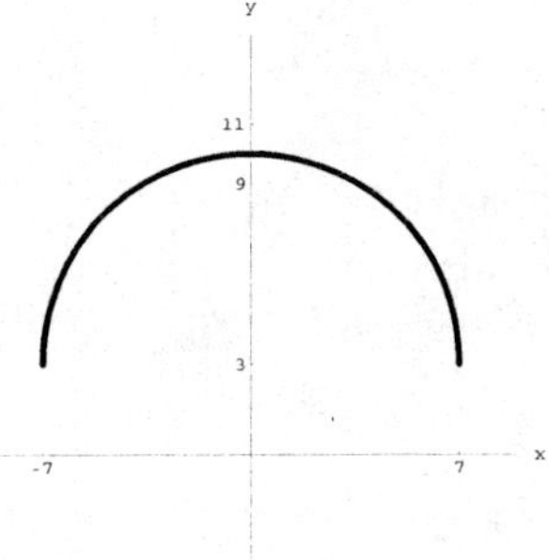

21. Since $2x > 3$ or $x > 3/2$, the solution set is $(3/2, \infty)$.

22. Since $-2x \leq -6$ or $x \geq 3$, the solution set is $[3, \infty)$.

23. Based on the portion of the graph of $y = |x| - 100$ above the x-axis, and its x-intercepts, the solution set of $|x| - 100 \geq 0$ is $(-\infty, -100] \cup [100, \infty)$.

24. Based on the part of the graph of $y = 1 - 2|x + 90|$ above the x-axis, and its x-intercepts, the solution set of $1 - 2|x+90| > 0$ is $(-90.5, -89.5)$.

25. Based on the part of the graph of $y = 3 - 2\sqrt{x + 30}$ below the x-axis, and its x-intercepts, the solution set of $3 - 2\sqrt{x + 30} \leq 0$ is $[-27.75, \infty)$.

26. Solution set is $[3, \infty)$ since the graph is entirely above the x-axis.

27. Based on the portion of the graph of $y = -2(x-2)^2 + 1$ below the x-axis, and its x-intercepts, the solution set of $-2(x-2)^2+1<0$ is

$$\left(-\infty, \frac{4-\sqrt{2}}{2}\right) \cup \left(\frac{4+\sqrt{2}}{2}, \infty\right).$$

28. Based on the part of the graph of $y=4(x+2)^2-1$ above the x-axis, and its x-intercepts, the solution set of $4(x+2)^2-1<0$ is $(-\infty,-5/2] \cup [-3/2,\infty)$.

29. Based on the part of the graph of $y=\sqrt{9-x^2}-2$ above the x-axis, and its x-intercepts, the solution set of $\sqrt{9-x^2}-2 \geq 0$ is $[-\sqrt{5},\sqrt{5}]$.

30. Since the graph of $y=\sqrt{49-x^2}+3$ is entirely above the line $y=3$, the solution to $\sqrt{49-x^2}+3 \leq 0$ is the empty set $\emptyset$.

31. $f(2)=3(2+1)^3-24=3(27)-24=57$

32. The graph of $f(x)=3(x+1)^3-24$ is given.

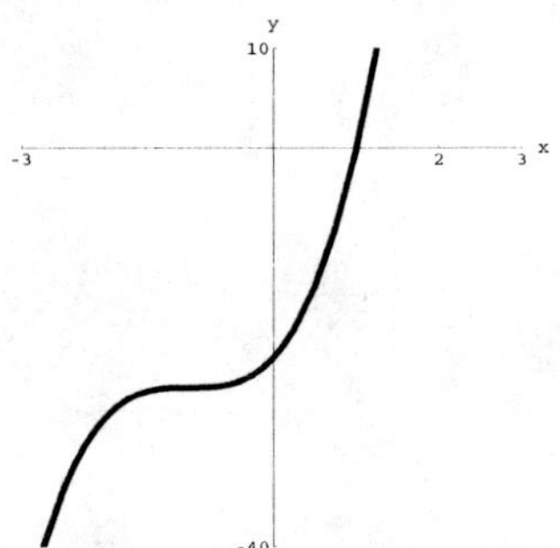

33. $f=K \circ F \circ H \circ G$

34. Solving for x, we get

$$\begin{aligned} 3(x+1)^3 &= 24 \\ (x+1)^3 &= 8 \\ x+1 &= 2 \\ x &= 1. \end{aligned}$$

The solution set is $\{1\}$.

35. Based on the part of the graph of $y=3(x+1)^3-24$ above the x-axis, and its x-intercept $(1,0)$, the solution set of $3(x+1)^3-24 \geq 0$ is $[1,\infty)$.

36. Solving for x, we obtain

$$\begin{aligned} y+24 &= 3(x+1)^3 \\ \frac{y+24}{3} &= (x+1)^3 \\ \sqrt[3]{\frac{y+24}{3}} &= x+1 \\ \sqrt[3]{\frac{y+24}{3}}-1 &= x. \end{aligned}$$

37. Based on the answer from Exercise 36, the inverse is $f^{-1}(x)=\sqrt[3]{\frac{x+24}{3}}-1$

38. The graph of f^{-1} is given below.

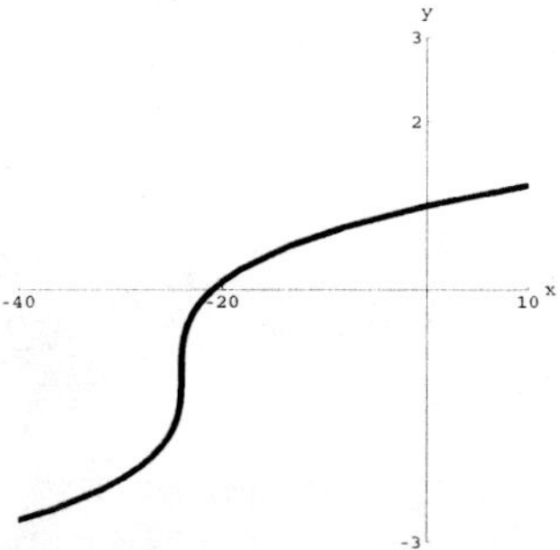

39. Based on the part of the graph of f^{-1} above the x-axis, and its x-intercept $(-21,0)$, the solution set of $f^{-1}(x)>0$ is $(-21,\infty)$.

40. Since $f=K \circ F \circ H \circ G$, we get

$$f^{-1}=G^{-1} \circ H^{-1} \circ F^{-1} \circ K^{-1}.$$

41. a, $-a$

42. exponential, exponent, base

43. domain

44. term

45. product

46. power of a power

47. one

48. square root

49. radical, index, radicand

50. product rule for radicals

For Thought

1. False, the range of $y = x^2$ is $[0, \infty)$.

2. False, the vertex is the point $(3, -1)$.

3. True **4.** True **5.** True, since $\frac{-b}{2a} = \frac{6}{2 \cdot 3} = 1$.

6. True, the x-intercept of $y = (3x+2)^2$ is the vertex $(-2/3, 0)$ and the y-intercept is $(0, 4)$.

7. True

8. True, since $(x - \sqrt{3})^2$ is always nonnegative.

9. True, since if x and $\frac{p-2x}{2}$ are the length and the width, respectively, of a rectangle with perimeter p, then the area is $y = x \cdot \frac{p-2x}{2}$. This is a parabola opening down with vertex $\left(\frac{p}{4}, \frac{p^2}{16}\right)$. Thus, the maximum area is $\frac{p^2}{16}$.

10. False, $f(x) = (x-3)^2$ is increasing on $[3, \infty)$.

3.1 Exercises

1. upward

3. vertex

5. minimum

7. axis of symmetry

9. Completing the square, we get

$$\begin{aligned} y &= \left(x^2 + 4x + \left(\frac{4}{2}\right)^2\right) - \left(\frac{4}{2}\right)^2 \\ y &= \left(x^2 + 4x + 4\right) - 4 \\ y &= (x+2)^2 - 4. \end{aligned}$$

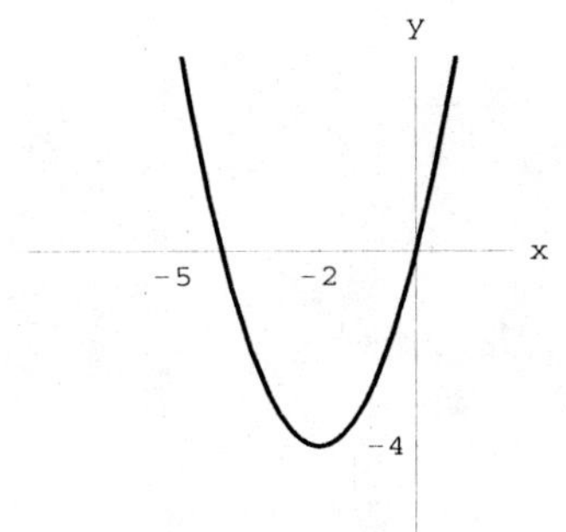

11. $y = \left(x^2 - 3x + \frac{9}{4}\right) - \frac{9}{4} = \left(x - \frac{3}{2}\right)^2 - \frac{9}{4}$

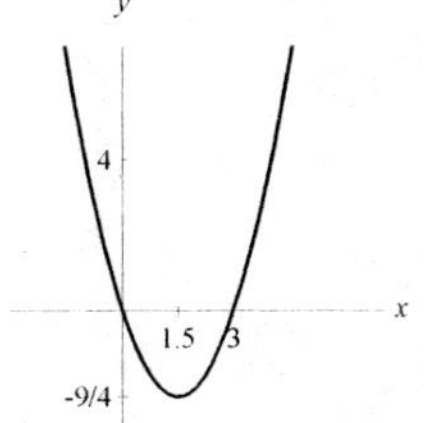

13. Completing the square, we get

$$\begin{aligned} y &= 2\left(x^2 - 6x + \left(\frac{6}{2}\right)^2\right) - 2\left(\frac{6}{2}\right)^2 + 22 \\ y &= 2(x-3)^2 + 4. \end{aligned}$$

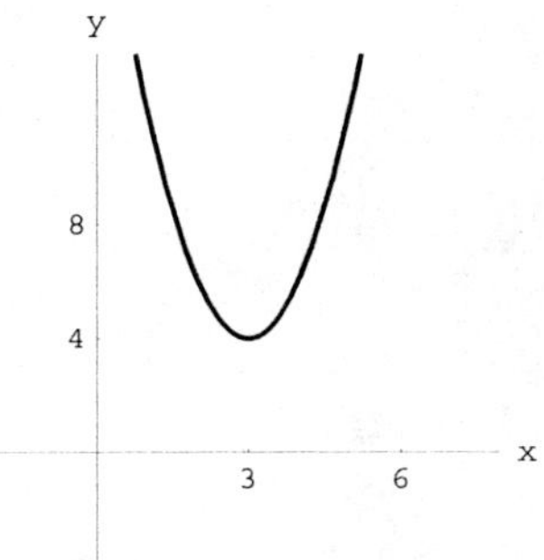

15. Completing the square, we find

$$\begin{aligned} y &= -3\left(x^2 - 2x + \left(\frac{2}{2}\right)^2\right) + 3\left(\frac{2}{2}\right)^2 - 3 \\ y &= -3(x-1)^2. \end{aligned}$$

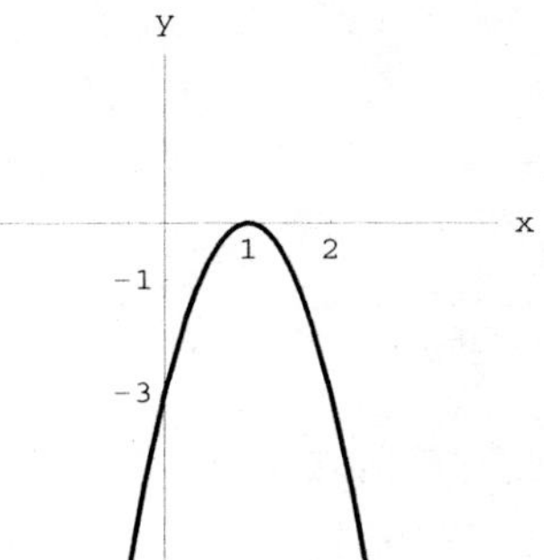

17. $y = \left(x^2 + 3x + \frac{9}{4}\right) - \frac{9}{4} + \frac{5}{2} = \left(x + \frac{3}{2}\right)^2 + \frac{1}{4}$

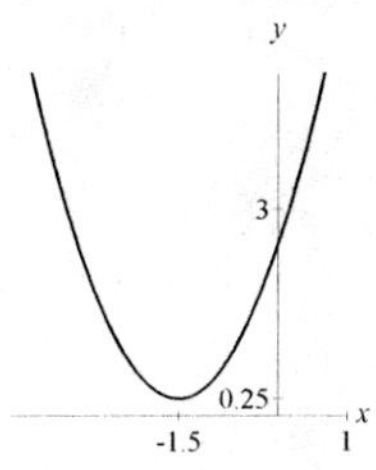

19. $y = -2\left(x^2 - \frac{3}{2}x + \frac{9}{16}\right) + \frac{9}{8} - 1 = -2\left(x - \frac{3}{4}\right)^2 + \frac{1}{8}$

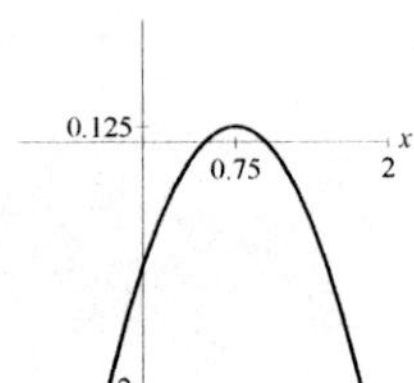

21. Let $a = 1$ and $b = -1$. Since $\frac{-b}{2a} = \frac{1}{2}$ and $f\left(\frac{1}{2}\right) = \left(\frac{1}{2}\right)^2 - \frac{1}{2} = -\frac{1}{4}$, the vertex is $\left(\frac{1}{2}, \frac{1}{4}\right)$.

23. Since $\frac{-b}{2a} = \frac{12}{6} = 2$ and $f(2) = 12 - 24 + 1 = -11$, the vertex is $(2, -11)$.

25. Vertex: $(4, 1)$

27. Since $\frac{-b}{2a} = \frac{1/3}{-1} = -\frac{1}{3}$ and $f(-1/3) = -\frac{1}{18} + \frac{1}{9} = \frac{1}{18}$, the vertex is $\left(-\frac{1}{3}, \frac{1}{18}\right)$.

29. Up, vertex $(1, -4)$, axis of symmetry $x = 1$, range $[-4, \infty)$, minimum value -4, decreasing on $(-\infty, 1)$, inreasing on $(1, \infty)$.

31. Since it opens down with vertex $(0, 3)$, the range is $(-\infty, 3]$, maximum value is 3, decreasing on $[0, \infty)$, and increasing on $(-\infty, 0]$.

33. Since it opens up with vertex $(1, -1)$, the range is $[-1, \infty)$, minimum value is -1, decreasing on $(-\infty, 1]$, and increasing on $[1, \infty)$.

35. Since it opens up with vertex $(-4, -18)$, range is $[-18, \infty)$, minimum value is -18, decreasing on $(-\infty, -4]$, and increasing on $[-4, \infty)$.

37. Since it opens up with vertex is $(3, 4)$, the range is $[4, \infty)$, minimum value is 4, decreasing on $(-\infty, 3]$, and increasing on $[3, \infty)$.

39. Since it opens down with vertex $(3/2, 27/2)$, the range is $(-\infty, 27/2]$, maximum value is $27/2$, decreasing on $[3/2, \infty)$, and increasing on $(-\infty, 3/2]$.

41. Since it opens down with vertex is $(1/2, 9)$, the range is $(-\infty, 9]$, maximum value is 9, decreasing on $[1/2, \infty)$, and increasing on $(-\infty, 1/2]$.

43. Vertex $(0, -3)$, axis $x = 0$, y-intercept $(0, -3)$, x-intercepts $(\pm\sqrt{3}, 0)$, opening up

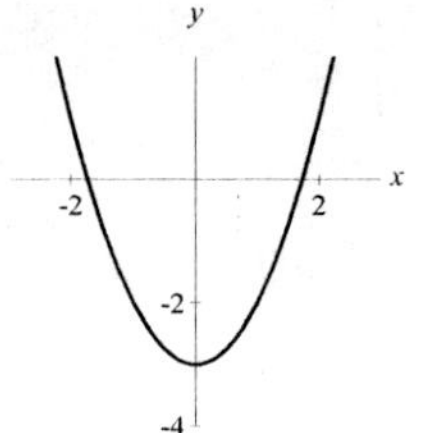

45. Vertex $(1/2, -1/4)$, axis $x = 1/2$, y-intercept $(0, 0)$, x-intercepts $(0, 0), (1, 0)$, opening up

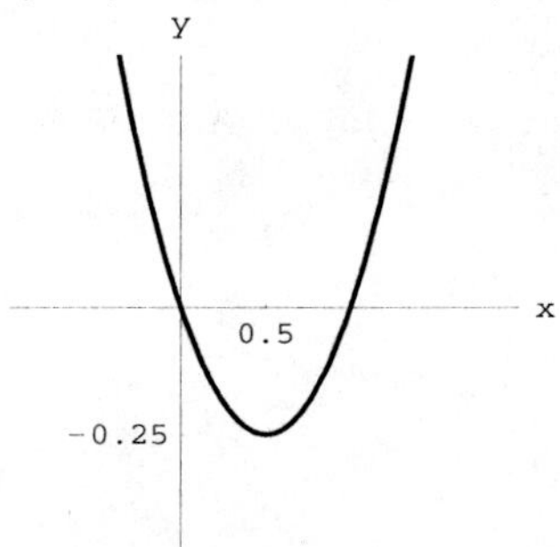

47. Vertex $(-3, 0)$, axis $x = -3$, y-intercept $(0, 9)$, x-intercept $(-3, 0)$, opening up

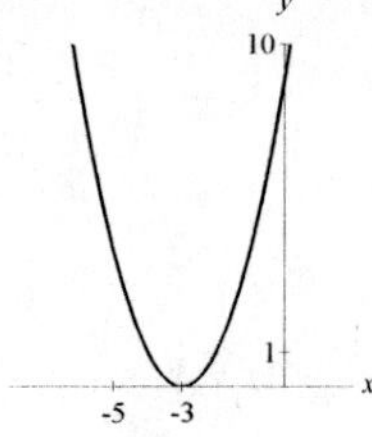

49. Vertex $(3, -4)$, axis $x = 3$, y-intercept $(0, 5)$, x-intercepts $(1, 0), (5, 0)$, opening up

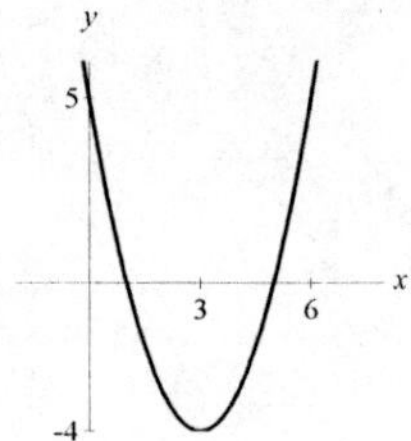

51. Vertex $(2, 12)$, axis $x = 2$, y-intercept $(0, 0)$, x-intercepts $(0, 0), (4, 0)$, opening down

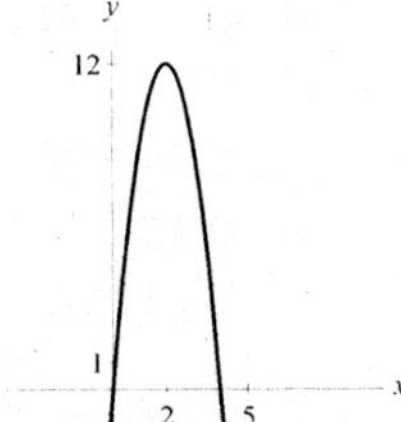

53. Vertex $(1, 3)$, axis $x = 1$, y-intercept $(0, 1)$, x-intercepts $\left(1 \pm \frac{\sqrt{6}}{2}, 0\right)$, opening down

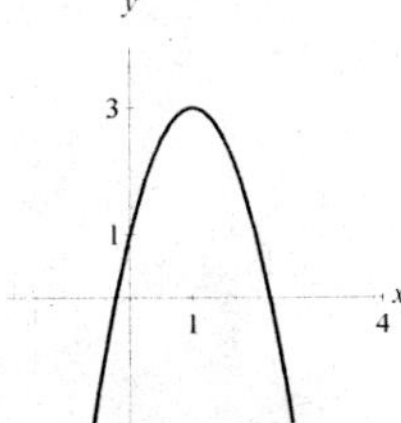

55. The x-intercepts are $x = -1, 3$. Since the parabola opens upward, the solution set for $x^2 - 2x - 3 > 0$ is $(-\infty, -1) \cup (3, \infty)$.

57. Since $x^2 - 4x + 1 = (x-2)^2 - 3$, the x-intercepts are $x = 2 \pm \sqrt{3}$. Since the parabola opens upward, the solution set for $x^2 - 4x + 1 < 0$ is $(2 - \sqrt{3}, 2 + \sqrt{3})$.

59. Since $6x^2 - x - 1 = (3x + 1)(2x - 1)$, the x-intercepts are $x = -1/3, 1/2$. Then the solution set for $x + 1 < 6x^2$ or $6x^2 - x - 1 > 0$ is $(-\infty, -1/3) \cup (1/2, \infty)$.

61. $(-\infty, -1] \cup [3, \infty)$

63. $(-3, 1)$

65. $[-3, 1]$

67. The roots of $x^2 - 4x + 2 = 0$ are $x_1 = 2 - \sqrt{2}$ and $x_2 = 2 + \sqrt{2}$.

If $x = -5$, then $(-5)^2 - 4(-5) + 2 = 47 > 0$.
If $x = 2$, then $(2)^2 - 4(2) + 2 = -2 < 0$.
If $x = 5$, then $(5)^2 - 4(5) + 2 = 7 > 0$.

+ 0 − 0 +

-5 x_1 0 x_2 5

The solution set of $x^2 - 4x + 2 < 0$ is $(2 - \sqrt{2}, 2 + \sqrt{2})$ and its graph follows.

$2 - \sqrt{2}$ $2 + \sqrt{2}$

69. The roots of $x^2 - 10 = 0$ are $x_1 = -\sqrt{10}$ and $x_2 = \sqrt{10}$.

If $x = -4$, then $(-4)^2 - 10 = 6 > 0$.
If $x = 0$, then $(0)^2 - 10 = -10 < 0$.
If $x = 4$, then $(4)^2 - 10 = 6 > 0$.

+ 0 − 0 +

−4 x_1 0 x_2 4

The solution set of $x^2 - 9 > 1$ is $(-\infty, -\sqrt{10}) \cup (\sqrt{10}, \infty)$ and its graph is

$-\sqrt{10}$ $\sqrt{10}$

71. The roots of $y^2 - 10y + 18 = 0$ are $y_1 = 5 - \sqrt{7}$ and $y_2 = 5 + \sqrt{7}$.

If $y = 2$, then $(2)^2 - 10(2) + 18 = 2 > 0$.
If $y = 5$, then $(5)^2 - 10(5) + 18 = -7 < 0$.
If $y = 8$, then $(8)^2 - 10(8) + 18 = 2 > 0$.

+ 0 − 0 +

2 y_1 5 y_2 8

The solution set of $y^2 - 10y + 18 > 0$ is $(-\infty, 5 - \sqrt{7}) \cup (5 + \sqrt{7}, \infty)$ and its graph is

$5 - \sqrt{7}$ $5 + \sqrt{7}$

73. Note, $p^2 + 9 = 0$ has no real roots. If $p = 0$, then $(0)^2 + 9 > 0$. The signs of $p^2 + 9$ are shown below.

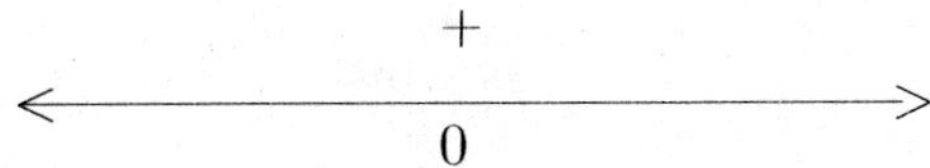

The solution set of $p^2 + 9 > 0$ is $(-\infty, \infty)$ and its graph follows.

75. Note, $a^2 - 8a + 20 = 0$ has no real roots. If $a = 0$, then $(0)^2 - 8(0) + 20 > 0$. The signs of $a^2 - 8a + 20$ are shown below.

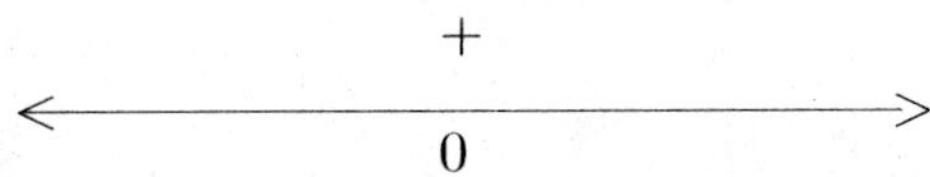

The solution set of $a^2 - 8a + 20 \leq 0$ is $\emptyset$.

77. Note, $2w^2 - 5w + 6 = 0$ has no real roots. If $w = 0$, then $2(0)^2 - 5(0) + 6 > 0$. The signs of $2w^2 - 5w + 6$ are shown below.

+

0

The solution set of $2w^2 - 5w + 6 > 0$ is $(-\infty, \infty)$

and its graph follows.

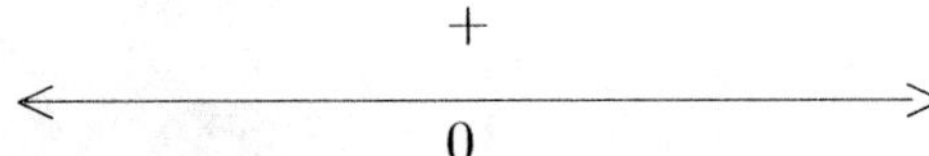

79. The zeros of $f(x) = (2x - 3)(x + 1)$ are $x_1 = -1$ and $x_2 = 3/2$.

If $x = -2$, then $f(-2) = 7 > 0$.
If $x = 0$, then $f(0) = -3 < 0$.
If $x = 2$, then $f(2) = 3 > 0$.

+ 0 − 0 +

-2 −1 0 3/2 2

The solution set is $(-1, 3/2)$

81. The zeros of $f(x) = x^2 - 2x - 15 = (x+3)(x-5)$ are $x_1 = -3$ and $x_2 = 5$.

If $x = -4$, then $f(-4) = 9 > 0$.
If $x = 0$, then $f(0) = -15 < 0$.
If $x = 6$, then $f(6) = 9 > 0$.

+ 0 − 0 +

-4 -3 0 5 6

The solution set is $(-\infty, -3) \cup (5, \infty)$.

83. The zeros of $f(w) = w^2 - 4w - 12 = (w + 2)(w - 6)$ are $w_1 = -2$ and $w_2 = 6$.

If $w = -3$, then $f(-3) = 9 > 0$.
If $w = 0$, then $f(0) = -12 < 0$.
If $w = 7$, then $f(7) = 9 > 0$.

+ 0 − 0 +

-3 -2 0 6 7

The solution set is $(-\infty, -2] \cup [6, \infty)$.

85. The zeros of $f(t) = t^2 - 16 = (t + 4)(t - 4)$ are $t_1 = -4$ and $t_2 = 4$.

If $t = -5$, then $f(-5) = 9 > 0$.
If $t = 0$, then $f(0) = -16 < 0$.
If $t = 5$, then $f(5) = 9 > 0$.

+ 0 − 0 +

-5 -4 0 4 5

The solution set is $[-4, 4]$.

87. The zero of $f(a) = (a + 3)^2$ is $a_1 = -3$.

If $a = -4$, then $f(-4) = 1 > 0$.
If $a = 0$, then $f(0) = 9 > 0$.

+ 0 +

-4 -3 0

The solution set is $\{-3\}$.

89. The zero of $f(z) = (2z - 3)^2$ is $z_1 = 3/2$.

If $z = 0$, then $f(0) = 9 > 0$.
If $z = 2$, then $f(2) = 1 > 0$.

+ 0 +

0 3/2 2

The solution set is $(-\infty, 3/2) \cup (3/2, \infty)$.

91. **a)** Since $x^2 - 3x - 10 = (x - 5)(x + 2) = 0$, the solution set is $\{-2, 5\}$.

b) Since $x^2 - 3x - 10 = -10$, we get $x^2 - 3x = 0$ or $x(x - 3) = 0$. The solution set is $\{0, 3\}$.

c) If $x = -3$, then $(-3)^2 - 3(-3) - 10 = 8 > 0$.
If $x = 0$, then $(0)^2 - 3(0) - 10 = -10 < 0$.
If $x = 6$, then $(6)^2 - 3(6) - 10 = 8 > 0$.
The signs of $x^2 - 3x - 10$ are shown below.

+ 0 − 0 +

−3 −2 0 5 6

The solution set of $x^2 - 3x - 10 > 0$. is $(-\infty, -2) \cup (5, \infty)$

d) Using the sign graph of $x^2 - 3x - 10$ given in part c), the solution set of $x^2 - 3x - 10 \leq 0$ is $[-2, 5]$.

e) By using the method of completing the square, one obtains

$$x^2 - 3x - 10 = \left(x - \frac{3}{2}\right)^2 - 10 - \frac{9}{4}$$
$$x^2 - 3x - 10 = \left(x - \frac{3}{2}\right)^2 - \frac{49}{4}.$$

The graph of f is obtained from the graph of $y = x^2$ by shifting to the right $\frac{3}{2}$ units, and down $\frac{49}{4}$ units.

f) Domain is $(-\infty, \infty)$, range is $\left[-\frac{49}{4}, \infty\right)$, minimum y-value is $-\frac{49}{4}$

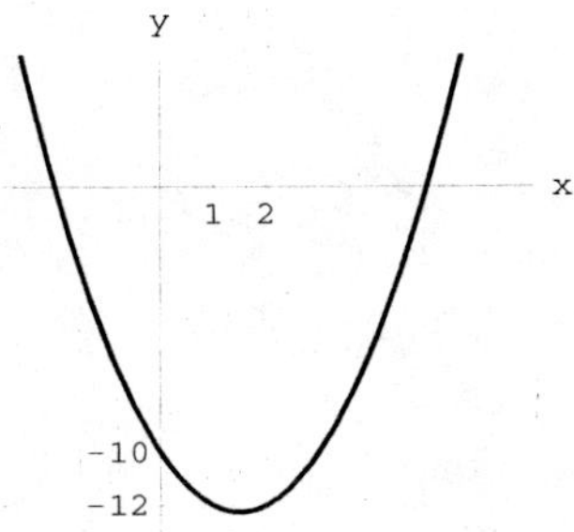

g) The solution to $f(x) > 0$ may be obtained by considering the part of the parabola that is above the x-axis, i.e., when x lies in $(-\infty, -2) \cup (5, \infty)$.
While, the solution to $f(x) \leq 0$ may be obtained by considering the part of the parabola on or below the x-axis. i.e., when x is in $[-2, 5]$.

h) x-intercepts are $(5, 0)$ and $(-2, 0)$, the y-intercept is $(0, -10)$, axis of symmetry $x = \frac{3}{2}$, vertex $\left(\frac{3}{2}, -\frac{49}{4}\right)$, opens upward, increasing on $\left[\frac{3}{2}, \infty\right)$, decreasing on $\left(-\infty, \frac{3}{2}\right]$

93. Since $\frac{-b}{2a} = \frac{-128}{-2(-16)} = 4$, the maximum height is $h(4) = 261$ ft.

95. a) Finding the vertex involves the number

$$\frac{-b}{2a} = \frac{-160}{-32} = 5.$$

Thus, the maximum height is

$$h(5) = -16(5)^2 + 160(5) + 8 = 408 \text{ ft}.$$

b) When the arrow reaches the ground, one has $h(t) = 0$.

$$-16t^2 + 160t + 8 = 0$$
$$t^2 - 10t = \frac{1}{2}$$
$$(t-5)^2 = 25 + \frac{1}{2}$$
$$t = 5 \pm \sqrt{\frac{51}{2}}$$

Since $t \geq 0$, the arrow reaches the ground in

$$5 + \frac{\sqrt{102}}{2} = \frac{10 + \sqrt{102}}{2} \approx 10.05 \text{ sec}.$$

97. a) About 100 mph

b) The value of A that would maximize M is $A = \frac{-b}{2a} = \frac{-0.127}{-0.001306} \approx 97.24$ mph.

c) Lindbergh flying at 97 mph would use $\frac{97}{1.2} \approx 80.83$ lbs. of fuel or

$$\frac{80.83 \text{ lbs}}{6.12 \text{ lbs per gal}} \approx 13.2 \text{ gal/hr}.$$

99. Let x and y be the length and width, respectively. Since $2x + 2y = 200$, we find $y = 100 - x$. The area as a function of x is

$$f(x) = x(100 - x) = 100x - x^2.$$

The graph of f is a parabola and its vertex is $(50, 2500)$. Thus, the maximum area is 2500 yd^2. Using $x = 50$ from the vertex, we get $y = 100 - x = 100 - 50 = 50$. The dimensions are 50 yd by 50 yd.

101. Let the length of the sides be x, x, x, y, y. Then $3x + 2y = 120$ and $y = \frac{120 - 3x}{2}$.
The area of rectangular enclosure is

$$A(x) = xy = x\left(\frac{120 - 3x}{2}\right) = \frac{1}{2}(120x - 3x^2).$$

This is a parabola opening down. Since $-\frac{b}{2a} = 20$ and $y = \frac{120 - 3(20)}{2} = 30$, the optimal dimensions are 20 ft by 30 ft.

103. Let the length of the sides be x, x, and $30 - 2x$. The area of rectangular enclosure is

$$A(x) = x(30 - 2x) = 30x - 2x^2.$$

We have a parabola opening down. Since $-\dfrac{b}{2a} = 7.5$ and $y = 30 - 2(7.5) = 15$, the optimal dimensions are 15 ft by 7.5 ft.

105. Let x be the length of a folded side. The area of the cross-section is $A = x(10 - 2x)$. This is a parabola opening down with $-b/2a = 2.5$. The dimensions of the cross-section are 2.5 in. high and 5 in. wide.

107. Let n and p be the number of persons and the price of a tour per person, respectively.

a) The function expressing p as a function of n is $p = 50 - n$.

b) The revenue is $R = (50 - n)n$ or

$$R = 50n - n^2.$$

c) Since the graph of R is a parabola opening down with vertex $(25, 625)$, we find that 25 persons will give her the maximum revenue of \$625.

109. Since $v = 50p - 50p^2$ is a parabola opening down, to maximize v choose

$$p = \frac{-b}{2a} = \frac{50}{100} = 1/2.$$

111. a) A graph of atmospheric pressure versus altitude is given below.

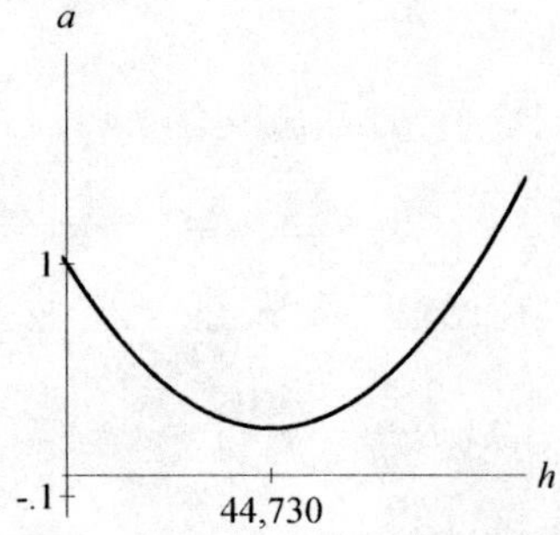

b) From the graph, one finds the atmospheric pressure is decreasing as they went to $h = 29,029$ feet.

c) Since $\dfrac{-b}{2a} = \dfrac{3.48 \times 10^{-5}}{2(3.89 \times 10^{-10})} \approx 44,730$ feet, the function is decreasing on the interval $[0, 44730]$ and increasing on $[44730, \infty)$.

d) No, it does not make sense to speak of atmospheric pressure at heights that lie in $(44730, \infty)$ since it is higher than the summit of 29,029 ft.

e) It is valid for altitudes less than 30,000 feet which is less than the height of the summit.

113. a) Using a graphing calculator, we find that the equation of the regression line is

$$y = -3135x + 43,624.$$

The equation of the quadratic regression curve is $y = 148x^2 - 4612x + 46,331$

b) The quadratic regression curve seems seems to fit the data more.

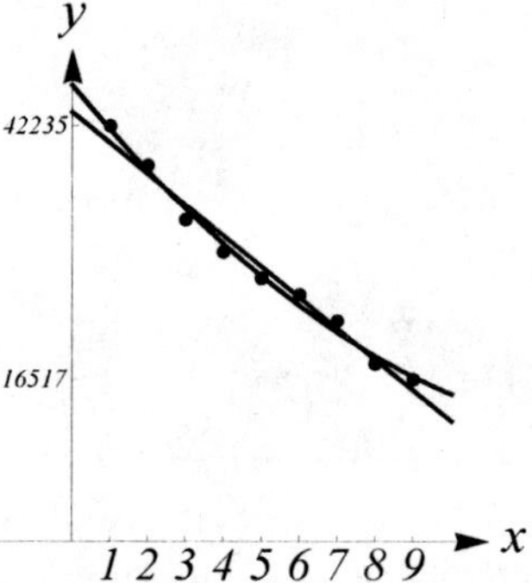

c) We substitute $x = 11$ into the regression equations in part (a). Using the linear function, the price of an 11-year old car is \$9139.

Using the quadratic curve, we obtain \$13,507.

115. Let $y = k/x$. If $y = 3$ and $x = 12$, we find $k = 36$. Thus, if $x = 40$ then

$$y = \frac{k}{x} = \frac{36}{40} = \frac{9}{10}.$$

117. Since $f(9) = 2(9) - 9 = 9$ and $g(9) = \sqrt{9} = 3$, we obtain $(f + g)(9) = 9 + 3 = 12$.

119. Domain $\{-3, 1, 3, 4\}$, range $\{2, 5\}$

121. The vertices of the quadrilateral where the triangle and square overlap are $A(81/30, 81/40)$, $B(81/30, 19/10)$, $C(143/40, 19/10)$, and $D(96/30, 24/10)$.

Subdividing the quadrilateral into two triangles and a rectangle, we find that the area of the quadrilateral is 1/4.

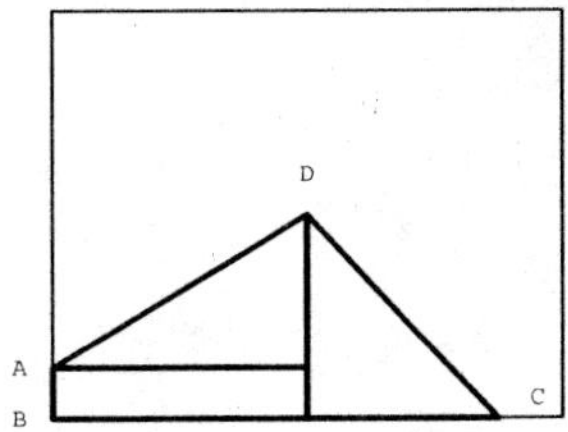

For Thought

1. False, the function is not a polynomial.

2. True, by the Remainder Theorem.

3. True, by the Factor Theorem.

4. True, since $2^5 - 1 = 31$.

5. False, since $P(x) = 1$ has no zero.

6. False, rather $c^3 - c^2 + 4c - 5 = b$.

7. False, since $P(4) = -15$. **8.** True

9. True, since 1 is a root.

10. False, since -3 is not a root.

3.2 Exercises

1. zero

3. zero, factor

5. x-coordinates

7. Quotient $x - 3$, remainder 1

$$\begin{array}{r@{\,}l} & x - 3 \\ x - 2 & \overline{)x^2 - 5x + 7} \\ & \underline{x^2 - 2x} \\ & \quad -3x + 7 \\ & \quad \underline{-3x + 6} \\ & \qquad\quad 1 \end{array}$$

9. Quotient $-2x^2 + 6x - 14$, remainder 33

$$\begin{array}{r@{\,}l} & -2x^2 + 6x - 14 \\ x + 3 & \overline{)-2x^3 + 0x^2 + 4x - 9} \\ & \underline{-2x^3 - 6x^2} \\ & \qquad 6x^2 + 4x \\ & \qquad \underline{6x^2 + 18x} \\ & \qquad\qquad -14x - 9 \\ & \qquad\qquad \underline{-14x - 42} \\ & \qquad\qquad\qquad 33 \end{array}$$

11. Quotient $s^2 + 2$, remainder 16

$$\begin{array}{r@{\,}l} & s^2 + 2 \\ s^2 - 5 & \overline{)s^4 - 3s^2 + 6} \\ & \underline{s^4 - 5s^2} \\ & \qquad 2s^2 + 6 \\ & \qquad \underline{2s^2 - 10} \\ & \qquad\quad 16 \end{array}$$

13. Quotient $x + 6$, remainder 13

2	1	4	1
		2	12
	1	6	13

15. Quotient $-x^2 + 4x - 16$, remainder 57

−3	−1	1	−4	9
		3	−12	48
	−1	4	-16	57

17. Quotient $4x^2 + 2x - 4$, remainder 0

1/2	4	0	-5	2
		2	1	-2
	4	2	-4	0

19. Quotient $2a^2 - 4a + 6$, remainder 0

-1/2	2	-3	4	3
		-1	2	-3
	2	-4	6	0

21. Quotient $x^3 + x^2 + x + 1$, remainder -2

1	1	0	0	0	-3
		1	1	1	1
	1	1	1	1	-2

23. Quotient $x^4 + 2x^3 - 2x^2 - 4x - 4$, remainder -13

$$\begin{array}{r|rrrrrr} 2 & 1 & 0 & -6 & 0 & 4 & -5 \\ & & 2 & 4 & -4 & -8 & -8 \\ \hline & 1 & 2 & -2 & -4 & -4 & -13 \end{array}$$

25. $f(1) = 0$

$$\begin{array}{r|rrrrrr} 1 & 1 & 0 & 0 & 0 & 0 & -1 \\ & & 1 & 1 & 1 & 1 & 1 \\ \hline & 1 & 1 & 1 & 1 & 1 & 0 \end{array}$$

27. $f(-2) = -33$

$$\begin{array}{r|rrrrrr} -2 & 1 & 0 & 0 & 0 & 0 & -1 \\ & & -2 & 4 & -8 & 16 & -32 \\ \hline & 1 & -2 & 4 & -8 & 16 & -33 \end{array}$$

29. $g(1) = 5$

$$\begin{array}{r|rrrr} 1 & 1 & -4 & 0 & 8 \\ & & 1 & -3 & -3 \\ \hline & 1 & -3 & -3 & 5 \end{array}$$

31. $g(-1/2) = 55/8$

$$\begin{array}{r|rrrr} -1/2 & 1 & -4 & 0 & 8 \\ & & -1/2 & 9/4 & -9/8 \\ \hline & 1 & -9/2 & 9/4 & 55/8 \end{array}$$

33. $h(-1) = 0$

$$\begin{array}{r|rrrrr} -1 & 2 & 1 & -1 & 3 & 3 \\ & & -2 & 1 & 0 & -3 \\ \hline & 2 & -1 & 0 & 3 & 0 \end{array}$$

35. $h(1) = 8$

$$\begin{array}{r|rrrrr} 1 & 2 & 1 & -1 & 3 & 3 \\ & & 2 & 3 & 2 & 5 \\ \hline & 2 & 3 & 2 & 5 & 8 \end{array}$$

37. Yes, $(x+3)(x^2+x-2) = (x+3)(x+2)(x-1)$

$$\begin{array}{r|rrrr} -3 & 1 & 4 & 1 & -6 \\ & & -3 & -3 & 6 \\ \hline & 1 & 1 & -2 & 0 \end{array}$$

39. Yes, $(x-4)(x^2+8x+15) = (x-4)(x+5)(x+3)$

$$\begin{array}{r|rrrr} 4 & 1 & 4 & -17 & -60 \\ & & 4 & 32 & 60 \\ \hline & 1 & 8 & 15 & 0 \end{array}$$

41. Yes, since the remainder below is zero

$$\begin{array}{r|rrrr} 3 & 2 & -5 & -4 & 3 \\ & & 6 & 3 & -3 \\ \hline & 2 & 1 & -1 & 0 \end{array}$$

43. No, since the remainder below is not zero

$$\begin{array}{r|rrrr} -2 & 1 & 2 & 3 & 1 \\ & & -2 & 0 & -6 \\ \hline & 1 & 0 & 3 & -5 \end{array}$$

45. Yes, since the remainder below is zero

$$\begin{array}{r|rrrrr} -1 & 1 & 2 & 4 & 6 & 3 \\ & & -1 & -1 & -3 & -3 \\ \hline & 1 & 1 & 3 & 3 & 0 \end{array}$$

47. No, since the remainder below is not zero

$$\begin{array}{r|rrrr} 1/2 & 1 & 3 & -5 & 7 \\ & & 1/2 & 7/4 & -13/8 \\ \hline & 1 & 7/2 & -13/4 & 43/8 \end{array}$$

49. $\pm\{1, 2, 3, 4, 6, 8, 12, 24\}$

51. $\pm\{1, 3, 5, 15\}$

53. $\pm\left\{1, 3, 5, 15, \frac{1}{2}, \frac{1}{4}, \frac{1}{8}, \frac{3}{2}, \frac{3}{4}, \frac{3}{8}, \frac{5}{2}, \frac{5}{4}, \frac{5}{8}, \frac{15}{2}, \frac{15}{4}, \frac{15}{8}\right\}$

55. $\pm\left\{1, 2, \frac{1}{2}, \frac{1}{3}, \frac{1}{6}, \frac{1}{9}, \frac{1}{18}, \frac{2}{3}, \frac{2}{9}\right\}$

57. Zeros are $2, 3, 4$ since

2	1	-9	26	-24
		2	-14	24
	1	-7	12	0

and $x^2 - 7x + 12 = (x - 4)(x - 3)$

59. Zeros are $-3, 2 \pm i$ since

-3	1	-1	-7	15
		-3	12	-15
	1	-4	5	0

$x^2 - 4x + 5 = (x - 2)^2 + 1 = 0$ or $x - 2 = \pm i$

61. Zeros are $1/2, 3/2, 5/2$ since

1/2	8	-36	46	-15
		4	-16	15
	8	-32	30	0

and $8a^2 - 32a + 30 = 2(4a^2 - 16a + 15) = 2(2a - 3)(2a - 5)$

63. Zeros are $\frac{1}{2}, \frac{1 \pm i}{3}$ since

1/2	18	-21	10	-2
		9	-6	2
	18	-12	4	0

and the zeros of $18t^2 - 12t + 4$ are (by the quadratic formula) $\frac{1 \pm i}{3}$

65. Zeros are $-6, 3 \pm i$ since

-6	1	0	-26	60
		-6	36	-60
	1	-6	10	0

and the zeros of $x^2 - 6x + 10$ are (by the quadratic formula) $3 \pm i$

67. Zeros are $1, -2, \pm i$ since

1	1	1	-1	1	-2
		1	2	1	2
	1	2	1	2	0

-2	1	2	1	2
		-2	0	-2
	1	0	1	0

and the zeros of $w^2 + 1$ are $\pm i$

69. Zeros are $-1, \pm\sqrt{2}$ since

-1	1	2	-1	-4	-2
		-1	-1	2	2
	1	1	-2	-2	0

-1	1	1	-2	-2
		-1	0	2
	1	0	-2	0

and the zeros of $x^2 - 2$ are $\pm\sqrt{2}$

71. Zeros are $1/2, 1/3, 1/4$ since

1/2	24	-26	9	-1
		12	-7	1
	24	-14	2	0

and $2(12x^2 - 7x + 1) = 2(4x - 1)(3x - 1)$

73. Rational zero is $1/16$ since

1/16	16	-33	82	-5
		1	-2	5
	16	-32	80	0

and by the quadratic formula $16x^2 - 32x + 80$ has imaginary zeros $1 \pm 2i$.

75. Rational zeros are 7/3, $-6/7$ since

$$\begin{array}{r|rrrrr} 7/3 & 21 & -31 & -21 & -31 & -42 \\ & & 49 & 42 & 49 & 42 \\ \hline & 21 & 18 & 21 & 18 & 0 \end{array}$$

$$\begin{array}{r|rrrr} -6/7 & 21 & 18 & 21 & 18 \\ & & -18 & 0 & -18 \\ \hline & 21 & 0 & 21 & 0 \end{array}$$

and $21x^2 + 21 = 0$ has imaginary zeros $\pm i$.

77. Dividing $x^3 + 6x^2 + 3x - 10$ by $x - 1$, we find

$$\begin{array}{r|rrrr} 1 & 1 & 6 & 3 & -10 \\ & & 1 & 7 & 10 \\ \hline & 1 & 7 & 10 & 0 \end{array}$$

Moreover, $x^2 + 7x + 10 = (x + 5)(x + 2)$.
Note, the zeros of $x^2 + 9$ are $\pm 3i$.
Thus, the zeros of $f(x)$ are $x = -5, -2, 1, \pm 3i$.

79. Dividing $x^3 - 9x^2 + 23x - 15$ by $x - 1$, we find

$$\begin{array}{r|rrrr} 1 & 1 & -9 & 23 & -15 \\ & & 1 & -8 & 15 \\ \hline & 1 & -8 & 15 & 0 \end{array}$$

For the quotient, we find

$$x^2 - 8x + 15 = (x - 5)(x - 3).$$

Then the zeros of $x^3 - 9x^2 + 23x - 15$ are $x = 1, 3, 5$.

By using the method of completing the square, we find

$$x^2 - 4x + 1 = (x - 2)^2 - 3.$$

Then the zeros of $(x - 2)^2 - 3$ are $2 \pm \sqrt{3}$.
Thus, all the zeros are $x = 1, 3, 5, 2 \pm \sqrt{3}$.

81.

$$\frac{2x+1}{x-2} = 2 + \frac{5}{x-2} \text{ since}$$

$$\begin{array}{r|rr} 2 & 2 & 1 \\ & & 4 \\ \hline & 2 & 5 \end{array}$$

83.

$$\frac{a^2 - 3a + 5}{a - 3} = a + \frac{5}{a-3} \text{ since}$$

$$\begin{array}{r|rrr} 3 & 1 & -3 & 5 \\ & & 3 & 0 \\ \hline & 1 & 0 & 5 \end{array}$$

85.

$$\frac{c^2 - 3c - 4}{c^2 - 4} = 1 + \frac{-3c}{c^2 - 4} \text{ since}$$

$$\begin{array}{rl} & \quad 1 \\ c^2 - 4 & \overline{)c^2 - 3c - 4} \\ & \underline{\;c^2 + 0c - 4\;} \\ & \quad -3c \end{array}$$

87.

$$\frac{4t - 5}{2t + 1} = 2 + \frac{-7}{2t+1} \text{ since}$$

$$\begin{array}{rl} & \quad 2 \\ 2t + 1 & \overline{)4t - 5} \\ & \underline{\;4t + 2\;} \\ & \quad -7 \end{array}$$

89. a) Note, $\dfrac{P(t)}{t} = -t^3 + 12t^2 - 58t + 132$.

$$\begin{array}{r|rrrr} 6 & -1 & 12 & -58 & 132 \\ & & -6 & 36 & -132 \\ \hline & -1 & 6 & -22 & 0 \end{array}$$

The drug will be eliminated in $t = 6$ hr.

b) About 120 ppm

c) About 3 hours

d) Between 1 and 5 hours approximately, the concentration is above 80 ppm. Thus, the concentration is above 80 ppm for about 4 hours.

91. If w is the width, then $w(w + 4)(w + 9) = 630$. This can be re-written as

$$w^3 + 13w^2 + 36w - 630 = 0.$$

Using synthetic division, we find

$$\begin{array}{r|rrrr} 5 & 1 & 13 & 36 & -630 \\ & & 5 & 90 & 630 \\ \hline & 1 & 18 & 126 & 0 \end{array}$$

Since $w^2 + 18w + 126 = 0$ has non-real roots, the width of the HP box is $w = 5$ in. The dimensions are 5 in. by 9 in. by 14 in.

93. The quotient and remainder when

$$3x^3 + 4x^2 + 2x - 4 \text{ is divided by } 3x - 2$$

are the same as the quotient and remainder, respectively, when

$$x^3 + \frac{4}{3}x^2 + \frac{2}{3}x - \frac{4}{3} \text{ is divided by } x - \frac{2}{3}.$$

Then one can use sythetic division for the latter case since the divisor is $x - \dfrac{2}{3}$.

95. Using the method of completing the square, we find

$$\begin{aligned} f(x) &= 2\left(x^2 - \frac{3}{2}x\right) + 1 \\ &= 2\left(x - \frac{3}{4}\right)^2 - \frac{9}{8} + 1 \\ &= 2\left(x - \frac{3}{4}\right)^2 - \frac{1}{8}. \end{aligned}$$

97. Since an absolute value is always nonnegative, the solution set of $|3x+7| \geq 0$ is $(-\infty, \infty)$.

99. **a)** $6a(4a^2 + 3a - 10) = 6a(4a-5)(a+2)$
b) $x(x^4 - 16) = x(x^2+4)(x^2-4) = x(x^2+4)(x-2)(x+2)$

101. When a point on a circle with radius r is rotated through an angle of $\pi/2$, the distance the point rotates is $s = r\dfrac{\pi}{2}$. The sum of the distances traveled by point A is

$$\sqrt{45}\frac{\pi}{2} + 3\frac{\pi}{2} + 0 + 6\frac{\pi}{2} = \frac{(3\sqrt{5}+9)\pi}{2} \text{ ft.}$$

For Thought

1. False, since 1 has multiplicity 1. **2.** True

3. True **4.** False, it factors as $(x-5)^4(x+2)$.

5. False, rather $4+5i$ is also a solution. **6.** True

7. False, since they are solutions to a polynomial with real coefficients of degree at least 4.

8. False, 2 is not a solution.

9. True, since $-x^3 - 5x^2 - 6x - 1 = 0$ has no sign changes.

10. True

3.3 Exercises

1. multiplicity

3. $a - bi$

5. conjugate

7. Degree 2; 5 with multiplicity 2 since $(x-5)^2 = 0$

9. Degree 5; 0 with multiplicity 3, and ± 3 since $x^3(x-3)(x+3) = 0$

11. Degree 4; 0 and 1 each have multiplicity 2 since $x^2(x^2 - 2x + 1) = x^2(x-1)^2$

13. Degree 4; 3/2 and $-4/3$ each with multiplicity 2

15. Degree 3; the roots are $0, 2 \pm \sqrt{10}$ since $x(x^2 - 4x - 6) = x((x-2)^2 - 10) = 0$

17. $x^2 + 9$

19. $\left[(x-1) - \sqrt{2}\right]\left[(x-1) + \sqrt{2}\right] = (x-1)^2 - 2 = x^2 - 2x - 1$

21. $[(x-3) - 2i][(x-3) + 2i] = (x-3)^2 + 4 = x^2 - 6x + 13$

23. $(x-2)\left[(x-3) - 4i\right]\left[(x-3) + 4i\right] = (x-2)\left[(x-3)^2 + 16\right] = x^3 - 8x^2 + 37x - 50$

25. $(x+3)(x-5) = 0$ or $x^2 - 2x - 15 = 0$

27. $(x+4i)(x-4i) = 0$ or $x^2 + 16 = 0$

29. $(x-(3-i))(x-(3+i)) = 0$ or $x^2 - 6x + 10 = 0$

31. $(x+2)(x-i)(x+i) = 0$ or $x^3 + 2x^2 + x + 2 = 0$

33. $x(x - i\sqrt{3})(x + i\sqrt{3}) = 0$ or $x^3 + 3x = 0$

35. $(x-3)\left[x - (1-i)\right]\left[x - (1+i)\right] = 0$ or $x^3 - 5x^2 + 8x - 6 = 0$

37. $(x-1)(x-2)(x-3)=0$ or
$x^3-6x^2+11x-6=0$

39. $(x-1)[x-(2-3i)][x-(2+3i)]=0$ or
$x^3-5x^2+17x-13=0$

41. $(2x-1)(3x-1)(4x-1)=0$ or
$24x^3-26x^2+9x-1=0$

43. $(x-i)(x+i)[x-(1+i)][x-(1-i)]=0$ or $x^4-2x^3+3x^2-2x+2=0$

45. $P(x)=x^3+5x^2+7x+1$ has no sign change and $P(-x)=-x^3+5x^2-7x+1$ has 3 sign changes. There are (a) 3 negative roots, or (b) 1 negative root & 2 imaginary roots.

47. $P(x)=-x^3-x^2+7x+6$ has 1 sign change and $P(-x)=x^3-x^2-7x+6$ has 2 sign changes. There are (a) 1 positive root and 2 negative roots, or (b) 1 positive root and 2 imaginary roots.

49. $P(y)=y^4+5y^2+7=P(-y)$ has no sign change. There are 4 imaginary roots.

51. $P(t)=t^4-3t^3+2t^2-5t+7$ has 4 sign changes and $P(-t)=t^4+3t^3+2t^2+5t+7$ has no sign change. There are (a) 4 positive roots, or (b) 2 positive roots and 2 imaginary roots, or (c) 4 imaginary roots.

53. $P(x)=x^5+x^3+5x$ and $P(-x)=-x^5-x^3-5x$ have no sign changes; 4 imaginary roots and 0.

55. Best integral bounds are $-1<x<3$. One checks that 1, 2 are not upper bounds and that 3 is a bound.

$$\begin{array}{r|rrrr} 3 & 2 & -5 & 0 & 6 \\ & & 6 & 3 & 9 \\ \hline & 2 & 1 & 3 & 15 \end{array}$$

-1 is a lower bound since

$$\begin{array}{r|rrrr} -1 & 2 & -5 & 0 & 6 \\ & & -2 & 7 & -7 \\ \hline & 2 & -7 & 7 & -1 \end{array}$$

57. Best integral bounds are $-3<x<2$. One checks that 1 is not an upper bound and that 2 is a bound.

$$\begin{array}{r|rrrr} 2 & 4 & 8 & -11 & -15 \\ & & 8 & 32 & 42 \\ \hline & 4 & 16 & 21 & 27 \end{array}$$

$-1,-2$ are not lower bounds but -3 is a bound since

$$\begin{array}{r|rrrr} -3 & 4 & 8 & -11 & -15 \\ & & -12 & 12 & -3 \\ \hline & 4 & -4 & 1 & -18 \end{array}$$

59. Best integral bounds are $-1<w<5$. One checks that 1, 2, 3, 4 are not upper bounds and that 5 is a bound

$$\begin{array}{r|rrrrr} 5 & 1 & -5 & 3 & 2 & -1 \\ & & 5 & 0 & 15 & 85 \\ \hline & 1 & 0 & 3 & 17 & 84 \end{array}$$

-1 is a lower bound since

$$\begin{array}{r|rrrrr} -1 & 1 & -5 & 3 & 2 & -1 \\ & & -1 & 6 & -9 & 7 \\ \hline & 1 & -6 & 9 & -7 & 6 \end{array}$$

61. Best integral bounds are $-1<x<3$. Multiply equation by -1, this makes the leading coefficient positive. One checks that 1, 2 are not upper bounds and that 3 is a bound.

$$\begin{array}{r|rrrr} 3 & 2 & -5 & 3 & -9 \\ & & 6 & 3 & 18 \\ \hline & 2 & 1 & 6 & 9 \end{array}$$

-1 is a lower bound since

$$\begin{array}{r|rrrr} -1 & 2 & -5 & 3 & -9 \\ & & -2 & 7 & -10 \\ \hline & 2 & -7 & 10 & -19 \end{array}$$

63. Roots are $1, 5, -2$ since

$$\begin{array}{r|rrrr} 1 & 1 & -4 & -7 & 10 \\ & & 1 & -3 & -10 \\ \hline & 1 & -3 & -10 & 0 \end{array}$$

and $x^2 - 3x - 10 = (x-5)(x+2)$

65. Roots are $-3, \dfrac{3 \pm \sqrt{13}}{2}$ since

$$\begin{array}{r|rrrr} -3 & 1 & 0 & -10 & -3 \\ & & -3 & 9 & 3 \\ \hline & 1 & -3 & -1 & 0 \end{array}$$

and by the quadratic formula the roots of $x^2 - 3x - 1 = 0$ are $\dfrac{3 \pm \sqrt{13}}{2}$.

67. Roots are $2, -4, \pm i$ since

$$\begin{array}{r|rrrrr} 2 & 1 & 2 & -7 & 2 & -8 \\ & & 2 & 8 & 2 & 8 \\ \hline & 1 & 4 & 1 & 4 & 0 \end{array}$$

$$\begin{array}{r|rrrr} -4 & 1 & 4 & 1 & 4 \\ & & -4 & 0 & -4 \\ \hline & 1 & 0 & 1 & 0 \end{array}$$

and the root of $x^2 + 1 = 0$ are $\pm i$.

69. Roots are $1/3, 1/2, -5$ since

$$\begin{array}{r|rrrr} 1/3 & 6 & 25 & -24 & 5 \\ & & 2 & 9 & -5 \\ \hline & 6 & 27 & -15 & 0 \end{array}$$

and $6x^2 + 27x - 15 = 3(2x-1)(x+5)$.

71. $1, -2$ each have multiplicity 2 since

$$\begin{array}{r|rrrrr} 1 & 1 & 2 & -3 & -4 & 4 \\ & & 1 & 3 & 0 & -4 \\ \hline & 1 & 3 & 0 & -4 & 0 \end{array}$$

$$\begin{array}{r|rrrr} -2 & 1 & 3 & 0 & -4 \\ & & -2 & -2 & 4 \\ \hline & 1 & 1 & -2 & 0 \end{array}$$

and $x^2 + x - 2 = (x+2)(x-1)$.

73. Use synthetic division on the cubic factor in $x(x^3 - 6x^2 + 12x - 8) = 0$.

$$\begin{array}{r|rrrr} 2 & 1 & -6 & 12 & -8 \\ & & 2 & -8 & 8 \\ \hline & 1 & -4 & 4 & 0 \end{array}$$

Since $x^2 - 4x + 4 = (x-2)^2$, the roots are 2 (with multiplicity 3) and 0.

75. Use synthetic division on the 5th degree factor in $x(x^5 - x^4 - x^3 + x^2 - 12x + 12) = 0$.

$$\begin{array}{r|rrrrrr} 1 & 1 & -1 & -1 & 1 & -12 & 12 \\ & & 1 & 0 & -1 & 0 & -12 \\ \hline & 1 & 0 & -1 & 0 & -12 & 0 \end{array}$$

$$\begin{array}{r|rrrrr} 2 & 1 & 0 & -1 & 0 & -12 \\ & & 2 & 4 & 6 & 12 \\ \hline & 1 & 2 & 3 & 6 & 0 \end{array}$$

$$\begin{array}{r|rrrr} -2 & 1 & 2 & 3 & 6 \\ & & -2 & 0 & -6 \\ \hline & 1 & 0 & 3 & 0 \end{array}$$

Note, the roots of $x^2 + 3 = 0$ are $\pm i\sqrt{3}$. Thus, the roots are $x = 0, 1, \pm 2, \pm i\sqrt{3}$.

77. The roots are $\pm 1, -2, 1/4, 3/2$ since

$$\begin{array}{r|rrrrrr} -1 & 8 & 2 & -33 & 4 & 25 & -6 \\ & & -8 & 6 & 27 & -31 & 6 \\ \hline & 8 & -6 & -27 & 31 & -6 & 0 \end{array}$$

$$\begin{array}{r|rrrrr} -2 & 8 & -6 & -27 & 31 & -6 \\ & & -16 & 44 & -34 & 6 \\ \hline & 8 & -22 & 17 & -3 & 0 \end{array}$$

1	8	−22	17	−3
		8	−14	3
	8	−14	3	0

and the last quotient is

$$8x^2 - 14x + 4 = (4x - 1)(2x - 3).$$

79. By the Theorem of Bounds and the graph below

6	2	-3	-50	18
		12	54	24
	2	9	4	42

-5	2	-3	-50	18
		-10	65	-75
	2	-13	15	-57

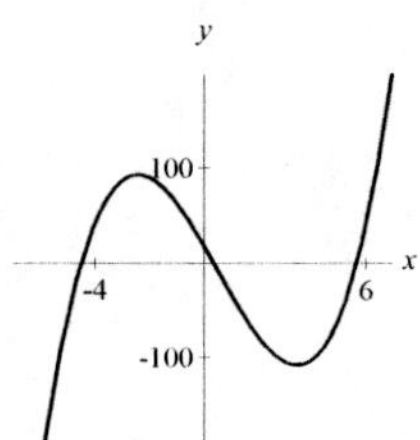

the intervals are $-5 < x < 6, -5 < x < 6$

81. By the Theorem of Bounds and the graph below

6	1	0	-26	0	153
		6	36	60	360
	1	6	10	60	513

-6	1	0	-26	0	153
		-6	36	-60	360
	1	-6	10	-60	513

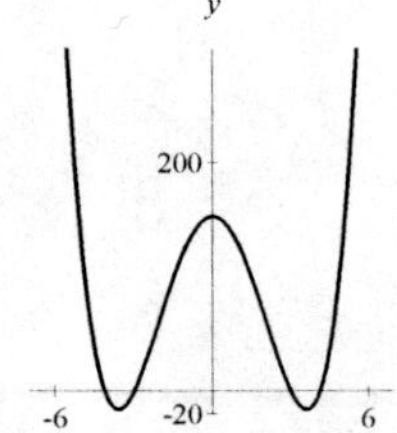

the intervals are $-6 < x < 6, -5 < x < 5$.

83. By the Theorem of Bounds and the graph below

23	4	-90	-2	45
		92	46	1012
	4	2	44	1057

-1	4	-90	-2	45
		-4	94	-92
	4	-94	92	-47

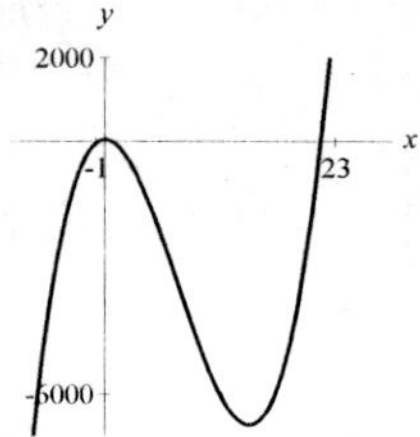

the intervals are $-1 < x < 23, -1 < x < 23$.

85. Multiplying by 100, $t^3 - 8t^2 + 11t + 20 = 0$.

4	1	-8	11	20
		4	-16	-20
	1	-4	-5	0

Since $t^2 - 4t - 5 = (t-5)(t+1) = 0$, we obtain $t = 4$ hr and $t = 5$ hr.

87. Let x be the radius of the cone. The volume of the cone is $\frac{\pi}{3}x^2 \cdot 2$ and the volume of the cylinder with height $4x$ is $\pi x^2 \cdot (4x)$. So $114\pi = \frac{\pi}{3}2x^2 + \pi x^2(4x)$. Divide by π, multiply by 3, and simplify to obtain $12x^3 + 2x^2 - 342 = 0$.

3	12	2	0	-342
		36	114	342
	12	38	114	0

The radius is $x = 3$ in. since $12x^2+38x+114 = 0$ has no real roots.

89. From the following

$$\begin{aligned}\overline{(a+bi)+(c+di)} &= \overline{(a+c)+i(b+d)}\\ &= (a+c) - i(b+d)\end{aligned}$$

and

$$\begin{aligned}\overline{a+bi}+\overline{c+di} &= (a-bi)+(c-di)\\ &= (a+c)-i(b+d)\end{aligned}$$

we obtain that the conjugate of the sum of two complex numbers is equal to the sum of their conjugates.

91. If a is real then $\overline{a}=\overline{a+0i}=a-0i=a$.

93. Let $f(x)=a(x-1)(x-2)(x-3)$.

Since $f(0)=-6a=3$, we find $a=-\frac{1}{2}$.

Then substitute $a=-\frac{1}{2}$ and multiply out

$f(x)$ to obtain $f(x)=-\frac{1}{2}x^3+3x^2-\frac{11}{2}x+3$.

95. The divisors of the constant term 6 are $p=\pm1,\pm2,\pm3,\pm6$. The divisors of the leading coefficient 2 are $q=\pm1,\pm2$. Then the possible rational zeros are $p/q=\pm1,\pm2,\pm3,\pm6,\pm1/2,\pm3/2$.

97. b is a function of a since there is a unique number of primes less than a.

a is not a function of b since the ordered pairs $(b,a)=(3,6)$ and $(a,b)=(3,7)$ have the same first coordinates and different second coordinates.

99. $(f\circ g)(x)=2(2x-4)^2-9=8x^2-32x+23$

101.

a) Consider the top layer of four balls and the ball that sits above it. Connecting the centers of the five balls gives a pyramid with a 2-by-2 square base and a slanted height of 2. The diagonal of the square base is $2\sqrt{2}$. By looking at a slanted height and half of the base, the height of the pyramid is $\sqrt{2}$. Then the height of the box is $4+\sqrt{2}$. Multiplying by the area of the 4-by-4 base of the box which is 16, we obtain the volume of the box. That is, the volume of the box is

$$16(4+\sqrt{2})=64+16\sqrt{2}\approx 86.63.$$

b) Using the same pyramid in part a), we obtain that the height of the box is $2+2\sqrt{2}$. Multiplying by the area of the 4-by-4 base of the box which is 16, we find that the volume of the box is

$$16(2+2\sqrt{2})=32+32\sqrt{2}\approx 77.25.$$

c) The distance between a vertex of the cubic box and the center of the ball that is closest to the vertex is $\sqrt{3}$. This is obtained by using the Pythagorean theorem.

Then the diagonal of the cube holding the five balls has a length of

$$4+2\sqrt{3}.$$

If x is the length of an edge of the cube then

$$x^2+x^2+x^2=(4+2\sqrt{3})^2.$$

Solving for x, we find $x=\dfrac{4+2\sqrt{3}}{\sqrt{3}}$. Then the volume of the cube is

$$\text{Volume}=x^3=\left(\frac{4+2\sqrt{3}}{\sqrt{3}}\right)^3.$$

Simplifying, we obtain

$$\text{Volume}=\frac{208\sqrt{3}}{9}+40\approx 80.03.$$

The box with the least volume is **b)**

For Thought

1. False, to be symmetric about the origin one must have $P(-x)=-P(x)$ for *all* x in the domain.

2. True **3.** True

4. False, since $f(-x)=-f(x)$.

5. True **6.** True **7.** False

8. False, y-intercept is $(0,38)$. **9.** True

10. False, only one x-intercept.

3.4 Exercises

1. y-axis

3. $-\dfrac{b}{2a}$

5. every

7. y-axis, since $f(-x) = f(x)$

9. $x = 3/2$, since $\dfrac{3}{2}$ is the x-coordinate of the vertex of a parabola

11. None

13. Origin, since $f(-x) = -f(x)$

15. $x = 5$, since 5 is the x-coordinate of the vertex of a parabola

17. Origin, since $f(-x) = -f(x)$

19. It does not cross $(4, 0)$ since $x - 4$ is raised to an even power.

21. It crosses $(1/2, 0)$.

23. It crosses $(1/4, 0)$

25. No x-intercepts since $x^2 - 3x + 10 = 0$ has no real root.

27. It crosses $(3, 0)$ and not $(0, 0)$ since $x^3 - 3x^2 = x^2(x - 3)$.

29. It crosses $(1/2, 0)$ and not $(1, 0)$ since

1	2	-5	4	-1
		2	-3	1
	2	-3	1	0

and $2x^3 - 5x^2 + 4x - 1 = (x - 1)(2x^2 - 3x + 1) = (x - 1)^2(2x - 1)$.

31. It crosses $(2, 0)$ and not $(-3, 0)$ since

-3	-2	-8	6	36
		6	6	-36
	-2	-2	12	0

and $(x + 3)(-2x^2 - 2x + 12) =$
$-2(x + 3)(x^2 + x - 6) =$
$-2(x + 3)(x + 3)(x - 2) = -2(x + 3)^2(x - 2)$.

33. $y \to \infty$ **35.** $y \to -\infty$ **37.** $y \to -\infty$

39. $y \to \infty$ **41.** $y \to \infty$

43. Neither symmetry, crosses $(-2, 0)$, does not cross $(1, 0)$, $y \to \infty$ as $x \to \infty$, $y \to -\infty$ as $x \to -\infty$

45. Symmetric about y-axis, no x-intercepts, $y \to \infty$ as $x \to \infty$, $y \to \infty$ as $x \to -\infty$

47. Applying the leading coefficient test, we find

$$\lim_{x\to\infty} (x^2 - 4) = \infty.$$

49. Applying the leading coefficient test, we obtain

$$\lim_{x\to\infty} (-x^5 - x^2) = -\infty.$$

51. Using the leading coefficient test, we get

$$\lim_{x\to-\infty} (-3x) = \infty.$$

53. Using the leading coefficient test, we get

$$\lim_{x\to-\infty} (-2x^2 + 1) = -\infty.$$

55. The graph of $f(x) = (x - 1)^2(x + 3)$ is shown below.

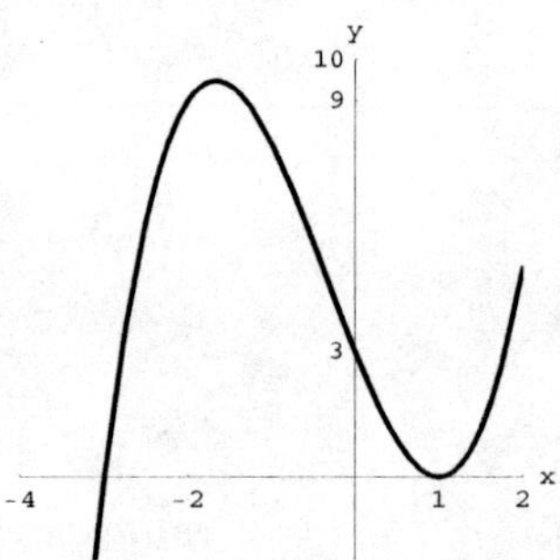

57. The graph of $f(x) = -2(2x - 1)^2(x + 1)^3$ is given below.

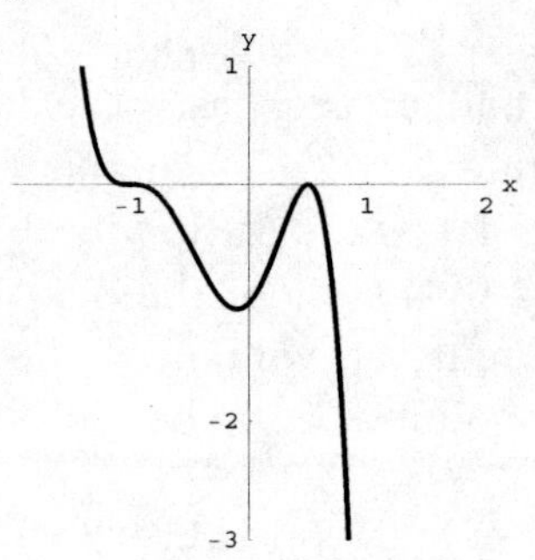

59. e, line

61. g, cubic polynomial and y-intercept $(0, 1)$

63. b, y-intercept $(0, 6)$, 4th degree polynomial

65. c, y-intercept $(0, -4)$, 3rd degree polynomial

67. $f(x) = x - 30$ has x-intercept $(30, 0)$ and y-intercept $(0, -30)$

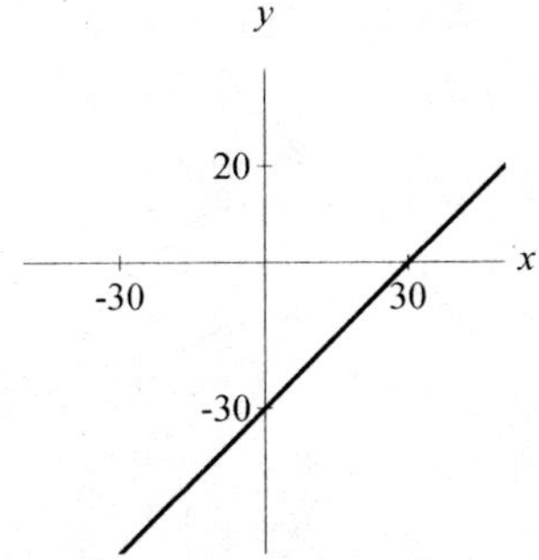

69. $f(x) = (x - 30)^2$ does not cross $(30, 0)$, y-intercept $(0, 900)$, $y \to \infty$ as $x \to \infty$ and as $x \to -\infty$

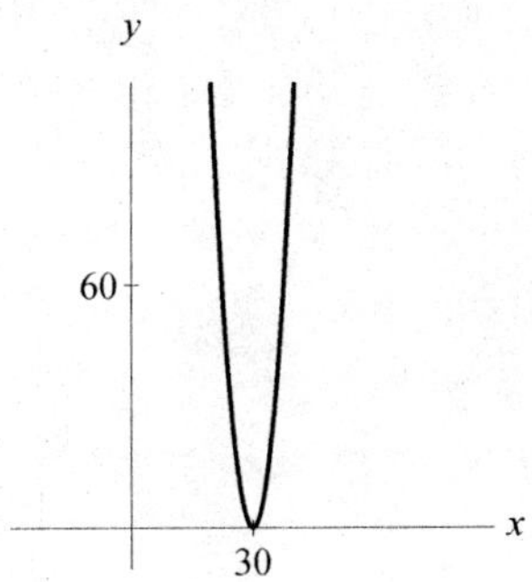

71. $f(x) = x^2(x - 40)$ crosses $(40, 0)$ but does not cross $(0, 0)$, $y \to \infty$ as $x \to \infty$, $y \to -\infty$ as $x \to -\infty$

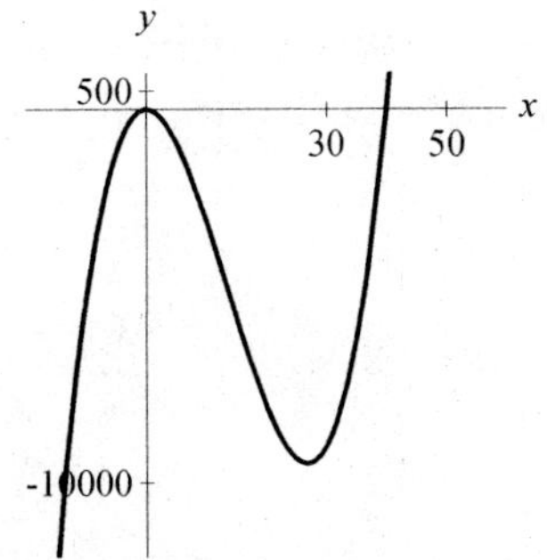

73. $f(x) = (x - 20)^2(x + 20)^2$ does not cross $(20, 0), (-20, 0)$, y-intercept $(0, 160000)$, $y \to \infty$ as $x \to \infty$ and as $x \to -\infty$

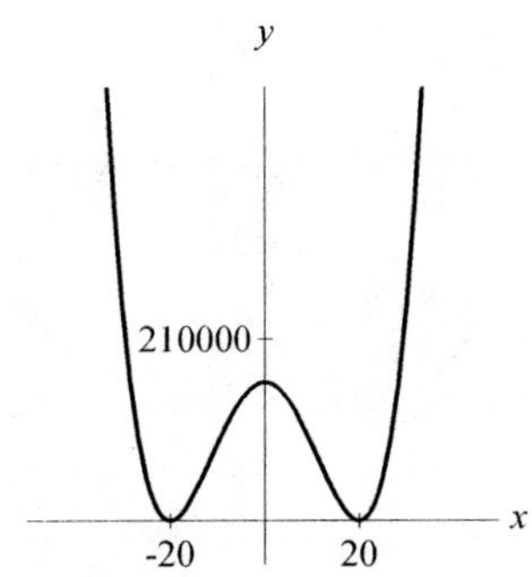

75. Since $f(x) = -x^3 - x^2 + 5x - 3$

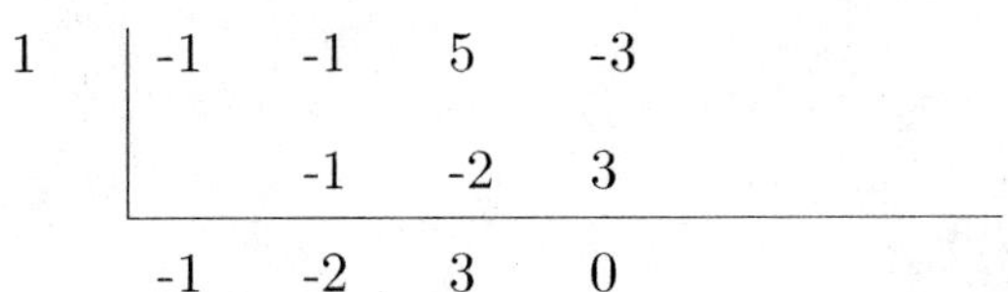

1	-1	-1	5	-3
		-1	-2	3
	-1	-2	3	0

$f(x) = (x - 1)(-x^2 - 2x + 3) = -(x-1)(x+3)(x-1)$, the graph crosses $(-3, 0)$ but not $(1, 0)$, y-intercept $(0, -3)$, $y \to -\infty$ as $x \to \infty$, and $y \to \infty$ as $x \to -\infty$

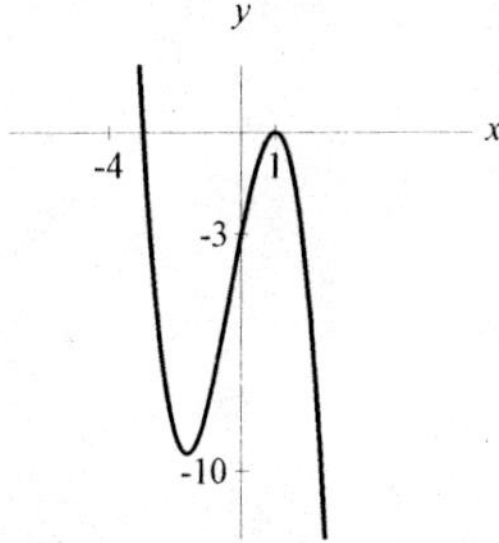

77. Since $x^3 - 10x^2 - 600x = x(x - 30)(x + 20)$, graph crosses $(0, 0), (30, 0), (-20, 0)$, $y \to \infty$ as $x \to \infty$, $y \to -\infty$ as $x \to -\infty$

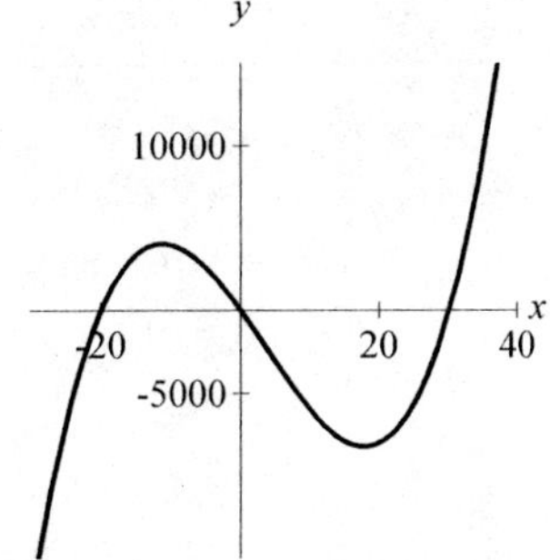

79. $f(x) = x^3 + 18x^2 - 37x + 60$ has only one x-intercept since

-20	1	18	-37	60
		-20	40	-60
	1	-2	3	0

and $x^2 - 2x + 3$ has no real root.
Graph crosses $(-20, 0)$, y-intercept $(0, 60)$, $y \to \infty$ as $x \to \infty$, $y \to -\infty$ as $x \to -\infty$

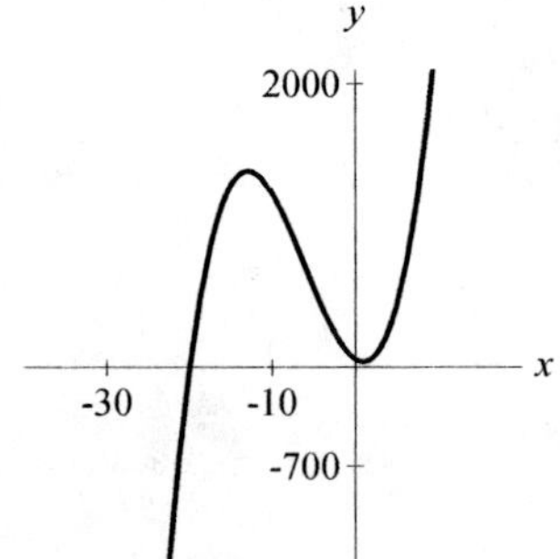

81. Since $-x^2(x^2 - 196) = -x^2(x - 14)(x + 14)$, graph crosses $(\pm 14, 0)$ and does not cross $(0, 0)$, $y \to -\infty$ as $x \to \infty$ and as $x \to -\infty$

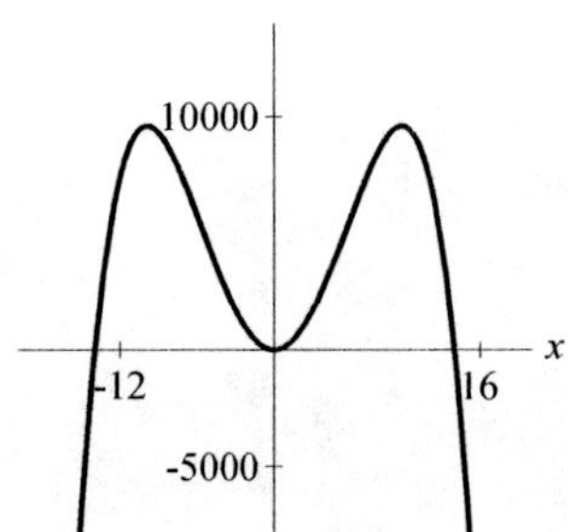

83. Use synthetic division on $f(x) = x^3 + 3x^2 + 3x + 1$ with $c = 1$.

-1	1	3	3	1
		-1	-2	-1
	1	2	1	0

Since $x^2 + 2x + 1 = (x + 1)^2$ then $f(x) = (x + 1)^3$. Graph crosses $(-1, 0)$, y-intercept $(0, 1)$, $y \to \infty$ as $x \to \infty$, and $y \to -\infty$ as $x \to -\infty$.

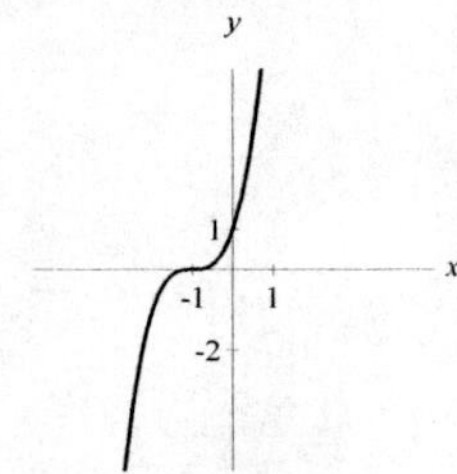

85. The graph crosses $(-7, 0)$, but it does not cross $(3, 0)$ and $(-5, 0)$, y-intercept is $(0, 1575)$, $y \to \infty$ as $x \to \infty$, and $y \to -\infty$ as $x \to -\infty$.

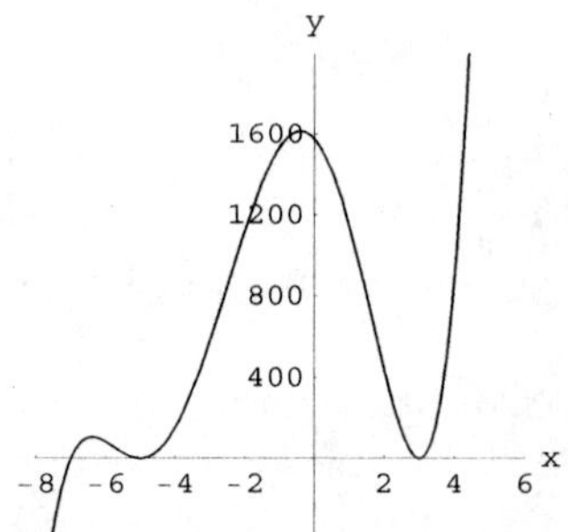

87. $(-2, 0) \cup (2, \infty)$, see the part of the graph above the x-axis

89. $[-2, 0] \cup [2, \infty)$, see the part of the graph above the x-axis including the x-intercepts

91. $(-2, -1) \cup (1, 2)$, see the part of the graph below the x-axis

93. $(-\infty - 2] \cup [-1, 1] \cup [2, \infty)$, see the part of the graph above the x-axis including the x-intercepts

95. From the graph of $y = x^3 - 1$ and the x-intercepts, the y coordinates are positive if x lies in $(1, \infty)$.

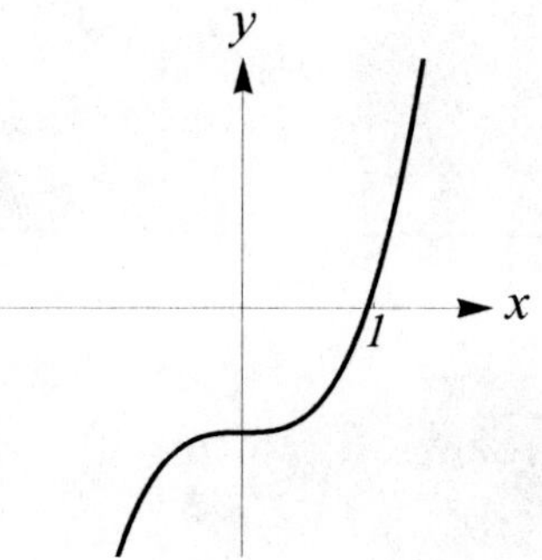

97. From the graph of $y = x^3 - 4x^2 - x + 4$, the y coordinates are negative if x belongs to $(-\infty, -1) \cup (1, 4)$.

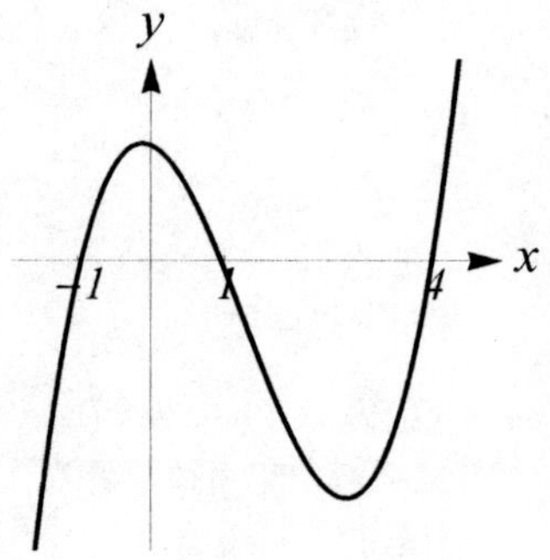

99. Using the graph of $y = x^4 - 4x^2 + 3$ and the x-intercepts, we find $y < 0$ if x lies in $(-\sqrt{3}, -1) \cup (1, \sqrt{3})$.

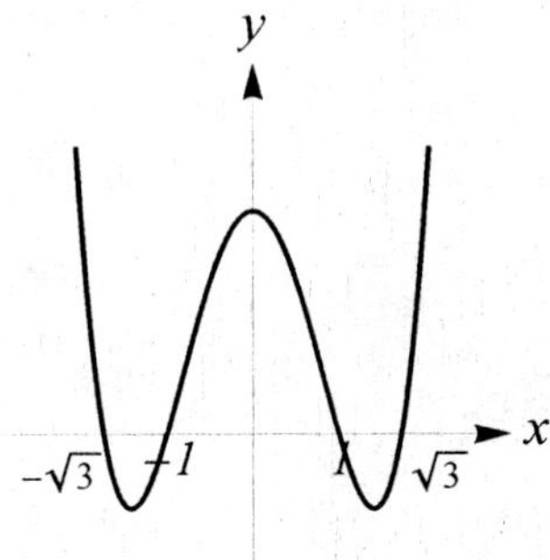

101. From the graph of $y = (x-0.01)(0.02-x)(x-0.03)$ and the x-intercepts, the y coordinates are nonnegative if x lies in

$$(-\infty, 0.01] \cup [0.02, 0.03].$$

103. The roots of $x(x^2 - 3) = 0$ are $x = 0, \pm\sqrt{3}$.

If $x = 2$, then $(2)^2 - 3(2) > 0$.
If $x = 1$, then $(1)^2 - 3(1) < 0$.
If $x = -1$, then $(-1)^2 - 3(-1) > 0$.
If $x = -2$, then $(-2)^2 - 3(-2) < 0$.

$-$ 0 $+$ 0 $-$ 0 $+$

-2 $-\sqrt{3}$ -1 0 1 $\sqrt{3}$ 2

The solution set is $(-\sqrt{3}, 0) \cup (\sqrt{3}, \infty)$.

105. The roots of $x^2(2 - x^2) = 0$ are $x = 0, \pm\sqrt{2}$.

If $x = 3$, then $2(3)^2 - (3)^4 < 0$.
If $x = 1$, then $2(1)^2 - (1)^4 > 0$.
If $x = -1$, then $2(-1)^2 - (-1)^4 > 0$.
If $x = -3$, then $2(-3)^2 - (-3)^4 < 0$.

$-$ 0 $+$ 0 $+$ 0 $-$

-3 $-\sqrt{2}$ -1 0 -1 $\sqrt{2}$ 3

The solution set is $(-\infty, -\sqrt{2}] \cup \{0\} \cup [\sqrt{2}, \infty)$.

107. Let $f(x) = x^3 + 4x^2 - x - 4$. Since

$$x^2(x+4) - (x+4) = (x+4)(x^2 - 1) = 0,$$

the zeros of $f(x)$ are $x = -4, \pm 1$.

If $x = 2$, then $f(2) > 0$.
If $x = 0$, then $f(0) < 0$.
If $x = -2$, then $f(-2) > 0$.
If $x = -5$, then $f(-5) < 0$.

$-$ 0 $+$ 0 $-$ 0 $+$

-5 -4 -2 -1 0 1 2

The solution set is $(-4, -1) \cup (1, \infty)$.

109. Let $f(x) = x^3 - 4x^2 - 20x + 48$.
Since $f(x) = (x+4)(x-2)(x-6)$, the zeros of $f(x)$ are $x = -4, 2, 6$.

If $x = 7$, then $f(7) > 0$.
If $x = 4$, then $f(4) < 0$.
If $x = 0$, then $f(0) > 0$.
If $x = -5$, then $f(-5) < 0$.

$-$ 0 $+$ 0 $-$ 0 $+$

-5 -4 0 2 4 6 7

The solution set is $[-4, 2] \cup [6, \infty)$.

111. Note, $f(x) = x^3 - x^2 + x - 1 =$
$x^2(x-1) + (x-1) = (x-1)(x^2+1) = 0$
has only one solution, namely, $x = 1$.

If $x = 2$, then $f(2) > 0$.
If $x = 0$, then $f(0) < 0$.

$-$ 0 $+$

0 1 2

The solution set is $(-\infty, 1)$.

113. Let $f(x) = x^4 - 19x^2 + 90$.
Since $f(x) = (x^2 - 10)(x^2 - 9)$, the zeros of $f(x)$ are $x = \pm\sqrt{10}, \pm 3$.

If $x = 4$, then $f(4) > 0$.
If $x = 3.1$, then $f(3.1) < 0$.
If $x = 0$, then $f(0) > 0$.
If $x = -3.1$, then $f(-3.1) < 0$.
If $x = -4$, then $f(-4) > 0$.

$$\begin{array}{ccccccccccc} + & 0 & - & 0 & + & 0 & - & 0 & + \\ & -\sqrt{10} & & -3 & & 3 & & \sqrt{10} & \end{array}$$

The solution set is $[-\sqrt{10}, -3] \cup [3, \sqrt{10}]$.

115. d, since the x-intercepts of

$$f(x) = \frac{1}{3}(x+3)(x-2)$$

are -3 and 2 and $f(0) = -2$

117. Since $f(-5) = 0 = f(4)$, the quadratic function f can be expressed as

$$f(x) = a(x+5)(x-4).$$

Also, $3 = f(0) = a(5)(-4)$ and so $a = -3/20$. Thus, we have

$$f(x) = -\frac{3}{20}(x+5)(x-4).$$

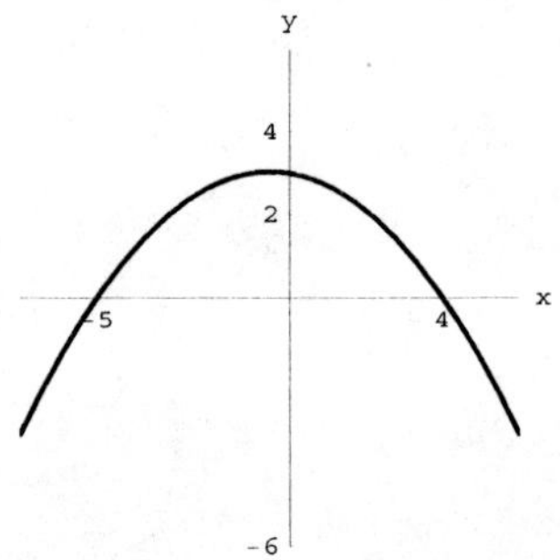

119. Since $f(-1) = 0 = f(3)$, the quadratic function f can be written as

$$f(x) = a(x+1)(x-3).$$

Also, $2 = f(1) = a(2)(-2)$ and we find $a = -1/2$. Thus, we obtain

$$f(x) = -\frac{1}{2}(x+1)(x-3).$$

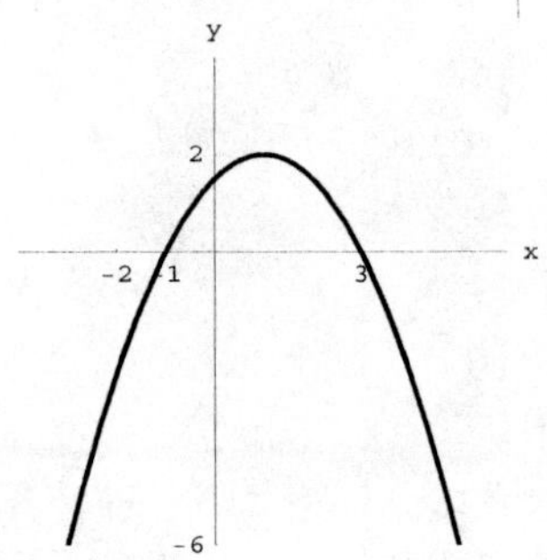

121. Since $f(2) = f(-3) = f(4) = 0$, the cubic function f can be expressed as

$$f(x) = a(x-2)(x+3)(x-4).$$

Also, $6 = f(0) = a(-2)(3)(-4)$ and we get $a = 1/4$. Thus, we obtain

$$f(x) = \frac{1}{4}(x-2)(x+3)(x-4).$$

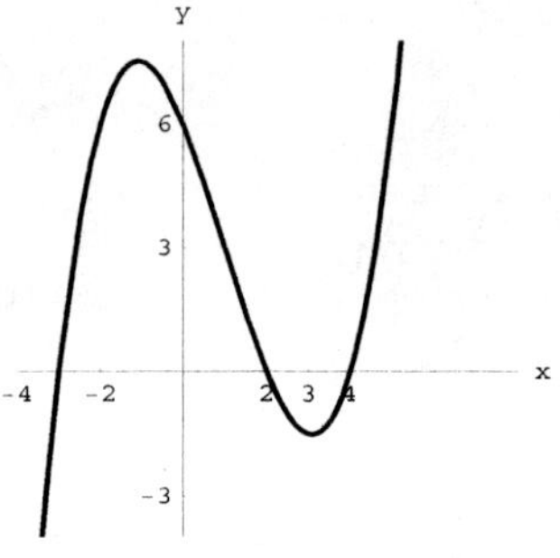

123. Since $f(\pm 2) = 0 = f(\pm 4)$, the quartic function f can be written as

$$f(x) = a(x^2-4)(x^2-16).$$

Also, $3 = f(1) = a(-3)(-15)$ and we get $a = 1/15$. Thus, we obtain

$$f(x) = \frac{1}{15}(x^2-4)(x^2-16).$$

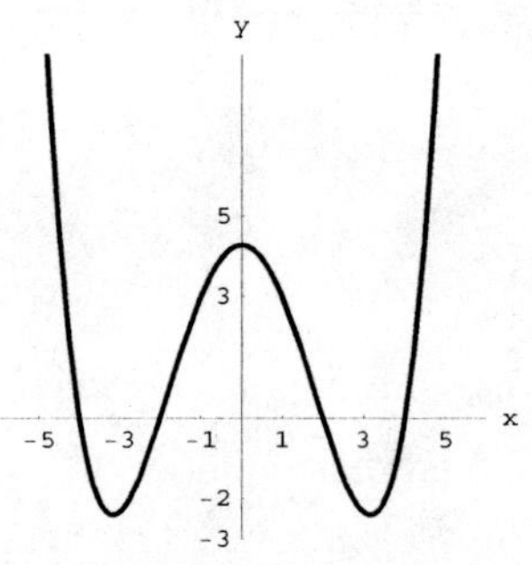

125. The area A of the rectangle is a function of the base x, i.e.,

$$A(x) = x(x^2 - 4x + 4)$$

Using a graphing calculator, the maximum of $A(x)$ where $0 < x < 2$ occurs when $x = 2/3$. Then the maximum area is

$$A(2/3) \approx 1.185$$

127. A graph of $P = \frac{x}{10}(x^2 - 60x + 900)$ is given below. If $x = 10$ stores are in operation, the profit decreases to \$0 at $x = 30$ stores. Then the profit increases when $x > 30$ stores.

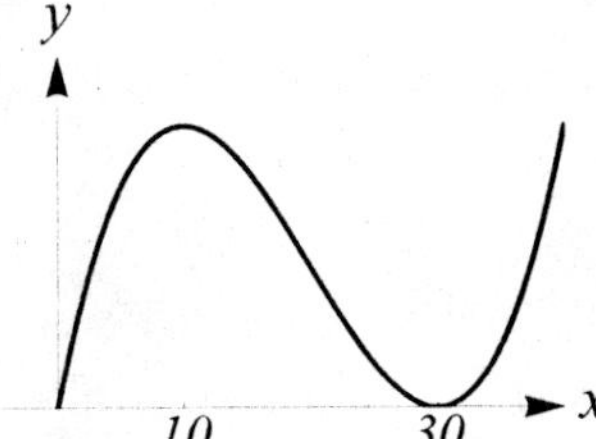

129. Since $3x + 2y = 12$, we get $y = \frac{12 - 3x}{2}$.

The volume V of the block is given by

$$\begin{aligned} V &= xy\frac{4y}{3} \\ &= \frac{4x}{3}y^2 \\ &= \frac{4x}{3}\left(\frac{12-3x}{2}\right)^2 \\ V &= 3x^3 - 24x^2 + 48x. \end{aligned}$$

One finds from the graph of

$$V = \frac{4x}{3}\left(\frac{12-3x}{2}\right)^2$$

that the optimal dimensions of the block are $x = \frac{4}{3}$ in. by $y = \frac{12-4}{2} = 4$ in. by $\frac{16}{3}$ in.

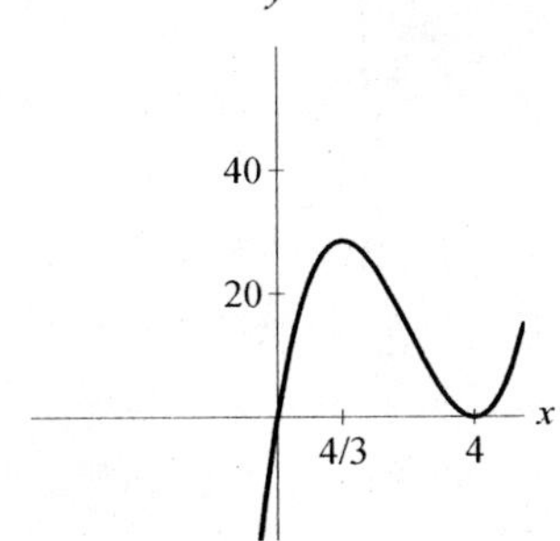

131. Isolate one radical and square both sides.

$$\begin{aligned} x + 12 &= 9 + 6\sqrt{x-9} + (x-9) \\ 12 &= 6\sqrt{x-9} \\ 2 &= \sqrt{x-9} \\ 4 &= x - 9 \\ 13 &= x \end{aligned}$$

The solution set is $\{13\}$.

133. Quotient $x^3 + x^2 - 11x + 24$ and remainder -53 since

-2	1	3	-9	2	-5
		-2	-2	22	-48
	1	1	-11	24	-53

135. Since the absolute value of a number is nonnegative and $x = |y - 5|$, the domain or set of all x-values is $[0, \infty)$. The range or set of all permissible y-values in $|y - 5|$ is $(-\infty, \infty)$.

137. Using similar triangles, we find

$$\frac{7}{d} = \frac{x}{1}.$$

By the Pythagorean Theorem, we obtain

$$1 + (d-1)^2 = (7-x)^2.$$

Then we obtain

$$1 + (d-1)^2 = 49\left(1 - \frac{1}{d}\right)^2.$$

Using a computer algebra system, we find

$$d = \frac{\sqrt{47 - 10\sqrt{2}} + 5\sqrt{2} + 1}{2} \approx 6.9016 \text{ ft.}$$

For Thought

1. False, $\sqrt{x} - 3$ is not a polynomial.

2. False, domain is $(-\infty, 2) \cup (2, \infty)$.

3. False

4. False, it has three vertical asymptotes.

5. True

6. False, $y = 5$ is the horizontal asymptote.

7. True **8.** False

9. True, it is an even function.

10. True, since $x = -3$ is not a vertical asymptote.

3.5 Exercises

1. rational

3. vertical asymptote

5. domain

7. $(-\infty, -2) \cup (-2, \infty)$

9. $(-\infty, -2) \cup (-2, 2) \cup (2, \infty)$

11. $(-\infty, 3) \cup (3, \infty)$

13. $(-\infty, 0) \cup (0, \infty)$

15. $(-\infty, -1) \cup (-1, 0) \cup (0, 1) \cup (1, \infty)$ since

$$f(x) = \frac{3x^2 - 1}{x(x^2 - 1)}$$

17. $(-\infty, -3) \cup (-3, -2) \cup (-2, \infty)$ since

$$f(x) = \frac{-x^2 + x}{(x+3)(x+2)}$$

19. Domain $(-\infty, 2) \cup (2, \infty)$, asymptotes $y = 0$ and $x = 2$

21. Domain $(-\infty, 0) \cup (0, \infty)$, asymptotes $y = x$ and $x = 0$

23. Asymptotes $x = 2$, $y = 0$

25. Asymptotes $x = \pm 3$, $y = 0$

27. Asymptotes $x = 1$, $y = 2$

29. Asymptotes $x = 0$, $y = x - 2$ since

$$f(x) = x - 2 + \frac{1}{x}$$

31. Asymptotes $x = -1$, $y = 3x - 3$ since

$$\begin{array}{r|l} & \quad 3x - 3 \\ x+1 & \overline{)3x^2 + 0x + 4} \\ & \underline{3x^2 + 3x} \\ & \quad -3x + 4 \\ & \quad \underline{-3x - 3} \\ & \qquad\quad 7 \end{array}$$

and $f(x) = 3x - 3 + \dfrac{7}{x+1}$

and $f(x) = x + 9 + \dfrac{81}{x - 9}$

33. Asymptotes $x = -2$, $y = -x + 6$ since

$$\begin{array}{r|l} & \quad -x + 6 \\ x+2 & \overline{)-x^2 + 4x + 0} \\ & \underline{-x^2 - 2x} \\ & \qquad 6x + 0 \\ & \qquad \underline{6x + 12} \\ & \qquad\quad -12 \end{array}$$

and $f(x) = -x + 6 + \dfrac{-12}{x+2}$

35. Asymptotes $x = 0$, $y = 0$, no x or y-intercept

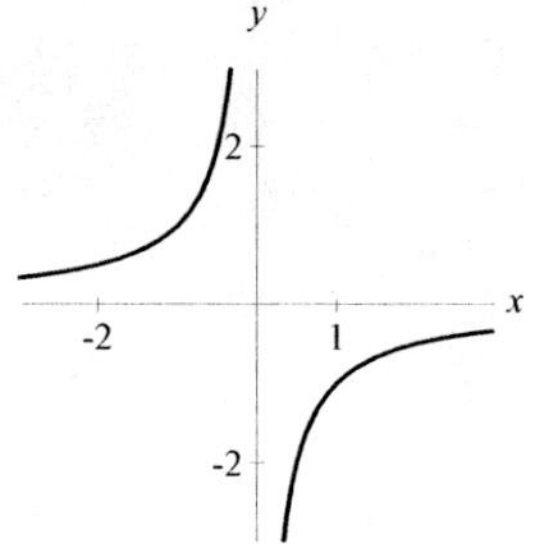

37. Asymptotes $x = 2$, $y = 0$, no x-intercept, y-intercept $(0, -1/2)$

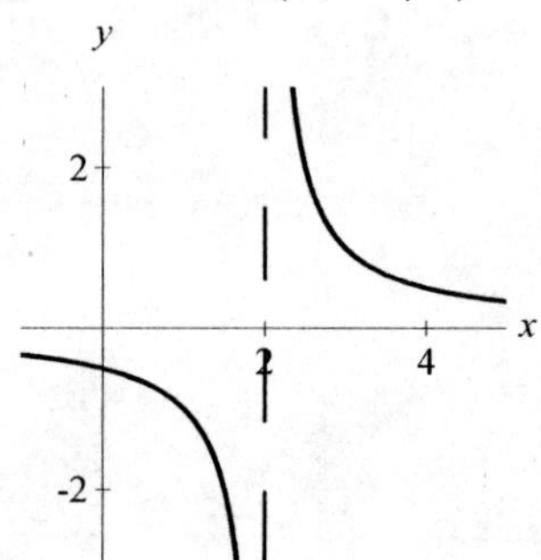

39. Asymptotes $x = \pm 2$, $y = 0$, no x-intercept, y-intercept $(0, -1/4)$

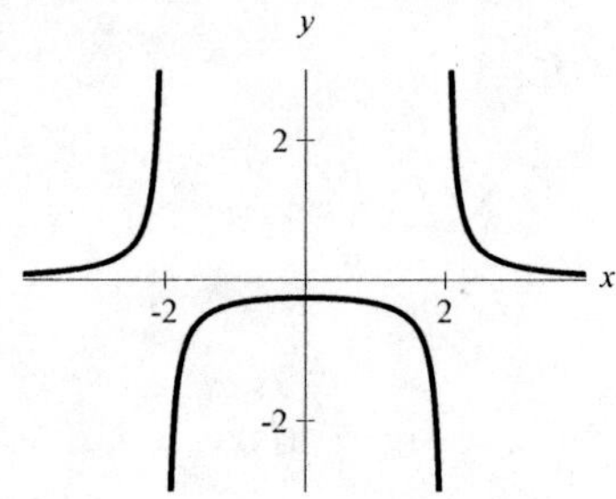

41. Asymptotes $x = -1$, $y = 0$, no x-intercept, y-intercept $(0, -1)$

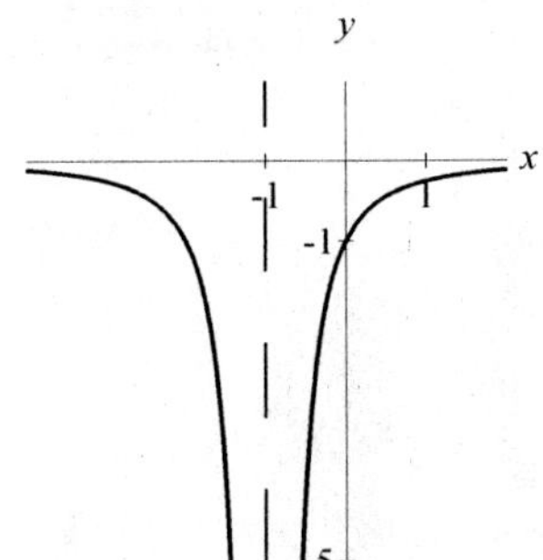

43. Asymptotes $x = 1$, $y = 2$, x-intercept $(-1/2, 0)$, y-intercept $(0, -1)$

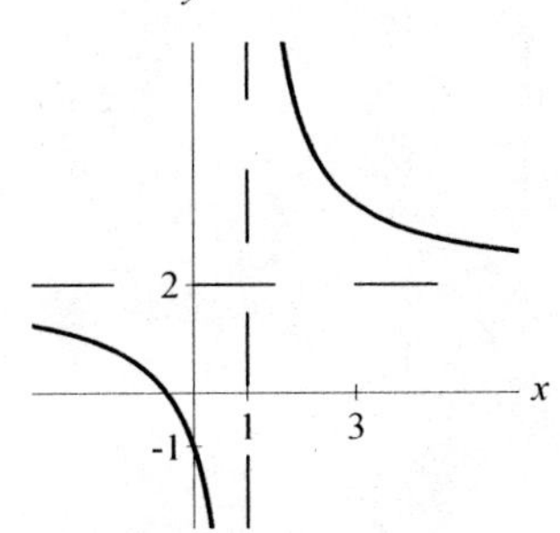

45. Asymptotes $x = -2$, $y = 1$, x-intercept $(3, 0)$, y-intercept $(0, -3/2)$

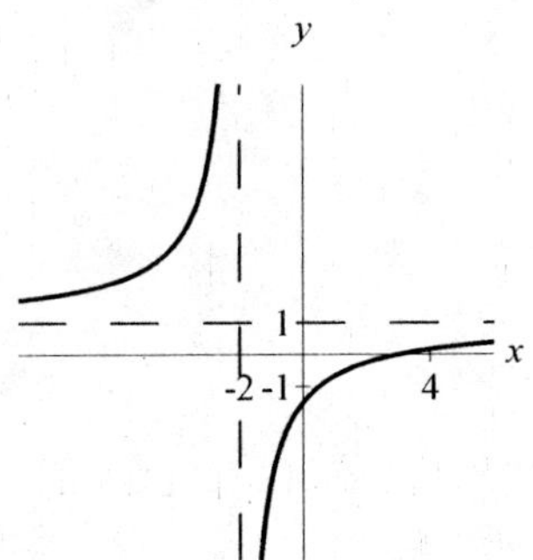

47. Asymptotes $x = \pm 1$, $y = 0$, x- and y-intercept is $(0, 0)$

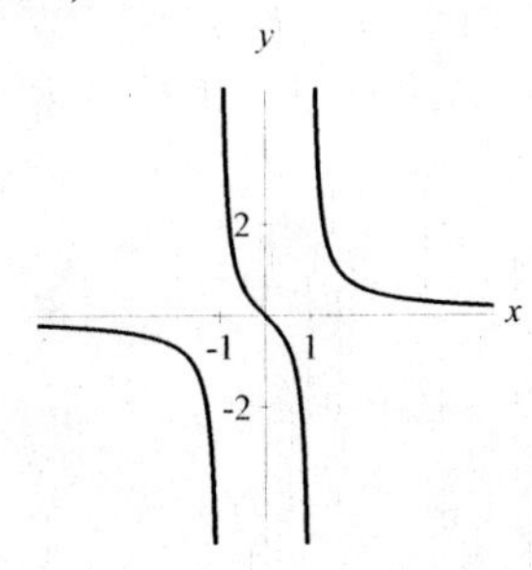

49. Since $f(x) = \dfrac{4x}{(x-1)^2}$, asymptotes are $x = 1$, $y = 0$, x- and y-intercept is $(0, 0)$

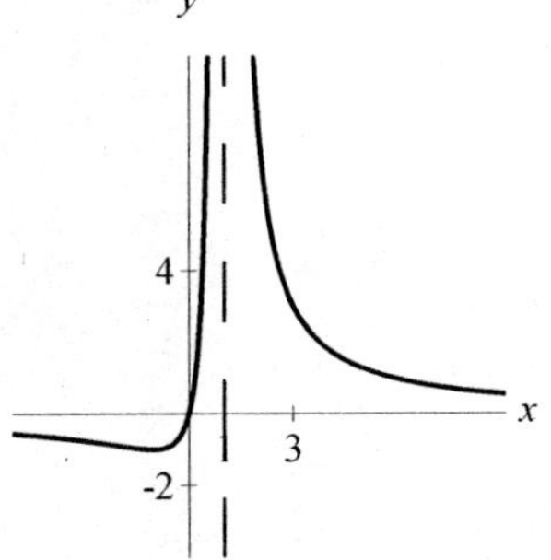

51. Asymptotes are $x = \pm 3$, $y = -1$, x-intercept $(\pm 2\sqrt{2}, 0)$, y-intercept $(0, -8/9)$

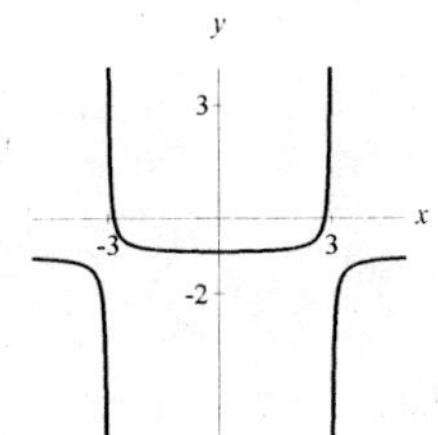

53. Since $f(x) = \dfrac{2x^2 + 8x + 2}{(x+1)^2}$,

asymptotes are $x = -1$, $y = 2$, by solving $2x^2 + 8x + 2 = 0$ one gets the x-intercepts $\left(-2 \pm \sqrt{3}, 0\right)$, y-intercept $(0, 2)$

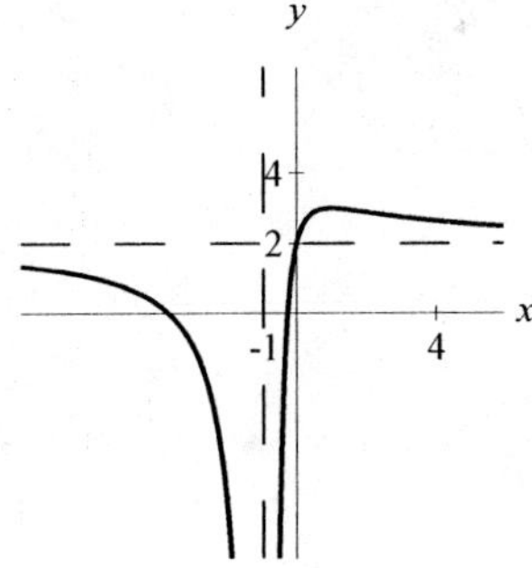

55. We see from the graph that $\lim\limits_{x\to\infty} \dfrac{1}{x^2} = 0$

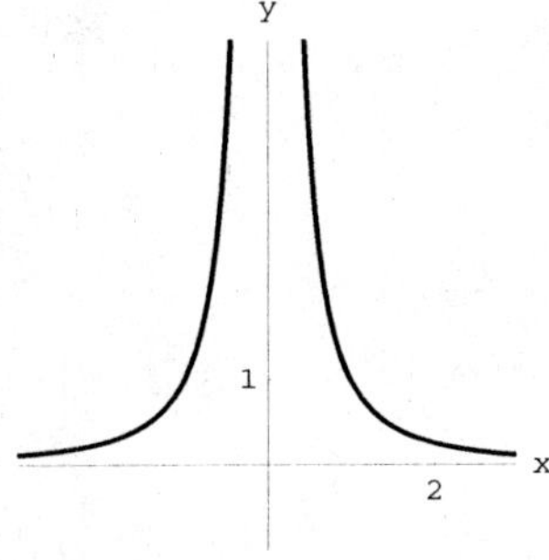

57. From the graph, we find $\lim_{x \to \infty} \dfrac{2x-3}{x-1} = 2$

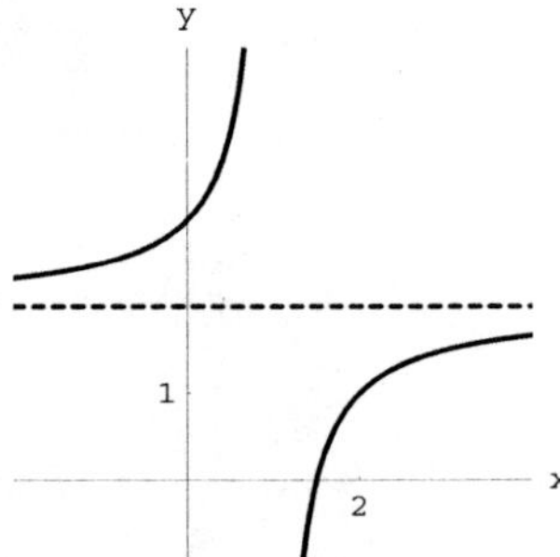

59. From the graph, we find $\lim_{x \to 0^+} \dfrac{1}{x^2} = \infty$

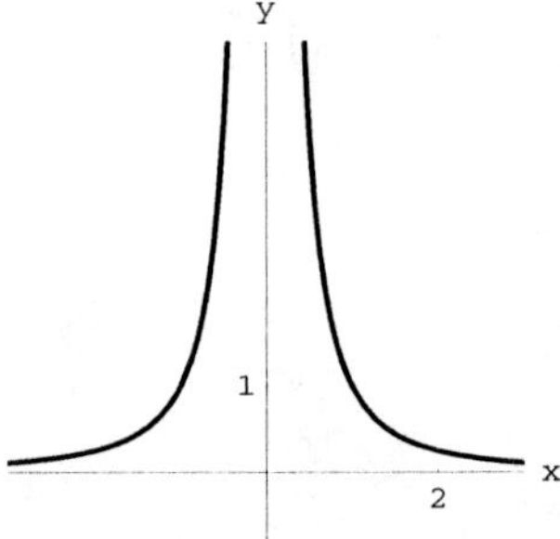

61. We see from the graph that $\lim_{x \to 1^+} \dfrac{2}{x-1} = \infty$

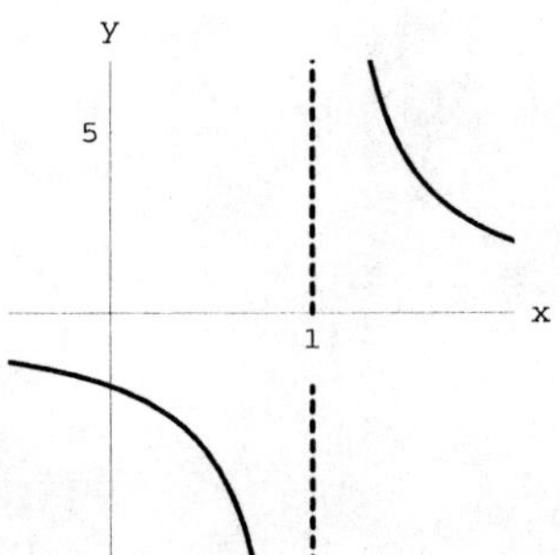

63. Since $f(x) = x + \dfrac{1}{x}$, oblique asymptote is $y = x$, asymptote $x = 0$, no x-intercept, no y-intercept, graph goes through $(1, 2), (-1, -2)$

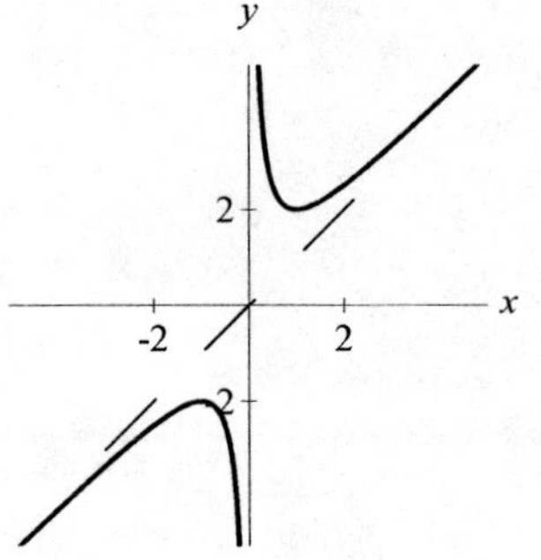

65. Since $f(x) = x - \dfrac{1}{x^2}$, oblique asymptote is $y = x$, asymptote $x = 0$, x-intercept $(1, 0)$, no y-intercept, graph goes through $(-1, -2)$

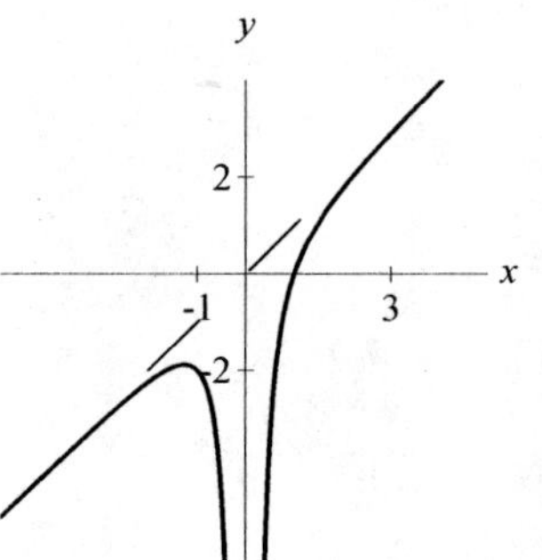

67. $f(x) = x - 1 + \dfrac{1}{x+1}$ since

$$\begin{array}{r|l} & x - 1 \\ x+1 & x^2 + 0x \\ & \underline{x^2 + x} \\ & -x + 0 \\ & \underline{-x - 1} \\ & 1 \end{array}$$

Oblique asymptote $y = x - 1$, asymptote $x = -1$, x-intercept $(0, 0)$, graph goes through $(-2, -4)$

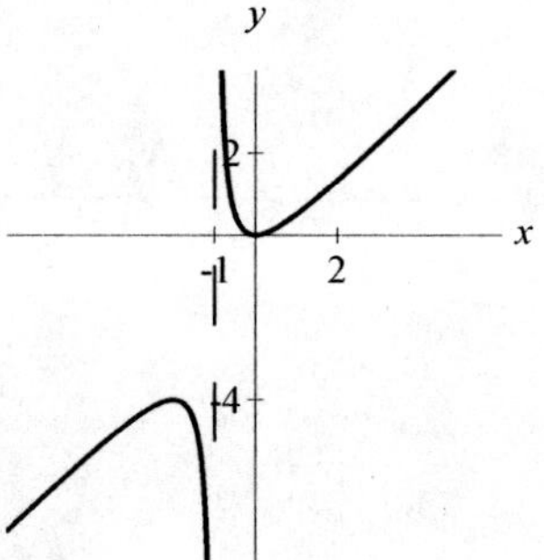

69. $f(x) = 2x + 1 + \dfrac{1}{x-1}$ since

$$\begin{array}{r|l} & 2x + 1 \\ x-1 & 2x^2 - x + 0 \\ & \underline{2x^2 - 2x} \\ & x + 0 \\ & \underline{x - 1} \\ & 1 \end{array}$$

Oblique asymptote $y = 2x + 1$, asymptote $x = 1$, x-intercepts $(0, 0), (1/2, 0)$, graph goes through $(2, 6)$

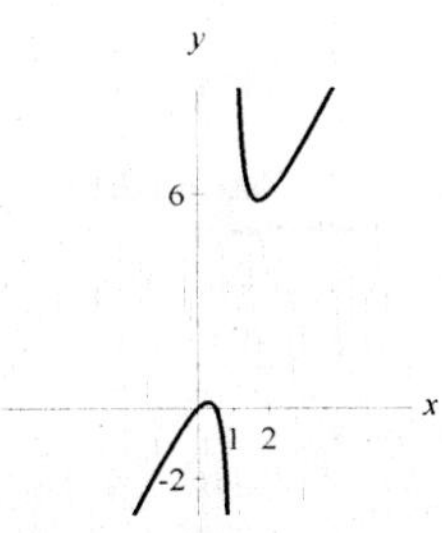

71. Using long division, we find

$$\frac{x^3 - x^2 - 4x + 5}{x^2 - 4} = (x - 1) + \frac{1}{x^2 - 4}.$$

The oblique asymptote is $y = x - 1$.

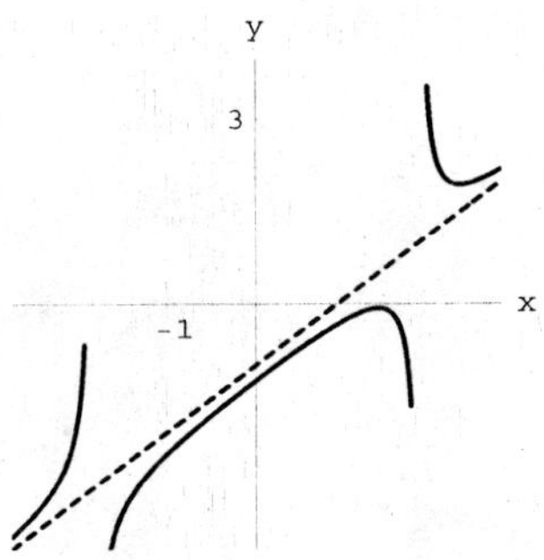

73. Using long division, we find

$$\frac{-x^3 + x^2 + 5x - 4}{x^2 + x - 2} = (-x + 2) + \frac{x}{x^2 + x - 2}.$$

The oblique asymptote is $y = -x + 2$.

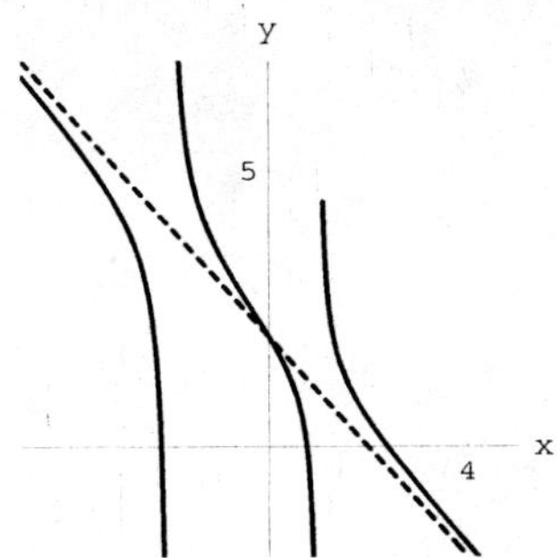

75. e **77.** a

79. b **81.** c

83. Domain is $\{x : x \neq \pm 1\}$ since

$$f(x) = \frac{x + 1}{(x + 1)(x - 1)} = \frac{1}{x - 1} \text{ if } x \neq -1,$$

a 'hole' at $(-1, -1/2)$, asymptotes $x = 1$, $y = 0$, no x-intercept, y-intercept $(0, -1)$

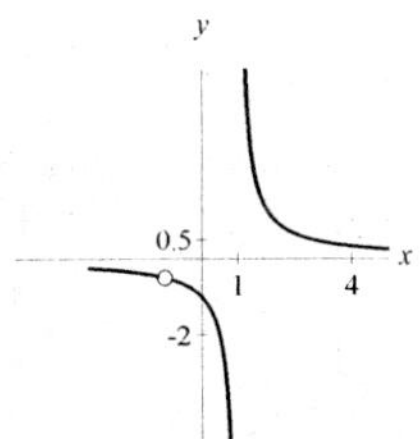

85. The domain is $\{x : x \neq 1\}$ since

$$f(x) = \frac{(x - 1)(x + 1)}{x - 1} = x + 1 \text{ where } x \neq 1,$$

a line with a 'hole' at $(1, 2)$

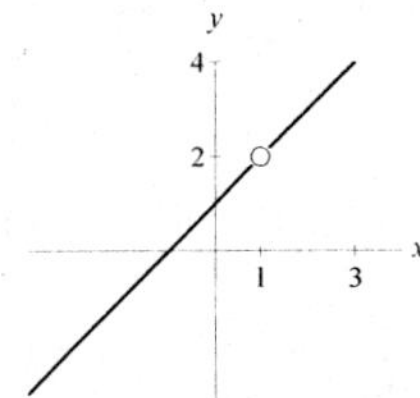

87. Parabola $y = x^2$ with domain $x \neq \pm 1$.

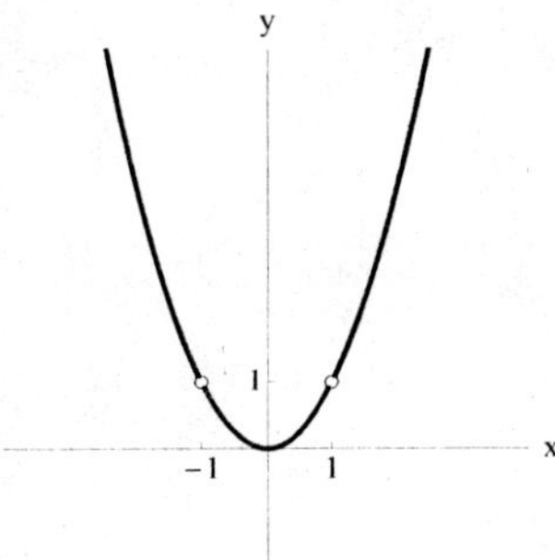

89. The graph of $y = \dfrac{1}{x^2}$ with domain $x \neq 0, \pm 2$

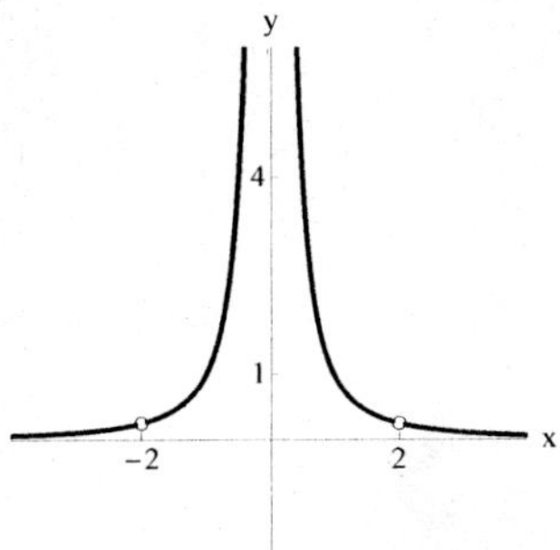

91. Symmetric about y-axis, asymptote $x = 0$, goes through $(0, 2), (1, 1)$

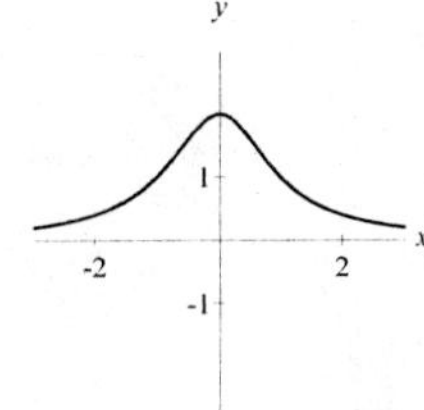

93. Since $f(x) = \dfrac{x-1}{x(x^2-9)}$, asymptotes are

$x = \pm 3, x = 0, y = 0$, crosses $(1,0)$

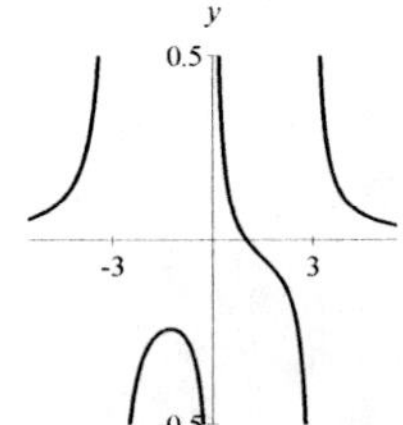

95. Asymptotes $x = 0, y = 0$, x-intercept $(-1,0)$, no y-intercept, graph goes through $(1,2)$

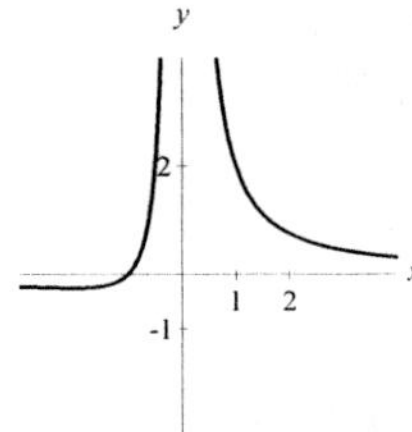

97. In the graph of Exercise 91, the y-coordinate is positive for all x in $(-\infty,\infty)$.

99. In the graph of Exercise 93, the y-coordinate is nonpositive if x lies in $(-3,0) \cup [1,3)$.

101. In the graph of Exercise 95, the y-coordinate is nonnegative if x lies in $[-1,0) \cup (0,\infty)$.

103. Let $f(x) = \dfrac{x-4}{x+2}$.

If $x = -3$, then $f(-3) > 0$.
If $x = 0$, then $f(0) < 0$.
If $x = 5$, then $f(5) > 0$.

+ U − 0 +

-3 -2 0 4 5

The solution set is $(-2,4]$.

105. Let $f(q) = \dfrac{q+8}{q+3} > 0$.

If $q = -9$, then $f(-9) > 0$.
If $q = -6$, then $f(-6) < 0$.
If $q = 0$, then $f(0) > 0$.

+ 0 − U +

-9 -8 -6 -3 0

The solution set is $(-\infty,-8) \cup (-3,\infty)$.

107. Let $f(w) = \dfrac{(w-3)(w+2)}{w-6} \geq 0$.

If $w = -3$, then $f(-3) < 0$.
If $w = 0$, then $f(0) > 0$.
If $w = 4$, then $f(4) < 0$.
If $w = 7$, then $f(7) > 0$.

− 0 + 0 − U +

-3 -2 0 3 4 6 7

The solution set is $[-2,3] \cup (6,\infty)$.

109. Let $f(x) = \dfrac{-5}{(x+2)(x-3)} > 0$.

If $x = -3$, then $f(-3) < 0$.
If $x = 0$, then $f(0) > 0$.
If $x = 4$, then $f(4) < 0$.

− U + U −

-3 -2 0 3 4

The solution set is $(-2,3)$.

111. Let $f(x) = \dfrac{(x-4)(x+2)}{5-x} > 0$.

If $x = -3$, then $f(-3) > 0$.
If $x = 0$, then $f(0) < 0$.
If $x = 4.5$, then $f(4.5) > 0$.
If $x = 6$, then $f(6) < 0$.

+ 0 − 0 + U −

-3 -2 0 4 4.5 5 6

The solution set is $(-\infty,-2) \cup (4,5)$.

113. Let $f(x) = \dfrac{(x-3)(x+1)}{x-5} \geq 0$.

If $x = -2$, then $f(-2) < 0$.
If $x = 0$, then $f(0) > 0$.
If $x = 4$, then $f(4) < 0$.
If $x = 7$, then $f(7) > 0$.

− 0 + 0 − U +

-2 -1 0 3 4 5 7

The solution set is $[-1,3] \cup (5,\infty)$.

115. Let $f(x) = \dfrac{(x-\sqrt{7})(x+\sqrt{7})}{(\sqrt{2}-x)(\sqrt{2}+x)}$.

If $x = 3$, then $f(3) < 0$.
If $x = 2$, then $f(2) > 0$.
If $x = 0$, then $f(0) < 0$.
If $x = -2$, then $f(-2) > 0$.
If $x = -3$, then $f(-3) < 0$.

− 0 + U − U + 0 −

$-\sqrt{7}$ $-\sqrt{2}$ $\sqrt{2}$ $\sqrt{7}$

The solution set is

$$(-\infty, -\sqrt{7}] \cup (-\sqrt{2}, \sqrt{2}) \cup [\sqrt{7}, \infty).$$

117. Let $R(x) = \dfrac{(x+1)^2}{(x-5)(x+3)} \geq 0$.

Since $R(-4) > 0$, $R(-2) < 0$, $R(0) < 0$, and $R(6) > 0$, we get

+ U − 0 − U +

-4 -3 -2 -1 0 5 6

The solution set is $(-\infty, -3) \cup \{-1\} \cup (5, \infty)$.

119. Let $f(w) = \dfrac{w-1}{w^2} > 0$.

If $w = -1$, then $f(-1) < 0$.
If $w = 0.5$, then $f(0.5) < 0$.
If $w = 2$, then $f(2) > 0$.

− U − 0 +

-1 0 0.5 1 2

The solution set is $(1, \infty)$.

121. Let $f(w) = \dfrac{w^2 - 4w + 5}{w - 3} > 0$. The numerator has no real zero.

If $w = 0$, then $f(0) < 0$.
If $w = 4$, then $f(4) > 0$.

− U +

0 3 4

The solution set is $(3, \infty)$.

123. $(-\infty, 1) \cup (3, \infty)$, see the part of the graph above the x-axis

125. $(-\infty, 1] \cup (3, \infty)$, see the part of the graph above the x-axis including the x-intercepts

127. $(-\infty, -2) \cup (0, 2)$, see the part of the graph below the x-axis

129. $(-2, 0] \cup (2, \infty)$, see the part of the graph above the x-axis including the x-intercepts

131. Since $y = 0$ and $x = 1$ are asymptotes, the rational function can be written as

$$y = \frac{a}{x-1}.$$

Since $(3, 1)$ satifies the equation above, we find $1 = a/2$ or $a = 2$. Thus, the function is

$$y = \frac{2}{x-1}.$$

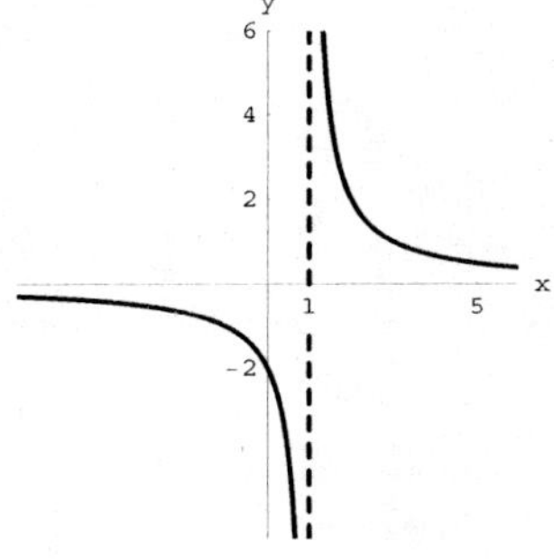

133. Since $y = 1$ and $x = 2$ are asymptotes, the function can be expressed as

$$y = \frac{x-a}{x-2}.$$

Since $(0, 5)$ satifies the equation above, we obtain $5 = (-a)/(-2)$ or $a = 10$. Thus, the function is

$$y = \frac{x-10}{x-2} \text{ or } y = \frac{-8}{x-2} + 1.$$

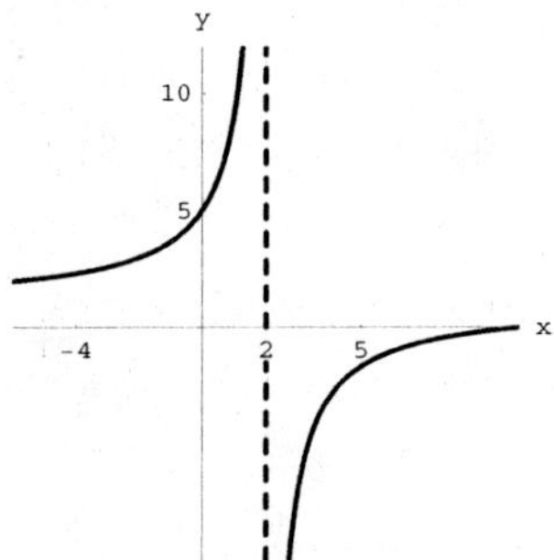

135. Since $y = 0$ and $x = \pm 3$ are asymptotes, the function can be written as

$$y = \frac{ax + b}{x^2 - 9}.$$

Since $(0, 0)$ and $(4, 8/7)$ satifies the equation above, we obtain $0 = (b)/(-9)$ and $8/7 = (4a + b)/7$, respectively. Then $b = 0$ and $8/7 = 4a/7$ or $a = 2$. Thus, the function is

$$y = \frac{2x}{x^2 - 9}.$$

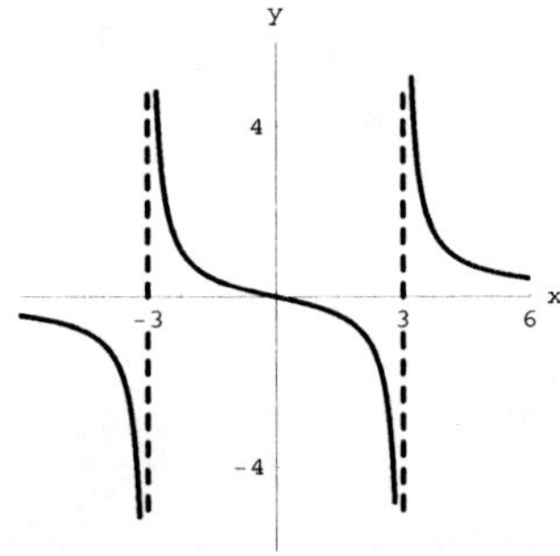

137. Since $y = 2x + 1$ and $x = 1$ are asymptotes, the function can be written as

$$y = (2x + 1) + \frac{a}{x - 1}.$$

Since $(0, 3)$ satifies the equation above, we find $3 = 1 + a/(-1)$ or $a = -2$. Thus, the function is

$$y = 2x + 1 + \frac{-2}{x - 1}.$$

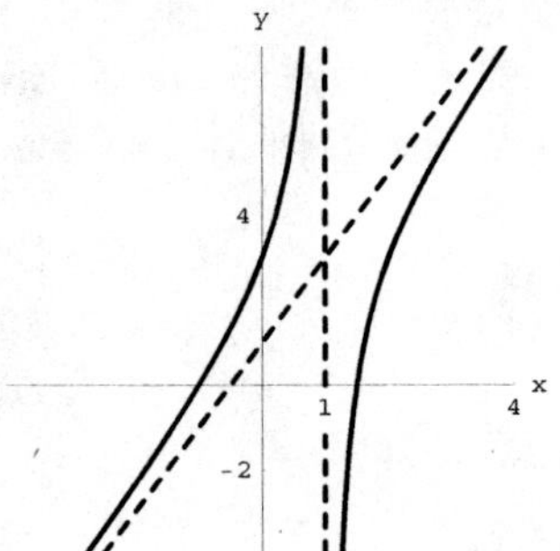

139. Average cost is $C = \frac{100 + x}{x}$. If she visits the zoo 100 times, then her average cost per visit is $C = \frac{200}{100} = \$2$. Since $C \to \$1$ as $x \to \infty$, over a long period her average cost per visit is \$1.

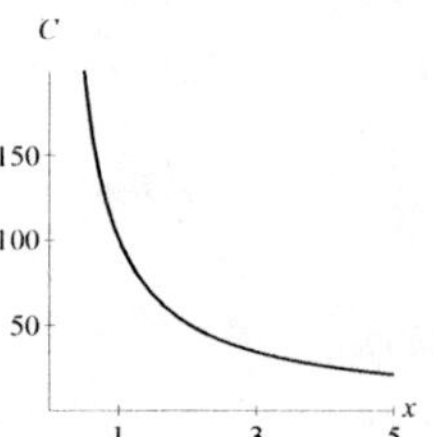

141. Since the second half is 100 miles long and it must be completed in $4 - x$ hours, the average speed for the second half is $S(x) = \frac{100}{4 - x}$. The asymptote $x = 4$ implies $S \to \infty$ as the completion time in the first half shortens to four hours.

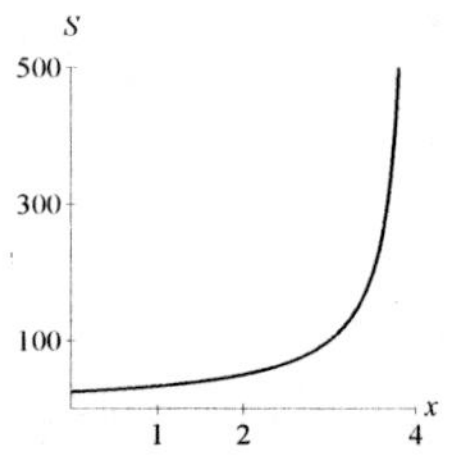

143. a) About 15 minutes

b) About 200 PPM

c) The horizontal asymptote is PPM $= 0$ and the vertical asymptote is $t = 0$. In simple terms, a long exposure to low level carbon monoxide still leads to permanent brain damage, and a very high level carbon monoxide can lead to permanent brain damage in a few minutes.

145.

(a) $h = \frac{500}{\pi r^2}$ since $500 = \pi r^2 h$

(b) $S = 2\pi r^2 + 2\pi r h = 2\pi r^2 + 2\pi r \frac{500}{\pi r^2}$

or equivalently $S = 2\pi r^2 + \frac{1000}{r}$

(c) As can be seen from the graph of

$S = 2\pi r^2 + \frac{1000}{r}$, S is minimized

when $r \approx 4.3$ feet.

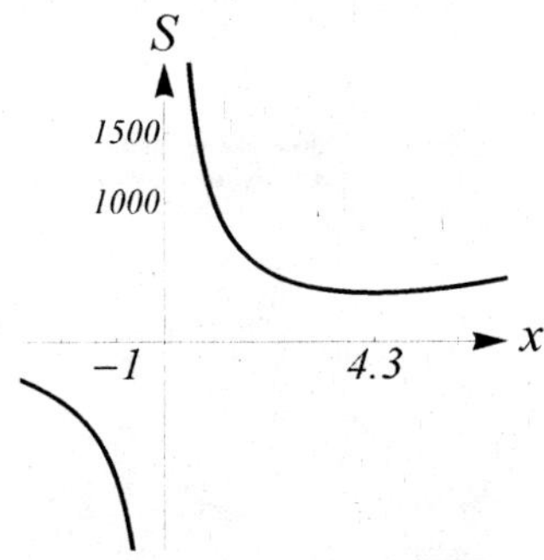

(d) If it costs \$8 per ft^2 to construct the tank in part (c), then the cost of the tank is given by

$$8\left(2\pi(4.3)^2 + \frac{1000}{4.3}\right) \approx \$2789.87.$$

149. Let $f(x) = (x-1)^2(x+5) > 0$.
If $x = -6$, then $f(-6) < 0$.
If $x = 0$, then $f(0) > 0$.
If $x = 2$, then $f(2) > 0$.

−	0	+	0	+
-6	−5	0	1	2

The solution set is $(-5, 1) \cup (1, \infty)$.

151. Factoring, we find

$$x(x^4 + 13x^2 + 36) = x(x^2+4)(x^2+9)$$

The solution set is $\{0, \pm 2i, \pm 3i\}$.

153. Since $-\frac{b}{2a} = \frac{3}{2}$ and $f\left(\frac{3}{2}\right) = -\frac{31}{4}$, the vertex is $\left(\frac{3}{2}, -\frac{31}{4}\right)$. Since the parabola opens downward, the function is increasing on $\left(-\infty, \frac{3}{2}\right]$.

155. Suppose there are n rows of circles with n circles in the bottom row, $n-1$ circles in the next, ..., and one in the top row. Then the number of circles is

$$1 + 2 + 3 + \ldots + n = \frac{n(n+1)}{2}.$$

a) Look at the n circles in the bottom row. The distance between the points of tangency to the bottom side of the equilateral triangle of the first and last circle is $2r(n-1)$. Also, the distance from the last point of tangency to the closest vertex of the triangle is $r\sqrt{3}$. Since the side of the triangle has length 1, we have

$$2r(n-1) + 2r\sqrt{3} = 1.$$

Solving for r, we obtain

$$r = \frac{1}{2(n + \sqrt{3} - 1)}.$$

Then the sum of the areas of all the circles is

$$\begin{aligned} A(n) &= \frac{n(n+1)}{2}\pi r^2 \\ &= \frac{n(n+1)}{2}\pi\left(\frac{1}{2(n+\sqrt{3}-1)}\right)^2 \\ A(n) &= \frac{\pi(n^2+n)}{8(n+\sqrt{3}-1)^2}. \end{aligned}$$

b) By dividing the leading coefficients of the numerator and denominator of the rational function $A(n)$, we find that $A(n)$ approaches

$$\frac{\pi}{8} \approx 0.39$$

as n approaches infinity.
Since the area of the equilateral triangle is

$$\frac{\sqrt{3}}{4} \approx 0.43$$

the triangular pipe will not be filled by the circular cables.

Chapter 3 Review Exercises

1. By using the method of completing the square, we obtain

$$\begin{aligned} f(x) &= 3\left(x^2 - \frac{2}{3}x + \frac{1}{9}\right) - \frac{1}{3} + 1 \\ &= 3\left(x - \frac{1}{3}\right)^2 + \frac{2}{3} \end{aligned}$$

3. Since $y = 2(x^2 - 2x + 1) - 2 - 1 = 2(x-1)^2 - 3$, the vertex is $(1, -3)$ and axis of symmetry is $x = 1$. Setting $y = 0$, we get $x - 1 = \pm\frac{\sqrt{6}}{2}$ and the x-intercepts are $\left(\frac{2 \pm \sqrt{6}}{2}, 0\right)$. The y-intercept is $(0, -1)$.

5. From the x-intercepts, $y = a(x+1)(x-3)$. Substitute $(0,6)$, so $6 = a(-3)$ or $a = -2$. The equation of the parabola is $y = -2(x^2 - 2x - 3)$ or $y = -2x^2 + 4x + 6$.

7. $1/3$ **9.** $\pm 2\sqrt{2}$

11. Factoring, we obtain

$$m(x) = (2x-1)(4x^2 + 2x + 1).$$

By using the quadratic formula, we obtain the zeros to the second factor. Namely,

$$x = \frac{-2 \pm \sqrt{-12}}{8} = \frac{-2 \pm 2i\sqrt{3}}{8}.$$

The zeros are $\frac{1}{2}, \frac{-1 \pm i\sqrt{3}}{4}$.

13. Since $P(t) = (t^2 - 10)(t^2 + 10)$, the zeros are $\pm\sqrt{10}, \pm i\sqrt{10}$

15. Factoring, we obtain $R(s) = 4s^2(2s-1) - (2s-1) = (4s^2-1)(2s-1)$. The zeros are $\frac{1}{2}$ (with multiplicity 2) and $-\frac{1}{2}$.

17. Find the roots of the second factor of $f(x) = x(x^2 + 2x - 6)$. Since $x^2 + 2x - 6 = (x^2 + 2x + 1) - 6 - 1 = (x+1)^2 - 7$, the zeros are $0, -1 \pm \sqrt{7}$.

19. $P(3) = 108 - 27 + 3 - 1 = 83$. By synthetic division, we get

3	4	-3	1	-1
		12	27	84
	4	9	28	83

Remainder is $P(3) = 83$.

21. $P(-1/2) = \frac{1}{4} - \frac{1}{4} + 3 + 2 = 5$. By synthetic division, we obtain

-1/2	-8	0	2	0	-6	2
		4	-2	0	0	3
	-8	4	0	0	-6	5

So the remainder is $P(-1/2) = 5$.

23. $\pm\left\{1, \frac{1}{3}, 2, \frac{2}{3}\right\}$

25. $\pm\left\{1, \frac{1}{2}, \frac{1}{3}, \frac{1}{6}, 3, \frac{3}{2}\right\}$

27.

$$\begin{aligned} 2\left(x + \frac{1}{2}\right)(x-3) &= (2x+1)(x-3) \\ &= 2x^2 - 5x - 3 \end{aligned}$$

An equation is $2x^2 - 5x - 3 = 0$.

29. $(x - (3-2i))(x - (3+2i)) =$

$$\begin{aligned} &= ((x-3)+2i)((x-3)-2i) \\ &= (x-3)^2 + 4 \\ &= x^2 - 6x + 13 \end{aligned}$$

An equation is $x^2 - 6x + 13 = 0$.

31. $(x-2)(x-(1-2i))(x-(1+2i)) =$

$$\begin{aligned} &= (x-2)((x-1)+2i)((x-1)-2i) \\ &= (x-2)\left((x-1)^2 + 4\right) \\ &= (x-2)(x^2 - 2x + 5) \\ &= x^3 - 4x^2 + 9x - 10 \end{aligned}$$

Thus, an equation is $x^3 - 4x^2 + 9x - 10 = 0$.

33. $\left(x - (2-\sqrt{3})\right)\left(x - (2+\sqrt{3})\right) =$

$$\begin{aligned} &= \left((x-2)+\sqrt{3}\right)\left((x-2)-\sqrt{3}\right) \\ &= \left((x-2)^2 - 3\right) \\ &= x^2 - 4x + 1 \end{aligned}$$

Thus, an equation is $x^2 - 4x + 1 = 0$.

35. $P(x) = P(-x) = x^8 + x^6 + 2x^2$ has no sign variation. There are 6 imaginary roots and 0 has multiplicity 2.

37. $P(x) = 4x^3 - 3x^2 + 2x - 9$ has 3 sign variations and $P(-x) = -4x^3 - 3x^2 - 2x - 9$ has no sign variation. There are
(a) 3 positive roots, or
(b) 1 positive and 2 imaginary roots.

39. $P(x) = x^3 + 2x^2 + 2x + 1$ has no sign variation and $P(-x) = -x^3 + 2x^2 - 2x + 1$ has 3 sign variations. There are
(a) 3 negative roots, or
(b) 1 negative root and 2 imaginary roots.

41. Best integral bounds: $-4 < x < 3$. One checks that $1, 2$ are not upper bounds and that 3 is a bound.

3	6	5	-50
		18	69
	6	23	19

One checks that $-1, -2, -3$ are not lower bounds and that -4 is a bound since

-4	6	5	-50
		-24	76
	6	-19	26

43. Best integral bounds: $-1 < x < 8$. One checks that $1, 2, 3, 4, 5, 6, 7$ are not upper bounds and that 8 is a bound.

8	2	-15	31	-12
		16	8	312
	2	1	39	300

-1 is a lower bound since

-1	2	-15	31	-12
		-2	17	-48
	2	-17	48	-60

45. Best integral bounds: $-1 < x < 1$

1	12	-4	-3	1
		12	8	5
	12	8	5	6

-1	12	-4	-3	1
		-12	16	-13
	12	-16	13	-12

47. Roots are $1, 2, 3$ since

1	1	-6	11	-6
		1	-5	6
	1	-5	6	0

and $x^2 - 5x + 6 = (x - 3)(x - 2)$.

49. Roots are $\pm i, 1/3, 1/2$ since

1/2	6	-5	7	-5	1
		3	-1	3	-1
	6	-2	6	-2	0

1/3	6	-2	6	-2
		2	0	2
	6	0	6	0

and the zeros of $6x^2 + 6 = 6(x^2 + 1) = 0$ are $\pm i$.

51. Roots are $3, 3 \pm i$ since

3	1	-9	28	-30
		3	-18	30
	1	-6	10	0

and the zeros of $x^2 - 6x + 10 = (x - 3)^2 + 1 = 0$ are $3 \pm i$.

53. Roots are $2, 1 \pm i\sqrt{2}$ since

2	1	-4	7	-6
		2	-4	6
	1	-2	3	0

and the zeros of $x^2 - 2x + 3 = (x - 1)^2 + 2 = 0$ are $1 \pm i\sqrt{2}$.

55. Apply synthetic division to the second factor in $x(2x^3 - 5x^2 - 2x + 2) = 0$.

1/2	2	-5	-2	2
		1	-2	-2
	2	-4	-4	0

By completing the square, the zeros of $2x^2-4x-4 = 2(x^2-2x)-4 = 2(x-1)^2-6 = 0$ are $1\pm\sqrt{3}$. All the roots are $x = 0, 1/2, 1\pm\sqrt{3}$.

57. Symmetric about $x = 3/4$ since $\dfrac{-b}{2a} = \dfrac{3}{4}$.

59. Symmetric about y-axis since $f(-x) = f(x)$.

61. Symmetric about the origin for $f(-x) = -f(x)$.

63. $(-\infty, -2.5) \cup (-2.5, \infty)$

65. $(-\infty, \infty)$

67. Since $f(x) = (x-2)(x+1)$, the x-intercepts are $(2,0), (-1,0)$, y-intercept is $(0,-2)$.

Since $\dfrac{-b}{2a} = \dfrac{1}{2}$, the vertex is $(1/2, -9/4)$.

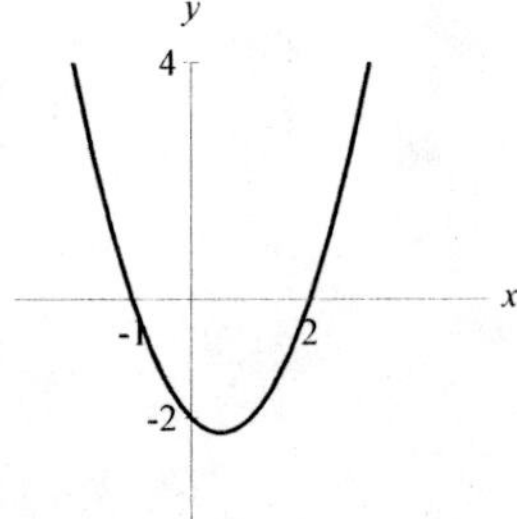

69. Use synthetic division on $f(x) = x^3 - 3x - 2$.

-1	1	0	-3	-2
		-1	1	2
	1	-1	-2	0

Since $x^2 - x - 2 = (x-2)(x+1)$, $f(x) = (x+1)^2(x-2)$. Graph crosses $(2,0)$ but does not cross $(-1,0)$. y-intercept is $(0,-2)$.

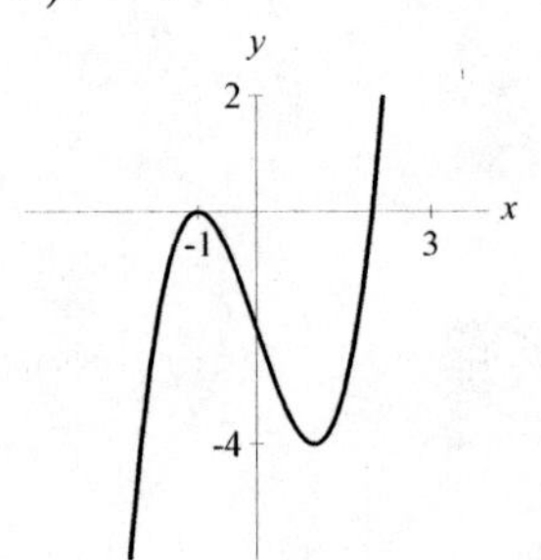

71. Use synthetic division on

$$f(x) = \frac{1}{2}x^3 - \frac{1}{2}x^2 - 2x + 2.$$

1	1/2	-1/2	-2	2
		1/2	0	-2
	1/2	0	-2	0

Since the roots of $\dfrac{1}{2}x^2 - 2 = 0$ are ± 2, the graph crosses x-intercepts $(\pm 2, 0), (1, 0)$, y-intercept is $(0, 2)$.

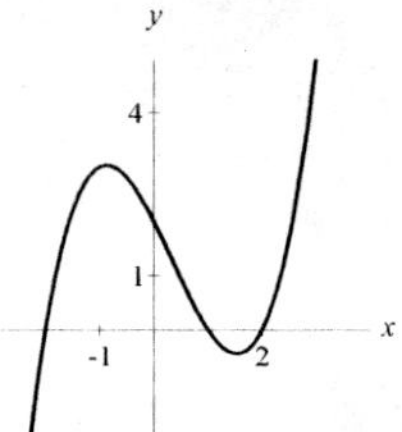

73. Factoring, we get

$$f(x) = \frac{1}{4}(x^4 - 8x^2 + 16) = \frac{1}{4}(x^2 - 4)^2.$$

The graph does not cross through x-intercepts $(\pm 2, 0)$, y-intercept is $(0, 4)$.

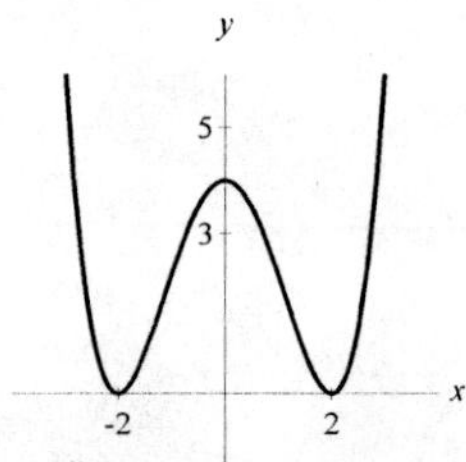

75. $f(x) = \dfrac{2}{x+3}$ has no x-intercept, y-intercept is $(0, 2/3)$, asymptotes are $x = -3$, $y = 0$

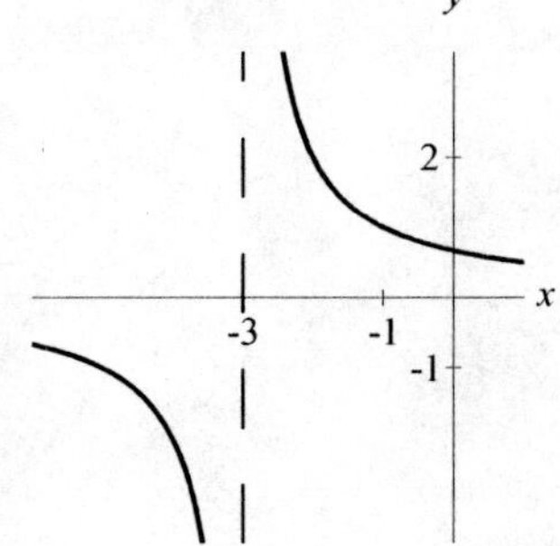

77. $f(x) = \dfrac{2x}{x^2-4}$ has x-intercept $(0,0)$, asymptotes are $x = \pm 2$, $y = 0$. Symmetric about the origin. Graph goes through $(1, -2/3)$, $(3, 6/5)$, and $(-3, -6/5)$.

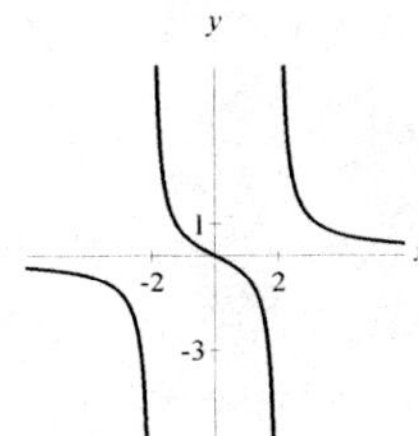

79. $f(x) = \dfrac{(x-1)^2}{x-2}$ has x-intercept $(1,0)$, y-intercept is $(0, -1/2)$, asymptote $x = 2$, and oblique asymptote $y = x$ since

2	1	-2	1
		2	0
	1	0	1

and $f(x) = x + \dfrac{1}{x-2}$.

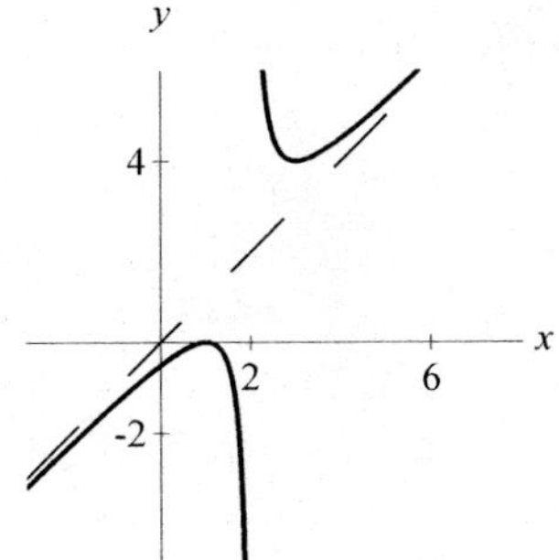

81. $f(x) = \dfrac{2x-1}{2-x}$ has x-intercept $(1/2, 0)$, y-intercept is $(0, -1/2)$, asymptotes are $x = 2$ and $y = -2$, graph goes through $(1,1)$, $(3,-5)$.

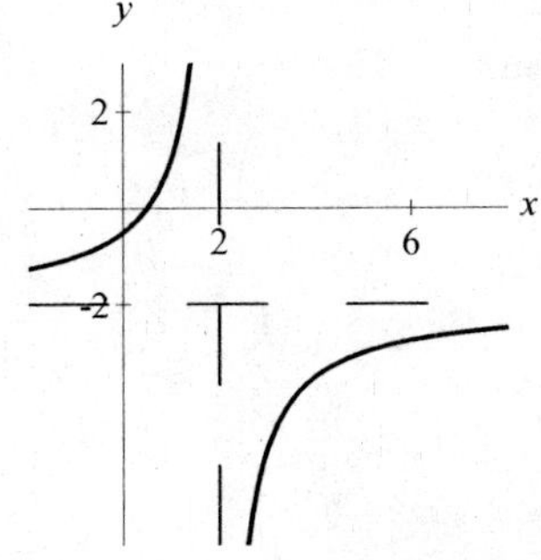

83. Since $f(x) = \dfrac{x^2-4}{x-2} = x + 2$ if $x \neq 2$, x-intercept is $(-2, 0)$, y-intercept is $(0, 2)$, no asymptotes

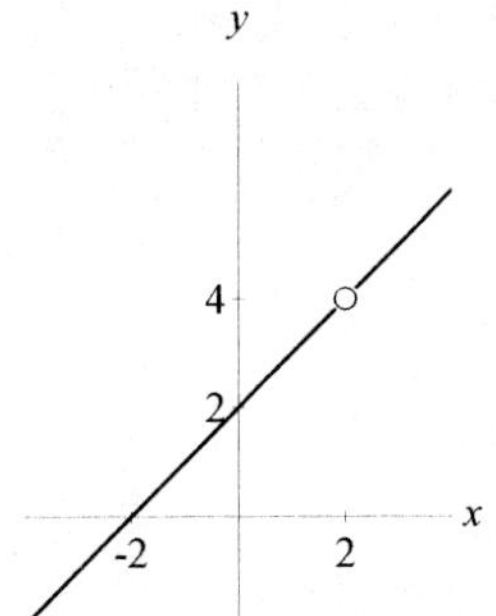

85. Let $f(x) = 8x^2 - 6x + 1 = (4x-1)(2x-1) < 0$. Using the test-point method, we find $f(0) > 0$, $f(3/4) < 0$, and $f(1) > 0$.

$$\begin{array}{ccccccccc} & + & 0 & - & 0 & + & \\ 0 & & \frac{1}{4} & \frac{3}{4} & \frac{1}{2} & & 1 \end{array}$$

The solution is $(1/4, 1/2)$.

87. Let $f(x) = (3-x)(x+5) \geq 0$. Using the test-point method, we find $f(-6) < 0$, $f(0) > 0$, and $f(4) < 0$.

$$\begin{array}{ccccc} - & 0 & + & 0 & - \\ -6 & -5 & 0 & 3 & 4 \end{array}$$

The solution is $[-5, 3]$.

89. Using the Rational Zero Theorem and synthetic division, we obtain

1/2	4	-400	-1	100
		2	-199	-100
	4	-398	-200	0

Note, $4x^2 - 398x - 200 = (4x+2)(x-100)$. Let $f(x) = 4x^3 - 400x^2 - x + 100 =$ $(x - \frac{1}{2})(4x+2)(x-100) \geq 0$.

Apply the test-point method as follows:

$$\begin{array}{cccccccc} - & 0 & + & 0 & - & 0 & + \\ -1 & -\frac{1}{2} & 0 & \frac{1}{2} & 1 & 100 & 101 \end{array}$$

The solution set is $\left[-\frac{1}{2}, \frac{1}{2}\right] \cup [100, \infty)$.

91. Let $R(x) = \dfrac{x+10}{x+2} - 5 = \dfrac{-4x}{x+2} < 0$.
Using the test-point method, we find $R(-3) < 0, R(-1) > 0$, and $R(1) < 0$.

−	U	+	0	−
-3	-2	-1	0	1

The solution is $(-\infty, -2) \cup (0, \infty)$.

93. $R(x) = \dfrac{12-7x}{x^2} + 1 = \dfrac{(x-3)(x-4)}{x^2} > 0$.
Using the test-point method, we get $R(-1) > 0, R(1) > 0, R(3.5) < 0$, and $R(5) > 0$.

+	U	+	0	−	0	+
-1	0	1	3	3.5	4	5

the solution is $(-\infty, 0) \cup (0, 3) \cup (4, \infty)$.

95. Let $R(x) = \dfrac{(x-1)(x-2)}{(x-3)(x-4)}$. We will use the test-point method. Note, $R(0) > 0$, $R(1.5) < 0, R(2.5) > 0$, $R(3.5) < 0$, and $R(5) > 0$.

+	0	−	0	+	U	−	U	+
0	1	1.5	2	2.5	3	3.5	4	5

The solution is $(-\infty, 1] \cup [2, 3) \cup (4, \infty)$.

97. Quotient $x^2 - 3x$, remainder -15

3	1	-6	9	-15
		3	-9	0
	1	-3	0	-15

99. Since $\dfrac{-b}{2a} = \dfrac{-156}{-32} = 4.875$, the maximum height is $-16(4.875)^2 + 156(4.875) = 380.25$ ft.

101. Let x and $36 - x$ be the lengths of the two pieces of wood. The square of the length of the hypotenuse is

$$S = x^2 + (36 - x)^2.$$

Rewriting, we find $S = 2x^2 - 72x + 36^2$. The x-coordinate of the vertex of the parabola is $x = -\frac{b}{2a} = \frac{72}{4} = 18$. The square of the length of the blade is minimized when $x = 18$ in.

103. Since $b = a^2 - 16$, the area A of the triangle satisfies

$$\begin{aligned} A &= \frac{1}{2}(\text{base})(\text{height}) \\ A &= \frac{1}{2}(2a)(-b) \\ A &= -ab \\ A &= -a(a^2 - 16) \\ A &= -a^3 + 16a. \end{aligned}$$

Note, A is maximized when $a = \dfrac{4}{\sqrt{3}} \approx 2.3$ as seen from the given graph of $A = -a^3 + 16a$.

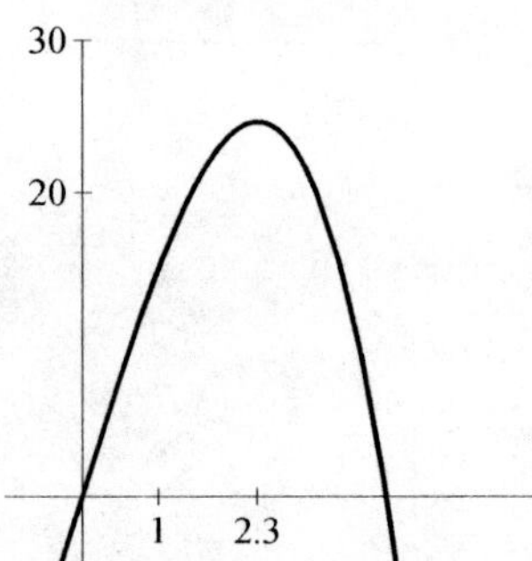

The point $(a, b) = \left(\dfrac{4}{\sqrt{3}}, -\dfrac{32}{3}\right) \approx (2.3, -10.7)$ maximizes the area.

105. $(x - \sqrt{3} - \sqrt{5})(x - \sqrt{3} + \sqrt{5})(x + \sqrt{3} - \sqrt{5}) \cdot (x + \sqrt{3} + \sqrt{5}) = x^4 - 16x^2 + 4 = 0$

Chapter 3 Test

1. Use the method of completing the square.

$$y = 3(x^2 - 4x + 4) + 1 - 12 = 3(x-2)^2 - 11$$

2. $y = 3(x-2)^2 - 11$ has vertex $(2, -11)$, axis of symmetry $x = 2$, y-intercept $(0, 1)$,

the x-intercepts are

$$\left(\frac{6 \pm \sqrt{33}}{3}, 0\right),$$

and the range is $[-11, \infty)$

3. Minimum value is -11 since vertex is $(2, -11)$

4. Quotient $2x^2 - 6x + 14$, remainder -37

-3	2	0	-4	5
		-6	18	-42
	2	-6	14	-37

5. By the Remainder Theorem, remainder is -14.

6. $\pm\left\{1, \frac{1}{3}, 2, \frac{2}{3}, 3, 6\right\}$

7.

$$\begin{aligned}(x+3)(x-4i)(x+4i) &= (x+3)(x^2+16) \\ &= x^3 + 3x^2 + 16x + 48\end{aligned}$$

An equation is $x^3 + 3x^2 + 16x + 48 = 0$.

8. $P(x) = x^3 - 3x^2 + 5x + 7$ has 2 sign variations and $P(-x) = -x^3 - 3x^2 - 5x + 7$ has 1 sign variation. There are
(a) 2 positive roots and 1 negative root, or
(b) 1 negative root and 2 imaginary roots.

9. Since $\frac{-b}{2a} = \frac{-128}{-32} = 4$,

the maximum height is $S(4) = 256$ feet.

10. ± 3

11. $\pm 2, \pm 2i$

12.

2	1	-4	-1	10
		2	-4	-10
	1	-2	-5	0

Since $x^2 - 2x - 5 = (x-1)^2 - 6$, the zeros are $2, 1 \pm \sqrt{6}$.

13. Zeros are $x = \pm i$, each with multiplicity 2 since $f(x) = (x^2+1)^2 = (x-i)^2(x+i)^2$

14. The zeros are 2, 0 with multiplicity 2, and $-3/2$ with mulitiplicity 3 since the equation can be written as $x^2(x-2)(2x+3)^3 = 0$.

15.

1/2	2	-9	14	-5
		1	-4	5
	2	-8	10	0

By completing the square, we find that the roots of $2(x^2 - 4x) + 10 = 2(x-2)^2 + 2 = 0$ are $2 \pm i$. The zeros are $x = 2 \pm i, 1/2$.

16. Parabola $y = 2(x-3)^2 + 1$ with vertex $(3, 1)$

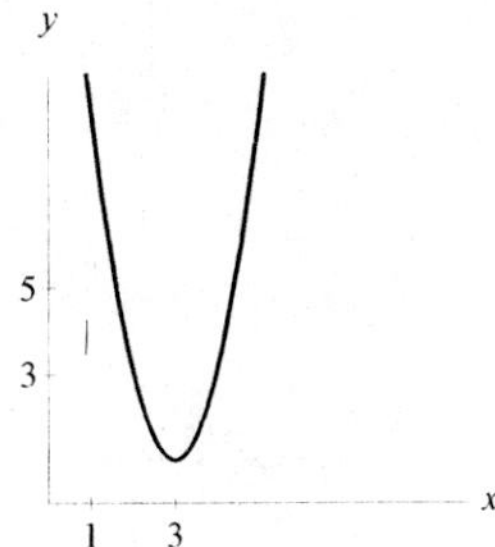

17. $y = (x-2)^2(x+1)$ crosses $(-1, 0)$ but does not cross $(2, 0)$, y-intercept is $(0, 4)$

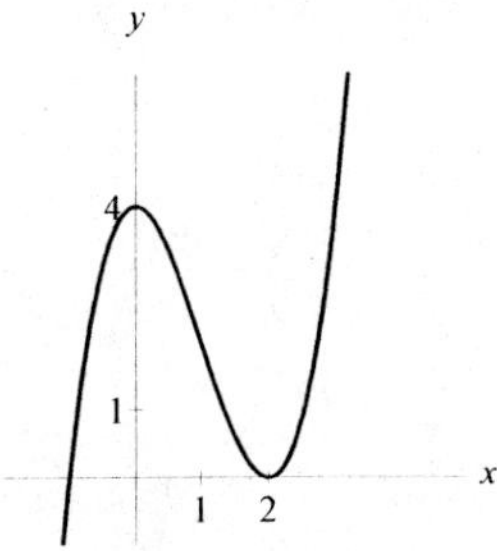

18. $y = x(x-2)(x+2)$ crosses $(0,0), (\pm 2, 0)$, and goes through $(1, -3)$

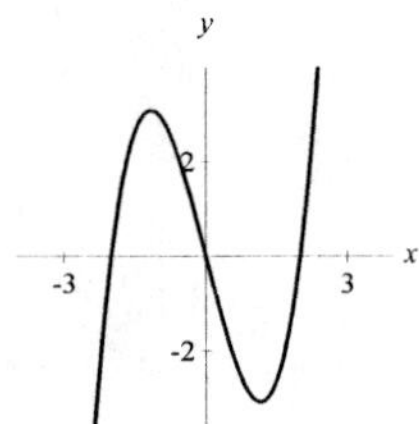

19. $y = \dfrac{1}{x-2}$ has asymptotes $x = 2$, $y = 0$, and goes through $(1, -1), (3, 1), (4, 1/2)$

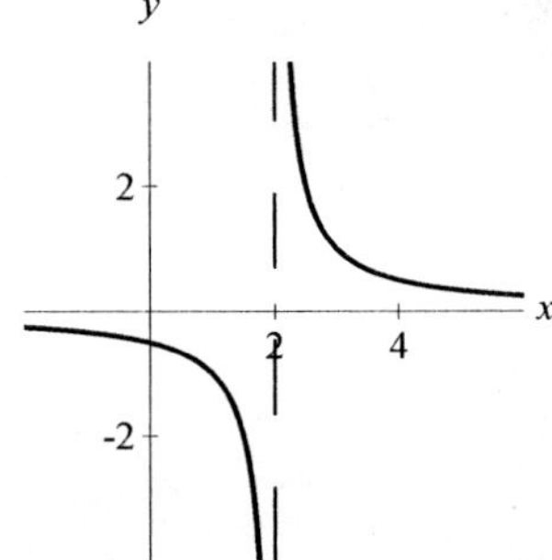

20. $y = \dfrac{2x-3}{x-2}$ has asymptotes $x = 2, y = 2$, x-intercept $(3/2, 0)$, y-intercept $(0, 3/2)$

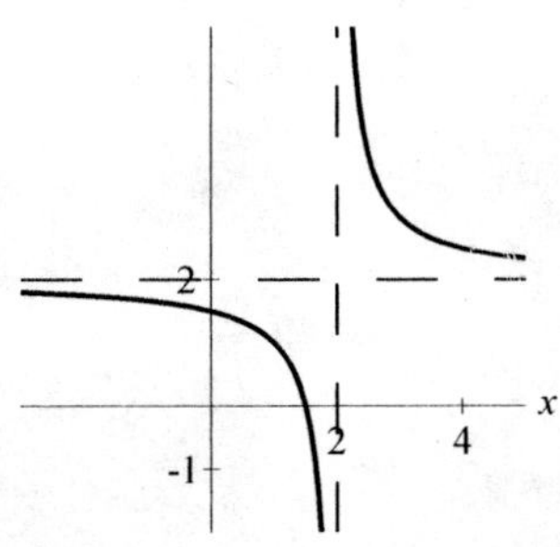

21. $f(x) = x + \dfrac{1}{x}$ has oblique asymptote $y = x$, asymptote $x = 0$, goes through $(1, 2), (-1, -2)$

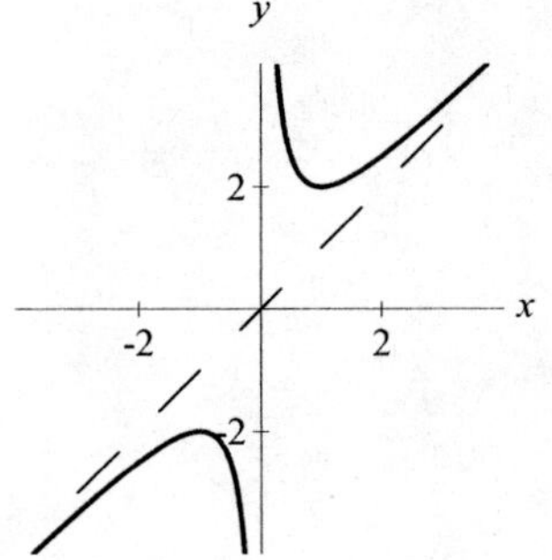

22. $y = \dfrac{4}{x^2 - 4}$ has asymptotes $x = \pm 2, y = 0$, y-intercept $(0, -1)$

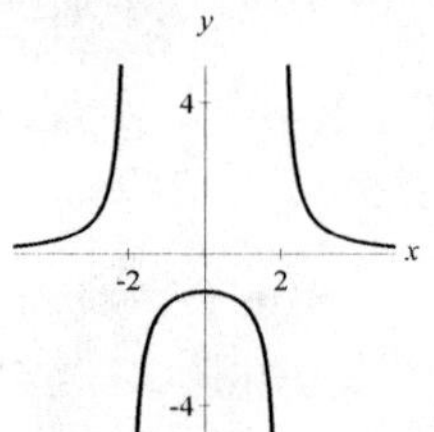

23. Note, $\dfrac{x^2 - 2x + 1}{x - 1} = x - 1$ provided $x \neq 1$; no x-intercept, y-intercept $(0, -1)$

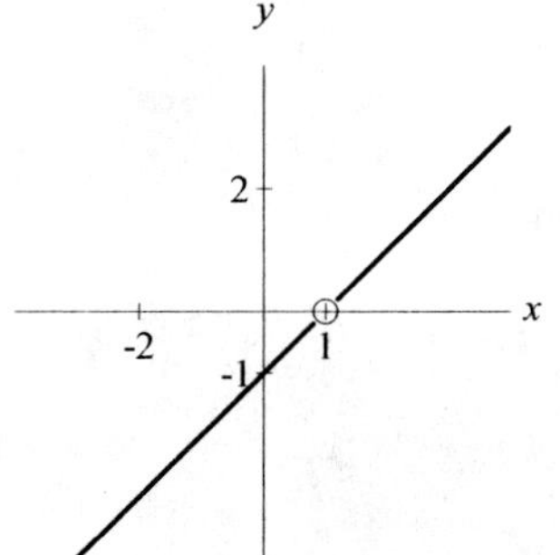

24. Let $f(x) = (x-4)(x+2) < 0$.

Using the test-point method, we get $f(-3) > 0, f(0) < 0, f(5) > 0$.

+	0	−	0	+
-3	-2	0	4	5

The solution set is the interval $(-2, 4)$.

25. Let $f(x) = \dfrac{2x-1}{x-3} > 0$.

Using the test-point method we find

+	0	−	U	+
0	$\frac{1}{2}$	1	3	4

The solution set is $(-\infty, 1/2) \cup (3, \infty)$.

26. Let $f(x) = \dfrac{x+3}{(4-x)(x+1)} \geq 0$.

Using the test-point method we find

+	0	−	U	+	U	−
-4	-3	-2	-1	2	4	5

The solution set is $(-\infty,-3]\cup(-1,4)$.

27. Let $f(x)=x(x^2-7)>0$.

Applying the test-point method, we obtain $f(-4)<0$, $f(-2)>0$, $f(2)<0$, $f(5)>0$.

$$\begin{array}{ccccccccccccc} & - & & 0 & + & 0 & & - & 0 & & + & \\ \leftarrow & & & & & & & & & & & \rightarrow \\ & -4 & & -\sqrt{7} & -2 & 0 & & 2 & \sqrt{7} & & 5 & \end{array}$$

The solution set is $(-\sqrt{7},0)\cup(\sqrt{7},\infty)$.

Tying It All Together

1. Since $2x-3=0$ is equivalent to $2x=3$, the solution is $x=\dfrac{3}{2}$.

2. Factoring, we obtain $2x^2-3x=x(2x-3)=0$.

Thus, the solutions are $x=0,\dfrac{3}{2}$.

3. Using synthetic division, we obtain

$$\begin{array}{r|rrrr} -1 & 1 & 3 & 3 & 1 \\ & & -1 & -2 & -1 \\ \hline & 1 & 2 & 1 & 0 \end{array}$$

Since $x^2+2x+1=(x+1)^2$, we find
$x^3+3x^2+3x+1=(x+1)^3$.
Thus, the only solution is $x=-1$.

4. Multiply both sides of the equation of $\dfrac{x-1}{x-2}=0$ by $x-2$. Then we obtain $x-1=0$.
The solution is $x=1$.

5. If we raise each side of $x^{2/3}=9$ to the power 3, then we obtain $x^2=729$.

Thus, $x=\pm\sqrt{729}$ and the solution set is $\{\pm27\}$.

6. Since $x^2=-1$, the solution set is $\{\pm i\}$.

7. x-intercept $(2,0)$, y-intercept $(0,-4)$, increasing $(-\infty,\infty)$ domain $(-\infty,\infty)$, range $(-\infty,\infty)$, solution set to $f(x)=0$ is $\{2\}$, solution set to $f(x)>0$ is $(2,\infty)$, solution set to $f(x)\le 0$ is $(-\infty,2]$,

8. x-intercepts $(-1,0)$ and $(3,0)$, y-intercept $(0,3)$, increasing $(-\infty,1]$, decreasing $[1,\infty)$, domain $(-\infty,\infty)$, range $(-\infty,4]$, solution set to $f(x)=0$ is $\{-1,3\}$, solution set to $f(x)>0$ is $(-1,3)$, solution set to $f(x)\le 0$ is $(-\infty,-1]\cup[3,\infty)$, vertex $(1,4)$

9. x-intercepts $(-3,0)$ and $(3,0)$, y-intercept $(0,27)$, increasing $(-\infty,-1]$ and $[3,\infty)$, decreasing $[-1,3]$, domain $(-\infty,\infty)$, range $(-\infty,\infty)$, solution set to $f(x)=0$ is $\{-3,3\}$, solution set to $f(x)>0$ is $(-3,3)\cup(3,\infty)$, solution set to $f(x)\le 0$ is $(-\infty,-3]\cup\{3\}$

10. x-intercepts $(-1,0)$ and $(1,0)$, y-intercept $(0,1/4)$, increasing $(-\infty,-2)$ and $(-2,0]$, decreasing $[0,2)$ and $(2,\infty)$, domain $(-\infty,-2)\cup(-2,2)\cup(2,\infty)$, range $(-\infty,\frac{1}{4}]\cup(1,\infty)$, solution set to $f(x)=0$ is $\{-1,1\}$, solution set to $f(x)>0$ is $(-\infty,-2)\cup(-1,1)\cup(2,\infty)$, solution set to $f(x)\le 0$ is $(-2,1]\cup[1,2)$.

Vertical asymptotes $x=-2$, $x=2$, and the horizontal asymptote is $y=1$.

11. The line $y=2x-3$ is increasing and has x-intercept at $x=3/2$. The set of all x's of the part of the line below the x-axis satisfies $2x-3<0$. Thus, the solution set is $\left(-\infty,\dfrac{3}{2}\right)$.

12. The x-intercepts of $y=2x^2-3x$ are at $x=0,\frac{3}{2}$. Since the parabola opens upward, the solution set of $2x^2-3x<0$ is $\left(0,\dfrac{3}{2}\right)$.

13. Since $(x+1)^3=x^3+3x^2+3x+1$, the solution set of $(x+1)^3>0$ is $(-1,\infty)$.

14. Let $f(x)=\dfrac{x-1}{x-2}>0$.

Using the test-point method we find

$$\begin{array}{cccccccc} & + & & 0 & - & & \text{U} & + \\ \leftarrow & & & & & & & \rightarrow \\ & 0 & & 1 & \frac{1}{2} & & 2 & 2 \end{array}$$

The solution set is $(-\infty,1)\cup(2,\infty)$.

15. Let $f(x)=(\sqrt[3]{x}-3)(\sqrt[3]{x}-3)<0$.

Using the test-point method we find

+	0	−	0	+
-30	-27	0	27	30

The solution set is $(-27, 27)$.

16. Since $x^2 < -1$, the solution set is $\emptyset$.

17. Let $y = f(x)$, and interchange x and y.

Then $x = 2y - 3$. Solving for y, we obtain $y = \dfrac{x+3}{2}$. The inverse is

$$f^{-1}(x) = \frac{x+3}{2}.$$

18. Since the graph of $f(x) = 2x^2 - 3x$ does not satisfy the Horizontal Line Test, the function is not one-to-one. Thus, the inverse does not exist.

19. From the solution of number 3, we see that $f(x) = (x+1)^3$. Let $y = f(x)$, and interchange x and y. Then $x = (y+1)^3$. Taking the cube root of both sides, we get $\sqrt[3]{x} = y + 1$. Thus, the inverse is

$$f^{-1}(x) = \sqrt[3]{x} - 1.$$

20. Let $y = f(x)$, and interchange x and y.

Then $x = \dfrac{y-1}{y-2}$. Solving for y, we obtain

$$\begin{aligned} x(y-2) &= y-1 \\ xy - 2x &= y-1 \\ xy - y &= 2x-1 \\ y(x-1) &= 2x-1 \\ y &= \frac{2x-1}{x-1}. \end{aligned}$$

The inverse is

$$f^{-1}(x) = \frac{2x-1}{x-1}.$$

21. Since the graph of $f(x) = x^{2/3} - 9$ does not satisfy the Horizontal Line Test, the function is not one-to-one. Thus, the inverse does not exist.

22. Not invertible since the graph of $f(x) = x^2 + 1$ does not satisfy the horizontal line test.

23. trinomial

24. binomial

25. conjugates

26. dividend, quotient, divisor, remainder

27. greatest common factor

28. prime

29. real part, imaginary part

30. formula

31. ordered pair, interval

32. Cartesian coordinate system

For Thought

1. False, the base of an exponential function is positive. **2.** True

3. True, since $2^{-3} = \frac{1}{8}$.

4. True **5.** True **6.** True **7.** True

8. False, since it is decreasing.

9. True, since $0.25 = 4^{-1}$.

10. True, since $\sqrt[100]{2^{173}} = \left(2^{173}\right)^{1/100}$.

4.1 Exercises

1. algebraic

3. exponential

5. increasing, decreasing

7. range

9. 27

11. $-(2^0) = -1$

13. $\frac{1}{2^3} = \frac{1}{8}$

15. $\left(\frac{2}{1}\right)^4 = 2^4 = 16$

17. $\left(8^{1/3}\right)^2 = 2^2 = 4$

19. $-\left(9^{1/2}\right)^{-3} = -(3)^{-3} = -\frac{1}{3^3} = -\frac{1}{27}$

21. $3^2 = 9$ **23.** $3^{-2} = 1/9$ **25.** $2^{-1} = 1/2$

27. $2^3 = 8$

29. $(1/4)^{-1} = 4$

31. $4^{1/2} = 2$

33. $f(x) = 5^x$ goes through $(-1, 1/5), (0, 1), (1, 5)$, domain is $(-\infty, \infty)$, range is $(0, \infty)$, increasing

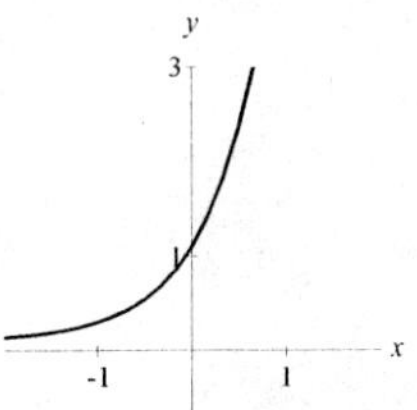

35. $f(x) = 10^{-x}$ goes through $(-1, 10), (0, 1)$, $(1, 1/10)$, domain is $(-\infty, \infty)$, range is $(0, \infty)$, decreasing

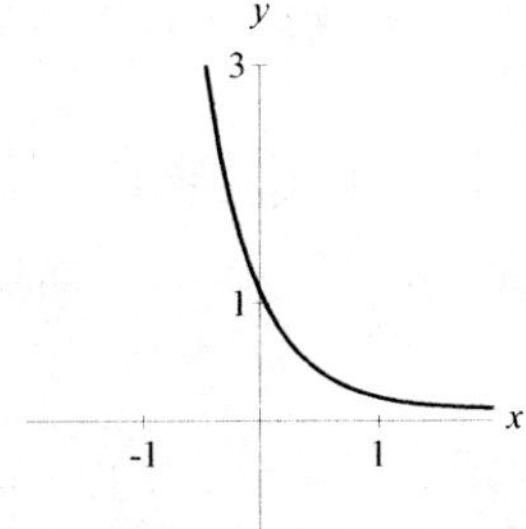

37. $f(x) = (1/4)^x$ goes through $(-1, 4), (0, 1)$, $(1, 1/4)$, domain is $(-\infty, \infty)$, range is $(0, \infty)$, decreasing

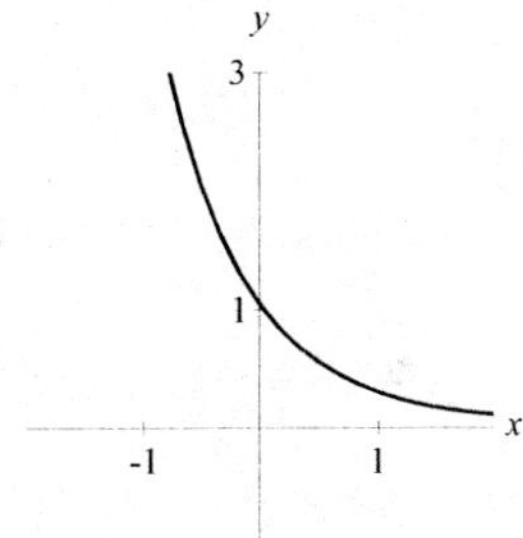

39. From the graph, we find $\lim_{x\to\infty} 3^x = \infty$.

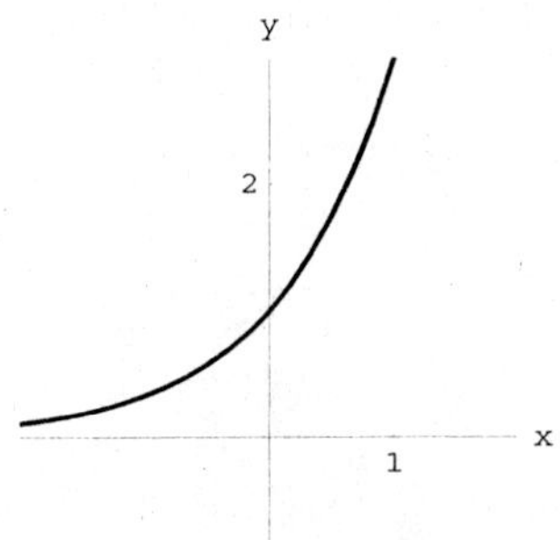

41. Using the graph, we obtain $\lim_{x\to\infty} 5^{-x} = 0$.

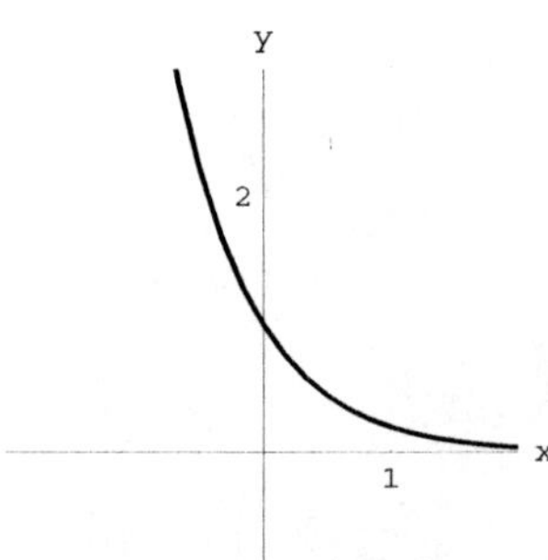

43. We see from the graph that $\lim_{x\to\infty} \left(\frac{1}{3}\right)^x = 0$.

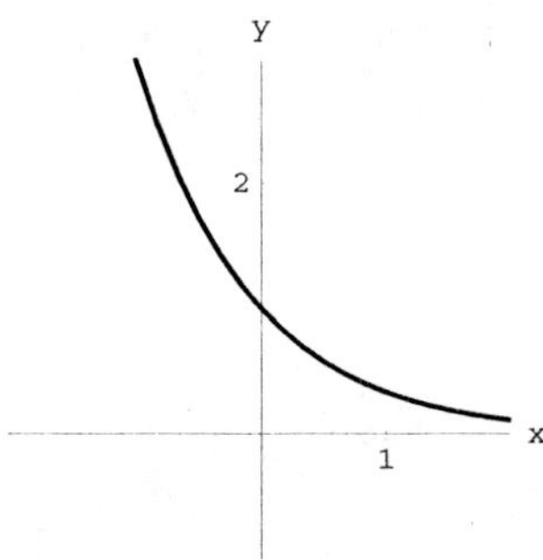

45. Using the graph, we get $\lim_{x\to-\infty} e^{-x} = \infty$.

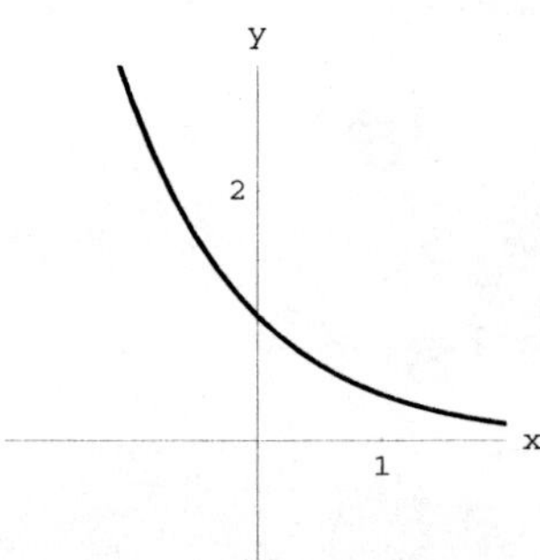

47. Shift $y = 2^x$ down by 3 units; $f(x) = 2^x - 3$ goes through $(-1, -2.5), (0, -2), (2, 1)$, domain $(-\infty, \infty)$, range $(-3, \infty)$, asymptote $y = -3$, increasing

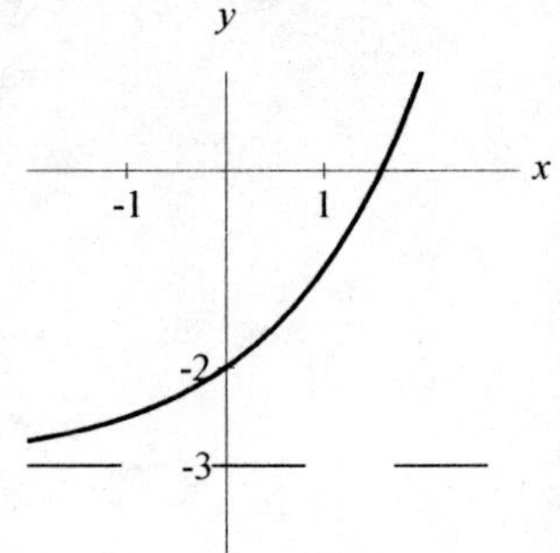

49. Shift $y = 2^x$ to left by 3 units and down by 5 units; $f(x) = 2^{x+3} - 5$ goes through $(-4, -4.5)$,$(-3, -4)$,$(0, 3)$, domain $(-\infty, \infty)$, range $(-5, \infty)$, asymptote $y = -5$, increasing

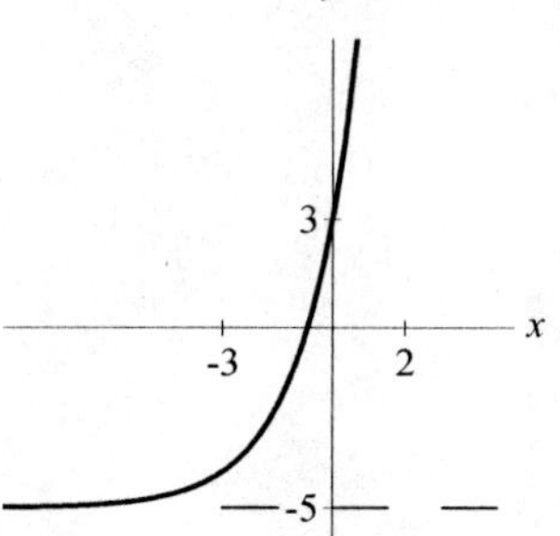

51. Reflect $y = 2^{-x}$ about x-axis, $f(x) = -2^{-x}$ goes through $(-1, -2)$,$(0, -1)$,$(1, -1/2)$, domain $(-\infty, \infty)$, range $(-\infty, 0)$, asymptote $y = 0$, increasing

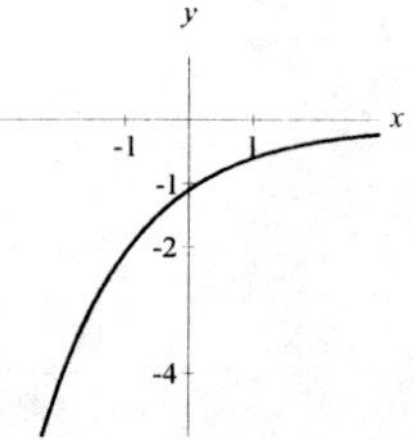

53. Reflect $y = 2^x$ about x-axis and shift up by 1 unit, $f(x) = 1 - 2^x$ goes through $(-1, 0.5)$,$(0, 0)$,$(1, -1)$, domain $(-\infty, \infty)$, range $(-\infty, 1)$, asymptote $y = 1$, decreasing

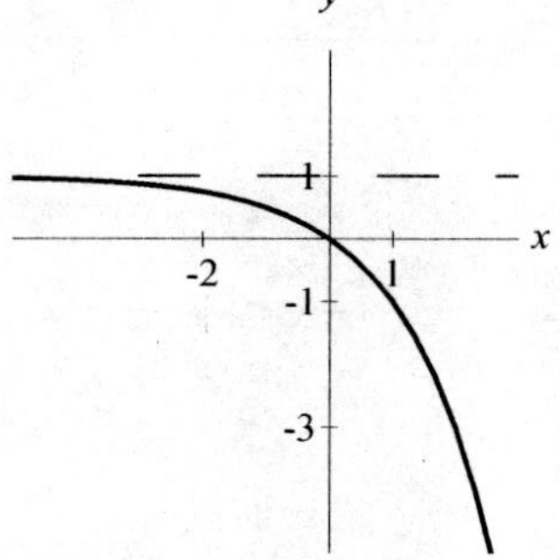

55. Shift $y = 3^x$ to right by 2 and shrink by a factor of 0.5, $f(x) = 0.5 \cdot 3^{x-2}$ goes through $(0, 1/18)$,$(2, 0.5)$,$(3, 1.5)$, domain $(-\infty, \infty)$, range $(0, \infty)$, asymptote $y = 0$, increasing

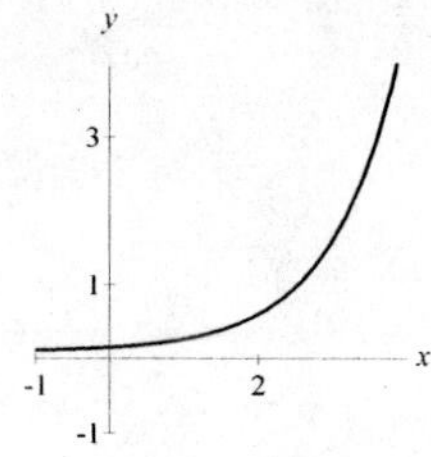

57. Stretch $y = (0.5)^x$ by a factor of 500, $f(x) = 500 \cdot (0.5)^x$ goes through $(0, 500)$, $(1, 250)$, domain $(-\infty, \infty)$, range $(0, \infty)$, asymptote $y = 0$, decreasing

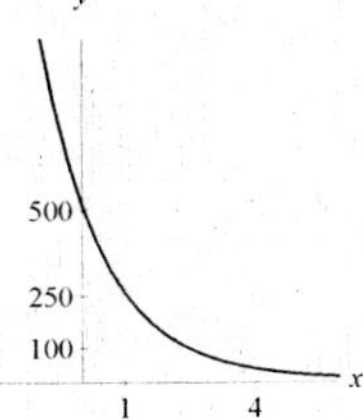

59. $y = 2^{x-5} - 2$

61. $y = -\left(\frac{1}{4}\right)^{x-1} - 2$

63. Since $2^x = 2^6$, solution set is $\{6\}$.

65. $\{-1\}$

67. Multiplying the equation by -1, $3^x = 3^3$ and the solution set is $\{3\}$.

69. $\{-2\}$

71. Since $(2^3)^x = 2^{3x} = 2$, $3x = 1$.

The solution set is $\left\{\frac{1}{3}\right\}$.

73. $\{-2\}$

75. Since $(2^{-1})^x = 2^{-x} = 2^3$, $-x = 3$. The solution set is $\{-3\}$.

77. Since $10^{x-1} = 10^{-2}$, $x - 1 = -2$. The solution set is $\{-1\}$.

79. Since $2^{2x} = 4^x$, the original equation may be written as $4^x = 64$. Then the solution set is $\{3\}$.

81. Since $2^x = 4$, we find $x = 2$.

83. Since $2^x = \frac{1}{2}$, we obtain $x = -1$.

85. Since $\left(\frac{1}{3}\right)^x = 1$, we get $x = 0$.

87. Since $\left(\frac{1}{3}\right)^x = 3^3$, we find $x = -3$.

89. Since $10^x = 1000$, we obtain $x = 3$.

91. Since $10^x = 0.1 = 10^{-1}$, we find $x = -1$.

93. 1 **95.** -1

97. $(2, 9), (1, 3), (-1, 1/3), (-2, 1/9)$

99. $(0, 1), (-2, 25), (-1, 5), (1, 1/5)$

101. $(4, -16), (-2, -1/4), (-1, -1/2), (5, -32)$

103. When interest is compounded n times a year, the amount at the end of 6 years is

$$A(n) = 5000\left(1 + \frac{0.08}{n}\right)^{6n}$$

and the interest earned after 6 years is

$$I(n) = A(n) - 5000.$$

a) If $n = 1$, then $A(1) = \$7934.37$ and $I(1) = \$2934.37$.

b) If $n = 4$, then $A(4) = \$8042.19$ and $I(4) = \$3042.19$.

c) If $n = 12$, then $A(12) = \$8067.51$ and $I(12) = \$3067.51$.

d) If $n = 365$, then $A(365) = \$8079.95$ and $I(365) = \$3079.95$.

105. After t years, a deposit of \$5000 will amount to

$$A(t) = 5000e^{0.03t}.$$

We use 30 days per month, and 365 days for a full year.

a) After 6 years, the amount is

$$A(6) = \$5986.09.$$

b) After 8 years and 3 months, the amount is

$$A\left(8 + \frac{3}{12}\right) = \$6404.10.$$

c) After 5 years, 4 months, and 22 days, the amount is

$$A\left(5 + \frac{4(30) + 22}{365}\right) = \$5877.37.$$

d) After 20 years, 321 days, the amount is

$$A\left(20 + \frac{321}{365}\right) = \$9354.16.$$

107. Assume there are 365 days in a year and 30 days in a month. The present value is

$$3000\left(1+\frac{0.065}{365}\right)^{-(365)(5+120/365)} = \$2121.82.$$

109. $20,000e^{-0.0542(30)} = \$3934.30$

111. a) The interest for first hour is

$$10^6 \cdot e^{0.06\left(\frac{1}{(24)(365)}\right)} - 10^6 = \$6.85.$$

b) The interest for 500th hour is the difference between the amounts in the account at the end of the 500th and 499th hours i.e.

$$10^6 e^{0.06\left(\frac{500}{(24)(365)}\right)} - 10^6 e^{0.06\left(\frac{499}{(24)(365)}\right)} = \$6.87$$

113. If $t = 0$, then $A = 200e^0 = 200$ g.
If $t = 500$, then $A = 200e^{-0.001(500)} \approx 121.3$ g.

115. a) Using a calculator, we find that the exponential regression curve is

$$y = 8.93(1.22)^x$$

b) Yes

c) In 2015 when $t = 25$, the number of subscribers is

$$8.93(1.22)^{25} \approx 1288 \text{ million.}$$

117. $P = 10\left(\frac{1}{2}\right)^n$

119. Let M be the number of members on the nth day.

$$M = 12(3)^{n-1}$$

In July 2013, the world population is about 7 billion. Since $M(19) \approx 4.7$ billion and $M(20) \approx 14$ billion, the membership will exceed the world population in 19 days.

121. When $t = 31$, the number of damaged O-rings is $n = 644e^{-0.15(31)} \approx 6$.

125. $y = 2$, $x = -7$

127. Rewrite $|2x - 5| > 7$ as follows:

$$\begin{array}{ccc} 2x - 5 > 7 & \text{or} & 2x - 5 < -7 \\ 2x > 12 & \text{or} & 2x < -2 \\ x > 6 & \text{or} & x < -1 \end{array}$$

The solution set is $(-\infty, -1) \cup (6, \infty)$.

129. Solve for x:

$$\begin{aligned} 2x - 5 &= y \\ 2x &= y + 5 \\ x &= \frac{y+5}{2} \end{aligned}$$

131. Let x be the rate of the swimmer who after 40 feet passes the other swimmer. Let y be the rate of the other swimmer. If x is the length of the pool, then

$$\frac{40}{x} = \frac{d-40}{y}$$

and

$$\frac{d+45}{x} = \frac{2d-45}{y}.$$

Since we may solve for the ratio y/x from both equations, we find

$$\frac{y}{x} = \frac{d-40}{40} = \frac{2d-45}{d+45}.$$

Solving for d, we obtain $d = 75$ or $d = 0$. Thus, the length of the pool is 75 feet.

For Thought

1. True **2.** False, since $\log_{100}(10) = 1/2$.

3. True **4.** True

5. False, the domain is $(0, \infty)$. **6.** True

7. True

8. False, since $\log_a(0)$ is undefined.

9. True **10.** True

4.2 Exercises

1. logarithmic

3. natural

5. vertical asymptote

7. logarithmic family

9. 6 **11.** -4

13. $\frac{1}{4}$ **15.** -3

17. Since $2^6 = 64$, $\log_2(64) = 6$.

19. Since $3^{-4} = \frac{1}{81}$, $\log_3\left(\frac{1}{81}\right) = -4$.

21. Since $16^{1/4} = 2$, $\log_{16}(2) = \frac{1}{4}$.

23. Since $\left(\frac{1}{5}\right)^{-3} = 125$, $\log_{1/5}(125) = -3$.

25. Since $10^{-1} = 0.1$, $\log(0.1) = -1$

27. Since $10^0 = 1$, $\log(1) = 0$.

29. Since $e^1 = e$, $\ln(e) = 1$.

31. -5

33. $y = \log_3(x)$ goes through $(1/3, -1)$, $(1, 0)$, $(3, 1)$, domain $(0, \infty)$, range $(-\infty, \infty)$

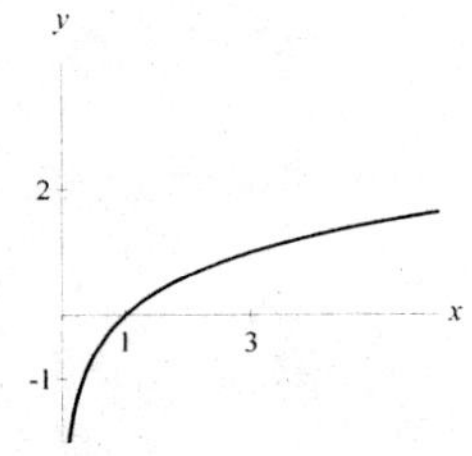

35. $f(x) = \log_5(x)$ goes through $(1/5, -1)$,$(1, 0)$, and $(5, 1)$, domain $(0, \infty)$, range $(-\infty, \infty)$

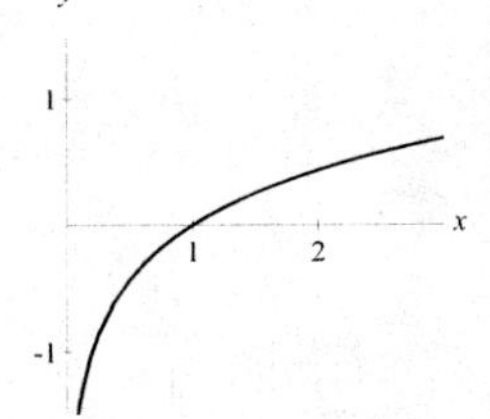

37. $y = \log_{1/2}(x)$ goes through $(2, -1)$,$(1, 0)$, $(1/2, 1)$, domain $(0, \infty)$, range $(-\infty, \infty)$

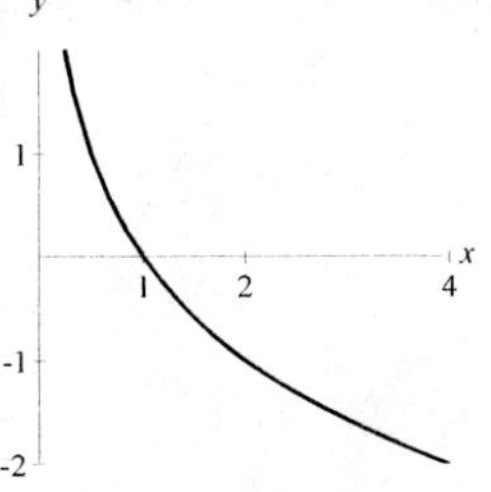

39. $h(x) = \log_{1/5}(x)$ goes through $(5, -1)$, $(1, 0)$,$(1/5, 1)$, domain $(0, \infty)$, range $(-\infty, \infty)$

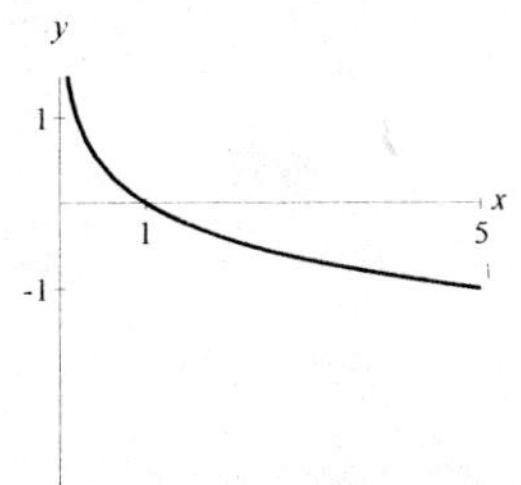

41. $f(x) = \ln(x - 1)$ goes through $\left(1 + \frac{1}{e}, -1\right)$, $(2, 0)$, $(1 + e, 1)$, domain $(1, \infty)$, range $(-\infty, \infty)$

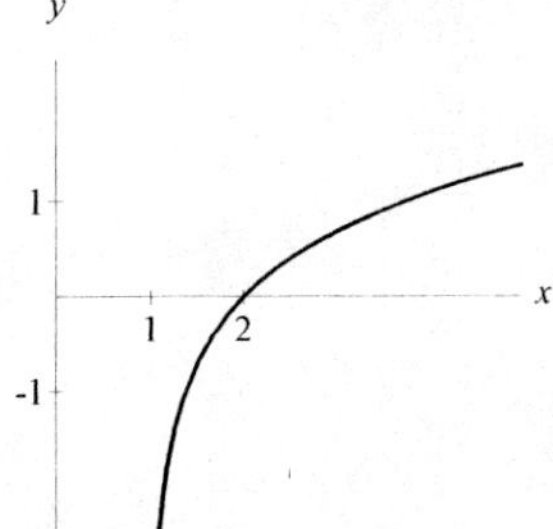

43. $f(x) = -3 + \log(x + 2)$ goes through $(-1.9, -4)$, $(-1, -3)$, $(8, -2)$, domain $(-2, \infty)$, range $(-\infty, \infty)$

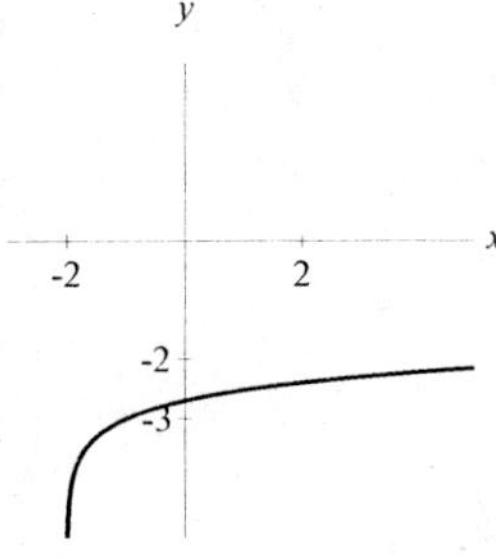

45. $f(x) = -\frac{1}{2}\log(x-1)$ goes through $(1.1, 0.5)$, $(2, 0)$, $(11, -0.5)$, domain $(1, \infty)$, range $(-\infty, \infty)$

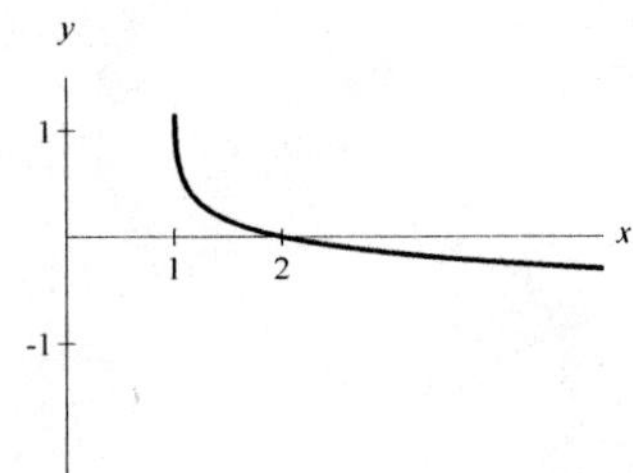

47. From the graph, we find $\lim_{x\to\infty} \log_3 x = \infty$.

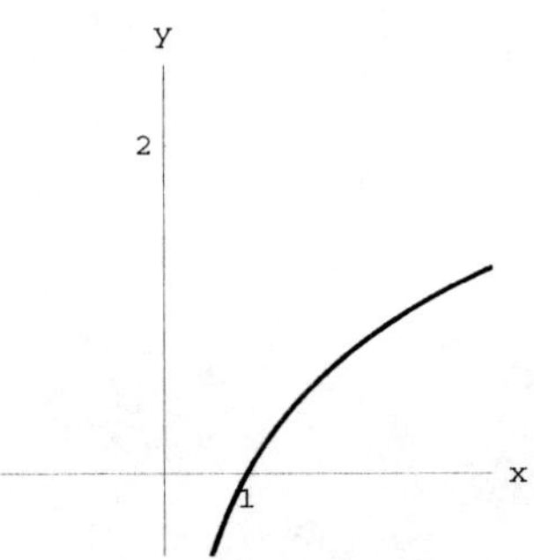

49. Using the graph, we obtain $\lim_{x\to 0^+} \log_{1/2} x = \infty$.

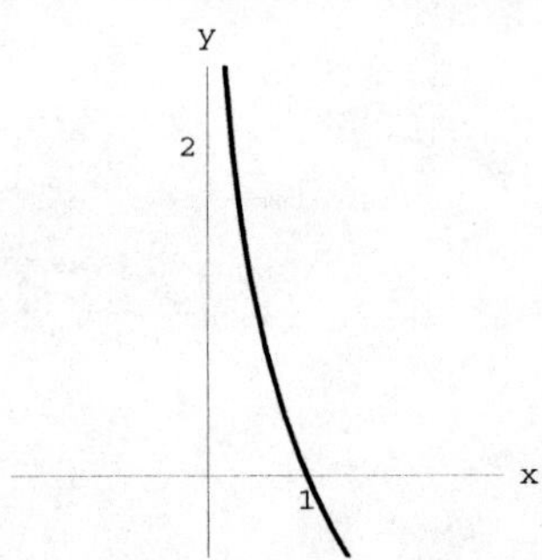

51. From the graph, we get $\lim_{x\to 0^+} \ln x = -\infty$.

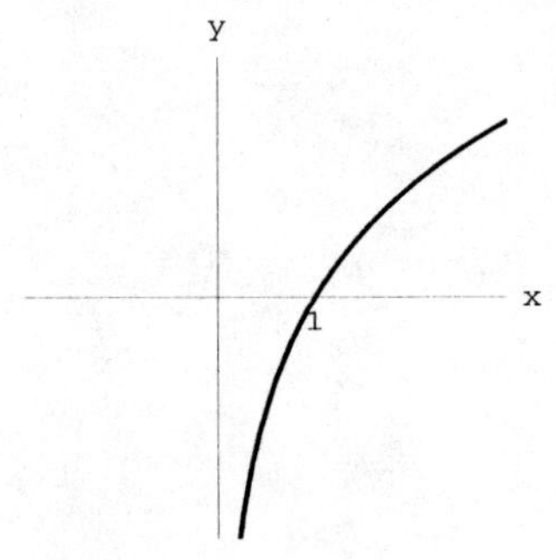

53. From the graph, we find $\lim_{x\to\infty} \log x = \infty$.

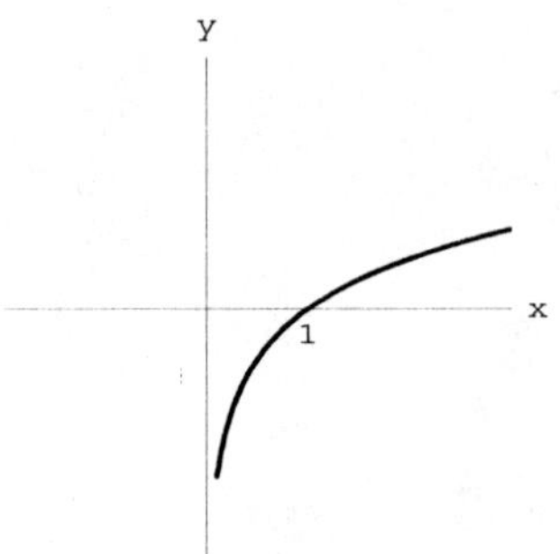

55. $y = \ln(x-3) - 4$

57. $y = -\log_2(x-5) - 1$

59. $2^5 = 32$ **61.** $5^y = x$ **63.** $10^z = 1000$

65. $e^x = 5$

67. $a^m = x$

69. $\log_5(125) = 3$

71. $\ln(y) = 3$

73. $\log(y) = m$

75. $\log_a(y) = z$

77. $\log_a(n) = x - 1$

79. $f^{-1}(x) = \log_2(x)$

81. $f^{-1}(x) = 7^x$

83. Replace $f(x)$ by y, interchange x and y, solve for y, and replace y by $f^{-1}(x)$.

$$\begin{aligned} y &= \ln(x-1) \\ x &= \ln(y-1) \\ e^x &= y - 1 \\ y = f^{-1}(x) &= e^x + 1 \end{aligned}$$

85. Replace $f(x)$ by y, interchange x and y, solve for y, and replace y by $f^{-1}(x)$.

$$\begin{aligned} y &= 3^{x+2} \\ x &= 3^{y+2} \\ y + 2 &= \log_3(x) \\ y = f^{-1}(x) &= \log_3(x) - 2 \end{aligned}$$

87. Replace $f(x)$ by y, interchange x and y, solve for y, and replace y by $f^{-1}(x)$.

$$\begin{aligned} x &= \frac{1}{2}10^{y-1}+5 \\ x-5 &= \frac{1}{2}10^{y-1} \\ 2x-10 &= 10^{y-1} \\ \log(2x-10) &= y-1 \\ \log(2x-10)+1 &= y \\ f^{-1}(x) &= \log(2x-10)+1 \end{aligned}$$

89. Since $2^8 = x$, the solution is $x = 256$.

91. Since $3^{1/2} = x$, the solution is $x = \sqrt{3}$.

93. Since $x^2 = 16$ and the base of a logarithm is positive, $x = 4$.

95. By using the definition of the logarithm, we get $x = \log_3(77)$.

97. Since $y = \ln(x)$ is one-to-one, $x - 3 = 2x - 9$. The solution is $x = 6$.

99. Since $x^2 = 18$, we obtain $x = \sqrt{18} = 3\sqrt{2}$. The base of a logarithm is positive.

101. Since $x + 1 = \log_3(7)$, $x = \log_3(7) - 1$.

103. Since $y = \log(x)$ is one-to-one, we get

$$\begin{aligned} x &= 6-x^2 \\ x^2+x-6 &= 0 \\ (x+3)(x-2) &= 0 \\ x &= -3, 2. \end{aligned}$$

But $\log(-3)$ is undefined, so the solution is $x = 2$.

105. Since $x^{-2/3} = \dfrac{1}{9}$, we find

$$x = \pm\left(\frac{1}{9}\right)^{-3/2} = \pm\left(\frac{1}{3}\right)^{-3} = \pm 27.$$

The solution is $x = 27$ since a negative number cannot be a base.

107. Since $(2^2)^{2x-1} = 2^{4x-2} = 2^{-1}$, we get $4x - 2 = -1$. The solution is $x = \dfrac{1}{4}$.

109. Since $32^x = 64$, we get $2^{5x} = 2^6$. Thus, the solution is $x = \dfrac{6}{5}$.

111. Since $\log_3(\log_4(x)) = 1$, we obtain $\log_4 x = 3$. Then $x = 4^3 = 64$.

113. $x = \log 25 \approx 1.3979$

115. Solving for x, we obtain

$$\begin{aligned} e^{2x} &= 3 \\ 2x &= \ln(3) \\ x &= \frac{1}{2}\ln(3) \\ x &\approx 0.5493. \end{aligned}$$

117. Solving for x, we get

$$\begin{aligned} e^{x} &= \frac{4}{5} \\ x &= \ln\left(\frac{4}{5}\right) \\ x &\approx -0.2231. \end{aligned}$$

119. Solving for x, we find

$$\begin{aligned} \frac{1}{10^x} &= 2 \\ \frac{1}{2} &= 10^x \\ x &= \log\left(\frac{1}{2}\right) \\ x &\approx -0.3010. \end{aligned}$$

121. Solving for the year t when \$10 grows to \$20, we find

$$\begin{aligned} 20 &= 10e^{rt} \\ 2 &= e^{rt} \\ \ln 2 &= rt \\ \frac{\ln 2}{r} &= t. \end{aligned}$$

a) If $r = 2\%$, then $t = \dfrac{\ln 2}{0.02} \approx 34.7$ years.

b) If $r = 4\%$, then $t = \dfrac{\ln 2}{0.04} \approx 17.3$ years.

c) If $r = 8\%$, then $t = \dfrac{\ln 2}{0.08} \approx 8.7$ years.

d) If $r = 16\%$, then $t = \dfrac{\ln 2}{0.16} \approx 4.3$ years.

123. Solving for the annual percentage rate r when \$10 becomes \$30, we find

$$\begin{aligned} 30 &= 10e^{rt} \\ 3 &= e^{rt} \\ \ln 3 &= rt \\ \frac{\ln 3}{t} &= r. \end{aligned}$$

a) If $t = 5$ years, then $r = \dfrac{\ln 3}{5} \approx 0.2197 \approx$ 22%.

b) If $t = 10$ years, then $r = \dfrac{\ln 3}{10} \approx 0.10986 \approx$ 11%.

c) If $t = 20$ years, then $r = \dfrac{\ln 3}{20} \approx 0.0549 \approx$ 5.5%.

d) If $t = 40$ years, then $r = \dfrac{\ln 3}{40} \approx 0.027465 \approx$ 2.7%.

125. Let t be the number of years.

$$\begin{aligned} 1000 \cdot e^{0.14t} &= 10^6 \\ e^{0.14t} &= 1000 \\ 0.14t &= \ln(1000) \\ t &\approx 49.341 \text{ years} \end{aligned}$$

Note, $0.341(365) \approx 125$.
It will take 49 years and 125 days.

127. Since $e^{rt} = A/P$, $rt = \ln(A/P)$ and

$$r = \frac{\ln(A/P)}{t}.$$

Thus, \$1000 will double in 3 years if the rate is $r = \dfrac{\ln(2000/1000)}{3} \approx 0.231$ or 23.1%.

129.

(a) Let t be the number of years.

$$\begin{aligned} P \cdot e^{0.1t} &= 2P \\ e^{0.1t} &= 2 \\ 0.1t &= \ln(2) \\ t = \frac{\ln(2)}{0.1} &\approx 6.9 \end{aligned}$$

An investment at 10% doubles every 6.9 years.

(b) If t is the number of years it takes before an investment doubles, then

$$\begin{aligned} P \cdot e^{rt} &= 2P \\ e^{rt} &= 2 \\ rt &= \ln(2) \\ t &= \frac{\ln(2)}{r} \\ t &\approx \frac{0.70}{r}. \end{aligned}$$

In particular, if $r = 0.07$ then

$$t \approx \frac{0.70}{0.07} = 10 \text{ years.}$$

That is, at 10%, an investment will double in about 10 years.

131. Let r be the interest rate.

$$\begin{aligned} 4,000 \cdot e^{200r} &= 4,500,000 \\ e^{200r} &= 1,125 \\ 200r &= \ln(1,125) \\ r = \frac{\ln(1,125)}{200} &\approx 0.035 \end{aligned}$$

The rate is 3.5% .

133. Let t be the number of years.

$$\begin{aligned} F_o \cdot e^{-0.052t} &= 0.6F_o \\ e^{-0.052t} &= 0.6 \\ -0.052t &= \ln(0.6) \\ t = \frac{\ln(0.6)}{-0.052} &\approx 9.8 \end{aligned}$$

Only 60% of the present forest will remain after 9.8 years.

135. Let r be the annual rate from 1950 to 1987.

$$\begin{aligned} 2.5 \cdot e^{37r} &= 5 \\ e^{37r} &= 2 \\ 37r &= \ln(2) \\ r = \frac{\ln(2)}{37} &\approx 0.0187 \end{aligned}$$

The annual rate is 1.87%.

If the annual rate is 1.63% and the initial population is 5 billion in 1987, the world population in year 2010 will be

$$5 \cdot e^{0.0163(23)} \approx 7.3 \text{ billion.}$$

137.

(a) $30e^{1.2(7)} \approx 133,412$ acres

(b) Solve for t:

$$\begin{aligned} 30e^{1.2t} &= 53,480,960 \\ 1.2t &= \ln\left(\frac{53,480,960}{30}\right) \\ t &\approx 11.9 \end{aligned}$$

It will take about 12 days.

139.

a) The function is given by

$$\begin{aligned} p - 100 &= \frac{100-10}{2-4}(x-2) \\ p - 100 &= -45(x-2) \\ p &= -45x + 90 + 100 \\ p &= -45x + 190 \\ p &= -45\log(I) + 190. \end{aligned}$$

b) $p = -45\log(100,000) + 190 = -35\%$ or 0% of the population is expected to be without safe drinking water, i.e., everyone is expected to have safe water.

141. $pH = -\log\left(10^{-4.1}\right) = 4.1$

143. $pH = -\log\left(10^{-3.7}\right) = 3.7$

145. By substituting $x = 1$, we find a formula for c.

$$\begin{aligned} a \cdot b^x &= a \cdot e^{cx} \\ b^x &= e^{cx} \\ b &= e^c \\ c &= \ln b. \end{aligned}$$

By using this formula, we find

$$y = 500(1.036)^x = 500e^{\ln(1.036)x}$$

and the continuous growth rate is

$$\ln(1.036) \cdot 100 \approx 3.54\%.$$

149. Domain $(-\infty, \infty)$, range $(-\infty, 7)$

151. $(8 \times 10^{-27})(25 \times 10^6) = 200 \times 10^{-21} = 2 \times 10^{-19}$

153. Rewrite the equation:

$$\begin{aligned} x(x^2 - 4x + 13) &= 0 \\ x((x-2)^2 + 9) &= 0 \end{aligned}$$

Then $x = 0$ or $x - 2 = \pm 3i$.

The solution set is $\{0, 2 \pm 3i\}$.

155. First, \$59 is not a possible total, i.e., for all whole numbers x and y we have

$$7x + 11y \neq 59.$$

But for each $60 \leq n \leq 69$, there are whole numbers x and y satisfying

$$n = 7x + 11y.$$

Suppose

$$m = 7a + 11b \geq 69$$

and a, b are whole numbers. Since

$$m + 1 = 7(a-3) + 11(b+2)$$

we see that $m+1$ is a possible total if $a - 3 \geq 0$.

However, if $a - 3 < 0$ then a is either 0, 1, or 2 for $a \geq 0$. Since $m - 7a = 11b$ and $m \geq 69$, we find

$$\begin{aligned} b - 5 &= \frac{m - 7a - 55}{11} \\ &\geq \frac{69 - 7a - 55}{11} \\ &= \frac{14 - 7a}{11} \\ &\geq 0 \end{aligned}$$

for $a = 0, 1, 2$. In any case, $m + 1$ is a possible total since

$$m + 1 = 7(a+8) + 11(b-5).$$

Since $m \geq 69$, all whole numbers greater than 59 is a possible total. Hence, 59 is the largest integer that is not a possible total.

For Thought

1. False, since $\log(8)-\log(3)=\log(8/3)\neq\dfrac{\log(8)}{\log(3)}$.

2. True, since $\ln(3^{1/2})=\dfrac{1}{2}\cdot\ln(3)=\dfrac{\ln(3)}{2}$.

3. True, since $\dfrac{\log_{19}(8)}{\log_{19}(2)}=\log_2(8)=3=\log_3(27)$.

4. True, because of the base-change formula.

5. False, $\ln x$ is defined for $x>0$.

6. False, since $\log(x)-\log(2)=\log(x/2)$.

7. False, since the solution of the first equation is $x=-2$ and the second equation is not defined when $x=-2$.

8. True **9.** False, since x can be negative.

10. False, since a can be negative and so $\ln(a)$ will not be a real number.

4.3 Exercises

1. sum

3. power

5. $\sqrt{y}$ **7.** $y+1$ **9.** 999

11. $\log(15)$

13. $\log_2((x-1)x)=\log_2(x^2-x)$

15. $\log_4(6)$ **17.** $\ln\left(\dfrac{x^8}{x^3}\right)=\ln(x^5)$

19. $\log_2(3x)=\log_2(3)+\log_2(x)$

21. $\log\left(\dfrac{x}{2}\right)=\log(x)-\log(2)$

23. $\log((x-1)(x+1))=\log(x-1)+\log(x+1)$

25. $\ln\left(\dfrac{x-1}{x}\right)=\ln(x-1)-\ln(x)$

27. $\log_a(5^3)=3\log_a(5)$

29. $\log_a(5^{1/2})=\dfrac{1}{2}\cdot\log_a(5)$

31. $\log_a(5^{-1})=-\log_a(5)$

33. $\log_a(2)+\log_a(5)$

35. $\log_a(5/2)=\log_a(5)-\log_a(2)$

37. $\log_a(\sqrt{2^2\cdot 5})=\dfrac{1}{2}(2\log_a(2)+\log_a(5))=$ $\log_a(2)+\dfrac{1}{2}\log_a(5)$

39. $\log_a(4)-\log_a(25)=\log_a(2^2)-\log_a(5^2)=$ $2\log_a(2)-2\log_a(5)$

41. $\log_3(5)+\log_3(x)$

43. $\log_2(5)-\log_2(2y)=\log_2(5)-\log_2(2)-\log_2(y)$

45. $\log(3)+\dfrac{1}{2}\log(x)$

47. $\log(3)+(x-1)\log(2)$

49. $\dfrac{1}{3}\cdot\ln(xy)-\dfrac{4}{3}\ln(t)=\dfrac{1}{3}\cdot\ln(x)+\dfrac{1}{3}\cdot\ln(y)-\dfrac{4}{3}\ln(t)$

51. $\ln(6\sqrt{x-1})-\ln(5x^3)=$ $\ln(6)+\dfrac{1}{2}\cdot\ln(x-1)-\ln(5)-3\cdot\ln(x)$

53. $\log_2(5x^3)$

55. $\log_7(x^5)-\log_7(x^8)=\log(x^5/x^8)=\log_7(x^{-3})$

57. $\log(2xy/z)$

59. $\log\left(\dfrac{\sqrt{x}}{y}\right)+\log\left(\dfrac{z}{\sqrt[3]{w}}\right)=\log\left(\dfrac{z\sqrt{x}}{y\sqrt[3]{w}}\right)$

61. $\log_4(x^6)+\log_4(x^{12})+\log_4(x^2)=\log_4(x^{20})$

63. Since $2^x=9$, we get $x=\dfrac{\log 9}{\log 2}\approx 3.1699$.

65. Since $0.56^x=8$, we get $x=\dfrac{\log 8}{\log 0.56}\approx -3.5864$.

67. Since $1.06^x=2$, we get $x=\dfrac{\log 2}{\log 1.06}\approx 11.8957$.

69. Since $0.73^x=0.5$, we get $x=\dfrac{\log 0.5}{\log 0.73}\approx 2.2025$.

71. $\dfrac{\ln(9)}{\ln(4)}\approx\dfrac{2.1972246}{1.3862944}\approx 1.5850$

73. $\frac{\ln(2.3)}{\ln(9.1)} \approx \frac{0.8329091}{2.2082744} \approx 0.3772$

75. $\frac{\ln(12)}{\ln(1/2)} \approx -3.5850$

77. Since $4t = \log_{1.02}(3) = \frac{\ln(3)}{\ln(1.02)}$,

we find $t = \frac{\ln(3)}{4 \cdot \ln(1.02)} \approx 13.8695$.

79. Since $365t = \log_{1.0001}(3.5) = \frac{\ln(3.5)}{\ln(1.0001)}$,

we get $t = \frac{\ln(3.5)}{365 \cdot \ln(1.0001)} \approx 34.3240$.

81. $1 + r = \sqrt[3]{2.3}$, so $r = \sqrt[3]{2.3} - 1 \approx 0.3200$

83.

$$\begin{aligned} \left(1 + \frac{r}{12}\right)^{360} &= 4.2 \\ 1 + \frac{r}{12} &= \pm \sqrt[360]{4.2} \\ r &= 12\left(\pm \sqrt[360]{4.2} - 1\right) \\ r &\approx 0.0479, -24.0479 \end{aligned}$$

85. Since $x^5 = 33.4$, we get $x = \sqrt[5]{33.4} \approx 2.0172$.

87. Since $x^{-1.3} = 0.546$, we have $x = 0.546^{1/(-1.3)}$ or $x \approx 1.5928$.

89. Let t be the number of years.

$$\begin{aligned} 800\left(1 + \frac{0.08}{365}\right)^{365t} &= 2000 \\ \left(1 + \frac{0.08}{365}\right)^{365t} &= 2.5 \\ (1.0002192)^{365t} &\approx 2.5 \\ 365t &\approx \log_{1.0002192}(2.5) \\ t &\approx \frac{1}{365} \cdot \frac{\ln(2.5)}{\ln(1.0002192)} \\ t &\approx 11.454889 \\ t &\approx 11 \text{ years, } 166 \text{ days} \end{aligned}$$

91. Let t be the number of years.

$$\begin{aligned} W\left(1 + \frac{0.1}{4}\right)^{4t} &= 3W \\ (1.025)^{4t} &= 3 \\ 4t &\approx \log_{1.025}(3) \\ t &\approx \frac{1}{4} \cdot \frac{\ln(3)}{\ln(1.025)} \\ t &\approx 11.123 \text{ years} \\ t &\approx 11.123(4) \approx 44 \text{ quarters} \end{aligned}$$

93. Let t be the number of years.

$$\begin{aligned} 500\,(1 + r)^{25} &= 2000 \\ (1 + r)^{25} &= 4 \\ 1 + r &\approx \sqrt[25]{4} \\ r &= \sqrt[25]{4} - 1 \\ r &\approx 0.057 \text{ or } 5.7\% \end{aligned}$$

95. Let t be the number of years.

$$\begin{aligned} 4000\,(1 + r)^{200} &= 4.5 \times 10^6 \\ (1 + r)^{200} &= 1125 \\ r &= \sqrt[200]{1125} - 1 \\ r &\approx 0.035752 \text{ or } 3.58\% \end{aligned}$$

97. Let r be the annual growth rate.

$$\begin{aligned} 1995\,(1 + r)^{42} &= 11,750 \\ r &= \left(\frac{11,750}{1995}\right)^{1/42} - 1 \\ r &\approx 0.043 \\ r &\approx 4.3\% \end{aligned}$$

99. The Richter scale rating is

$$\log(I) - \log(I_o) = \log\left(\frac{I}{I_o}\right).$$

When $I = 1000 \cdot I_o$, the Richter scale rating is

$$\log\left(\frac{1000 \cdot I_o}{I_o}\right) = \log(1000) = 3.$$

101. $t = \frac{1}{r}\ln(P/P_o) = \frac{1}{r}\ln(P) - \frac{1}{r}\ln(P_o)$

103.

(a) p decreases as n increases

(b) Solving for n, one can take the logarithm of both sides and to note that $y = \log(x)$ is an increasing function.

$$\begin{aligned}
\left(\frac{7,059,051}{7,059,052}\right)^n &> \frac{1}{2} \\
\log\left(\left(\frac{7,059,051}{7,059,052}\right)^n\right) &> \log\left(\frac{1}{2}\right) \\
n\log\left(\frac{7,059,051}{7,059,052}\right) &> \log\left(\frac{1}{2}\right) \\
n &< \frac{\log(1/2)}{\log(7,059,051/7,059,052)} \\
n &< 4,892,962
\end{aligned}$$

If at most 4,892,961 tickets are purchsed, then the probability of a rollover is greater than 50%.

105. We note that $MR(x) = R(x+1) - R(x) = 500 \cdot \log(x+2) - 500 \cdot \log(x+1) =$

$$500 \cdot \log\left(\frac{x+2}{x+1}\right) = \log\left(\left(\frac{x+2}{x+1}\right)^{500}\right).$$

Hence, as $x \to \infty$, then $\dfrac{x+2}{x+1} \to 1$ and $MR(x) \to 0$.

107.

a) Let x be the number of years since 1990 and let y be the number of computers per 1000 people. With the aid of a calculator, we find that an exponential regression curve is

$$y = 217.9(1.084)^x$$

b) $y = 217.9(e^{\ln 1.084})^x \approx 217.9e^{0.0807x}$

c) From part b), the continuous growth is 8.07%

d) Substitute $y = 1500$ into the equation in part a):

$$\begin{aligned}
1500 &= 217.9(1.084)^x \\
\frac{1500}{217.9} &= 1.084^x \\
\ln\left(\frac{1500}{217.9}\right) &= x\ln 1.084 \\
23.9 &\approx x
\end{aligned}$$

In 2014 (= 1990 + 24), there will be 1500 computers per 1000 people.

e) No, the data does not look exponential, rather it looks linear.

109. Note, $x^2 \geq 0$ for all real numbers x. Then the domain of $y = \log(x^2)$ is $(-\infty, 0) \cup (0, \infty)$ and the domain of $y = 2\log(x)$ is $(0, \infty)$. Thus, these two functions are not the same. The graph of $y = \log(x^2)$ is shown below

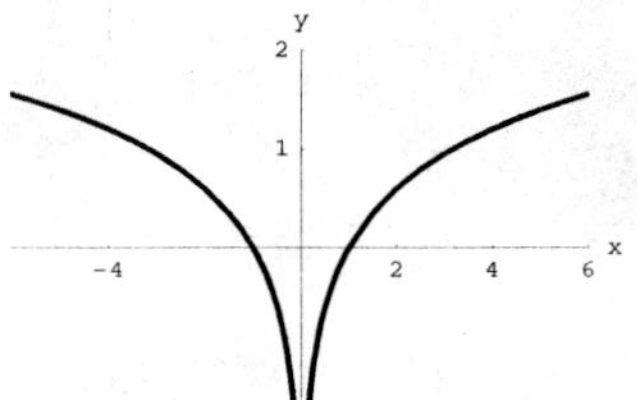

and the graph of $y = 2\log(x)$ is given next.

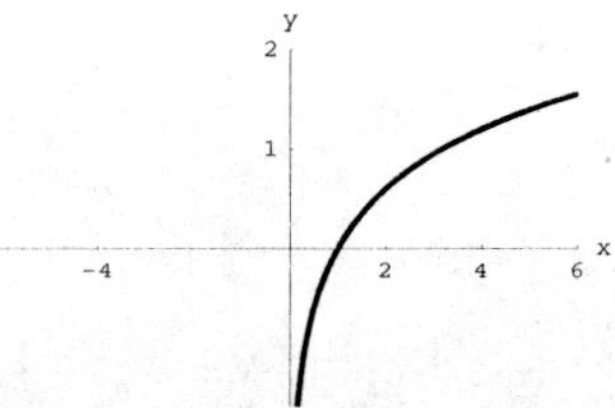

Note, the domain $y = \log(x(x-1))$ and the domain of $y = \log(x) + \log(x-1)$ are not the same.

111. Let $a > 0$, $a \neq 1$. Using the definition of a logarithm and properties of exponents, we have

$$a^{\log_a(M^p)} = M^p = \left(a^{\log_a(M)}\right)^p = a^{p \cdot \log_a(M)}.$$

Then $a^{\log_a(M^p)} = a^{p \cdot \log_a(M)}$. Since $a^s = a^t$ implies $s = t$, we have $\log_a(M^p) = p \cdot \log_a(M)$.

113. Domain $(1, \infty)$, range $(-\infty, \infty)$

115. $10,000e^{0.043(5.25)} = \$12,532.62$

117. Solve for x:

$$\begin{aligned}
\frac{3}{8}x - \frac{15}{8} &= \frac{3}{2}x + \frac{5}{6} \\
-\frac{15}{8} - \frac{5}{6} &= \frac{3}{2}x - \frac{3}{8}x \\
-\frac{65}{24} &= \frac{9}{8}x
\end{aligned}$$

$$-\frac{65}{24}\cdot\frac{8}{9} = x$$
$$-\frac{65}{27} = x$$

The solution set is $\{-\frac{65}{27}\}$.

119.

a) $\frac{6}{23} = \frac{1}{4} + \frac{1}{92}$

b) $\frac{14}{15} = \frac{1}{2} + \frac{1}{3} + \frac{1}{10}$

c) $\frac{7}{11} = \frac{1}{2} + \frac{1}{11} + \frac{1}{22}$

For Thought

1. True, since $(1.02)^x = 7$ is equivalent to $x = \log_{1.02}(7)$.

2. True, for $x(1-\ln(3)) = 8$ implies $x = \frac{8}{1-\ln(3)}$.

3. False, $\ln(1-\sqrt{6})$ is undefined.

4. True, by the definition of a logarithm.

5. False, the exact solution is $x = \log_3(17)$ which is not the same as 2.5789.

6. False, since $x = -2$ is not a solution of the first equation but is a solution of the second one.

7. True, since $4^x = 2^{2x} = 2^{x-1}$ is equivalent to $2x = x - 1$.

8. True, since we may take the ln of both sides of $1.09^x = 2.3$.

9. True, since $\frac{\ln(2)}{\ln(7)} = \frac{\log(2)}{\log(7)}$.

10. True, since $\log(e)\cdot\ln(10) = \ln\left(10^{\log(e)}\right) = \ln(e) = 1$.

4.4 Exercises

1. 10^t

3. $\log_5 9$

5. Since $x = 2^3$, we get $x = 8$.

7. Since $10^2 = x + 20$, $x = 80$.

9. Since $10^1 = x^2 - 15$, we get $x^2 = 25$. The solutions are $x = \pm 5$.

11. Note, $x^2 = 9$ and $x = \pm 3$. Since the base of a logarithm is positive, the solution is $x = 3$.

13. Note, $x^{-2} = 4$ or $\frac{1}{4} = x^2$. Since the base of a logarithm is positive, the solution is $x = \frac{1}{2}$.

15. Since $x^3 = 10$, the solution is $x = \sqrt[3]{10}$.

17. Since $x = \left(8^{-1/3}\right)^2$, the solution is $x = \frac{1}{4}$.

19.

$$\begin{aligned} \log_2(x^2-4) &= 5 \\ x^2 - 4 &= 2^5 \\ x^2 &= 36 \\ x &= \pm 6 \end{aligned}$$

Checking $x = -6$, one gets $\log_2(-6+2)$ which is undefined. The solution is $x = 6$.

21.

$$\begin{aligned} \log_6\left(\frac{x^2-x-6}{14}\right) &= 0 \\ \frac{x^2-x-6}{14} &= 6^0 = 1 \\ x^2-x-6 &= 14 \\ x^2-x-20 &= 0 \\ (x-5)(x+4) &= 0 \end{aligned}$$

Note, if $x = -4$, then $\log\frac{x-3}{2}$ is undefined. Thus, $x = 5$.

23.

$$\begin{aligned} \log\left(\frac{x+1}{x}\right) &= 3 \\ \frac{x+1}{x} &= 10^3 \\ x+1 &= 1000x \\ 1 &= 999x \\ x &= \frac{1}{999} \end{aligned}$$

25.

$$\begin{aligned}\log_4\left(\frac{x}{x+2}\right) &= 2\\ \frac{x}{x+2} &= 16\\ x &= 16x+32\\ -\frac{32}{15} &= x\end{aligned}$$

Note, if $x = -32/15$, then $\log_4 x$ is undefined. The solution set is $\emptyset$.

27.

$$\begin{aligned}\log(5)+\log(x) &= 2\\ \log(5x) &= 2\\ 5x &= 10^2\\ x &= 20\end{aligned}$$

29. Since $\ln(x(x+2)) = \ln(8)$ and $y = \ln(x)$ is one-to-one, we get

$$\begin{aligned}x^2+2x &= 8\\ x^2+2x-8 &= 0\\ (x+4)(x-2) &= 0\\ x &= -4, 2\end{aligned}$$

Since $\ln(-4)$ is undefined, the solution is $x = 2$.

31. Since $\log(4x) = \log\left(\frac{5}{x}\right)$ and $y = \log(x)$ is one-to-one, we obtain

$$\begin{aligned}4x &= \frac{5}{x}\\ 4x^2 &= 5\\ x^2 &= \frac{5}{4}\\ x &= \pm\frac{\sqrt{5}}{2}.\end{aligned}$$

But $\log\left(-\frac{\sqrt{5}}{2}\right)$ is undefined, so $x = \frac{\sqrt{5}}{2}$.

33. Since $\log_2\left(\frac{x}{3x-1}\right) = 0$, we get

$$\begin{aligned}\frac{x}{3x-1} &= 1\\ x &= 3x-1\\ 1 &= 2x\\ x &= \frac{1}{2}.\end{aligned}$$

35.

$$\begin{aligned}x\cdot\ln(3)+x\cdot\ln(2) &= 2\\ x(\ln(3)+\ln(2)) &= 2\\ &= \frac{2}{\ln(3)+\ln(2)}\\ x &= \frac{2}{\ln(6)}\end{aligned}$$

37. Since $x-1 = \log_2(7)$, $x = \frac{\ln(7)}{\ln(2)} + 1 \approx 3.8074$.

39. Since $4x = \log_{1.09}(3.4)$, we find

$$x = \frac{1}{4}\cdot\frac{\ln(3.4)}{\ln(1.09)} \approx 3.5502.$$

41. Since $-x = \log_3(30)$, we obtain

$$x = -\frac{\ln(30)}{\ln(3)} \approx -3.0959.$$

43. Note, $-3x^2 = \ln(9)$. There is no solution since the left-hand side is non-negative and the right-hand side is positive.

45.

$$\begin{aligned}\ln(6^x) &= \ln(3^{x+1})\\ x\cdot\ln(6) &= (x+1)\cdot\ln(3)\\ x\cdot\ln(6) &= x\cdot\ln(3)+\ln(3)\\ x(\ln(6)-\ln(3)) &= \ln(3)\\ x &= \frac{\ln(3)}{\ln(6)-\ln(3)}\\ x &\approx 1.5850\end{aligned}$$

47.

$$\begin{aligned}\ln(e^{x+1}) &= \ln(10^x)\\ (x+1)\cdot\ln(e) &= x\cdot\ln(10)\\ x+1 &= x\cdot\ln(10)\\ 1 &= x(\ln(10)-1)\\ x &= \frac{1}{\ln(10)-1}\\ x &\approx 0.7677\end{aligned}$$

49.

$$\begin{aligned} 2^{x-1} &= (2^2)^{3x} \\ 2^{x-1} &= 2^{6x} \\ x-1 &= 6x \\ -1 &= 5x \\ x &= -0.2 \end{aligned}$$

51.

$$\begin{aligned} \ln(6^{x+1}) &= \ln(12^x) \\ (x+1)\cdot\ln(6) &= x\cdot\ln(12) \\ x\cdot\ln(6)+\ln(6) &= x\cdot\ln(12) \\ \ln(6) &= x(\ln(12)-\ln(6)) \\ x &= \frac{\ln(6)}{\ln(12)-\ln(6)} \\ x &\approx 2.5850 \end{aligned}$$

53. Since $3 = e^{-\ln(w)} = e^{\ln(1/w)} = 1/w$, we have $\frac{1}{w} = 3$ and $w = \frac{1}{3}$.

55.

$$\begin{aligned} (\log(z))^2 &= 2\cdot\log(z) \\ (\log(z))^2 - 2\cdot\log(z) &= 0 \\ \log(z)\cdot(\log(z)-2) &= 0 \\ \log(z)=0 \quad &\text{or} \quad \log(z)=2 \\ z=10^0 \quad &\text{or} \quad z=10^2 \\ z=1 \quad &\text{or} \quad z=100 \end{aligned}$$

The solutions are $z = 1, 100$.

57. Divide the equation by $4(1.03)^x$.

$$\begin{aligned} \left(\frac{1.02}{1.03}\right)^x &= \frac{3}{4} \\ \ln\left(\left(\frac{1.02}{1.03}\right)^x\right) &= \ln\left(\frac{3}{4}\right) \\ x\cdot\ln\left(\frac{1.02}{1.03}\right) &= \ln\left(\frac{3}{4}\right) \\ x &= \frac{\ln\left(\frac{3}{4}\right)}{\ln\left(\frac{1.02}{1.03}\right)} \\ x &\approx 29.4872 \end{aligned}$$

59. Note that $e^{\ln((x^2)^3)-\ln(x^2)} = e^{\ln(x^6)-\ln(x^2)} = e^{\ln(x^6/x^2)} = e^{\ln(x^4)} = x^4$.
Thus, $x^4 = 16$ and $x = \pm 2$.
But $\ln(-2)$ is undefined, so $x = 2$.

61. Since $\left(\frac{1}{2}\right)^2 = \frac{1}{4}$, we find

$$\begin{aligned} \left(\frac{1}{2}\right)^{2x-1} &= \left(\frac{1}{2}\right)^{6x+4} \\ 2x-1 &= 6x+4 \\ -5 &= 4x \\ x &= -\frac{5}{4}. \end{aligned}$$

63. By approximating the x-intercepts of the graph $y = 2^x - 3^{x-1} - 5^{-x}$, we find that the solutions are $x \approx 0.194, 2.70$.

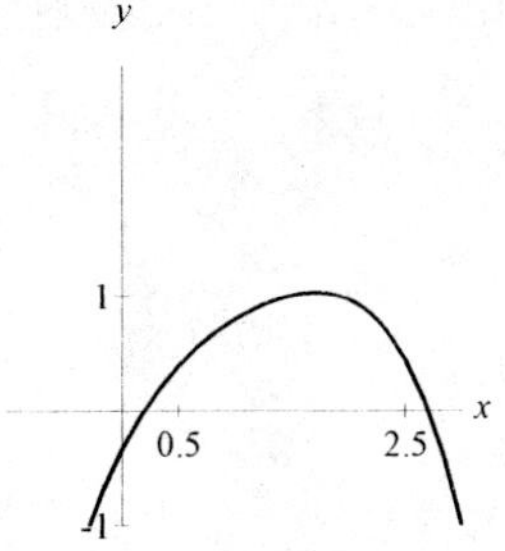

65. By approximating the x-intercept of the graph $y = \ln(x+51) - \log(-48-x)$, we obtain that the solution is $x \approx -49.73$.

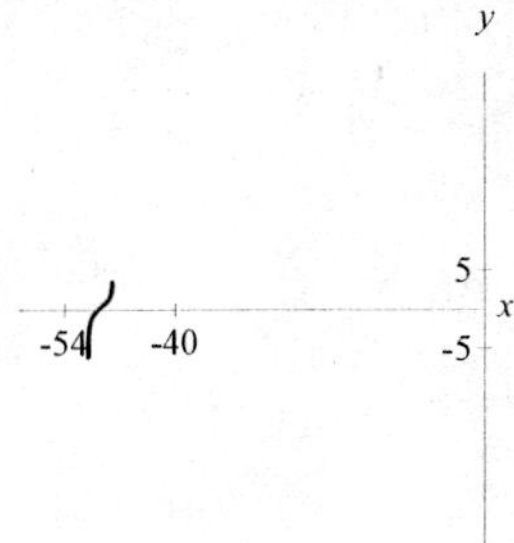

67. By approximating the x-intercepts of the graph $y = x^2 - 2^x$, we find that the solutions are $x \approx -0.767, 2, 4$.

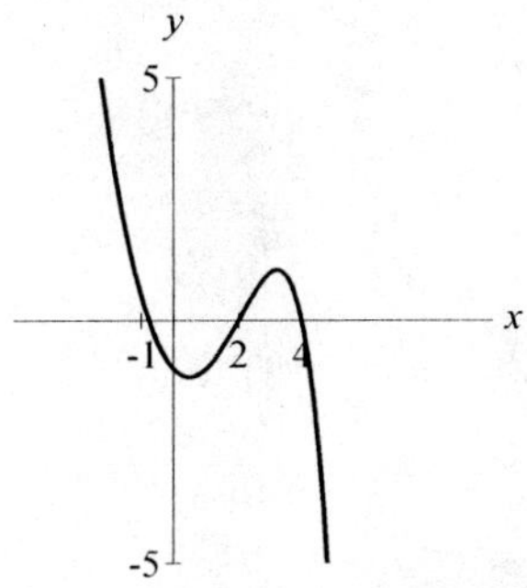

69. Solving for the rate of decay r, we find

$$\begin{aligned} A_0/2 &= A_0 e^{10{,}000r} \\ 1/2 &= e^{10{,}000r} \\ \ln(1/2) &= 10,000r \\ \frac{\ln(1/2)}{10,000} &= r \end{aligned}$$

Approximately, $r \approx -6.93 \times 10^{-5}$.

71. Using $A = A_o e^{rt}$ with $A_o = 1$ and the half-life, we obtain $\frac{1}{2} = e^{5730r}$. Thus, $5730r = \ln\left(\frac{1}{2}\right)$ and $r \approx -0.000120968$. When $A = 0.1$, we get

$$\begin{aligned} 0.1 &= e^{-0.000120968t} \\ \ln(0.1) &= -0.000120968t \\ t = \frac{\ln(0.1)}{-0.000120968} &\approx 19,035 \text{ years} \end{aligned}$$

73. From Number 71, $r \approx -0.000120968$. When $A = 10$ and $A_o = 12$, we obtain

$$\begin{aligned} 10 &= 12 \cdot e^{-0.000120968t} \\ \ln(10/12) &= -0.000120968t \\ t = \frac{\ln(5/6)}{-0.000120968} &\approx 1507 \text{ years.} \end{aligned}$$

75. Let $A = A_o e^{rt}$ where $A_o = 25$, $A = 20$, and $t = 8000$. Then

$$\begin{aligned} 20 &= 25 \cdot e^{8000r} \\ \ln(0.8) &= 8000r \\ r &\approx -0.000027893. \end{aligned}$$

To find the half-life, let $A = 12.5$. Thus, we have

$$\begin{aligned} 12.5 &= 25 \cdot e^{-0.000027893t} \\ \ln(0.5) &= -0.000027893t \\ t &\approx 24,850 \text{ years.} \end{aligned}$$

77. $\frac{2.5(0.5)^{24/14}}{2.5} \times 100 \approx 30.5\%$, the percentage of the last dosage that remains before the next dosage is taken

79. Since the half-life is $t = 5730$ years and $A_o = 1$ and $A = 0.5$, we get

$$\begin{aligned} 0.5 &= e^{(5730)\cdot r} \\ \ln(0.5) &= 5730 \cdot r \\ r &\approx -0.000121. \end{aligned}$$

If 79.3% of the carbon is still present, then

$$\begin{aligned} 0.793 &= e^{(-0.000121)\cdot t} \\ \ln(0.793) &= -0.000121 \cdot t \\ t &\approx 1917. \end{aligned}$$

The scrolls were made in the year $1951 - 1917 = 34$ AD.

81.

a) Solve for r:

$$\begin{aligned} 11,981 &= 21,075e^{3r} \\ \ln\left(\frac{11,981}{21,075}\right) &= 3r \\ -0.188255 &= r \\ -18.8\% &\approx r \end{aligned}$$

b) If $t = 5$ years, then

$$P = 21,075e^{5(-0.188)} = \$8,200$$

83. In year $2009 + x$, the number of blog sites is $C(x) = 0.1e^{rx}$ for some r. Since $C = 4.8$ when $x = 2$, we find

$$\begin{aligned} 4.8 &= 0.1e^{2r} \\ r &= \frac{\ln 48}{2}. \end{aligned}$$

In year 2014, we have $x = 5$ and the number of blog sites is

$$\begin{aligned} C &= 0.1e^{5\ln(48)/2} \\ &\approx 1596 \text{ million.} \end{aligned}$$

85. The initial difference in temperature is $325 - 35 = 290$, and after $t = 3$ hours the difference is $325 - 140 = 185$. Then

$$\begin{aligned} 185 &= 290 \cdot e^{3k} \\ \ln\left(\frac{185}{290}\right) &= 3k \\ k &\approx -0.1498417. \end{aligned}$$

The difference in temperature when the roast is well-done is $325 - 170 = 155$. Thus,

$$\begin{aligned} 155 &= 290 \cdot e^{(-0.1498417) \cdot t} \\ \ln\left(\frac{155}{290}\right) &= (-0.1498417) \cdot t \\ t &\approx 4.18 \text{ hr} \\ t &\approx 4 \text{ hr and } 11 \text{ min.} \end{aligned}$$

James must wait 1 hour, 11 minutes longer.

If the oven temperature is set at 170°, then the initial and final differences are 135 and 0, respectively. Since $0 = 135 \cdot e^{(-0.1498417) \cdot t}$ has no solution, James has to wait forever.

87. At 7:00 a.m., the difference in temperature is $80 - 40 = 40$, and $t = 1$ hour later the difference in temperature is $72 - 40 = 32$. Then

$$\begin{aligned} 32 &= 40 \cdot e^{1 \cdot k} \\ \ln\left(\frac{32}{40}\right) &= k \\ k &\approx -0.2231436. \end{aligned}$$

Let n be the number of hours before 7:00 a.m. when death occured. At the time of death, the difference in temperature is $98 - 40 = 58$. Then

$$\begin{aligned} 40 &= 58 \cdot e^{-0.2231436 \cdot n} \\ \ln\left(\frac{40}{58}\right) &= -0.2231436 \cdot n \\ n &\approx 1.665 \text{ hr} \\ n &\approx 1 \text{ hr and } 40 \text{ min.} \end{aligned}$$

The death occured at 5:20 a.m.

89. Since $R = P\dfrac{i}{1-(1+i)^{-nt}}$, we obtain

$$\begin{aligned} 1-(1+i)^{-nt} &= Pi/R \\ 1 - Pi/R &= (1+i)^{-nt} \\ \ln(1 - Pi/R) &= -nt\ln(1+i) \\ \frac{-\ln(1-Pi/R)}{n\ln(1+i)} &= t. \end{aligned}$$

Let $i = 0.09/12 = 0.0075$, $P = 100,000$, and $R = 1250$. If $n = 12$, then

$$t = \frac{-\ln(1-Pi/R)}{n\ln(1+i)} \approx 10.219.$$

It will take 10 yr, 3 mo to pay off the loan.

91. The future values of the \$1000 and \$1100 investments are equal. Then

$$\begin{aligned} 1000 \cdot e^{0.06t} &= 1100\left(1+\frac{0.06}{365}\right)^{365t} \\ e^{0.06t} &= 1.1\left(1+\frac{0.06}{365}\right)^{365t} \\ 0.06t &= \ln(1.1) + 365t\ln\left(1+\frac{0.06}{365}\right) \end{aligned}$$

$$0.06t - 365t\ln\left(1+\frac{0.06}{365}\right) = \ln(1.1)$$

$$\begin{aligned} t &= \frac{\ln(1.1)}{0.06 - 365\ln\left(1+\dfrac{0.06}{365}\right)} \\ t &\approx 19,328.84173 \text{ years} \end{aligned}$$

They will be equal after 19, 328 yr, 307 days.

93. **a)** Let $t = 0$. The present number of rabbits is

$$P = 12,300 + 1000 \cdot \ln(1) = 12,300 + 0 = 12,300.$$

b) In about 15 years.

c) The number of years before there will be 15, 000 rabbits is given by

$$\begin{aligned} 12,300 + 1000 \cdot \ln(t+1) &= 15,000 \\ 1000 \cdot \ln(t+1) &= 2,700 \\ \ln(t+1) &= 2.7 \\ t+1 &= e^{2.7} \\ t &\approx 13.9 \text{ yr.} \end{aligned}$$

95. **a)** Let $n = 2500$ and $A = 400$. Since $n = k\log(A)$, $2500 = k\log(400)$. Then $k = \dfrac{2500}{\log(400)}$. When $A = 200$, the number of species left is

$$n = \frac{2500}{\log(400)}\log(200) \approx 2211 \text{ species.}$$

b) Let $n = 3500$ and $A = 1200$. Since $n = k\log(A)$, $3500 = k\log(1200)$. Then $k = \dfrac{3500}{\log(1200)}$. When $n = 1000$, the remaining forest area is given by

$$\begin{aligned} 1000 &= \frac{3500}{\log(1200)}\log(A) \\ \frac{1000}{3500}\log(1200) &= \log(A) \\ \frac{2}{7}\log(1200) &= \log(A) \\ \log\left(1200^{2/7}\right) &= \log(A) \\ 1200^{2/7} &= A \end{aligned}$$

Thus, the percentage of forest that has been destroyed is $100 - \frac{A}{1200}(100) \approx 99\%$.

97. Since $m = 0$ and $M_v = 4.39$, the distance to Alpha Centauri is given by

$$\begin{aligned} 4.39 - 5 + 5 \cdot \log(d) &= 0 \\ 5 \cdot \log(d) &= 0.61 \\ \log(d) &= 0.122 \\ d = 10^{0.122} &\approx 1.32 \text{ parsecs.} \end{aligned}$$

99. Let P be the price for a gigabyte of hard drive storage in year $1982 + x$.

a) Let $y = \log(P)$. A formula for P is

$$\begin{aligned} y - 4.5 &= \frac{4.5+1}{0-30}(x-0) \\ y &= -\frac{11}{60}x + 4.5 \\ \log(P) &= -\frac{11}{60}x + 4.5 \\ P &= 10^{(-11x/60+4.5)}. \end{aligned}$$

b) In 2002, $x = 20$ and

$$P = 10^{(-11(20)/60+4.5)} \approx \$6.81.$$

c) Solving for x, we derive

$$\begin{aligned} 0.01 &= 10^{(-11x/60+4.5)} \\ \log(0.01) &= -\frac{11}{60}x + 4.5 \\ \frac{\log(0.01) - 4.5}{-11/60} &= x \\ x &\approx 35. \end{aligned}$$

The cost will be $P = \$0.01$ in year $2017 = 1982 + 35$.

101. If the sound level is 90 db, then the intensity of the sound is given by

$$\begin{aligned} 10 \cdot \log(I \times 10^{12}) &= 90 \\ \log(I) + \log(10^{12}) &= 9 \\ \log(I) + 12 &= 9 \\ \log(I) &= -3 \\ I &= 10^{-3} \text{ watts/m}^2. \end{aligned}$$

103. Since $P \cdot e^{(0.06)18} = 20,000$, the investment will grow to $P = \frac{20,000}{e^{(0.06)18}} \approx \$6791.91$.

105.

a) The logarithmic regression line is

$$y = 114.0 - 12.5\ln(x)$$

where the year is $1960 + x$.

b) Let $x = 55$ for year 2015. Since

$$114.0 - 12.5\ln(55) \approx 64$$

the percentage of two-parent families in 2015 is 64%.

c) If $y = 60$, then

$$\begin{aligned} 60 &= 114.0 - 12.5\ln(x) \\ \ln(x) &= \frac{54}{12.5} \\ x &\approx 75. \end{aligned}$$

Since $1960 + 75 = 2035$, the percentage of two-parent families will reach 60% in year 2035.

d) The data looks like a quadratic or exponential model.

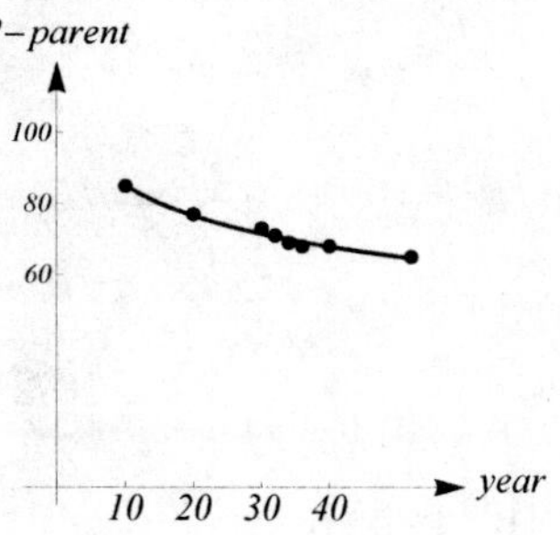

107. By using the first five terms of the formula,

$$\begin{aligned} e^{0.1} &\approx 1 + 0.1 + \frac{(0.1)^2}{2} + \frac{(0.1)^3}{6} + \frac{(0.1)^4}{24} \\ &\approx 1.105170833 \end{aligned}$$

From a calculator, $e^{0.1} \approx 1.105170918$.

109. $\log_6(4 \cdot 9) = \log_6(36) = 2$

111. Convert to exponential form:

$$\begin{aligned} x^{2.4} &= 9.8 \\ x &= 9.8^{1/2.4} \\ x &\approx 2.5883 \end{aligned}$$

The solution set is $\{2.5883\}$.

113. By the identity $\log_3(3^w) = w$, we obtain

$$\begin{aligned} (g \circ f)(x) &= \log_3\left(3^{(x-5)}\right) + 5 \\ &= (x-5) + 5 \\ &= x \end{aligned}$$

115. Consider the point of intersection between the two left-most circles on the bottom row. The (vertical) distance between this point of intersection and the center of the left-most circle in the middle row is $\sqrt{5}$.

Let x be the radius of the circle at the top row. We draw another right triangle but this time its hypotenuse is the line segment that joins the center of the left-most circle in the middle row and the center of the circle in the top row. The length of the hypotenuse of this triangle is $x + 1$.

From the center of the left-most circle in the middle row, draw a horizontal side that is 2-units long in such a way that the endpoint of this side should be directly below the center of the circle on the top row. Then a right triangle is formed when this endpoint is joined to the center of the top-most circle by a line segment.

Applying the Pythagorean Theorem, we find

$$2^2 + (x + 2 - \sqrt{5})^2 = (x+1)^2.$$

Solving for x, we find

$$x = \sqrt{5} - 1.$$

Chapter 4 Review Exercises

1. 64 **3.** 6

5. 0 **7.** 17

9. $4^{1/\log_3 4 + \log_9 3} = 4^{\log_4 3 + 1/2} =$

$4^{\log_4 3} 4^{1/2} = 3 \cdot 2 = 6$

11. $2^5 = 32$

13. $\log(10^3) = 3$

15. $\log_2(2^9) = 9$

17. $\log(1000) = 3$

19. $\log_2(1) - \log_2(8) = 0 - 3 = -3$

21. $\log_2(8) = 3$

23. $\log((x-3)x) = \log(x^2 - 3x)$

25. $\ln(x^2) + \ln(3y) = \ln(3x^2 y)$

27. $\log(3) + \log(x^4) = \log(3) + 4 \cdot \log(x)$

29. $\log_3(5) + \log_3(x^{1/2}) - \log_3(y^4) =$
$\log_3(5) + \frac{1}{2} \cdot \log_3(x) - 4 \cdot \log_3(y)$

31. $\ln(2 \cdot 5) = \ln(2) + \ln(5)$

33. $\ln(5^2 \cdot 2) = \ln(5^2) + \ln(2) = 2 \cdot \ln(5) + \ln(2)$

35. Since $\log_{10}(x) = 10$, we get $x = 10^{10}$.

37. Since $x^4 = 81$ and $x > 0$, $x = 3$.

39. Since $\log_{1/3}(27) = -3 = x + 2$, we get $x = -5$.

41. Since $3^{x+2} = 3^{-2}$, $x + 2 = -2$. So $x = -4$.

43. Since $x - 2 = \ln(9)$, we obtain $x = 2 + \ln(9)$.

45. Since $(2^2)^{x+3} = 2^{2x+6} = 2^{-x}$, we get $2x + 6 = -x$. Then $6 = -3x$ and so $x = -2$.

47.

$$\begin{aligned} \log(2x^2) &= 5 \\ 2x^2 &= 10^5 \\ x^2 &= 50,000 \\ x &= \pm 100\sqrt{5} \end{aligned}$$

Since $\log(-100\sqrt{5})$ is undefined, $x = 100\sqrt{5}$.

49.

$$\begin{aligned} \log_2\left(x^2 - 4x\right) &= \log_2(x+24) \\ x^2 - 4x &= x + 24 \\ x^2 - 5x - 24 &= 0 \\ (x-8)(x+3) &= 0 \\ x &= 8, -3 \end{aligned}$$

Since $\log(-3)$ is undefined, $x = 8$.

51. Since $\ln((x+2)^2) = \ln(4^3)$ and $y = \ln(x)$ is a one-to-one function, we obtain

$$\begin{aligned} (x+2)^2 &= 64 \\ x + 2 &= \pm 8 \\ x &= -2 \pm 8 \\ x &= 6, -10. \end{aligned}$$

Checking $x = -10$ one gets $2\ln(-8)$ which is undefined. So $x = 6$.

53.

$$\begin{aligned} x \cdot \log(4) + x \cdot \log(25) &= 6 \\ x(\log(4) + \log(25)) &= 6 \\ x \cdot \log(100) &= 6 \\ x \cdot 2 &= 6 \\ x &= 3 \end{aligned}$$

55. The missing coordinates are

(i) 3 since $\left(\frac{1}{3}\right)^{-1} = 3$,

(ii) -3 since $\left(\frac{1}{3}\right)^{-3} = 27$,

(iii) $\sqrt{3}$ since $\left(\frac{1}{3}\right)^{-1/2} = \sqrt{3}$, and

(iv) 0 since $\left(\frac{1}{3}\right)^{0} = 1$.

57. c

59. b

61. d

63. e

65. Domain $(-\infty, \infty)$, range $(0, \infty)$, increasing, asymptote $y = 0$

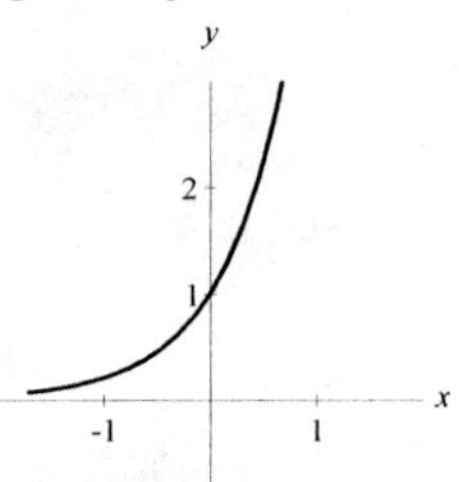

67. Domain $(-\infty, \infty)$, range $(0, \infty)$, decreasing, asymptote $y = 0$

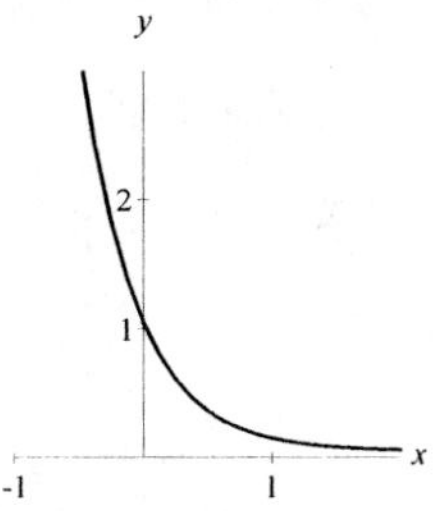

69. Domain $(0, \infty)$, range $(-\infty, \infty)$, increasing, asymptote $x = 0$

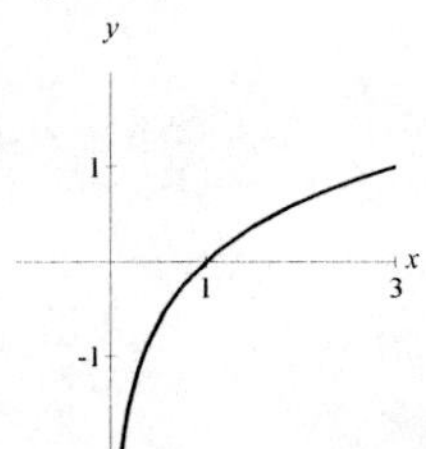

71. Domain $(-3, \infty)$, range $(-\infty, \infty)$, increasing, asymptote $x = -3$

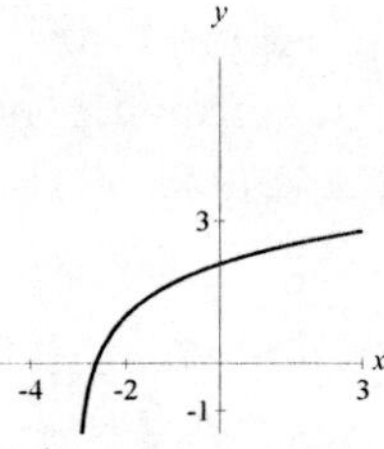

73. Domain $(-\infty, \infty)$, range $(1, \infty)$, increasing, asymptote $y = 1$

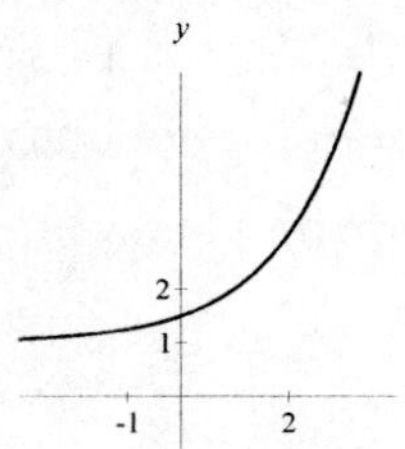

75. Domain $(-\infty, 2)$, range $(-\infty, \infty)$, decreasing, asymptote $x = 2$

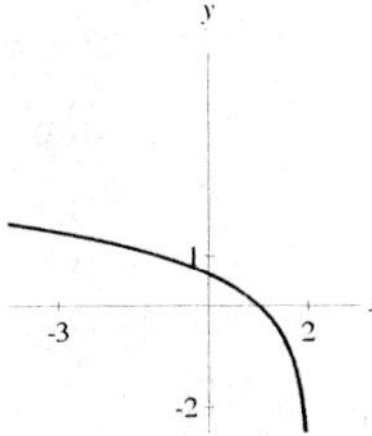

77. $f^{-1}(x) = \log_7(x)$

79. $f^{-1}(x) = 5^x$

81. Replace $f(x)$ by y, interchange x and y, solve for y, and replace y by $f^{-1}(x)$.

$$\begin{aligned} y &= 3\log(x-1) \\ x &= 3\log(y-1) \\ \frac{x}{3} &= \log(y-1) \\ 10^{x/3} &= y-1 \\ 10^{x/3}+1 &= y \\ f^{-1}(x) &= 10^{x/3}+1 \end{aligned}$$

83. Replace $f(x)$ by y, interchange x and y, solve for y, and replace y by $f^{-1}(x)$.

$$\begin{aligned} y &= e^{x+2}-3 \\ x &= e^{y+2}-3 \\ x+3 &= e^{y+2} \\ \ln(x+3) &= y+2 \\ \ln(x+3)-2 &= y \\ f^{-1}(x) &= \ln(x+3)-2 \end{aligned}$$

85. Since $3^x = 10$, we have

$$x = \log_3(10) = \frac{\ln(10)}{\ln(3)} \approx 2.0959.$$

87. Since $\log_3(x) = 1.876$, $x = 3^{1.876} \approx 7.8538$.

89. After taking the natural logarithm of both sides, we have

$$\begin{aligned} \ln(5^x) &= \ln(8^{x+1}) \\ x \cdot \ln(5) &= (x+1)\cdot\ln(8) \\ x \cdot \ln(5) &= x\cdot\ln(8)+\ln(8) \\ x\cdot(\ln(5)-\ln(8)) &= \ln(8) \\ x &= \frac{\ln(8)}{\ln(5)-\ln(8)} \\ x &\approx -4.4243. \end{aligned}$$

91. True, since $\log_3(81) = 4$ and $2 = \log_3(9)$.

93. False, since $\ln(3^2) = 2 \cdot \ln(3) \neq (\ln(3))^2$.

95. True, since $4 \cdot \log_2(8) = 4 \cdot 3 = 12$.

97. False, since $3 + \log(6) = \log(10^3) + \log(6) = \log(6000)$.

99. False, since $\log_2(16) = 4$, $\log_2(8) = 3$, and $\frac{3}{4} \neq 3 - 4$.

101. False, since $\log_2(25) = 2\log_2(5) = 2 \cdot \frac{\log(5)}{\log(2)} \neq 2 \cdot \log(5)$.

103. True, because of the base-changing formula.

105. If H_B^+ is the hydrogen ion concentration of liquid B then $10 \cdot H_B^+$ is the hydrogen ion concentration of liquid A. The *pH* of A is

$$-\log(10 \cdot H_B^+) = -1 - \log(H_B^+)$$

i.e. the pH of A is one less than the pH of B.

107. The value at the end of 18 years is

$$50,000\left(1+\frac{0.05}{4}\right)^{18\cdot 4} \approx \$122,296.01.$$

109. Let t be the number of years.

$$\begin{aligned} 50,000\left(1+\frac{0.05}{4}\right)^{4t} &= 100,000 \\ (1.0125)^{4t} &= 2 \\ 4t &= \log_{1.0125}(2) \\ t &= \frac{1}{4}\cdot\frac{\ln(2)}{\ln(1.0125)} \\ t &\approx 13.9 \text{ years} \end{aligned}$$

It doubles in $4 \cdot 13.9 \approx 56$ quarters.

111. The present amount is $A = 25 \cdot e^0 = 25$ g. After $t = 1000$ years, the amount left is $A = 25 \cdot e^{-0.32} \approx 18.15$ g.

To find the half-life, let $A = 12.5$.

$$\begin{aligned} 25 \cdot e^{-0.00032t} &= 12.5 \\ e^{-0.00032t} &= 0.5 \\ -0.00032t &= \ln(0.5) \\ t &\approx 2166 \end{aligned}$$

The half-life is 2166 years.

113. Let $f(t) = 10,000$.

$$\begin{aligned} 40,000 \cdot (1 - e^{-0.0001t}) &= 10,000 \\ 1 - e^{-0.0001t} &= 0.25 \\ 0.75 &= e^{-0.0001t} \\ \ln(0.75) &= -0.0001t \\ t &\approx 2877 \text{ hr} \end{aligned}$$

It takes 2877 hours to learn 10,000 words.

115. In the following equations, we solve for x.

$$\begin{aligned} 1026\left(\frac{25,005}{64}\right)^x &= 19.2 \\ \left(\frac{25,005}{64}\right)^x &= \frac{19.2}{1026} \\ x\ln\left(\frac{25,005}{64}\right) &= \ln\left(\frac{19.2}{1026}\right) \\ x &= \frac{\ln\left(\frac{19.2}{1026}\right)}{\ln\left(\frac{25,005}{64}\right)} \end{aligned}$$

$$x \approx -0.667 \quad \text{or} \quad x \approx -\frac{2}{3}$$

In the following equations, we obtain y.

$$\begin{aligned} \frac{25005}{2240}\left(\frac{35+\frac{1}{12}}{100}\right)^y &= 258.51 \\ \left(\frac{35+\frac{1}{12}}{100}\right)^y &= \frac{258.51(2240)}{25005} \\ y &= \frac{\ln\left(\frac{258.51(2240)}{25005}\right)}{\ln\left(\frac{35+\frac{1}{12}}{100}\right)} \\ y &\approx -3 \end{aligned}$$

Next, we solve for z.

$$\begin{aligned} 13.5\left(\frac{25005}{64}\right)^z &= 1.85 \\ \left(\frac{25005}{64}\right)^z &= \frac{1.85}{13.5} \\ z\ln\left(\frac{25005}{64}\right) &= \ln\left(\frac{1.85}{13.5}\right) \\ z &= \frac{\ln\left(\frac{1.85}{13.5}\right)}{\ln\left(\frac{25005}{64}\right)} \end{aligned}$$

$$z \approx -0.333 \quad \text{or } z \approx -\frac{1}{3}$$

117. We begin by noting that

$$\left(\begin{array}{c}\text{area of}\\ \text{two crescents}\end{array}\right) + \left(\begin{array}{c}\text{area of}\\ \text{largest semicircle}\end{array}\right) =$$

$$\left(\begin{array}{c}\text{area of two}\\ \text{smaller circles}\end{array}\right) + \left(\begin{array}{c}\text{area of the}\\ \text{right triangle}\end{array}\right) =$$

Let x and y be the sides of the right triangle where crescents A and B intersect the triangle, respectively. Then the hypotenuse of the right triangle is $\sqrt{x^2 + y^2}$.

Recall, the area of a semicircle with diameter d is $\pi d^2/8$. Thus, the sum of the areas of the two smaller semicircles with diameters x and y is equal to the area of the largest semicircle with diameter $\sqrt{x^2 + y^2}$.

If we subtract the areas of the circles from the equation above, then we obtain

$$\left(\begin{array}{c}\text{area of}\\ \text{two crescents}\end{array}\right) = \left(\begin{array}{c}\text{area of the}\\ \text{right triangle}\end{array}\right).$$

Hence, the ratio of the total area of the two crescents to the area of the triangle is 1.

Chapter 4 Test

1. 3 **2.** -2 **3.** 6.47 **4.** $\sqrt{2}$

5. $f^{-1}(x) = e^x$

6. Replace $f(x)$ by y, interchange x and y, solve for y, and replace y by $f^{-1}(x)$.

$$\begin{aligned} y &= 8^{x+1} - 3 \\ x &= 8^{y+1} - 3 \end{aligned}$$

$$\begin{aligned} x+3 &= 8^{y+1} \\ \log_8(x+3) &= y+1 \\ \log_8(x+3)-1 &= y \\ f^{-1}(x) &= \log_8(x+3)-1 \end{aligned}$$

7. $\log(x)+\log(y^3)=\log(xy^3)$

8. $\ln(\sqrt{x-1})-\ln(33)=\ln\left(\dfrac{\sqrt{x-1}}{33}\right)$

9. $\log_a(2^2\cdot 7)=\log_a(2^2)+\log_a(7)=$
$2\log_a(2)+\log_a(7)$

10. $\log_a\left(\dfrac{7}{2}\right)=\log_a(7)-\log_a(2)$

11.

$$\begin{aligned} \log_2(x^2-2x) &= 3 \\ x^2-2x &= 2^3 \\ x^2-2x-8 &= 0 \\ (x-4)(x+2) &= 0 \\ x &= 4,-2 \end{aligned}$$

But $\log_2(-2)$ is undefined, so $x=4$.

12.

$$\begin{aligned} \log\left(\frac{10x}{x+2}\right) &= \log(3^2) \\ \frac{10x}{x+2} &= 9 \\ 10x &= 9x+18 \\ x &= 18 \end{aligned}$$

13.

$$\begin{aligned} \ln(3^x) &= \ln(5^{x-1}) \\ x\cdot\ln(3) &= (x-1)\cdot\ln(5) \\ x\cdot\ln(3) &= x\cdot\ln(5)-\ln(5) \\ x(\ln(3)-\ln(5)) &= -\ln(5) \\ x(\ln(5)-\ln(3)) &= \ln(5) \\ x &= \frac{\ln(5)}{\ln(5)-\ln(3)} \\ x &\approx 3.1507 \end{aligned}$$

14. By the definition of a logarithm, we obtain $x-1=3^{5.46}$. Then $x=1+3^{5.46}\approx 403.7931$.

15. Domain $(-\infty,\infty)$, range $(1,\infty)$, increasing, asymptote $y=1$

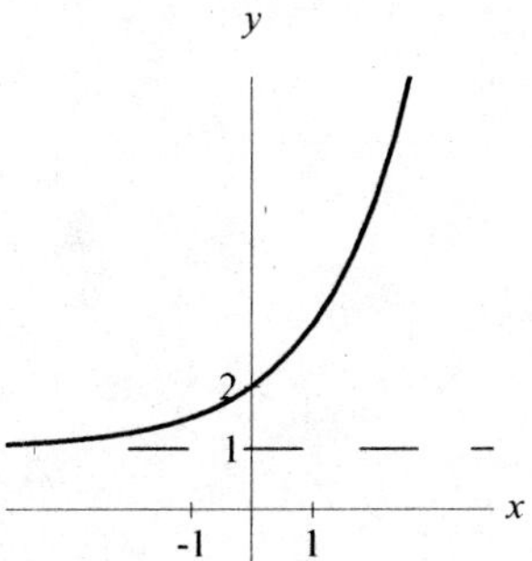

16. Domain $(1,\infty)$, range $(-\infty,\infty)$, decreasing, asymptote $x=1$

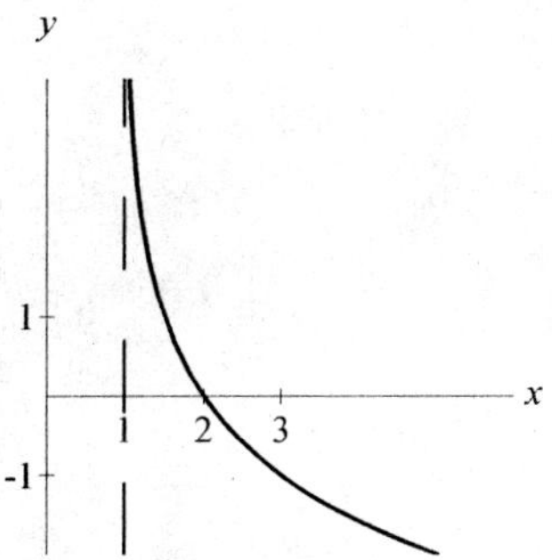

17. $(1,0)$

18. Compounded quarterly, the investment is worth

$$2000\left(1+\frac{0.08}{4}\right)^{80}\approx \$9750.88.$$

Compounded continuously, the investment is worth

$$2000\cdot e^{0.08(20)}\approx \$9906.06.$$

19. The amount of power at the end of $t=200$ days is $P=50\cdot e^{-200/250}\approx 22.5$ watts.

To find the half-life, let $P=25$.

$$\begin{aligned} 50\cdot e^{-t/250} &= 25 \\ e^{-t/250} &= 0.5 \\ -\frac{t}{250} &= \ln(0.5) \\ t &\approx 173.3 \end{aligned}$$

The half-life is 173.3 days.

The operational life for a power of $P = 9$ watts is given by

$$\begin{aligned} 50 \cdot e^{-t/250} &= 9 \\ e^{-t/250} &= \frac{9}{50} \\ -\frac{t}{250} &= \ln\left(\frac{9}{50}\right) \\ t &\approx 428.7 \text{ days.} \end{aligned}$$

20.

$$\begin{aligned} 4,000 \cdot \left(1 + \frac{0.06}{4}\right)^{4t} &= 10,000 \\ (1.015)^{4t} &= 2.5 \\ 4t &= \log_{1.015}(2.5) \\ t &\approx 15.38576 \text{ years} \\ t &\approx 61.5 \text{ quarters} \end{aligned}$$

21. Substituting $t = 100$, we obtain

$$\begin{aligned} -50 \cdot \ln(1-p) &= 100 \\ \ln(1-p) &= -2 \\ 1-p &= e^{-2} \\ p &= 1-e^{-2} \\ p &\approx 0.86. \end{aligned}$$

The level reached after 100 hr is $p = 0.86$. When $p = 1$, then $t = -50\ln(0)$ which is undefined, so it is impossible to master MGM.

Tying It All Together

1. Since $x - 3 = \pm 2$, we get $x = 3 \pm 2 = 1, 5$.

2. Since $\log((x-3)^2) = \log(4)$, we obtain

$$\begin{aligned} (x-3)^2 &= 4 \\ x-3 &= \pm 2 \\ x &= 3 \pm 2 \\ x &= 5, 1. \end{aligned}$$

Checking $x = 1$, one gets $2\log(-2)$ which is undefined, so $x = 5$.

3. Using the definition of a logarithm, we obtain $x - 3 = 2^4$. The solution is $x = 16 + 3 = 19$.

4. Since $2^{x-3} = 2^2$, we have $x - 3 = 2$. So $x = 5$.

5. Square both sides of $\sqrt{x-3} = 4$. Then $x - 3 = 16$. The solution is $x = 19$.

6. An equivalent equation is $x - 3 = \pm 4$. Thus, $x = 3 \pm 4 = -1, 7$.

7. Completing the square, we obtain

$$\begin{aligned} x^2 - 4x + 4 &= -2 + 4 \\ (x-2)^2 &= 2 \\ x - 2 &= \pm\sqrt{2} \\ x &= 2 \pm \sqrt{2}. \end{aligned}$$

8. Since $2^{x-3} = 2^{2x}$, $x - 3 = 2x$. So $x = -3$.

9. Raise both sides to the third power.

$$\begin{aligned} (\sqrt[3]{x-5})^3 &= 5^3 \\ x - 5 &= 125 \\ x &= 130 \end{aligned}$$

10. By using the definition of a logarithm, we have $x = \log_2(3)$.

11.

$$\begin{aligned} \log(4x - 12) &= \log(x) \\ 4x - 12 &= x \\ 3x &= 12 \\ x &= 4 \end{aligned}$$

12. Use synthetic division with $c = -1$.

-1	1	-4	1	6
		-1	5	-6
	1	-5	6	0

The quotient factors as $x^2 - 5x + 6 = (x-3)(x-2)$. The solutions are $x = -1, 2, 3$.

13. Parabola $y = x^2$ goes through $(0,0)$,$(\pm 1, 1)$

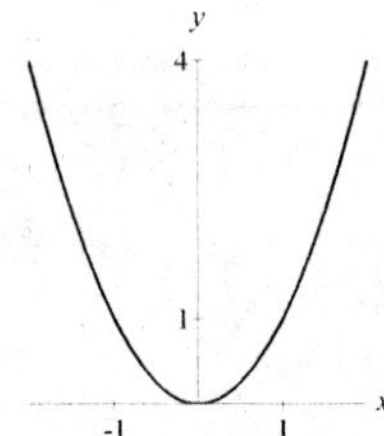

14. Parabola $y = (x-2)^2$ goes through $(0,4)$, $(1,1)$, $(3,1)$

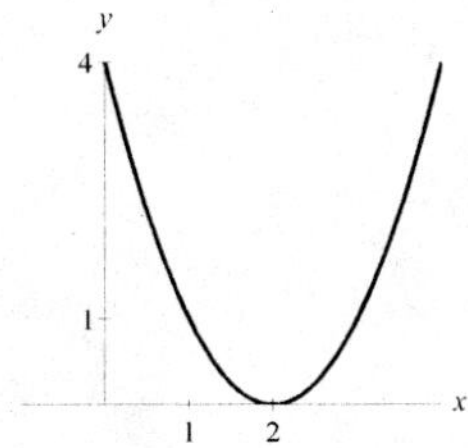

15. $y = 2^x$ goes through $\left(-1, \dfrac{1}{2}\right)$, $(0,1)$, $(1,2)$

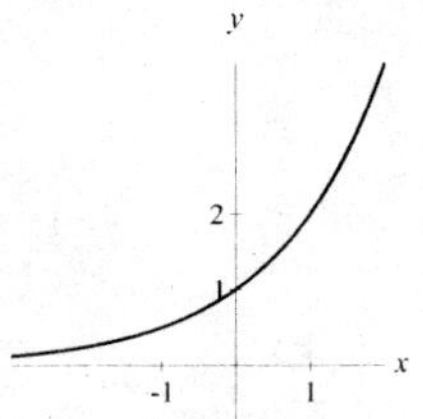

16. $y = x^{-2}$ goes through $(\pm 1, 1)$, $\left(\pm 2, \dfrac{1}{4}\right)$, and $\left(\pm\dfrac{1}{2}, 4\right)$

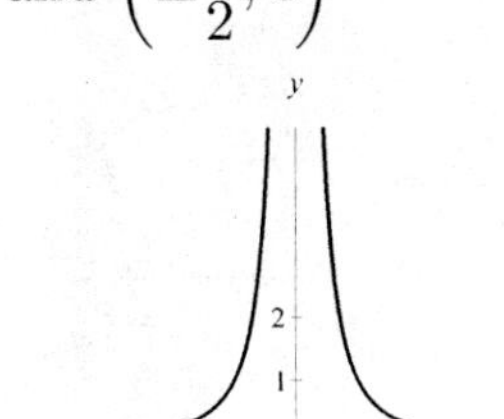

17. $y = \log_2(x-2)$ goes through $(3,0)$, $(4,1)$, and $(2.5,-1)$

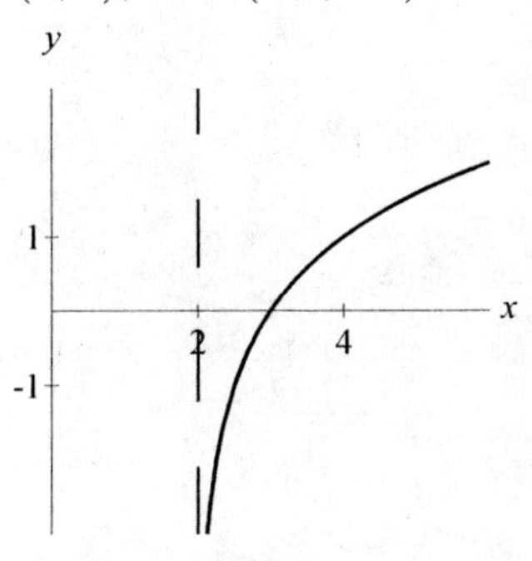

18. Line $y = x - 2$ goes through $(0,-2)$, $(2,0)$

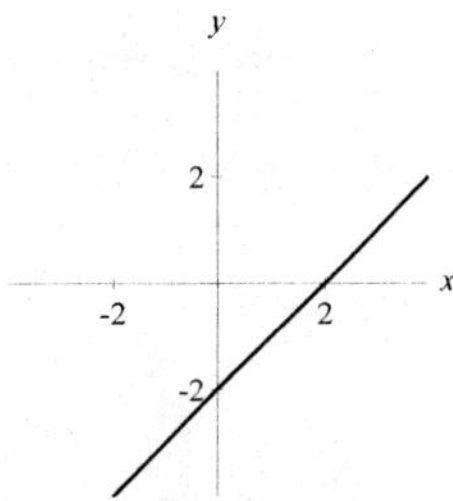

19. Line $y = 2x$ goes through $(0,0)$, $(1,2)$

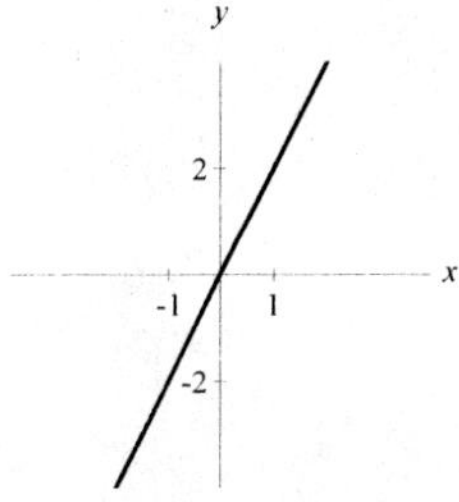

20. Line $y = x \cdot \log(2)$ goes through $(0,0)$, $(1, \log(2))$

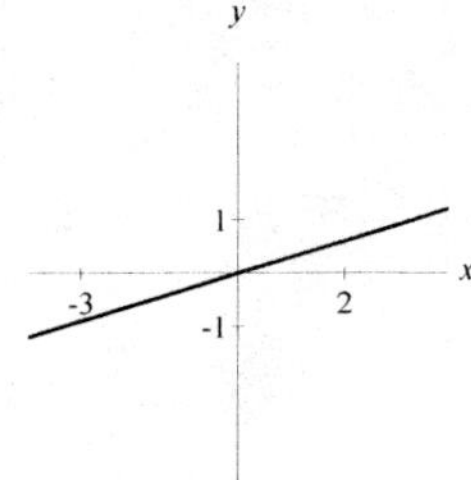

21. Horizontal line $y = e^2$ goes through $(0, e^2)$

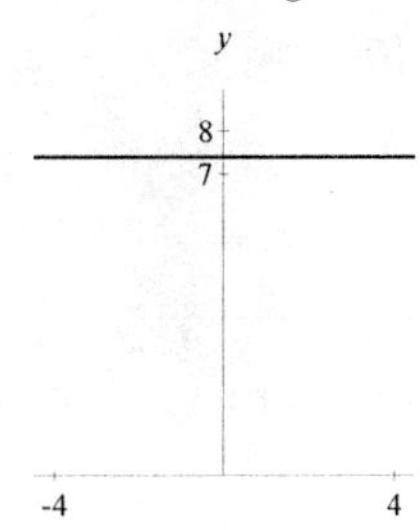

22. Parabola $y = 2 - x^2$ goes through $(0,2)$, $(\pm 1, 1)$

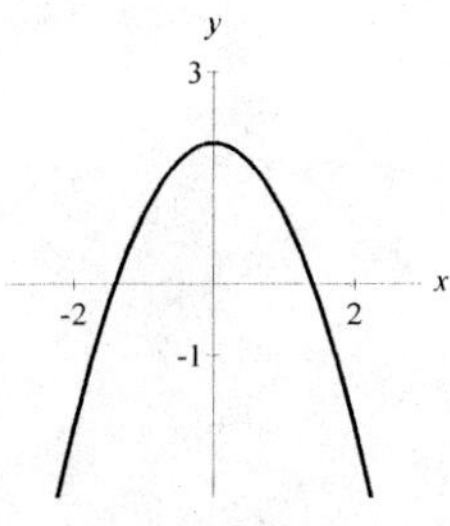

23. $y = \dfrac{2}{x}$ is symmetric about the origin and goes through $(1, 2)$, $(2, 1)$, $\left(\frac{1}{2}, 4\right)$, $\left(4, \frac{1}{2}\right)$

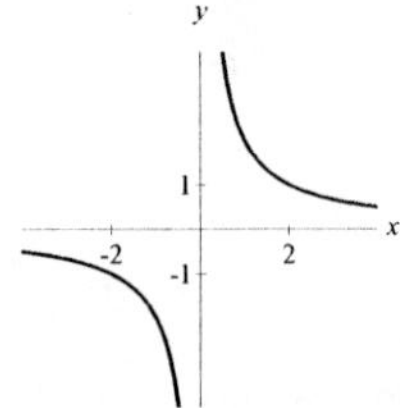

24. The graph of
$$y = \frac{1}{x-2}$$
is obtained by shifting the graph
$$y = \frac{1}{x}$$
to the right by 2 units.

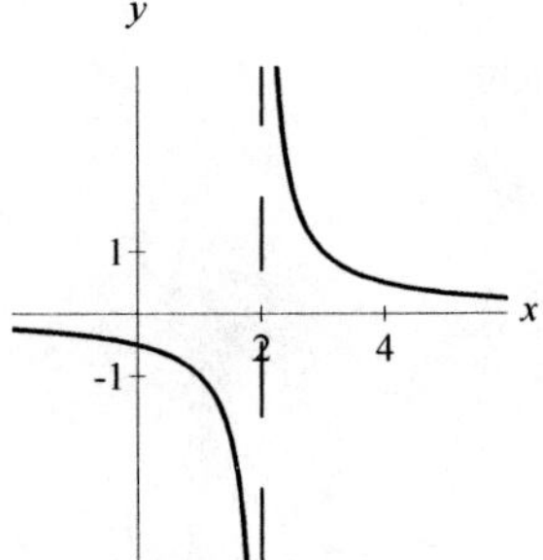

25. $f^{-1}(x) = 3x$

26. $f^{-1}(x) = -\log_3(x)$ since $f(x) = 3^{-x}$

27. $f^{-1}(x) = x^2 + 2 \quad \text{for } x \geq 0$

28. $f^{-1}(x) = \sqrt[3]{x-2} + 5$

29. $f^{-1}(x) = (10^x + 3)^2$ **30.** $\{(1, 3), (4, 5)\}$

31. $f^{-1}(x) = 5 + \dfrac{1}{x-3}$

32. $f^{-1}(x) = (\ln(3-x))^2$ for $x \leq 2$

33. $(p \circ m)(x) = e^{x+5}$, domain $(-\infty, \infty)$, range $(0, \infty)$

34. $(p \circ q)(x) = e^{\sqrt{x}}$, domain $[0, \infty)$, range $[1, \infty)$

35. $(q \circ p \circ m)(x) = q\left(e^{x+5}\right) = \sqrt{e^{x+5}}$, domain $(-\infty, \infty)$, range $(0, \infty)$

36. $(m \circ r \circ q)(x) = m\left(\ln(\sqrt{x})\right) = \ln(\sqrt{x}) + 5$, domain $(0, \infty)$, range $(-\infty, \infty)$

37. $(p \circ r \circ m)(x) = p\left(\ln(x+5)\right) = e^{\ln(x+5)}$, domain $(-5, \infty)$, range $(0, \infty)$

38. $(r \circ q \circ p)(x) = r\left(\sqrt{e^x}\right) = \ln\left(\sqrt{e^x}\right) = \frac{1}{2}x$, domain $(-\infty, \infty)$, range $(-\infty, \infty)$

39. $F(x) = (f \circ g \circ h)(x)$

40. $H(x) = (g \circ h \circ f)(x)$

41. $G(x) = (h \circ g \circ f)(x)$

42. $M(x) = (h \circ f \circ g)(x)$

43. origin

44. circle

45. rise, run

46. point-slope form

47. slope-intercept form

48. perpendicular

49. parallel

50. quadratic

51. quadratic

52. equivalent

For Thought

1. True **2.** False

3. False, $5°$ is coterminal with $-355°$.

4. False **5.** True, since $\frac{38\pi}{4} = \frac{19\pi}{2}$.

6. False, $210° = \frac{7\pi}{6}$.

7. False, since $25°20'40'' \neq 25.34°$; $25.34°$ is an approximation to $25°20'40''$.

8. True, since Seattle makes an angle of 2π every 24 hours, then the angular velocity is $\frac{2\pi}{24} = \frac{\pi}{12}$ radians per hour.

9. False, Seattle has a smaller linear velocity since its orbit about the axis of the earth is smaller than the orbit of Los Angeles.

10. True

5.1 Exercises

1. angle

3. standard position

5. obtuse

7. coterminal

9. minute

11. unit

13. Substitute $k = 1, 2, -1, -2$ into $60° + k \cdot 360°$ to obtain the coterminal angles

$$420°, 780°, -300°, -660°.$$

There are other coterminal angles.

15. Substitute $k = 1, 2, -1, -2$ into $-16° + k \cdot 360°$ to find the coterminal angles

$$344°, 704°, -376°, -736°.$$

There are other coterminal angles.

17. Yes, since $123.4° - (-236.6°) = 360°$ is an integral multiple of $360°$.

19. No, since $1055° - (155°) = 900° = k \cdot 360°$ does not have an integral solution for any k.

21. Quadrant I

23. $-125°$ lies in Quadrant III since $-125° + 360° = 235°$ and $180° < 235° < 270°$

25. Quadrant IV

27. $750°$ lies in Quadrant I since $750° - 720° = 30°$

29. $45°$

31. $60°$

33. $120°$

35. $400° - 360° = 40°$

37. $-340° + 360° = 20°$

39. $-1100° + 4 \cdot 360° = 340°$

41. $13° + \frac{12}{60}° = 13.2°$

43. $-8° - \frac{30}{60}° - \frac{18}{3600}° = -8.505°$

45. $28° + \frac{5}{60}° + \frac{9}{3600}° \approx 28.0858°$

47. $75.5° = 75°30'$ since $0.5(60) = 30$

49. $-17.33° = -17°19'48''$ since $0.33(60) = 19.8$ and $0.8(60) = 48$

51. $18.123° \approx 18°7'23''$ since $0.123(60) = 7.38$ and $0.38(60) \approx 23$

53. $\frac{\pi}{6}$

55. $18° \cdot \frac{\pi}{180} = \frac{\pi}{10}$

57. $-67.5° \cdot \frac{\pi}{180} = -\frac{135\pi}{360} = -\frac{3\pi}{8}$

59. $630° \cdot \frac{\pi}{180} = \frac{7\pi}{2}$

61. $37.4° \cdot \frac{\pi}{180} \approx 0.653$

63. $\left(-13 - \frac{47}{60}\right) \cdot \frac{\pi}{180} \approx -0.241$

65. $\left(-53 - \frac{37}{60} - \frac{6}{3600}\right) \cdot \frac{\pi}{180} \approx -0.936$

67. $\frac{5\pi}{12} \cdot \frac{180}{\pi} = 75^\circ$

69. $\frac{7\pi}{4} \cdot \frac{180}{\pi} = 315^\circ$

71. $-6\pi \cdot \frac{180}{\pi} = -1080^\circ$

73. $2.39 \cdot \frac{180}{\pi} \approx 136.937^\circ$

75. Substitute $k = 1, 2, -1, -2$ into $\frac{\pi}{3} + k \cdot 2\pi$ to obtain the coterminal angles

$$\frac{7\pi}{3}, \frac{13\pi}{3}, -\frac{5\pi}{3}, -\frac{11\pi}{3}.$$

There are other coterminal angles.

77. Substitute $k = 1, 2, -1, -2$ into $-\frac{\pi}{6} + k \cdot 2\pi$ to find the coterminal angles

$$\frac{11\pi}{6}, \frac{23\pi}{6}, -\frac{13\pi}{6}, -\frac{25\pi}{6}.$$

There are other coterminal angles.

79. $3\pi - 2\pi = \pi$

81. $\frac{9\pi}{2} - 4\pi = \frac{\pi}{2}$

83. $-\frac{5\pi}{3} + 2\pi = \frac{\pi}{3}$

85. $-\frac{13\pi}{3} + 6\pi = \frac{5\pi}{3}$

87. $8.32 - 2\pi \approx 2.04$

89. No, since $\frac{29\pi}{4} - \frac{3\pi}{4} = \frac{26\pi}{4} = k \cdot 2\pi$ does not have an integral solution for any k.

91. Yes, since $\frac{7\pi}{6} - \frac{-5\pi}{6} = \frac{12\pi}{6} = 2\pi$.

93. Quadrant I

95. Quadrant III

97. $\frac{13\pi}{8}$ lies in Quadrant IV since
$\frac{3\pi}{2} = \frac{12\pi}{8} < \frac{13\pi}{8} < 2\pi$

99. Note $2\pi \approx 6.28$ and $3\pi/2 \approx 4.71$. Since -7.3 is coterminal with $-7.3 + 2(6.28) = 5.26$ and $4.71 < 5.26 < 6.28$, it follows that -7.3 lies in Quadrant IV.

101. $30^\circ = \frac{\pi}{6}$, $45^\circ = \frac{\pi}{4}$, $60^\circ = \frac{\pi}{3}$, $90^\circ = \frac{\pi}{2}$,
$120^\circ = \frac{2\pi}{3}$, $135^\circ = \frac{3\pi}{4}$, $150^\circ = \frac{5\pi}{6}$, $180^\circ = \pi$,
$210^\circ = \frac{7\pi}{6}$, $225^\circ = \frac{5\pi}{4}$, $240^\circ = \frac{4\pi}{3}$,
$270^\circ = \frac{3\pi}{2}$, $300^\circ = \frac{5\pi}{3}$, $315^\circ = \frac{7\pi}{4}$,
$330^\circ = \frac{11\pi}{6}$, $360^\circ = 2\pi$

103. $s = 12 \cdot \frac{\pi}{4} = 3\pi$ ft

105. $s = 4000 \cdot \frac{3\pi}{180} \approx 209.4$ miles

107. radius is $r = \frac{s}{\alpha} = \frac{1}{1} = 1$ mile.

109. radius is $r = \frac{s}{\alpha} = \frac{10}{\pi} \approx 3.18$ km

111. Distance from Peshtigo to the North Pole is

$$s = r\alpha = 3950\left(45 \cdot \frac{\pi}{180}\right) \approx 3102 \text{ miles.}$$

113. central angle is $\alpha = \frac{2000}{3950} \approx 0.506329$ radians
$\approx 0.506329 \cdot \frac{180}{\pi} \approx 29.0^\circ$

115. Linear velocity is $v = \frac{s}{t} = \frac{r\alpha}{t} =$
$\frac{6 \cdot (10,350) \cdot 2\pi}{1} \approx 390,185.8$ cm/min.

117. The radius of the blade is
10 in. $= 10 \cdot \frac{1}{12 \cdot 5280} \approx 0.0001578$ miles.
Since the angle rotated in one hour is

$$2800 \cdot 2\pi \cdot 60 = 336,000\pi$$

the linear velocity is

$$v = \frac{r\alpha}{t} \approx \frac{(0.0001578)(336,000\pi)}{1} \approx 166.6 \text{ mph.}$$

119. In 1 hr,the saw rotates through an angle of $3450(60)\cdot 2\pi$. After converting the radii into miles, the linear velocity is

$$3450(60)\cdot 2\pi\left(\frac{6}{12(5280)}-\frac{5}{12(5280)}\right)\approx$$

20.5 mph.

121. The angular velocity of any point on the surface of the earth is $w=\dfrac{\pi}{12}$ rad/hr.

A point 1 mile from the North Pole is approximately 1 mile from the axis of the earth. The linear velocity of that point is $v=w\cdot r=\dfrac{\pi}{12}\cdot 1\approx 0.26$ mph.

123. Since $7^\circ\approx 0.12217305$, the radius of the earth according to Eratosthenes is

$$r=\frac{s}{\alpha}\approx\frac{800}{0.12217305}\approx 6548.089 \text{ km}.$$

The circumference is

$$2\pi r\approx 41,143 \text{ km}.$$

Using $r = 6378$ km, the circumference is 40,074 km.

125. The area of a 16-inch diameter pizza is $\pi r^2=\pi\cdot 8^2=64\pi$. The area of one slice is $\dfrac{64\pi}{6}\approx 33.5 \text{ in}^2$.

127. Since the velocity at point A is 10 ft/sec, the linear velocity at B and C are both 10 ft/sec. The angular velocity at B is

$$\omega=\frac{v}{r}=\frac{10 \text{ ft/sec}}{5/12 \text{ ft}}=24 \text{ rad/sec}.$$

While, the angular velocity at C is

$$\omega=\frac{v}{r}=\frac{10 \text{ ft/sec}}{3/12 \text{ ft}}=40 \text{ rad/sec}.$$

129. Let t be the number of minutes after 12 noon. The number of degrees spanned by the minute hand and hour hand since 12 noon are $6t$ and $t/2$, respectively. If the minute hand and the hour hand form a 90-degree angle, then

$$6t-\frac{t}{2}=90 \text{ or } 6t-\frac{t}{2}=270.$$

The solutions to this equation are

$$y=\frac{180}{11}\text{ min}\approx 16 \text{ min}, 22 \text{ sec}.$$

and

$$y=\frac{540}{11}\text{ min}\approx 49 \text{ min}, 5 \text{ sec.},$$

respectively. Thus, the two hands form a 90-degree angle at 12:16:22 and 12:49:05.

131. A region S bounded by a chord and a circle of radius 10 meters is shaded below. The central angle is 60°. The area A_s of S may be obtained by subtracting the area of an equilateral triangle from the area of a sector. That is,

$$A_s=100\left(\frac{\pi}{6}-\frac{\sqrt{3}}{4}\right).$$

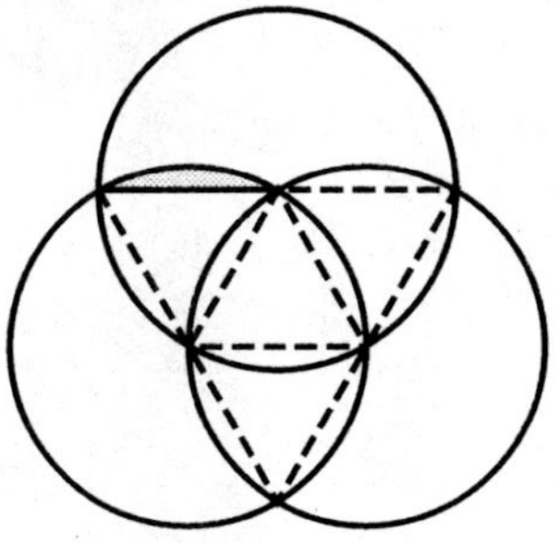

The common region bounded by two or three circles consists of 4 equilateral triangles and six regions like S.

The area of the region inside the higher circle but outside the common region is $100(\pi/2-2A_s)$. There are two other such regions for the two circles on the left and right.

Then the total area A watered by the three circular sprinklers is

$$\begin{aligned} A &= \left[100\sqrt{3}+6A_s+3\left(\frac{100\pi}{2}-2A_s\right)\right] \\ &= 100\left[\frac{3\pi}{2}+\sqrt{3}\right] \text{ m}^2 \\ &\approx 644.4 \text{ m}^2. \end{aligned}$$

133. a) Given an angle α (in degrees) as in the problem, the radius r of the cone must satisfy

$$2\pi r=8\pi-4\alpha\frac{\pi}{180}$$

Note, 8π inches is the circumference of a circle with radius 4 inches. Then we obtain

$$r = 4 - \frac{\alpha}{90}.$$

Note, $h = \sqrt{16 - r^2}$ by the Pythagorean theorem. Since the volume $V(\alpha)$ of the cone is $\frac{\pi}{3}r^2h$,

$$V(\alpha) = \frac{\pi}{3}\left(4 - \frac{\alpha}{90}\right)^2\sqrt{16 - \left(4 - \frac{\alpha}{90}\right)^2}.$$

This reduces to

$$V(\alpha) = \frac{\pi(360 - \alpha)^2\sqrt{720\alpha - \alpha^2}}{2,187,000}$$

If $\alpha = 30°$, then $V(30°) \approx 22.5$ inches^3.

b) As shown in part a), the volume of the cone obtained by an overlapping angle α is

$$V(\alpha) = \frac{\pi(360 - \alpha)^2\sqrt{720\alpha - \alpha^2}}{2,187,000}$$

135. If s is the arc length, the radius is $r = \frac{6-s}{2}$. Let θ be the central angle such that

$$s = r\theta = \frac{(6 - s)\theta}{2}.$$

Solving for s, we find

$$s = \frac{6\theta}{\theta + 2}.$$

The area of the sector is

$$A = \frac{\theta r^2}{2} = \frac{\theta}{2}\left(\frac{s}{\theta}\right)^2 = \frac{s^2}{2\theta}.$$

Substituting, we obtain

$$\begin{aligned} A &= \frac{1}{2\theta}\left(\frac{6\theta}{\theta + 2}\right)^2 \\ &= \frac{18\theta}{(\theta + 2)^2}. \end{aligned}$$

137. Since $x^2 + 3x - 10 = (x + 5)(x - 2) = 0$, the solution set is $\{-5, 2\}$.

139. Writing in exponential form, we have

$$\begin{aligned} \frac{x}{x + 3} &= 2^{-3} \\ 8x &= x + 3 \\ 7x &= 3 \end{aligned}$$

The solution set is $\{3/7\}$.

141. Set one side to zero:

$$\frac{2x - 1}{x + 3} - 1 = \frac{x - 4}{x + 3} = f(x) \geq 0$$

If $x = -2$, then $f(-2) > 0$.
If $x = 0$, then $f(0) < 0$.
If $x = 5$, then $f(5) > 0$.

$$\begin{array}{ccccc} + & \text{U} & - & 0 & + \\ \hline -2 & -3 & 0 & 4 & 5 \end{array}$$

The solution set is $(-\infty, -3) \cup [4, \infty)$.

143. Let r be the radius of the circle. Let p be the distance from the lower left most corner of the 1-by-1 square to the point of tangency of the left most, lower most circle at the base of the 1-by-1 square. By the Pythagorean Theorem,

$$(p + r)^2 + (r + 1 - p)^2 = 1$$

or equivalently

$$p^2 + r^2 + r - p = 0. \quad (1)$$

Since the area of the four triangles plus the area of the small square in the middle is 1, we obtain

$$4r^2 + 4\left[\frac{1}{2}(p + r)(r + 1 - p)\right] = 1$$

or

$$6r^2 - 2p^2 + 2p + 2r = 1. \quad (2)$$

Multiply (1) by two and add the result to (2). Then $8r^2 + 4r = 1$. The solution is $r = \dfrac{\sqrt{3} - 1}{4}$.

For Thought

1. False, since $\cos 90° = 0$.

2. False, since $\cos 90 \approx -0.4$.

3. True, since $\sin(45°) = \dfrac{\sqrt{2}}{2} = \dfrac{1}{\sqrt{2}}$.

4. False, since

$$\sin\left(-\frac{\pi}{3}\right) = -\frac{\sqrt{3}}{2}$$

and

$$\sin\left(\frac{\pi}{3}\right) = \frac{\sqrt{3}}{2}.$$

5. False, since

$$\cos\left(-\frac{\pi}{3}\right) = \frac{1}{2}$$

and

$$-\cos\left(\frac{\pi}{3}\right) = -\frac{1}{2}.$$

6. True, since the reference arc of 390° is 30° and 390° lies in Quadrant I.

7. False, since α lies in Quadrant IV.

8. False, since $\sin(\alpha) = -\frac{1}{2}$.

9. False, since possibly $\alpha = \frac{13\pi}{6}$.

10. True, since $(1 - \sin\alpha)(1 + \sin\alpha) = 1 - \sin^2\alpha = \cos^2\alpha$.

5.2 Exercises

1. $\sin\alpha$, $\cos\alpha$

3. domain

5. $(1, 0), \left(\frac{\sqrt{2}}{2}, \frac{\sqrt{2}}{2}\right), (0, 1), \left(-\frac{\sqrt{2}}{2}, \frac{\sqrt{2}}{2}\right),$
$(-1, 0), \left(-\frac{\sqrt{2}}{2}, -\frac{\sqrt{2}}{2}\right), (0, -1), \left(\frac{\sqrt{2}}{2}, -\frac{\sqrt{2}}{2}\right)$

7. 0 **9.** 0

11. 0 **13.** 0

15. $\frac{\sqrt{2}}{2}$ **17.** $-\frac{\sqrt{2}}{2}$

19. $\frac{1}{2}$ **21.** $\frac{1}{2}$

23. $-\frac{\sqrt{3}}{2}$ **25.** $\frac{\sqrt{3}}{2}$

27. $\sin(390°) = \sin(30°) = \frac{1}{2}$

29. $\cos(-420°) = \cos(300°) = \frac{1}{2}$

31. $\cos\left(\frac{13\pi}{6}\right) = \cos\left(\frac{\pi}{6}\right) = \frac{\sqrt{3}}{2}$

33. $30°, \pi/6$

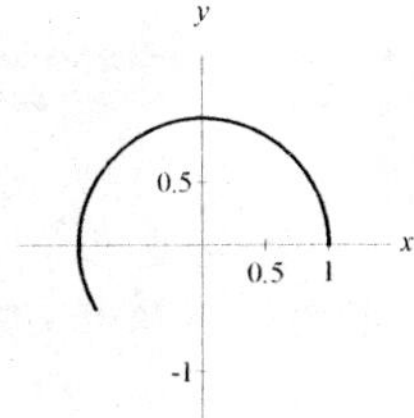

35. $60°, \pi/3$

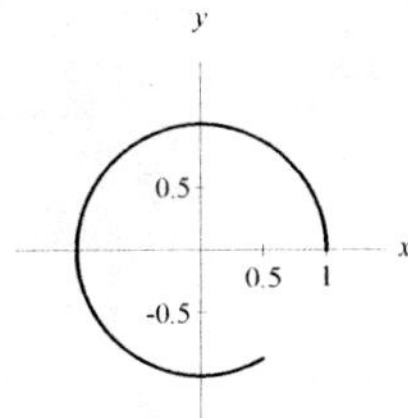

37. $60°, \pi/3$

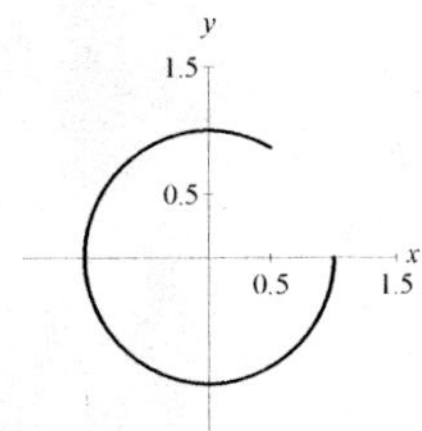

39. $30°, \pi/6$

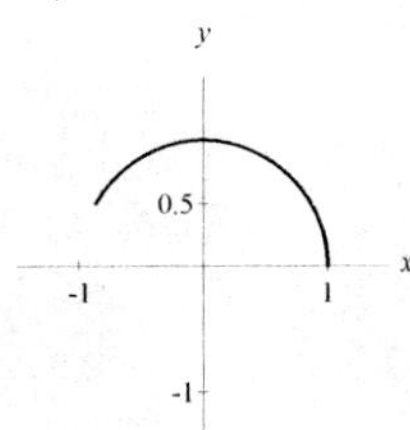

41. $45°, \pi/4$

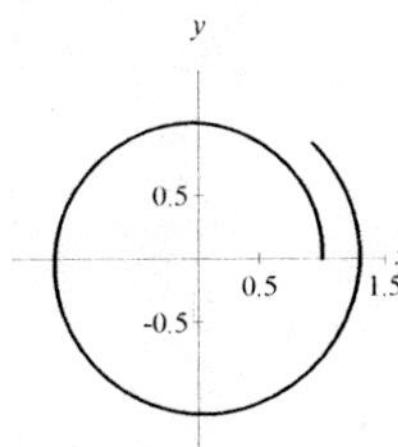

43. $45°, \pi/4$

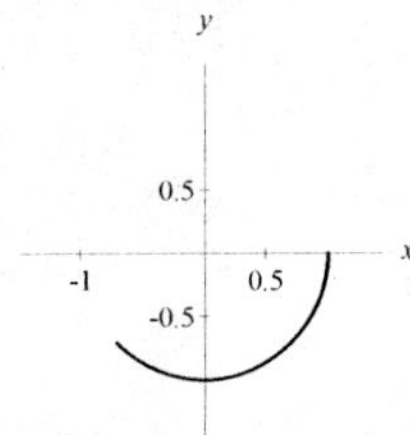

45. +, since sine is positive in the 2nd quadrant

47. +, since cosine is positive in the 4th quadrant

49. −, since sine is negative in the 3rd quadrant

51. −, since cosine is negative in the 3rd quadrant

53. $\sin(135°) = \sin(45°) = \dfrac{\sqrt{2}}{2}$

55. $\cos\left(\dfrac{5\pi}{3}\right) = \cos\left(\dfrac{\pi}{3}\right) = \dfrac{1}{2}$

57. $\sin\left(\dfrac{7\pi}{4}\right) = -\sin\left(\dfrac{\pi}{4}\right) = -\dfrac{\sqrt{2}}{2}$

59. $\cos\left(-\dfrac{17\pi}{6}\right) = -\cos\left(\dfrac{\pi}{6}\right) = -\dfrac{\sqrt{3}}{2}$

61. $\sin(-45°) = -\sin(45°) = -\dfrac{\sqrt{2}}{2}$

63. $\cos(-240°) = -\cos(60°) = -\dfrac{1}{2}$

65. $\dfrac{\cos(\pi/3)}{\sin(\pi/3)} = \dfrac{1/2}{\sqrt{3}/2} = \dfrac{1}{\sqrt{3}} = \dfrac{\sqrt{3}}{3}$

67. $\dfrac{\sin(7\pi/4)}{\cos(7\pi/4)} = \dfrac{-\sqrt{2}/2}{\sqrt{2}/2} = -1$

69. $\sin\left(\dfrac{\pi}{3} + \dfrac{\pi}{6}\right) = \sin\left(\dfrac{\pi}{2}\right) = 1$

71. $\dfrac{1 - \cos(5\pi/6)}{\sin(5\pi/6)} = \dfrac{1 - (-\sqrt{3}/2)}{1/2} \cdot \dfrac{2}{2} = 2 + \sqrt{3}$

73. $\dfrac{\sqrt{2}}{2} + \dfrac{\sqrt{2}}{2} = \sqrt{2}$

75. 0.9999

77. 0.4035

79. −0.7438

81. 1.0000

83. −0.2588

85. $\sin\left(\dfrac{\pi}{2}\right) = 1$

87. $\cos\left(\dfrac{\pi}{3}\right) = \dfrac{1}{2}$

89. $\sin\left(\dfrac{3\pi}{4}\right) = \dfrac{\sqrt{2}}{2}$

91. $\cos\left(\dfrac{\pi}{6}\right) = \dfrac{\sqrt{3}}{2}$

93. Use the Fundamental Identity.

$$\begin{aligned} \left(\frac{5}{13}\right)^2 + \cos^2(\alpha) &= 1 \\ \frac{25}{169} + \cos^2(\alpha) &= 1 \\ \cos^2(\alpha) &= \frac{144}{169} \\ \cos(\alpha) &= \pm\frac{12}{13} \end{aligned}$$

Since α is in quadrant II, $\cos(\alpha) = -12/13$.

95. Use the Fundamental Identity.

$$\begin{aligned} \left(\frac{3}{5}\right)^2 + \sin^2(\alpha) &= 1 \\ \frac{9}{25} + \sin^2(\alpha) &= 1 \\ \sin^2(\alpha) &= \frac{16}{25} \\ \sin(\alpha) &= \pm\frac{4}{5} \end{aligned}$$

Since α is in quadrant IV, $\sin(\alpha) = -4/5$.

97. Use the Fundamental Identity.

$$\begin{aligned} \left(\frac{1}{3}\right)^2 + \cos^2(\alpha) &= 1 \\ \frac{1}{9} + \cos^2(\alpha) &= 1 \\ \cos^2(\alpha) &= \frac{8}{9} \\ \cos(\alpha) &= \pm\frac{2\sqrt{2}}{3} \end{aligned}$$

Since $\cos(\alpha) > 0$, $\cos(\alpha) = \dfrac{2\sqrt{2}}{3}$.

99. Since $x(t) = 4\sin(t) - 3\cos(t)$, the location of the weight after 3 seconds is $x(3) = 4\sin(3) - 3\cos(3) \approx 3.53$ cm. It is below its equilibrium position.

101. The angle between the tips of two adjacent teeth is $\dfrac{2\pi}{22} = \dfrac{\pi}{11}$. The actual distance is

$c = 6\sqrt{2 - 2\cos(\pi/11)} \approx 1.708$ in.

The length of the arc is $s = 6 \cdot \frac{\pi}{11} \approx 1.714$ in.

103. Note, $\cos^2\alpha = 1 - \sin^2\alpha$ or $\cos(\alpha) = \pm\sqrt{1-\sin^2\alpha}$. Then $\cos\alpha = \sqrt{1-\sin^2\alpha}$ if the terminal side of α lies in quadrant I or IV, while $\cos\alpha = -\sqrt{1-\sin^2\alpha}$ if the terminal side of α lies in quadrant II or III.

105. If $R = 3$, $r = 14$, and $\theta = 60°$, then

$$\begin{aligned} T &= \frac{14\cos 60° - 3}{\sin 60°} \\ &= \frac{14(1/2) - 3}{\sqrt{3}/2} \\ &\approx 4.6 \text{ in.} \end{aligned}$$

107. $48°13.8' = 48°13'48''$

109. The factors of 6 are $p = \pm1 \pm 2, \pm3, \pm6$. The factors of 4 are $q = \pm1, \pm2, \pm4$.

Then the possible rational solutions are

$\frac{p}{q} = \pm1, \pm2, \pm3, \pm6, \pm\frac{1}{4}, \pm\frac{1}{2}, \pm\frac{3}{2}, \pm\frac{3}{4}$

111. Using $3x - 9 \geq 0$ or $x \geq 3$, it follows that the domain is $[3, \infty)$.

Note, $-\sqrt{w}$ takes all the values in $(-\infty, 0]$ as w assumes all the nonnegative values. Then the range of $y = 1 - \sqrt{3x-9}$ is $(-\infty, 1]$.

113.

a) Let t be a fraction of an hour, i.e., $0 \leq t \leq 1$. If the angle between the hour and minute hands is 120°, then

$$\begin{aligned} 360t - 30t &= 120 \\ 330t &= 120 \\ t &= \frac{12}{33}\text{hr} \\ t &\approx 21 \text{ min}, 49.1 \text{ sec.} \end{aligned}$$

Thus, the hour and minute hands will be 120° apart when the time is 12:21:49.1. Moreover, this is the only time between 12 noon and 1 pm that the angle is 120°.

b) We measure angles clockwise from the 12 0'clock position.

At 12:21:49.1, the hour hand is pointing to the 10.9°-angle, the minute hand is pointing to the 130.9°-angle, and the second hand is pointing to the 294.5°-angle. Thus, the three hands of the clock cannot divide the face of the clock into thirds.

c) No, as discussed in part b).

For Thought

1. False, the period is 1.

2. False, the range is $[-1, 7]$.

3. False, the phase shift is $-\pi/12$.

4. True

5. True

6. True, $\frac{2\pi}{0.1\pi} = 20$.

7. False

8. True, since $\frac{2\pi}{b} = \frac{2\pi}{4} = \frac{\pi}{2}$.

9. True

10. True

5.3 Exercises

1. sine wave

3. periodic

5. amplitude

7. period

9. $y = -2\sin(x)$, amplitude 2

11. $y = 3\cos(x)$, amplitude 3

13. Amplitude 2, period 2π, phase shift 0

15. Amplitude 1, period 2π, phase shift $\pi/2$

17. Amplitude 2, period 2π, phase shift $-\pi/3$

19. Amplitude 1, phase shift 0, some points are $(0,0)$, $\left(\frac{\pi}{2},-1\right)$, $(\pi,0)$, $\left(\frac{3\pi}{2},1\right)$,$(2\pi,0)$

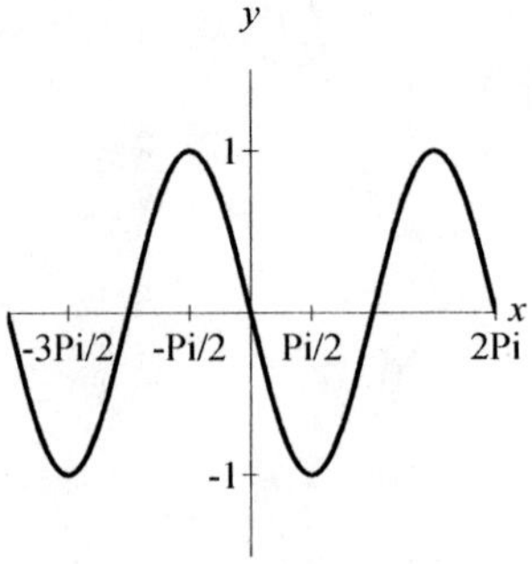

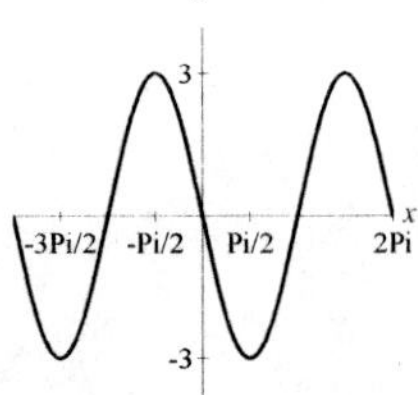

21. Amplitude 3, phase shift 0, some points are $(0,0)$, $\left(\frac{\pi}{2},-3\right)$, $(\pi,0)$, $\left(\frac{3\pi}{2},3\right)$, $(2\pi,0)$

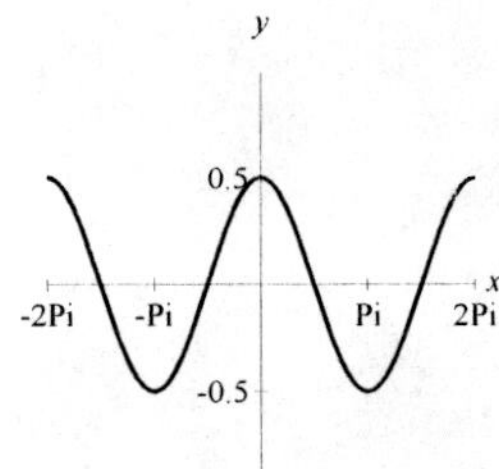

23. Amplitude 1/2, phase shift 0, some points are $(0,1/2)$, $(\pi/2,0)$, $(\pi,-1/2)$ $(3\pi/2,0)$, $(2\pi,1/2)$

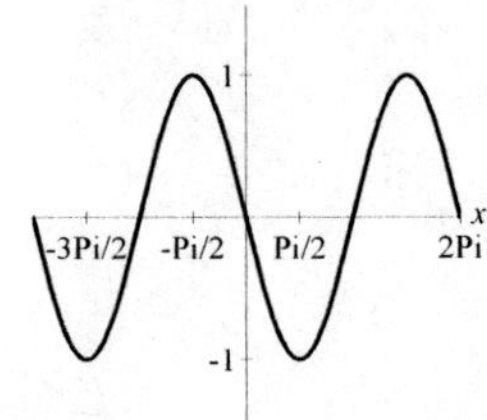

25. Amplitude 1, phase shift $-\pi$, some points are $(0,0)$, $\left(\frac{\pi}{2},-1\right)$, $(\pi,0)$, $\left(\frac{3\pi}{2},1\right)$, $(2\pi,0)$

27. Amplitude 1, phase shift $\pi/3$, some points are $\left(-\frac{2\pi}{3},-1\right)$, $\left(-\frac{\pi}{6},0\right)$, $\left(\frac{\pi}{3},1\right)$, $\left(\frac{5\pi}{6},0\right)$, $\left(\frac{4\pi}{3},-1\right)$

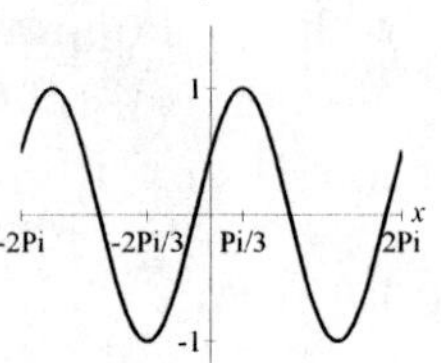

29. Amplitude 1, phase shift 0, some points are $(0,3)$, $\left(\frac{\pi}{2},2\right)$, $(\pi,1)$ $\left(\frac{3\pi}{2},2\right)$, $(2\pi,3)$

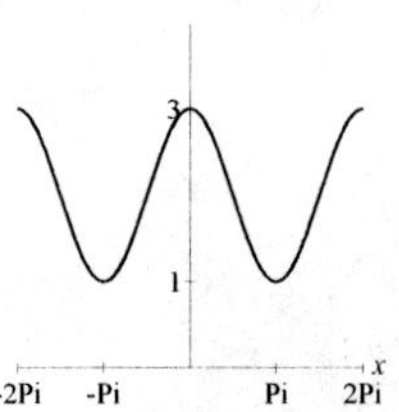

31. Amplitude 1, phase shift 0, some points are $(0,-1)$, $\left(\frac{\pi}{2},-2\right)$, $(\pi,-1)$,$\left(\frac{3\pi}{2},0\right)$, $(2\pi,-1)$

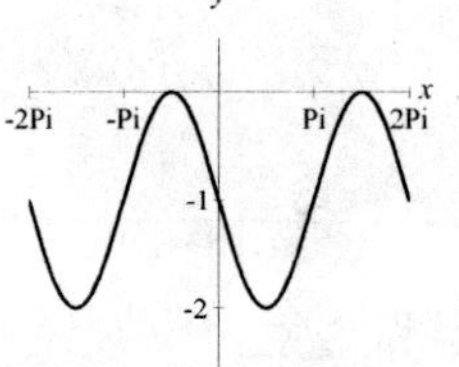

33. Amplitude 1, phase shift $-\pi/4$, some points are $\left(-\frac{\pi}{4},2\right)$, $\left(\frac{\pi}{4},3\right)$, $\left(\frac{3\pi}{4},2\right)$, $\left(\frac{5\pi}{4},1\right)$, $\left(\frac{7\pi}{4},2\right)$

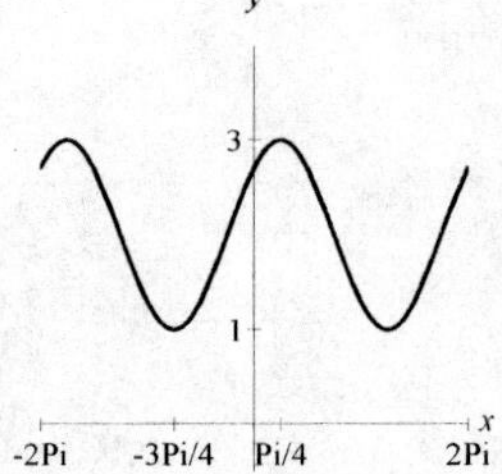

35. Amplitude 2, phase shift $-\pi/6$, some points are $\left(-\frac{\pi}{6}, 3\right)$, $\left(\frac{\pi}{3}, 1\right)$, $\left(\frac{5\pi}{6}, -1\right)$, $\left(\frac{4\pi}{3}, 1\right)$, $\left(\frac{11\pi}{6}, 3\right)$

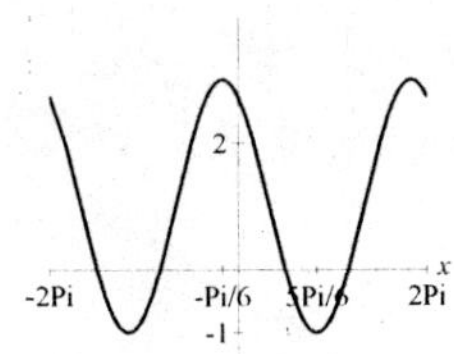

37. Amplitude 2, phase shift $\pi/3$, some points are $\left(-\frac{\pi}{6}, 3\right)$, $\left(\frac{\pi}{3}, 1\right)$, $\left(\frac{5\pi}{6}, -1\right)$, $\left(\frac{4\pi}{3}, 1\right)$, $\left(\frac{11\pi}{6}, 3\right)$

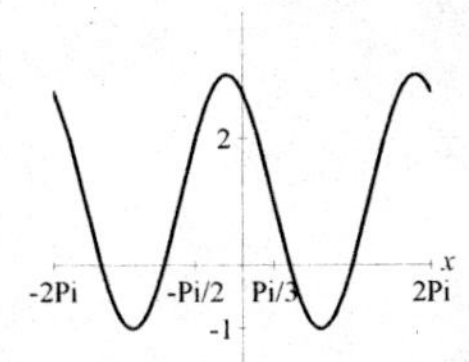

39. Amplitude 3, period $\pi/2$, phase shift 0

41. Amplitude 1, period $\frac{2\pi}{1/2}$ or 4π, phase shift 0

43. Amplitude 2, period 2π, phase shift π

45. Amplitude 2; since $y = -2\cos(2(x+\pi/4))$, the period is π and the phase shift is $-\pi/4$

47. Amplitude 2; since $y = -2\cos\left(\frac{\pi}{2}(x+2)\right)$, the period is $\frac{2\pi}{\pi/2}$ or 4, and the phase shift is -2

49. Note, $A = (7-3)/2 = 2$. Since $A + D = 2 + D = 7$, we find that $D = 5$. Since $C = -\pi/2$ and $2\pi/B = \pi$, we obtain $B = 2$. Thus,

$$y = 2\sin\left(2\left(x + \frac{\pi}{2}\right)\right) + 5.$$

51. Note, $A = (9 - (-1))/2 = 5$. Since $A + D = 5 + D = 9$, we obtain that $D = 4$. Since $C = 2$ and $2\pi/B = 2$, we find $B = \pi$. Thus,

$$y = 5\sin(\pi(x-2)) + 4.$$

53. Note, $A = (3 - (-9))/2 = 6$. Since

$$A + D = 6 + D = 3$$

we find that $D = -3$. Since $C = -\pi$ and $2\pi/B = 1/2$, we find $B = 4\pi$. Hence,

$$y = 6\sin(4\pi(x+\pi)) - 3.$$

55. $y = -\sin\left(x - \frac{\pi}{4}\right) + 1$

57. $y = -[3\cos(x-\pi) - 2]$ or equivalently $y = -3\cos(x-\pi) + 2$

59. $F(x) = \sin\left(3x - \frac{\pi}{4}\right)$

61. $F(x) = \sin\left(3x - \frac{3\pi}{4}\right)$

63. Period $2\pi/3$, phase shift 0, range $[-1, 1]$, labeled points are $(0,0)$, $\left(\frac{\pi}{6}, 1\right)$, $\left(\frac{\pi}{3}, 0\right)$, $\left(\frac{\pi}{2}, -1\right)$, $\left(\frac{2\pi}{3}, 0\right)$

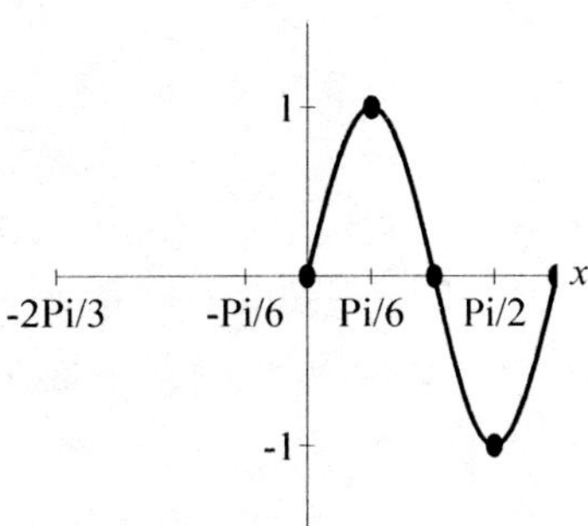

65. Period π, phase shift 0, range $[-1, 1]$, labeled points are $(0,0)$, $\left(\frac{\pi}{4}, -1\right)$, $\left(\frac{\pi}{2}, 0\right)$, $\left(\frac{3\pi}{4}, 1\right)$, $(\pi, 0)$

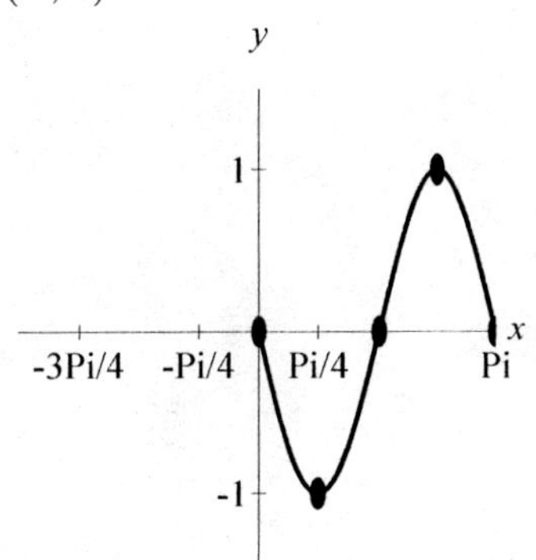

67. Period $\pi/2$, phase shift 0, range $[1, 3]$, labeled points are $(0, 3)$, $\left(\frac{\pi}{8}, 2\right)$, $\left(\frac{\pi}{4}, 1\right)$, $\left(\frac{3\pi}{8}, 2\right)$, $\left(\frac{\pi}{2}, 3\right)$

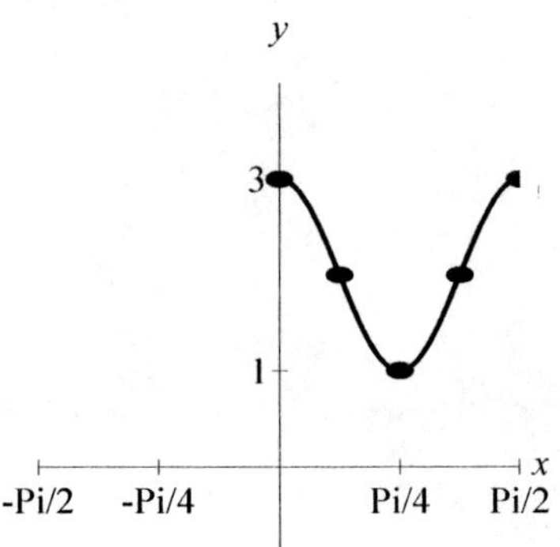

69. Period 8π, phase shift 0, range $[1, 3]$, labeled points are $(0, 2)$, $(2\pi, 1)$, $(4\pi, 2)$, $(6\pi, 3)$, $(8\pi, 2)$

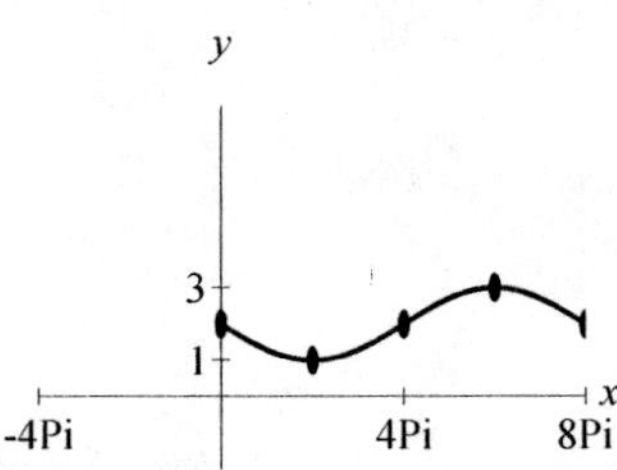

71. Period 6, phase shift 0, range $[-1, 1]$, labeled points are $(0, 0)$, $(1.5, 1)$, $(3, 0)$, $(4.5, -1)$, $(6, 0)$

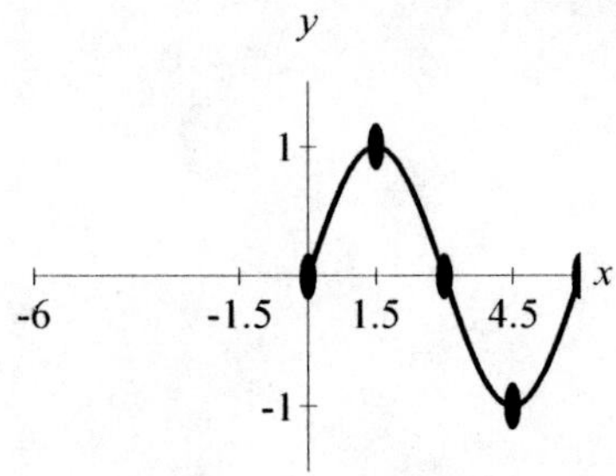

73. Period π, phase shift $\pi/2$, range $[-1, 1]$, labeled points are $\left(\frac{\pi}{2}, 0\right)$, $\left(\frac{3\pi}{4}, 1\right)$, $(\pi, 0)$, $\left(\frac{5\pi}{4}, -1\right)$, $\left(\frac{3\pi}{2}, 0\right)$

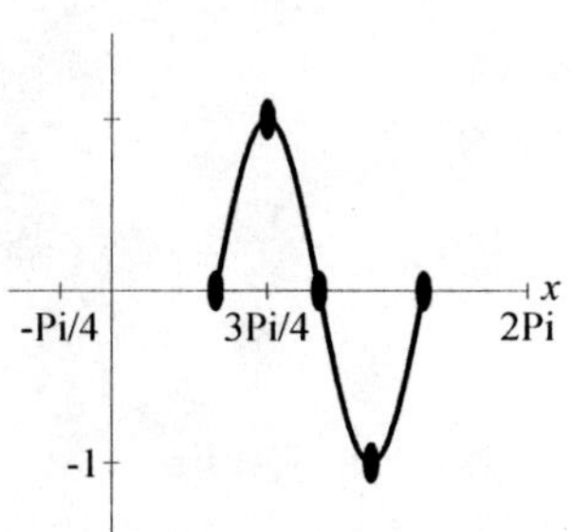

75. Period 4, phase shift -3, range $[-1, 1]$, labeled points are $(-3, 0)$, $(-2, 1)$, $(-1, 0)$, $(0, -1)$, $(1, 0)$

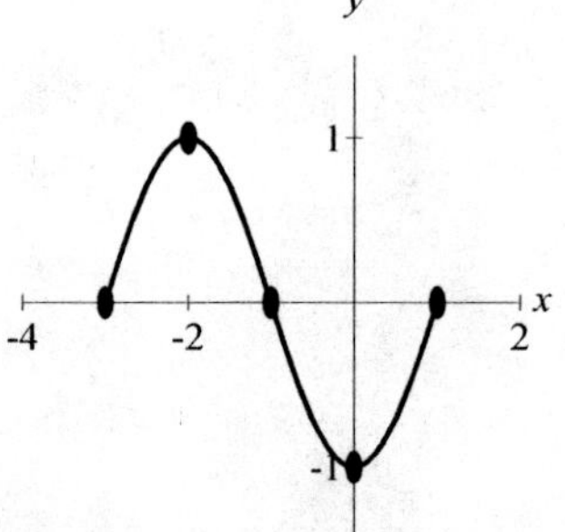

77. Period π, phase shift $-\pi/6$, range $[-1, 3]$, labeled points are $\left(-\frac{\pi}{6}, 3\right)$, $\left(\frac{\pi}{12}, 1\right)$, $\left(\frac{\pi}{3}, -1\right)$, $\left(\frac{7\pi}{12}, 1\right)$, $\left(\frac{5\pi}{6}, 3\right)$

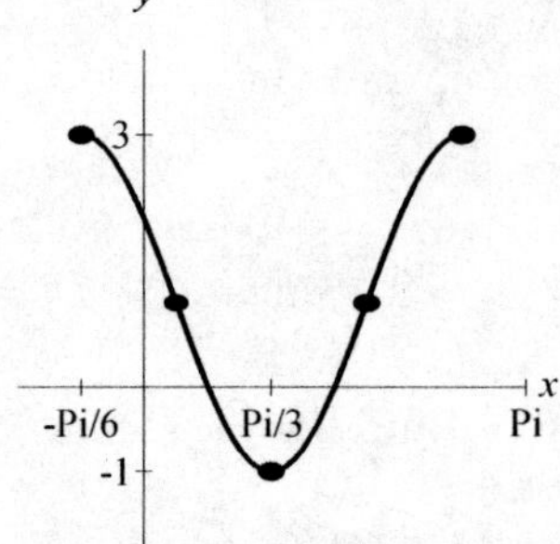

79. Period $\frac{2\pi}{3}$, phase shift $\frac{\pi}{6}$, range $\left[-\frac{3}{2}, -\frac{1}{2}\right]$,

labeled points are $\left(\frac{\pi}{6}, -1\right)$, $\left(\frac{\pi}{3}, -\frac{3}{2}\right)$, $\left(\frac{\pi}{2}, -1\right)$, $\left(\frac{2\pi}{3}, -\frac{1}{2}\right)$, $\left(\frac{5\pi}{6}, -1\right)$

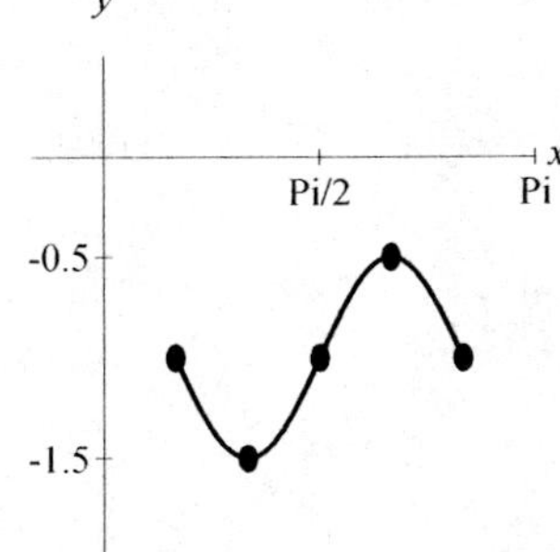

81. $y = 2\sin\left(2\left(x - \frac{\pi}{4}\right)\right)$

83. $y = 3\sin\left(\frac{3}{2}\left(x + \frac{\pi}{3}\right)\right) + 3$

85. 100 cycles per second

87. Frequency is $\frac{1}{0.025} = 40$ cycles per hour

89. Substitute $v_o = 6$, $\omega = 2$, and $x_o = 0$ into $x(t) = \frac{v_o}{\omega} \cdot \sin(\omega t) + x_o \cdot \cos(\omega t)$. Then

$$x(t) = 3\sin(2t).$$

The amplitude is 3 and period is π.

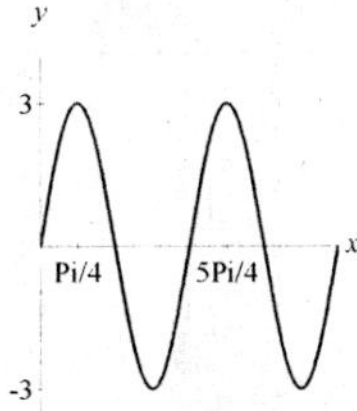

91. 11 years

93. Note that the maximum and minimum values of $\sin(t)$ are 1 and -1.
(a) Maximum volume is 1300 cc and minimum volume is 500 cc
(b) The runner takes a breath every 1/30 (which is the period) of a minute.Thus, the runner makes 30 breaths/minute.

95. Period is 12, amplitude is 15,000, phase shift is -3, vertical translation is 25,000, a formula for the curve is

$$y = 15,000\sin\left(\frac{\pi}{6}x + \frac{\pi}{2}\right) + 25,000.$$

For April (when $x = 4$), the revenue is $15,000\sin\left(\frac{\pi}{6}x + \frac{\pi}{2}\right) + 25,000 \approx \$17,500$.

97.

a) period is 40, amplitude is 65, an equation for the sine wave is $y = 65\sin\left(\frac{\pi}{20}x\right)$

b) 40 days

c) $65\sin\left(\frac{\pi}{20}(36)\right) \approx -38.2$ meters/second

d) The planet is between the Earth and Rho.

99. Since the period is 20, the amplitude is 1, and the vertical translation is 1, an equation for the swell is $y = \sin\left(\frac{\pi}{10}x\right) + 1$.

101. The sine regression curve is

$$y = 50\sin(0.214x - 0.615) + 48.8$$

or approximately

$$y = 50\sin(0.21x - 0.62) + 48.8$$

The period is

$$\frac{2\pi}{b} = \frac{2\pi}{0.214} \approx 29.4 \text{ days}.$$

When $x = 39$, we find

$$y = 50\sin(0.214(39) - 0.615) + 48.8 \approx 98\%$$

On February 8, 2020, 98% of the moon is illuminated. Shown below is a graph of the regression equation and the data points.

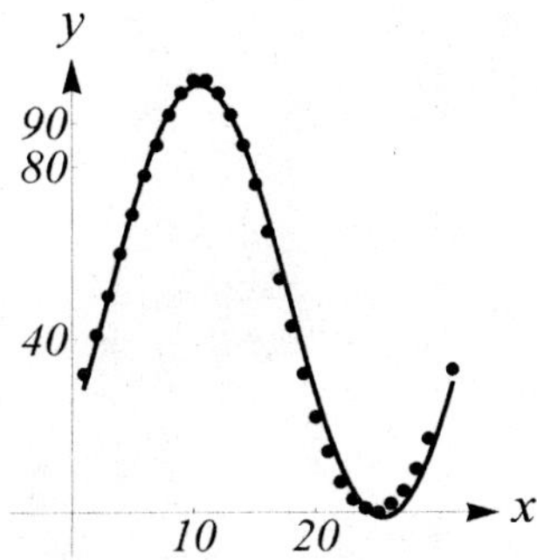

105. $225° \cdot \dfrac{\pi}{180°} = \dfrac{5\pi}{4}$

107. Since the radius is $r = 15$ inches, the linear velocity of the tip of the blade that is rotating at 2000 revolutions/minute is given by

$$v = 15(2000)2\pi \cdot \frac{60}{12(5280)} = 178.5 \text{ mph}$$

109. a) $\sin 45° = \dfrac{\sqrt{2}}{2}$, **b)** $\cos\left(\dfrac{-\pi}{4}\right) = \dfrac{\sqrt{2}}{2}$

111.

a) The contestants that leave the table are # 1, # 3, # 5, # 7, # 9, # 11, # 13, # 4, # 8, # 12, # 2, #6 (in this order). Then contestant # 10 is the unlucky contestant.

b) Let $n = 8$. The contestants that leave the table are # 1, # 3, # 5, # 7, # 2, # 6, # 4 (in this order). Thus, contestant # 8 is the unlucky contestant.

Let $n = 16$. The contestants that leave the table are # 1, # 3, # 5, # 7, # 9, # 11, # 13, # 15, # 2, # 6, # 10, # 14, # 4, # 12, # 8 (in this order). Thus, contestant # 16 is the unlucky contestant.

Let $n = 41$. The contestants that leave the table are # 1, # 3, # 5, # 7, # 9, # 11, # 13, # 15, # 17, # 19, # 21, # 23, # 25, # 27, # 29, # 31, # 33, # 35, # 37, # 39, # 4, # 8 # 12, # 16, # 20, # 24, # 28, # 32, # 36, # 40, # 6, # 14, # 22, # 30, # 38, # 10, # 26, # 2, # 34 (in this order).
Thus, contestant # 18 is the unlucky contestant.

c) Let $m \geq 1$ and let n satisfy

$$2^m < n \leq 2^{m+1} \tag{3}$$

where

$$n = 2^m + k. \tag{4}$$

We claim the unlucky number is $2k$. One can check the claim is true for all n when $m = 1$. Suppose the claim is true for $m - 1$ and all such n.

Consider the case when k is an even integer satisfying (3) and (4). After selecting the survivors in round 1, the remaining contestants are

$$2, 4, 6, ..., 2^m + k.$$

Renumber, the above contestant by the rule

$$f(x) = x/2$$

so that the remaining contestant are renumbered as

$$1, 2, 3, ..., 2^{m-1} + k/2.$$

Note,

$$2^{m-1} < 2^{m-1} + k/2 \leq 2^m$$

Since the claim is true for $m - 1$, the unlucky contestant in the renumbering is

$$2\left(\frac{k}{2}\right) = k.$$

But by the renumbering, we find

$$f(x) = \frac{x}{2} = k$$

or the unlucky contestant is $2k$.
Finally, let k be an odd integer satisfying (3) and (4). After selecting the survivors in round 1, the remaining contestants are

$$2, 4, 6, ..., 2^m + (k - 1).$$

Note, the next survivor is 4 since the last survivor chosen is $2^m + k$.
Renumber, the above contestants by the rule

$$g(x) = \frac{x}{2} - 1$$

so the remaining contestant are renumbered as follows:

$$0, 1, 2, 3, ..., [2^{m-1} + (k - 1)/2 - 1].$$

Since the claim is true for $m - 1$, the unlucky contestant using the previous renumbering is

$$2\left(\frac{k-1}{2}\right) \quad \text{or} \quad k - 1$$

Using the original numbering, the unlucky contestant is

$$g(x) = \frac{x}{2} - 1 = k - 1 \text{ or } x = 2k.$$

Equivalently, the unlucky number is

$$2k = 2(n - 2^m).$$

For Thought

1. True, since $\sin(\pi/4) = \cos(\pi/4)$.

2. False, since $\cot(\pi/2) = 0$ and $\dfrac{1}{\tan(\pi/2)}$ is undefined for $\tan(\pi/2)$ is undefined.

3. True, $\csc(60°) = \dfrac{2}{\sqrt{3}} \cdot \dfrac{\sqrt{3}}{\sqrt{3}} = \dfrac{2\sqrt{3}}{3}$.

4. False, since $\tan(5\pi/2)$ is undefined.

5. False, $\sec(95°) < 0$.

6. True, since $\sin 120° = \dfrac{\sqrt{3}}{2}$ and $\csc 120° = 1/\sin 120°$.

7. False, since the ranges are $(-\infty, -2] \cup [2, \infty)$ and $(-\infty, -0.5] \cup [0.5, \infty)$, respectively.

8. True, since $|\csc x| \geq 1$ and $|0.5 \csc x| \geq 0.5$.

9. True, since $\tan\left(3 \cdot \dfrac{\pm\pi}{6}\right) = \tan\left(\pm\dfrac{\pi}{2}\right)$ is undefined.

10. True, since $\cot\left(4 \cdot \dfrac{\pm\pi}{4}\right) = \cot(\pm\pi)$ is undefined.

5.4 Exercises

1. $\tan\alpha$, $\sec\alpha$

3. range

5. $\tan(0) = 0$, $\tan(\pi/4) = 1$, $\tan(\pi/2)$ undefined, $\tan(3\pi/4) = -1$, $\tan(\pi) = 0$, $\tan(5\pi/4) = 1$, $\tan(3\pi/2)$ undefined, $\tan(7\pi/4) = -1$

7. $\sqrt{3}$ **9.** -1

11. 0

13. $-\sqrt{3}/3$

15. $\dfrac{2\sqrt{3}}{3}$

17. Undefined

19. Undefined

21. $\sqrt{2}$

23. -1

25. $\sqrt{3}$

27. -2

29. $-\sqrt{2}$

31. 0

33. 48.0785 **35.** −2.8413 **37.** 500.0003

39. 1.0353 **41.** 636.6192 **43.** −1.4318

45. 71.6221 **46.** 1.0000 **47.** −0.9861

49. $\sec^2\left(2\left(\dfrac{\pi}{6}\right)\right) = \sec^2\left(\dfrac{\pi}{3}\right) = 2^2 = 4$

51. $\tan\left(\dfrac{\frac{\pi}{3}}{2}\right) = \tan\left(\dfrac{\pi}{6}\right) = \dfrac{\sqrt{3}}{3}$

53. $\sec\left(\dfrac{\frac{3\pi}{2}}{2}\right) = \sec\left(\dfrac{3\pi}{4}\right) = -\sqrt{2}$

55. $y = \tan(3x)$ has period $\pi/3$

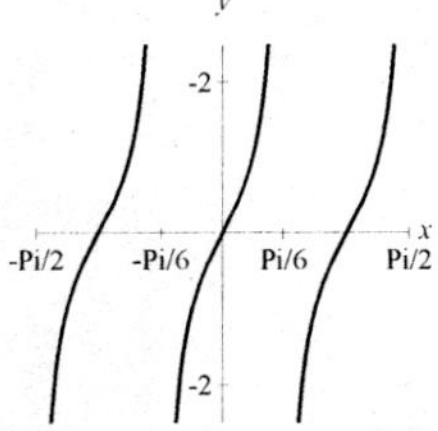

57. $y = \cot(x + \pi/4)$ has period π

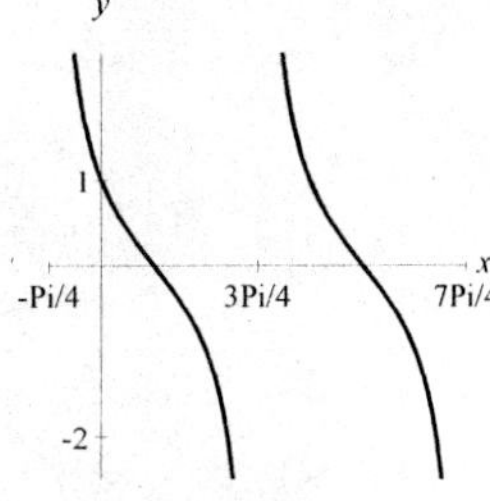

59. $y = \cot(x/2)$ has period 2π

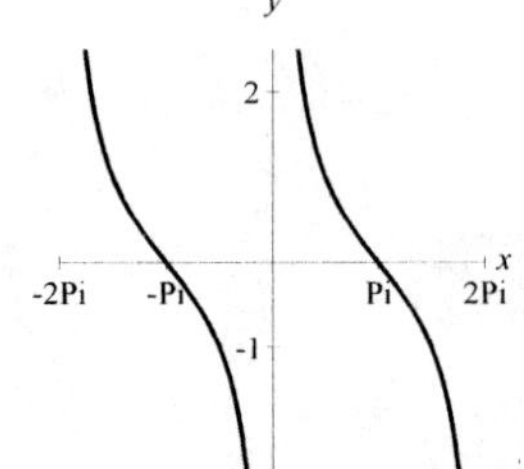

61. $y = \tan(\pi x)$ has period 1

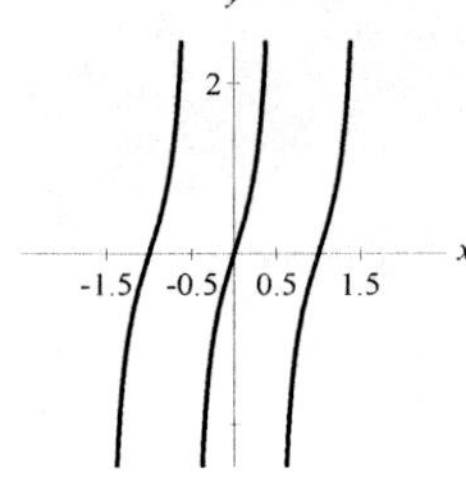

63. $y = -2\tan(x)$ has period π

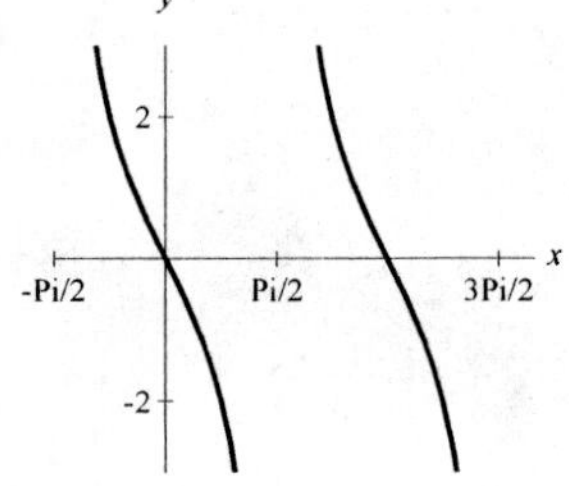

65. $y = -\cot(x + \pi/2)$ has period π

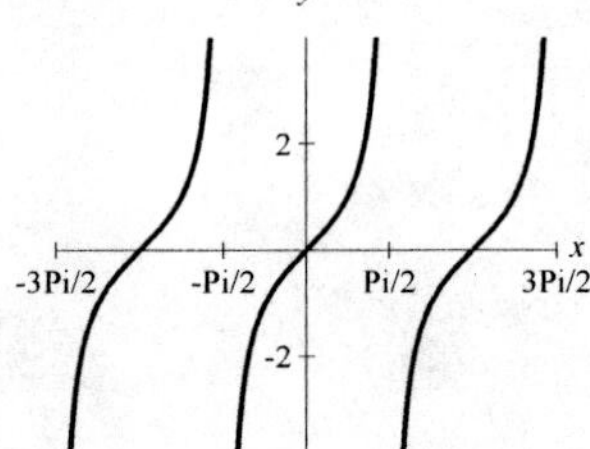

67. $y = \cot(2x - \pi/2)$ has period $\pi/2$

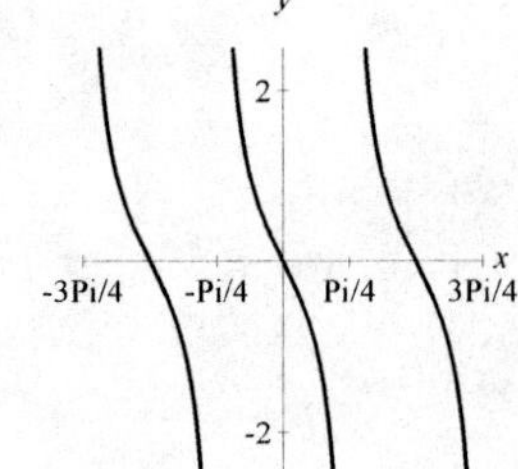

69. $y = \tan\left(\frac{\pi}{2} \cdot x - \frac{\pi}{2}\right)$ has period 2

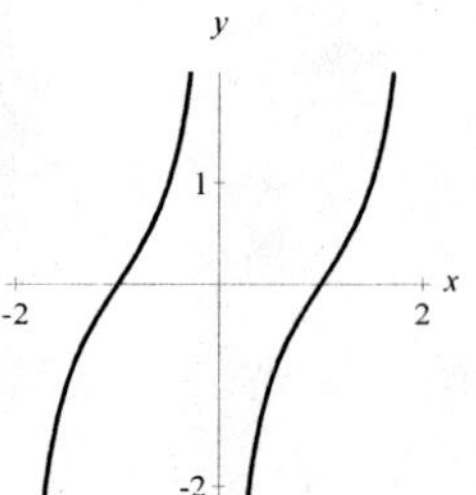

71. period π, range $(-\infty, -1] \cup [1, \infty)$

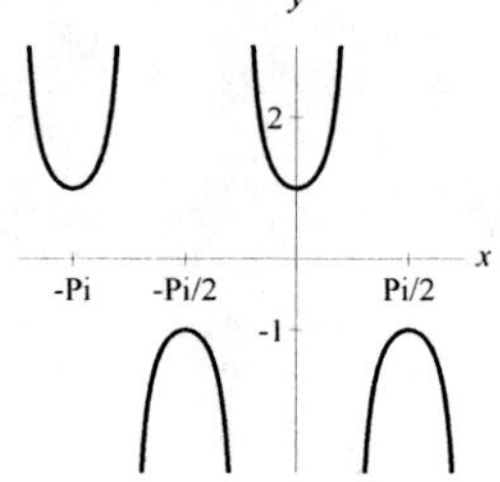

73. period 2π, $(-\infty, -1] \cup [1, \infty)$

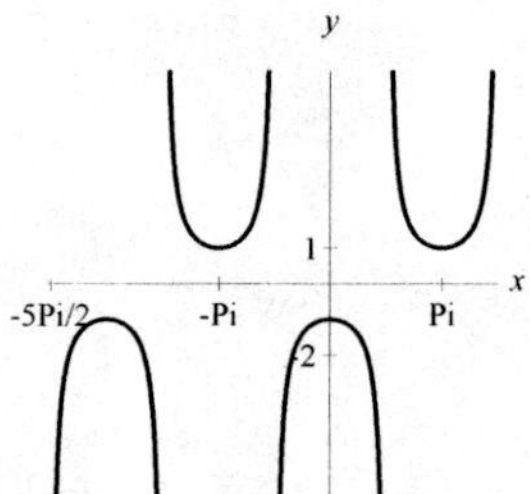

75. period 4π, $(-\infty, -1] \cup [1, \infty)$

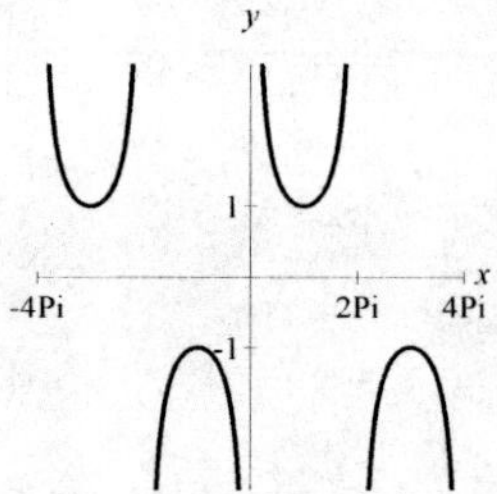

77. period 4, $(-\infty, -1] \cup [1, \infty)$

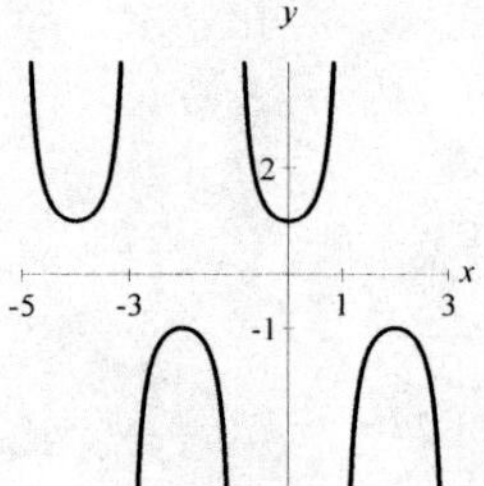

79. period 2π, $(-\infty, -2] \cup [2, \infty)$

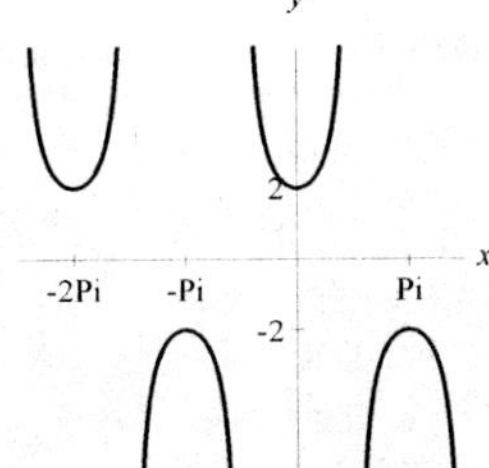

81. period π, $(-\infty, -1] \cup [1, \infty)$

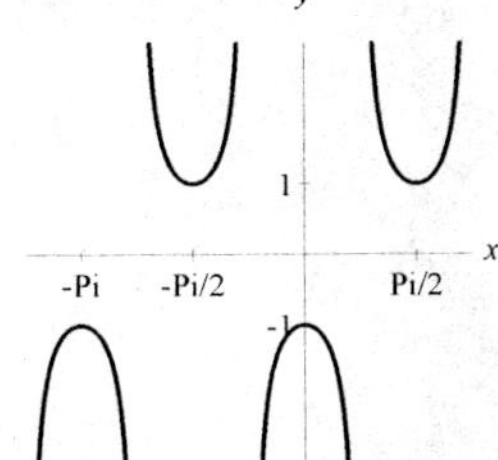

83. period 4, $(-\infty, -1] \cup [1, \infty)$

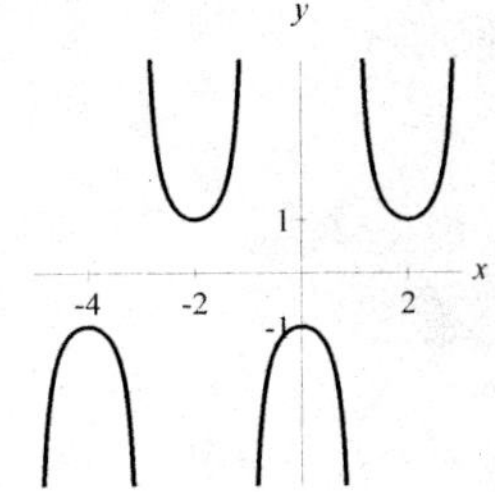

85. period π, $(-\infty, 0] \cup [4, \infty)$

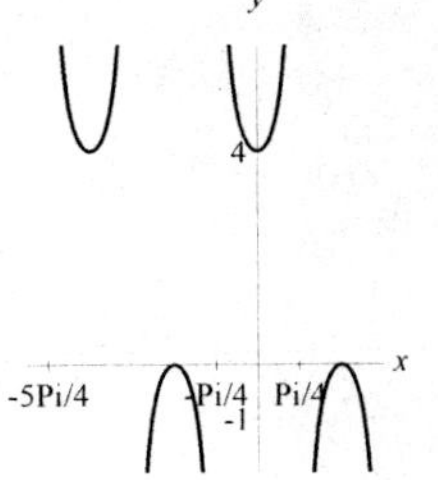

87. Period $\pi/B = \pi/2$, range $(-\infty, \infty)$

89. Period $2\pi/B = 2\pi/(1/2) = 4\pi$, range $(-\infty, -2-1] \cup [2-1, \infty)$ or $(-\infty, -3] \cup [1, \infty)$

91. Period $2\pi/B = 2\pi/2 = \pi$, range $(-\infty, -3-4] \cup [3-4, \infty)$ or $(-\infty, -7] \cup [-1, \infty)$

93. $y = 3\tan\left(x - \frac{\pi}{4}\right) + 2$

95. $y = -\sec(x + \pi) + 2$

97. By adding the ordinates of $y = x$ and $y = \sin(x)$, we obtain the graph of $y = x + \sin(x)$ (which is given below).

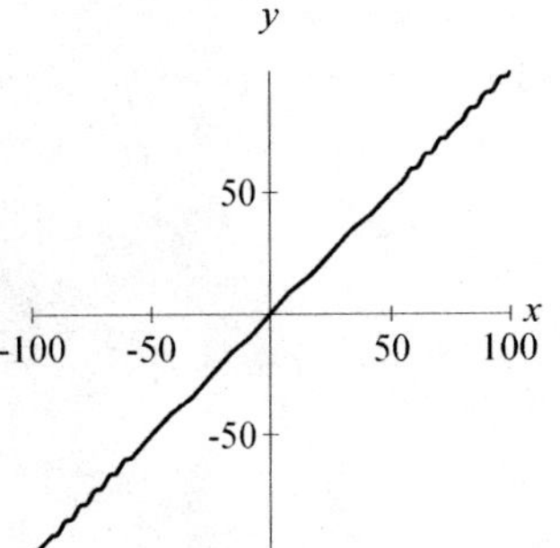

99. a) The graph of y_2 (as shown) looks like the graph of $y = \cos(x)$where $y_1 = \sin(x)$.

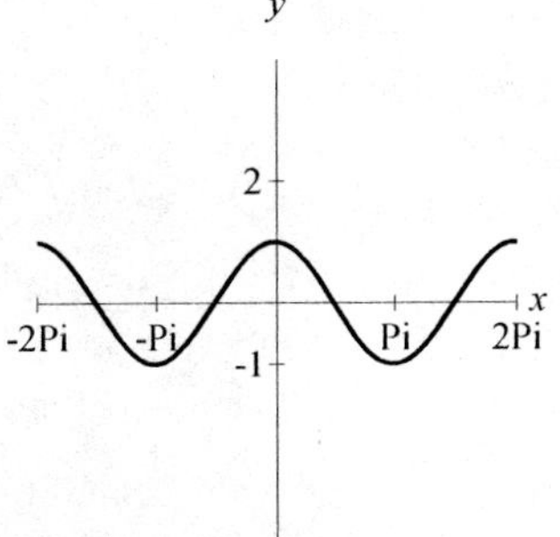

b) The graph of y_2 (as shown) looks like the graph of $y = -\sin(x)$ where $y_1 = \cos(x)$

.

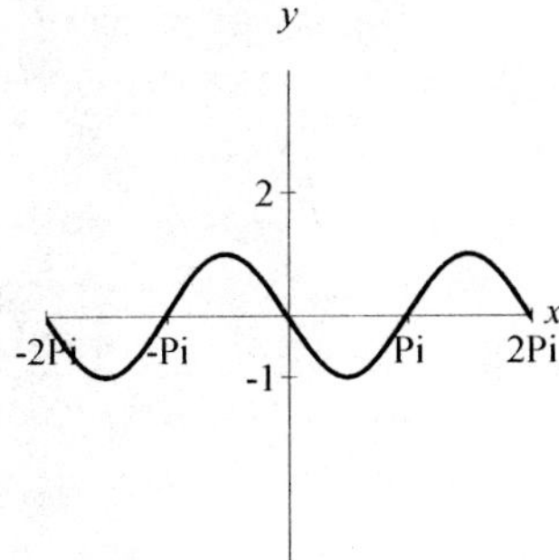

The graph of y_2 (as shown) where $y_1 = e^x$ looks like the graph of $y = e^x$.

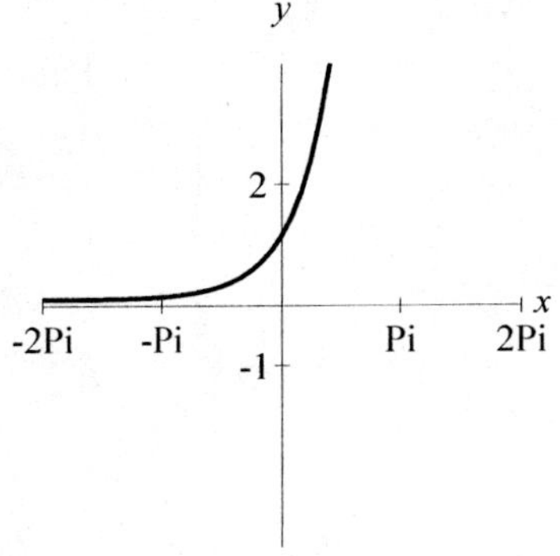

The graph of y_2 (as shown) looks like the graph of $y = 1/x$ where $y_1 = \ln(x)$.

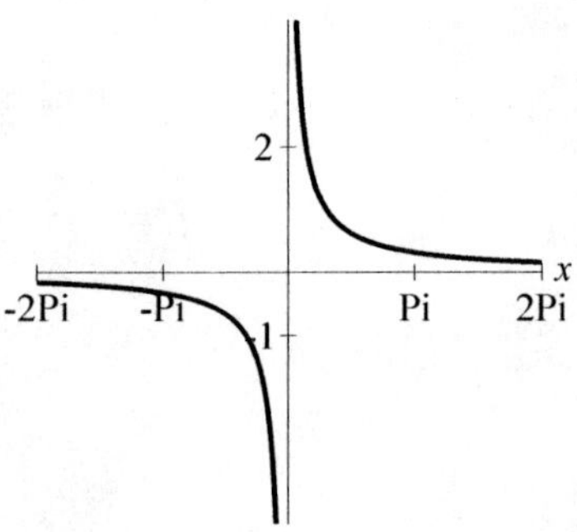

The graph of y_2 (as shown) looks like the graph of $y = 2x$ where $y_1 = x^2$.

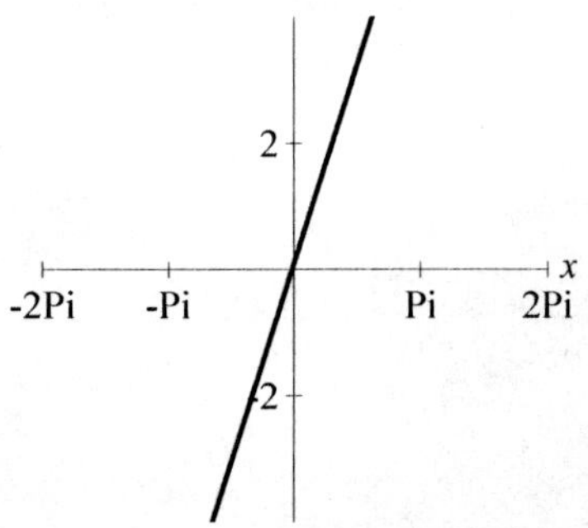

101. $-210°$

103. Using $s = r\alpha$, the linear velocity is

$$3950(2\pi) \cdot \frac{5280}{(24)60^2} \approx 1517 \text{ ft/sec}$$

105. $y = -\cos(x + \pi) + 2$

107. Let P stand for Porsche, N for Nissan, and C for Chrysler. The fifteen preferences could be the following:

Six preferences: P, N, C (1st, 2nd, 3rd respectively)

Five preferences: C, N, P

Three preferences: N, C, P

One preference: N, P, C

For Thought

1. True, $\sin^{-1}(0) = 0 = \sin(0)$.

2. True, since $\sin(3\pi/4) = \dfrac{\sqrt{2}}{2} = \dfrac{1}{\sqrt{2}}$.

3. False, $\cos^{-1}(0) = \pi/2$.

4. False, $\sin^{-1}(\sqrt{2}/2) = 45°$.

5. False, since it equals $\tan^{-1}(1/5)$.

6. True, since $1/5 = 0.2$.

7. True, $\sin(\cos^{-1}(\sqrt{2}/2)) = \sin(\pi/4) = 1/\sqrt{2}$.

8. True by definition of $y = \sec^{-1}(x)$.

9. False, since $f^{-1}(x) = \sin(x)$ where $-\pi/2 \le x \le \pi/2$.

10. False, the secant and cosecant functions are not one-to-one functions and therefore do not have inverse functions.

5.5 Exercises

1. domain

3. domain

5. $\sin^{-1} y$

7. $-\pi/6$ **9.** $\pi/6$

11. $\pi/4$ **13.** $-45°$

15. $30°$ **17.** $0°$

19. $-19.5°$

21. $34.6°$

23. $3\pi/4$ **25.** $\pi/3$

27. π **29.** $135°$

31. $180°$ **33.** $120°$

35. $173.2°$

37. $89.9°$

39. $-\pi/4$ **41.** $\pi/3$

43. $\pi/4$ **45.** $-\pi/6$

47. 0 **49.** $\pi/2$

51. $3\pi/4$ **53.** $2\pi/3$

55. 0.60 **57.** 3.02 **59.** -0.14

61. 1.87 **63.** 1.15 **65.** −0.36

67. 3.06 **69.** 0.06

71. $\tan(\pi/3) = \sqrt{3}$

73. $\sin^{-1}(-1/2) = -\pi/6$

75. $\cot^{-1}(\sqrt{3}) = \pi/6$

77. $\arcsin(\sqrt{2}/2) = \pi/4$

79. $\tan(\pi/4) = 1$

81. $\cos^{-1}(0) = \pi/2$

83. $\cos(2 \cdot \pi/4) = \cos(\pi/2) = 0$

85. $\sin^{-1}(2 \cdot 1/2) = \sin^{-1}(1) = \pi/2$

87. 0.8930

89. Undefined **91.** −0.9802 **93.** −0.4082

95. 3.4583 **97.** 1.0183

99. Solving for y, one finds

$$\begin{aligned} x &= \sin(2y) \\ \sin^{-1}(x) &= 2y \\ y &= \frac{\sin^{-1}(x)}{2}. \end{aligned}$$

Then

$$f^{-1}(x) = 0.5\sin^{-1}(x).$$

As x takes the values in $\left[-\frac{\pi}{4}, \frac{\pi}{4}\right]$, $2x$ takes all the values in $\left[-\frac{\pi}{2}, \frac{\pi}{2}\right]$, and $f(x) = \sin(2x)$ takes all the values in $[-1, 1]$. Thus, the range of f is $[-1, 1]$ which is the domain of f^{-1}.

101. Solving for y, one obtains

$$\begin{aligned} x &= 3 + \tan(\pi y) \\ x - 3 &= \tan(\pi y) \\ \tan^{-1}(x-3) &= \pi y \\ \frac{\tan^{-1}(x-3)}{\pi} &= y. \end{aligned}$$

Thus, we find

$$f^{-1}(x) = \frac{\tan^{-1}(x-3)}{\pi}.$$

As x takes the values in $\left(-\frac{1}{2}, \frac{1}{2}\right)$, πx takes all the values in $\left(-\frac{\pi}{2}, \frac{\pi}{2}\right)$, $\tan(\pi x)$ takes all the values in $(-\infty, \infty)$, and $f(x) = 3 + \tan(\pi x)$ will take all the values in $(-\infty, \infty)$. Thus, the range of f is $(-\infty, \infty)$, which is the domain of f^{-1}.

103. Solving for y, one obtains

$$\begin{aligned} x &= \sin^{-1}\left(\frac{y}{2}\right) + 3 \\ x - 3 &= \sin^{-1}\left(\frac{y}{2}\right) \\ \sin(x-3) &= \frac{y}{2} \\ y &= 2\sin(x-3). \end{aligned}$$

Thus, we obtain

$$f^{-1}(x) = 2\sin(x-3).$$

As x takes the values in $[-2, 2]$, $\frac{x}{2}$ takes all the values in $[-1, 1]$, $\sin^{-1}\left(\frac{x}{2}\right)$ takes all the values in $\left[-\frac{\pi}{2}, \frac{\pi}{2}\right]$, and $f(x) = \sin^{-1}\left(\frac{x}{2}\right) + 3$ will take all the values in $\left[-\frac{\pi}{2} + 3, \frac{\pi}{2} + 3\right]$. Thus, the range of f is $\left[3 - \frac{\pi}{2}, 3 + \frac{\pi}{2}\right]$, which is the domain of f^{-1}.

105. Consider the right triangle with hypotenuse 2400, altitude 2000, and the angle between the hypotenuse and the altitude is $\frac{\theta}{2}$. Since

$$\cos\left(\frac{\theta}{2}\right) = \frac{2000}{2400},$$

we obtain

$$\begin{aligned} \theta &= 2\cos^{-1}\left(\frac{2000}{2400}\right) \\ &\approx 67.1°. \end{aligned}$$

Thus, the airplane is within the range of the gun for $\theta \approx 67.1°$.

107. If $t = 0.25$, then

$$u = 1 - 8(0.25) + 8(0.25)^2 = -0.5$$

Consequently,

$$\begin{aligned} P &= \frac{\cos^{-1}(-0.5) - \sin(\cos^{-1}(-0.5))}{\pi} \\ &= \frac{2\pi/3 - \sqrt{3}/2}{\pi} \\ &\approx 39\% \end{aligned}$$

109. Note that

$$y = \sin\left(\sin^{-1} x\right) = x$$

and the graph is a segment of the line $y = x$.

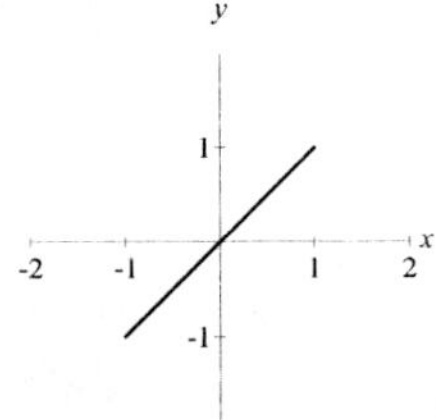

111. Note, $\sin^{-1}(1/x) = \csc^{-1}(x)$ for $-1 \le x \le 1$ and $x \ne 0$.

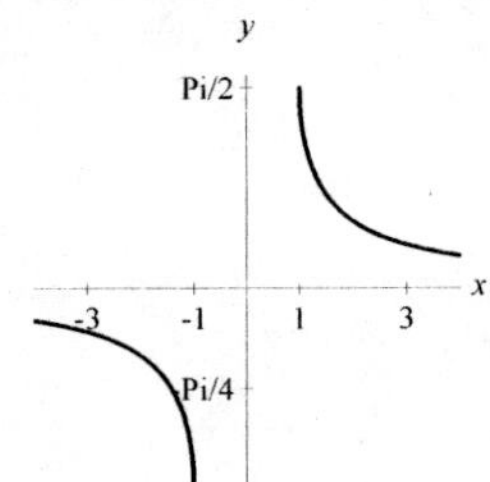

113. **a)** -1 **b)** $\dfrac{\sqrt{2}}{2}$ **c)** $\sqrt{3}$
d) Undefined **e)** -2
f) Undefined **g)** $-\sqrt{3}$ **h)** $-\dfrac{\sqrt{2}}{2}$

115. $\dfrac{\pi}{6}$

117. Period π

Solve for x:

$$\begin{aligned} 2x - \pi &= \frac{\pi}{2} + k\pi \\ 2x &= \frac{\pi}{2} + (k+1)\pi \\ x &= \frac{\pi}{4} + \frac{(k+1)\pi}{2} \end{aligned}$$

Since $k + 1$ is an integer, the vertical asymptotes are

$$x = \frac{\pi}{4} + \frac{k\pi}{2}$$

where k is an integer

The range is $(-\infty, -3] \cup [3, \infty)$.

119.

a) Draw a line segment along a diameter of the porthole with one endpoint at the point where the wiper blade is attached. Then form a right triangle such that this line segment-diameter is a hypotenuse of length 2. Let θ be the angle of the sector that is formed by the wiper blades. Then the angle of the right triangle at the point of attachment is $\theta/2$. Using right triangle trigonometry, we obtain

$$\theta = 2\cos^{-1}\left(\frac{x}{2}\right).$$

Since the area of the sector is $A = \dfrac{\theta x^2}{2}$, we find

$$A = x^2 \cos^{-1}\left(\frac{x}{2}\right).$$

b) The area of a unit circle, i.e., radius 1, is π. If the area of the sector in part a) is one-half of the area of a unit circle, then

$$x^2 \cos^{-1}\left(\frac{x}{2}\right) = \frac{\pi}{2}.$$

Note, the graph of

$$y = x^2 \cos^{-1}\left(\frac{x}{2}\right) - \frac{\pi}{2}, \quad 0 < x < 2$$

has only two x-intercepts. By inspection, these are $x = \sqrt{2}$ ft or $x = \sqrt{3}$ ft.

c) Using a graphing calculator, we find that the area

$$A = x^2 \cos^{-1}\left(\frac{x}{2}\right)$$

is maximized when $x \approx 1.5882$ ft.

For Thought

1. False, since $\sin\alpha = -10/\sqrt{125}$.

2. True, since

$$r = \sqrt{(-1)^2 + 2^2} = \sqrt{5}$$

we get

$$\sec\alpha = \frac{r}{x} = \frac{\sqrt{5}}{-1} = -\sqrt{5}.$$

3. False, α may not lie in $[-\pi/2, \pi/2]$.

4. False, α may not lie in $[0, \pi]$.

5. True, since the side opposite α is the side adjacent to β.

6. False, $c = \sqrt{20}$.

7. False, since $\tan\beta = 1/3$.

8. True, since $\tan(55°) = 8/b$.

9. True, the smallest angle is $\cos^{-1}(4/5)$.

10. False, $\sin(90°) = 1 \neq \text{hyp}/\text{adj}$.

5.6 Exercises

1. opposite side, hypotenuse

3. opposite side, adjacent side

5. elevation

7. $\sin(\alpha) = 4/5, \cos(\alpha) = 3/5, \tan(\alpha) = 4/3,$
$\csc(\alpha) = 5/4, \sec(\alpha) = 5/3, \cot(\alpha) = 3/4$

9. $\sin(\alpha) = 3\sqrt{10}/10, \cos(\alpha) = -\sqrt{10}/10,$
$\tan(\alpha) = -3, \csc(\alpha) = \sqrt{10}/3,$
$\sec(\alpha) = -\sqrt{10}, \cot(\alpha) = -1/3$

11. $\sin(\alpha) = -\sqrt{3}/3, \cos(\alpha) = -\sqrt{6}/3,$
$\tan(\alpha) = \sqrt{2}/2, \csc(\alpha) = -\sqrt{3},$
$\sec(\alpha) = -\sqrt{6}/2, \cot(\alpha) = \sqrt{2}$

13. $\sin(\alpha) = -1/2, \cos(\alpha) = \sqrt{3}/2,$
$\tan(\alpha) = -\sqrt{3}/3, \csc(\alpha) = -2,$
$\sec(\alpha) = 2\sqrt{3}/3, \cot(\alpha) = -\sqrt{3}$

15. $\sin(\alpha) = \sqrt{5}/5, \cos(\alpha) = 2\sqrt{5}/5, \tan(\alpha) = 1/2,$
$\sin(\beta) = 2\sqrt{5}/5, \cos(\beta) = \sqrt{5}/5, \tan(\beta) = 2$

17. $\sin(\alpha) = 3\sqrt{34}/34, \cos(\alpha) = 5\sqrt{34}/34,$
$\tan(\alpha) = 3/5, \sin(\beta) = 5\sqrt{34}/34,$
$\cos(\beta) = 3\sqrt{34}/34, \tan(\beta) = 5/3$

19. $\sin(\alpha) = 4/5, \cos(\alpha) = 3/5, \tan(\alpha) = 4/3,$
$\sin(\beta) = 3/5, \cos(\beta) = 4/5, \tan(\beta) = 3/4$

21. $\tan^{-1}(9/1.5) \approx 80.5°$

23. $\tan^{-1}(\sqrt{3}) = 60°$

25. $\tan^{-1}(6.3/4) \approx 1.0$

27. $\tan^{-1}(1/\sqrt{5}) \approx 0.4$

29. Form the right triangle with $\alpha = 60°$, $c = 20$.

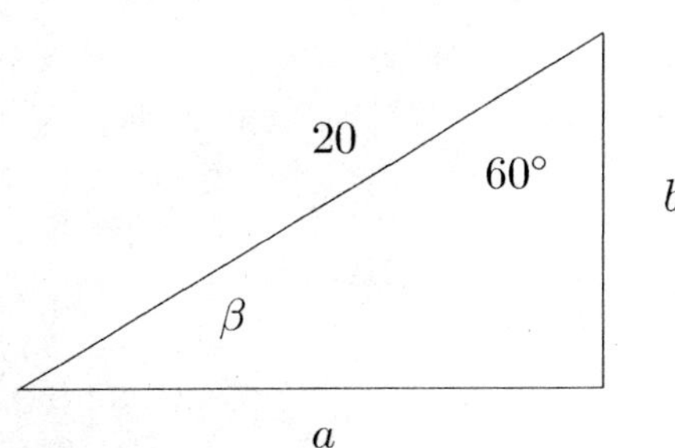

Since $\sin 60° = \dfrac{a}{20}$, we get

$$a = 20 \cdot \frac{\sqrt{3}}{2} = 10\sqrt{3}.$$

Since $\cos 60° = \dfrac{b}{20}$, we find

$$b = 20 \cdot \frac{1}{2} = 10.$$

Also, $\beta = 90° - 60° = 30°$.

31. Form the right triangle with $a = 6$, $b = 8$.

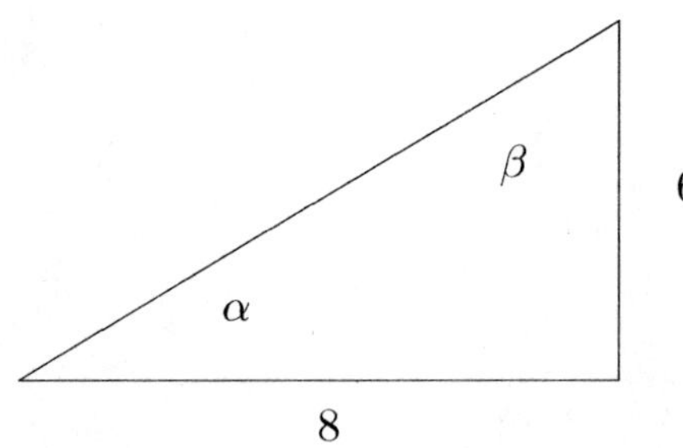

Note: $c = \sqrt{6^2 + 8^2} = 10$, $\tan(\alpha) = 6/8$, so $\alpha = \tan^{-1}(6/8) \approx 36.9^\circ$ and $\beta \approx 53.1^\circ$.

33. Form the right triangle with $b = 6$, $c = 8.3$.

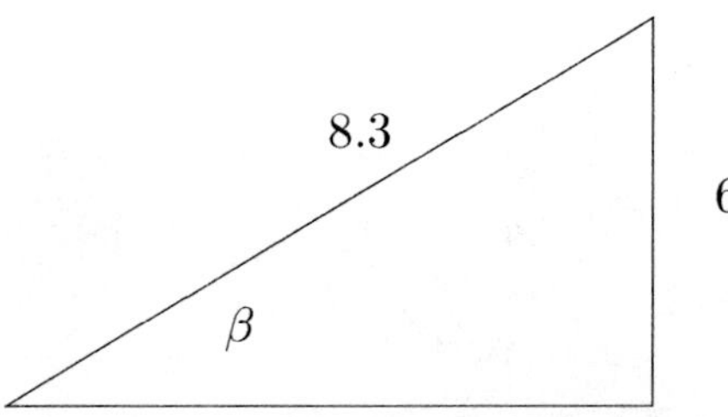

Note: $a = \sqrt{8.3^2 - 6^2} \approx 5.7$, $\sin(\beta) = 6/8.3$, so $\beta = \sin^{-1}(6/8.3) \approx 46.3^\circ$ and $\alpha \approx 43.7^\circ$.

35. Form the right triangle with $\alpha = 16^\circ$, $c = 20$.

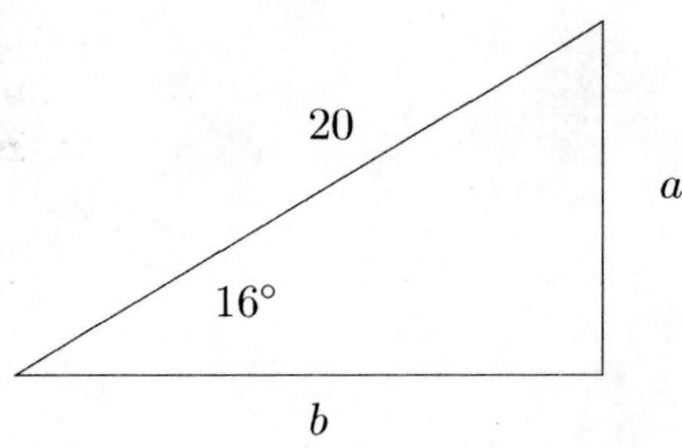

Since $\sin(16^\circ) = a/20$ and $\cos(16^\circ) = b/20$,
$a = 20\sin(16^\circ) \approx 5.5$ and
$b = 20\cos(16^\circ) \approx 19.2$. Also $\beta = 74^\circ$.

37. Form the right triangle with $\alpha = 39^\circ 9'$, $a = 9$.

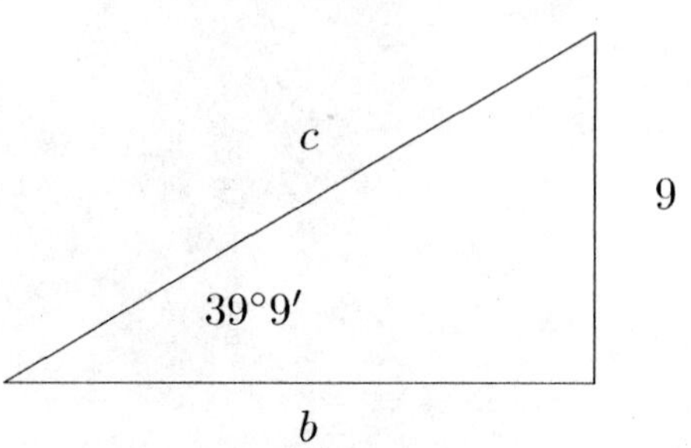

Since $\sin(39^\circ 9') = 9/c$ and $\tan(39^\circ 9') = 9/b$, then $c = 9/\sin(39^\circ 9') \approx 14.3$ and $b = 9/\tan(39^\circ 9') \approx 11.1$. Also $\beta = 50^\circ 51'$.

39. 25, the least number of significant digits is 2.

41. 0.831, the least number of significant digits is 3.

43. 18.8, the least number of significant digits is 3.

45. −289, least number of significant digits is 3.

47. Let h be the height of the buliding.

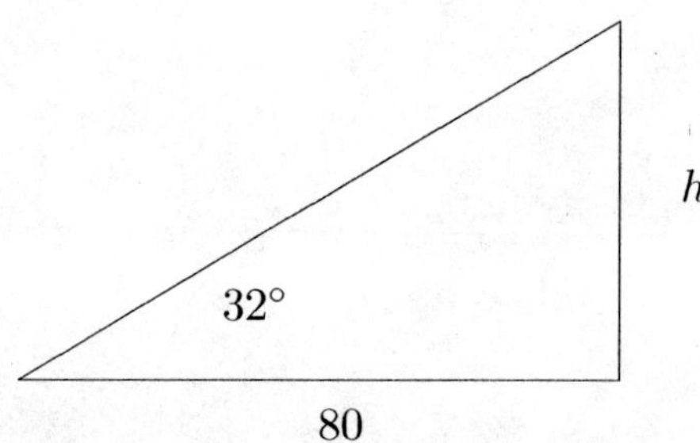

Since $\tan(32^\circ) = h/80$, we obtain

$$h = 80 \cdot \tan(32^\circ) \approx 50 \text{ ft.}$$

49. Let x be the distance between Muriel and the road at the time she encountered the swamp.

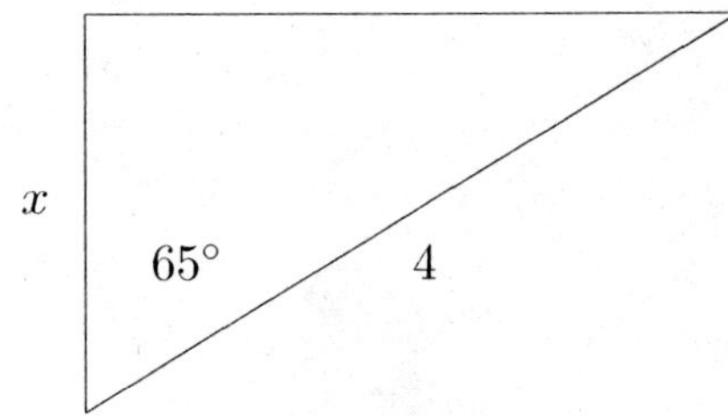

Since $\cos(65°) = x/4$, we find

$$x = 4 \cdot \cos(65°) \approx 1.7 \text{ miles.}$$

51. Let x be the distance between the car and a point on the highway directly below the observer.

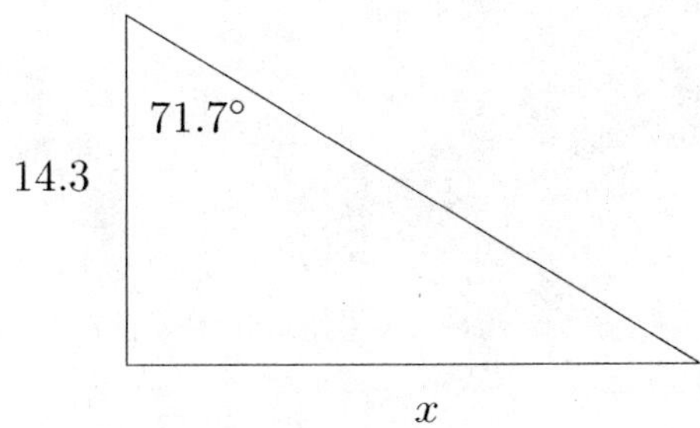

Since $\tan(71.7°) = x/14.3$, we obtain $x = 14.3 \cdot \tan(71.7°) \approx 43.2$ meters.

53. Let h be the height as in the picture below.

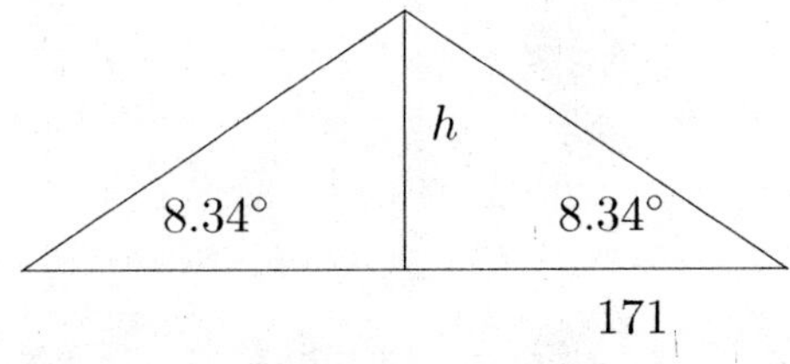

Since $\tan(8.34°) = h/171$, we obtain $h = 171 \cdot \tan(8.34°) \approx 25.1$ ft.

55. Let α be the angle the guy wire makes with the ground.

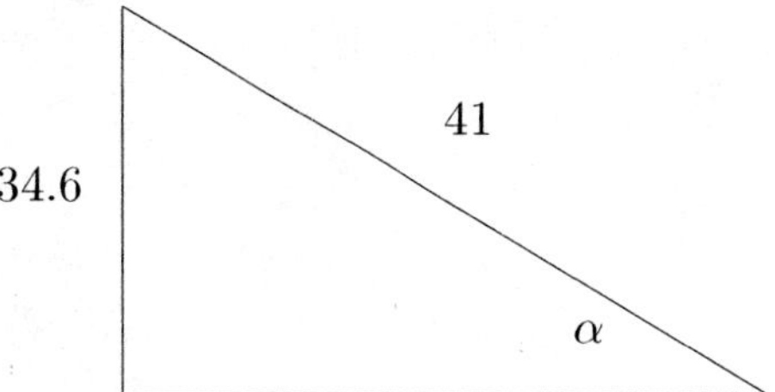

From the Pythagorean Theorem, the distance of the point to the base of the antenna is

$$\sqrt{41^2 - 34.6^2} \approx 22 \text{ meters.}$$

Also, $\alpha = \sin^{-1}(34.6/41) \approx 57.6°$.

57. Consider a right triangle with a height of 13.3 in., and the base is the length of the trail. Let 64° be the angle opposite the height. If x is the length of the trail, then

$$\tan 64° = \frac{13.3}{x}.$$

Thus, $x \approx 6.5$ in.

59. Choose the ball in the left corner of the triangle. From the center A of the ball draw the perpendicular line segment to the point B below and on the outside perimeter of the rack.

Since the diameter of the ball is 2.25 in., and the sides of the triangular rack is 0.25 in., we obtain $AB = 2.25/2 + 0.25 = 1.375$ in.

Let x be the length from the left vertex of the rack to B. By right triangle trigonometry,

$$\tan 30° = \frac{1.375}{x}$$

$$x = \frac{1.375}{\tan 30°} \approx 2.38.$$

Then the length of the horizontal outside perimeter of the rack is $2.25 \times 4 + 2x$. The total length of the outside perimeter is

$$3\,(2.25(4) + 2x) \approx 41.3 \text{ in.}$$

61. Note, 1.75 sec. = 1.75/3600 hour.

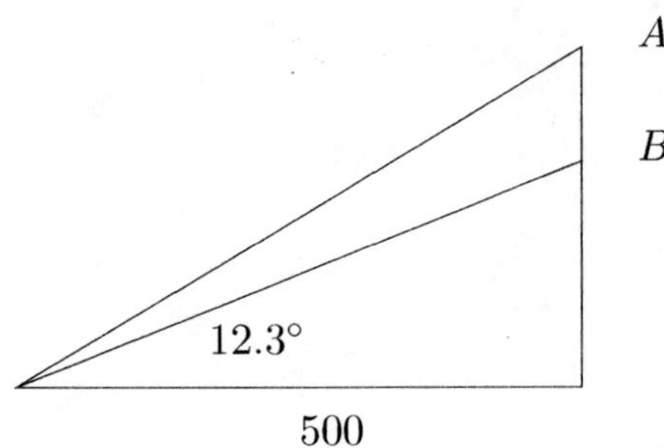

The distance in miles between A and B is

$$\frac{500\,(\tan(15.4°) - \tan(12.3°))}{5280} \approx 0.0054366$$

The speed is

$$\frac{0.0054366}{(1.75/3600)} \approx 11.2 \text{ mph}$$

and the car is not speeding.

63. Let h be the height.

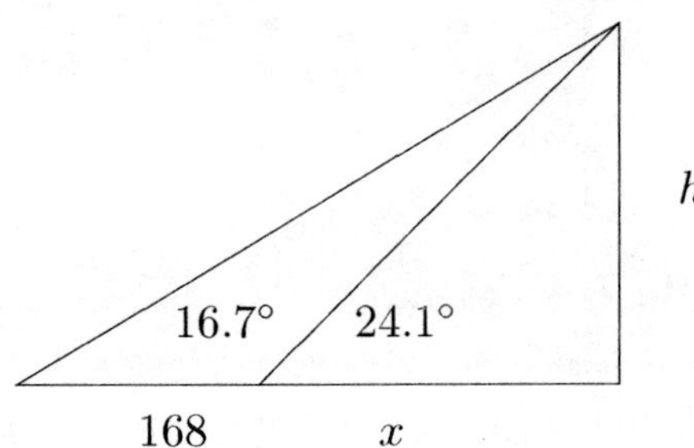

Note, $\tan 24.1° = \frac{h}{x}$ and $\tan 16.7° = \frac{h}{168+x}$.
Solve for h in the second equation and substitute $x = \frac{h}{\tan 24.1°}$.

$$h = \tan(16.7°) \cdot \left(168 + \frac{h}{\tan 24.1°}\right)$$

$$h - \frac{h\tan(16.7°)}{\tan 24.1°} = \tan(16.7°) \cdot 168$$

$$h = \frac{168 \cdot \tan(16.7°)}{1 - \tan(16.7°)/\tan(24.1°)}$$

$$h \approx 153.1\text{meters}$$

The height is 153.1 meters.

65. Let x be the closest distance the boat can come to the lighthouse LH.

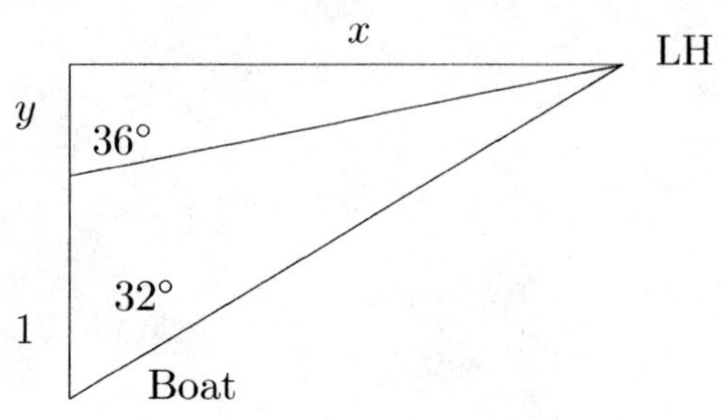

Since $\tan(36°) = x/y$ and $\tan(32°) = x/(1+y)$, we obtain

$$\tan(32°) = \frac{x}{1 + x/\tan(36°)}$$

$$\tan(32°) + \frac{\tan(32°)x}{\tan(36°)} = x$$

$$\tan(32°) = x\left(1 - \frac{\tan(32°)}{\tan(36°)}\right)$$

$$x = \frac{\tan(32°)}{1 - \tan(32°)/\tan(36°)}$$

$$x \approx 4.5 \text{ km.}$$

The closest the boat will come to the lighthouse is 4.5 km.

67. Let p be the number of miles in one parsec.

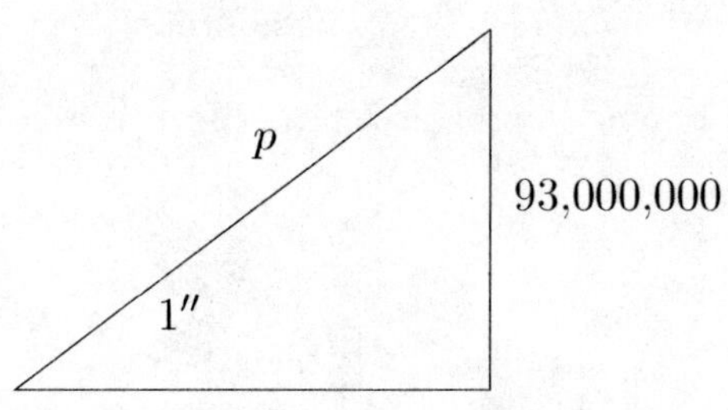

Since $\sin(1'') = \frac{93,000,000}{p}$, we obtain

$$p = \frac{93,000,000}{\sin(1/3600°)} \approx 1.9 \times 10^{13} \text{ miles.}$$

Light travels one parsec in 3.3 years since

$$\frac{p}{186,000 \text{ miles/sec.}} \cdot \frac{1}{365} \cdot \frac{1}{24} \cdot \frac{1}{3600} \approx 3.3$$

69. In the triangle below CE stands for the center of the earth and PSE is a point on the surface of the earth on the horizon of the cameras of Landsat.

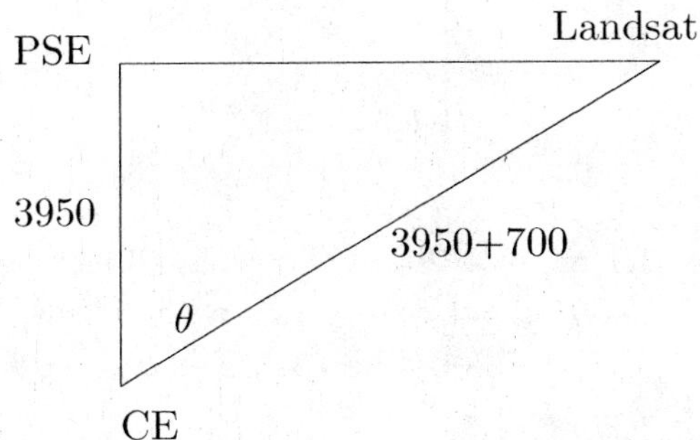

Since $\cos(\theta) = \dfrac{3950}{3950+700}$, we have

$\theta = \cos^{-1}\left(\dfrac{3950}{3950+700}\right) \approx 0.5558$ radian.

But 2θ is the central angle, with vertex at CE, intercepted by the path on the surface of the earth as can be seen by Landsat. The width of this path is the arclength subtended by 2θ, i.e., $s = r \cdot 2\theta = 3950 \cdot 2 \cdot 0.5558 \approx 4391$ miles

71. Let h be the height of the building, and let x be the distance between C and the building. Using right triangle trigonometry, we obtain

$$\frac{1}{\sqrt{3}} = \tan 30^\circ = \frac{h}{40+x}$$

and

$$1 = \tan 45^\circ = \frac{h}{20+x}.$$

Solving simultaneously, we find $x = 10(\sqrt{3}-1)$ and $h = 10(\sqrt{3}+1)$. However,

$$\tan C = \frac{h}{x} = \frac{10(\sqrt{3}+1)}{10(\sqrt{3}-1)} = 2+\sqrt{3}$$

Since $\tan 75^\circ = 2+\sqrt{3}$ by the addition formula for tangent, we obtain

$$C = 75^\circ$$

73. a) The distance from a center to the nearest vertex of the square is $6\sqrt{2}$ by the Pythagorean theorem. Then the diagonal of the square is $12+12\sqrt{2}$. From which, the side of the square is $12+6\sqrt{2}$ by the Pythagorean theorem as shown below.

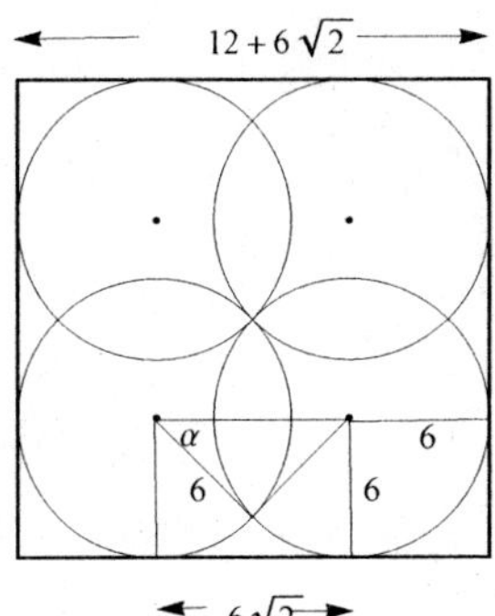

The area A_c of the region in one corner of the box that is *not* watered is obtained by subtracting one-fourth of the area of a circle of radius 6 from the area of a 6-by-6 square:

$$A_c = 36 - 9\pi.$$

Note, the distance between two horizontal centers is $6\sqrt{2}$ as shown above. Then the angle α between the line joining the centers and the line to the intersection of the circles is $\alpha = \pi/4$.

Thus, the area A_b of the region between two adjacent circles that is not watered is the area of a 6-by-$6\sqrt{2}$ square minus the area, 18, of the isosceles triangle with base angle $\alpha = \pi/4$, and minus the combined area 9π of two sectors with central angle $\pi/4$:

$$A_b = 36\sqrt{2} - 18 - 9\pi$$

Hence, the total area not watered is

$$\begin{aligned} 4(A_c + A_b) &= 4\left(36 - 9\pi + 36\sqrt{2} - 18 - 9\pi\right) \\ &= 72\left(1 + 2\sqrt{2} - \pi\right) \end{aligned}$$

b) Since the side of the square is $12+6\sqrt{2}$, the area that is watered by at least one sprinkler is

$$100\% - \frac{4(A_c + A_b)}{(12+6\sqrt{2})^2} \cdot 100\% \approx 88.2\%$$

75. First, consider the figure below.

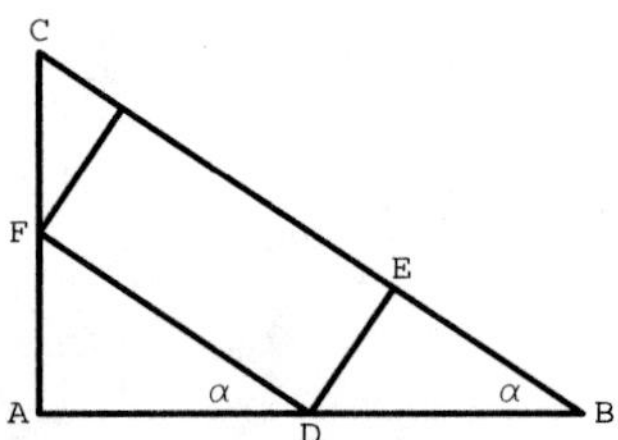

Suppose $AC = s$, $AB = \dfrac{3s}{2}$, $BD = h$, $DE = w$, and $DF = L$. Note,

$$\tan\alpha = \frac{AC}{AB} = \frac{2}{3}.$$

Since $\sin\alpha = \dfrac{2}{\sqrt{13}}$ and $\sin\alpha = \dfrac{DE}{BD}$, we obtain $w = \dfrac{2h}{\sqrt{13}}$. Since $\cos\alpha = \dfrac{AD}{DF}$, we get

$$\cos\alpha = \frac{\frac{3s}{2} - h}{L}.$$

Solving for L. we find $L = \dfrac{\frac{3s}{2} - h}{\cos\alpha}$

and since $\cos\alpha = \dfrac{3}{\sqrt{13}}$, we get

$$L = \frac{\sqrt{13}}{3}\left(\frac{3s}{2} - \frac{w\sqrt{13}}{2}\right).$$

So, the area of the house is

$$wL = \frac{w\sqrt{13}}{3}\left(\frac{3s}{2} - \frac{w\sqrt{13}}{2}\right).$$

Since this area represents a quadratic function of w, one can find the vertex of its graph and conclude that the maximum area of the rectangle is obtained if one chooses $w = \dfrac{3s}{2\sqrt{13}}$.

Correspondingly, we obtain $L = \dfrac{s\sqrt{13}}{4}$.

Finally, given $s = 100$ feet, the dimensions of the house with maximum area are

$$w = \frac{3(100)}{2\sqrt{13}} \approx 41.60 \text{ ft and } L = \frac{100\sqrt{13}}{4} \approx 90.14 \text{ ft}.$$

77. In the figure below,

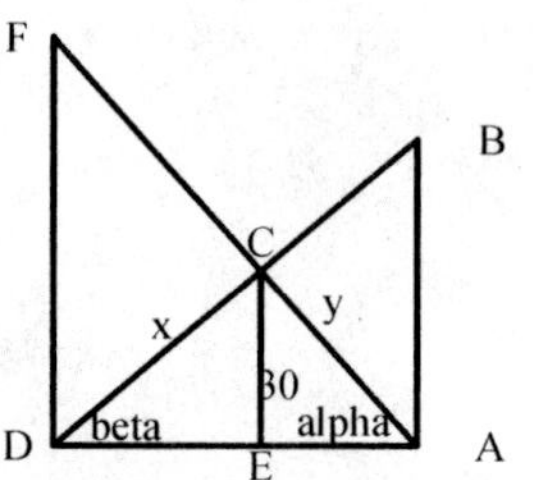

let $AF = 120$, $AC = y$, $BD = 90$, and $CD = x$. Also, let α be the angle formed by AE and AC, and let β be the angle formed by CD and DE.

Choose the point Q in DF so that DF is perpendicular to CQ. In $\triangle CQF$, we have $\sin(\pi/2 - \alpha) = \dfrac{QC}{120 - y}$. In $\triangle CQD$, we get $\sin(\pi/2 - \beta) = \dfrac{QC}{x}$. Therefore,

$$\begin{aligned} \sin(\pi/2 - \alpha)(120 - y) &= x\sin(\pi/2 - \beta) \\ \frac{x}{\sin(\pi/2 - \alpha)} &= \frac{120 - y}{\sin(\pi/2 - \beta)} \\ \frac{x}{\cos\alpha} &= \frac{120 - y}{\cos\beta}. \end{aligned}$$

Similarly, $\dfrac{90 - x}{\cos\alpha} = \dfrac{y}{\cos\beta}$. Thus, $\dfrac{x}{120 - y} = \dfrac{\cos\alpha}{\cos\beta} = \dfrac{90 - x}{y}$. Consequently, $4x + 3y = 360$ or equivalently

$$y = \frac{360 - 4x}{3}.$$

From the right triangles $\triangle$ ACE and $\triangle$ CDE, one finds $\sin\alpha = \dfrac{30}{y}$ and $\sin\beta = \dfrac{30}{x}$.

Consequently,

$$\cos\alpha = \frac{\sqrt{y^2 - 30^2}}{y} \quad \text{and} \quad \cos\beta = \frac{\sqrt{x^2 - 30^2}}{x}.$$

Moreover, using triangles $\triangle$ ABD and $\triangle$ ADF, one obtains $\cos\beta = \dfrac{AD}{90}$ and $\cos\alpha = \dfrac{AD}{120}$. Combining all of these, one derives

$$\begin{aligned} 90\cos\beta &= 120\cos\alpha \\ \frac{3}{4} &= \frac{\cos\alpha}{\cos\beta} \end{aligned}$$

$$\frac{3}{4} = \frac{\frac{\sqrt{y^2-30^2}}{y}}{\frac{\sqrt{x^2-30^2}}{x}}$$

$$\frac{3}{4} = \frac{\sqrt{y^2-30^2}}{\sqrt{x^2-30^2}} \cdot \frac{x}{y}$$

$$\frac{3}{4} = \frac{\sqrt{\left(\frac{360-4x}{3}\right)^2 - 30^2}}{\sqrt{x^2-30^2}} \cdot \frac{x}{\frac{360-4x}{3}}$$

$$\frac{3}{4} = \frac{3x}{360-4x} \cdot \frac{\sqrt{\left(\frac{360-4x}{3}\right)^2 - 30^2}}{\sqrt{x^2-30^2}}$$

$$\frac{1}{16} = \frac{x^2}{(360-4x)^2} \cdot \frac{\left(\frac{360-4x}{3}\right)^2 - 30^2}{x^2-30^2}$$

$$(360-4x)^2(x^2-30^2) = \\ = 16x^2\left[\left(\frac{360-4x}{3}\right)^2 - 30^2\right]$$

Using a graphing calculator, for $0 < x < 90$, one finds $x \approx 60.4$. Working backwards, one derives $\sin\beta = \frac{30}{60.4}$ or $\beta \approx 29.8°$, and $\cos(29.8) = \frac{AD}{90}$. Hence, the width of the property is $AD = 78.1$ feet.

79. Consider the right triangle formed by the hook, the center of the circle, and a point on the circle where the chain is tangent to the circle.

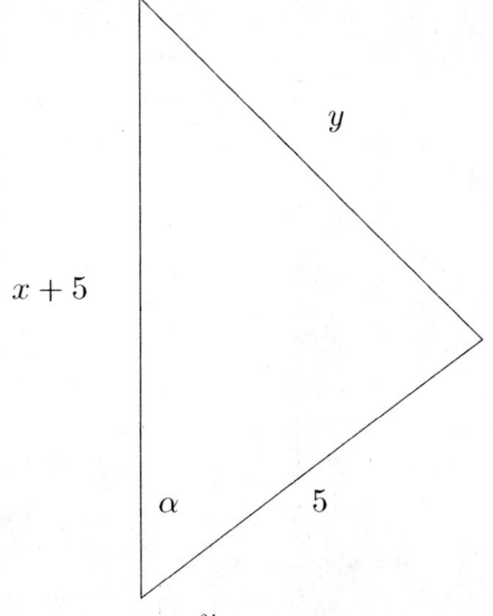

Then $\tan\alpha = \frac{y}{5}$ or $y = 5\tan\alpha$. Since the chain is 40 ft long and the angle $2\pi - 2\alpha$ intercepts an arc around the pipe where the chain wraps around the circle, we obtain

$$2y + 5(2\pi - 2\alpha) = 40.$$

By substitution, we get

$$10\tan\alpha + 10\pi - 10\alpha = 40.$$

With a graphing calculator, we obtain $\alpha \approx 1.09835$ radians. From the figure above, we get $\cos\alpha = \frac{5}{5+x}$. Solving for x, we obtain

$$x = \frac{5 - 5\cos\alpha}{\cos\alpha} \approx 5.987 \text{ ft.}$$

81. Assume the circle is given by

$$x^2 + (y-r)^2 = r^2$$

where r is the radius, see figure below.

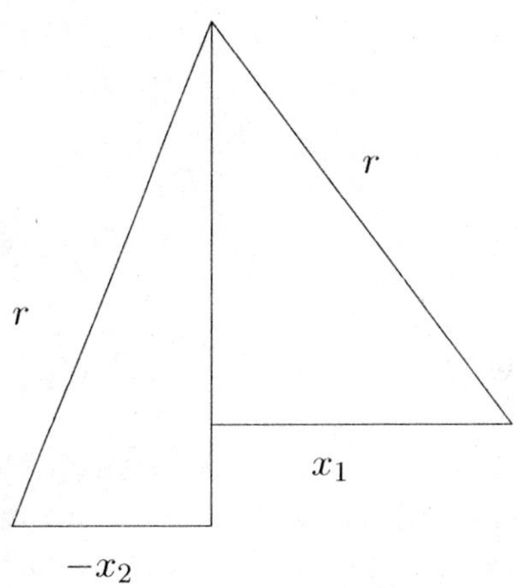

Using the Pythagorean Theorem, we find

$$x_1^2 + (r-2)^2 = r^2 \text{ and } x_2^2 + (r-1)^2 = r^2.$$

From which we obtain

$$x_1^2 + 4 - 4r = 0 \text{ and } x_2^2 + 1 - 2r = 0.$$

Then

$$x_1 = \sqrt{4r-4} \text{ and } x_2 = \sqrt{2r-1}.$$

Note, the central angle of the arc joining the points $(x_1, 2)$ and $(-x_2, 1)$ on the blocks is

$$\alpha = \arcsin\frac{x_1}{r} + \arcsin\frac{x_2}{r}$$

Since the length $r\alpha$ of the arc is 6 ft, we obtain

$$r\left(\arcsin\frac{\sqrt{4r-4}}{r}+\arcsin\frac{\sqrt{2r-1}}{r}\right)=6$$

Using a calculator, we find $r \approx 2.768$ ft.

85. **a)** $-30°$ **b)** $120°$ **c)** $-45°$

87. Amplitude 3; period is $\dfrac{2\pi}{B}=\dfrac{2\pi}{\pi/2}=4$

Since $\dfrac{\pi x}{2}-\dfrac{\pi}{2}=\dfrac{\pi}{2}(x-1)$, the phase shift is 1

If we add 7 to the interval $[-3,3]$, we find that the range is $[4,10]$

89. period is $\dfrac{2\pi}{B}=\dfrac{2\pi}{\pi}=2$

range $(-\infty,-5]\cup[5,\infty)$

91. We begin by writing the length of a diagonal of a rectangular box. If the dimensions of a box are a-by-b-by-c, then the length of a diagonal is $\sqrt{a^2+b^2+c^2}$.

Suppose the ball is placed at the center of the field and 60 feet from the goal line. Then the distance between the ball and the right upright of the goal is

$$A=\sqrt{90^2+10^2+9.25^2}\approx 91.025.$$

Consider the triangle formed by the ball, and the left and right uprights of the goal. Opposite the angle θ_1 is the 10-ft horizontal bar. Using the cosine law in Chapter 7, we find

$$18.5^2=2A^2-2A^2\cos\theta$$

and $\theta_1 \approx 11.66497°$.

Now, place the ball on the right hash mark which is 9.25 ft from the centerline. The ball is also 60 feet from the goal line. Then the distance between the ball and the right upright of the goal is

$$B=\sqrt{90^2+10^2}\approx 90.554.$$

And, the distance between the ball and the left upright of the goal is

$$C=\sqrt{90^2+10^2+18.5^2}\approx 92.424.$$

Similarly, consider the triangle formed by the ball, and the left and right uprights of the goal. Opposite the angle θ is the 10-ft horizontal bar. Using the cosine law in Chapter 7, we find

$$18.5^2=B^2+C^2-2BC\cos\theta_2$$

and $\theta_2 \approx 11.54654°$.

Thus, the difference between the values of θ is

$$\theta_1-\theta_2\approx 11.66497°-11.54654°\approx 0.118°$$

Review Exercises

1. $388°-360°=28°$

3. $-153°14'27''+359°59'60''=206°45'33''$

5. $180°$

7. $13\pi/5-2\pi=3\pi/5=3\cdot 36°=108°$

9. $5\pi/3=5\cdot 60°=300°$

11. $270°$ **13.** $11\pi/6$ **15.** $-5\pi/3$

17.

θ deg	0	30	45	60	90	120	135	150	180
θ rad	0	$\frac{\pi}{6}$	$\frac{\pi}{4}$	$\frac{\pi}{3}$	$\frac{\pi}{2}$	$\frac{2\pi}{3}$	$\frac{3\pi}{4}$	$\frac{5\pi}{6}$	π
$\sin\theta$	0	$\frac{1}{2}$	$\frac{\sqrt{2}}{2}$	$\frac{\sqrt{3}}{2}$	1	$\frac{\sqrt{3}}{2}$	$\frac{\sqrt{2}}{2}$	$\frac{1}{2}$	0
$\cos\theta$	1	$\frac{\sqrt{3}}{2}$	$\frac{\sqrt{2}}{2}$	$\frac{1}{2}$	0	$-\frac{1}{2}$	$-\frac{\sqrt{2}}{2}$	$-\frac{\sqrt{3}}{2}$	-1
$\tan\theta$	0	$\frac{\sqrt{3}}{3}$	1	$\sqrt{3}$	NA	$-\sqrt{3}$	-1	$-\frac{\sqrt{3}}{3}$	0

19. $-\sqrt{2}/2$ **21.** $\sqrt{3}$ **23.** $-2\sqrt{3}/3$

25. 0 **27.** 0

29. -1 **31.** $\cot(60°)=\sqrt{3}/3$ **33.** $-\sqrt{2}/2$

35. -2 **37.** $-\sqrt{3}/3$

39. $\sin(\alpha)=5/13, \cos(\alpha)=12/13, \tan(\alpha)=5/12,$ $\csc(\alpha)=13/5, \sec(\alpha)=13/12, \cot(\alpha)=12/5$

41. 0.6947 **43.** -0.0923 **45.** 0.1869

47. 1.0356 **49.** $-\pi/6$ **51.** $-\pi/4$

53. $\pi/4$ **55.** $\pi/6$ **57.** $90°$

59. $135°$ **61.** $30°$ **63.** $90°$

65. Form the right triangle with $a = 2$, $b = 3$.

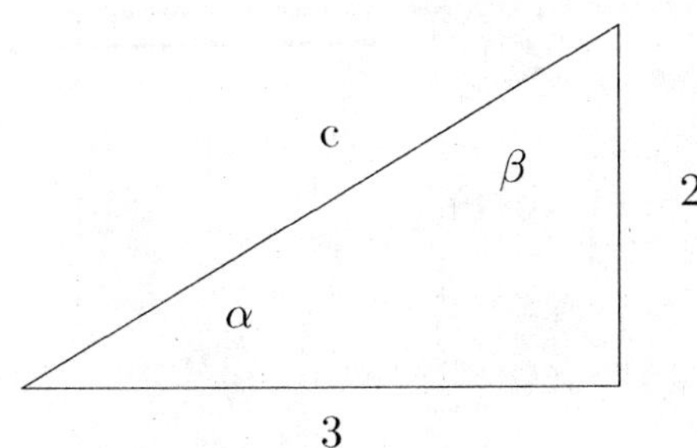

Note that $c = \sqrt{2^2 + 3^2} = \sqrt{13}$, $\tan(\alpha) = 2/3$, so $\alpha = \tan^{-1}(2/3) \approx 33.7^\circ$ and $\beta \approx 56.3^\circ$.

67. Form the right triangle with $a = 3.2$, $\alpha = 21.3^\circ$.

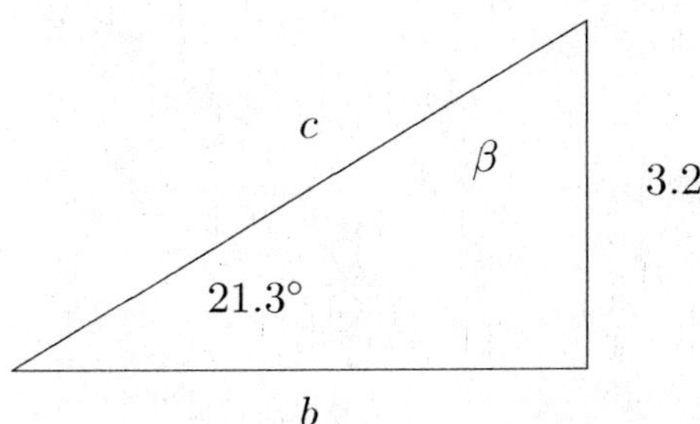

Since $\sin 21.3^\circ = \dfrac{3.2}{c}$ and $\tan 21.3^\circ = \dfrac{3.2}{b}$, $c = \dfrac{3.2}{\sin 21.3^\circ} \approx 8.8$ and $b = \dfrac{3.2}{\tan 21.3^\circ} \approx 8.2$

Also, $\beta = 90^\circ - 21.3^\circ = 68.7^\circ$

69. $f(x) = 2\sin(3x)$ has period $2\pi/3$, range $[-2, 2]$

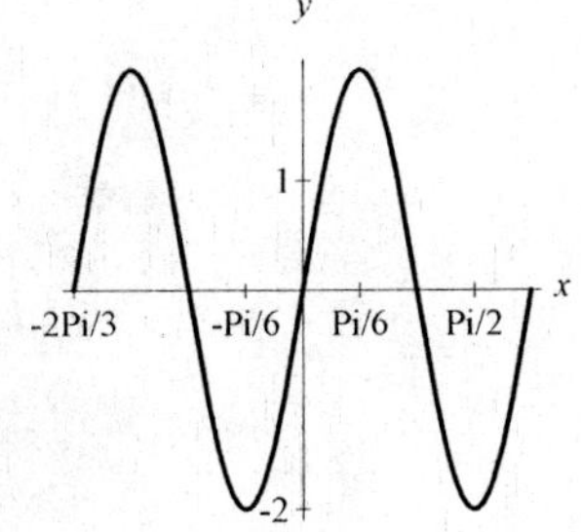

71. $f(x) = \tan(2x + \pi)$ has period $\pi/2$, range $(-\infty, \infty)$

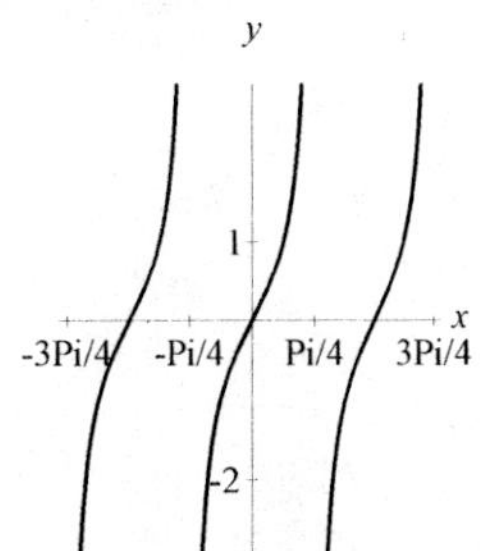

73. $f(x) = \sec(x/2)$ has period 4π, range $(-\infty, -1] \cup [1, \infty)$

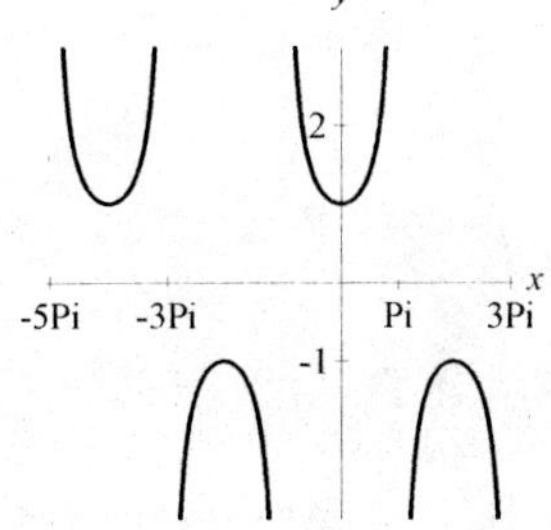

75. $f(x) = \dfrac{1}{2} \cdot \cos(2x)$ has period π, range $[-1/2, 1/2]$

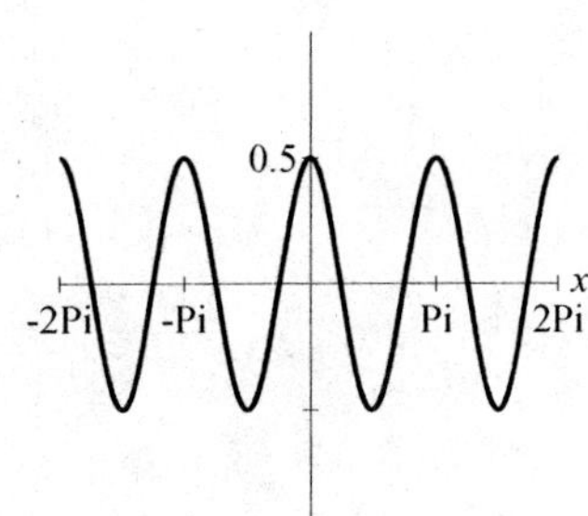

77. $f(x) = \cot\left(2x + \dfrac{\pi}{3}\right)$ has period $\dfrac{\pi}{2}$, range $(-\infty, \infty)$

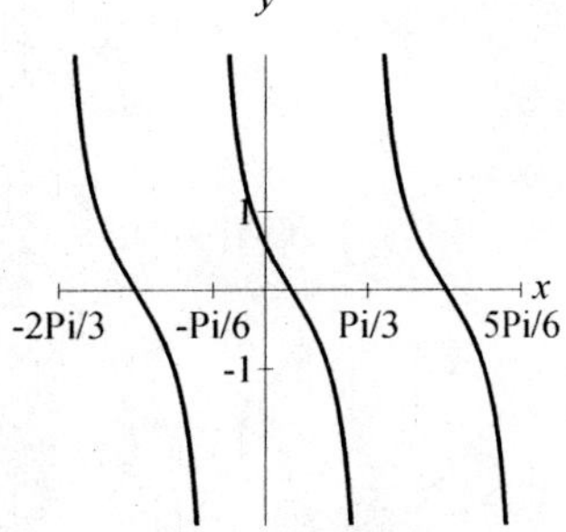

79. $f(x) = \frac{1}{3}\csc(2x + \pi)$ has period π, range $(-\infty, -1/3] \cup [1/3, \infty)$

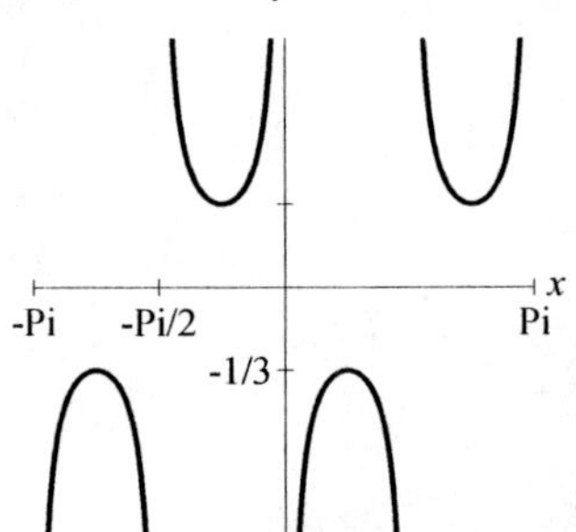

81. One full period is between $2 \le x \le 6$. So $\frac{2\pi}{b} = 4$ or $b = \frac{\pi}{2}$ and there is a right shift of 2 units. An equation is $y = 2\sin\left(\frac{\pi}{2}(x-2)\right)$.

83. Note, one full period is between $8\pi \le x \le 10\pi$, there is a vertical upward shift of 40 units, and $a = 20$. An equation is $y = 20\sin(x) + 40$.

85. $\alpha = 150°$

87. $\sin(\alpha) = -\sqrt{1-(1/5)^2} = -\sqrt{24/25} = -2\sqrt{6}/5$

89. In the given right triangle, the hypotenuse is 24 feet and the side adjacent to the 16° angle is 18 feet. Since $\sin(16°) = \frac{x}{24}$, the shortest side is $x = 24\sin(16°) \approx 6.6$ feet.

91. Since the largest angle α must be opposite the 8 cm leg (which is the longest leg) and the leg adjacent to α is 6 cm long, we obtain

$$\alpha = \tan^{-1}\left(\frac{8}{6}\right) \approx 53.1°.$$

93. Period is $\frac{1}{92.3 \times 10^6} \approx 1.08 \times 10^{-8}$ sec

95. In one hour the nozzle revolves through an angle of $2\pi/8$. The linear velocity, $v = r \cdot \alpha = 120 \cdot 2\pi/8 \approx 94.2$ ft/hr.

97. The height of the man is

$$s = r \cdot \alpha = 1000(0.4) \cdot \frac{\pi}{180} \approx 6.9813 \text{ ft.}$$

99. Since the period is 20 minutes, $\frac{2\pi}{b} = 20$ or $b = \frac{\pi}{10}$. Since the depth is between 12 ft and 16 ft, the vertical upward shift is 14 and $a = 2$. Since the depth is 16 ft at time $t = 0$, one can assume there is a left shift of 5 minutes. An equation is

$$y = 2\sin\left(\frac{\pi}{10}(x+5)\right) + 14$$

and its graph is given.

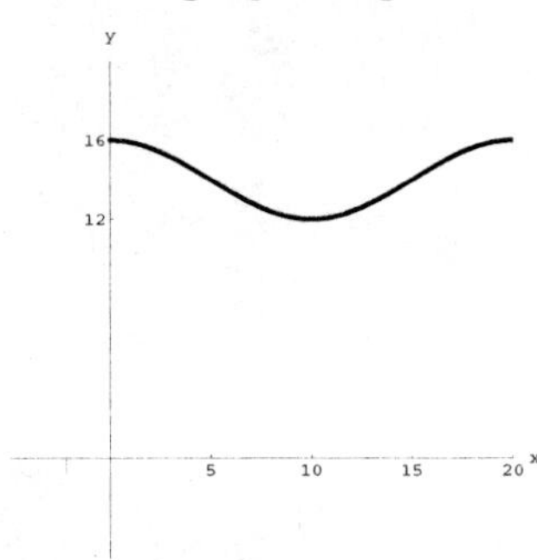

101. Form the right triangle below.

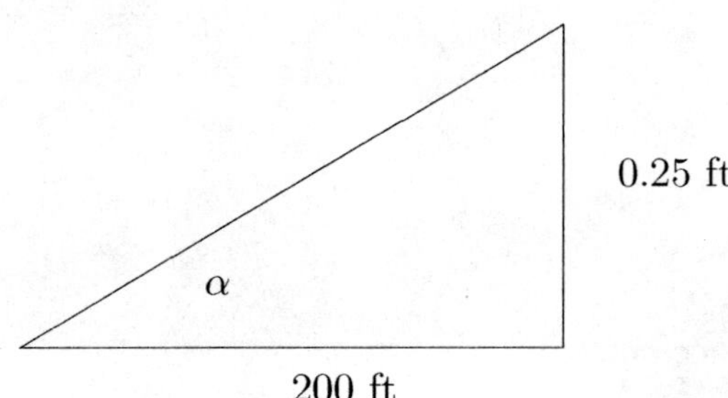

Since $\tan(\alpha) = 0.25/200$, $\alpha = \tan^{-1}(0.25/200) \approx 0.0716°$. She will not hit the target if she deviates by 0.1° from the center of the circle.

103. If $r = 90$ meters, then the time for one revolution is given by

$$\begin{aligned} T^2(9.8) &= 4\pi^2(90) \\ T &\approx 19.04 \text{ sec.} \end{aligned}$$

Angular velocity when the period $T = 19.04$ sec is $\omega = 2\pi/19.04 \approx 0.33$ radians/sec.

105. Consider the right triangle below where vertex A represents the center of the earth, the cargo ship is located at vertex B, and d is the distance to the horizon.

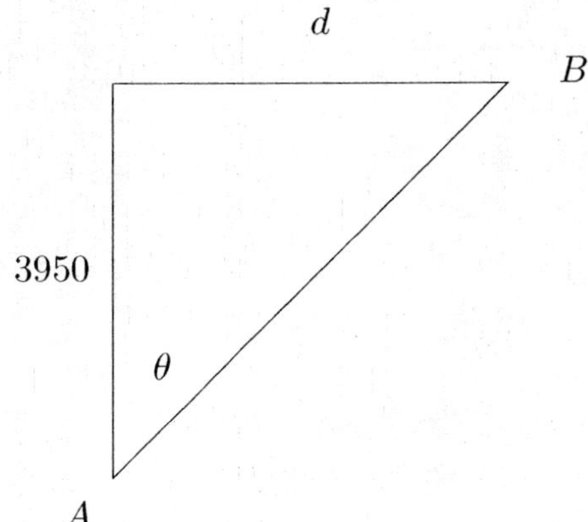

From the triangle, we find $\tan\theta = \dfrac{d}{3950}$ and since $AB = 3950 + \dfrac{90}{5280}$ we obtain

$$\sin\theta = \frac{d}{3950 + \frac{90}{5280}}.$$

Also, we have

$$\sin\theta = \frac{\text{opp}}{\text{hyp}} = \frac{d}{\sqrt{d^2 + 3950^2}}.$$

Then the distance to the horizon is obtained as follows:

$$\begin{aligned}
\frac{d}{\sqrt{d^2+3950^2}} &= \frac{d}{3950 + \frac{90}{5280}} \\
\frac{1}{\sqrt{d^2+3950^2}} &= \frac{1}{3950 + \frac{90}{5280}} \\
3950 + \frac{90}{5280} &= \sqrt{d^2 + 3950^2} \\
\left(3950 + \frac{90}{5280}\right)^2 &= d^2 + 3950^2 \\
\left(3950 + \frac{90}{5280}\right)^2 - 3950^2 &= d^2 \\
\sqrt{\left(3950 + \frac{90}{5280}\right)^2 - 3950^2} &= d \\
11.6 \text{ miles} &= d
\end{aligned}$$

107. Draw an isosceles triangle containing the two circles as shown below. The indicated radii are perpendicular to the sides of the triangle.

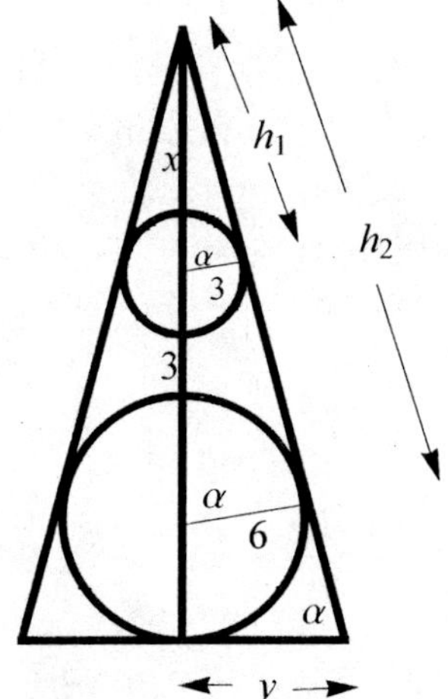

Using similar triangles, we obtain

$$\frac{x+3}{3} = \frac{x+15}{6}$$

from which we solve $x = 9$. Applying the Pythagorean theorem, we find

$$h_1 = 3\sqrt{15}, \quad h_2 = 6\sqrt{15}$$

Note, the height of the triangle is 30 inches. By similar triangles,

$$\frac{3\sqrt{15}}{3} = \frac{30}{y}$$

or $y = 2\sqrt{15}$. Thus, the base angle α of the isosceles triangle is $\alpha = \arctan\sqrt{15}$.

Then the arc lengths of the belt that wrap around the two pulleys are (using $s = r\alpha$)

$$s_1 = 6 \cdot (2\pi - 2\alpha)$$

and

$$s_2 = 3 \cdot 2\alpha = 6\alpha.$$

Hence, the length of the belt is

$$\begin{aligned}
s_1 + s_2 + 2(h_2 - h_1) &= \\
12\pi - 6\arctan(\sqrt{15}) + 6\sqrt{15} &\approx \\
53.0 \text{ in.}
\end{aligned}$$

109. Consider a right triangle whose height is h, the hypotenuse is 2 ft, and the angle between h and the hypotenuse is 18°. Then

$$h = 2\cos 18° \approx 1.9 \text{ ft} = 22.8 \text{ in.}$$

111. Draw a line segment from the center A of the pentagon to the midpoint C of one of the sides of the regular pentagon. This line segment should be perpendicular to the side. Consider a right triangle with vertices at A, C, and an adjacent vertex B of the hexagon near C. The angle at A of the right triangle is 36°, and the hypotenuse r is the distance from the center A to vertex B. Using right triangle trigonometry, we obtain

$$r = \frac{1}{\sin 36°} \approx 1.701.$$

Furthermore, the side adjacent to 36° is $h - r$. Applying right triangle trigonometry, we have

$$\cos 36° = \frac{h-r}{r}$$

Solving for h, we find

$$h = r(1 + \cos 36°) \approx 3.08 \text{ m}$$

113. We assume that the center of the bridge moves straight upward to the center of the arc where the bridge expands. The arc has length $100 + \frac{1}{12}$ feet. To find the central angle α, we use $s = r\alpha$ and $\sin(\alpha/2) = 50/r$. Then $\alpha = 2\sin^{-1}(50/r)$. Solving the equation

$$100 + \frac{1}{12} = 2r\sin^{-1}\left(\frac{50}{r}\right)$$

with a graphing calculator, we find

$$r \approx 707.90241 \text{ ft.}$$

The distance from the chord to the center of the circle can be found by using the Pythagorean theorem, and it is

$$d = \sqrt{r^2 - 50^2} \approx 706.1344221 \text{ ft.}$$

Then the distance from the center of the arc to the cord is

$$r - d \approx 1.767989 \text{ ft} \approx 21.216 \text{ inches.}$$

Chapter 5 Test

1. $\cos(60°) = 1/2$ **2.** $\sin(-30°) = -1/2$

3. -1 **4.** 2 **5.** -2 **6.** $\sqrt{3}/3$

7. $-\pi/6$ **8.** $2\pi/3$ **9.** $-\pi/4$

10. Undefined since $0 < \frac{\sqrt{3}}{2} < 1$

11. Undefined since $-1 < -\frac{\sqrt{2}}{2} < 0$ **12.** $2\sqrt{2}/3$

13. $y = \sin(3x) - 2$ has period $2\pi/3$, range $[-3, -1]$, amplitude 1

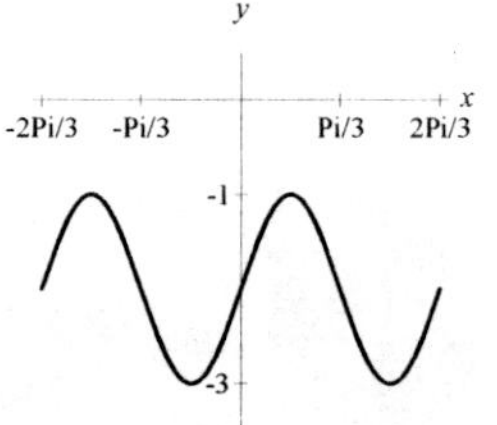

14. $y = \cos(x + \pi/2)$ has period 2π, range $[-1, 1]$, amplitude 1

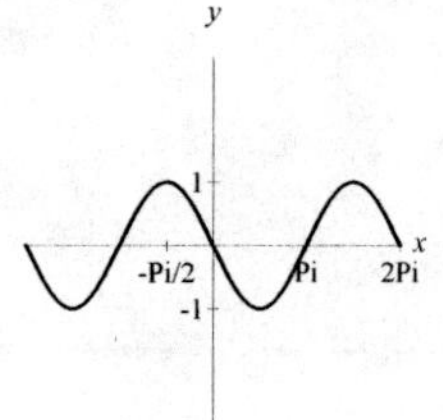

15. $y = \tan(\pi x/2)$ has period 2, range $(-\infty, \infty)$

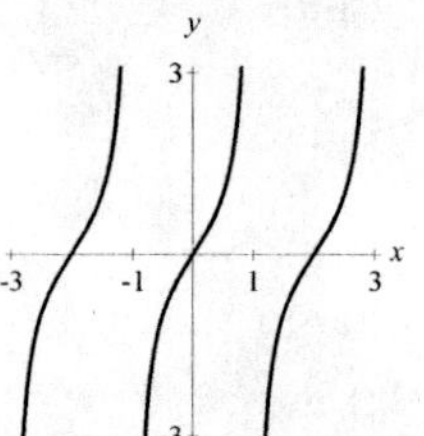

16. $y = 2\sin(2x + \pi)$ has period π, range $[-2, 2]$, amplitude 2

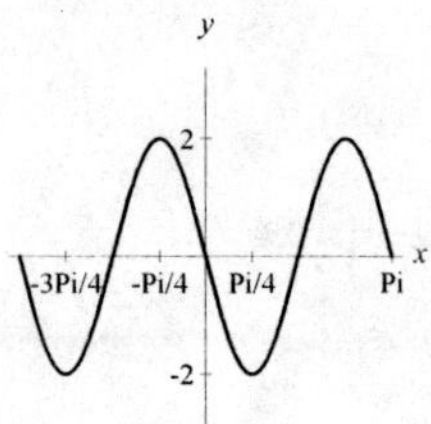

17. $y = 2\sec(x - \pi)$ has period 2π, range $(-\infty, -2] \cup [2, \infty)$

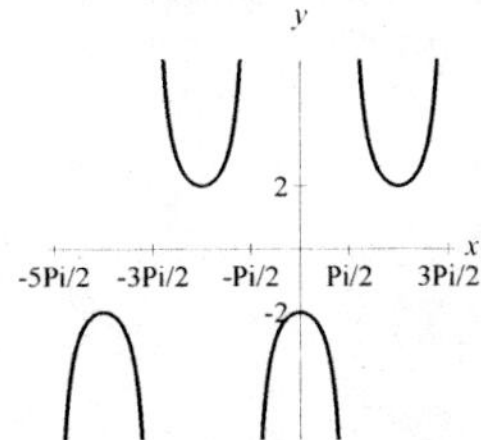

18. $y = \csc(x - \pi/2)$ has period 2π, range $(-\infty, -1] \cup [1, \infty)$

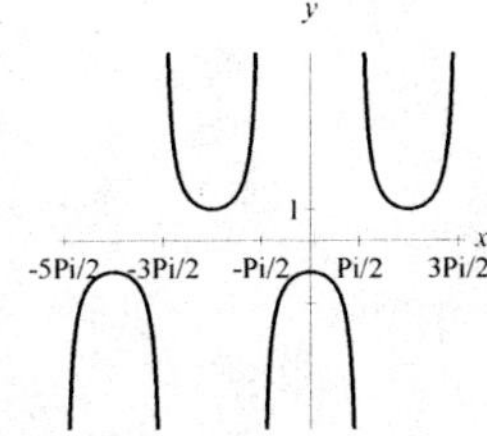

19. $y = \cot(2x)$ has period $\pi/2$, range $(-\infty, \infty)$

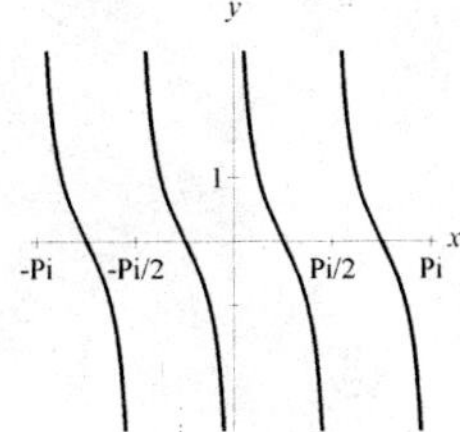

20. $y = -\cos(x - \pi/2)$ has period 2π, range $[-1, 1]$, amplitude 1

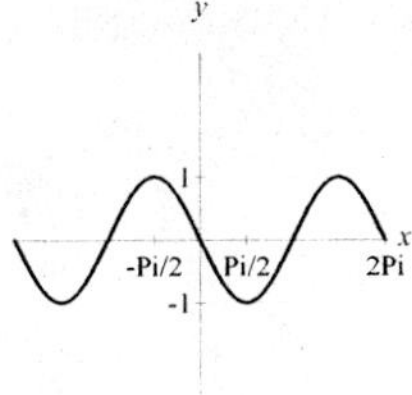

21. Since $46°24'6'' \approx 0.8098619$, the arclength is $s = r\alpha = 35.62(0.8098619) \approx 28.85$ meters.

22. $2.34 \cdot \dfrac{180°}{\pi} \approx 134.1°$

23. $\cos(\alpha) = -\sqrt{1 - (1/4)^2} = -\sqrt{15}/4$

24. $r = \sqrt{5^2 + (-2)^2} = \sqrt{29}$
$\sin(\alpha) = -2/\sqrt{29}$, $\cos(\alpha) = 5/\sqrt{29}$,
$\tan(\alpha) = -2/5$, $\csc(\alpha) = -\sqrt{29}/2$,
$\sec(\alpha) = \sqrt{29}/5$, $\cot(\alpha) = -5/2$

25. $\omega = 103 \cdot 2\pi \approx 647.2$ radians/minute

26. In one minute, the wheel turns through an arclength of $13(103 \cdot 2\pi)$ inches.

Multiplying this by $\dfrac{60}{12 \cdot 5280}$ results in the speed in mph which is 7.97 mph.

27. Let h be the height of the head.

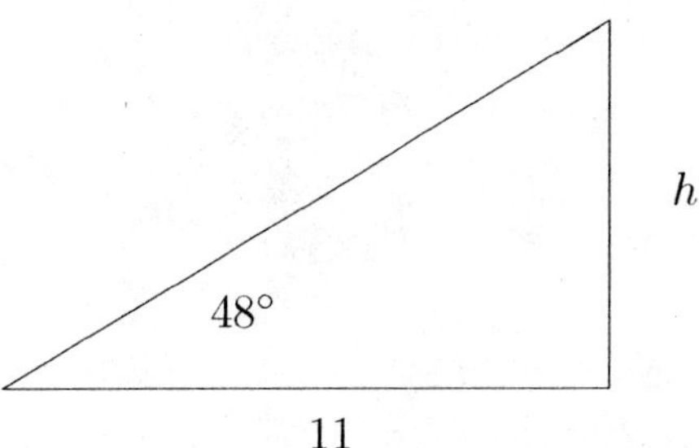

Since $\tan(48°) = h/11$, we obtain $h = 11 \tan 48° \approx 12.2$ m.

28. Let h be the height of the building.

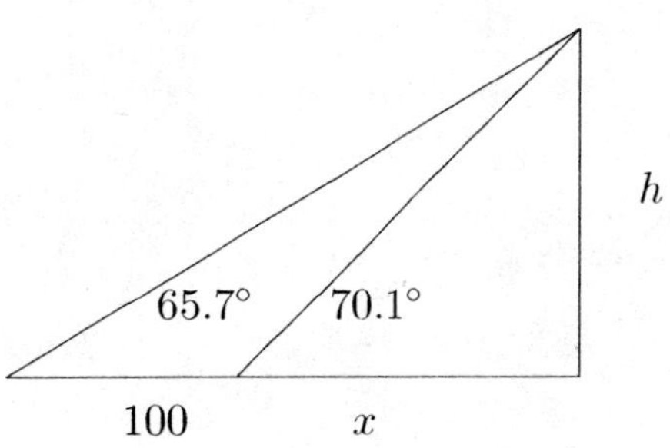

Since $\tan 70.1° = \dfrac{h}{x}$ and $\tan 65.7° = \dfrac{h}{100 + x}$, we obtain

$$\tan(65.7°) = \frac{h}{100 + h/\tan(70.1°)}$$

$$100 \cdot \tan(65.7°) + h \cdot \frac{\tan(65.7°)}{\tan(70.1°)} = h$$

$$100 \cdot \tan(65.7°) = h\left(1 - \frac{\tan(65.7°)}{\tan(70.1°)}\right)$$

$$h = \frac{100 \cdot \tan(65.7°)}{1 - \tan(65.7°)/\tan(70.1°)}$$

$$h \approx 1117 \text{ ft.}$$

29. Since the pH oscillates between 7.2 and 7.8, there is a vertical upward shift of 7.5 and $a = 0.3$. Note, the period is 4 days. Thus, $\frac{2\pi}{b} = 4$ or $b = \frac{\pi}{2}$. Since the pH is 7.2 on day 13, the pH is 7.5 on day 14. Using the period, the pH is also 7.5 on day 2. We can assume a right shift of 2 days. Hence, an equation is

$$y = 0.3\sin\left(\frac{\pi}{2}(x-2)\right) + 7.5.$$

A graph of one cycle is given.

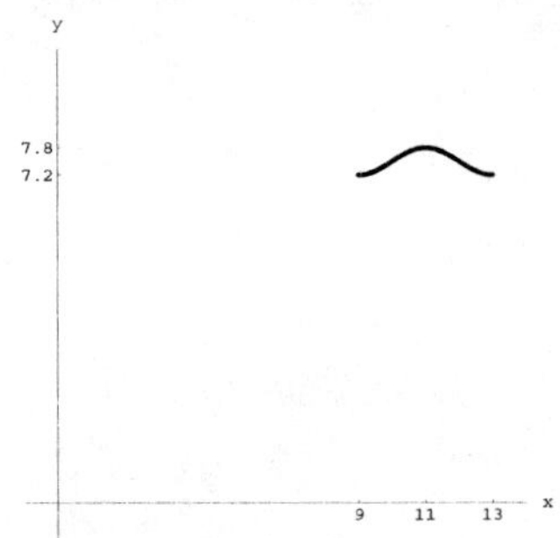

Tying It All Together

1. $y = 2 + e^x$ has domain $(-\infty, \infty)$, range $(2, \infty)$

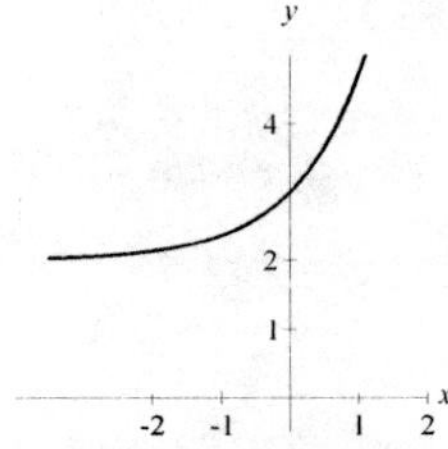

2. $y = 2 + x^2$ has domain $(-\infty, \infty)$, range $[2, \infty)$

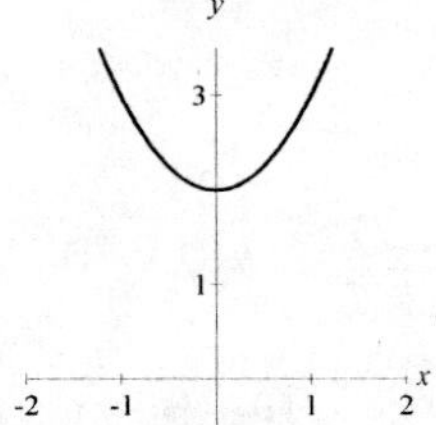

3. $y = 2 + \sin(x)$ has domain $(-\infty, \infty)$, range $[1, 3]$

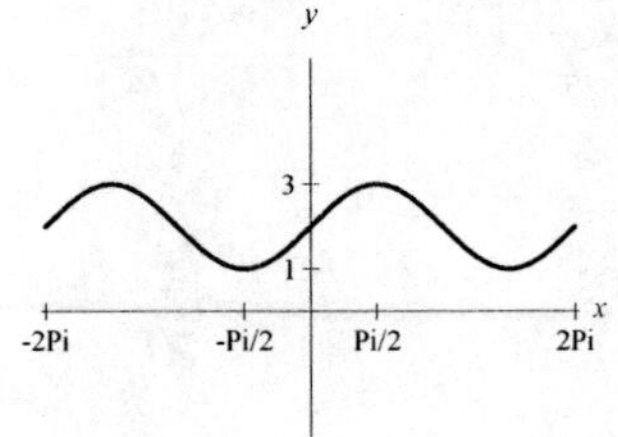

4. $y = 2 + \ln(x)$ has domain $(0, \infty)$, range $(-\infty, \infty)$

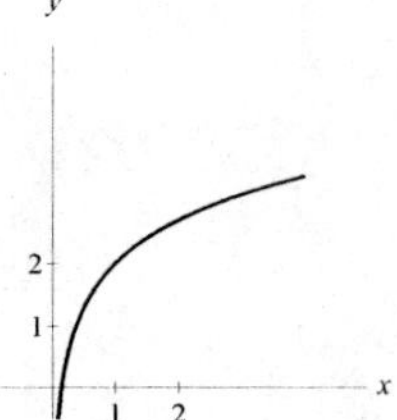

5. $y = \ln(x - \pi/4)$ has domain $(\pi/4, \infty)$, range $(-\infty, \infty)$

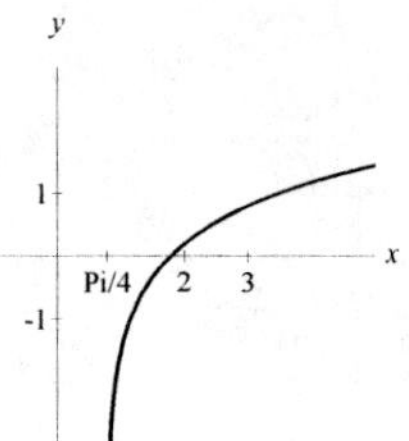

6. $y = \sin(x - \pi/4)$ has domain $(-\infty, \infty)$, range $[-1, 1]$

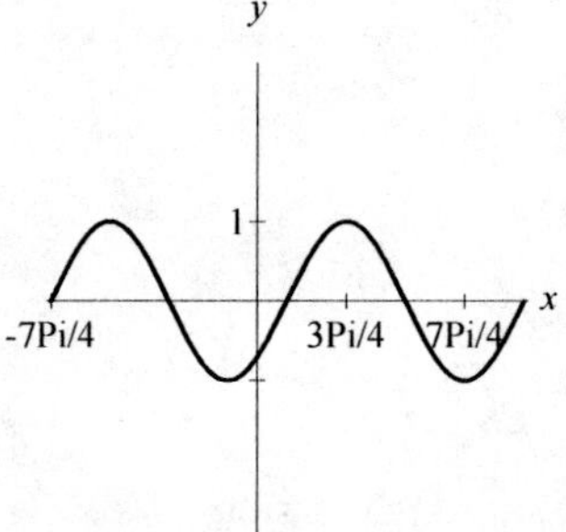

7. $y = \log_2(2x)$ has domain $(0, \infty)$, range $(-\infty, \infty)$

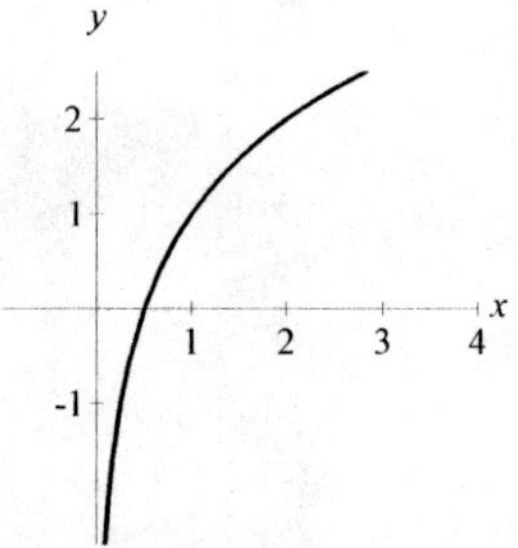

8. $y = \cos(2x)$ has domain $(-\infty, \infty)$, range $[-1, 1]$

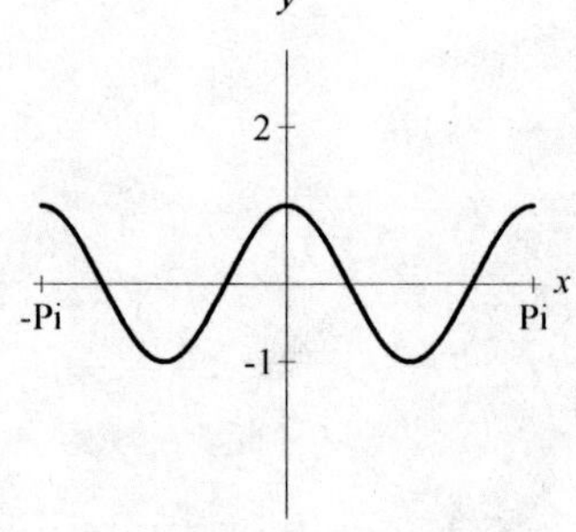

9. Neither

10. Odd

11. Even

12. Even

13. Odd

14. Even

15. Neither

16. Odd

17. Increasing

18. Increasing

19. Increasing

20. Decreasing

21. Decreasing

22. Increasing

23. perfect square

24. exponential

25. logarithmic

26. common

27. natural

28. angle

29. central

30. acute

31. quadrantal

32. coterminal

For Thought

1. False, for example $\sin 0 = 0$ but $\cos 0 = 1$.

2. True, since $\dfrac{\sin(x)}{\cos(x)} \cdot \dfrac{\cos(x)}{\sin(x)} = 1$.

3. True, since $f(-x) = (\sin(-x))^2 = (-\sin(x))^2 = \sin^2(x) = f(x)$.

4. False, it is even since $f(-x) = (\cos(-x))^3 = (\cos(x))^3 = f(x)$.

5. True **6.** False, since $(\sin x + \cos x)^2 = 1 + 2\sin(x)\cos(x) \neq 1 = \sin^2 x + \cos^2 x$.

7. False, since $\tan(x) = \pm\sqrt{\sec^2(x) - 1}$.

8. True, since $[\sin(-3)\csc(-3)] \cdot [\cos(-3)\sec(3)] \cdot [\tan(-3)\cot(3)] = [1] \cdot [1] \cdot [-1] = -1$.

9. False, $\sin^2(-\pi/9) + \cos^2(-\pi/9) = 1$.

10. True, $1 - \sin^2(-\pi/7) = \cos^2(-\pi/7) = \cos^2(\pi/7)$.

6.1 Exercises

1. even

3. Pythagorean

5. odd

7. $\dfrac{\sin x}{\cos x} \cdot \cos x = \sin x$

9. $\dfrac{1}{\cos x} \cdot \cos x = 1$

11. $\dfrac{1/\cos(x)}{\sin(x)/\cos(x)} = \dfrac{1}{\cos(x)} \cdot \dfrac{\cos(x)}{\sin(x)} = \dfrac{1}{\sin(x)} = \csc(x)$

13. $\dfrac{\sin x}{1/\sin x} + \cos^2 x = \sin^2 x + \cos^2 x = 1$

15. $\dfrac{\sin(x)}{1/\sin(x)} + \dfrac{\cos(x)}{1/\cos(x)} = \sin^2(x) + \cos^2(x) = 1$

17. $1 - \sin^2 \alpha = \cos^2 \alpha$

19. $(\sin\beta + 1)(\sin\beta - 1) = \sin^2\beta - 1 = -\cos^2\beta$

21.
$$\frac{1 + \cos\alpha \cdot \dfrac{\sin\alpha}{\cos\alpha} \cdot \dfrac{1}{\sin\alpha}}{1/\sin\alpha} = \frac{1+1}{1/\sin\alpha} = 2\sin\alpha$$

23. Since $\cot^2 x = \csc^2 x - 1$, we obtain $\cot x = \pm\sqrt{\csc^2 x - 1}$.

25. $\sin(x) = \dfrac{1}{\csc(x)} = \dfrac{1}{\pm\sqrt{1 + \cot^2(x)}}$

27. Since $\cot^2(x) = \csc^2(x) - 1$, we obtain

$$\tan(x) = \frac{1}{\cot(x)} = \frac{1}{\pm\sqrt{\csc^2(x) - 1}}$$

29. Since $\sec\alpha = \sqrt{1 + (1/2)^2} = \sqrt{5}/2$, $\cos\alpha = 2/\sqrt{5}$, $\sin\alpha = \sqrt{1 - (2/\sqrt{5})^2} = 1/\sqrt{5}$. So $\csc\alpha = \sqrt{5}$ and $\cot\alpha = 2$.

31. Since $\sin\alpha = -\sqrt{1 - (-\sqrt{3}/5)^2} = -\sqrt{1 - 3/25} = -\sqrt{22}/5$, $\csc\alpha = -5/\sqrt{22}$, $\sec\alpha = -5/\sqrt{3}$, $\tan\alpha = \dfrac{-\sqrt{22}/5}{-\sqrt{3}/5} = \sqrt{22}/\sqrt{3}$, and $\cot\alpha = \sqrt{3}/\sqrt{22}$.

33. Since α is in Quadrant IV, we get $\csc\alpha = -\sqrt{1 + (-1/3)^2} = -\sqrt{10}/3$, $\sin\alpha = -3/\sqrt{10}$, $\cos\alpha = \sqrt{1 - (-3/\sqrt{10})^2} = \sqrt{1 - 9/10} = 1/\sqrt{10}$, $\sec\alpha = \sqrt{10}$, and $\tan\alpha = -3$.

35. Let $\theta = \arccos x$. Then $\cos\theta = x$ and θ lies in quadrant 1 or 2. Since $\sin^2\theta = 1 - \cos^2\theta = 1 - x^2$, we obtain $\sin(\arccos x) = \sin\theta = \pm\sqrt{1 - x^2}$. Since sine is positive in both quadrants 1 and 2, we have $\sin(\arccos x) = \sqrt{1 - x^2}$.

37. Note, by Exercise 26, $\cos(x) = \dfrac{\pm 1}{\sqrt{\tan^2(x) + 1}}$.

Since $\arctan x$ is an angle in quadrant 1 or 4, and cosine is positive in both quadrants 1 and 4, we get

$$\cos(\arctan x) = \frac{1}{\sqrt{\tan^2(\arctan x) + 1}} = \frac{1}{\sqrt{x^2 + 1}}.$$

39. Note, $\tan x = \pm\sqrt{\sec^2 x - 1}$ and $\cos(\arcsin x) = \sqrt{1-x^2}$. Then

$$\begin{aligned}\tan(\arcsin x) &= \pm\sqrt{\sec^2(\arcsin x) - 1} \\ &= \pm\sqrt{\left(\frac{1}{\sqrt{1-x^2}}\right)^2 - 1} \\ &= \pm\sqrt{\frac{1}{1-x^2} - 1} \\ &= \pm\sqrt{\frac{x^2}{1-x^2}} \\ &= \pm\frac{\sqrt{x^2}}{\sqrt{1-x^2}} \\ &= \pm\frac{\pm x}{\sqrt{1-x^2}} \\ &= \pm\frac{x}{\sqrt{1-x^2}}.\end{aligned}$$

Note, $\tan(\arcsin x)$ is positive exactly when $x > 0$, and $\tan(\arcsin x)$ is negative exactly when $x < 0$. Thus, $\tan(\arcsin x) = \frac{x}{\sqrt{1-x^2}}$.

41. Note, $\arctan x$ is an angle in quadrant 1 or 4, and secant is positive in both quadrants 1 and 4. Since $\sec(\theta) = \pm\sqrt{\tan^2(\theta)+1}$, we have $\sec(\arctan x) = \sqrt{\tan^2(\arctan x)+1} = \sqrt{x^2+1}$.

43. $(-\sin x)\cdot(-\cot x) = \sin(x)\cdot\frac{\cos x}{\sin x} = \cos(x)$

45. $\sin(y) + (-\sin(y)) = 0$

47. $\frac{\sin(x)}{\cos(x)} + \frac{-\sin(x)}{\cos(x)} = 0$

49. $(1+\sin\alpha)(1-\sin\alpha) = 1-\sin^2\alpha = \cos^2\alpha$

51. $(-\sin\beta)(\cos\beta)(1/\sin\beta) = -\cos\beta$

53. Odd, since $\sin(-y) = -\sin(y)$ for any y, including $y = 2x$.

55. Neither, since $f(-\pi/6) \neq f(\pi/6)$ and $f(-\pi/6) \neq -f(\pi/6)$.

57. Even, since $\sec^2(-t) - 1 = \sec^2(t) - 1$.

59. Even, $f(-\alpha) = 1 + \sec(-\alpha) = 1 + \sec(\alpha) = f(\alpha)$

61. Even, $f(-x) = \frac{\sin(-x)}{-x} = \frac{-\sin(x)}{-x} = f(x)$

63. Odd, $f(-x) = -x + \sin(-x) = -x - \sin(x) = -f(x)$

65. h, since $\frac{1}{\csc(x)} = \frac{1}{1/\sin(x)} = 1\cdot\frac{\sin(x)}{1} = \sin(x)$

67. n **69.** m

71. k, since $\sin^2(x) + \cos^2(x) = 1$

73. l, since $\sec^2(x) = 1 + \tan^2(x)$

75. g, since $\cos(-x) = \cos(x)$ and $\cos(x) = \frac{1}{\sec(x)}$

77. b, since $\cot(-x) = \frac{\cos(-x)}{\sin(-x)} = \frac{\cos(x)}{-\sin(x)} = -\cot(x)$

79. f, since $\csc(-x) = \frac{1}{\sin(-x)} = \frac{1}{-\sin(x)} = -\csc(x)$

81. d, since $\sec^2(x) = 1 + \tan^2(x)$

83. It is not an identity. If $\gamma = \pi/3$ then $(\sin(\pi/3) + \cos(\pi/3))^2 = (\sqrt{3}/2 + 1/2)^2 = \frac{(\sqrt{3}+1)^2}{4} = \frac{4+2\sqrt{3}}{4} \neq 1 = \sin^2(\pi/3) + \cos^2(\pi/3)$.

85. It is not an identity. If $\beta = \pi/6$ then $(1+\sin(\pi/6))^2 = (1+1/2)^2 = (3/2)^2 = 9/4$ and $1 + \sin^2(\pi/6) = 1 + (1/2)^2 = 5/4$.

87. It is not an identity. If $\alpha = 7\pi/6$ then $\sin(7\pi/6) = -1/2$ while $\sqrt{1-\cos^2(7\pi/6)}$ is a positive number.

89. It is not an identity. If $y = \pi/6$ then $\sin(\pi/6) = 1/2$ and $\sin(-\pi/6) = -1/2$.

91. It is not an identity. If $y = \pi/6$ then $\cos^2(\pi/6) - \sin^2(\pi/6) = (\sqrt{3}/2)^2 - (1/2)^2 = 3/4 - 1/4 = 1/2$ and $\sin(2\cdot\pi/6) = \sin\pi/3 = \sqrt{3}/2$.

93. $1 - \frac{1}{\cos^2(x)} = 1 - \sec^2(x) = -\tan^2(x)$

95. $\dfrac{-(\tan^2 t + 1)}{\sec^2 t} = \dfrac{-\sec^2 t}{\sec^2 t} = -1$

97. $\dfrac{(1-\cos^2 w) - \cos^2 w}{1 - 2\cos^2 w} = \dfrac{1 - 2\cos^2 w}{1 - 2\cos^2 w} = 1$

99. $\dfrac{\tan x(\tan^2 x - \sec^2 x)}{-\cot x} = \dfrac{(\tan x)(-1)}{-\cot x} = \tan^2 x$

101.
$\dfrac{1}{\sin^3 x} - \dfrac{\cos^2(x)/\sin^2(x)}{\sin x} = \dfrac{1}{\sin^3 x} - \dfrac{\cos^2(x)}{\sin^3 x} =$
$\dfrac{1-\cos^2 x}{\sin^3 x} = \dfrac{\sin^2 x}{\sin^3 x} = \dfrac{1}{\sin x} = \csc x$

103. $(\sin^2 x - \cos^2 x)(\sin^2 x + \cos^2 x) =$
$(\sin^2 x - \cos^2 x)(1) = \sin^2 x - \cos^2 x$

105. $\cos\theta = \pm\sqrt{1-\sin^2\theta} = \pm\sqrt{1-(1/3)^2} =$
$\pm\sqrt{1-1/9} = \pm\sqrt{\dfrac{8}{9}} = \pm\dfrac{2\sqrt{2}}{3}$

107. $\cos\theta = \pm\sqrt{1-\sin^2\theta} = \pm\sqrt{1-u^2}$

109. Note, $\tan x = \dfrac{\sin x}{\cos x}$ is not valid if $\cos x = 0$.
Thus, the identity is not valid if $x = \dfrac{\pi}{2} + k\pi$ where k is an integer.

113. Let h be the height of the building. Using right triangle trigonometry, we find

$$h = 2000\tan 30^\circ \approx 1155 \text{ ft.}$$

115. The amplitude is 5.

Since $B = 2$, the period is $\dfrac{2\pi}{B} = \dfrac{2\pi}{2} = \pi$.

Since $2x - \pi = 2\left(x - \dfrac{\pi}{2}\right)$, phase shift is $\dfrac{\pi}{2}$.

The range is the interval $[-5+3, 5+3] = [-2, 8]$

117. $\cos\beta = 0$, for $\cos^2\beta + \sin^2\beta = 1$

119. Let r be the radius of the small circle, and let x be the distance from the center of the small circle to the point of tangency of any two circles with radius 1.

By the Pythagorean theorem, we find

$$1 + (x+r)^2 = (1+r)^2$$

and

$$1 + (1 + 2r + x)^2 = 2^2.$$

The second equation may be written as

$$1 + (r+1)^2 + 2(r+1)(r+x) + (r+x)^2 = 4.$$

Using the first equation, the above equation simplifies to

$$(r+1)^2 + 2(r+1)(r+x) + (1+r)^2 = 4$$

or

$$(r+1)^2 + (r+1)(r+x) = 2.$$

Since (from first equation, again)

$$x + r = \sqrt{(1+r)^2 - 1}$$

we obtain

$$(r+1)^2 + (r+1)\left(\sqrt{(1+r)^2 - 1}\right) = 2.$$

Solving for r, we find

$$r = \frac{2\sqrt{3} - 3}{3}.$$

For Thought

1. True, $\dfrac{\sin x}{1/\sin x} = \sin x \cdot \dfrac{\sin x}{1} = \sin^2 x$.

2. False, if $x = \pi/3$ then $\dfrac{\cot(\pi/3)}{\tan(\pi/3)} =$
$\dfrac{\sqrt{3}/3}{\sqrt{3}} = \dfrac{1}{3}$ and $\tan^2(\pi/3) = (\sqrt{3})^2 = 3$.

3. True, $\dfrac{1/\cos x}{1/\sin x} = \dfrac{1}{\cos x} \cdot \dfrac{\sin x}{1} = \dfrac{\sin x}{\cos x} = \tan x$.

4. True, $\sin x \cdot \dfrac{1}{\cos x} = \dfrac{\sin x}{\cos x} = \tan x$.

5. True, $\dfrac{\cos x}{\cos x} + \dfrac{\sin x}{\cos x} = 1 + \tan x$.

6. False, if $x = \pi/4$ then
$\sec(\pi/4) + \dfrac{\sin(\pi/4)}{\cos(\pi/4)} = \sqrt{2} + 1$ and
$\dfrac{1 + \sin(\pi/4)\cos(\pi/4)}{\cos(\pi/4)} = \dfrac{1 + (\sqrt{2}/2)(\sqrt{2}/2)}{\sqrt{2}/2} =$
$\dfrac{1 + (1/2)}{\sqrt{2}/2} = (3/2)(2/\sqrt{2}) = 3/\sqrt{2}$.

7. True, $\frac{1+\sin x}{1-\sin^2 x} = \frac{1+\sin x}{(1-\sin x)(1+\sin x)} = \frac{1}{1-\sin x}$.

8. True, since $\tan x \cdot \cot x = \tan x \cdot \frac{1}{\tan x} = 1$.

9. False, if $x = \pi/3$ then $(1-\cos(\pi/3))^2 = (1-1/2)^2 = (1/2)^2 = 1/4$ and $\sin^2(\pi/3) = (\sqrt{3}/2)^2 = 3/4$.

10. False, if $x = \pi/6$ then $(1-\csc(\pi/6))(1+\csc(\pi/6)) = (1-2)(1+2) = -3$ and $\cot^2(\pi/6) = (\sqrt{3})^2 = 3$.

6.2 Exercises

1. complicated

3. numerator, denominator

5. D, $\cos x \tan x = \cos x \cdot \frac{\sin x}{\cos x} = \sin x$.

7. A, $\csc^2 x - \cot^2 x = 1$.

9. B, $1 - \sec^2 x = -\tan^2 x$.

11. H, $\frac{\csc x}{\csc x} - \frac{\sin x}{\csc x} = 1 - \sin^2 x = \cos^2 x$.

13. G, $\csc^2 x = 1 + \cot^2 x$.

15. $2\cos^2\beta - \cos\beta - 1$

17. $\csc^2 x + 2\csc x \sin x + \sin^2 x = \csc^2 x + 2 + \sin^2 x$

19. $4\sin^2\theta - 1$

21. $9\sin^2\theta + 12\sin\theta + 4$

23. $4\sin^4 y - 4\sin^2 y \csc^2 y + \csc^4 y = 4\sin^4 y - 4 + \csc^4 y$

25. Note the factorization of a difference of two squares: $(1-\sin\alpha)(1+\sin\alpha) = 1 - \sin^2\alpha = \cos^2\alpha$.

27. Note the factorization of a difference of two squares: $(\csc\alpha - 1)(\csc\alpha + 1) = \csc^2\alpha - 1 = \cot^2\alpha$.

29. Note the factorization of a difference of two squares: $(\tan\alpha - \sec\alpha)(\tan\alpha + \sec\alpha) = \tan^2\alpha - \sec^2\alpha = -1$.

31. $(2\sin\gamma + 1)(\sin\gamma - 3)$

33. $(\tan\alpha - 4)(\tan\alpha - 2)$

35. $(2\sec\beta + 1)^2$

37. $(\tan\alpha - \sec\beta)(\tan\alpha + \sec\beta)$

39. $\cos\beta(\sin^2\beta + \sin\beta - 2) = \cos\beta(\sin\beta + 2)(\sin\beta - 1)$

41. $(2\sec^2 x - 1)^2$

43. $\cos\alpha(\sin\alpha + 1) + (\sin\alpha + 1) = (\sin\alpha + 1)(\cos\alpha + 1)$

45. Combining, we get $\frac{1-\cos^2 x}{a} = \frac{\sin^2 x}{a}$.

47. We obtain $\frac{\sin(2x)}{2} + \frac{2\sin(2x)}{2} = \frac{3\sin(2x)}{2}$.

49. Since 6 is the LCD, we get $\frac{2\tan x}{6} + \frac{3\tan x}{6} = \frac{5\tan x}{6}$.

51. Separating the fraction, we obtain $\frac{\sin x}{\sin x} - \frac{\sin^2 x}{\sin x} = 1 - \sin x$.

53. Factoring: $\frac{(\sin x - \cos x)(\sin x + \cos x)}{\sin x - \cos x} = \sin x + \cos x$.

55. Factoring: $\frac{(\sin x - 2)(\sin x + 1)}{(\sin x - 2)(\sin x + 2)} = \frac{\sin x + 1}{\sin x + 2}$.

57. Note, $\sin(-x) = -\sin(x)$. Factoring, we obtain $\frac{\sin^2 x + \sin x}{1 + \sin x} = \frac{\sin x(\sin x + 1)}{1 + \sin x} = \sin x$.

59.

$$\begin{aligned} \sin x \cot x &= \\ \sin x \frac{\cos x}{\sin x} &= \\ \cos x & \end{aligned}$$

61.

$$\begin{aligned} 1-\sec x\cos^3 x &= \\ 1-\frac{1}{\cos x}\cos^3 x &= \\ 1-\cos^2 x &= \\ \sin^2 x & \end{aligned}$$

63.

$$\begin{aligned} 1+\sec^2 x\sin^2 x &= \\ 1+\frac{1}{\cos^2 x}\sin^2 x &= \\ 1+\tan^2 x &= \\ \sec^2 x & \end{aligned}$$

65.

$$\begin{aligned} \frac{\sin^3 x+\sin x\cos^2 x}{\cos x} &= \\ \frac{\sin x(\sin^2 x+\cos^2 x)}{\cos x} &= \\ \frac{(\sin x)(1)}{\cos x} &= \\ \tan x & \end{aligned}$$

67.

$$\begin{aligned} \frac{\sin x}{\csc x}+\frac{\cos x}{\sec x} &= \\ \frac{\sin x}{1/\sin x}+\frac{\cos x}{1/\cos x} &= \\ \sin^2 x+\cos^2 x &= \\ 1 & \end{aligned}$$

69.

$$\begin{aligned} \frac{1}{\csc\theta-\cot\theta}\cdot\frac{\sin\theta}{\sin\theta} &= \\ \frac{\sin\theta}{1-\cos\theta}\cdot\frac{1+\cos\theta}{1+\cos\theta} &= \\ \frac{\sin\theta(1+\cos\theta)}{1-\cos^2\theta} &= \\ \frac{\sin\theta(1+\cos\theta)}{\sin^2\theta} &= \\ \frac{1+\cos\theta}{\sin\theta} & \end{aligned}$$

71.

$$\begin{aligned} \frac{\sec x-\cos x}{\sec x} &= \\ 1-\frac{\cos x}{\sec x} &= \\ 1-\cos^2 x &= \\ \sin^2 x & \end{aligned}$$

73.

$$\begin{aligned} &= \frac{1-(-\sin x)^2}{1+\sin x} \\ &= \frac{1-\sin^2 x}{1+\sin x} \\ &= \frac{(1-\sin x)(1+\sin x)}{1+\sin x} \\ &= 1-\sin(x) \end{aligned}$$

75.

$$\begin{aligned} &= \frac{1-\cot^2 w\,(1-\cos^2 w)}{\csc^2 w} \\ &= \frac{1-\cot^2 w\sin^2 w}{\csc^2 w} \\ &= \frac{1-\cos^2 w}{\csc^2 w} \\ &= \frac{\sin^2 w}{\csc^2 w} \\ &= \sin^4 w \end{aligned}$$

77.

$$\begin{aligned} &= \frac{\cos x+\csc x}{\cos x} \\ &= \frac{\cos x}{\cos x}+\frac{\csc x}{\cos x} \\ &= 1+\csc x\sec x \end{aligned}$$

79. Rewrite the left side of the equation.

$$\begin{aligned} \tan(x)\cos(x)+\csc(x)\sin^2(x) &= \\ \sin x+\sin x &= \\ 2\sin x & \end{aligned}$$

81.

$$\begin{aligned}(1+\sin\alpha)^2+\cos^2\alpha &= \\ 1+2\sin\alpha+\sin^2\alpha+\cos^2\alpha &= \\ 2+2\sin\alpha\end{aligned}$$

83.

$$\begin{aligned}\frac{\sin^2\beta+\sin\beta-2}{2\sin\beta-2} &= \\ \frac{(\sin\beta+2)(\sin\beta-1)}{2(\sin\beta-1)} &= \\ \frac{\sin\beta+2}{2}\end{aligned}$$

85.

$$\begin{aligned}2-\csc(\beta)\sin(\beta) &= \\ 2-1 &= \\ 1 &= \\ \sin^2(\beta)+\cos^2(\beta)\end{aligned}$$

87.

$$\begin{aligned}\frac{\sin x}{\cos x}+\frac{\cos x}{\sin x} &= \\ \frac{\sin^2 x+\cos^2 x}{\sin(x)\cos(x)} &= \\ \frac{1}{\sin(x)\cos(x)} &= \\ \sec(x)\csc(x)\end{aligned}$$

89.

$$\begin{aligned}\frac{\sec(x)}{\tan(x)}-\frac{\tan(x)}{\sec(x)} &= \\ \frac{\sec^2(x)-\tan^2(x)}{\tan(x)\sec(x)} &= \\ \frac{1}{\tan(x)\sec(x)} &= \\ \cot(x)\cos(x)\end{aligned}$$

91. Rewrite the right side of the equation.

$$\begin{aligned}&= \frac{\csc x}{\csc x-\sin x}\cdot\frac{\sin x}{\sin x} \\ &= \frac{1}{1-\sin^2 x} \\ &= \frac{1}{\cos^2 x} \\ &= \sec^2 x\end{aligned}$$

93.

$$\begin{aligned}&= \frac{1+\sin(y)}{1-\sin(y)}\cdot\frac{\csc(y)}{\csc(y)} \\ &\quad \frac{\csc(y)+1}{\csc(y)-1}\end{aligned}$$

95.

$$\begin{aligned}\ln(\sec\theta) &= \\ \ln((\cos\theta)^{-1}) &= \\ -\ln(\cos\theta)\end{aligned}$$

97.

$$\begin{aligned}\ln\left|(\sec\alpha+\tan\alpha)\cdot\frac{\sec\alpha-\tan\alpha}{\sec\alpha-\tan\alpha}\right| &= \\ \ln\left|\frac{\sec^2\alpha-\tan^2\alpha}{\sec\alpha-\tan\alpha}\right| &= \\ \ln\left|\frac{1}{\sec\alpha-\tan\alpha}\right| &= \\ -\ln|\sec\alpha-\tan\alpha|\end{aligned}$$

99. It is an identity since

$$\begin{aligned}\frac{\sin\theta}{\sin\theta}+\frac{\cos\theta}{\sin\theta} &= \\ 1+\cot\theta.\end{aligned}$$

The graphs of $y=\frac{\sin\theta+\cos\theta}{\sin\theta}$ and $y=1+\cot\theta$ are shown to be identical.

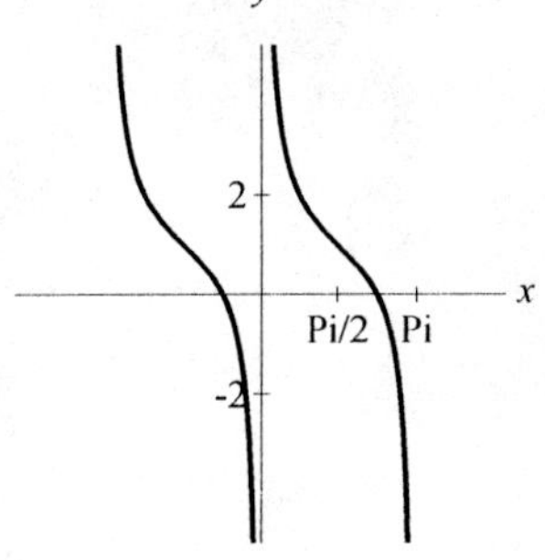

101. It is not an identity since the graphs of $y = (\sin x + \csc x)^2$ and $y = \sin^2 x + \csc^2 x$ do not coincide as shown.

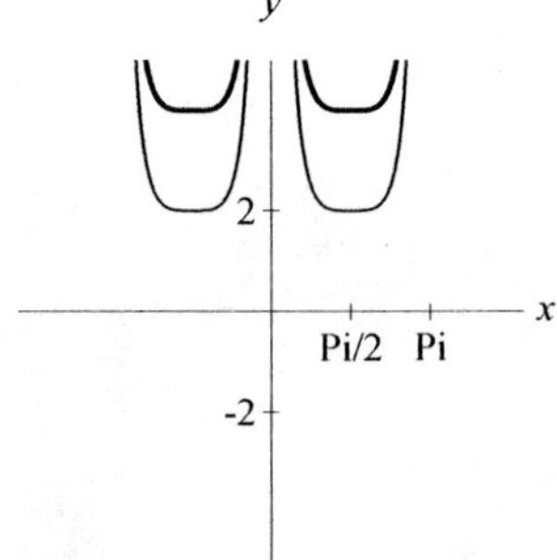

103. It is an identity. Re-arranging the numerator of the right-hand side one finds

$$\begin{aligned} &= \frac{1 - \cos^2 x + \cos x}{\sin x} \\ &= \frac{\sin^2 x + \cos x}{\sin x} \\ &= \frac{\sin^2 x}{\sin x} + \frac{\cos x}{\sin x} \\ &= \sin x + \cot x. \end{aligned}$$

The graphs of $y = \cot x + \sin x$ and $y = \dfrac{1 + \cos x - \cos^2 x}{\sin x}$ are shown to be identical.

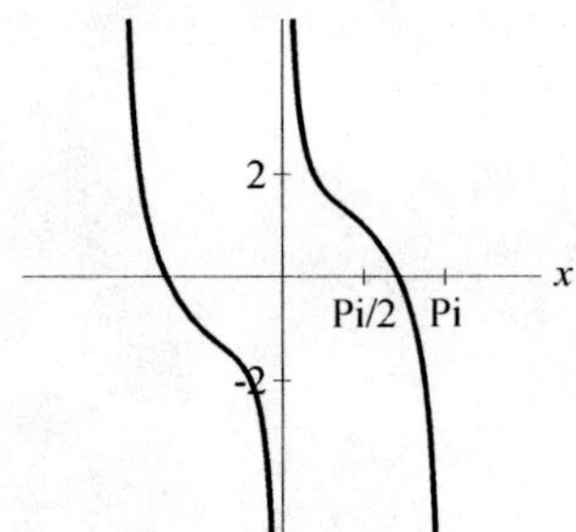

105. It is not an identity since the graphs of $y = \dfrac{\sin x}{\cos x} - \dfrac{\cos x}{\sin x}$ and $y = \dfrac{2\cos^2 x - 1}{\sin x \cos x}$ are not the same as shown.

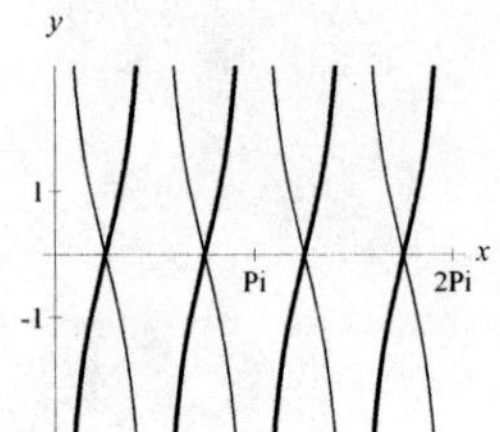

107. It is an identity.

$$\begin{aligned} \frac{\cos x}{1 - \sin(x)} \cdot \frac{1 + \sin x}{1 + \sin x} &= \\ \frac{\cos x(1 + \sin x)}{1 - \sin^2 x} &= \\ \frac{\cos x(1 + \sin x)}{\cos^2 x} &= \\ \frac{1 + \sin x}{\cos x} &= \\ \frac{1 - \sin(-x)}{\cos x} &= \end{aligned}$$

The graphs of $y = \dfrac{\cos(-x)}{1 - \sin x}$ and $y = \dfrac{1 - \sin(-x)}{\cos x}$ are shown to be identical.

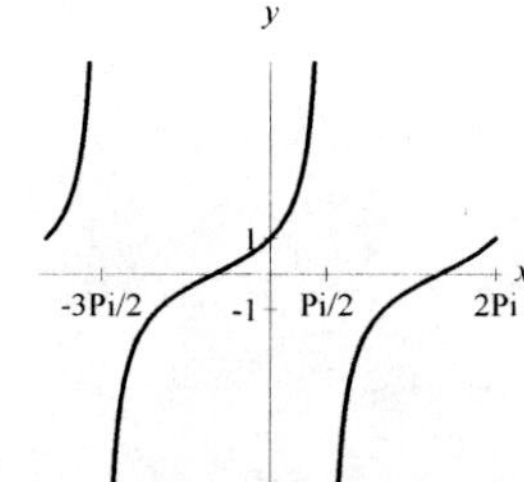

111. $\sin^2 x + \cos^2 x = 1$, $1 + \cot^2 x = \csc^2 x$, $\tan^2 x + 1 = \sec^2 x$

113. $\dfrac{\csc x}{\sec x} = \dfrac{1/\sin x}{1/\cos x} = \dfrac{\cos x}{\sin x} = \cot x$

115. Amplitude is 4. Since $B = 2\pi/3$, the period is $\dfrac{2\pi}{B} = \dfrac{2\pi}{2\pi/3} = 3$.

Since $\dfrac{2\pi x}{3} - \dfrac{\pi}{3} = \dfrac{2\pi}{3}\left(x - \dfrac{1}{2}\right)$, the phase shift is $\dfrac{1}{2}$.

117. Solve for v_0 in ft/sec:

$$\begin{aligned} \frac{1}{32} v_0^2 \sin 2(33°) &= 200 \\ v_0 &= \sqrt{\frac{200(32)}{\sin 66°}} \text{ ft/sec} \\ v_0 &= \sqrt{\frac{200(32)}{\sin 66°}} \cdot \frac{3600}{5280} \text{ mph} \\ v_0 &\approx 57.1 \text{ mph}. \end{aligned}$$

119. Since $\sin x = 3\cos x$, we find

$$\begin{aligned} 1 &= \sin^2 x + \cos^2 x \\ 1 &= 9\cos^2 x + \cos^2 x \\ \cos^2 x &= \frac{1}{10}. \end{aligned}$$

Then $\sin x \cos x = 3\cos^2 x = \frac{3}{10}$.

For Thought

1. False, the right-hand side should be $\cos(5°)$.

2. True, by the sum identity for cosine.

3. True, $\cos(t - \pi/2) = \cos(\pi/2 - t) = \sin t$.

4. False, $\sin(\alpha - \pi/2) = -\sin(\pi/2 - \alpha) = -\cos\alpha$.

5. True, $\sec(\pi/3) = \sec(\pi/2 - \pi/6) = \csc(\pi/6)$.

6. False, since $\sin(5\pi/6) = 1/2$ and $\sin(2\pi/3) + \sin(\pi/6) = \sqrt{3}/2 + 1/2$.

7. True, since the sum identity for sine is applied to $5\pi/12 = \pi/6 + \pi/4$.

8. True, since the cofunction identity for tangent is applied to $90° - 68°29'55" = 21°30'5"$.

9. False, the equation fails when $x = \pi/2$.

10. True, since both sides of the equation (by the sum identity for tangent) are equal to $\tan(-7°)$.

6.3 Exercises

1. cosine

3. cotangent

5. $\cos(\pi/3 + \pi/4) =$
$\cos(\pi/3)\cos(\pi/4) - \sin(\pi/3)\sin(\pi/4) =$

$$\frac{1}{2}\cdot\frac{\sqrt{2}}{2} - \frac{\sqrt{3}}{2}\cdot\frac{\sqrt{2}}{2} = \frac{\sqrt{2}-\sqrt{6}}{4}$$

7. $\cos(60° - 45°) =$
$\cos(60°)\cos(45°) + \sin(60°)\sin(45°) =$

$$\frac{1}{2}\cdot\frac{\sqrt{2}}{2} + \frac{\sqrt{3}}{2}\cdot\frac{\sqrt{2}}{2} = \frac{\sqrt{2}+\sqrt{6}}{4}$$

9. Since $\sin(20°) = \cos(90° - 20°) = \cos(70°)$, the answer is $70°$.

11. Since $\tan\left(\frac{\pi}{6}\right) = \cot\left(\frac{\pi}{2} - \frac{\pi}{6}\right) = \cot\left(\frac{\pi}{3}\right)$, the answer is $\pi/3$.

13. Since $\sec(90° - 6°) = \csc(6°)$, the answer is $6°$.

15. $\sin(\pi/3 + \pi/4) =$
$\sin(\pi/3)\cos(\pi/4) + \cos(\pi/3)\sin(\pi/4) =$

$$\frac{\sqrt{3}}{2}\cdot\frac{\sqrt{2}}{2} + \frac{1}{2}\cdot\frac{\sqrt{2}}{2} = \frac{\sqrt{6}+\sqrt{2}}{4}$$

17. $\sin(60° - 45°) =$
$\sin(60°)\cos(45°) - \cos(60°)\sin(45°) =$

$$\frac{\sqrt{3}}{2}\cdot\frac{\sqrt{2}}{2} - \frac{1}{2}\cdot\frac{\sqrt{2}}{2} = \frac{\sqrt{6}-\sqrt{2}}{4}$$

19. $\tan\left(\frac{3\pi}{4} + \frac{\pi}{3}\right) = \frac{\tan(3\pi/4) + \tan(\pi/3)}{1 - \tan(3\pi/4)\tan(\pi/3)} =$

$$\frac{-1+\sqrt{3}}{1-(-1)(\sqrt{3})} = \frac{\sqrt{3}-1}{\sqrt{3}+1} = \frac{\sqrt{3}-1}{\sqrt{3}+1}\cdot\frac{\sqrt{3}-1}{\sqrt{3}-1} =$$

$$\frac{4-2\sqrt{3}}{2} = 2 - \sqrt{3}$$

21. $\tan(210° - 45°) =$

$$\frac{\tan(210°) - \tan(45°)}{1 + \tan(210°)\tan(45°)} =$$

$$\frac{\sqrt{3}/3 - 1}{1 + (\sqrt{3}/3)(1)} = \frac{\sqrt{3}/3 - 1}{1 + (\sqrt{3}/3)(1)}\cdot\frac{3}{3} =$$

$$\frac{\sqrt{3}-3}{3+\sqrt{3}} = \frac{\sqrt{3}-3}{3+\sqrt{3}}\cdot\frac{3-\sqrt{3}}{3-\sqrt{3}} = \frac{-12+6\sqrt{3}}{6} =$$

$$\sqrt{3} - 2$$

23. $\frac{7\pi}{12}$

25. $\frac{13\pi}{12}$

27. $\sin(23° + 67°) = \sin(90°) = 1$

29. $\cos\left(\frac{\pi}{6} + \frac{\pi}{3}\right) = \cos\left(\frac{\pi}{2}\right) = 0$

31. $\tan\left(\frac{\pi}{12} + \frac{\pi}{6}\right) = \tan\left(\frac{\pi}{4}\right) = 1$

33. $\sin(2k + k) = \sin(3k)$

35. $30° + 45°$

37. $120° + 45°$

39. $\cos(2\pi/3 - \pi/4) =$

$\cos(2\pi/3)\cos(\pi/4) + \sin(2\pi/3)\sin(\pi/4) =$

$-\frac{1}{2}\cdot\frac{\sqrt{2}}{2} + \frac{\sqrt{3}}{2}\cdot\frac{\sqrt{2}}{2} = \frac{\sqrt{6}-\sqrt{2}}{4}$

41. $\sin(\pi/3 + \pi/4) =$
$\sin(\pi/3)\cos(\pi/4) + \cos(\pi/3)\sin(\pi/4) =$

$\frac{\sqrt{3}}{2}\cdot\frac{\sqrt{2}}{2} + \frac{1}{2}\cdot\frac{\sqrt{2}}{2} = \frac{\sqrt{6}+\sqrt{2}}{4}$

43. $\tan(45° + 30°) = \frac{\tan(45°) + \tan(30°)}{1 - \tan(45°)\tan(30°)} =$

$\frac{1 + \sqrt{3}/3}{1 - 1\cdot\sqrt{3}/3}\cdot\frac{3}{3} = \frac{3+\sqrt{3}}{3-\sqrt{3}}\cdot\frac{3+\sqrt{3}}{3+\sqrt{3}} =$

$\frac{12 + 6\sqrt{3}}{9-3} = 2 + \sqrt{3}$

45. $\sin(30° - 45°) =$
$\sin(30°)\cos(45°) - \cos(30°)\sin(45°) =$

$\frac{1}{2}\cdot\frac{\sqrt{2}}{2} - \frac{\sqrt{3}}{2}\cdot\frac{\sqrt{2}}{2} = \frac{\sqrt{2}-\sqrt{6}}{4}$

47. $\cos(135° + 60°) =$
$\cos(135°)\cos(60°) - \sin(135°)\sin(60°) =$

$\frac{-\sqrt{2}}{2}\cdot\frac{1}{2} - \frac{\sqrt{2}}{2}\cdot\frac{\sqrt{3}}{2} = \frac{-\sqrt{2}-\sqrt{6}}{4}$

49. $\tan(-13\pi/12) = -\tan(13\pi/12) =$

$-\tan\left(\frac{3\pi}{4} + \frac{\pi}{3}\right) =$

$-\frac{\tan(3\pi/4) + \tan(\pi/3)}{1 - \tan(3\pi/4)\tan(\pi/3)} =$

$-\frac{-1+\sqrt{3}}{1-(-1)\sqrt{3}} = \frac{1-\sqrt{3}}{1+\sqrt{3}}\cdot\frac{1-\sqrt{3}}{1-\sqrt{3}} =$

$\frac{4-2\sqrt{3}}{-2} = -2 + \sqrt{3}$

51. $\sin(3°)\cos(-87°) + \cos(3°)\sin(87°) =$
$\sin(3°)\cos(87°) + \cos(3°)\sin(87°) =$
$\sin(3° + 87°) = \sin(90°) = 1$

53. $\cos(\pi/2)\cos(\pi/5) + \sin(\pi/2)\sin(\pi/5) =$

$\cos\left(\frac{\pi}{2} - \frac{\pi}{5}\right) = \cos(3\pi/10)$

55. $\frac{\tan(\pi/7) + \tan(\pi/6)}{1 - \tan(\pi/7)\tan(\pi/6)} = \tan\left(\frac{\pi}{7} + \frac{\pi}{6}\right) =$
$\tan(13\pi/42)$

57. $\sin(14°)\cos(35°) + \cos(14°)\sin(35°) =$
$\sin(14° + 35°) = \sin(49°)$

59. G, $\cos(44°) = \sin(90° - 44°) = \sin(46°)$

61. H, $\cos(46°) = \sin(90° - 46°) = \sin(44°)$

63. F, $\sec(1) = \csc\left(\frac{\pi}{2} - 1\right) = \csc\left(\frac{\pi-2}{2}\right)$

65. A, $\csc(\pi/2) = 1 = \cos(0)$

67. Since α is in quadrant II and β is in quadrant I, $\cos\alpha = -\sqrt{1 - \left(\frac{3}{5}\right)^2} =$

$-\sqrt{1 - \frac{9}{25}} = -\sqrt{\frac{16}{25}} = -\frac{4}{5}$ and $\cos\beta =$

$\sqrt{1 - \left(\frac{5}{13}\right)^2} = \sqrt{1 - \frac{25}{169}} = \sqrt{\frac{144}{169}} = \frac{12}{13}.$

So $\sin(\alpha + \beta) = \sin\alpha\cos\beta + \cos\alpha\sin\beta =$

$\frac{3}{5}\cdot\frac{12}{13} + \frac{-4}{5}\cdot\frac{5}{13} = \frac{16}{65}.$

69. Since α is in quadrant I and β is in quadrant III, we obtain

$\cos\alpha = \sqrt{1 - \left(\frac{2}{3}\right)^2} = \sqrt{1 - \frac{4}{9}} =$

$\sqrt{\frac{5}{9}} = \frac{\sqrt{5}}{3}$ and $\cos\beta = -\sqrt{1 - \left(\frac{-1}{2}\right)^2} =$

$-\sqrt{1 - \frac{1}{4}} = -\sqrt{\frac{3}{4}} = -\frac{\sqrt{3}}{2}.$

So $\cos(\alpha + \beta) = \cos\alpha\cos\beta - \sin\alpha\sin\beta =$

$\frac{\sqrt{5}}{3}\cdot\frac{-\sqrt{3}}{2} - \frac{2}{3}\cdot\frac{-1}{2} = \frac{2-\sqrt{15}}{6}.$

71. Since α is in quadrant III and β is in quadrant II, we find

$\cos\alpha = -\sqrt{1 - \left(\frac{-24}{25}\right)^2} = -\frac{7}{25}$

and $\sin\beta = \sqrt{1 - \left(\frac{-8}{17}\right)^2} = \frac{15}{17}.$

Then $\sin(\alpha - \beta) = \sin\alpha\cos\beta - \cos\alpha\sin\beta =$

$\frac{-24}{25}\cdot\frac{-8}{17} - \frac{-7}{25}\cdot\frac{15}{17} = \frac{297}{425}.$

73. Since α is in quadrant II and β is in quadrant IV, we find

$$\cos\alpha = -\sqrt{1-\left(\frac{24}{25}\right)^2} = -\frac{7}{25}$$

and $\sin\beta = -\sqrt{1-\left(\frac{8}{17}\right)^2} = -\frac{15}{17}$.

Then $\cos(\alpha-\beta) = \cos\alpha\cos\beta + \sin\alpha\sin\beta =$
$\frac{-7}{25}\cdot\frac{8}{17} + \frac{24}{25}\cdot\frac{-15}{17} = -\frac{416}{425}$.

75. $\cos(\pi/2 - (-\alpha)) = \sin(-\alpha) = -\sin\alpha$

77. $\cos 180^\circ \cos\alpha + \sin 180^\circ \sin\alpha =$
$(-1)\cdot\cos\alpha + 0\cdot\sin\alpha = -\cos\alpha$

79. The period is 360°, so $\sin(360^\circ - \alpha) =$
$\sin(-\alpha) = -\sin\alpha$

81. $\sin(90^\circ - (-\alpha)) = \cos(-\alpha) = \cos\alpha$

83.

$$\begin{aligned}
\sin(180^\circ - \alpha) &= \\
\sin(180^\circ)\cos\alpha - \cos(180^\circ)\ \sin\alpha &= \\
\sin\alpha &= \\
\frac{\sin^2\alpha}{\sin\alpha} &= \\
\frac{1-\cos^2\alpha}{\sin\alpha}
\end{aligned}$$

85.

$$\begin{aligned}
\frac{\cos(x+y)}{\cos(x)\cos(y)} &= \\
\frac{\cos(x)\cos(y) - \sin(x)\sin(y)}{\cos(x)\cos(y)} &= \\
\frac{\cos(x)\cos(y)}{\cos(x)\cos(y)} - \frac{\sin(x)\sin(y)}{\cos(x)\cos(y)} &= \\
1 - \tan(x)\tan(y)
\end{aligned}$$

87. Substitute the sum and difference sine identities into the left-hand side to get a difference of two squares.

$$\begin{aligned}
\sin(\alpha+\beta)\sin(\alpha-\beta) &= \\
(\sin\alpha\cos\beta)^2 - (\cos\alpha\sin\beta)^2 &= \\
\sin^2\alpha(1-\sin^2\beta) - (1-\sin^2\alpha)\sin^2\beta &= \\
\sin^2\alpha - \sin^2\alpha\sin^2\beta - \sin^2\beta + \sin^2\alpha\sin^2\beta &= \\
\sin^2\alpha - \sin^2\beta
\end{aligned}$$

89. Using the sum identity for cosine, we obtain

$$\begin{aligned}
\cos(x+x) &= \\
\cos x\cos x - \sin x\sin x &= \\
\cos^2 x - \sin^2 x
\end{aligned}$$

91.

$$\begin{aligned}
\sin(x-y) - \sin(y-x) &= \\
\sin(x-y) + \sin(x-y) &= \\
2\sin(x-y) &= \\
2(\sin x\cos y - \cos x\sin y) &= \\
2\sin x\cos y - 2\cos x\sin y
\end{aligned}$$

93.

$$\begin{aligned}
\tan(s+t)\tan(s-t) &= \\
\frac{\tan s + \tan t}{1-\tan(s)\tan(t)}\cdot\frac{\tan s - \tan t}{1+\tan(s)\tan(t)} &= \\
\frac{\tan^2 s - \tan^2 t}{1+\tan^2(s)\tan^2(t)}
\end{aligned}$$

95. In the proof, divide each term by $\cos\alpha\cos\beta$.

$$\begin{aligned}
\frac{\cos(\alpha+\beta)}{\sin(\alpha-\beta)} &= \\
\frac{\frac{\cos\alpha\cos\beta}{\cos\alpha\cos\beta} - \frac{\sin\alpha\sin\beta}{\cos\alpha\cos\beta}}{\frac{\sin\alpha\cos\beta}{\cos\alpha\cos\beta} - \frac{\cos\alpha\sin\beta}{\cos\alpha\cos\beta}} &= \\
\frac{1-\tan(\alpha)\tan(\beta)}{\tan(\alpha)-\tan(\beta)}
\end{aligned}$$

97. In the proof, multiply each term by $\cos(v-t)$. Also, the sum and difference identities for cosine expresses $\cos(v+t)\cos(v-t)$ as a difference of two squares.

$$\begin{aligned}
\sec(v+t) &= \\
\frac{1}{\cos(v+t)} &= \\
\frac{\cos(v-t)}{\cos(v+t)\cos(v-t)} &= \\
\frac{\cos(v-t)}{\cos^2(v)\cos^2(t) - \sin^2(v)\sin^2(t)} &=
\end{aligned}$$

$$\frac{\cos(v-t)}{\cos^2(v)\cos^2(t)-(1-\cos^2 v)(1-\cos^2 t)} =$$

$$\cos(v-t) \div \Big[\cos^2(v)\cos^2(t) -$$

$$(1-\cos^2 v-\cos^2 t+\cos^2(v)\cos^2(t))\Big] =$$

$$\frac{\cos(v-t)}{-1+\cos^2 v+\cos^2 t} =$$

$$\frac{\cos(v-t)}{\cos^2 v-\sin^2 t} =$$

$$\frac{\cos(v)\cos(t)+\sin(v)\sin(t)}{\cos^2 v-\sin^2 t}$$

99. In the proof, divide each term by $\cos x \sin y$.

$$\frac{\cos(x+y)}{\cos(x-y)} =$$

$$\frac{\cos(x)\cos(y)-\sin(x)\sin(y)}{\cos(x)\cos(y)+\sin(x)\sin(y)} =$$

$$\frac{\dfrac{\cos(x)\cos(y)}{\cos(x)\sin(y)}-\dfrac{\sin(x)\sin(y)}{\cos(x)\sin(y)}}{\dfrac{\cos(x)\cos(y)}{\cos(x)\sin(y)}+\dfrac{\sin(x)\sin(y)}{\cos(x)\sin(y)}} =$$

$$\frac{\cot(y)-\tan(x)}{\cot(y)+\tan(x)}$$

101. In the proof, multiply each term by $\sin(\alpha-\beta)$. Also, the sum and difference identities for sine expresses $\sin(\alpha+\beta)\sin(\alpha-\beta)$ as a difference of two squares.

$$\frac{\sin(\alpha+\beta)}{\sin\alpha+\sin\beta} =$$

$$\frac{\sin(\alpha+\beta)}{\sin\alpha+\sin\beta}\cdot\frac{\sin(\alpha-\beta)}{\sin(\alpha-\beta)} =$$

$$\frac{\sin^2\alpha\cos^2\beta-\cos^2\alpha\sin^2\beta}{(\sin\alpha+\sin\beta)\sin(\alpha-\beta)} =$$

$$\frac{\sin^2\alpha(1-\sin^2\beta)-(1-\sin^2\alpha)\sin^2\beta}{(\sin\alpha+\sin\beta)\sin(\alpha-\beta)} =$$

$$\frac{\sin^2\alpha-\sin^2\alpha\sin^2\beta-\sin^2\beta+\sin^2\alpha\sin^2\beta}{(\sin\alpha+\sin\beta)\sin(\alpha-\beta)} =$$

$$\frac{\sin^2\alpha-\sin^2\beta}{(\sin\alpha+\sin\beta)\sin(\alpha-\beta)} =$$

$$\frac{(\sin\alpha-\sin\beta)(\sin\alpha+\sin\beta)}{(\sin\alpha+\sin\beta)\sin(\alpha-\beta)} =$$

$$\frac{\sin\alpha-\sin\beta}{\sin(\alpha-\beta)}$$

105. If $\alpha=\beta=\pi/6$, then $\sin(\alpha+\beta)\neq\sin\alpha+\sin\beta$

106. The following formulas will be useful

$$\sin(90^\circ-\alpha)=\cos\alpha$$

and

$$\cos(90^\circ-\alpha)=\sin\alpha.$$

In particular, $\cos(89^\circ)=\sin(1^\circ)$, $\cos(88^\circ)=\sin(2^\circ)$, $\sin(89^\circ)=\cos(1^\circ)$, and so on. Thus, for $k=1^\circ,...,44^\circ$ we have

$$\sin^2(k^\circ)+\sin^2((90-k)^\circ)=1.$$

Since $\sin^2(45^\circ)=1/2$, we find

$$\begin{aligned}\sin^2(1^\circ)+\sin^2(2^\circ)+...+\sin^2(90^\circ) &=\\ 44+\sin^2(45^\circ)+\sin^2(90^\circ) &=\\ 45+\frac{1}{2}.\end{aligned}$$

Similarly, we obtain

$$\cos^2(1^\circ)+\cos^2(2^\circ)+...+\cos^2(90^\circ)=44+\frac{1}{2}.$$

Finally, we obtain

$$\frac{\sin^2(1^\circ)+...+\sin^2(90^\circ)}{\cos^2(1^\circ)+...+\cos^2(90^\circ)} =$$

$$\frac{45+1/2}{44+1/2} = \frac{91}{89}.$$

107. $1-\sin^2\alpha=\cos^2\alpha.$

109. Since $B=2$, the period is $\dfrac{\pi}{B}=\dfrac{\pi}{2}$.

Solve $2x=k\pi$ where k is an integer. Then the asymptotes are $x=\dfrac{k\pi}{2}$.

111. $\cos\alpha=-\sqrt{1-(1/4)^2}=-\frac{\sqrt{15}}{4}$

$$\tan\alpha=\frac{\sin\alpha}{\cos\alpha}=\frac{1/4}{-\sqrt{15}/4}=-\frac{1}{\sqrt{15}}=-\frac{\sqrt{15}}{15}$$

$$\csc\alpha=\frac{1}{\sin\alpha}=4$$

$$\sec\alpha=\frac{1}{\cos\alpha}=-\frac{4}{\sqrt{15}}=-\frac{4\sqrt{15}}{15}$$

$$\cot\alpha=\frac{1}{\tan\alpha}=-\frac{15}{\sqrt{15}}=-\sqrt{15}$$

113. The angle spanned by the first seventeen triangles is

$$\tan^{-1}\left(\frac{1}{1}\right)+\tan^{-1}\left(\frac{1}{\sqrt{2}}\right)+...+\tan^{-1}\left(\frac{1}{\sqrt{17}}\right) \approx 365°$$

while the angle spanned by the first sixteen triangles is

$$\tan^{-1}\left(\frac{1}{1}\right)+\tan^{-1}\left(\frac{1}{\sqrt{2}}\right)+...+\tan^{-1}\left(\frac{1}{\sqrt{16}}\right) \approx 351°.$$

Thus, the 17th triangle is the first triangle that overlaps with the first triangle.

For Thought

1. True, $\dfrac{\sin(2\cdot 21°)}{2} = \dfrac{2\sin(21°)\cos(21°)}{2}$
$= \sin(21°)\cos(21°)$.

2. True, by a cosine double angle identity $\cos(2\sqrt{2}) = 2\cos^2(\sqrt{2}) - 1$.

3. False, $\sin\left(\dfrac{300°}{2}\right) = \sqrt{\dfrac{1-\cos(300°)}{2}}$.

4. True, $\sin\left(\dfrac{400°}{2}\right) = -\sqrt{\dfrac{1-\cos(400°)}{2}}$
$= -\sqrt{\dfrac{1-\cos(40°)}{2}}$.

5. False, $\tan\left(\dfrac{7\pi/4}{2}\right) = -\sqrt{\dfrac{1-\cos(7\pi/4)}{1+\cos(7\pi/4)}}$.

6. True, $\tan\left(\dfrac{-\pi/4}{2}\right) = \dfrac{1-\cos(-\pi/4)}{\sin(-\pi/4)} =$
$\dfrac{1-\cos(\pi/4)}{\sin(-\pi/4)}$

7. False, if $x = \pi/4$ then $\dfrac{\sin(2\cdot\pi/4)}{2} =$
$\dfrac{\sin(\pi/2)}{2} = \dfrac{1}{2}$ and $\sin(\pi/4) = \sqrt{2}/2$.

8. False, since $\cos(2\pi/3) = -1/2$ while
$\sqrt{\dfrac{1+\cos(2x)}{2}}$ is a non-negative number.

9. True, since $1-\cos x \geq 0$ we find $\sqrt{(1-\cos x)^2} = |1-\cos x| = 1-\cos x$

10. True, α is in quadrant III or IV, while $\alpha/2$ is in quadrant II.

6.4 Exercises

1. $\sin(2\cdot 45°) = 2\sin(45°)\cos(45°) =$
$2\cdot\dfrac{\sqrt{2}}{2}\cdot\dfrac{\sqrt{2}}{2} = 2\cdot\dfrac{2}{4} = 1.$

3. $\tan(2\cdot 30°) = \dfrac{2\tan(30°)}{1-\tan^2(30°)} = \dfrac{2(\sqrt{3}/3)}{1-(\sqrt{3}/3)^2} =$
$\dfrac{2\sqrt{3}/3}{1-1/3} = \dfrac{2\sqrt{3}/3}{2/3} = \sqrt{3}$

5. $\sin\left(2\cdot\dfrac{3\pi}{4}\right) = 2\sin(3\pi/4)\cos(3\pi/4) =$
$2\cdot\dfrac{\sqrt{2}}{2}\cdot\dfrac{-\sqrt{2}}{2} = 2\cdot\dfrac{-2}{4} = -1$

7. $\tan\left(2\cdot\dfrac{2\pi}{3}\right) = \dfrac{2\tan(2\pi/3)}{1-\tan^2(2\pi/3)} = \dfrac{2(-\sqrt{3})}{1-(-\sqrt{3})^2}$
$= \dfrac{-2\sqrt{3}}{1-3} = \dfrac{-2\sqrt{3}}{-2} = \sqrt{3}$

9. $\cos\left(\dfrac{30°}{2}\right) = \sqrt{\dfrac{1+\cos(30°)}{2}} =$
$\sqrt{\dfrac{1+\sqrt{3}/2}{2}\cdot\dfrac{2}{2}} = \sqrt{\dfrac{2+\sqrt{3}}{4}} = \dfrac{\sqrt{2+\sqrt{3}}}{2}$

11. $\sin\left(\dfrac{30°}{2}\right) = \sqrt{\dfrac{1-\cos(30°)}{2}} =$
$\sqrt{\dfrac{1-\sqrt{3}/2}{2}\cdot\dfrac{2}{2}} = \sqrt{\dfrac{2-\sqrt{3}}{4}} = \dfrac{\sqrt{2-\sqrt{3}}}{2}$

13. $\tan\left(\dfrac{30°}{2}\right) = \dfrac{1-\cos(30°)}{\sin(30°)} =$
$\dfrac{1-\sqrt{3}/2}{1/2}\cdot\dfrac{2}{2} = 2-\sqrt{3}$

15. $\sin\left(\dfrac{45°}{2}\right) = \sqrt{\dfrac{1-\cos(45°)}{2}} =$
$\sqrt{\dfrac{1-\sqrt{2}/2}{2}\cdot\dfrac{2}{2}} = \sqrt{\dfrac{2-\sqrt{2}}{4}} = \dfrac{\sqrt{2-\sqrt{2}}}{2}$

17. Positive, 118.5° is in quadrant II

19. Negative, 100° is in quadrant II

21. Negative, $-5\pi/12$ is in quadrant IV

23. $\sin(2\cdot 13°) = \sin 26°$

25. $\cos(2 \cdot 22.5°) = \cos 45° = \sqrt{2}/2$

27. $\frac{1}{2} \cdot \frac{2\tan 15°}{1 - \tan^2 15°} = \frac{1}{2} \cdot \tan(2 \cdot 15°) =$
$\frac{1}{2} \cdot \tan 30° = \frac{1}{2} \cdot \frac{\sqrt{3}}{3} = \frac{\sqrt{3}}{6}$

29. $\tan\left(\frac{12°}{2}\right) = \tan 6°$

31. $2\sin\left(\frac{\pi}{9} - \frac{\pi}{2}\right)\cos\left(\frac{\pi}{9} - \frac{\pi}{2}\right) =$
$\sin\left(2 \cdot \left(\frac{\pi}{9} - \frac{\pi}{2}\right)\right) = \sin\left(\frac{2\pi}{9} - \pi\right) =$
$\sin(-7\pi/9) = -\sin(7\pi/9)$.

33. $\cos(2 \cdot (\pi/9)) = \cos(2\pi/9)$

35. c, since $\sin^2 x = 1 - \cos^2 x$

37. g, for $\cos\left(\frac{x}{2}\right) = \pm\sqrt{\frac{1 + \cos(x)}{2}}$

39. a, for $\sin(2x) = 2\sin x \cos x$

41. h, for $\tan\left(\frac{x}{2}\right) = \pm\sqrt{\frac{1 - \cos x}{1 + \cos x}}$

43. f, since $\sin\left(\frac{x}{2}\right) = \pm\sqrt{\frac{1 - \cos(x)}{2}}$

45. Since $\cos(2\alpha) = 2\cos^2\alpha - 1$, we get

$$\begin{aligned} 2\cos^2\alpha - 1 &= \frac{3}{5} \\ 2\cos^2\alpha &= \frac{8}{5} \\ \cos^2\alpha &= \frac{4}{5} \\ \cos\alpha &= \pm\frac{2}{\sqrt{5}}. \end{aligned}$$

But $0° < \alpha < 45°$, so $\cos\alpha = \frac{2}{\sqrt{5}}$ and

$\sin\alpha = \sqrt{1 - \left(\frac{2}{\sqrt{5}}\right)^2} = \sqrt{1 - \frac{4}{5}} =$
$\sqrt{\frac{1}{5}} = \frac{1}{\sqrt{5}}$.

Furthermore, $\sec\alpha = \frac{\sqrt{5}}{2}$, $\csc\alpha = \sqrt{5}$,

$\tan\alpha = \frac{1/\sqrt{5}}{2/\sqrt{5}} = \frac{1}{2}$, $\cot\alpha = 2$.

47. Since $2\alpha = \sin^{-1}(5/13) \approx 22.6°$, we find

$\cos 2\alpha = \sqrt{1 - \left(\frac{5}{13}\right)^2} = \frac{12}{13}$. Then

$\sin\alpha = \sqrt{\frac{1 - \cos 2\alpha}{2}} = \sqrt{\frac{1 - 12/13}{2}} = \frac{\sqrt{26}}{26}$,

$\cos\alpha = \sqrt{\frac{1 + \cos 2\alpha}{2}} = \sqrt{\frac{1 + 12/13}{2}} = \frac{5\sqrt{26}}{26}$,

$\tan\alpha = \frac{\sin\alpha}{\cos\alpha} = \frac{\sqrt{26}/26}{5\sqrt{26}/26} = \frac{1}{5}$

$\csc\alpha = \frac{1}{\sin\alpha} = \sqrt{26}$,

$\sec\alpha = \frac{1}{\cos\alpha} = \frac{\sqrt{26}}{5}$,

and $\cot\alpha = \frac{1}{\tan\alpha} = 5$.

49. By a half-angle identity, we have

$$\begin{aligned} -\sqrt{\frac{1 + \cos\alpha}{2}} &= -\frac{1}{4} \\ \frac{1 + \cos\alpha}{2} &= \frac{1}{16} \\ 1 + \cos\alpha &= \frac{1}{8} \\ \cos\alpha &= -\frac{7}{8}. \end{aligned}$$

But $\pi \le \alpha \le 3\pi/2$,

so $\sin\alpha = -\sqrt{1 - \left(-\frac{7}{8}\right)^2} =$

$-\sqrt{1 - \frac{49}{64}} = -\sqrt{\frac{15}{64}} = -\frac{\sqrt{15}}{8}$.

Furthermore, $\sec\alpha = -\frac{8}{7}$, $\csc\alpha = -\frac{8}{\sqrt{15}}$,

$\tan\alpha = \frac{-\sqrt{15}/8}{-7/8} = \frac{\sqrt{15}}{7}$, $\cot\alpha = \frac{7}{\sqrt{15}}$.

51. By a half-angle identity, we find

$$\begin{aligned} \sqrt{\frac{1 - \cos\alpha}{2}} &= \frac{4}{5} \\ \frac{1 - \cos\alpha}{2} &= \frac{16}{25} \\ 1 - \cos\alpha &= \frac{32}{25} \\ \cos\alpha &= -\frac{7}{25}. \end{aligned}$$

Since $(\pi/2+2k\pi) \le \alpha/2 \le (\pi+2k\pi)$ for some integer k, $(\pi+4k\pi) \le \alpha \le (2\pi+4k\pi)$.
So α is in quadrant III because $\cos\alpha < 0$.

$\sin\alpha = -\sqrt{1-\left(-\frac{7}{25}\right)^2} = -\sqrt{1-\frac{49}{625}} =$
$-\sqrt{\frac{576}{625}} = -\frac{24}{25}$.

Furthermore, $\sec\alpha = -\frac{25}{7}$, $\csc\alpha = -\frac{25}{24}$,
$\tan\alpha = \frac{-24/25}{-7/25} = \frac{24}{7}$, and $\cot\alpha = \frac{7}{24}$.

53.

$$\begin{aligned} \cos^4 s - \sin^4 s &= \\ (\cos^2 s - \sin^2 s)(\cos^2 s + \sin^2 s) &= \\ \cos(2s)\cdot(1) &= \\ \cos(2s) & \end{aligned}$$

55.

$$\begin{aligned} \cos(2t+t) &= \\ \cos(2t)\cos(t) - \sin(2t)\sin(t) &= \\ \left[\cos^2 t - \sin^2 t\right]\cos t - [2\sin t\cos t]\sin t &= \\ \cos^3 t - \sin^2 t\cos t - 2\sin^2 t\cos t &= \\ \cos^3 t - 3\sin^2 t\cos t & \end{aligned}$$

57.

$$\begin{aligned} \frac{\cos(2x)+\cos(2y)}{\sin(x)+\cos(y)} &= \\ \frac{1-2\sin^2 x + 2\cos^2 y - 1}{\sin x + \cos y} &= \\ 2\frac{\cos^2 y - \sin^2 x}{\sin x + \cos y} &= \\ 2\frac{(\cos y - \sin x)(\cos y + \sin x)}{\sin x + \cos y} &= \\ 2\cos(y) - 2\sin(x) & \end{aligned}$$

59.

$$\begin{aligned} \frac{\cos 2x}{\sin^2 x} &= \\ \frac{1-2\sin^2 x}{\sin^2 x} &= \\ \frac{1}{\sin^2 x} - 2\cdot\frac{\sin^2 x}{\sin^2 x} &= \\ \csc^2 x - 2 & \end{aligned}$$

61.

$$\begin{aligned} &= \frac{\sin^2 u}{1+\cos u} \\ &= \frac{1-\cos^2 u}{1+\cos u} \\ &= \frac{(1-\cos u)(1+\cos u)}{1+\cos u} \\ &= (1-\cos u)\cdot\frac{2}{2} \\ &= 2\cdot\frac{1-\cos u}{2} \\ &= 2\sin^2(u/2) \end{aligned}$$

63. Multiply and divide by $\cos x$.

$$\begin{aligned} &= \frac{\sec x + \cos x - 2}{\sec x - \cos x}\cdot\frac{\cos x}{\cos x} \\ &= \frac{1+\cos^2 x - 2\cos x}{1-\cos^2 x} \\ &= \frac{\cos^2 x - 2\cos x + 1}{1-\cos^2 x} \\ &= \frac{(1-\cos x)^2}{(1+\cos x)(1-\cos x)} \\ &= \frac{1-\cos x}{1+\cos x} \\ &= \tan^2(x/2) \end{aligned}$$

65.

$$\begin{aligned} \frac{1-\sin^2(x/2)}{1+\sin^2(x/2)} &= \\ \frac{1-\left(\frac{1-\cos x}{2}\right)}{1+\left(\frac{1-\cos x}{2}\right)}\cdot\frac{2}{2} &= \\ \frac{2-(1-\cos x)}{2+(1-\cos x)} &= \\ \frac{1+\cos x}{3-\cos x} & \end{aligned}$$

67. It is not an identity. If $x = \pi/4$, then $\sin(2\cdot\pi/4) = \sin(\pi/2) = 1$ and $2\sin(\pi/4) = 2\cdot(\sqrt{2}/2) = \sqrt{2}$.

69. It is not an identity. If $x = 2\pi/3$, then $\tan\left(\frac{2\pi/3}{2}\right) = \tan(\pi/3) = \sqrt{3}$ and $\frac{1}{2}\cdot\tan(2\pi/3) = \frac{1}{2}\cdot(-\sqrt{3})$.

71. It is not an identity. If $x = \pi/2$, then

$$\sin(2 \cdot \pi/2)\sin\left(\frac{\pi/2}{2}\right) = \sin(\pi)\sin(\pi/4)$$

$$= 0 \cdot \frac{\sqrt{2}}{2} = 0 \text{ and } \sin^2(\pi/2) = 1.$$

73. It is an identity. The proof below uses the double-angle identity for tangent.

$$\begin{aligned}
\cot(x/2) - \tan(x/2) &= \\
\frac{1}{\tan(x/2)} - \tan(x/2) &= \\
\frac{1 - \tan^2(x/2)}{\tan(x/2)} &= \\
2 \cdot \frac{1 - \tan^2(x/2)}{2 \cdot \tan(x/2)} &= \\
2 \cdot \frac{1}{\tan x} &= \\
2 \cdot \frac{\cos x}{\sin x} \cdot \frac{\sin x}{\sin x} &= \\
\frac{2\sin x \cos x}{\sin^2 x} &= \\
\frac{\sin(2x)}{\sin^2 x}
\end{aligned}$$

75. Note, $\cos\alpha = -\sqrt{1 - \left(\frac{3}{5}\right)^2} = -\frac{4}{5}$.

Then $\sin 2\alpha = 2\sin\alpha\cos\alpha =$
$2 \cdot \frac{3}{5} \cdot \frac{-4}{5} = -\frac{24}{25}$.

77. $\cos 2\alpha = 1 - 2\sin^2\alpha = 1 - 2\left(\frac{8}{17}\right)^2 = \frac{161}{289}$

79. Since $\tan\alpha = \frac{3}{5}$, $\sin\alpha = \frac{3}{\sqrt{34}}$ and

$\cos\alpha = \frac{5}{\sqrt{34}}$. By a half-angle identity, we obtain

$$\begin{aligned}
\tan\frac{\alpha}{2} &= \frac{\sin\alpha}{1 + \cos\alpha} \\
&= \frac{\frac{3}{\sqrt{34}}}{1 + \frac{5}{\sqrt{34}}} \\
&= \frac{3}{5 + \sqrt{34}}
\end{aligned}$$

and since $\tan\frac{\alpha}{2} = \frac{BD}{5}$ then

$$\begin{aligned}
BD &= \frac{15}{5 + \sqrt{34}} \\
&= \frac{15(5 - \sqrt{34})}{25 - 34} \\
&= \frac{15(\sqrt{34} - 5)}{9} \\
BD &= \frac{5\sqrt{34} - 25}{3}.
\end{aligned}$$

81. Since the base of the TV screen is $b = d\cos\alpha$ and its height is $h = d\sin\alpha$, then the area A is given by

$$\begin{aligned}
A &= bh \\
&= (d\cos\alpha)(d\sin\alpha) \\
&= d^2\cos\alpha\sin\alpha \\
A &= \frac{d^2}{2}\sin(2\alpha).
\end{aligned}$$

85. $\frac{1}{1 - \sin x} + \frac{1}{1 + \sin x} = \frac{1 + \sin x + 1 - \sin x}{1 - \sin^2 x} =$

$\frac{2}{\cos^2 x} = 2\sec^2 x$

$\frac{\cos x\,(\cos^2 x + \sin^2 x)}{\sin x} = \frac{\cos x \cdot 1}{\sin x} = \cot x$

87. **a)** $\cos x\cos y - \sin x\sin y$

b) $\cos x\cos y + \sin x\sin y$

89. a) $\frac{1}{2}$ b) -1 c) Undefined

d) 2 e) -1 f) 1

91. An eighth of the region that gets watered by all sprinklers is region R_a below with vertices B, C, and D.

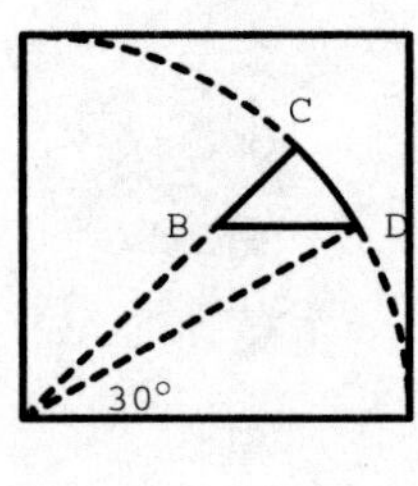

The area of R_a is the area of the sector determined by C, A, and D minus the area of triangle ΔABD. In the figure above, we have $AB = \sqrt{2}$, $BC = 2 - \sqrt{2}$, $BD = \sqrt{3} - 1$, angle $\langle ABD = 135°$, and $\langle CBD = 45°$.

The area of the sector is

$$A_s = \frac{1}{2}\left(2^2 \frac{\pi}{12}\right) = \frac{\pi}{6}$$

and the area of ΔABD is

$$\begin{aligned} A_t &= \frac{1}{2}(AB)(BD)\sin 135° \\ &= \frac{1}{2}\sqrt{2}(\sqrt{3}-1)\frac{\sqrt{2}}{2} \\ A_t &= \frac{\sqrt{3}-1}{2}. \end{aligned}$$

Thus, the area watered by all sprinklers is

$$\begin{aligned} \text{Area} &= 8\,(A_s - A_t) \\ &= 8\left(\frac{\pi}{6} - \frac{\sqrt{3}-1}{2}\right) \\ \text{Area} &= 4\pi/\sqrt{3} + 4 - 4\sqrt{3}\ \text{m}^2. \end{aligned}$$

For Thought

1. True, $\sin 45° \cos 15° =$
$(1/2)\left[\sin(45° + 15°) + \sin(45° - 15°)\right] =$
$0.5\left[\sin 60° + \sin 30°\right]$.

2. False, $\cos(\pi/8)\sin(\pi/4) =$
$(1/2)\left[\sin(\pi/8 + \pi/4) - \sin(\pi/8 - \pi/4)\right] =$
$0.5\left[\sin(3\pi/8) - \sin(-\pi/8)\right] =$
$0.5\left[\sin(3\pi/8) + \sin(\pi/8)\right]$.

3. True, $2\cos(6°)\cos(8°) =$
$\cos(6° - 8°) + \cos(6° + 8°) =$
$\cos(-2°) + \cos(14°) = \cos(2°) + \cos(14°)$.

4. False, $\sin(5°) - \sin(9°) =$
$2\cos\left(\frac{5° + 9°}{2}\right)\sin\left(\frac{5° - 9°}{2}\right) =$
$2\cos(7°)\sin(-2°) = -2\cos(7°)\sin(2°)$.

5. True, $\cos(4) + \cos(12) =$
$2\cos\left(\frac{4+12}{2}\right)\cos\left(\frac{4-12}{2}\right) =$
$2\cos(8)\cos(-4) = 2\cos(8)\cos(4)$.

6. False, $\cos(\pi/3) - \cos(\pi/2) =$
$-2\sin\left(\frac{\pi/3 + \pi/2}{2}\right)\sin\left(\frac{\pi/3 - \pi/2}{2}\right) =$
$-2\sin(5\pi/12)\sin(-\pi/12) =$
$2\sin(5\pi/12)\sin(\pi/12)$.

7. True, $\sqrt{2}\sin(\pi/6 + \pi/4) =$
$\sqrt{2}\left[\sin(\pi/6)\cos(\pi/4) + \cos(\pi/6)\sin(\pi/4)\right] =$
$\sqrt{2}\left[\sin(\pi/6)\cdot\frac{1}{\sqrt{2}} + \cos(\pi/6)\cdot\frac{1}{\sqrt{2}}\right] =$
$\sin(\pi/6) + \cos(\pi/6)$.

8. True, $\frac{1}{2}\sin(\pi/6) + \frac{\sqrt{3}}{2}\cos(\pi/6) =$
$\frac{1}{2}\cdot\frac{1}{2} + \frac{\sqrt{3}}{2}\cdot\frac{\sqrt{3}}{2} = \frac{1}{4} + \frac{3}{4} = 1 = \sin(\pi/2)$.

9. True, $y = \cos(\pi/3)\sin x + \sin(\pi/3)\cos x =$
$\sin(x + \pi/3)$.

10. True, since $y = \cos(\pi/4)\sin x + \sin(\pi/4)\cos x$
$= \sin(x + \pi/4)$ by the sum identity for sine.

6.5 Exercises

1.
$\frac{1}{2}\left[\cos(13° - 9°) - \cos(13° + 9°)\right] =$
$0.5\left[\cos 4° - \cos 22°\right]$

3.
$\frac{1}{2}\left[\sin(16° + 20°) + \sin(16° - 20°)\right] =$
$0.5\left[\sin 36° + \sin(-4°)\right] = 0.5\left[\sin 36° - \sin 4°\right]$

5.
$\frac{1}{2}\left[\sin(5° + 10°) + \sin(5° - 10°)\right] =$
$0.5\left[\sin 15° + \sin(-5°)\right] = 0.5\left[\sin 15° - \sin 5°\right]$

7.
$\frac{1}{2}\left[\cos\left(\frac{\pi}{6} - \frac{\pi}{5}\right) + \cos\left(\frac{\pi}{6} + \frac{\pi}{5}\right)\right] =$
$0.5\left[\cos\left(\frac{-\pi}{30}\right) + \cos\left(\frac{11\pi}{30}\right)\right] =$
$0.5\left[\cos\left(\frac{\pi}{30}\right) + \cos\left(\frac{11\pi}{30}\right)\right]$

9.
$\frac{1}{2}\left[\cos(5y^2 - 7y^2) + \cos(5y^2 + 7y^2)\right] =$
$0.5\left[\cos(-2y^2) + \cos(12y^2)\right] =$
$0.5\left[\cos(2y^2) + \cos(12y^2)\right]$

11.
$$\frac{1}{2}\Big[\sin((2s-1)+(s+1))+ \sin((2s-1)-(s+1))\Big] =$$
$$0.5\Big[\sin(3s)+\sin(s-2)\Big]$$

13.
$$\frac{1}{2}\left[\cos(52.5^\circ-7.5^\circ)-\cos(52.5^\circ+7.5^\circ)\right] =$$
$$\frac{1}{2}\left[\cos 45^\circ - \cos 60^\circ\right] =$$
$$\frac{1}{2}\left[\frac{\sqrt{2}}{2}-\frac{1}{2}\right] = \frac{\sqrt{2}-1}{4}$$

15.
$$\frac{1}{2}\left[\sin\left(\frac{13\pi}{24}+\frac{5\pi}{24}\right)+\sin\left(\frac{13\pi}{24}-\frac{5\pi}{24}\right)\right] =$$
$$\frac{1}{2}\left[\sin(18\pi/24)+\sin(8\pi/24)\right] =$$
$$\frac{1}{2}\left[\sin(3\pi/4)+\sin(\pi/3)\right] =$$
$$\frac{1}{2}\left[\frac{\sqrt{2}}{2}+\frac{\sqrt{3}}{2}\right] = \frac{\sqrt{2}+\sqrt{3}}{4}$$

17.
$$2\cos\left(\frac{12^\circ+8^\circ}{2}\right)\sin\left(\frac{12^\circ-8^\circ}{2}\right) =$$
$$2\cos 10^\circ \sin 2^\circ$$

19.
$$-2\sin\left(\frac{80^\circ+87^\circ}{2}\right)\sin\left(\frac{80^\circ-87^\circ}{2}\right) =$$
$$-2\sin 83.5^\circ \sin(-3.5^\circ) = 2\sin 83.5^\circ \sin 3.5^\circ$$

21.
$$2\cos\left(\frac{3.6+4.8}{2}\right)\sin\left(\frac{3.6-4.8}{2}\right) =$$
$$2\cos(4.2)\sin(-0.6) = -2\cos(4.2)\sin(0.6)$$

23.
$$-2\sin\left(\frac{\pi/3+\pi/5}{2}\right)\sin\left(\frac{\pi/3-\pi/5}{2}\right) =$$
$$-2\sin(4\pi/15)\sin(\pi/15) =$$

25.
$$-2\sin\left(\frac{(5y-3)+(3y+9)}{2}\right)\cdot \sin\left(\frac{(5y-3)-(3y+9)}{2}\right) =$$
$$-2\sin(4y+3)\sin(y-6)$$

27.
$$2\cos\left(\frac{5\alpha+8\alpha}{2}\right)\sin\left(\frac{5\alpha-8\alpha}{2}\right) =$$
$$2\cos(6.5\alpha)\sin(-1.5\alpha) =$$
$$-2\cos(6.5\alpha)\sin(1.5\alpha)$$

29.
$$2\sin\left(\frac{75^\circ+15^\circ}{2}\right)\cos\left(\frac{75^\circ-15^\circ}{2}\right) =$$
$$2\sin 45^\circ\cos(30^\circ) = 2\cdot\frac{\sqrt{2}}{2}\frac{\sqrt{3}}{2} = \frac{\sqrt{6}}{2}$$

31.
$$-2\sin\left(\frac{\frac{-\pi}{24}+\frac{7\pi}{24}}{2}\right)\sin\left(\frac{\frac{-\pi}{24}-\frac{7\pi}{24}}{2}\right) =$$
$$-2\sin(3\pi/24)\sin(-4\pi/24) =$$
$$-2\sin(\pi/8)\sin(-\pi/6) =$$
$$-2\sin\left(\frac{\pi/4}{2}\right)\cdot\frac{-1}{2} = -2\sqrt{\frac{1-\cos(\pi/4)}{2}}\cdot\frac{-1}{2}$$
$$= \sqrt{\frac{1-\sqrt{2}/2}{2}\cdot\frac{2}{2}} = \sqrt{\frac{2-\sqrt{2}}{4}} = \frac{\sqrt{2-\sqrt{2}}}{2}$$

33. Since $a = 1$ and $b = -1$, we obtain
$$r = \sqrt{1^2+(-1)^2} = \sqrt{2}.$$
If the terminal side of α passes through $(1,-1)$, then $\cos\alpha = a/r = 1/\sqrt{2}$ and $\sin\alpha = b/r = -1/\sqrt{2}$. Choose $\alpha = -\pi/4$.

Thus, $\sin x - \cos x = r\sin(x+\alpha) = \sqrt{2}\sin(x-\pi/4)$.

35. Since $a = -1/2$ and $b = \sqrt{3}/2$, we obtain $r = \sqrt{(-1/2)^2+(\sqrt{3}/2)^2} = 1$. If the terminal side of α passes through $(-1/2, \sqrt{3}/2)$, then $\cos\alpha = a/r = a/1 = a = -1/2$ and $\sin\alpha = b/r = b/1 = b = \sqrt{3}/2$. Choose $\alpha = 2\pi/3$. So $-\frac{1}{2}\sin x + \frac{\sqrt{3}}{2}\cos x = r\sin(x+\alpha) = \sin(x+2\pi/3)$.

37. Since $a = \sqrt{3}/2$ and $b = -1/2$, we have
$$r = \sqrt{(\sqrt{3}/2)^2+(-1/2)^2} = 1.$$
If the terminal side of α passes through $(\sqrt{3}/2, -1/2)$, then
$$\cos\alpha = a/r = a/1 = a = \sqrt{3}/2$$

and

$$\sin\alpha = b/r = b/1 = b = -1/2.$$

Choose $\alpha = -\pi/6$. Thus,

$$\frac{\sqrt{3}}{2}\sin x - \frac{1}{2}\cos x = r\sin(x+\alpha) =$$

$$\sin(x - \pi/6).$$

39. Since $a = -1$ and $b = 1$, we obtain $r = \sqrt{(-1)^2 + 1^2} = \sqrt{2}$. If the terminal side of α passes through $(-1, 1)$, then $\cos\alpha = a/r = -1/\sqrt{2}$ and $\sin\alpha = b/r = 1/\sqrt{2}$. Choose $\alpha = 3\pi/4$. Then $y = -\sin x + \cos x = r\sin(x+\alpha) = \sqrt{2}\sin(x + 3\pi/4)$. Amplitude is $\sqrt{2}$, period is 2π, and phase shift is $-3\pi/4$.

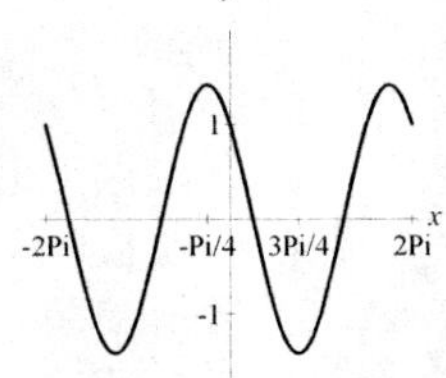

41. Since $a = \sqrt{2}$ and $b = -\sqrt{2}$, we obtain $r = \sqrt{\sqrt{2}^2 + (-\sqrt{2})^2} = 2$. If the terminal side of α passes through $(\sqrt{2}, -\sqrt{2})$, then $\cos\alpha = a/r = \sqrt{2}/2$ and $\sin\alpha = b/r = -\sqrt{2}/2$. Choose $\alpha = -\pi/4$. So $y = \sqrt{2}\sin x - \sqrt{2}\cos x = r\sin(x+\alpha) = 2\sin(x - \pi/4)$. Amplitude is 2, period is 2π, and phase shift is $\pi/4$.

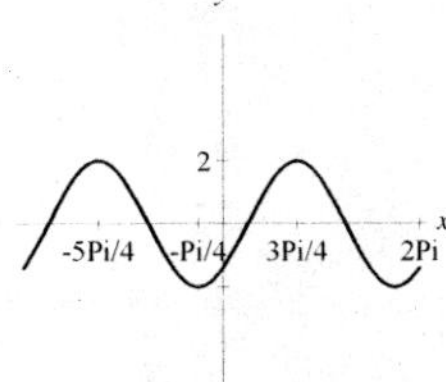

43. Since $a = -\sqrt{3}$ and $b = -1$, we find $r = \sqrt{(-\sqrt{3})^2 + (-1)^2} = 2$. If the terminal side of α passes through $(-\sqrt{3}, -1)$, then $\cos\alpha = a/r = -\sqrt{3}/2$ and $\sin\alpha = b/r = -1/2$. Choose $\alpha = 7\pi/6$. Then $y = -\sqrt{3}\sin x - \cos x = r\sin(x+\alpha) = 2\sin(x + 7\pi/6)$. Amplitude is 2, period is 2π, and phase shift is $-7\pi/6$.

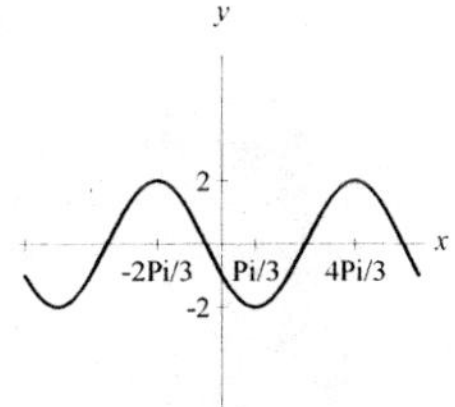

45. Since $a = 3$ and $b = 4$, the amplitude is $\sqrt{3^2 + 4^2} = \sqrt{25} = 5$. If the terminal side of α passes through $(3, 4)$, then $\tan\alpha = 4/3$ and $\alpha = \tan^{-1}(4/3) \approx 0.9$. Phase shift is -0.9.

47. Since $a = -6$ and $b = 1$, amplitude is

$$\sqrt{(-6)^2 + 1^2} = \sqrt{37}.$$

If the terminal side of α passes through $(-6, 1)$, then $\tan\alpha = -1/6$. Using a calculator, one gets $\tan^{-1}(-1/6) \approx -0.165$ which is an angle in quadrant IV. Since $(-6, 1)$ is in quadrant II and π is the period of $\tan x$,

$$\alpha \approx -0.165 + \pi \approx 3.0.$$

The phase shift is -3.0.

49. Since $a = -3$ and $b = -5$, amplitude is $\sqrt{(-3)^2 + (-5)^2} = \sqrt{34}$. If the terminal side of α passes through $(-3, -5)$, then $\tan\alpha = 5/3$. Using a calculator, one gets $\tan^{-1}(5/3) \approx 1.03$ which is an angle in quadrant I. Since $(-3, -5)$ is in quadrant III and π is the period of $\tan x$, $\alpha \approx 1.03 + \pi \approx 4.2$. Phase shift is -4.2.

51. By using a sum-to-product identity, we get

$$\frac{\sin(3t) - \sin(t)}{\cos(3t) + \cos(t)} =$$

$$\frac{2\cos\left(\frac{3t+t}{2}\right)\sin\left(\frac{3t-t}{2}\right)}{2\cos\left(\frac{3t+t}{2}\right)\cos\left(\frac{3t-t}{2}\right)} =$$

$$\frac{2\cos(2t)\sin t}{2\cos(2t)\cos t} =$$

$$\tan t$$

53. By using a sum-to-product identity, we find

$$\frac{\cos x - \cos(3x)}{\cos x + \cos(3x)} =$$

$$\frac{-2\sin\left(\frac{x+3x}{2}\right)\sin\left(\frac{x-3x}{2}\right)}{2\cos\left(\frac{x+3x}{2}\right)\cos\left(\frac{x-3x}{2}\right)} =$$

$$\frac{-2\sin(2x)\sin(-x)}{2\cos(2x)\cos(-x)} =$$

$$\frac{2\sin(2x)\sin x}{2\cos(2x)\cos x} =$$

$$\tan(2x)\tan(x)$$

55. By using a product-to-sum identity, we get

$$\begin{aligned}
&= -\sin(x+y)\sin(x-y) \\
&= -\frac{1}{2}\Big[\cos\left((x+y)-(x-y)\right) - \\
&\qquad \cos\left((x+y)+(x-y)\right)\Big] \\
&= -\frac{1}{2}\Big[\cos(2y) - \cos(2x)\Big] \\
&= -\frac{1}{2}\Big[(2\cos^2 y - 1) - (2\cos^2 x - 1)\Big] \\
&= -\frac{1}{2}\Big[2\cos^2 y - 2\cos^2 x\Big] \\
&= \cos^2 x - \cos^2 y
\end{aligned}$$

57. Let $A = \dfrac{x+y}{2}$ and $B = \dfrac{x-y}{2}$.
Note, $A + B = x$ and $A - B = y$. Expand the left-hand side and use product-to-sum identities.

$$(\sin A + \cos A)(\sin B + \cos B) =$$

$$\sin A \sin B + \sin A \cos B + \cos A \sin B + \cos A \cos B =$$

$$\frac{1}{2}\Big[\cos(A-B) - \cos(A+B)\Big] + \frac{1}{2}\Big[\sin(A+B) + \sin(A-B)\Big] + \frac{1}{2}\Big[\sin(A+B) - \sin(A-B)\Big] + \frac{1}{2}\Big[\cos(A-B) + \cos(A+B)\Big] =$$

$$\frac{1}{2}\Big[\cos y - \cos x\Big] + \frac{1}{2}\Big[\sin x + \sin y\Big] + \frac{1}{2}\Big[\sin x - \sin y\Big] + \frac{1}{2}\Big[\cos y + \cos x\Big] =$$

$$\frac{1}{2}\Big[2\cos y + 2\sin x\Big] =$$

$$\sin x + \cos y$$

59. Use a sum-to-product identity in the 2nd line, and a product-to-sum identity in the 5th line.

$$\begin{aligned}
&= \sin^2(A+B) - \sin^2(A-B) \\
&= \sin(2A)\sin(2B) \\
&= (2\sin A\cos A)(2\sin B\cos B) \\
&= [2\cos A\cos B]\cdot[2\sin A\sin B] \\
&= \Big[\cos(A-B) + \cos(A+B)\Big]\cdot \\
&\qquad \Big[\cos(A-B) - \cos(A+B)\Big] \\
&= \cos^2(A-B) - \cos^2(A+B)
\end{aligned}$$

61. Note that x can be written in the form $x = a\sin(t+\alpha)$. The maximum displacement of $x = \sqrt{3}\sin t + \cos t$ is

$$a = \sqrt{\sqrt{3}^2 + 1^2} = 2.$$

Thus, 2 meters is the maximum distance between the block and its resting position. Since the terminal side of α goes through $(\sqrt{3}, 1)$, we get $\tan\alpha = 1/\sqrt{3}$ and one can choose $\alpha = \pi/6$. Then $x = 2\sin(t + \pi/6)$.

67. a) $\dfrac{\tan x + \tan y}{1 - \tan x \tan y}$ b) $\dfrac{\tan x - \tan y}{1 + \tan x \tan y}$

69. $\dfrac{1-\cos x}{\sin x} = \dfrac{1-(-1/3)}{-\sqrt{8}/3} =$

$$\frac{4/3}{-2\sqrt{2}/3} = \frac{2}{-\sqrt{2}} = -\sqrt{2}$$

71. Since $\sin y = -\dfrac{4}{5}$ and $\cos y = \dfrac{3}{5}$, we obtain

$$\sin 2y = 2\sin y\cos y = 2\left(-\frac{4}{5}\right)\frac{3}{5} = -\frac{24}{25}$$

73. Given below are two circles with radii a. Consider the triangle with a vertex at the top point of intersection and with the centers of the circles as the other two vertices. This triangle is an equilateral triangle.

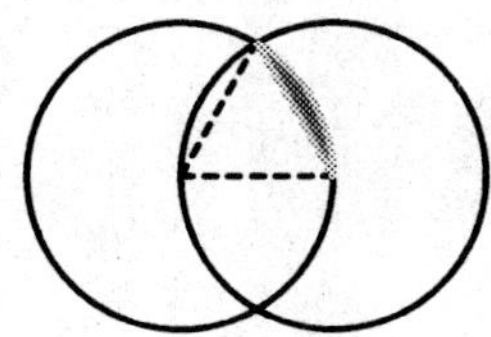

By subtracting the area of a triangle from the area of a sector, we find that the area of the shaded region above is

$$\frac{1}{2}a^2\frac{\pi}{3} - \frac{1}{2}a^2\frac{\sqrt{3}}{2}$$

or

$$\frac{a^2(2\pi - 3\sqrt{3})}{12}.$$

Then the area of the region inside the circle on the right that is outside the circle on the left is

$$2\left[\frac{1}{2}a^2\frac{2\pi}{3}\right] - 2\left[\frac{a^2(2\pi - 3\sqrt{3})}{12}\right]$$

or equivalently

$$\frac{a^2(2\pi + 3\sqrt{3})}{6}.$$

If we add the area of the circle on the left to the above expression we obtain the total area sprinkled, that is,

$$\pi a^2 + \frac{a^2(2\pi + 3\sqrt{3})}{6} = \frac{a^2(8\pi + 3\sqrt{3})}{6}.$$

For Thought

1. False, the only solutions are 45° and 315°.

2. False, there is no solution in $[0, \pi)$.

3. True, since −29° and 331° are coterminal angles.

4. True **5.** True, since the right-side is a factorization of the left-side.

6. False, $x = 0$ is a solution to the first equation and not to the second equation.

7. False, $\cos^{-1} 2$ is undefined. **8.** True

9. False, $x = 3\pi/4$ is not a solution to the first equation but is a solution to the second equation.

10 . False, rather $\left\{x | 3x = \frac{\pi}{2} + 2k\pi\right\} = \left\{x | x = \frac{\pi}{6} + \frac{2k\pi}{3}\right\}$.

6.6 Exercises

1. $\{x \mid x = \pi + 2k\pi, k \text{ an integer}\}$

3. $\{x \mid x = k\pi, k \text{ an integer}\}$

5. $\left\{x \mid x = \frac{3\pi}{2} + 2k\pi, k \text{ an integer}\right\}$

7. Solutions in $[0, 2\pi)$ are $x = \frac{\pi}{3}, \frac{5\pi}{3}$. So solution set is $\left\{x \mid x = \frac{\pi}{3} + 2k\pi \text{ or } x = \frac{5\pi}{3} + 2k\pi\right\}$.

9. Solutions in $[0, 2\pi)$ are $x = \frac{\pi}{4}, \frac{3\pi}{4}$. So solution set is $\left\{x \mid x = \frac{\pi}{4} + 2k\pi \text{ or } x = \frac{3\pi}{4} + 2k\pi\right\}$.

11. Solution in $[0, \pi)$ is $x = \frac{\pi}{4}$.

The solution set is $\left\{x \mid x = \frac{\pi}{4} + k\pi\right\}$.

13. Solutions in $[0, 2\pi)$ are $x = \frac{5\pi}{6}, \frac{7\pi}{6}$.

Then the solution set is

$\left\{x \mid x = \frac{5\pi}{6} + 2k\pi \text{ or } x = \frac{7\pi}{6} + 2k\pi\right\}$.

15. Solutions in $[0, 2\pi)$ are $x = \frac{5\pi}{4}, \frac{7\pi}{4}$.

The solution set is

$\left\{x \mid x = \frac{5\pi}{4} + 2k\pi \text{ or } x = \frac{7\pi}{4} + 2k\pi\right\}$.

17. Solution in $[0, \pi)$ is $x = \dfrac{3\pi}{4}$. The solution set is $\left\{x \mid x = \dfrac{3\pi}{4} + k\pi\right\}$.

19. Solutions in $[0, 360°)$ are $\alpha = 90°, 270°$. So solution set is $\{\alpha \mid \alpha = 90° + k \cdot 180°\}$.

21. Solution in $[0, 360°)$ is $\alpha = 90°$. So the solution set is $\{\alpha \mid \alpha = 90° + k \cdot 360°\}$.

23. Solution in $[0, 180°)$ is $\alpha = 0°$. The solution set is $\{\alpha \mid \alpha = k \cdot 180°\}$.

25. One solution is $\cos^{-1}(0.873) \approx 29.2°$. Another solution is $360° - 29.2° = 330.8°$. Solution set is $\{\alpha \mid \alpha = 29.2° + k360° \text{ or } \alpha = 330.8° + k360°\}$.

27. One solution is $\sin^{-1}(-0.244) \approx -14.1°$. This is coterminal with $345.9°$. Another solution is $180° + 14.1° = 194.1°$. Solution set is $\{\alpha \mid \alpha = 345.9° + k360° \text{ or } \alpha = 194.1° + k360°\}$.

29. One solution is $\tan^{-1}(5.42) \approx 79.5°$. Solution set is $\{\alpha \mid \alpha = 79.5° + k \cdot 180°\}$.

31. Values of $x/2$ in $[0, 2\pi)$ are $\pi/3$ and $5\pi/3$. Then we get

$$\frac{x}{2} = \frac{\pi}{3} + 2k\pi \text{ or } \frac{x}{2} = \frac{5\pi}{3} + 2k\pi$$

$$x = \frac{2\pi}{3} + 4k\pi \text{ or } x = \frac{10\pi}{3} + 4k\pi.$$

The solution set is

$$\left\{x \mid x = \frac{2\pi}{3} + 4k\pi \text{ or } x = \frac{10\pi}{3} + 4k\pi\right\}.$$

33. Value of $3x$ in $[0, 2\pi)$ is 0. Thus, $3x = 2k\pi$. The solution set is $\left\{x \mid x = \dfrac{2k\pi}{3}\right\}$.

35. Since $\sin(x/2) = 1/2$, values of $x/2$ in $[0, 2\pi)$ are $\pi/6$ and $5\pi/6$. Then

$$\frac{x}{2} = \frac{\pi}{6} + 2k\pi \text{ or } \frac{x}{2} = \frac{5\pi}{6} + 2k\pi$$

$$x = \frac{\pi}{3} + 4k\pi \text{ or } x = \frac{5\pi}{3} + 4k\pi.$$

The solution set is

$$\left\{x \mid x = \frac{\pi}{3} + 4k\pi \text{ or } x = \frac{5\pi}{3} + 4k\pi\right\}.$$

37. Since $\sin(2x) = -\sqrt{2}/2$, values of $2x$ in $[0, 2\pi)$ are $5\pi/4$ and $7\pi/4$. Thus,

$$2x = \frac{5\pi}{4} + 2k\pi \text{ or } 2x = \frac{7\pi}{4} + 2k\pi$$

$$x = \frac{5\pi}{8} + k\pi \text{ or } x = \frac{7\pi}{8} + k\pi.$$

The solution set is

$$\left\{x \mid x = \frac{5\pi}{8} + k\pi \text{ or } x = \frac{7\pi}{8} + k\pi\right\}.$$

39. Value of $2x$ in $[0, \pi)$ is $\pi/3$. Then

$$2x = \frac{\pi}{3} + k\pi.$$

The solution set is $\left\{x \mid x = \dfrac{\pi}{6} + \dfrac{k\pi}{2}\right\}$.

41. Value of $4x$ in $[0, \pi)$ is 0. Then

$$4x = k\pi.$$

The solution set is $\left\{x \mid x = \dfrac{k\pi}{4}\right\}$.

43. The values of πx in $[0, 2\pi)$ are $\pi/6$ and $5\pi/6$. Then

$$\pi x = \frac{\pi}{6} + 2k\pi \text{ or } \pi x = \frac{5\pi}{6} + 2k\pi$$

$$x = \frac{1}{6} + 2k \text{ or } x = \frac{5}{6} + 2k.$$

The solution set is

$$\left\{x \mid x = \frac{1}{6} + 2k \text{ or } x = \frac{5}{6} + 2k\right\}.$$

45. Values of $2\pi x$ in $[0, 2\pi)$ are $\pi/2$ and $3\pi/2$. So

$$2\pi x = \frac{\pi}{2} + 2k\pi \text{ or } 2\pi x = \frac{3\pi}{2} + 2k\pi$$

$$x = \frac{1}{4} + k \text{ or } x = \frac{3}{4} + k.$$

The solution set is

$$\left\{x \mid x = \frac{1}{4} + \frac{k}{2}\right\}.$$

47. Since $\sin\alpha = -\sqrt{3}/2$, the solution set is $\{240°, 300°\}$.

49. Since $\cos 2\alpha = 1/\sqrt{2}$, values of 2α in $[0, 360°)$ are $45°$ and $315°$. Thus,

$$2\alpha = 45° + k \cdot 360° \text{ or } 2\alpha = 315° + k \cdot 360°$$

$$\alpha = 22.5° + k \cdot 180° \text{ or } \alpha = 157.5° + k \cdot 180°.$$

Then let $k = 0, 1$. The solution set is

$$\{22.5°, 157.5°, 202.5°, 337.5°\}.$$

51. Values of 3α in $[0, 360°)$ are $135°$ and $225°$. Then

$$3\alpha = 135° + k \cdot 360° \text{ or } 3\alpha = 225° + k \cdot 360°$$

$$\alpha = 45° + k \cdot 120° \text{ or } \alpha = 75° + k \cdot 120°.$$

By choosing $k = 0, 1, 2$, one obtains the solution set $\{45°, 75°, 165°, 195°, 285°, 315°\}$.

53. The value of $\alpha/2$ in $[0, 180°)$ is $30°$. Then

$$\frac{\alpha}{2} = 30° + k \cdot 180°$$

$$\alpha = 60° + k \cdot 360°.$$

By choosing $k = 0$, the solution set is $\{60°\}$.

55. A solution is $3\alpha = \sin^{-1}(0.34) \approx 19.88°$. Another solution is $3\alpha = 180° - 19.88° = 160.12°$. Then

$$3\alpha = 19.88° + k \cdot 360° \text{ or } 3\alpha = 160.12° + k \cdot 360°$$

$$\alpha \approx 6.6° + k \cdot 120° \text{ or } \alpha \approx 53.4° + k \cdot 120°.$$

Solution set is

$$\{\alpha \mid \alpha = 6.6° + k \cdot 120° \text{ or } \alpha = 53.4° + k \cdot 120°\}.$$

57. A solution is $3\alpha = \sin^{-1}(-0.6) \approx -36.87°$. This is coterminal with $323.13°$. Another solution is $3\alpha = 180° + 36.87° = 216.87°$. Then

$$3\alpha = 323.13° + k \cdot 360° \text{ or } 3\alpha = 216.87° + k \cdot 360°$$

$$\alpha \approx 107.7° + k \cdot 120° \text{ or } \alpha \approx 72.3° + k \cdot 120°.$$

The solution set is

$$\{\alpha \mid \alpha = 107.7° + k120° \text{ or } \alpha = 72.3° + k120°\}.$$

59. A solution is $2\alpha = \cos^{-1}(1/4.5) \approx 77.16°$. Another solution is $2\alpha = 360° - 77.16° = 282.84°$. Thus,

$$2\alpha = 77.16° + k \cdot 360° \text{ or } 2\alpha = 282.84° + k \cdot 360°$$

$$\alpha \approx 38.6° + k \cdot 180° \text{ or } \alpha \approx 141.4° + k \cdot 180°.$$

The solution set is

$$\{\alpha \mid \alpha = 38.6° + k180° \text{ or } \alpha = 141.4° + k180°\}.$$

61. A solution is $\alpha/2 = \sin^{-1}(-1/2.3) \approx -25.77°$. This is coterminal with $334.23°$. Another solution is $\alpha/2 = 180° + 25.77° = 205.77°$. Thus,

$$\frac{\alpha}{2} = 334.23° + k \cdot 360° \text{ or } \frac{\alpha}{2} = 205.77° + k \cdot 360°$$

$$\alpha \approx 668.5° + k \cdot 720° \text{ or } \alpha \approx 411.5° + k \cdot 720°.$$

The solution set is

$$\{\alpha \mid \alpha = 668.5° + k720° \text{ or } \alpha = 411.5° + k720°\}.$$

63. Set the right-hand side to zero and factor.

$$\begin{aligned} 3\sin^2 x - \sin x &= 0 \\ \sin x(3\sin x - 1) &= 0 \end{aligned}$$

Set each factor to zero.

$$\begin{aligned} \sin x = 0 \quad &\text{or} \quad \sin x = 1/3 \\ x = 0, \pi \quad &\text{or} \quad x = \sin^{-1}(1/3) \approx 0.3 \end{aligned}$$

Another solution to $\sin x = 1/3$ is $x = \pi - 0.3 \approx 2.8$.
The solution set is $\{0, 0.3, 2.8, \pi\}$.

65. Set the right-hand side to zero and factor.

$$\begin{aligned} 2\cos^2 x + 3\cos x + 1 &= 0 \\ (2\cos x + 1)(\cos x + 1) &= 0 \end{aligned}$$

Set the factors to zero.

$$\begin{aligned} \cos x = -1/2 \quad &\text{or} \quad \cos x = -1 \\ x = 2\pi/3, 4\pi/3 \quad &\text{or} \quad x = \pi \end{aligned}$$

The solution set is $\{\pi, 2\pi/3, 4\pi/3\}$.

67. Substitute $\cos^2 x = 1 - \sin^2 x$.

$$\begin{aligned} 5\sin^2 x - 2\sin x &= 1 - \sin^2 x \\ 6\sin^2 x - 2\sin x - 1 &= 0 \end{aligned}$$

Apply the quadratic formula.

$$\begin{aligned} \sin x &= \frac{2 \pm \sqrt{28}}{12} \\ \sin x &= \frac{1 \pm \sqrt{7}}{6} \end{aligned}$$

Then

$$x = \sin^{-1}\left(\frac{1+\sqrt{7}}{6}\right) \quad \text{or} \quad x = \sin^{-1}\left(\frac{1-\sqrt{7}}{6}\right)$$

$$x \approx 0.653 \quad \text{or} \quad x \approx -0.278.$$

Another solution is $\pi - 0.653 \approx 2.5$. An angle coterminal with -0.278 is $2\pi - 0.278 \approx 6.0$. Another solution is $\pi + 0.278 \approx 3.4$. The solution set is $\{0.7, 2.5, 3.4, 6.0\}$.

69. Squaring both sides of the equation, we obtain

$$\begin{aligned} \tan^2 x &= \sec^2 x - 2\sqrt{3}\sec x + 3 \\ \sec^2 x - 1 &= \sec^2 x - 2\sqrt{3}\sec x + 3 \\ -4 &= -2\sqrt{3}\sec x \\ \sec x &= 2/\sqrt{3} \\ x &= \pi/6, 11\pi/6. \end{aligned}$$

Checking $x = \pi/6$, one gets $\tan(\pi/6) = 1/\sqrt{3}$ and $\sec(\pi/6) - \sqrt{3} = 2/\sqrt{3} - \sqrt{3} = -1/\sqrt{3}$. Then $x = \pi/6$ is an extraneous root and the solution set is $\{11\pi/6\}$.

71. Square both sides of the equation.

$$\begin{aligned} \sin^2 x + 2\sqrt{3}\sin x + 3 &= 27\cos^2 x \\ \sin^2 x + 2\sqrt{3}\sin x + 3 &= 27(1 - \sin^2 x) \\ 28\sin^2 x + 2\sqrt{3}\sin x - 24 &= 0 \\ 14\sin^2 x + \sqrt{3}\sin x - 12 &= 0 \end{aligned}$$

By the quadratic formula, we get

$$\begin{aligned} \sin x &= \frac{-\sqrt{3} \pm \sqrt{675}}{28} \\ \sin x &= \frac{-\sqrt{3} \pm 15\sqrt{3}}{28} \\ \sin x &= \frac{\sqrt{3}}{2}, \frac{-4\sqrt{3}}{7}. \end{aligned}$$

Thus,

$$x = \frac{\pi}{3}, \frac{2\pi}{3} \quad \text{or} \quad x = \sin^{-1}\left(\frac{-4\sqrt{3}}{7}\right)$$

$$x = \frac{\pi}{3}, \frac{2\pi}{3} \quad \text{or} \quad x \approx -1.427.$$

Checking $x = 2\pi/3$, one finds $\sin(2\pi/3) + \sqrt{3} = \sqrt{3}/2 + \sqrt{3}$ and $3\sqrt{3}\cos(2\pi/3)$ is a negative number. Then $x = 2\pi/3$ is an extraneous root.

An angle coterminal with -1.427 is $2\pi - 1.427 \approx 4.9$. In a similar way, one checks that $\pi + 1.427 \approx 4.568$ is an extraneous root. Thus, the solution set is $\{\pi/3, 4.9\}$.

73. Express the equation in terms of $\sin x$ and $\cos x$.

$$\begin{aligned} \frac{\sin x}{\cos x} \cdot 2\sin x \cos x &= 0 \\ 2\sin^2 x &= 0 \\ \sin x &= 0 \end{aligned}$$

Solution set is $\{0, \pi\}$.

75. Substitute the double-angle identity for $\sin x$.

$$\begin{aligned} 2\sin x \cos x - \sin x \cos x &= \cos x \\ \sin x \cos x - \cos x &= 0 \\ \cos x(\sin x - 1) &= 0 \end{aligned}$$

$$\cos x = 0 \quad \text{or} \quad \sin x = 1$$

$$x = \pi/2, 3\pi/2 \quad \text{or} \quad x = \pi/2$$

Solution set is $\{\pi/2, 3\pi/2\}$.

77. Use the sum identity for sine.

$$\sin(x + \pi/4) = 1/2$$

$$x + \frac{\pi}{4} = \frac{\pi}{6} + 2k\pi \quad \text{or} \quad x + \frac{\pi}{4} = \frac{5\pi}{6} + 2k\pi$$

$$x = \frac{-\pi}{12} + 2k\pi \quad \text{or} \quad x = \frac{7\pi}{12} + 2k\pi$$

By choosing $k = 1$ in the first case and $k = 0$ in the second case, one finds the solution set is $\{23\pi/12, 7\pi/12\}$.

79. Apply the difference identity for sine.

$$\begin{aligned} \sin(2x - x) &= -1/2 \\ \sin x &= -1/2 \end{aligned}$$

The solution set is $\{7\pi/6, 11\pi/6\}$.

81. Since $4 \cdot 4^{2\sin^2 x} = 4^{3\sin x}$, we set the exponents equal to each other. Then

$$\begin{aligned} 2\sin^2 x + 1 &= 3\sin x \\ 2\sin^2 x - 3\sin x + 1 &= 0 \\ (2\sin x - 1)(\sin x - 1) &= 0 \\ \sin x &= 1, \frac{1}{2}. \end{aligned}$$

Thus, the solution set is $\{\pi/6, \pi/2, 5\pi/6\}$.

83. Use a half-angle identity for cosine and express equation in terms of $\cos\theta$.

$$\begin{aligned} \frac{1+\cos\theta}{2} &= \frac{1}{\cos\theta} \\ \cos\theta + \cos^2\theta &= 2 \\ \cos^2\theta + \cos\theta - 2 &= 0 \\ (\cos\theta + 2)(\cos\theta - 1) &= 0 \\ \cos\theta = -2 \quad &\text{or} \quad \cos\theta = 1 \\ \text{no solution} \quad &\text{or} \quad \theta = 0^\circ \end{aligned}$$

The solution set is $\{0^\circ\}$.

85. Dividing the equation by $2\cos\theta$, we get

$$\begin{aligned} \frac{\sin\theta}{\cos\theta} &= \frac{1}{2} \\ \tan\theta &= 0.5 \\ \theta &= \tan^{-1}(0.5) \approx 26.6^\circ. \end{aligned}$$

Another solution is $180^\circ + 26.6^\circ = 206.6^\circ$. Solution set is $\{26.6^\circ, 206.6^\circ\}$.

87. Express equation in terms of $\sin 3\theta$.

$$\begin{aligned} \sin 3\theta &= \frac{1}{\sin 3\theta} \\ \sin^2 3\theta &= 1 \\ \sin 3\theta &= \pm 1 \end{aligned}$$

Then

$$\begin{aligned} 3\theta = 90^\circ + k\cdot 360^\circ \quad &\text{or} \quad 3\theta = 270^\circ + k \cdot 360^\circ \\ \theta = 30^\circ + k \cdot 120^\circ \quad &\text{or} \quad \theta = 90^\circ + k\cdot 120^\circ. \end{aligned}$$

By choosing $k = 0, 1, 2$, one finds that the solution set is $\{30^\circ, 90^\circ, 150^\circ, 210^\circ, 270^\circ, 330^\circ\}$.

89. By the method of completing the square, we get

$$\begin{aligned} \tan^2\theta - 2\tan\theta &= 1 \\ \tan^2\theta - 2\tan\theta + 1 &= 2 \\ (\tan\theta - 1)^2 &= 2 \\ \tan\theta - 1 &= \pm\sqrt{2} \\ \theta = \tan^{-1}(1+\sqrt{2}) \quad &\text{or} \quad \theta = \tan^{-1}(1-\sqrt{2}) \\ \theta \approx 67.5^\circ \quad &\text{or} \quad \theta = -22.5^\circ. \end{aligned}$$

Other solutions are $180^\circ + 67.5^\circ = 247.5^\circ$, $180^\circ - 22.5^\circ = 157.5^\circ$, and $180^\circ + 157.5^\circ = 337.5^\circ$. The solution set is $\{67.5^\circ, 157.5^\circ, 247.5^\circ, 337.5^\circ\}$.

91. Factor as a perfect square.

$$\begin{aligned} (3\sin\theta + 2)^2 &= 0 \\ \sin\theta &= -2/3 \\ \theta &= \sin^{-1}(-2/3) \approx -41.8^\circ \end{aligned}$$

An angle coterminal with -41.8° is $360^\circ - 41.8^\circ = 318.2^\circ$. Another solution is $180^\circ + 41.8^\circ = 221.8^\circ$. The solution set is $\{221.8^\circ, 318.2^\circ\}$.

93. By using the sum identity for tangent, we get

$$\begin{aligned} \tan(3\theta - \theta) &= \sqrt{3} \\ 2\theta &= 60^\circ + k\cdot 180^\circ \\ \theta &= 30^\circ + k \cdot 90^\circ. \end{aligned}$$

By choosing $k = 1, 3$, one obtains that the solution set is $\{120^\circ, 300^\circ\}$. Note, 30° and 210° are not solutions.

95. Factoring, we get

$$(4\cos^2\theta - 3)(2\cos^2\theta - 1) = 0.$$

Then

$$\begin{aligned} \cos^2\theta = 3/4 \quad &\text{or} \quad \cos^2\theta = 1/2 \\ \cos\theta = \pm\sqrt{3}/2 \quad &\text{or} \quad \cos\theta = \pm 1/\sqrt{2}. \end{aligned}$$

The solution set is

$$\{30^\circ, 45^\circ, 135^\circ, 150^\circ, 210^\circ, 225^\circ, 315^\circ, 330^\circ\}.$$

97. Factoring, we obtain

$$\begin{aligned}(\sec^2\theta - 1)(\sec^2\theta - 4) &= 0\\ \sec^2\theta = 1 \quad &\text{or} \quad \sec^2\theta = 4\\ \sec\theta = \pm 1 \quad &\text{or} \quad \sec\theta = \pm 2.\end{aligned}$$

Solution set is $\{0°, 60°, 120°, 180°, 240°, 300°\}$.

99. Multiplying the equation by LCD, we get

$$\begin{aligned}13.7\sin 33.2° &= a\cdot\sin 45.6°\\ \frac{13.7\sin 33.2°}{\sin 45.6°} &= a\\ 10.5 &\approx a.\end{aligned}$$

101. Multiplying by the LCD, we get

$$\begin{aligned}25.9\sin\alpha &= 23.4\sin 67.2°\\ \sin\alpha &= \frac{23.4\sin 67.2°}{25.9}\\ \sin\alpha &\approx 0.833\\ \alpha &\approx \sin^{-1}(0.833)\\ \alpha &\approx 56.4°.\end{aligned}$$

103. Isolate $\cos\alpha$ on one side.

$$\begin{aligned}2(5.4)(8.2)\cos\alpha &= 5.4^2 + 8.2^2 - 3.6^2\\ \cos\alpha &= \frac{5.4^2 + 8.2^2 - 3.6^2}{2(5.4)(8.2)}\\ \cos\alpha &\approx 0.942\\ \alpha &\approx \cos^{-1}(0.942)\\ \alpha &\approx 19.6°\end{aligned}$$

105. Given below is the graph of

$$y = \sin(x/2) - \cos(3x).$$

The intercepts or solutions on $[0, 2\pi)$ are approximately $\{0.4, 1.9, 2.2, 4.0, 4.4, 5.8\}$.

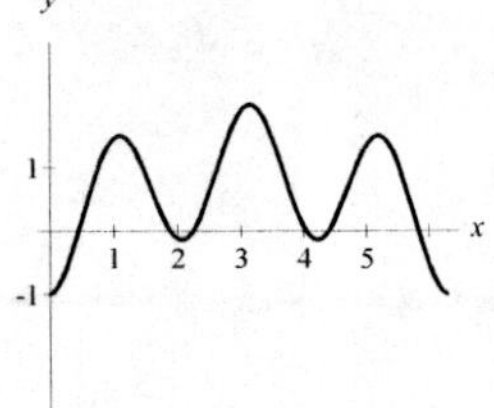

107. The graph of $y = \frac{x}{2} - \frac{\pi}{6} + \frac{\sqrt{3}}{2} - \sin x$ is shown. The solution set is $\{\pi/3\}$.

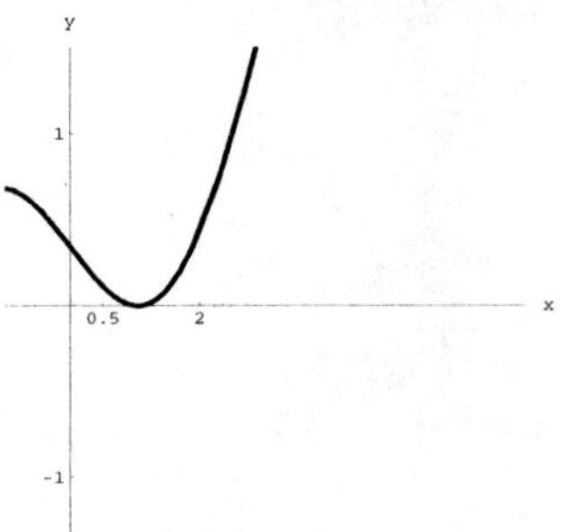

109. Since $a = \sqrt{3}$ and $b = 1$, we obtain $r = \sqrt{\sqrt{3}^2 + 1^2} = 2$. If the terminal side of α goes through $(\sqrt{3}, 1)$, then $\tan\alpha = 1/\sqrt{3}$. Then one can choose $\alpha = \pi/6$ and $x = 2\sin(2t + \pi/6)$. The times when $x = 0$ are given by

$$\begin{aligned}\sin\left(2t + \frac{\pi}{6}\right) &= 0\\ 2t + \frac{\pi}{6} &= k\cdot\pi\\ 2t &= -\frac{\pi}{6} + k\cdot\pi\\ t &= -\frac{\pi}{12} + \frac{k\cdot\pi}{2}\\ t &= -\frac{\pi}{12} + \frac{\pi}{2} + \frac{k\cdot\pi}{2}\\ t &= \frac{5\pi}{12} + \frac{k\cdot\pi}{2}\end{aligned}$$

where k is a nonnegative integer.

111. First, find the values of t when $x = \sqrt{3}$.

$$\begin{aligned}2\sin\left(\frac{\pi t}{3}\right) &= \sqrt{3}\\ \sin\left(\frac{\pi t}{3}\right) &= \frac{\sqrt{3}}{2}\end{aligned}$$

$$\begin{aligned}\frac{\pi t}{3} = \frac{\pi}{3} + 2k\pi \quad &\text{or} \quad \frac{\pi t}{3} = \frac{2\pi}{3} + 2k\pi\\ \pi t = \pi + 6k\pi \quad &\text{or} \quad \pi t = 2\pi + 6k\pi\\ t = 1 + 6k \quad &\text{or} \quad t = 2 + 6k\end{aligned}$$

Then the ball is $\sqrt{3}$ ft above sea level for the values of t satisfying

$$1 + 6k < t < 2 + 6k$$

where k is a nonnegative integer.

113. Since $v_o = 325$ and $d = 3300$, we have

$$\begin{aligned} 325^2 \sin 2\theta &= 32(3300) \\ \sin 2\theta &= \frac{32(3300)}{325^2} \\ \sin 2\theta &\approx 0.99976 \\ 2\theta &\approx \sin^{-1}(0.99976) \\ 2\theta &\approx 88.74^\circ \\ \theta &\approx 44.4^\circ. \end{aligned}$$

Another angle is given by $2\theta = 180^\circ - 88.74^\circ = 91.26^\circ$ or $\theta = 91.26^\circ/2 \approx 45.6^\circ$.
The muzzle was aimed at 44.4° or 45.6° .

115. Note, 90 mph$= 90 \cdot \dfrac{5280}{3600}$ ft/sec$=$ 132 ft/sec.
In $v_o^2 \sin 2\theta = 32d$, let $v_o = 132$ and $d = 230$.

$$\begin{aligned} 132^2 \sin 2\theta &= 32(230) \\ \sin 2\theta &= \frac{32(230)}{132^2} \\ \sin 2\theta &\approx 0.4224 \\ 2\theta = \sin^{-1}(0.4224) &\approx 25.0^\circ \text{ or } 155^\circ \\ \theta \approx 12.5^\circ \quad &\text{or} \quad 77.5^\circ \end{aligned}$$

The two possible angles are 12.5° and 77.5°. The time it takes the ball to reach home plate can be found by using $x = v_o t \cos\theta$.
(See Example 11). For the angle 12.5°, it takes

$$t = \frac{230}{132 \cos 12.5^\circ} \approx 1.78 \text{ sec}$$

while for 77.5° it takes

$$t = \frac{230}{132 \cos 77.5^\circ} \approx 8.05 \text{ sec.}$$

The difference in time is $8.05 - 1.78 \approx 6.3$ sec.

117. Observe,

$$\begin{aligned} y &= \sqrt{2}\left((\sin x)\frac{1}{\sqrt{2}} - (\cos x)\frac{1}{\sqrt{2}}\right) \\ &= \sqrt{2}\sin\left(x + \frac{7\pi}{4}\right). \end{aligned}$$

The amplitude is $\sqrt{2}$, period is 2π, and phase shift $-\frac{7\pi}{4}$

119. Using a cofunction relationship,

$$\sin\left(\frac{\pi}{2} - x\right) = \cos x = \frac{3}{4}.$$

121. Using a half-angle identity, we obtain

$$\begin{aligned} \sin\left(\frac{x}{2}\right) &= \sqrt{\frac{1-\cos x}{2}} \\ &= \sqrt{\frac{1-\frac{1}{4}}{2}} = \sqrt{\frac{3}{8}} \\ &= \sqrt{\frac{6}{16}} \\ \sin\left(\frac{x}{2}\right) &= \frac{\sqrt{6}}{4} \end{aligned}$$

123. The angles in the shaded triangle are 15°, 60°, and 105°. The side opposite the 105°-angle is one unit l long. Using the sine law (see Chapter 7), we find

$$x = \frac{\sin 60^\circ}{\sin 105^\circ}.$$

Recall, the area of a triangle is one-half times the product of the length of any two sides and the sine of the included angle of the two sides. Since the included angle between x and 1 is 15°, the area of the triangle is

$$\begin{aligned} \text{Area} &= \frac{1}{2}\frac{\sin 60^\circ}{\sin 105^\circ}\sin 15^\circ \\ &= \frac{\sqrt{3}}{4}\frac{\sin 15^\circ}{\sin 105^\circ} \\ &= \frac{\sqrt{3}}{4}\frac{\sqrt{(2-\sqrt{3}/2)/2}}{\sqrt{(2+\sqrt{3}/2)/2}} \\ &= \frac{\sqrt{3}}{4}\frac{\sqrt{2-\sqrt{3}}}{\sqrt{2+\sqrt{3}}} \\ &= \frac{\sqrt{3}}{4}(2-\sqrt{3}) \\ \text{Area} &= \frac{2\sqrt{3}-3}{4}. \end{aligned}$$

Review Exercises

1. $1 - \sin^2 \alpha = \cos^2 \alpha$

3. $(1 - \csc x)(1 + \csc x) = 1 - \csc^2 x = -\cot^2 x$

5.
$$\frac{1}{1+\sin\alpha} + \frac{\sin\alpha}{\cos^2\alpha} = \frac{1}{1+\sin\alpha} + \frac{\sin\alpha}{1-\sin^2\alpha} = \frac{(1-\sin\alpha)+\sin\alpha}{1-\sin^2\alpha} = \frac{1}{\cos^2\alpha} = \sec^2\alpha$$

7. $\tan(4s)$, by the double angle idenity for tangent

9. $\sin(3\theta - 6\theta) = \sin(-3\theta) = -\sin(3\theta)$

11. $\tan\left(\frac{2z}{2}\right) = \tan z$, by a double-angle identity for tangent

13. e **15.** c

17. a **19.** g

21. Note, $\sin\alpha = \sqrt{1 - \left(\frac{-5}{13}\right)^2} = \sqrt{1 - \frac{25}{169}} = \sqrt{\frac{144}{169}} = \frac{12}{13}$. Then $\tan\alpha = \frac{12/13}{-5/13} = -\frac{12}{5}$, $\cot\alpha = -\frac{5}{12}$, $\csc\alpha = \frac{13}{12}$, $\sec\alpha = -\frac{13}{5}$.

23. By using a cofunction identity, we get $\cos\alpha = \frac{-3}{5}$.

Then $\sin\alpha = -\sqrt{1 - \left(\frac{-3}{5}\right)^2} = -\sqrt{1 - \frac{9}{25}} = -\sqrt{\frac{16}{25}} = -\frac{4}{5}$, $\sec\alpha = -\frac{5}{3}$, $\csc\alpha = -\frac{5}{4}$, $\tan\alpha = \frac{-4/5}{-3/5} = \frac{4}{3}$, $\cot\alpha = \frac{3}{4}$.

25. By the half-angle identity for sine, we find

$$\begin{aligned} \sqrt{\frac{1-\cos\alpha}{2}} &= \frac{3}{5} \\ \frac{1-\cos\alpha}{2} &= \frac{9}{25} \\ 1-\cos\alpha &= \frac{18}{25} \\ \cos\alpha &= \frac{7}{25}. \end{aligned}$$

Since $\frac{3\pi}{2} < \alpha < 2\pi$, α is in quadrant IV and $\sin\alpha = -\sqrt{1 - \left(\frac{7}{25}\right)^2} = -\sqrt{1 - \frac{49}{625}} = -\sqrt{\frac{576}{625}} = -\frac{24}{25}$.

Then $\tan\alpha = \frac{-24/25}{7/25} = -\frac{24}{7}$, $\cot\alpha = -\frac{7}{24}$, $\sec\alpha = \frac{25}{7}$, $\csc\alpha = -\frac{25}{24}$.

27. It is an identity as shown below.

$$\begin{aligned} (\sin x + \cos x)^2 &= \\ \sin^2 x + 2\sin x\cos x + cos^2 x &= \\ 1 + 2\sin x\cos x &= \\ 1 + \sin(2x) & \end{aligned}$$

29. It is not an identity since $\csc^2 x - \cot^2 x = 1$ and $\tan^2 - \sec^2 x = -1$.

31. Odd, since $f(-x) = \frac{\sin(-x) - \tan(-x)}{\cos(-x)} = \frac{-\sin x + \tan x}{\cos x} = -\frac{\sin x - \tan x}{\cos x} = -f(x)$

33. It is neither even nor odd. Since $f(\pi/4) = \frac{\cos(\pi/4) - \sin(\pi/4)}{\sec(\pi/4)} = \frac{\sqrt{2}/2 - \sqrt{2}/2}{\sqrt{2}} = 0$ and $f(-\pi/4) = \frac{\cos(-\pi/4) - \sin(-\pi/4)}{\sec(-\pi/4)} = \frac{\sqrt{2}/2 + \sqrt{2}/2}{\sqrt{2}} = \frac{\sqrt{2}}{\sqrt{2}} = 1$, we see that $f(\pi/4) \neq \pm f(-\pi/4)$

35. Even, since $f(-x) = \frac{\sin(-x)\tan(-x)}{\cos(-x) + \sec(-x)} = \frac{(-\sin x)(-\tan x)}{\cos x + \sec x} = \frac{\sin x \tan x}{\cos x + \sec x} = f(x)$

37. f, since $\sin(\pi/2 - \alpha) = \cos\alpha$

39. e, for $\sin x + \sin y = 2\sin\left(\frac{x+y}{2}\right)\cos\left(\frac{x-y}{2}\right)$

41. b, since $\sin 2x = 2\sin x\cos x$

43. h, for $\cos x + \cos y = 2\cos\left(\frac{x+y}{2}\right)\cos\left(\frac{x-y}{2}\right)$

45. c, for $\tan\left(\frac{x}{2}\right) = \frac{\sin x}{1+\cos x}$

47. Rewrite the right side.

$$\begin{aligned} &= \frac{1+\tan^2\theta}{1-\tan^2\theta} \\ &= \frac{\sec^2\theta}{1-\frac{\sin^2\theta}{\cos^2\theta}} \cdot \frac{\cos^2\theta}{\cos^2\theta} \\ &= \frac{1}{\cos^2\theta - \sin^2\theta} \\ &= \frac{1}{\cos 2\theta} \\ &= \sec 2\theta \end{aligned}$$

49. Rewrite the right side as follows:

$$\begin{aligned} &= \frac{\csc^2 x - \cot^2 x}{2\csc^2 x + 2\csc x \cot x} \\ &= \frac{1}{\frac{2}{\sin^2 x} + 2 \cdot \frac{1}{\sin x} \cdot \frac{\cos x}{\sin x}} \\ &= \frac{1}{\frac{2}{\sin^2 x} + \frac{2\cos x}{\sin^2 x}} \cdot \frac{\sin^2 x}{\sin^2 x} \\ &= \frac{\sin^2 x}{2+2\cos x} \\ &= \frac{1-\cos^2 x}{2(1+\cos x)} \\ &= \frac{(1-\cos x)(1+\cos x)}{2(1+\cos x)} \\ &= \frac{1-\cos x}{2} \\ &= \sin^2\left(\frac{x}{2}\right) \end{aligned}$$

51. Rewrite the left side as follows:

$$\begin{aligned} \cot(\alpha - 45^\circ) &= \\ (\tan(\alpha - 45^\circ))^{-1} &= \\ \left(\frac{\tan\alpha - \tan 45^\circ}{1+\tan\alpha\tan 45^\circ}\right)^{-1} &= \\ \left(\frac{\tan\alpha - 1}{1+\tan\alpha}\right)^{-1} &= \\ \frac{1+\tan\alpha}{\tan\alpha - 1} & \end{aligned}$$

53. Rewrite the left side.

$$\begin{aligned} \frac{\sin 2\beta}{2\csc\beta} &= \\ \frac{2\sin\beta\cos\beta}{2/\sin\beta} \cdot \frac{\sin\beta}{\sin\beta} &= \\ \sin^2\beta\cos\beta & \end{aligned}$$

55. Factor the numerator on the left-hand side as a difference of two cubes.
Note, $\cot w \tan w = 1$.

$$\begin{aligned} \frac{\cot^3 y - \tan^3 y}{\sec^2 y + \cot^2 y} &= \\ \frac{(\cot y - \tan y)(\cot^2 y + 1 + \tan^2 y)}{\sec^2 y + \cot^2 y} &= \\ \frac{(\cot y - \tan y)(\cot^2 y + \sec^2 y)}{\sec^2 y + \cot^2 y} &= \\ \cot y - \tan y &= \\ \frac{1}{\tan y} - \tan y &= \\ \frac{1-\tan^2 y}{\tan y} &= \\ 2 \cdot \frac{1-\tan^2 y}{2\tan y} &= \\ 2 \cdot (\tan 2y)^{-1} &= \\ 2\cot(2y) & \end{aligned}$$

57. By using double-angle identities, we obtain

$$\begin{aligned} \cos(2 \cdot 2x) &= \\ 1 - 2\sin^2(2x) &= \\ 1 - 2(2\sin x\cos x)^2 &= \\ 1 - 8\sin^2 x\cos^2 x &= \\ 1 - 8\sin^2 x(1-\sin^2 x) &= \\ 8\sin^4 x - 8\sin^2 x + 1. & \end{aligned}$$

59. By the double-angle identity for sine, we get

$$\begin{aligned} \sin^4(2x) &= \\ (2\sin x\cos x)^4 &= \end{aligned}$$

$$
\begin{aligned}
16\sin^4 x\cos^4 x &= \\
16\sin^4 x(1-\sin^2 x)^2 &= \\
16\sin^4 x(1-2\sin^2 x+\sin^4 x) &= \\
16\sin^4 x - 32\sin^6 x + 16\sin^8 x &
\end{aligned}
$$

61. $\tan\left(\frac{-\pi/6}{2}\right) = \frac{1-\cos(-\pi/6)}{\sin(-\pi/6)} =$
$\frac{1-\sqrt{3}/2}{-1/2}\cdot\frac{2}{2} = \frac{2-\sqrt{3}}{-1} = \sqrt{3}-2$

63. $\sin\left(\frac{-150°}{2}\right) = -\sqrt{\frac{1-\cos(-150°)}{2}} =$
$-\sqrt{\frac{1-(-\sqrt{3}/2)}{2}\cdot\frac{2}{2}} = -\sqrt{\frac{2+\sqrt{3}}{4}} =$
$-\frac{\sqrt{2+\sqrt{3}}}{2}$

65. Let $a = 4$, $b = 4$, and $r = \sqrt{4^2+4^2} = 4\sqrt{2}$. If the terminal side of α goes through $(4,4)$, then $\tan\alpha = 4/4 = 1$ and one can choose $\alpha = \pi/4$. So $y = 4\sqrt{2}\sin(x+\pi/4)$, amplitude is $4\sqrt{2}$, and phase shift is $-\pi/4$.

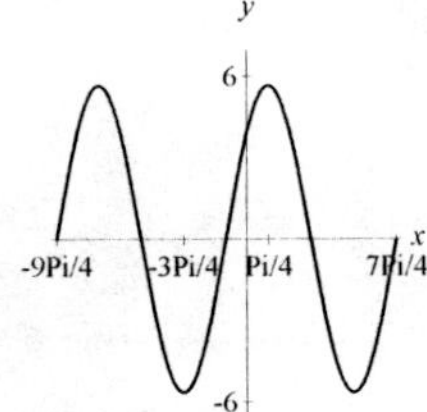

67. Let $a = -2$, $b = 1$, and $r = \sqrt{(-2)^2+1^2} = \sqrt{5}$. If the terminal side of α goes through $(-2,1)$, then $\tan\alpha = -1/2$. Since $\tan^{-1}(-1/2) \approx -0.46$ and $(-2,1)$ is in quadrant II, one can choose $\alpha = \pi - 0.46 = 2.68$. So $y = \sqrt{5}\sin(x+2.68)$, amplitude is $\sqrt{5}$, and phase shift is -2.68.

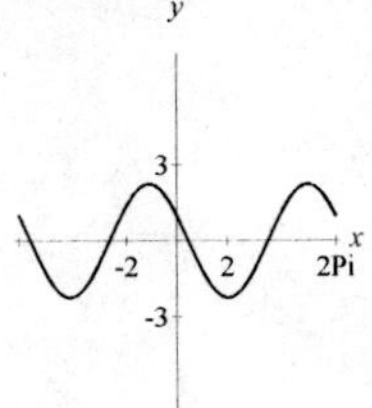

69. Isolate $\cos 2x$ on one side.

$$
\begin{aligned}
2\cos 2x &= -1 \\
\cos 2x &= -\frac{1}{2}
\end{aligned}
$$

$$2x = \frac{2\pi}{3} + 2k\pi \quad \text{or} \quad 2x = \frac{4\pi}{3} + 2k\pi$$

$$x = \frac{\pi}{3} + k\pi \quad \text{or} \quad x = \frac{2\pi}{3} + k\pi$$

The solution set is

$$\left\{x \mid x = \frac{\pi}{3} + k\pi \text{ or } x = \frac{2\pi}{3} + k\pi\right\}.$$

71. Set each factor to zero.

$$(\sqrt{3}\csc x - 2)(\csc x - 2) = 0$$

$$\csc x = \frac{2}{\sqrt{3}} \quad \text{or} \quad \csc x = 2$$

Thus, $x = \frac{\pi}{3}, \frac{2\pi}{3}, \frac{\pi}{6}, \frac{5\pi}{6}$ plus multiples of 2π.
The solution set is

$$\left\{x \mid x = \frac{\pi}{3} + 2k\pi, \frac{2\pi}{3} + 2k\pi, \frac{\pi}{6} + 2k\pi, \frac{5\pi}{6} + 2k\pi\right\}$$

73. Set the right-hand side to zero and factor.

$$
\begin{aligned}
2\sin^2 x - 3\sin x + 1 &= 0 \\
(2\sin x - 1)(\sin x - 1) &= 0
\end{aligned}
$$

$$\sin x = \frac{1}{2} \quad \text{or} \quad \sin x = 1$$

The $x = \frac{\pi}{6}, \frac{5\pi}{6}, \frac{\pi}{2}$ plus multiples of 2π.
The solution set is

$$\left\{x \mid x = \frac{\pi}{6} + 2k\pi, \frac{5\pi}{6} + 2k\pi, \frac{\pi}{2} + 2k\pi\right\}.$$

75. Isolate $\sin\frac{x}{2}$ on one side.

$$
\begin{aligned}
\sin\frac{x}{2} &= \frac{12}{8\sqrt{3}} \\
\sin\frac{x}{2} &= \frac{3}{2\sqrt{3}} \\
\sin\frac{x}{2} &= \frac{\sqrt{3}}{2}
\end{aligned}
$$

$$\frac{x}{2} = \frac{\pi}{3} + 2k\pi \quad \text{or} \quad \frac{x}{2} = \frac{2\pi}{3} + 2k\pi$$

$$x = \frac{2\pi}{3} + 4k\pi \quad \text{or} \quad x = \frac{4\pi}{3} + 4k\pi$$

The solution set is

$$\left\{x \mid x = \frac{2\pi}{3} + 4k\pi \text{ or } x = \frac{4\pi}{3} + 4k\pi\right\}.$$

77. By using the double-angle identity for sine, we get

$$\begin{aligned} \cos\frac{x}{2} - \sin\left(2\cdot\frac{x}{2}\right) &= 0 \\ \cos\frac{x}{2} - 2\sin\frac{x}{2}\cos\frac{x}{2} &= 0 \\ \cos\frac{x}{2}(1 - 2\sin\frac{x}{2}) &= 0 \end{aligned}$$

$$\cos\frac{x}{2} = 0 \quad \text{or} \quad \sin\frac{x}{2} = \frac{1}{2}.$$

Then $\frac{x}{2} = \frac{\pi}{2}, \frac{3\pi}{2}, \frac{\pi}{6}, \frac{5\pi}{6}$ plus mutiples of 2π.

Or $x = \pi, 3\pi, \frac{\pi}{3}, \frac{5\pi}{3}$ plus multiples of 4π.

The solution set is

$$\left\{x \mid x = \pi + 2k\pi, \frac{\pi}{3} + 4k\pi, \frac{5\pi}{3} + 4k\pi\right\}.$$

79. By the double-angle identity for cosine, we find

$$\begin{aligned} \cos 2x + \sin^2 x &= 0 \\ \cos^2 x - \sin^2 x + \sin^2 x &= 0 \\ \cos^2 x &= 0 \\ x &= \frac{\pi}{2} + k\pi. \end{aligned}$$

The solution set is $\left\{x \mid x = \frac{\pi}{2} + k\pi\right\}$.

81. By factoring, we obtain

$$\begin{aligned} \sin x(\cos x + 1) + (\cos x + 1) &= 0 \\ (\sin x + 1)(\cos x + 1) &= 0. \end{aligned}$$

Then

$$\sin x = -1 \quad \text{or} \quad \cos x = -1$$
$$x = \frac{3\pi}{2} + 2k\pi \quad \text{or} \quad x = \pi + 2k\pi.$$

The solution set is

$$\left\{x \mid x = \frac{3\pi}{2} + 2k\pi \text{ or } x = \pi + 2k\pi\right\}.$$

83. Since $\sin 2x = \pm\frac{1}{\sqrt{2}}$, we have

$$\begin{aligned} 2x &= \frac{\pi}{4} + k\pi \quad \text{or} \\ 2x &= \frac{3\pi}{4} + k\pi. \end{aligned}$$

Then

$$\begin{aligned} x &= \frac{(4k+1)\pi}{8} \quad \text{or} \\ x &= \frac{(4k+3)\pi}{8}. \end{aligned}$$

The solution set is

$$\{\frac{\pi}{8}, \frac{3\pi}{8}, \frac{5\pi}{8}, \frac{7\pi}{8}, \frac{9\pi}{8}, \frac{11\pi}{8}, \frac{13\pi}{8}, \frac{15\pi}{8}\}$$

85. Since $\cos x(2\cos x + 1) = 0$, we have $\cos x = 0$ or $\cos x = -\frac{1}{2}$. The solution set is

$$\{\frac{\pi}{2}, \frac{3\pi}{2}, \frac{2\pi}{3}, \frac{4\pi}{3}\}$$

87. Since $1 - \cos^2 2x + \cos 2x = 1$, we have $\cos 2x(\cos 2x - 1) = 0$. Then $\cos 2x = 0, 1$. Consequently, $2x = 2k\pi$ or $2x = \frac{\pi}{2} + k\pi$. That is, , $x = k\pi$ or $x = \frac{(2k+1)\pi}{4}$.

The solution set is $\{0, \pi, \frac{\pi}{4}, \frac{3\pi}{4}, \frac{5\pi}{4}, \frac{7\pi}{4}\}$.

89. Factoring we have

$$(2\sin x - 1)(\sin x - 1) = 0.$$

Then $\sin x = \frac{1}{2}, 1$. The solution set is $\{\frac{\pi}{6}, \frac{5\pi}{6}, \frac{\pi}{2}\}$.

91. Factoring we have

$$(2\cos x + 1)(\cos x + 1) = 0.$$

Then $\cos x = -\frac{1}{2}, -1$. The solution set is $\{\frac{2\pi}{3}, \frac{4\pi}{3}, \pi\}$.

93. Factoring we have

$$(\sin x + 1)(3\sin x - 1) = 0.$$

Then $\sin x = -1, \frac{1}{3}$. Notice, $\sin^{-1}(1/3) \approx 0.3$ and $\pi - \sin^{-1}(1/3) \approx 2.8$ The solution set is $\{\frac{3\pi}{2}, 0.3, 2.8\}$.

95. Factoring we have

$$(\cos 2x + 1)(3\cos 2x - 1) = 0.$$

Then $\cos 2x = -1, \frac{1}{3}$. Consequently, we have $2x = \pi + 2k\pi$, or

$2x = \cos^{-1}(1/3) + 2k\pi$, or

$2x = 2\pi - \cos^{-1}(1/3) + 2k\pi$.

The solution set is $\{\frac{\pi}{2}, \frac{3\pi}{2}, 0.6, 2.5, 3.8, 5.7\}$.

97. Factoring we have

$$(2\sin\frac{x}{2} - 1)(3\sin\frac{x}{2} - 1) = 0.$$

Then $\sin\frac{x}{2} = \frac{1}{2}, \frac{1}{3}$.

Consequently, either $\frac{x}{2} = \frac{\pi}{6} + 2k\pi$,

$\frac{x}{2} = \frac{5\pi}{6} + 2k\pi$,

$\frac{x}{2} = \sin^{-1}\frac{1}{3} + 2k\pi$, or

$\frac{x}{2} = \pi - \sin^{-1}\frac{1}{3} + 2k\pi$.

Let $k = 0$. The solution set is $\{\frac{\pi}{3}, \frac{5\pi}{3}, 0.7, 5.6\}$.

99. Since $|\sin\theta| \leq 1$ and by the quadratic formula, we find

$$\sin\theta = \frac{3 - \sqrt{13}}{2}.$$

Then $360° + \sin^{-1}\left(\frac{3-\sqrt{13}}{2}\right) \approx 342.4°$ and $180° - \sin^{-1}\left(\frac{3-\sqrt{13}}{2}\right) \approx 197.6°$

The solution set is $\{197.6°, 342.4°\}$.

101. Square both sides.

$$\begin{aligned} \tan^2\theta &= \sec^2\theta - 2\sqrt{3}\sec\theta + 3 \\ -1 &= -2\sqrt{3}\sec\theta + 3 \\ 2\sqrt{3}\sec\theta &= 4 \\ \sec\theta &= \frac{2}{\sqrt{3}}. \end{aligned}$$

Then $\theta = 30°, 330°$. But, $\theta = 30°$ is an extraneous root. The solution set is $\{330°\}$.

103. Applying a difference identity, we obtain

$$\sin(\theta - 45°) = \frac{0.8}{\sqrt{2}}$$

Then $\theta = 45° + \sin^{-1}\left(\frac{0.8}{\sqrt{2}}\right) \approx 79.5°$, or

$\theta = 45° + 180° - \sin^{-1}\left(\frac{0.8}{\sqrt{2}}\right) \approx 190.6°$

The solution set is $\{79.5°, 190.6°\}$.

105. By multiplying the equation by 2, we obtain

$$\begin{aligned} 2\sin\alpha\cos\alpha &= 1 \\ \sin 2\alpha &= 1 \\ 2\alpha &= 90° + k360° \\ \alpha &= 45° + k180°. \end{aligned}$$

By choosing $k = 0, 1$, one gets the solution set $\{45°, 225°\}$.

107. Suppose $1 + \cos\alpha \neq 0$. Dividing the equation by $1 + \cos\alpha$, we get

$$\begin{aligned} \frac{\sin\alpha}{1 + \cos\alpha} &= 1 \\ \tan\frac{\alpha}{2} &= 1 \\ \frac{\alpha}{2} &= 45° + k180° \\ \alpha &= 90° + k360°. \end{aligned}$$

One solution is 90°. On the other hand if $1 + \cos\alpha = 0$, then $\cos\alpha = -1$ and $\alpha = 180°$. Note $\alpha = 180°$ satisfies the given equation. The solution set is $\{90°, 180°\}$.

109. No solution since the left-hand side is equal to 1 by an identity. The solution set is $\emptyset$.

111. Isolate $\sin 2\alpha$ on one side.

$$\begin{aligned} \sin^4 2\alpha &= \frac{1}{4} \\ \sin 2\alpha &= \pm\sqrt[4]{\frac{1}{4}} \\ \sin 2\alpha &= \pm\frac{1}{\sqrt{2}} \\ 2\alpha &= 45° + k90° \\ \alpha &= 22.5° + k45° \end{aligned}$$

By choosing $k = 0, 1, ..., 7$, one gets the solution set $\{22.5°, 67.5°, 112.5°, 157.5°, 202.5°, 247.5°, 292.5°, 337.5°\}$.

113. Suppose $\tan\alpha \neq 0$. Divide the equation by $\tan\alpha$.

$$\begin{aligned} \frac{2\tan\alpha}{1 - \tan^2\alpha} &= \tan\alpha \\ \frac{2}{1 - \tan^2\alpha} &= 1 \\ 2 &= 1 - \tan^2\alpha \\ \tan^2\alpha &= -1 \end{aligned}$$

The last equation is inconsistent since $\tan^2\alpha$ is nonnegative. But if $\tan\alpha = 0$, then $\alpha = 0°, 180°$ and these two values of α satisfy the given equation.
The solution set is $\{0°, 180°\}$.

115. By using the sum identity for sine, we obtain

$$\begin{aligned} \sin(2\alpha+\alpha) &= \cos 3\alpha \\ \sin 3\alpha &= \cos 3\alpha \\ \tan 3\alpha &= 1 \\ 3\alpha &= 45° + k180° \\ \alpha &= 15° + k60°. \end{aligned}$$

By choosing $k = 0, 1, ..., 5$, one gets the solution set $\{15°, 75°, 135°, 195°, 255°, 315°\}$.

117. $\cos 15° + \cos 19° =$
$2\cos\left(\frac{15°+19°}{2}\right)\cos\left(\frac{15°-19°}{2}\right) =$
$2\cos 17° \cos(-2°) = 2\cos 17° \cos 2°$

119. $\sin(\pi/4) - \sin(-\pi/8) = \sin(\pi/4) + \sin(\pi/8) =$
$2\sin\left(\frac{\pi/4+\pi/8}{2}\right)\cos\left(\frac{\pi/4-\pi/8}{2}\right) =$
$2\sin\left(\frac{3\pi/8}{2}\right)\cos\left(\frac{\pi/8}{2}\right) =$
$2\sin(3\pi/16)\cos(\pi/16)$

121. $2\sin 11° \cos 13° =$
$\sin(11° + 13°) + \sin(11° - 13°) =$
$\sin 24° + \sin(-2°) = \sin 24° - \sin 2°$

123. $2\cos\frac{x}{4}\cos\frac{x}{3} =$
$\cos\left(\frac{x}{4} - \frac{x}{3}\right) + \cos\left(\frac{x}{4} + \frac{x}{3}\right) =$
$\cos\left(-\frac{x}{12}\right) + \cos\left(\frac{7x}{12}\right) =$
$\cos\left(\frac{x}{12}\right) + \cos\left(\frac{7x}{12}\right)$

125. Since α is in quadrant II and β is in quadrant I, $\cos\alpha = -\sqrt{1-\left(\frac{\sqrt{3}}{2}\right)^2} = -\frac{1}{2}$ and $\sin\beta = \sqrt{1-\left(\frac{\sqrt{2}}{2}\right)^2} = \frac{\sqrt{2}}{2}$.

So $\sin(\alpha-\beta) = \sin\alpha\cos\beta - \cos\alpha\sin\beta =$ $\frac{\sqrt{3}}{2}\cdot\frac{\sqrt{2}}{2} - \frac{-1}{2}\cdot\frac{\sqrt{2}}{2} = \frac{\sqrt{6}+\sqrt{2}}{4}$.

127. Since α is in quadrant I and β is in quadrant II, $\cos\alpha = \sqrt{1-\left(\frac{\sqrt{3}}{2}\right)^2} = \frac{1}{2}$ and $\sin\beta = \sqrt{1-\left(-\frac{\sqrt{2}}{2}\right)^2} = \frac{\sqrt{2}}{2}$.

So $\cos(\alpha-\beta) = \cos\alpha\cos\beta + \sin\alpha\sin\beta =$ $\frac{1}{2}\cdot\frac{-\sqrt{2}}{2} + \frac{\sqrt{3}}{2}\cdot\frac{\sqrt{2}}{2} = \frac{\sqrt{6}-\sqrt{2}}{4}$.

129. Let b be the hypotenuse of the smaller right triangle shown below. Let a be the side opposite the head angle θ. Then $a + b = r$, the radius of the wheel.

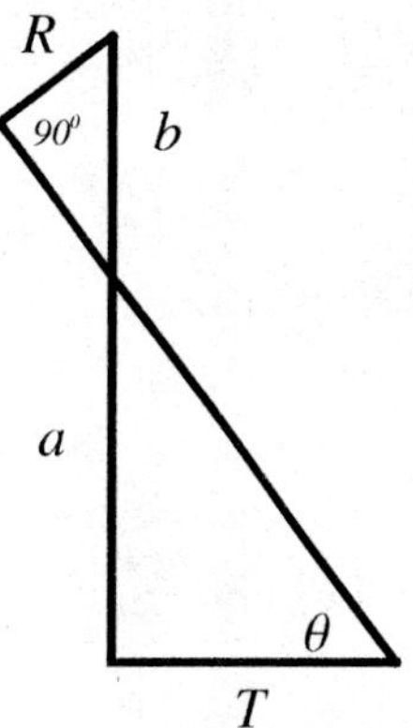

The two right triangles are similar triangles. Applying right triangle trigonometry,

$$b = \frac{R}{\cos\theta}.$$

Consequently,

$$a = r - b = r - \frac{R}{\cos\theta}.$$

Thus,

$$T = \frac{a}{\tan\theta} = \frac{r - \frac{R}{\cos\theta}}{\tan\theta}.$$

Simplifying, we obtain $T = \frac{r\cos\theta - R}{\sin\theta}$.

131. Let $a = 0.6$, $b = 0.4$, and $r = \sqrt{0.6^2 + 0.4^2} \approx 0.72$. If the terminal side of α goes through $(0.6, 0.4)$, then $\tan\alpha = 0.4/0.6$ and one can choose

$\alpha = \tan^{-1}(2/3) \approx 0.588$.
Thus, $x = 0.72\sin(2t + 0.588)$.
The values of t when $x = 0$ are given by

$$\begin{aligned} \sin(2t+0.588) &= 0 \\ 2t + 0.588 &= k\pi \\ 2t &= -0.588 + k\pi \\ t &= -0.294 + \frac{k\pi}{2} \end{aligned}$$

When $k = 1, 2$, one gets $t \approx 1.28, 2.85$.

133. WRONG = 25938 and RIGHT = 51876

Chapter 6 Test

1. $\dfrac{1}{\cos x} \cdot \dfrac{\cos x}{\sin x} \cdot 2\sin x \cos x = 2\cos x$

2. $\sin(2t + 5t) = \sin 7t$

3. $\dfrac{1}{1-\cos y} + \dfrac{1}{1+\cos y} = \dfrac{1+\cos y + 1 - \cos y}{1-\cos^2 y} =$

$\dfrac{2}{\sin^2 y} = 2\csc^2 y$

4. $\tan(\pi/5 + \pi/10) = \tan(3\pi/10)$

5.

$$\begin{aligned} \frac{\sin\beta\cos\beta}{\sin\beta/\cos\beta} &= \\ \sin\beta\cos\beta \cdot \frac{\cos\beta}{\sin\beta} &= \\ \cos^2\beta &= \\ 1 - \sin^2\beta \end{aligned}$$

6.

$$\begin{aligned} \frac{1}{\sec\theta - 1} - \frac{1}{\sec\theta + 1} &= \\ \frac{\sec\theta + 1 - (\sec\theta - 1)}{\sec^2\theta - 1} &= \\ \frac{2}{\tan^2\theta} &= \\ 2\cot^2\theta \end{aligned}$$

7. Using the cofunction identity for cosine, we get

$$\begin{aligned} \cos(\pi/2 - x)\cos(-x) &= \\ \sin x \cos x &= \\ \frac{2\sin x\cos x}{2} &= \\ \frac{\sin(2x)}{2} \end{aligned}$$

8. Factor the left-hand side and use a half-angle identity for tangent. Then

$$\begin{aligned} \tan(t/2)\cdot(\cos^2 t - 1) &= \\ \frac{1-\cos t}{\sin t}\cdot(-\sin^2 t) &= \\ (1-\cos t)\cdot(-\sin t) &= \\ (\cos t - 1)\sin t &= \\ \cos t \sin t - \sin t &= \\ \frac{\sin t}{\sec t} - \sin t. \end{aligned}$$

9. Since $-\sin\theta = 1$, we get $\sin\theta = -1$ and the solution set is $\left\{\theta \mid \theta = \dfrac{3\pi}{2} + 2k\pi\right\}$.

10. Since $\cos 3s = \dfrac{1}{2}$, we obtain

$$\begin{aligned} 3s &= \frac{\pi}{3} + 2k\pi \quad \text{or} \quad 3s = \frac{5\pi}{3} + 2k\pi \\ s &= \frac{\pi}{9} + \frac{2k\pi}{3} \quad \text{or} \quad s = \frac{5\pi}{9} + \frac{2k\pi}{3}. \end{aligned}$$

The solution set is

$$\left\{s \mid s = \frac{\pi}{9} + \frac{2k\pi}{3} \text{ or } s = \frac{5\pi}{9} + \frac{2k\pi}{3}\right\}.$$

11. Since $\tan 2t = -\sqrt{3}$, we have

$$\begin{aligned} 2t &= \frac{2\pi}{3} + k\pi \\ t &= \frac{\pi}{3} + \frac{k\pi}{2}. \end{aligned}$$

The solution set is $\left\{t \mid t = \dfrac{\pi}{3} + \dfrac{k\pi}{2}\right\}$.

12.

$$\begin{aligned} 2\sin\theta\cos\theta &= \cos\theta \\ \cos\theta(2\sin\theta - 1) &= 0 \\ \cos\theta = 0 \quad \text{or} & \quad \sin\theta = 1/2 \end{aligned}$$

The solution set is

$$\left\{\theta \mid \theta = \frac{\pi}{2} + k\pi, \frac{\pi}{6} + 2k\pi, \frac{5\pi}{6} + 2k\pi\right\}.$$

13. By factoring, we obtain

$$\begin{aligned} (3\sin\alpha - 1)(\sin\alpha - 1) &= 0 \\ \sin\alpha = 1/3 \quad \text{or} & \quad \sin\alpha = 1 \\ \alpha = \sin^{-1}(1/3) \approx 19.5^\circ \quad \text{or} & \quad \alpha = 90^\circ. \end{aligned}$$

Another solution is $\alpha = 180^\circ - 19.5^\circ = 160.5^\circ$. The solution set is $\{19.5^\circ, 90^\circ, 160.5^\circ\}$.

14.

$$\begin{aligned} \tan(2\alpha - 7\alpha) &= 1 \\ \tan(-5\alpha) &= 1 \\ -\tan 5\alpha &= 1 \\ \tan 5\alpha &= -1 \\ 5\alpha &= 135^\circ + k180^\circ \\ \alpha &= 27^\circ + k36^\circ \end{aligned}$$

The solution set is $\{27^\circ, 63^\circ, 99^\circ, 171^\circ, 207^\circ, 243^\circ, 279^\circ, 351^\circ\}$. Note, 135° and 315° are not solutions.

15. Let $a = 1$, $b = -\sqrt{3}$, $r = \sqrt{1^2 + (-\sqrt{3})^2} = 2$. If the terminal side of α goes through $(1, -\sqrt{3})$, then $\tan\alpha = -\sqrt{3}$ and one can choose $\alpha = 5\pi/3$. Then $y = 2\sin(x + 5\pi/3)$, the period is 2π, amplitude is 2, and phase shift is $-5\pi/3$.

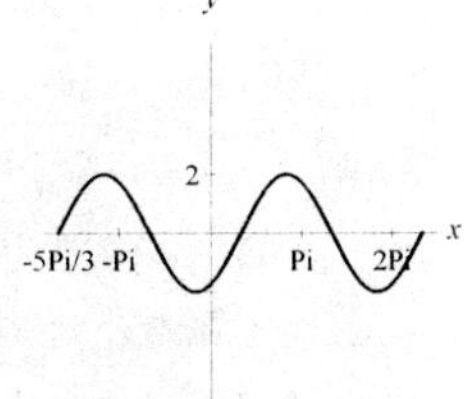

16. If $\csc\alpha = 2$, then $\sin\alpha = 1/2$. Since α is in quadrant II, we obtain $\cos\alpha = -\sqrt{1 - (1/2)^2} = -\sqrt{1 - 1/4} = -\sqrt{3/4} = -\sqrt{3}/2$, $\sec\alpha = -2/\sqrt{3}$, $\tan\alpha = \dfrac{1/2}{-\sqrt{3}/2} = -1/\sqrt{3}$, and $\cot\alpha = -\sqrt{3}$.

17. Even, $f(-x) = (-x)\sin(-x) = (-x)(-\sin x) = x\sin x = f(x)$.

18. By using a half-angle identity, we obtain

$$\sin\left(\frac{-\pi/6}{2}\right) = -\sqrt{\frac{1 - \cos(-\pi/6)}{2}} = -\sqrt{\frac{1 - \sqrt{3}/2}{2} \cdot \frac{2}{2}} = -\sqrt{\frac{2 - \sqrt{3}}{4}} = -\frac{\sqrt{2 - \sqrt{3}}}{2}.$$

19. If $x = y = \pi/6$, then $\tan x + \tan y = 2\tan(\pi/6) = 2 \cdot \dfrac{\sqrt{3}}{3} = \dfrac{2\sqrt{3}}{3}$ and $\tan(x+y) = \tan(\pi/6 + \pi/6) = \tan(\pi/3) = \sqrt{3}$. Thus, it is not an identity.

20. Let $a = 2$, $b = -4$, $r = \sqrt{2^2 + (-4)^2} = \sqrt{20}$. If the terminal side of α goes through $(2, -4)$, then one can choose $\alpha = \tan^{-1}(-4/2) \approx -1.107$. Then $d = \sqrt{20}\sin(3t - 1.107)$. The values of t when $d = 0$ are given by

$$\begin{aligned} \sin(3t - 1.107) &= 0 \\ 3t - 1.107 &= k\pi \\ 3t &= 1.107 + k\pi \\ t &\approx 0.4 + \frac{k\pi}{3}. \end{aligned}$$

By choosing $k = 0, 1, 2, 3$, one obtains the values of t in $[0, 4]$, namely, 0.4 sec, 1.4 sec, 2.5 sec, and 3.5 sec.

Tying It All Together

1. Odd, since $f(-x) = 3(-x)^3 - 2(-x) = -3x^3 + 2x = -f(x)$

2. Even, $f(-x) = 2|-x| = 2|x| = f(x)$

3. Odd, $f(-x) = (-x)^3 + \sin(-x) = -x^3 - \sin x = -f(x)$

4. Even, $f(-x) = (-x)^3 \sin(-x) = (-x^3)(-\sin x) = x^3 \sin x = f(x)$

5. Even, $f(-x) = (-x)^4 - (-x)^2 + 1 = x^4 - x^2 + 1 = f(x)$

6. Odd, $f(-x) = \dfrac{1}{-x} = -\dfrac{1}{x} = -f(x)$

7. Even, $f(-x) = \dfrac{\sin(-x)}{-x} = \dfrac{-\sin x}{-x} = \dfrac{\sin x}{x} = f(x)$

8. Even, $f(-x) = |\sin(-x)| = |-\sin x| = |\sin x| = f(x)$

9. It is not an identity. If $\alpha = \beta = \pi/6$, then $\sin(\alpha + \beta) = \sin(\pi/3) = \sqrt{3}/2$ and $\sin(\pi/6) + \sin(\pi/6) = 2 \cdot (1/2) = 1$.

10. It is not an identity. If $\alpha = \beta = 1$, then $(\alpha + \beta)^2 = 2^2 = 4$ and $\alpha^2 + \beta^2 = 1^2 + 1^2 = 2$.

11. Identity

12. Identity

13. It is not an identity; for if $x = 1$, then $\sin^{-1}(1) = \pi/2 \approx 1.57$ and $1/\sin 1 \approx 1.19$.

14. It is not an identity; for if $x = \sqrt{7\pi/6}$, then $\sin^2 \sqrt{7\pi/6}$ is a positive number and $\sin\left(\left(\sqrt{7\pi/6}\right)^2\right) = \sin(7\pi/6) = -1/2$.

15. Form a right triangle with $\alpha = 30°$, $a = 4$.

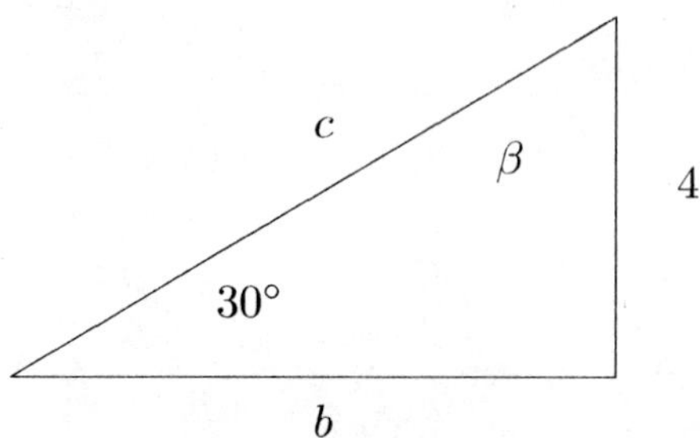

Since $\tan 30° = 4/b$ and $\sin 30° = 4/c$, we get $b = 4/\tan 30° = 4\sqrt{3}$ and $c = 4/\sin 30° = 8$. Also $\beta = 90° - 30° = 60°$.

16. Form a right triangle with $a = \sqrt{3}$, $b = 1$.

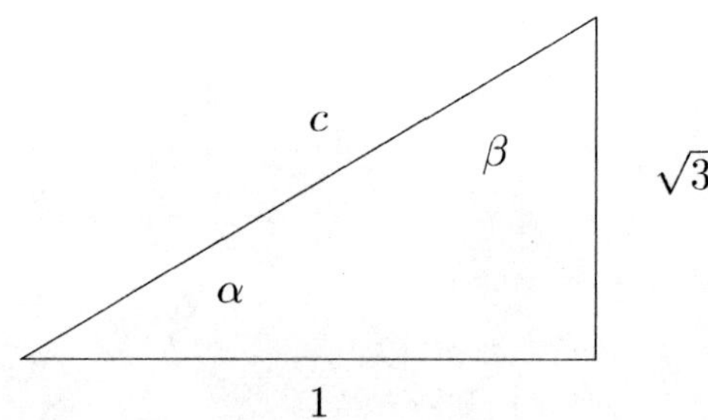

Then $c = \sqrt{(\sqrt{3})^2 + 1^2} = 2$ by the Pythagorean Theorem.
Since $\tan \alpha = \sqrt{3}/1 = \sqrt{3}$, we get $\alpha = 60°$ and $\beta = 30°$.

17. Form a right triangle with $b = 5$ as shown below.

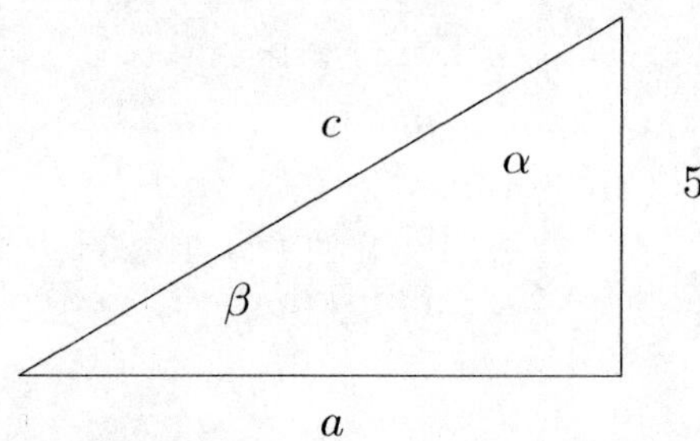

Since $\cos \beta = 0.3$, we find $\beta = \cos^{-1}(0.3) \approx 72.5°$ and $\alpha = 17.5°$. Since $\sin 72.5° = 5/c$ and $\tan 72.5° = 5/a$, we obtain $c = 5/\sin 72.5° \approx 5.2$ and $a = 5/\tan 72.5° \approx 1.6$.

18. Form a right triangle with $a = 2$.

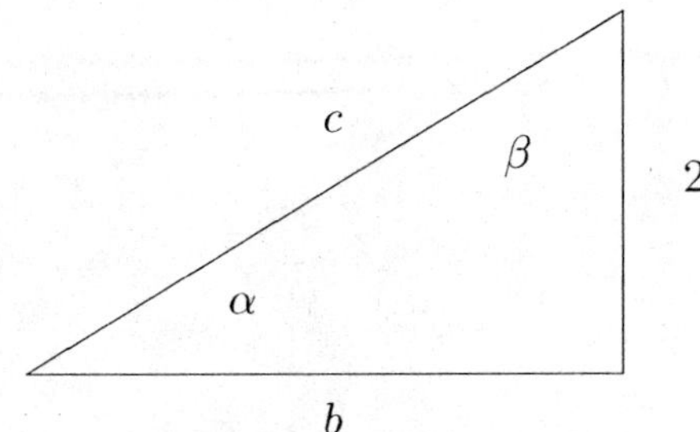

Since $\sin\alpha = 0.6$, we obtain
$\alpha = \sin^{-1}(0.6) \approx 36.9°$ and $\beta = 53.1°$.
Since $\sin 36.9° = 2/c$ and $\tan 36.9° = 2/b$,
we get $c = 2/\sin 36.9° \approx 3.3$ and
$b = 2/\tan 36.9° \approx 2.7$.

19. $\frac{1}{2} + i\frac{\sqrt{3}}{2}$

20. $2\left(-\frac{1}{2} + i \cdot \frac{\sqrt{3}}{2}\right) = -1 + i\sqrt{3}$

21. $\cos^2 225° + 2i\cos 225° \sin 225° - \sin^2 225° =$
$\cos^2 225° - \sin^2 225° + 2i\cos 225° \sin 225° =$
$\cos(2 \cdot 225°) + i \cdot \sin(2 \cdot 225°) =$
$\cos 450° + i \cdot \sin 450° =$
$\cos 90° + i \cdot \sin 90° = i$

22. $\cos^2 3° - (i\sin 3°)^2 = \cos^2 3° + \sin^2 3° = 1$

23. $(2+i)^2(2+i) = (4+4i-1)(2+i) =$
$(3+4i)(2+i) = 6+3i+8i-4 = 2+11i$

24. $(\sqrt{2}(1-i))^4 = \sqrt{2}^4(1-i)^4 = 4\left((1-i)^2\right)^2 =$
$4(1-2i-1)^2 = 4(-2i)^2 = 4(-4) = -16$

25. minutes

26. seconds

27. unit

28. π

29. αr

30. $r\omega$

31. y, x

32. fundamental

33. amplitude

34. phase shift

For Thought

1. True, for the sum of the measurements of the three angles is 180°.

2. False, $a \sin 17° = 88 \sin 9°$ and $a = \dfrac{88 \sin 9°}{\sin 17°}$.

3. False, since $\alpha = \sin^{-1}\left(\dfrac{5 \sin 44°}{18}\right) \approx 11°$ and $\alpha = 180 - 11° = 169°$.

4. True

5. True, since $\dfrac{\sin 60°}{\sqrt{3}} = \dfrac{\sqrt{3}/2}{\sqrt{3}} = \dfrac{1}{2}$ and $\dfrac{\sin 30°}{1} = \sin 30° = \dfrac{1}{2}$.

6. False, a triangle exists since $a = 500$ is bigger than $h = 10 \sin 60° \approx 8.7$.

7. True, since the triangle that exists is a right triangle.

8. False, there exists only one triangle and it is an obtuse triangle.

9. True **10.** True

7.1 Exercises

1. oblique

3. ambiguous

5. Note $\gamma = 180° - (64° + 72°) = 44°$.

   By the sine law $\dfrac{b}{\sin 72°} = \dfrac{13.6}{\sin 64°}$ and $\dfrac{c}{\sin 44°} = \dfrac{13.6}{\sin 64°}$. So $b = \dfrac{13.6}{\sin 64°} \cdot \sin 72°$ ≈ 14.4 and $c = \dfrac{13.6}{\sin 64°} \cdot \sin 44° \approx 10.5$.

7. Note $\beta = 180° - (12.2° + 33.6°) = 134.2°$.

   By the sine law $\dfrac{a}{\sin 12.2°} = \dfrac{17.6}{\sin 134.2°}$ and $\dfrac{c}{\sin 33.6°} = \dfrac{17.6}{\sin 134.2°}$.

   So $a = \dfrac{17.6}{\sin 134.2°} \cdot \sin 12.2° \approx 5.2$ and $c = \dfrac{17.6}{\sin 134.2°} \cdot \sin 33.6° \approx 13.6$.

9. Note $\beta = 180° - (10.3° + 143.7°) = 26°$.

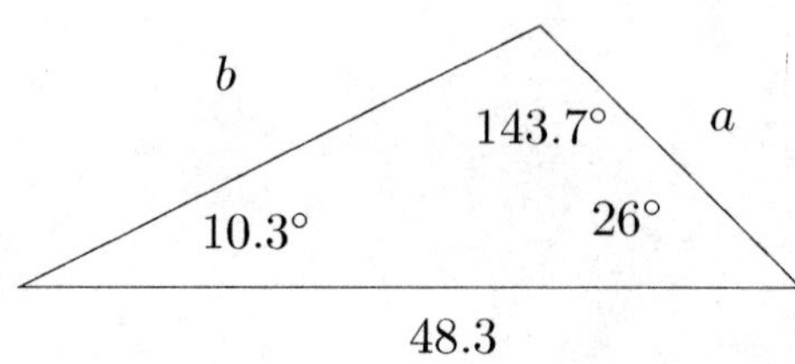

Since $\dfrac{a}{\sin 10.3°} = \dfrac{48.3}{\sin 143.7°}$ and $\dfrac{b}{\sin 26°} = \dfrac{48.3}{\sin 143.7°}$, we have $a = \dfrac{48.3}{\sin 143.7°} \cdot \sin 10.3° \approx 14.6$ and $b = \dfrac{48.3}{\sin 143.7°} \cdot \sin 26° \approx 35.8$

11. Note $\alpha = 180° - (120.7° + 13.6°) = 45.7°$.

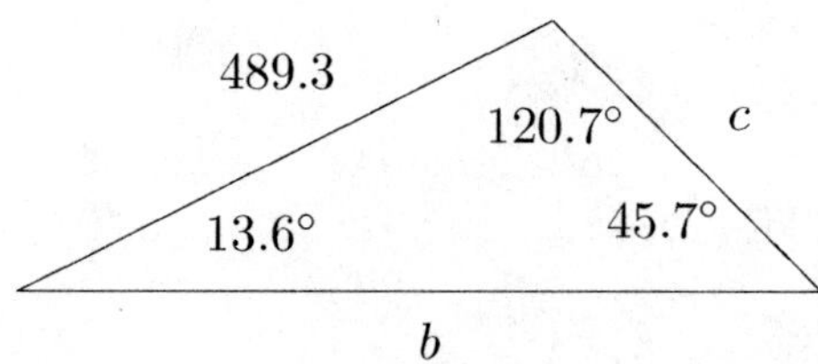

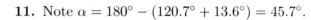

Since $\dfrac{c}{\sin 13.6°} = \dfrac{489.3}{\sin 45.7°}$ and $\dfrac{b}{\sin 120.7°} = \dfrac{489.3}{\sin 45.7°}$, we have $c = \dfrac{489.3}{\sin 45.7°} \cdot \sin 13.6° \approx 160.8$ and $b = \dfrac{489.3}{\sin 45.7°} \cdot \sin 120.7° \approx 587.9$

13. Draw angle $\alpha = 39.6°$ and let h be the height.

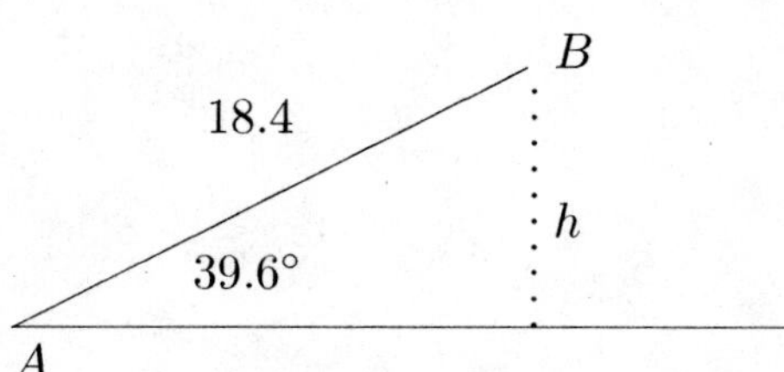

Since $\sin 39.6° = \dfrac{h}{18.4°}$, we have $h = 18.4 \sin 39.6° \approx 11.7$. There is no triangle since $a = 3.7$ is smaller than $h \approx 11.7$.

15. Draw angle $\gamma = 60°$ and let h be the height.

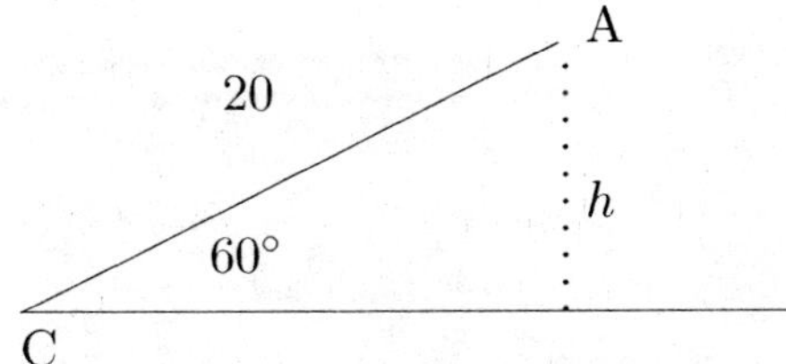

Since $h = 20 \sin 60° = 10\sqrt{3}$ and $c = h$, there is exactly one triangle and it is a right triangle. So $\beta = 90°$ and $\alpha = 30°$. By the Pythagorean Theorem,

$$a = \sqrt{20^2 - (10\sqrt{3})^2} = \sqrt{400 - 300} = 10.$$

17. Since β is an obtuse angle and $b > c$, there is exactly one triangle.

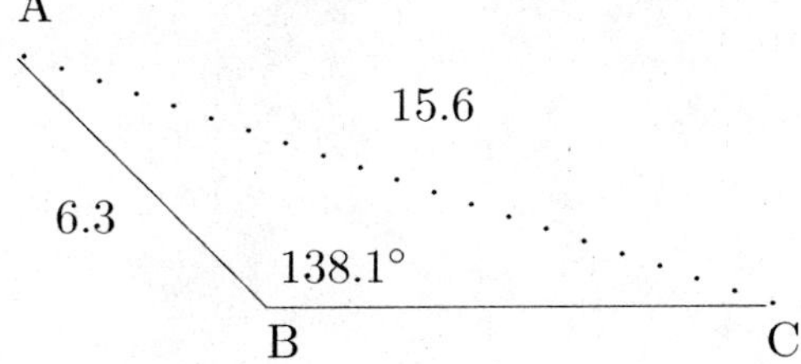

Apply the sine law.

$$\begin{aligned} \frac{15.6}{\sin 138.1°} &= \frac{6.3}{\sin \gamma} \\ \sin \gamma &= \frac{6.3 \sin 138.1°}{15.6} \\ \sin \gamma &\approx 0.2697 \\ \gamma = \sin^{-1}(0.2697) &\approx 15.6° \end{aligned}$$

So $\alpha = 180° - (15.6° + 138.1°) = 26.3°$. By the sine law, $a = \dfrac{15.6}{\sin 138.1°} \sin 26.3° \approx 10.3$.

19. Draw angle $\beta = 32.7°$ and let h be the height.

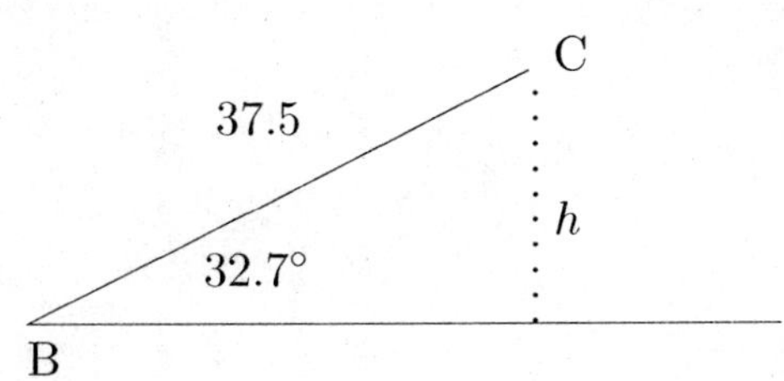

Since $h = 37.5 \sin 32.7° \approx 20.3$ and $20.3 < b < 37.5$, there are two triangles and they are given by

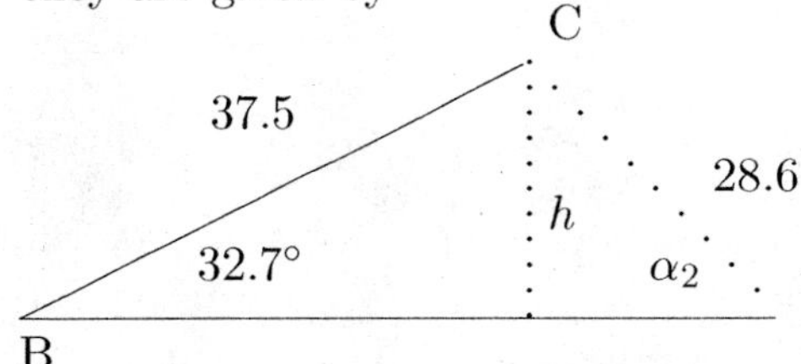

and

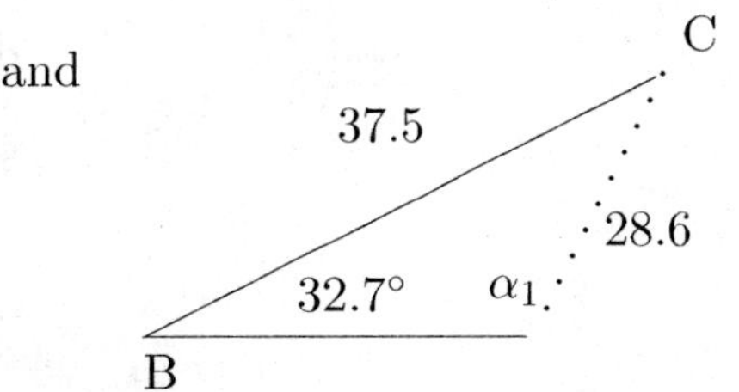

Apply the sine law to the acute triangle.

$$\begin{aligned} \frac{28.6}{\sin 32.7°} &= \frac{37.5}{\sin \alpha_2} \\ \sin \alpha_2 &= \frac{37.5 \sin 32.7°}{28.6} \\ \sin \alpha_2 &\approx 0.708 \\ \alpha_2 = \sin^{-1}(0.708) &\approx 45.1° \end{aligned}$$

So $\gamma_2 = 180° - (45.1° + 32.7°) = 102.2°$. By the sine law, $c_2 = \dfrac{28.6}{\sin 32.7°} \sin 102.2° \approx 51.7$.
On the obtuse triangle, we find $\alpha_1 = 180° - \alpha_2 = 134.9°$ and $\gamma_1 = 180° - (134.9° + 32.7°) = 12.4°$. By the sine law, $c_1 = \dfrac{28.6}{\sin 32.7°} \sin 12.4° \approx 11.4$.

21. Draw angle $\gamma = 99.6°$. Note, there is exactly one triangle since $12.4 > 10.3$.

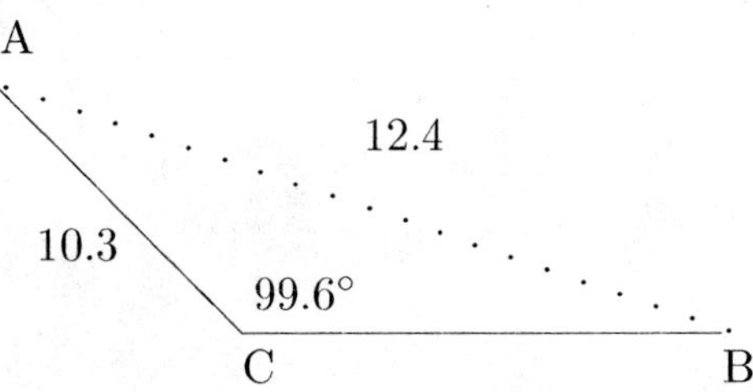

By the sine law, we obtain

$$\frac{12.4}{\sin 99.6°} = \frac{10.3}{\sin \beta}$$

$$\begin{aligned} \sin\beta &= \frac{10.3\sin 99.6^\circ}{12.4} \\ \sin\beta &\approx 0.819 \\ \beta = \sin^{-1}(0.819) &\approx 55.0^\circ. \end{aligned}$$

So $\alpha = 180^\circ - (55.0^\circ + 99.6^\circ) = 25.4^\circ$.

By the sine law, $a = \dfrac{12.4}{\sin 99.6^\circ}\sin 25.4^\circ \approx 5.4$.

23. Since two sides and an included angle are given, the area is
$A = \dfrac{1}{2}(12.9)(6.4)\sin 13.7^\circ \approx 9.8.$

25. Draw angle $\alpha = 39.4^\circ$.

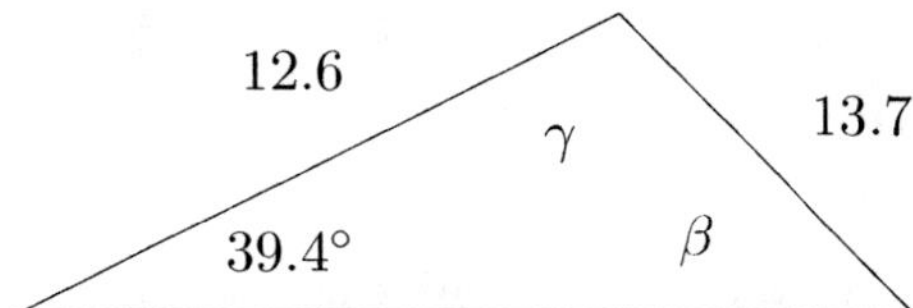

By the sine law, we obtain

$$\begin{aligned} \frac{12.6}{\sin\beta} &= \frac{13.7}{\sin 39.4^\circ} \\ \sin\beta &= \frac{12.6\sin 39.4^\circ}{13.7} \\ \sin\beta &\approx 0.5838 \\ \beta = \sin^{-1}(0.5838) &\approx 35.7^\circ. \end{aligned}$$

Then $\gamma = 180^\circ - (35.7^\circ + 39.4^\circ) = 104.9^\circ$.

The area is $A = \dfrac{1}{2}\cdot ab\sin\gamma =$
$\dfrac{1}{2}\cdot(13.7)(12.6)\sin 104.9^\circ \approx 83.4.$

27. Draw angle $\alpha = 42.3^\circ$.

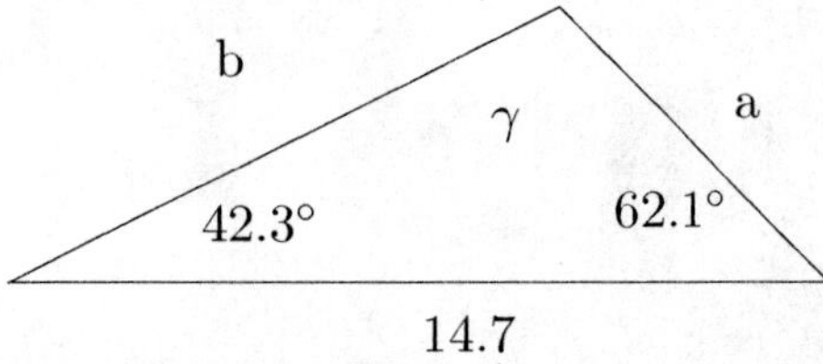

Note $\gamma = 180^\circ - (42.3^\circ + 62.1^\circ) = 75.6^\circ$.

By the sine law,

$$\begin{aligned} \frac{b}{\sin 62.1^\circ} &= \frac{14.7}{\sin 75.6^\circ} \\ b &= \frac{14.7}{\sin 75.6^\circ}\cdot\sin 62.1^\circ \\ b &\approx 13.41. \end{aligned}$$

The area is $A = \dfrac{1}{2}bc\sin\alpha =$
$\dfrac{1}{2}(13.41)(14.7)\sin 42.3^\circ \approx 66.3.$

29. Draw angle $\alpha = 56.3^\circ$.

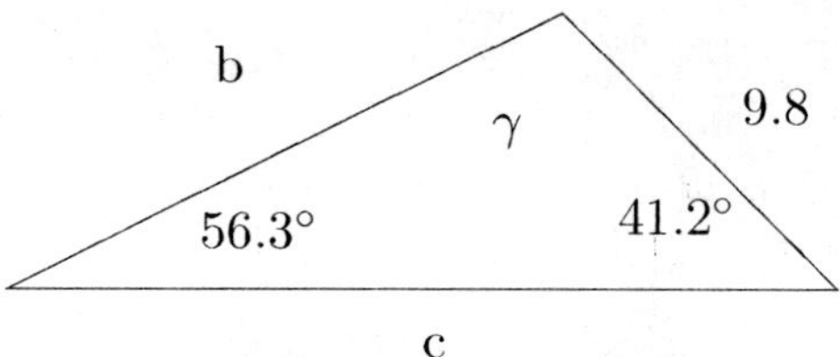

Note $\gamma = 180^\circ - (56.3^\circ + 41.2^\circ) = 82.5^\circ$.
By the sine law, we obtain

$$\begin{aligned} \frac{c}{\sin 82.5^\circ} &= \frac{9.8}{\sin 56.3^\circ} \\ c &= \frac{9.8}{\sin 56.3^\circ}\sin 82.5^\circ \\ c &\approx 11.679. \end{aligned}$$

The area is $A = \dfrac{1}{2}ac\sin\beta =$
$\dfrac{1}{2}(9.8)(11.679)\sin 41.2^\circ \approx 37.7.$

31. Divide the given 4-sided polygon into two triangles by drawing the diagonal that connects the 60° angle to the 135° angle. On each triangle two sides and an included angle are given. The area of the polygon is equal to the sum of the areas of the two triangles. Namely,

$$\frac{1}{2}(4)(10)\sin 120^\circ + \frac{1}{2}(12+2\sqrt{3})(2\sqrt{6})\sin 45^\circ =$$
$$20(\sqrt{3}/2) + \frac{1}{2}(24\sqrt{6} + 4\sqrt{18})(\sqrt{2}/2) =$$
$$10\sqrt{3} + \frac{1}{2}(12\sqrt{12} + 2\sqrt{36}) =$$
$$10\sqrt{3} + 6\sqrt{12} + \sqrt{36} = 10\sqrt{3} + 12\sqrt{3} + 6 =$$
$$22\sqrt{3} + 6.$$

33. Let x be the number of miles flown along I-20.

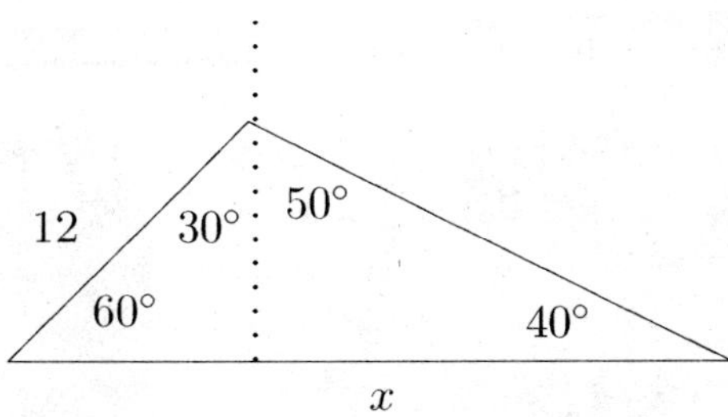

By the sine law, $\dfrac{x}{\sin 80°} = \dfrac{12}{\sin 40°}$.
Then $x \approx 18.4$ miles.

35. Let x be the length of the third side.

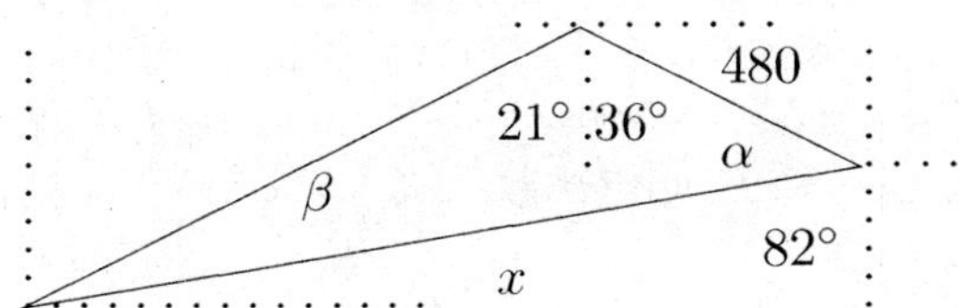

There is a 21° angle because of the $S21°W$ direction. There are 36° and 82° angles because opposite angles are equal and because of the directions $N36°W$ and $N82°E$.
Note $\alpha = 180° - (82° + 36°) = 62°$ and $\beta = 180 - (21° + 36° + 62°) = 61°$. By the sine law, $x = \dfrac{480}{\sin 61°} \sin 57° \approx 460.27$. The area is $\dfrac{1}{2}(460.27)(480)\sin 62° \approx 97,535$ sq ft.

37. Let h be the height of the tower.

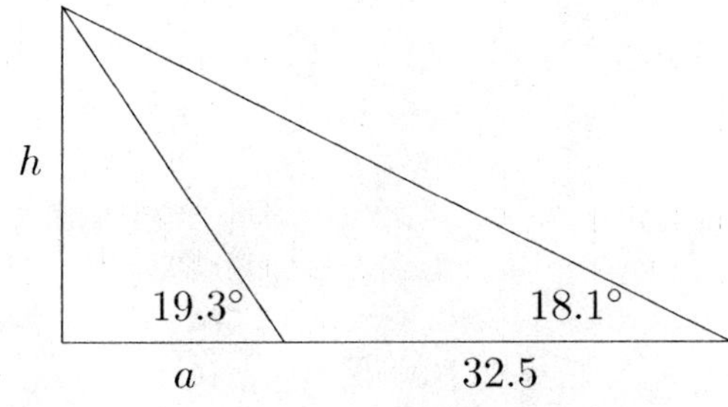

By using right triangle trigonometry, we get $\tan 19.3° = \dfrac{h}{a}$ or $a = \dfrac{h}{\tan 19.3°}$. Similarly, we have $\tan 18.1° = \dfrac{h}{a + 32.5}$. Then

$$\begin{aligned} \tan 18.1°(a + 32.5) &= h \\ a\tan 18.1° + 32.5\tan 18.1° &= h \\ \frac{h}{\tan 19.3°}\cdot\tan 18.1° + 32.5\tan 18.1° &= h \\ h\cdot\frac{\tan 18.1°}{\tan 19.3°} + 32.5\tan 18.1° &= h. \end{aligned}$$

Solving for h, we find that the height of the tower is $h \approx 159.4$ ft.

39. Note, $\tan\gamma = 6/12$ and $\gamma = \tan^{-1}(0.5) \approx 26.565°$. Also, $\tan\alpha = 3/12$ and $\alpha = \tan^{-1}(0.25) \approx 14.036°$.

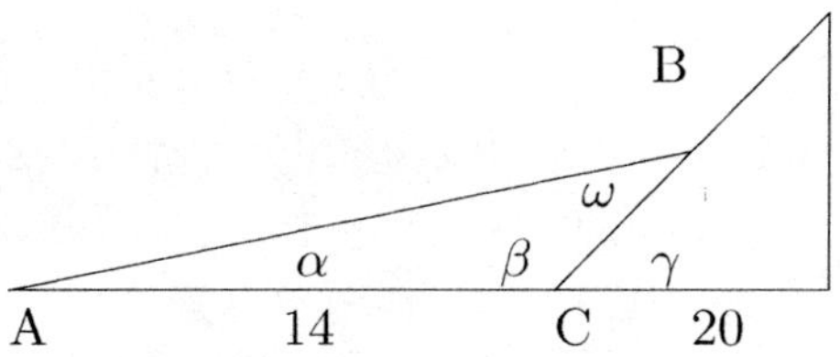

The remaining angles are $\beta = 153.435°$ and $\omega = 12.529°$.

By the sine law, $\dfrac{AB}{\sin 153.435°} = \dfrac{14}{\sin 12.529°}$

and $\dfrac{BC}{\sin 14.036°} = \dfrac{14}{\sin 12.529°}$.

Then $AB \approx 28.9$ ft and $BC \approx 15.7$ ft.

41. The kite consists of two equal triangles. The area of the kite is twice the area of the triangle.
It is $2\frac{1}{2}(24)(18)\sin 40° \approx 277.7$ in^2.

43. a) Consider the triangle where A is the center of the earth, B is a point on the surface of the earth, and C is a point on the atmosphere.

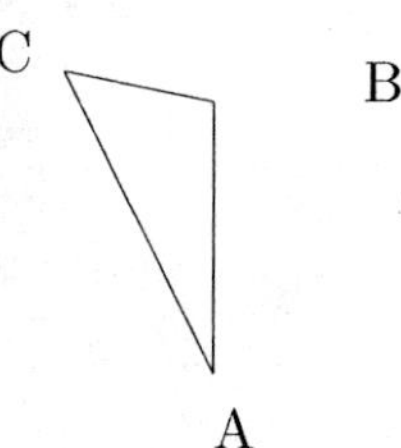

The angle at B is 120°. Let γ be the angle at C. Using the Sine Law, we obtain

$$\begin{aligned}\frac{3960}{\sin 120°} &= \frac{3950}{\sin\gamma}\\ \sin\gamma &= \frac{3950\sin 120°}{3960}\\ \gamma &\approx 59.75°.\end{aligned}$$

The angle at A is $\alpha \approx 0.25°$. Then d is given by

$$\begin{aligned}\frac{3960}{\sin 120°} &\approx \frac{d}{\sin 0.25°}\\ d &\approx \frac{3960\sin 0.25°}{\sin 120°}\\ d &\approx 19.9 \text{ miles.}\end{aligned}$$

b) In the picture given in part a), let θ be the angle at A. Using sine law, one finds

$$\begin{aligned}\frac{\sin\theta}{93{,}000{,}000} &= \frac{\sin(120°)}{93{,}003{,}950}\\ \theta &= \sin^{-1}\left(\frac{93{,}000{,}000\sin(120°)}{93{,}003{,}950}\right)\\ \theta &\approx 59.99579°.\end{aligned}$$

Suppose the sun is overhead at noon and the earth rotates 15° every hour. Since $\frac{59.99579}{15} \approx 3.999719$ (the number of hours since noon), when the angle of elevation is 30° the time is 1 second before 4:00 p.m.

c) In the triangle in part a), at sunset the angle at B is 90°. If d_s is the distance through the atmosphere at sunset, then $d_s^2 + 3950^2 = 3960^2$ or $d = \sqrt{3960^2 - 3950^2} \approx 281$ miles.

45. Let t be the number of seconds since the cruise missile was spotted.

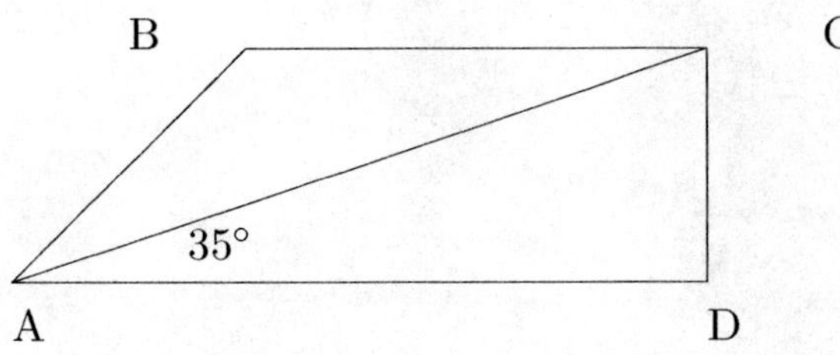

Let β be the angle at B. The angle formed by BAC is $180° - 35° - \beta$. After t seconds, the cruise missile would have traveled $548\frac{t}{3600}$ miles and the projectile $688\frac{t}{3600}$ miles. Using the law of sines, we have

$$\begin{aligned}\frac{\frac{548t}{3600}}{\sin(145° - \beta)} &= \frac{\frac{688t}{3600}}{\sin 35°}\\ \frac{548}{\sin(145° - \beta)} &= \frac{688}{\sin 35°}\\ \beta &= 145° - \sin^{-1}\left(\frac{548\sin 35°}{688}\right)\\ \beta &\approx 117.8°.\end{aligned}$$

Then angle BAC is 27.2°. The angle of elevation of the projectile must be the angle DAB which is 62.2° (= 35° + 27.2°).

47. Let t be the number of seconds it takes the fox to catch the rabbit. The distances travelled by the fox and rabbit are indicated below.

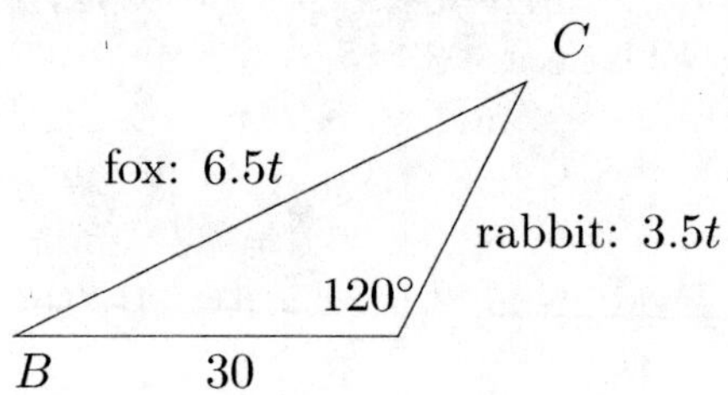

Apply the sine law as follows:

$$\begin{aligned}\frac{6.5t}{\sin 120°} &= \frac{3.5t}{\sin B}\\ \sin B &= \frac{3.5\sqrt{3}}{13}\end{aligned}$$

Note, $C = 60° - \arcsin\left(\frac{3.5\sqrt{3}}{13}\right)$. Then

$$\begin{aligned}\frac{30}{\sin C} &= \frac{3.5t}{\sin B}\\ t &= \frac{30\sin B}{3.5\sin C} = 7.5\end{aligned}$$

It will take 7.5 sec to catch the rabbit.

51. The shortest side is 7 cm. The angle opposite the shortest side is $90° - 64° = 26°$. If h is the hypotenuse, then

$$h = \frac{7}{\sin 26°} \approx 16.0 \text{ cm}.$$

If x is the longer leg, then

$$x = h \sin 64° \approx 14.4 \text{ cm}.$$

53. Rewrite as $y = 5\sin\left(4\left(x - \frac{\pi}{4}\right)\right) - 3$.

The amplitude is 5, period is $\frac{2\pi}{B} = \frac{2\pi}{4} = \frac{\pi}{2}$, phase shift is $\frac{\pi}{4}$, and the range is $[-5-3, 5-3]$ or $[-8, 2]$.

55. $3 - 3\sin^2 x = 3(1 - \sin^2 x) = 3\cos^2 x$

57. Triangle $\triangle ABC$ is an isosceles triangle. Also, $\triangle ACD$ and $\triangle ABD$ are isosceles triangles. Then $AC = AD = 2$. We apply the Angle Bisector Theorem in Exercise 50.

$$\frac{CD+2}{2} = \frac{2}{CD}.$$

Solving for CD, we find $CD = \sqrt{5} - 1$.

For Thought

1. True

2. False, $a = \sqrt{c^2 + b^2 - 2bc\cos\alpha}$.

3. False, $c^2 = a^2 + b^2 - 2ab\cos\gamma$.

4. True

5. False, it has only one solution.

6. True

7. True

8. True

9. True, since $\cos\gamma = \frac{3.4^2 + 4.2^2 - 8.1^2}{2(3.4)(4.2)} \approx -1.27$ has no real solution γ.

10. False, an obtuse triangle exists.

7.2 Exercises

1. law of cosines

3. cosines

5. By the cosine law, we obtain

$c = \sqrt{3.1^2 + 2.9^2 - 2(3.1)(2.9)\cos 121.3°}$

$\approx 5.23 \approx 5.2$. By the sine law, we find

$$\begin{aligned} \frac{3.1}{\sin\alpha} &= \frac{5.23}{\sin 121.3°} \\ \sin\alpha &= \frac{3.1\sin 121.3°}{5.23} \\ \sin\alpha &\approx 0.50647 \\ \alpha \approx \sin^{-1}(0.50647) &\approx 30.4°. \end{aligned}$$

Then $\beta = 180° - (30.4° + 121.3°) = 28.3°$.

7. By the cosine law, we find

$$\cos\beta = \frac{6.1^2 + 5.2^2 - 10.3^2}{2(6.1)(5.2)} \approx -0.6595$$

and so $\beta \approx \cos^{-1}(-0.6595) \approx 131.3°$.

By the sine law,

$$\begin{aligned} \frac{6.1}{\sin\alpha} &= \frac{10.3}{\sin 131.3°} \\ \sin\alpha &= \frac{6.1\sin 131.3°}{10.3} \\ \sin\alpha &\approx 0.4449 \\ \alpha \approx \sin^{-1}(0.4449) &\approx 26.4°. \end{aligned}$$

So $\gamma = 180° - (26.4° + 131.3°) = 22.3°$.

9. By the cosine law,

$b = \sqrt{2.4^2 + 6.8^2 - 2(2.4)(6.8)\cos 10.5°}$

$\approx 4.46167 \approx 4.5$ and

$$\cos\alpha = \frac{2.4^2 + 4.46167^2 - 6.8^2}{2(2.4)(4.46167)} \approx -0.96066.$$

So $\alpha = \cos^{-1}(-0.96066) \approx 163.9°$ and $\gamma = 180° - (163.9° + 10.5°) = 5.6°$

11. By the cosine law,

$$\cos\alpha = \frac{12.2^2 + 8.1^2 - 18.5^2}{2(12.2)(8.1)} \approx -0.6466.$$

Then $\alpha = \cos^{-1}(-0.6466) \approx 130.3°$.
By the sine law,

$$\begin{aligned} \frac{12.2}{\sin\beta} &= \frac{18.5}{\sin 130.3°} \\ \sin\beta &= \frac{12.2\sin 130.3°}{18.5} \\ \sin\beta &\approx 0.5029 \\ \beta \approx \sin^{-1}(0.5029) &\approx 30.2° \end{aligned}$$

So $\gamma = 180° - (30.2° + 130.3°) = 19.5°$

13. By the cosine law, we obtain
$a = \sqrt{9.3^2 + 12.2^2 - 2(9.3)(12.2)\cos 30°}$
$\approx 6.23 \approx 6.2$ and

$$\cos\gamma = \frac{6.23^2 + 9.3^2 - 12.2^2}{2(6.23)(9.3)} \approx -0.203.$$

So $\gamma = \cos^{-1}(-0.203) \approx 101.7°$ and
$\beta = 180° - (101.7° + 30°) = 48.3°$.

15. By the cosine law,

$$\cos\beta = \frac{6.3^2 + 6.8^2 - 7.1^2}{2(6.3)(6.8)} \approx 0.4146.$$

So $\beta = \cos^{-1}(0.4146) \approx 65.5°$.
By the sine law, we have

$$\begin{aligned} \frac{6.8}{\sin\gamma} &= \frac{7.1}{\sin 65.5°} \\ \sin\gamma &= \frac{6.8\sin 65.5°}{7.1} \\ \sin\gamma &\approx 0.8715 \\ \gamma \approx \sin^{-1}(0.8715) &\approx 60.6°. \end{aligned}$$

So $\alpha = 180° - (60.6° + 65.5°) = 53.9°$.

17. Note, $\alpha = 180° - 25° - 35° = 120°$.
Then by the sine law, we obtain

$$\frac{7.2}{\sin 120°} = \frac{b}{\sin 25°} = \frac{c}{\sin 35°}$$

from which we have

$$b = \frac{7.2\sin 25°}{\sin 120°} \approx 3.5$$

and

$$c = \frac{7.2\sin 35°}{\sin 120°} \approx 4.8.$$

19. There is no such triangle. Note, $a + b = c$ and in a triangle the sum of the lengths of two sides is greater than the length of the third side.

21. One triangle exists. The angles are uniquely determined by the law of cosines.

23. There is no such triangle since the sum of the angles in a triangle is 180°.

25. Exactly one triangle exists. This is seen by constructing a 179°-angle with two sides that have lengths 1 and 10. The third side is constructed by joining the endpoints of the first two sides.

27. Consider the figure below.

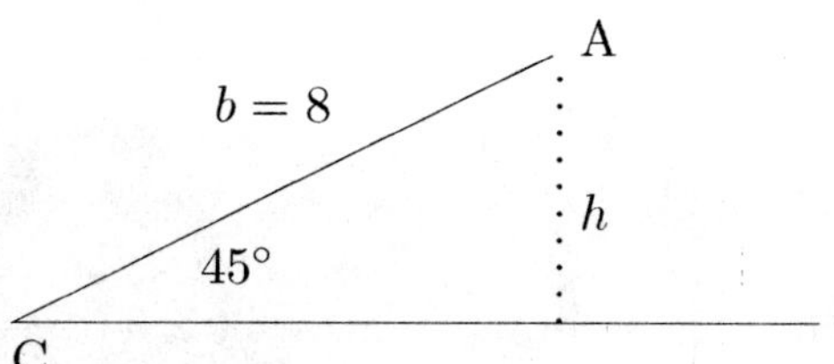

Note, $h = 8\sin 45° = 2\sqrt{2}$. Then the minimum value of c so that we will be able to make a triangle is $2\sqrt{2}$. Since $c = 2$, no such triangle is possible.

29. Note $S = \dfrac{16 + 9 + 10}{2} = 17.5$. The area is
$A = \sqrt{17.5(17.5 - 16)(17.5 - 9)(17.5 - 10)}$
$= \sqrt{17.5(1.5)(8.5)(7.5)} \approx 40.9$.

31. Note $S = \dfrac{3.6 + 9.8 + 8.1}{2} = 10.75$. The area is
$\sqrt{10.75(10.75 - 3.6)(10.75 - 9.8)(10.75 - 8.1)}$
$= \sqrt{10.75(7.15)(0.95)(2.65)} \approx 13.9$.

33. Note $S = \dfrac{346 + 234 + 422}{2} = 501$. The area is
$\sqrt{501(501 - 346)(501 - 234)(501 - 422)} =$
$\sqrt{501(155)(267)(79)} \approx 40,471.9$.

35. Since the base is 20 and the height is 10, the area is $\frac{1}{2}bh = \frac{1}{2}(20)(10) = 100$.

37. Since two sides and an included angle are given, the area is $\frac{1}{2}(6)(8)\sin 60° \approx 20.8$.

39. Note $S = \dfrac{9+5+12}{2} = 13$.
The area is $\sqrt{13(13-9)(13-5)(13-12)} = \sqrt{13(4)(8)(1)} \approx 20.4$

41. Recall, a central angle α in a circle of radius r intercepts a chord of length $r\sqrt{2-2\cos\alpha}$. Since $r = 30$ and $\alpha = 19°$, the length is $30\sqrt{2-2\cos 19°} \approx 9.90$ ft.

43. Note, a central angle α in a circle of radius r intercepts a chord of length $r\sqrt{2-2\cos\alpha}$. Since $921 = r\sqrt{2-2\cos 72°}$ (where $360 \div 5 = 72$), we get $r = \dfrac{921}{\sqrt{2-2\cos 72°}} \approx 783.45$ ft.

45. After 6 hours, Jan hiked a distance of 24 miles and Dean hiked 30 miles. Let x be the distance between them after 6 hrs.

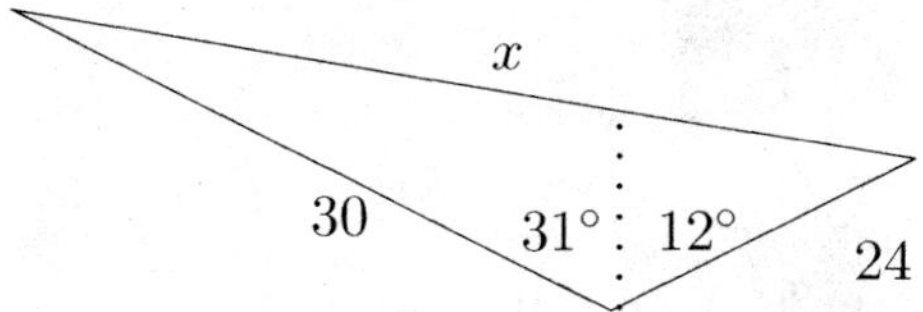

By the cosine law,
$x = \sqrt{30^2 + 24^2 - 2(30)(24)\cos 43°} = \sqrt{1476 - 1440\cos 43°} \approx 20.6$ miles.

47. By the cosine law, we find

$$\begin{aligned} \cos\alpha &= \frac{1.2^2 + 1.2^2 - 0.4^2}{2(1.2)(1.2)} \\ \cos\alpha &\approx 0.9444 \\ \alpha &\approx \cos^{-1}(0.9444) \\ \alpha &\approx 19.2°. \end{aligned}$$

49. Let α, β, and γ be the angles at gears A, B, and C. The length of the sides of the triangle are 5, 6, and 7. By the cosine law,

$$\begin{aligned} \cos\alpha &= \frac{5^2 + 6^2 - 7^2}{2(5)(6)} \\ \cos\alpha &= 0.2 \\ \alpha &= \cos^{-1}(0.2) \\ \alpha &\approx 78.5°. \end{aligned}$$

By the sine law,

$$\begin{aligned} \frac{6}{\sin\beta} &= \frac{7}{\sin 78.5°} \\ \sin\beta &\approx 0.8399 \\ \beta &\approx \sin^{-1}(0.8399) \\ \beta &\approx 57.1°. \end{aligned}$$

Then $\gamma = 180° - (57.1° + 78.5°) = 44.4°$.

51. By the cosine law,
$AB = \sqrt{5.3^2 + 7.6^2 - 2(5.3)(7.6)\cos 28°} = \sqrt{85.85 - 80.56\cos 28°} \approx 3.837 \approx 3.8$ miles.
Likewise,

$$\begin{aligned} \cos(\angle CBA) &= \frac{3.837^2 + 5.3^2 - 7.6^2}{2(3.837)(5.3)} \\ \cos(\angle CBA) &\approx -0.3675 \\ \angle CBA &\approx \cos^{-1}(-0.3675) \\ \angle CBA &\approx 111.6° \end{aligned}$$

and $\angle CAB = 180° - (111.6° + 28°) = 40.4°$.

53. The pentagon consists of 5 chords each of which intercepts a $\dfrac{360°}{5} = 72°$ angle.
By the cosine law, the length of a chord is given by

$$\sqrt{10^2 + 10^2 - 2(10)(10)\cos 72°} =$$

$\sqrt{200 - 200\cos 72°} \approx 11.76$ m.

55. Consider the figure below.

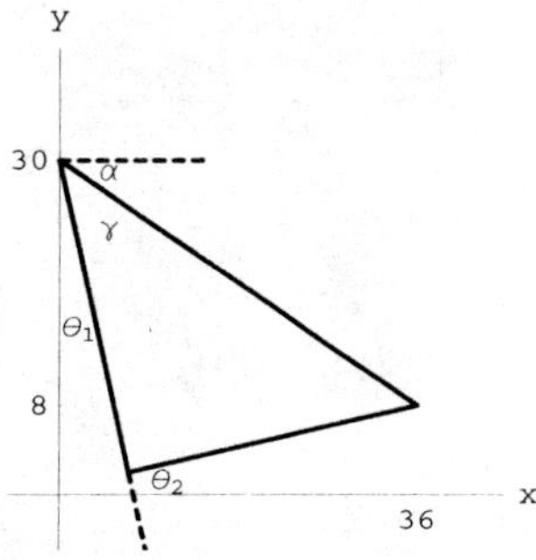

Note $\tan\alpha = 22/36$ and $\alpha = \tan^{-1}(22/36) \approx 31.4^\circ$. The distance between $(36, 8)$ and $(0, 30)$ is approximately 42.19. By the cosine law,

$$\begin{aligned} \cos\beta &= \frac{30^2 + 30^2 - 42.19^2}{2(30)(30)} \\ \cos\beta &\approx 0.011 \\ \beta \approx \cos^{-1}(0.011) &\approx 89.4^\circ. \end{aligned}$$

So $\theta_2 = 180^\circ - 89.4^\circ = 90.6^\circ$.

By the sine law, we find

$$\begin{aligned} \frac{30}{\sin\gamma} &= \frac{42.19}{\sin 89.4^\circ} \\ \sin\gamma &\approx 0.711 \\ \gamma \approx \sin^{-1}(0.711) &\approx 45.3^\circ. \end{aligned}$$

Then $\theta_1 = 90^\circ - (45.3^\circ + 31.4^\circ) = 13.3^\circ$.

57. a) Let α_m and α_M be the minimum and maximum values of α, respectively. By the law of cosines, we get

$$865,000^2 = 2(91,400,000)^2 - 2(91,400,000)^2 \cos\alpha_M.$$

Then

$$\begin{aligned} \alpha_M &= \cos^{-1}\left(\frac{2(91400000)^2 - 865000^2}{2(91400000)^2}\right) \\ \alpha_M &\approx 0.54^\circ. \end{aligned}$$

Likewise,

$$\begin{aligned} \alpha_m &= \cos^{-1}\left(\frac{2(94500000)^2 - 865000^2}{2(94500000)^2}\right) \\ \alpha_m &\approx 0.52^\circ. \end{aligned}$$

b) Let β_m and β_M be the minimum and maximum values of β, respectively. By the law of cosines, one obtains

$$2163^2 = 2(225,800)^2 - 2(225,800)^2 \cos\beta_M.$$

Then

$$\begin{aligned} \beta_M &= \cos^{-1}\left(\frac{2(225800)^2 - 2163^2}{2(225800)^2}\right) \\ \beta_M &\approx 0.55^\circ. \end{aligned}$$

Likewise,

$$\begin{aligned} \beta_m &= \cos^{-1}\left(\frac{2(252000)^2 - 2163^2}{2(252000)^2}\right) \\ \beta_m &\approx 0.49^\circ. \end{aligned}$$

c) Yes, even in perfect alignment a total eclipse may not occur, for instance when $\beta = 0.49^\circ$ and $\alpha = 0.52^\circ$.

59. a) The area of triangle is one-half the product of two sides and the sine of the included angle. If the sides have both length r, and the central angle is α, then the area of the triangle is

$$A_T = \frac{1}{2} r^2 \sin\alpha.$$

b) The area of a sector is proportional to the area of a circle. If the central angle is α and the radius is r, the area of the sector is

$$A_s = \frac{r^2\alpha}{2}.$$

c) The area A_L of a lens-shaped region is the difference of the area of a sector and the area of a triangle, see parts a) and b). Then

$$A_L = \frac{r^2\alpha}{2} - \frac{1}{2} r^2 \sin\alpha = \frac{r^2}{2}(\alpha - \sin\alpha).$$

61. Consider the lens-shaped region whose arc length is $s = 88.1$ ft. Since $s = r\alpha = 80\alpha$, the central angle is $\alpha = 88.1/80$.

The area A_L of the lens-shaped region is

$$A_L = \frac{80^2}{2}(\alpha - \sin\alpha) \approx 670.32$$

by Exercise 59c.

Join a line segment through the vertices of the lot that lie on the circle. This segment together with the other three sides of the lot form a trapezoid. The area A_T of the trapezoid is

$$A_T = \frac{80}{2}(102.5 + 127.1) = 9184.$$

Then the area of the property is

$$A_T - A_L \approx 9184 - 670.32 \approx 8513.68 \text{ ft}^2.$$

Multiplying by \$0.08, the property tax is

$$8513.68(0.08) \approx \$681.$$

63. Let d_b and d_h be the distance from the bear and hiker, respectively, to the base of the tower. Then $d_b = 150 \tan 80°$ and $d_h = 150 \tan 75°$.

Since the line segments joining the base of the tower to the bear and hiker form a 45° angle, by the cosine law the distance, d, between the bear and the hiker is

$$\begin{aligned} d &= \sqrt{d_b^2 + d_h^2 - 2(d_b)(d_h)\cos 45°} \\ &\approx \Big((850.69)^2 + (559.81)^2 - \\ &\qquad 2(850.69)(559.81)\cos 45°\Big)^{1/2} \\ &\approx 603 \text{ feet.} \end{aligned}$$

65. By using the cosine law, we obtain

$a = \sqrt{2r^2 - 2r^2\cos(\theta)} = \sqrt{4r^2\dfrac{1-\cos\theta}{2}} =$
$2r\sin(\theta/2)$.

67. If $\alpha = 90°$ in $a^2 = b^2 + c^2 - 2bc\cos\alpha$, then $a^2 = b^2 + c^2$ since $\cos 90° = 0°$. Thus the Pythagorean Theorem is a special case i.e., when the angle is 90°) of the law of cosines.

69. Note $S = \dfrac{31+87+56}{2} = 87$. By Heron's formula, the area of the triangle is suppose to be $\sqrt{87(87-31)(87-87)(87-56)}$. But this area is zero. Thus, no triangle exists with sides 31, 87, and 56.

71. Using the triangle below,

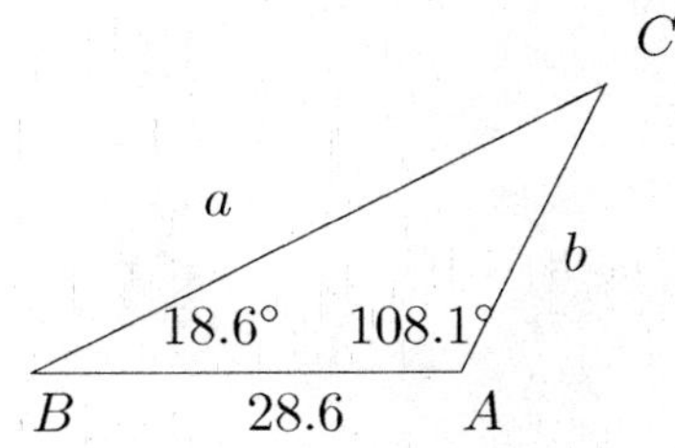

we find $C = 180° - A - B = 53.3°$. Using the sine law, we obtain

$$\begin{aligned} a &= \frac{28.6\sin 108.1°}{\sin 53.3°} \approx 33.9 \\ b &= \frac{28.6\sin 18.6°}{\sin 53.3°} \approx 11.4 \end{aligned}$$

73. Since $\sin x = \pm 1$, we find $x = \frac{\pi}{2}, \frac{3\pi}{2}$.

75. **a)** Domain $[-1,1]$, range $[-\pi/2, \pi/2]$

b) Domain $[-1,1]$, range $[0,\pi]$

c) Domain $(-\infty,\infty)$, range $(-\pi/2,\pi/2)$

77. Using Heron's formula, the area of the triangle is

$$A = \sqrt{15(6)4)(5)}.$$

Let α, β, and γ be the angles included by sides 9 & 10, 9 & 11, and 10 & 11, respectively. By the cosine law, we find

$$\alpha = \cos^{-1}\left(\frac{9^2+10^2-11^2}{2(9)(10)}\right)$$

$$\beta = \cos^{-1}\left(\frac{9^2+11^2-10^2}{2(9)(11)}\right)$$

$$\gamma = \cos^{-1}\left(\frac{10^2+11^2-9^2}{2(10)(11)}\right)$$

Draw a sector with central angle α and radius 4, and the area of this sector is

$$S_1 = \frac{1}{2}\left(4^2\alpha\right) \approx 9.847675339.$$

Similarly, let S_2 and S_3 be the areas of the sectors with central angle β and radius 5, and central angle γ and radius 6, respectively.

Thus, the area that is not sprayed by any of the three sprinklers is

$$A - (S_1 + S_2 + S_3) \approx 3.850 \text{ meters}^2.$$

For Thought

1. True

2. False, if $\boldsymbol{A} = \langle 1,0\rangle$ and $\boldsymbol{B} = \langle 0,1\rangle$ then $|\,\boldsymbol{A}+\boldsymbol{B}\,| = |\langle 1,1\rangle| = \sqrt{2}$ and $|\,\boldsymbol{A}\,| + |\,\boldsymbol{B}\,| = 2$.

3. True

4. True

5. False, rather the parallelogram law says that the magnitude of $\boldsymbol{A}+\boldsymbol{B}$ is the length of a diagonal of the parallelogram formed by $\boldsymbol{A}$ and $\boldsymbol{B}$.

6. False, the direction angle is formed with the positive x-axis.

7. True

8. True

9. True, the direction angle of a vector is unchanged when it is multipied by a positive scalar.

10. True

7.3 Exercises

1. vector

3. magnitude

5. parallelogram law

7. component

9. $\boldsymbol{A}+\boldsymbol{B} = 5\,\boldsymbol{j} + 4\,\boldsymbol{i} = 4\,\boldsymbol{i} + 5\,\boldsymbol{j}$ and $\boldsymbol{A}-\boldsymbol{B} = 5\,\boldsymbol{j} - 4\,\boldsymbol{i} = -4\,\boldsymbol{i} + 5\,\boldsymbol{j}$

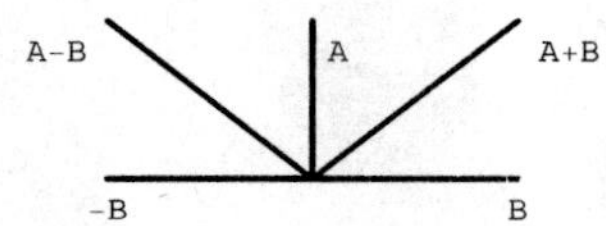

11. $\boldsymbol{A}+\boldsymbol{B} = (\,\boldsymbol{i} + 3\,\boldsymbol{j}\,) + (4\,\boldsymbol{i} + \boldsymbol{j}\,) = 5\,\boldsymbol{i} + 4\,\boldsymbol{j}$ and $\boldsymbol{A}-\boldsymbol{B} = (\,\boldsymbol{i} + 3\,\boldsymbol{j}\,) - (4\,\boldsymbol{i} + \boldsymbol{j}\,) = -3\,\boldsymbol{i} + 2\,\boldsymbol{j}$

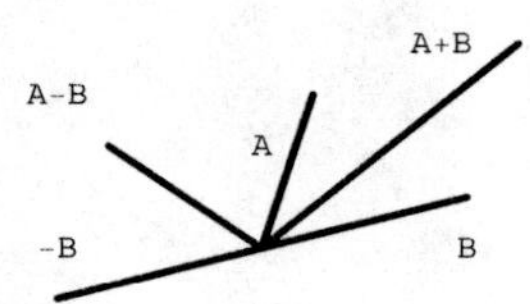

13. $\boldsymbol{A}+\boldsymbol{B} = (-\,\boldsymbol{i} + 4\,\boldsymbol{j}\,) + (4\,\boldsymbol{i}\,) = 3\,\boldsymbol{i} + 4\,\boldsymbol{j}$ and
$\boldsymbol{A}-\boldsymbol{B} = (-\,\boldsymbol{i} + 4\,\boldsymbol{j}\,) - (4\,\boldsymbol{i}\,) = -5\,\boldsymbol{i} + 4\,\boldsymbol{j}$

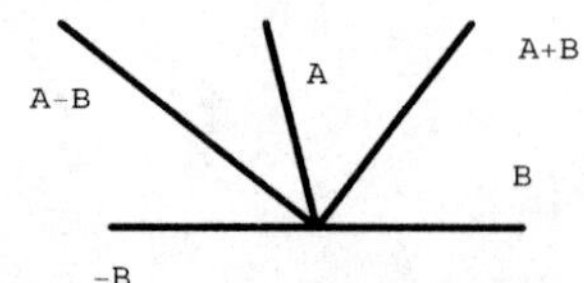

15. D **17.** E **19.** B

21. $|\,\boldsymbol{v_x}\,| = |4.5\cos 65.2°| = 1.9,$
$|\,\boldsymbol{v_y}\,| = |4.5\sin 65.2°| = 4.1$

23. $|\,\boldsymbol{v_x}\,| = |8000\cos 155.1°| \approx 7256.4,$
$|\,\boldsymbol{v_y}\,| = |8000\sin 155.1°| \approx 3368.3$

25. $|\,\boldsymbol{v_x}\,| = |234\cos 248°| \approx 87.7,$
$|\,\boldsymbol{v_y}\,| = |234\sin 248°| \approx 217.0$

27. The magnitude is $\sqrt{\sqrt{3}^2 + 1^2} = 2$.
Since $\tan\alpha = 1/\sqrt{3}$, the direction angle is $\alpha = 30°$.

29. The magnitude is $\sqrt{(-\sqrt{2})^2 + \sqrt{2}^2} = 2$.
Since $\tan\alpha = -\sqrt{2}/\sqrt{2} = -1$, the direction angle is $\alpha = 135°$.

31. The magnitude is $\sqrt{8^2 + (-8\sqrt{3})^2} = 16$.
Since $\tan\alpha = -8\sqrt{3}/8 = -\sqrt{3}$, the direction angle is $\alpha = 300°$.

33. The magnitude is $\sqrt{5^2 + 0^2} = 5$.
Since the terminal point is on the positive x-axis, the direction angle is $0°$.

35. The magnitude is $\sqrt{(-3)^2 + 2^2} = \sqrt{13}$.
Since $\tan^{-1}(-2/3) \approx -33.7°$, the direction angle is $180° - 33.7° = 146.3°$.

37. The magnitude is $\sqrt{3^2 + (-1)^2} = \sqrt{10}$.
Since $\tan^{-1}(-1/3) \approx -18.4°$, the direction angle is $360° - 18.4° = 341.6°$.

39. $\langle 8\cos 45°, 8\sin 45°\rangle = \langle 8(\sqrt{2}/2), 8(\sqrt{2}/2)\rangle$
$= \langle 4\sqrt{2}, 4\sqrt{2}\rangle$

41. $\langle 290\cos 145°, 290\sin 145°\rangle \approx \langle -237.6, 166.3\rangle$

43. $\langle 18\cos 347°, 18\sin 347°\rangle \approx \langle 17.5, -4.0\rangle$

45. $\langle 15, -10\rangle$

47. $\langle 6, -4\rangle + \langle 12, -18\rangle = \langle 18, -22\rangle$

49. $\langle -1, 5\rangle + \langle 12, -18\rangle = \langle 11, -13\rangle$

51. $\langle 3, -2\rangle - \langle 3, -1\rangle = \langle 0, -1\rangle$

53. $(3)(-1) + (-2)(5) = -13$

55. If $\boldsymbol{A} = \langle 2, 1\rangle$ and $\boldsymbol{B} = \langle 3, 5\rangle$, then the angle between these vectors is given by

$$\cos^{-1}\left(\frac{\boldsymbol{A}\cdot\boldsymbol{B}}{|\boldsymbol{A}|\cdot|\boldsymbol{B}|}\right) = \cos^{-1}\left(\frac{11}{\sqrt{5}\sqrt{34}}\right) \approx 32.5°$$

57. If $\boldsymbol{A} = \langle -1, 5\rangle$ and $\boldsymbol{B} = \langle 2, 7\rangle$, then the angle between these vectors is given by

$$\cos^{-1}\left(\frac{\boldsymbol{A}\cdot\boldsymbol{B}}{|\boldsymbol{A}|\cdot|\boldsymbol{B}|}\right) = \cos^{-1}\left(\frac{33}{\sqrt{26}\sqrt{53}}\right) \approx 27.3°$$

59. Since $\langle -6, 5\rangle \cdot \langle 5, 6\rangle = 0$, angle between them is $90°$.

61. Perpendicular since their dot product is zero

63. Parallel since $-2\langle 1, 7\rangle = \langle -2, -14\rangle$

65. Neither

67. $2\boldsymbol{i} + \boldsymbol{j}$ **69.** $-3\boldsymbol{i} + \sqrt{2}\boldsymbol{j}$ **71.** $-9\,\boldsymbol{j}$

73. $-7\,\boldsymbol{i} - \boldsymbol{j}$

75. The magnitude of $\boldsymbol{A} + \boldsymbol{B} = \langle 1, 4\rangle$ is $\sqrt{1^2 + 4^2} = \sqrt{17}$ and the direction angle is $\tan^{-1}(4/1) \approx 76.0°$

77. The magnitude of $-3\boldsymbol{A} = \langle -9, -3\rangle$ is $\sqrt{(-9)^2 + (-3)^2} = \sqrt{90} = 3\sqrt{10}$. Since $\tan^{-1}(3/9) \approx 18.4°$, the direction angle is $180° + 18.4° = 198.4°$

79. The magnitude of $\boldsymbol{B} - \boldsymbol{A} = \langle -5, 2\rangle$ is $\sqrt{(-5)^2 + 2^2} = \sqrt{29}$. Since $\tan^{-1}(-2/5) \approx -21.8°$, the direction angle is $180° - 21.8° = 158.2°$

81. Note $-\boldsymbol{A} + \dfrac{1}{2}\boldsymbol{B} = \langle -3 - 1, -1 + 3/2\rangle$ $= \langle -4, 1/2\rangle$. The magnitude is $\sqrt{(-4)^2 + (1/2)^2} = \sqrt{65}/2$.

Since $\tan^{-1}\left(\dfrac{1/2}{-4}\right) \approx -7.1°$, the direction angle is $180° - 7.1° = 172.9°$

83. The resultant is $\langle 2 + 6, 3 + 2\rangle = \langle 8, 5\rangle$. So the magnitude is $\sqrt{8^2 + 5^2} = \sqrt{89}$ and direction angle is $\tan^{-1}(5/8) = 32.0°$.

85. The resultant is $\langle -6 + 4, 4 + 2\rangle = \langle -2, 6\rangle$ and its magnitude is

$$\sqrt{(-2)^2 + 6^2} = 2\sqrt{10}.$$

Since $\tan^{-1}(-6/2) \approx -71.6°$, the direction angle is $180° - 71.6° = 108.4°$.

87. The resultant is $\langle -4 + 3, 4 - 6\rangle = \langle -1, -2\rangle$ and its magnitude is

$$\sqrt{(-1)^2 + (-2)^2} = \sqrt{5}.$$

Since $\tan^{-1}(2/1) \approx 63.4°$, the direction angle is $180° + 63.4° = 243.4°$.

89. Draw two perpendicular vectors whose magnitudes are 3 and 8.

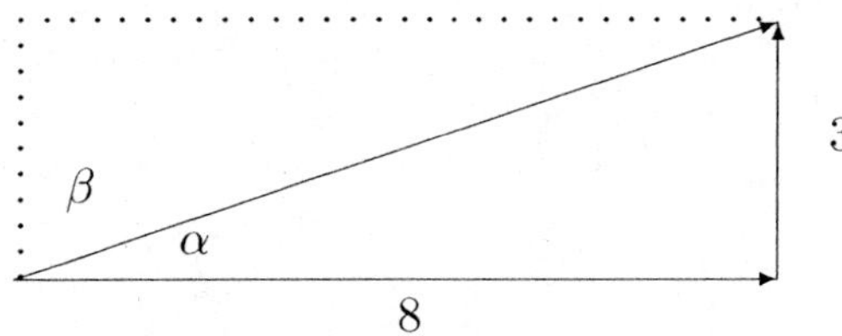

The magnitude of the resultant force is $\sqrt{8^2 + 3^2} = \sqrt{73}$ pounds by the Pythagorean Theorem. The angles between the resultant and each force are $\tan^{-1}(3/8) \approx 20.6°$ and $\beta = 90° - 20.6° = 69.4°$.

91. Draw two vectors with magnitudes 10.3 and 4.2 that act at an angle of $130°$ with each other.

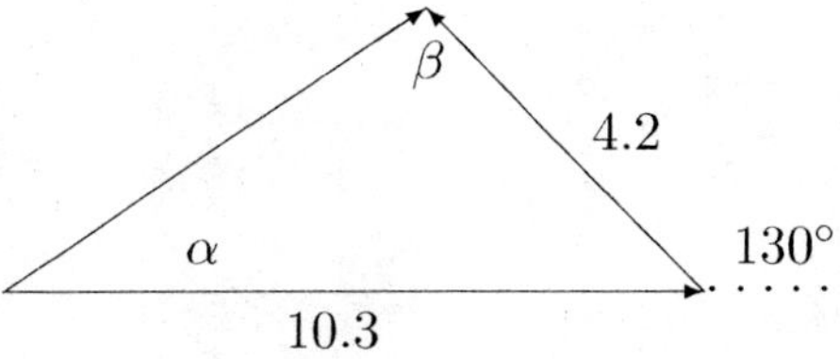

By using the cosine law, the magnitude

of the resultant force is

$\sqrt{10.3^2 + 4.2^2 - 2(10.3)(4.2)\cos 50°} \approx$ $8.253 \approx 8.25$ newtons.

By the sine law,

$$\begin{aligned} \frac{4.2}{\sin\alpha} &= \frac{8.253}{\sin 50°} \\ \sin\alpha &= \frac{4.2\sin 50°}{8.253} \\ \sin\alpha &\approx 0.3898 \\ \alpha \approx \sin^{-1}(0.3898) &\approx 22.9°. \end{aligned}$$

The angles between the resultant and each force are 22.9° and

$$\beta = 180° - 22.9° - 50° = 107.1°.$$

93. Draw two vectors with magnitudes 10 & 12.3 and whose angle between them is 23.4°.

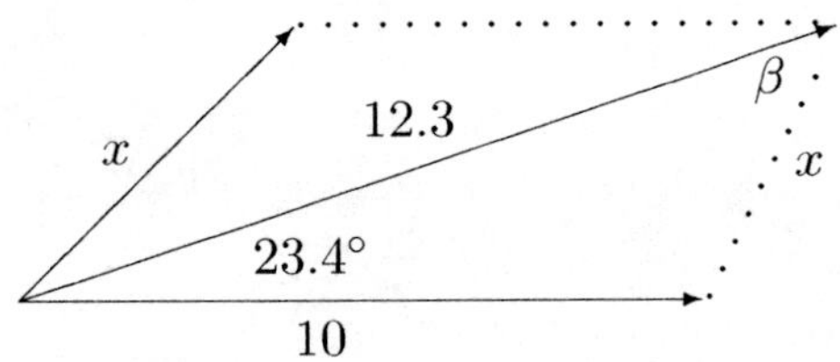

By the cosine law, the magnitude of the other force is $x = \sqrt{10^2 + 12.3^2 - 2(10)(12.3)\cos 23.4°}$ $\approx 5.051 \approx 5.1$ pounds. By the sine law,

$$\begin{aligned} \frac{10}{\sin\beta} &= \frac{5.051}{\sin 23.4°} \\ \sin\beta &= \frac{10\sin 23.4°}{5.051} \\ \sin\beta &\approx 0.7863 \\ \beta \approx \sin^{-1}(0.7863) &\approx 51.8°. \end{aligned}$$

The angle between the two forces is $51.8° + 23.4° = 75.2°$.

95. Since the angles in a parallelogram must add up to 360°, the angle formed by the

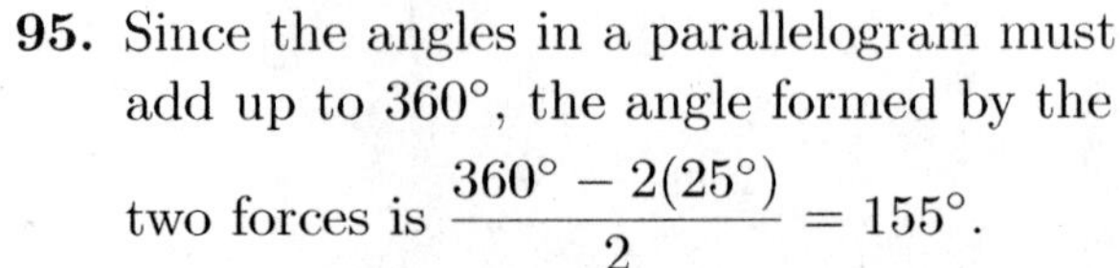

two forces is $\dfrac{360° - 2(25°)}{2} = 155°$.

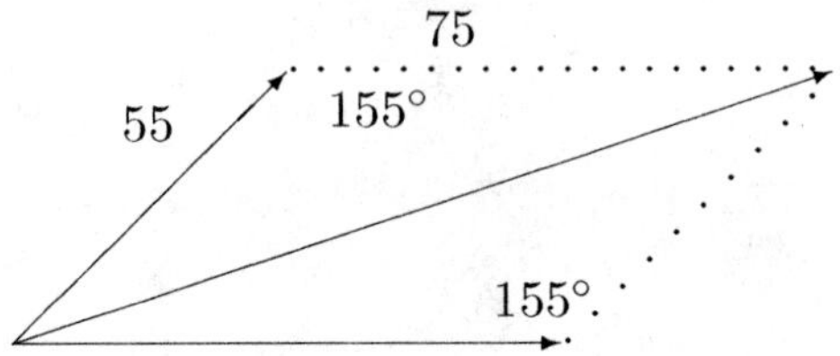

By the cosine law, the magnitude of the resultant force is

$\sqrt{55^2 + 75^2 - 2(55)(75)\cos 155°} \approx 127.0$ pounds.

So the donkey must pull a force of 127 pounds in the direction opposite that of the resultant's.

97. The magnitudes of the horizontal and vertical components are $|520\cos 30°| \approx 450.3$ mph and $|520\sin 30°| = 260$ mph, respectively.

99. Draw a vector pointing vertically down with magnitude 4000.

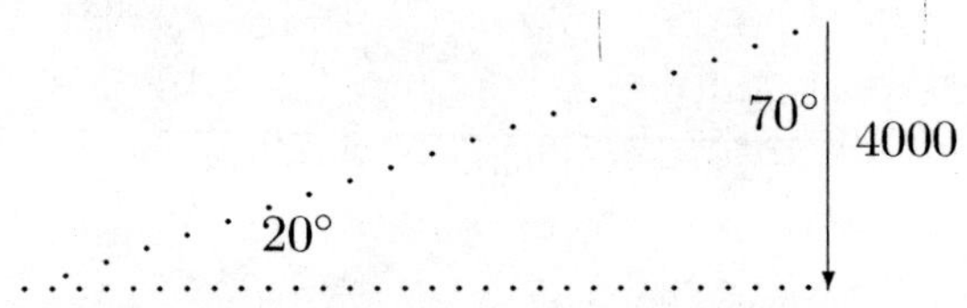

The magnitude of the component along the hill of the given vector is $4000\cos 70° \approx 1368.1$ pounds and this is the amount of force needed to prevent the rock from rolling.

101. In the diagram, $\alpha = 20°$ and x is the required force.

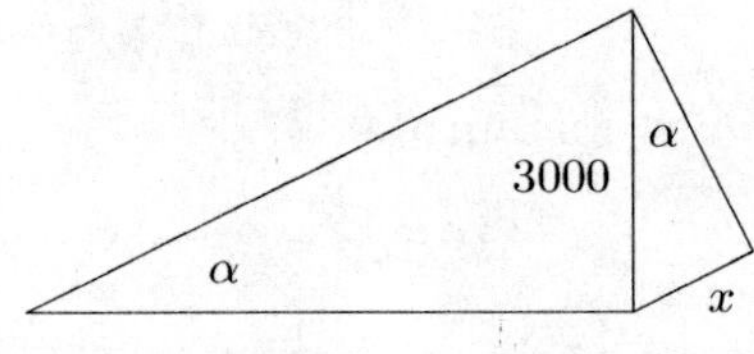

We find $x = 3000 \sin 20° \approx 1026.1$ lb.

103. Let β be the angle of inclination.

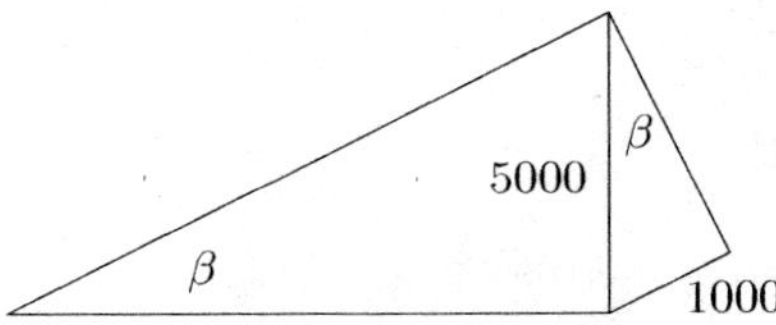

Since $\sin \beta = \dfrac{1000}{5000}$, we find

$$\beta = \sin^{-1} \frac{1000}{5000} \approx 11.5°.$$

105. Consider the figure below

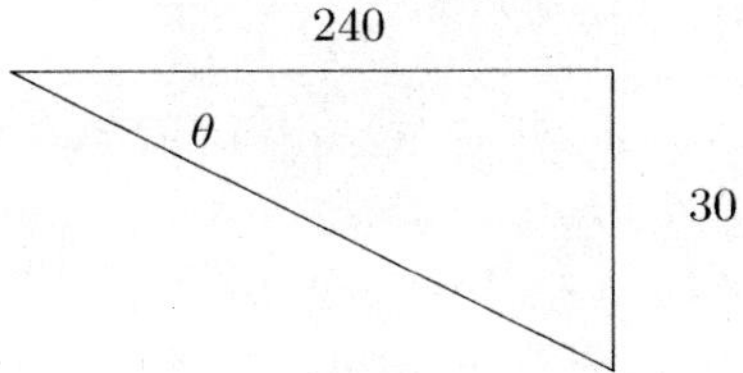

Since $\tan \theta = \dfrac{30}{240}$, we find

$$\theta = \tan^{-1} \frac{30}{240} \approx 7.1°.$$

Thus, the bearing of the plane's course is about 97.1° ($\approx 90° + \theta$).

Using the Pythagorean theorem, the ground speed is

$$\sqrt{240^2 + 30^2} \approx 241.9 \text{ mph}.$$

107. Let x be the ground speed in the figure below.

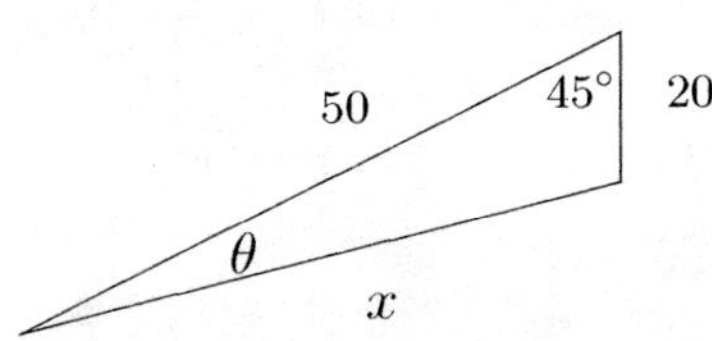

Using the cosine law, we find that the ground speed is

$$x = \sqrt{50^2 + 20^2 - 2(50)(20)\cos 45°} \approx 38.5 \text{ mph}.$$

Using the sine law, we obtain

$$\frac{x}{\sin 45°} = \frac{20}{\sin \theta}.$$

Solving for θ, we find

$$\theta = \sin^{-1}\left(\frac{20 \sin 45°}{x}\right) \approx 21.5°.$$

Thus, the bearing for the course is

$$45° + \theta \approx 66.5°.$$

109. Draw a vector $\boldsymbol{x}$ representing the airplane's course and ground speed. There is a 12° angle in the picture because of the plane's 102° heading.

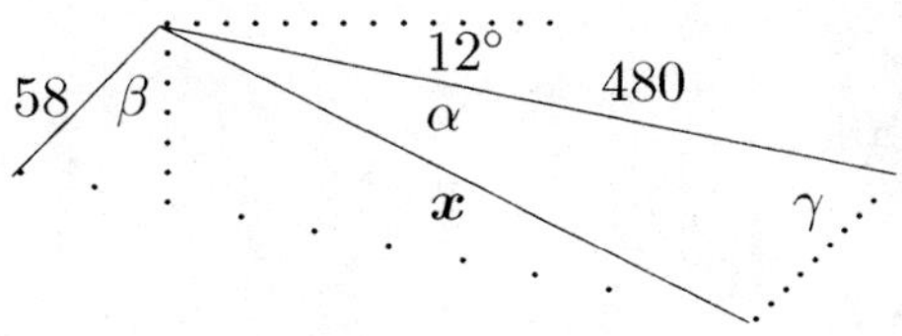

Note, $\beta = 45°$. Since the four angles in a parallelogram adds up to 360° and $45° + 78° = 123°$, we have

$$\gamma = \frac{360 - 2(123°)}{2} = 57°.$$

By the cosine law, the ground speed is $|\,\boldsymbol{x}\,| = \sqrt{480^2 + 58^2 - 2(480)(58)\cos 57°}$ ≈ 451.0 mph. By the sine law, we get

$$\begin{aligned}
\frac{58}{\sin \alpha} &= \frac{451}{\sin 57°} \\
\sin \alpha &= \frac{58 \sin 57°}{451} \\
\sin \alpha &\approx 0.107856 \\
\alpha &\approx \sin^{-1}(0.107856) \approx 6.2°.
\end{aligned}$$

The bearing of the airplane is

$$6.2° + 102° = 108.2°.$$

111. Draw two vectors representing the canoe and river current; the magnitudes of these vectors are 2 and 6, respectively.

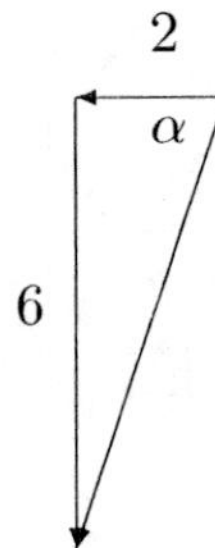

Since $\alpha = \tan^{-1}(6/2) \approx 71.6°$, the direction measured from the north is

$$270° - 71.6° = 198.4°.$$

Also, if d is the distance downstream from a point directly across the river to the point where she will land, then $\tan\alpha = d/2000$. Since $\tan\alpha = 6/2 = 3$, we get

$$d = 2000 \cdot 3 = 6000 \text{ ft}.$$

113. a) Assume we have a coordinate system where the origin is the point where the boat will start.

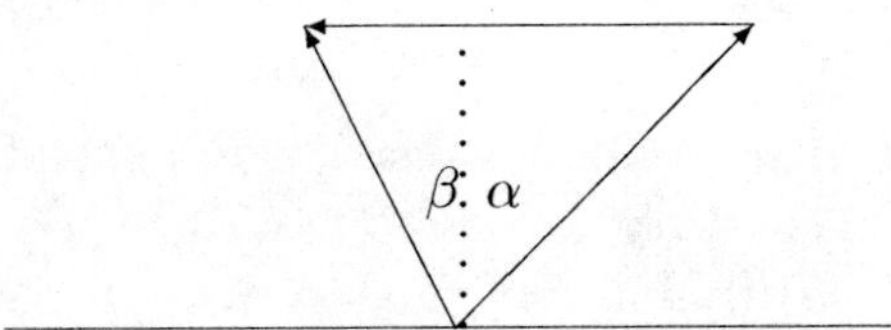

The intended direction and speed of the boat that goes 3 mph in still water is defined by the vector $3\sin\alpha\ \boldsymbol{i} + 3\cos\alpha\ \boldsymbol{j}$ and its actual direction and speed is determined by the vector

$$\boldsymbol{v} = (3\sin\alpha - 1)\ \boldsymbol{i} + 3\cos\alpha\ \boldsymbol{j}\ .$$

The number t of hours it takes the boat to cross the river is given by

$$t = \frac{0.2}{3\cos\alpha},$$

the solution to $3t\cos\alpha = 0.2$. Suppose $\beta > 0$ if $3\sin\alpha - 1 < 0$ and $\beta < 0$ if $3\sin\alpha - 1 > 0$. Using right triangle trigonometry, we find

$$\tan\beta = \frac{|3\sin\alpha - 1|}{3\cos\alpha}.$$

The distance d the boat travels as a function of β is given by

$$\begin{aligned} d &= |t\ \boldsymbol{v}\ | \\ &= \frac{0.2}{3\cos\alpha}|\ \boldsymbol{v}\ | \\ &= \frac{0.2}{3\cos\alpha}\sqrt{(3\sin\alpha - 1)^2 + (3\cos\alpha)^2} \\ &= 0.2\sqrt{\tan^2\beta + 1} \\ d &= 0.2|\sec\beta|. \end{aligned}$$

b) Since speed is distance divided by time, then by using the answer from part a) the speed s as a function of α and β is

$$\begin{aligned} s &= \frac{d}{t} \\ &= \frac{0.2|\sec\beta|}{0.2/(3\cos\alpha)} \\ &= 3\cos(\alpha)|\sec\beta|. \end{aligned}$$

c) As seen in the previous exercise, the number t of hours the trip will take as a function of α is

$$t = \frac{0.2}{3\cos\alpha} = \frac{1}{15}\sec\alpha.$$

The minimum value of t is attained when $\sec\alpha$ is the least, i.e., when $\alpha = 0°$.

117. Note, $\alpha < 90°$. By the sine law, we find

$$\begin{aligned} \sin\alpha &= \frac{19.4\sin 122.1°}{22.6} \\ \alpha &\approx 46.7° \end{aligned}$$

Then $\gamma = 180° - 122.1° - \alpha \approx 11.2°$.

By the sine law, we obtain

$$\begin{aligned} c &= \frac{22.6\sin\gamma}{\sin 122.1°} \\ c &\approx 5.2 \end{aligned}$$

119. Apply the cosine law to the given triangle.

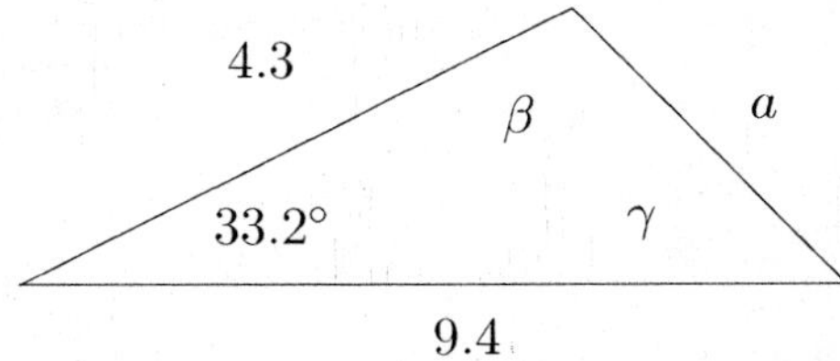

$$a^2 = 4.3^2 + 9.4^2 - 2(4.3)(9.4)\cos 33.2°$$
$$a \approx 6.3$$

Since $\gamma < \beta$, we find γ by the sine law.

$$\sin\gamma = \frac{4.3\sin 33.2°}{a}$$
$$\gamma \approx 22.1°$$

Then $\beta = 180° - 33.2° - \gamma \approx 124.7°$.

121. Since $\sin 3x = \pm\frac{\sqrt{3}}{2}$, we find

$$3x = \frac{\pi}{3}, \frac{2\pi}{3}, \frac{4\pi}{3}, \frac{5\pi}{3}, \frac{7\pi}{3}, \frac{8\pi}{3}, \frac{10\pi}{3}, \ldots$$

Since x lies in $(0, \pi)$, we obtain

$$x = \frac{\pi}{9}, \frac{2\pi}{9}, \frac{4\pi}{9}, \frac{5\pi}{9}, \frac{7\pi}{9}, \frac{8\pi}{9}$$

123. Let $\beta = 2\alpha$. By the cosine law, we find

$$a^2 = b^2 + c^2 - 2bc\cos\alpha$$

and

$$b^2 = a^2 + c^2 - 2ac\cos 2\alpha.$$

Since $\cos 2\alpha = 2\cos^2\alpha - 1$, we obtain

$$b^2 = a^2 + c^2 - 2ac\left[2\left(\frac{b^2 + c^2 - a^2}{2bc}\right)^2 - 1\right].$$

We may rewrite the above equation as

$$\frac{(a-b-c)(a+b-c)(a+c)(a^2-b^2+ac)}{b^2c} = 0.$$

Since the sum of any two sides of a triangle must be larger than the remaining side, we obtain $a^2 - b^2 + ac = 0$ or equivalently

$$c = \frac{b^2 - a^2}{a}.$$

a) Using the above equation, such a triangle with the smallest perimeter has sides

$$a = 4, b = 6, c = 5.$$

b) As seen in the above discussion, we have $\beta = 2\alpha$ and

$$c = \frac{b^2 - a^2}{a}.$$

c) The next two larger such triangles have sides

$$a = 9, b = 12, c = 7$$

and

$$a = 9, b = 15, c = 16$$

For Thought

1. True **2.** False, the absolute value is $\sqrt{(-2)^2 + (-5)^2} = \sqrt{29}$.

3. True **4.** False, $\tan\theta = 3/2$.

5. False, $i = 1(\cos 90° + i\sin 90°)$.

6. True **7.** True **8.** True

9. True, since $\frac{\pi}{4} + \frac{\pi}{2} = \frac{3\pi}{4}$.

10. False, since $\frac{3(\cos\pi/4 + i\sin\pi/4)}{3(\cos\pi/2 + i\sin\pi/2)} = 1.5(\cos(-\pi/4) + i\sin(-\pi/4)) = 1.5(\cos\pi/4 - i\sin\pi/4)$.

7.4 Exercises

1. complex

3. absolute value

5. modulus, argument

7. $\sqrt{0^2 + 8^2} = 8$

imaginary

8

real

8

9. $\sqrt{(-9)^2 + 0^2} = 9$

11. $\sqrt{2^2 + (-6)^2} = \sqrt{40} = 2\sqrt{10}$

13. $\sqrt{(-2)^2 + (2\sqrt{3})^2} = \sqrt{4 + 12} = 4$

15. $\sqrt{(1/\sqrt{2})^2 + (-1/\sqrt{2})^2} = \sqrt{1/2 + 1/2} = 1$

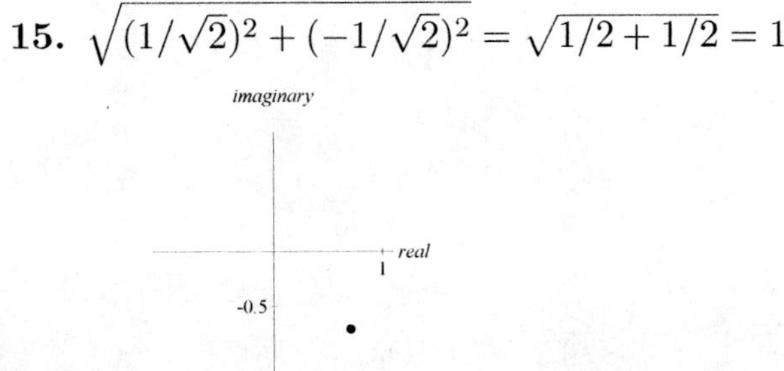

17. $\sqrt{3^2 + 3^2} = \sqrt{18} = 3\sqrt{2}$

19. Since terminal side of $0°$ goes through $(8, 0)$, the trigonomeric form is $8\,(\cos 0° + i \sin 0°)$

21. Since terminal side of $90°$ goes through $(0, \sqrt{3})$, the trigonomeric form is $\sqrt{3}\,(\cos 90° + i \sin 90°)$

23. Since $|-3 + 3i| = \sqrt{(-3)^2 + 3^2} = 3\sqrt{2}$ and if the terminal side of θ goes through $(-3, 3)$ then $\cos\theta = -3/(3\sqrt{2}) = -1/\sqrt{2}$. One can choose $\theta = 135°$ and the trigonometric form is $3\sqrt{2}\,(\cos 135° + i \sin 135°)$.

25. Since $\sqrt{(-3/\sqrt{2})^2 + (3/\sqrt{2})^2} = 3$ and if the terminal side of θ goes through $(-3/\sqrt{2}, 3/\sqrt{2})$ then $\cos\theta = (-3/\sqrt{2})/3 = -1/\sqrt{2}$. One can choose $\theta = 135°$. The trigonometric form is $3\,(\cos 135° + i \sin 135°)$.

27. Since $|-\sqrt{3} + i| = \sqrt{(-\sqrt{3})^2 + 1^2} = 2$ and if the terminal side of θ goes through $(-\sqrt{3}, 1)$ then $\cos\theta = -\sqrt{3}/2$. One can choose $\theta = 150°$. Trigonometric form is $2\,(\cos 150° + i \sin 150°)$.

29. Since $|3 + 4i| = \sqrt{3^2 + 4^2} = 5$ and if the terminal side of θ goes through $(3, 4)$ then $\cos\theta = 3/5$. One can choose $\theta = \cos^{-1}(3/5) \approx 53.1°$. The trigonometric form is $5\,(\cos 53.1° + i \sin 53.1°)$.

31. Since $|-3 + 5i| = \sqrt{(-3)^2 + 5^2} = \sqrt{34}$ and if the terminal side of θ goes through $(-3, 5)$ then $\cos\theta = -3/\sqrt{34}$. One can choose $\theta = \cos^{-1}(-3/\sqrt{34}) \approx 121.0°$. Trigonometric form is $\sqrt{34}\,(\cos 121.0° + i \sin 121.0°)$.

33. Note $|3 - 6i| = \sqrt{3^2 + (-6)^2} = \sqrt{45} = 3\sqrt{5}$. If the terminal side of θ goes through $(3, -6)$ then $\tan\theta = -6/3 = -2$. Since $\tan^{-1}(-2) \approx -63.4°$, one can choose $\theta = 360° - 63.4° = 296.6°$. The trigonometric form is $3\sqrt{5}\,(\cos 296.6° + i \sin 296.6°)$.

35. $\sqrt{2}\left(\frac{1}{\sqrt{2}} + i\frac{1}{\sqrt{2}}\right) = 1 + i$

37. $\frac{\sqrt{3}}{2}\left(-\frac{\sqrt{3}}{2} + \frac{1}{2}i\right) = -\frac{3}{4} + \frac{\sqrt{3}}{4}i$

39. $\frac{1}{2}(-0.848 - 0.53i) \approx -0.42 - 0.26i$

41. $3(0 + i) = 3i$

43. $\sqrt{3}(0 - i) = -i\sqrt{3}$

45. $\sqrt{6}\left(\frac{1}{2}+\frac{\sqrt{3}}{2}i\right)=$

$\frac{\sqrt{6}}{2}+\frac{\sqrt{18}}{2}i=\frac{\sqrt{6}}{2}+\frac{3\sqrt{2}}{2}i$

47. $6(\cos 450^\circ + i\sin 450^\circ)=$
$6(\cos 90^\circ + i\sin 90^\circ)=6(0+i)=6i$

49. $\sqrt{6}(\cos 30^\circ + i\sin 30^\circ)=$

$\sqrt{6}\left(\frac{\sqrt{3}}{2}+\frac{1}{2}i\right)=$

$\frac{\sqrt{18}}{2}+\frac{\sqrt{6}}{2}i=\frac{3\sqrt{2}}{2}+\frac{\sqrt{6}}{2}i$

51. $9(\cos 90^\circ + i\sin 90^\circ)=$
$9(0+i)=9i$

53. $2(\cos(\pi/6)+i\sin(\pi/6))=$

$2\left(\frac{\sqrt{3}}{2}+i\frac{1}{2}\right)=\sqrt{3}+i$

55. $0.5(\cos(-47.5^\circ)+i\sin(-47.5^\circ))\approx$
$0.5(0.6756-i\cdot 0.7373)\approx 0.34-0.37i$

57. Since $z_1=4\sqrt{2}(\cos 45^\circ + i\sin 45^\circ)$ and $z_2=5\sqrt{2}(\cos 225^\circ + i\sin 225^\circ)$, we have $z_1z_2=40(\cos 270^\circ + i\sin 270^\circ)=$ $40(0-i)=-40i$ and $\frac{z_1}{z_2}=0.8(\cos(-180^\circ)+i\sin(-180^\circ))=$ $0.8(-1+i\cdot 0)=-0.8$

59. Since $z_1=2(\cos 30^\circ + i\sin 30^\circ)$ and $z_2=4(\cos 60^\circ + i\sin 60^\circ)$, we have $z_1z_2=8(\cos 90^\circ + i\sin 90^\circ)=$ $8(0+i)=8i$ and $\frac{z_1}{z_2}=\frac{1}{2}(\cos(-30^\circ)+i\sin(-30^\circ))=$ $\frac{1}{2}\left(\frac{\sqrt{3}}{2}-\frac{1}{2}i\right)=\frac{\sqrt{3}}{4}-\frac{1}{4}i.$

61. Since $z_1=2\sqrt{2}(\cos 45^\circ + i\sin 45^\circ)$ and $z_2=2(\cos 315^\circ + i\sin 315^\circ)$, we have $z_1z_2=4\sqrt{2}(\cos 360^\circ + i\sin 360^\circ)=$ $4\sqrt{2}(1+i\cdot 0)=4\sqrt{2}$ and

$\frac{z_1}{z_2}=\sqrt{2}(\cos(-270^\circ)+i\sin(-270^\circ))=$

$\sqrt{2}(0+i)=\sqrt{2}i.$

63. Let α and β be angles whose terminal side goes through $(3,4)$ and $(-5,-2)$, respectively. Since $|3+4i|=5$ and $|-5-2i|=\sqrt{29}$, we have $\cos\alpha=3/5$, $\sin\alpha=4/5$, $\cos\beta=-5/\sqrt{29}$, and $\sin\beta=-2/\sqrt{29}$. From the sum and difference identities, we find

$$\cos(\alpha+\beta)=-\frac{7}{5\sqrt{29}},\ \sin(\alpha+\beta)=-\frac{26}{5\sqrt{29}},$$
$$\cos(\alpha-\beta)=-\frac{23}{5\sqrt{29}},\ \sin(\alpha-\beta)=-\frac{14}{5\sqrt{29}}.$$

Note $z_1=5(\cos\alpha+i\sin\alpha)$ and $z_2=\sqrt{29}(\cos\beta+i\sin\beta)$. Then

$$\begin{aligned} z_1z_2 &= 5\sqrt{29}(\cos(\alpha+\beta)+i\sin(\alpha+\beta)) \\ &= 5\sqrt{29}\left(-\frac{7}{5\sqrt{29}}-i\frac{26}{5\sqrt{29}}\right) \\ z_1z_2 &= -7-26i \end{aligned}$$

and

$$\begin{aligned} \frac{z_1}{z_2} &= \frac{5}{\sqrt{29}}(\cos(\alpha-\beta)+i\sin(\alpha-\beta)) \\ &= \frac{5}{\sqrt{29}}\left(-\frac{23}{5\sqrt{29}}-i\frac{14}{5\sqrt{29}}\right) \\ \frac{z_1}{z_2} &= -\frac{23}{29}-\frac{14}{29}i \end{aligned}$$

65. Let α and β be angles whose terminal sides goes through $(2,-6)$ and $(-3,-2)$, respectively. Since $|2-6i|=2\sqrt{10}$ and $|-3-2i|=\sqrt{13}$, we have $\cos\alpha=1/\sqrt{10}$, $\sin\alpha=-3/\sqrt{10}$, $\cos\beta=-3/\sqrt{13}$, and $\sin\beta=-2/\sqrt{13}$. From the sum and difference identities,

$$\cos(\alpha+\beta)=-\frac{9}{\sqrt{130}},\ \sin(\alpha+\beta)=\frac{7}{\sqrt{130}},$$
$$\cos(\alpha-\beta)=\frac{3}{\sqrt{130}},\ \sin(\alpha-\beta)=\frac{11}{\sqrt{130}}.$$

Note $z_1=2\sqrt{10}(\cos\alpha+i\sin\alpha)$ and $z_2=\sqrt{13}(\cos\beta+i\sin\beta)$. Thus,

$$\begin{aligned} z_1z_2 &= 2\sqrt{130}(\cos(\alpha+\beta)+i\sin(\alpha+\beta)) \\ &= 2\sqrt{130}\left(-\frac{9}{\sqrt{130}}+i\frac{7}{\sqrt{130}}\right) \\ z_1z_2 &= -18+14i \end{aligned}$$

and

$$\begin{aligned}\frac{z_1}{z_2} &= \frac{2\sqrt{10}}{\sqrt{13}}(\cos(\alpha-\beta)+i\sin(\alpha-\beta)) \\ &= \frac{2\sqrt{10}}{\sqrt{13}}\left(\frac{3}{\sqrt{130}}+i\frac{11}{\sqrt{130}}\right) \\ \frac{z_1}{z_2} &= \frac{6}{13}+\frac{22}{13}i\end{aligned}$$

67. Note $3i = 3(\cos 90° + i\sin 90°)$ and $1+i = \sqrt{2}(\cos 45° + i\sin 45°)$. Then we get

$$\begin{aligned}(3i)(1+i) &= 3\sqrt{2}(\cos 135° + i\sin 135°) \\ &= 3\sqrt{2}\left(-\frac{\sqrt{2}}{2}+i\cdot\frac{\sqrt{2}}{2}\right) \\ (3i)(1+i) &= -3+3i\end{aligned}$$

and

$$\begin{aligned}\frac{3i}{1+i} &= \frac{3}{\sqrt{2}}(\cos 45° + i\sin 45°) \\ &= \frac{3}{\sqrt{2}}\left(\frac{\sqrt{2}}{2}+\frac{\sqrt{2}}{2}i\right) \\ (3i)(1+i) &= \frac{3}{2}+\frac{3}{2}i.\end{aligned}$$

69. $3\left(\cos\left(-\frac{\pi}{4}\right)+i\sin\left(-\frac{\pi}{4}\right)\right)$

71. $2\sqrt{3}(\cos(20°)+i\sin(20°))$

73. $\left[3\left(\cos\frac{\pi}{6}+i\sin\frac{\pi}{6}\right)\right]\left[3\left(\cos\frac{\pi}{6}-i\sin\frac{\pi}{6}\right)\right]$
$= 9\left(\cos^2\frac{\pi}{6}+\sin^2\frac{\pi}{6}\right) = 9(1) = 9$

75. $[2(\cos 7° + i\sin 7°)][2(\cos 7° - i\sin 7°)]$
$= 4(\cos^2 7° + \sin^2 7°) = 4(1) = 4$

77. Since $4+4i = 4\sqrt{2}(\cos 45° + i\sin 45°)$,
$(4+4i)^2 = (4\sqrt{2})^2(\cos(2\cdot 45°)+i\sin(2\cdot 45°))$
$32(\cos 90° + i\sin 90°) =$
$32(0+i\cdot 1) = 32i.$

79. Since $3+3i = 3\sqrt{2}(\cos 45° + i\sin 45°)$,
$(3+3i)^3 = (3\sqrt{2})^3(\cos(3\cdot 45°)+i\sin(3\cdot 45°))$
$54\sqrt{2}(\cos 135° + i\sin 135°) =$
$54\sqrt{2}\left(-\frac{1}{\sqrt{2}}+i\frac{1}{\sqrt{2}}\right) = -54+54i.$

81. $\overline{z} = \overline{r(\cos\theta + i\sin\theta)} =$
$r(\cos\theta - i\sin\theta) = r(\cos(-\theta)+i\sin(-\theta))$

83. The reciprocal of z is $\frac{1}{z} = \frac{\cos 0 + i\sin 0}{r[\cos\theta + i\sin\theta]} =$
$r^{-1}[\cos(0-\theta)+i\sin(0-\theta)] =$
$r^{-1}[\cos\theta - i\sin\theta]$ provided $r \neq 0$.

85. The first sum is
$6(\cos 9° + i\sin 9°) + 3(\cos 5° + i\sin 5°) =$
$6\cos 9° + 3\cos 5° + i(6\sin 9° + 3\sin 5°)$.
Also, $(1+3i)+(5-7i) = 6-4i$. It is easier to add complex numbers in standard form.

87. $-2(3)+6(5) = 24$

89. The largest angle is γ and is opposite the longest side. By the cosine law,

$$\begin{aligned}10^2 &= 7^2+5^2-2(7)(5)\cos\gamma \\ 100 &= 74-70\cos\gamma \\ \gamma &= \cos^{-1}\left(-\frac{26}{70}\right) \\ \gamma &\approx 111.8°\end{aligned}$$

The remaining angles are less than 90°.

By the sine law,

$$\begin{aligned}\sin\beta &= \frac{7\sin\gamma}{10} \\ \beta &= \sin^{-1}\left(\frac{7\sin\gamma}{10}\right) \\ \beta &\approx 40.5°\end{aligned}$$

Finally, $\alpha = 180° - \gamma - \beta = 27.7°$.

91. If x is the height of the building, then

$$x = 230\tan 48° \approx 255.4 \text{ ft}$$

93. Note, $CT = CA = 5$ and $CS = CB = 6$. Then ΔABC is congruent to ΔTSC. We use Heron's formula, and let

$$s = \frac{5+6+7}{2} = 9.$$

The area of ΔTSC is

$$\text{Area} = \sqrt{9(9-5)(9-6)(9-7)} = 6\sqrt{6}.$$

For Thought

1. False, $(2+3i)^2 = 4+12i+9i^2$.

2. False, $z^3 = 8(\cos 360° + i\sin 360°) = 8$.

3. True **4.** False, the argument is 4θ.

5. False, since $\left[\frac{1}{2} + i\cdot\frac{1}{2}\right]^2 = \frac{i}{2}$ and

$\cos 2\pi/3 + i\sin \pi/3 = -\frac{1}{2} + i\cdot\frac{\sqrt{3}}{2}$.

6. False, since $\cos 5\pi/6 = \cos 7\pi/6$ and $5\pi/6 \neq 7\pi/6 + 2k\pi$ for any integer k. It is possible that $\alpha = 2k\pi - \beta$.

7. True **8.** True

9. True, $x = i$ is a solution not on $y = \pm x$.

10. False, it has four imaginary solutions.

7.5 Exercises

1. De Moivre's

3. $3^3(\cos 90° + i\sin 90°) = 27(0+i) = 27i$

5. $(\sqrt{2})^4(\cos 480° + i\sin 480°) =$
$4(\cos 120° + i\sin 120°) =$
$4\left(-\frac{1}{2} + i\cdot\frac{\sqrt{3}}{2}\right) = -2 + 2i\sqrt{3}$

7. $\cos(8\pi/12) + i\sin(8\pi/12) =$
$\cos(2\pi/3) + i\sin(2\pi/3) = -\frac{1}{2} + \frac{\sqrt{3}}{2}i$

9. $(\sqrt{6})^4\left[\cos(8\pi/3) + i\sin(8\pi/3)\right] =$
$36\left[\cos(2\pi/3) + i\sin(2\pi/3)\right] =$
$36\left[-\frac{1}{2} + i\frac{\sqrt{3}}{2}\right] = -18 + 18i\sqrt{3}$

11. $4.3^5\left[\cos 61.5° + i\sin 61.5°\right] \approx$
$1470.1\left[0.4772 + 0.8788i\right] \approx 701.5 + 1291.9i$

13. $\left(2\sqrt{2}\left[\cos 45° + i\sin 45°\right]\right)^3 =$
$16\sqrt{2}\left[\cos 135° + i\sin 135°\right] =$
$16\sqrt{2}\left[-\frac{1}{\sqrt{2}} + i\frac{1}{\sqrt{2}}\right] = -16 + 16i$

15. $(2\left[\cos(-30°) + i\sin(-30°)\right])^4 =$
$16\left[\cos(-120°) + i\sin(-120°)\right] =$
$16\left[-\frac{1}{2} - i\frac{\sqrt{3}}{2}\right] = -8 - 8i\sqrt{3}$

17. $(6\left[\cos 240° + i\sin 240°\right])^5 =$
$7776\left[\cos 1200° + i\sin 1200°\right] =$
$7776\left[\cos 120° + i\sin 120°\right] =$
$7776\left[-\frac{1}{2} + i\frac{\sqrt{3}}{2}\right] = -3888 + 3888i\sqrt{3}$

19. Note $|2+3i| = \sqrt{13}$. If the terminal side of α goes through $(2,3)$ then $\cos\alpha = 2/\sqrt{13}$ and $\sin\alpha = 3/\sqrt{13}$. By using the double-angle identities one can successively obtain
$\cos 2\alpha = -5/13$, $\sin 2\alpha = 12/13$,
$\cos 4\alpha = -119/169$, $\sin 4\alpha = -120/169$.

So $(2+3i)^4 = \left(\sqrt{13}\left[\cos\alpha + i\sin\alpha\right]\right)^4 =$
$169(\cos 4\alpha + i\sin 4\alpha) =$
$169\left(-119/169 - 120i/169\right) = -119 - 120i$

21. Note $|2-i| = \sqrt{5}$. If the terminal side of α goes through $(2,-1)$ then $\cos\alpha = 2/\sqrt{5}$ and $\sin\alpha = -1/\sqrt{5}$. By using the double-angle identities one can successively obtain
$\cos 2\alpha = 3/5$, $\sin 2\alpha = -4/5$,
$\cos 4\alpha = -7/25$, $\sin 4\alpha = -24/25$.

So $(2-4i)^4 = \left(\sqrt{5}\left[\cos\alpha + i\sin\alpha\right]\right)^4 =$
$25(\cos 4\alpha + i\sin 4\alpha) =$
$25\left(-7/25 - 24i/25\right) = -7 - 24i$.

23. Let $\omega = |1.2 + 3.6i|$. If the terminal side of α goes through $(1.2, 3.6)$ then $\cos\alpha = 1.2/\omega$ and $\sin\alpha = 3.6/\omega$. By using the double-angle identities one can obtain
$\cos 2\alpha = -11.52/\omega^2$ and $\sin 2\alpha = 8.64/\omega^2$.
By the sum identities one gets
$\cos 3\alpha = \cos(2\alpha + \alpha) = -44.928/\omega^3$ and
$\sin 3\alpha = \sin(2\alpha + \alpha) = -31.104/\omega^3$.

So $(1.2 + 3.6i)^3 = (\omega\left[\cos\alpha + i\sin\alpha\right])^3 =$
$\omega^3(\cos 3\alpha + i\sin 3\alpha) =$
$\omega^3\left(-44.928/\omega^3 - 31.104i/\omega^3\right) =$
$-44.928 - 31.104i$.

25. Square roots are given by

$$2\left[\cos\left(\frac{90° + k360°}{2}\right) + i\sin\left(\frac{90° + k360°}{2}\right)\right] =$$

$2\left[\cos(45° + k\cdot 180°) + i\sin(45° + k\cdot 180°)\right]$

When $k = 0, 1$ one gets
$2\left(\cos 45° + i\sin 45°\right)$ and
$2\left(\cos 225° + i\sin 225°\right)$.

27. Fourth roots are given by

$$\cos\left(\frac{120^\circ + k360^\circ}{4}\right) + i\sin\left(\frac{120^\circ + k360^\circ}{4}\right) =$$

$$\cos(30^\circ + k\cdot 90^\circ) + i\sin(30^\circ + k\cdot 90^\circ)$$

When $k = 0, 1, 2, 3$ one gets $\cos\alpha + i\sin\alpha$ where $\alpha = 30^\circ, 120^\circ, 210^\circ, 300^\circ$.

29. Sixth roots are given by

$$2\left[\cos\left(\frac{\pi + 2k\pi}{6}\right) + i\sin\left(\frac{\pi + 2k\pi}{6}\right)\right].$$

When $k = 0, 1, 2, 3, 4, 5$ one gets

$$2(\cos\alpha + i\sin\alpha)$$

where $\alpha = \frac{\pi}{6}, \frac{\pi}{2}, \frac{5\pi}{6}, \frac{7\pi}{6}, \frac{3\pi}{2}, \frac{11\pi}{6}$.

31. Cube roots of 1 are given by

$$\cos\left(\frac{k360^\circ}{3}\right) + i\sin\left(\frac{k360^\circ}{3}\right).$$

If $k = 0, 1, 2$ one obtains

$\cos 0^\circ + i\sin 0^\circ = 1$,

$\cos 120^\circ + i\sin 120^\circ = -\frac{1}{2} + i\frac{\sqrt{3}}{2}$, and

$\cos 240^\circ + i\sin 240^\circ = -\frac{1}{2} - i\frac{\sqrt{3}}{2}$.

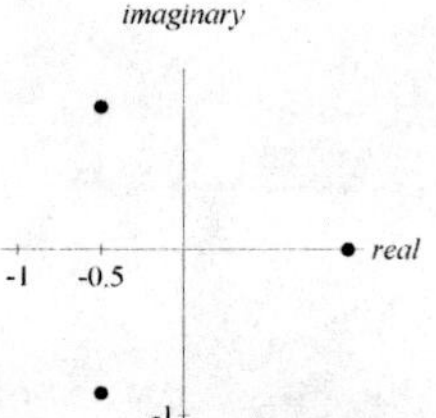

33. Fourth roots of 16 are given by

$$2\left[\cos\left(\frac{k360^\circ}{4}\right) + i\sin\left(\frac{k360^\circ}{4}\right)\right].$$

If $k = 0, 1, 2, 3$ one obtains

$2[\cos 0^\circ + i\sin 0^\circ] = 2$,

$2[\cos 90^\circ + i\sin 90^\circ] = 2i$,

$2[\cos 180^\circ + i\sin 180^\circ] = -2$, and

$2[\cos 270^\circ + i\sin 270^\circ] = -2i$.

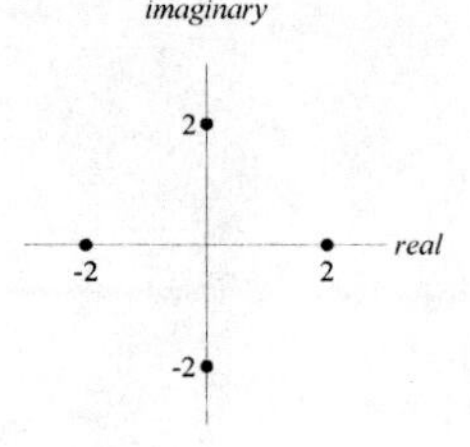

35. Fourth roots of -1 are given by

$$\cos\left(\frac{180^\circ + k360^\circ}{4}\right) + i\sin\left(\frac{180^\circ + k360^\circ}{4}\right)$$

$$= \cos(45^\circ + k90^\circ) + i\sin(45^\circ + k90^\circ).$$

If $k = 0, 1, 2, 3$ one obtains

$\cos 45^\circ + i\sin 45^\circ = \frac{\sqrt{2}}{2} + i\frac{\sqrt{2}}{2}$,

$\cos 135^\circ + i\sin 135^\circ = -\frac{\sqrt{2}}{2} + i\frac{\sqrt{2}}{2}$,

$\cos 225^\circ + i\sin 225^\circ = -\frac{\sqrt{2}}{2} - i\frac{\sqrt{2}}{2}$, and

$\cos 315^\circ + i\sin 315^\circ = \frac{\sqrt{2}}{2} - i\frac{\sqrt{2}}{2}$.

37. Cube roots of i are given by

$$\cos\left(\frac{90^\circ + k360^\circ}{3}\right) + i\sin\left(\frac{90^\circ + k360^\circ}{3}\right)$$

$$= \cos(30^\circ + k120^\circ) + i\sin(30^\circ + k120^\circ).$$

If $k = 0, 1, 2$ one obtains

$\cos 30^\circ + i\sin 30^\circ = \frac{\sqrt{3}}{2} + i\frac{1}{2}$,

$\cos 150^\circ + i\sin 150^\circ = -\frac{\sqrt{3}}{2} + i\frac{1}{2}$, and

$\cos 270^\circ + i\sin 270^\circ = -i$.

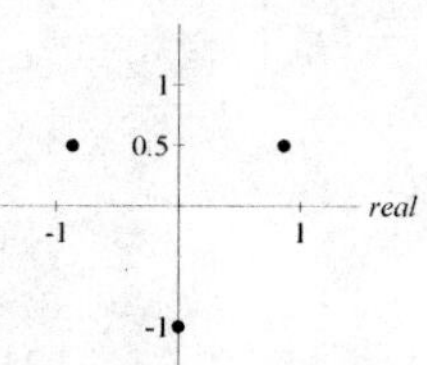

39. Since $|-2 + 2i\sqrt{3}| = 4$, the square roots are given by

$$2\left[\cos\left(\frac{120^\circ + k360^\circ}{2}\right) + i\sin\left(\frac{120^\circ + k360^\circ}{2}\right)\right]$$

$$= 2[\cos(60^\circ + k180^\circ) + i\sin(60^\circ + k180^\circ)].$$

If $k = 0, 1$ one obtains

$2[\cos 60^\circ + i\sin 60^\circ] = 1 + i\sqrt{3}$, and

$2[\cos 240^\circ + i\sin 240^\circ] = -1 - i\sqrt{3}$.

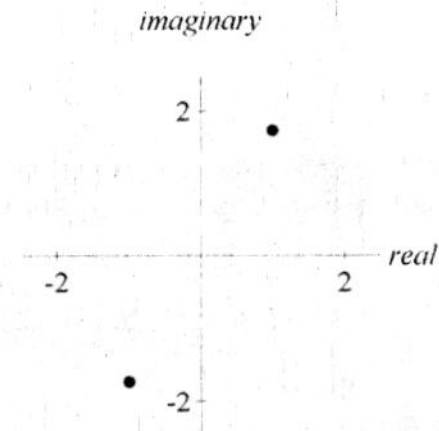

41. Note $|1+2i| = \sqrt{5}$. Since $\tan^{-1} 2 \approx 63.4°$ and

$$\frac{63.4° + k360°}{2} = 31.7° + k180°$$

the square roots are given by
$\sqrt[4]{5}\left[\cos(31.7° + k180°) + i\sin(31.7° + k180°)\right]$.

If $k = 0, 1$ one obtains
$\sqrt[4]{5}\left[\cos 31.7° + i\sin 31.7°\right] \approx 1.272 + 0.786i$,
and
$\sqrt[4]{5}\left[\cos 211.7° + i\sin 211.7°\right] \approx -1.272 - 0.786i$.

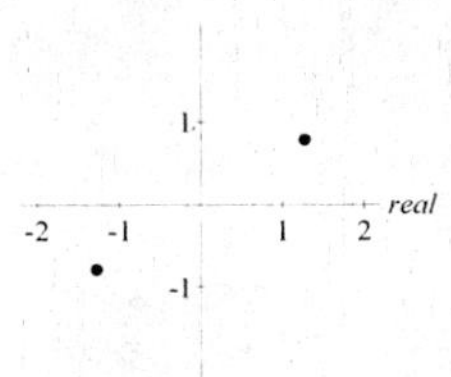

43. Solutions are the cube roots of -1. Namely,

$$\cos\left(\frac{180° + k360°}{3}\right) + i\sin\left(\frac{180° + k360°}{3}\right)$$
$$= \cos(60° + k120°) + i\sin(60° + k120°).$$

If $k = 0, 1, 2$ one obtains
$\cos 60° + i\sin 60° = \dfrac{1}{2} + i\dfrac{\sqrt{3}}{2}$,
$\cos 180° + i\sin 180° = -1$, and
$\cos 300° + i\sin 300° = \dfrac{1}{2} - i\dfrac{\sqrt{3}}{2}$.

45. Solutions are the fourth roots of 81. Namely,

$$3\left[\cos\left(\frac{k360°}{4}\right) + i\sin\left(\frac{k360°}{4}\right)\right]$$
$$= 3\left[\cos(k90°) + i\sin(k90°)\right].$$

If $k = 0, 1, 2, 3$ one obtains
$3[\cos 0° + i\sin 0°] = 3$,
$3[\cos 90° + i\sin 90°] = 3i$,
$3[\cos 180° + i\sin 180°] = -3$ and
$3[\cos 270° + i\sin 270°] = -3i$.

47. Solutions are the square roots of $-2i$.

If $k = 0, 1$ in $\alpha = \dfrac{-90° + k360°}{2}$ then
$\alpha = -45°, 135°$. These roots are
$\sqrt{2}\left[\cos(-45°) + i\sin(-45°)\right] = 1 - i$ and
$\sqrt{2}\left[\cos 135° + i\sin 135°\right] = -1 + i$.

49. Solutions of $x(x^6 - 64) = 0$ are $x = 0$ and the sixth roots of 64. The sixth roots are given by

$$2\left[\cos\left(\frac{k360°}{6}\right) + i\sin\left(\frac{k360°}{6}\right)\right]$$
$$= 2\left[\cos(k60°) + i\sin(k60°)\right].$$

If $k = 0, 1, 2, 3, 4, 5$ one obtains
$2[\cos 0° + i\sin 0°] = 2$,
$2[\cos 60° + i\sin 60°] = 1 + i\sqrt{3}$,
$2[\cos 120° + i\sin 120°] = -1 + i\sqrt{3}$,
$2[\cos 180° + i\sin 180°] = -2$,
$2[\cos 240° + i\sin 240°] = -1 - i\sqrt{3}$, and
$2[\cos 300° + i\sin 300°] = 1 - i\sqrt{3}$.

51. Factoring, we obtain

$$x^3(x^2 + 5) + 8(x^2 + 5) = (x^2 + 5)(x^3 + 8) = 0.$$

Then $x^2 = -5$ and $x^3 = -8$. Note, the square roots of -5 are $x = \pm i\sqrt{5}$. The cube roots of -8 are -2, $2(\cos 300° + i\sin 300°)$, and $2(\cos 60° + i\sin 60°)$. Thus, the solutions are $\pm i\sqrt{5}$, -2, and $1 \pm i\sqrt{3}$.

53. Solutions are the fifth roots of 2. Namely,

$$\sqrt[5]{2}\left[\cos\left(\frac{k360°}{5}\right) + i\sin\left(\frac{k360°}{5}\right)\right]$$
$$= \sqrt[5]{2}\left[\cos(k72°) + i\sin(k72°)\right].$$

Solutions are $x = \sqrt[5]{2}\left[\cos\alpha + i\sin\alpha\right]$ where $\alpha = 0°, 72°, 144°, 216°, 288°$.

55. Solutions are the fourth roots of $-3 + i$. Since $|-3 + i| = \sqrt{10}$, an argument of $-3 + i$ is $\cos^{-1}(-3/\sqrt{10}) \approx 161.6°$. Arguments of the fourth roots are given by

$$\frac{161.6° + k360°}{4} = 40.4° + k90°.$$

By choosing $k = 0, 1, 2, 3$, the solutions are $x = \sqrt[8]{10}\left[\cos\alpha + i\sin\alpha\right]$ where $\alpha = 40.4°, 130.4°, 220.4°, 310.4°$.

57. $[\cos\pi/3 + i\sin\pi/6]^3 =$
$[1/2 + i(1/2)]^3 = \left[\dfrac{1}{2}(1 + i)\right]^3 =$
$\dfrac{1}{8}\left[\sqrt{2}(\cos 45° + i\sin 45°)\right]^3 =$

$$\frac{1}{8}\left[2\sqrt{2}\,(\cos 135^\circ + i \sin 135^\circ)\right] =$$
$$\frac{\sqrt{2}}{4}\left(-\frac{\sqrt{2}}{2} + \frac{\sqrt{2}}{2}i\right) = -\frac{1}{4} + \frac{1}{4}i$$

59. By the quadratic formula, we get

$$x = \frac{-(-1+i) \pm \sqrt{(-1+i)^2 - 4(1)(-i)}}{2}$$
$$x = \frac{1 - i \pm \sqrt{(1 - 2i - 1) + 4i}}{2}$$
$$x = \frac{1 - i \pm \sqrt{2i}}{2}.$$

Note the square roots of $2i$ are given by $\sqrt{2}(\cos 45^\circ + i \sin 45^\circ) = 1 + i$ and $\sqrt{2}(\cos 225^\circ + i \sin 225^\circ) = -1 - i$. These two roots differ by a minus sign. So

$$x = \frac{1 - i \pm (1+i)}{2}$$
$$x = \frac{1 - i + 1 + i}{2} \quad \text{or} \quad x = \frac{1 - i - 1 - i}{2}$$
$$x = 1 \quad \text{or} \quad x = -i.$$

Solutions are $x = 1, -i$.

63. $\sqrt{3^2 + 5^2} = \sqrt{34}$

65. Since $\frac{\pi}{6} + \frac{\pi}{3} = \frac{\pi}{2} = 90^\circ$, the product is

$$6\,(\cos 90^\circ + i \sin 90^\circ) = 6i.$$

67. The magnitude is

$$\sqrt{(-3)^2 + (-9)^2} = \sqrt{90} = 3\sqrt{10}.$$

Note, $\alpha = \arctan\left(\frac{-9}{-3}\right) \approx 71.6^\circ$. Since the vector lies in the 3rd quadrant, the direction angle is

$$\alpha + 180^\circ \approx 251.6^\circ.$$

69. In the figure below, we have $AB = 1$, $AC = x$, and $BC = \sqrt{1 - x^2}$. Let y be the length of the statue.

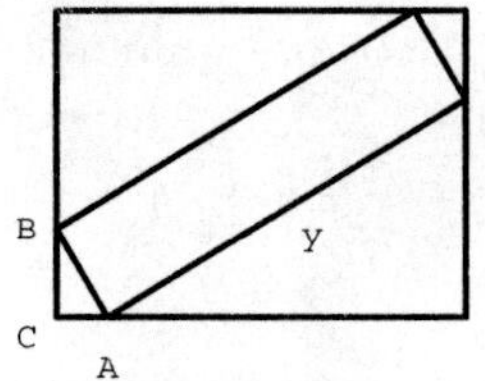

By the Pythagorean Theorem, we get

$$(4 - x)^2 + (3 - \sqrt{1 - x^2})^2 = y^2.$$

Using the area of the rectangle, we find

$$12 = x\sqrt{1 - x^2} + y + (4 - x)(3 - \sqrt{1 - x^2}).$$

The solutions of the system of two equations are

$$x \approx 0.5262 \text{ ft}, y \approx 4.0851 \text{ ft}.$$

The length of the statue is $y \approx 4.0851$ ft.

For Thought

1. True **2.** False, the distance is $|r|$.

3. False

4. False, $x = r\cos\theta$, $y = r\sin\theta$, and $x^2 + y^2 = r^2$.

5. True, since $x = -4\cos 225^\circ = 2\sqrt{2}$ and $y = -4\sin 225^\circ = 2\sqrt{2}$.

6. True, $\theta = \pi/4$ is a straight line through the origin which makes an angle of $\pi/4$ with the positive x-axis.

7. True, since each circle is centered at the origin with radius 5.

8. False, since upon substitution one gets $\cos 2\pi/3 = -1/2$ while $r^2 = 1/2$.

9. False, $r = 1/\sin\theta$ is the horizontal line $y = 1$.

10. False, since $r = \theta$ is a reflection of $r = -\theta$ about the origin.

7.6 Exercises

1. polar

3. $r\cos\theta$, $r\sin\theta$

5. $(3, \pi/2)$

7. Since $r = \sqrt{3^2 + 3^2} = 3\sqrt{2}$ and $\theta = \tan^{-1}(3/3) = \pi/4$, $(r, \theta) = (3\sqrt{2}, \pi/4)$.

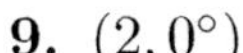

9. $(2, 0^\circ)$

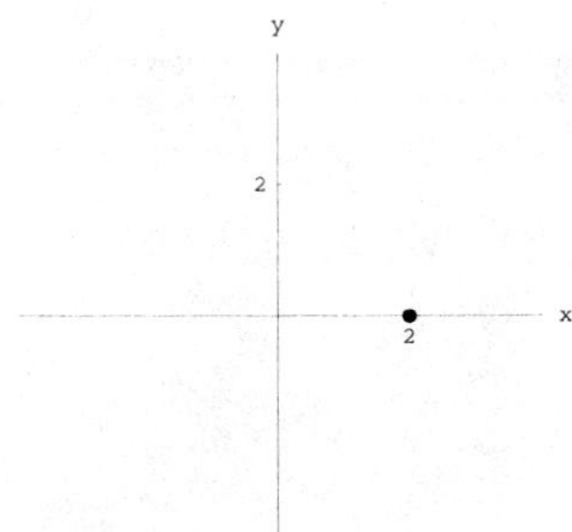

11. $(0, 35^\circ)$

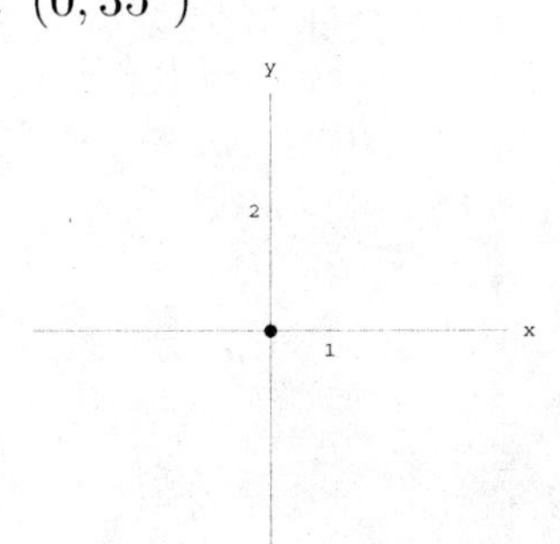

13. $(3, \pi/6)$

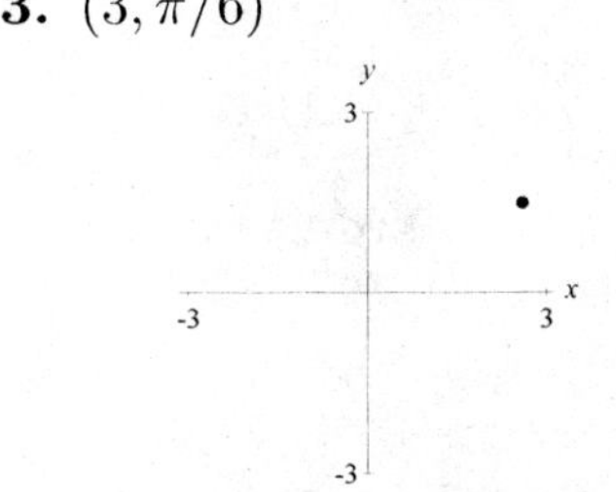

15. $(-2, 2\pi/3)$

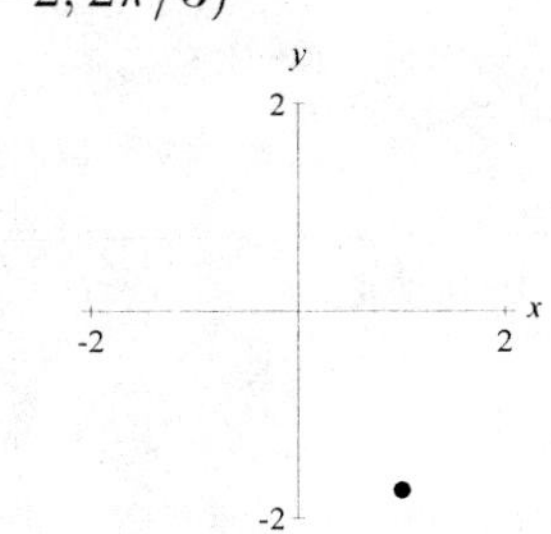

17. $(2, -\pi/4)$

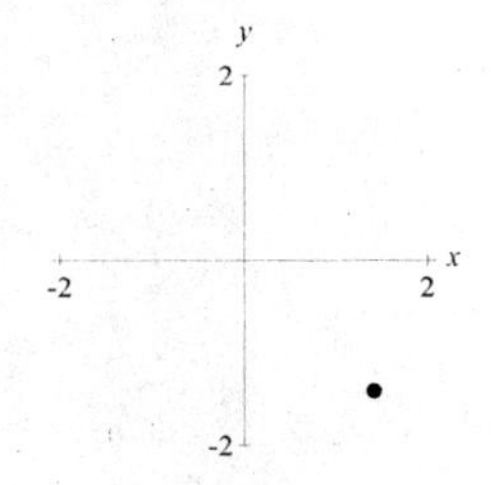

19. $(3, -225^\circ)$

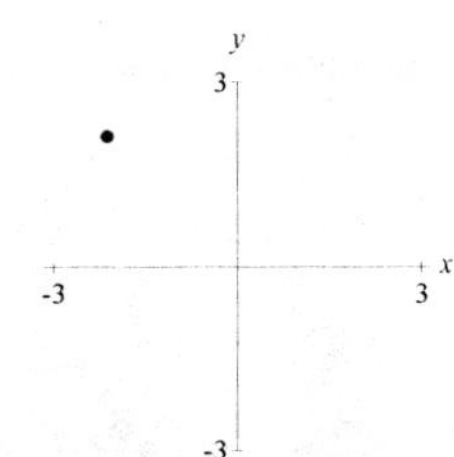

21. $(-2, 45^\circ)$

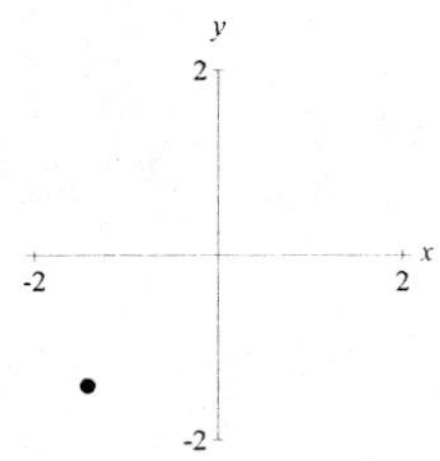

23. $(4, 390^\circ)$

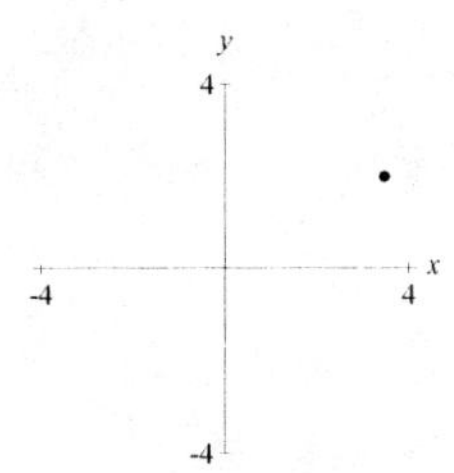

25. $(x, y) = (4 \cdot \cos(0^\circ), 4 \cdot \sin(0^\circ)) = (4, 0)$

27. $(x, y) = (0 \cdot \cos(\pi/4), 0 \cdot \sin(\pi/4)) = (0, 0)$

29. $(x, y) = (1 \cdot \cos(\pi/6), 1 \cdot \sin(\pi/6)) = \left(\frac{\sqrt{3}}{2}, \frac{1}{2}\right)$

31. $(x, y) = (-3\cos(3\pi/2), -3\sin(3\pi/2)) = (0, 3)$

33. $(x, y) = \left(\sqrt{2}\cos 135^\circ, \sqrt{2}\sin 135^\circ\right) = (-1, 1)$

35. $(x, y) = \left(-\sqrt{6}\cos(-60^\circ), -\sqrt{6}\sin(-60^\circ)\right) = \left(-\frac{\sqrt{6}}{2}, \frac{3\sqrt{2}}{2}\right)$

37. Since $r = \sqrt{(\sqrt{3})^2 + 3^2} = 2\sqrt{3}$ and $\cos\theta = \dfrac{x}{r} = \dfrac{\sqrt{3}}{2\sqrt{3}} = \dfrac{1}{2}$, one can choose $\theta = 60^\circ$. So $(r, \theta) = (2\sqrt{3}, 60^\circ)$.

39. Since $r = \sqrt{(-2)^2 + 2^2} = 2\sqrt{2}$ and $\cos\theta = \dfrac{x}{r} = \dfrac{-2}{2\sqrt{2}} = -\dfrac{1}{\sqrt{2}}$, one can choose $\theta = 135°$. So $(r, \theta) = (2\sqrt{2}, 135°)$.

41. $(r, \theta) = (2, 90°)$

43. Note $r = \sqrt{(-3)^2 + (-3)^2} = 3\sqrt{2}$.
Since $\tan\theta = \dfrac{y}{x} = \dfrac{-3}{-3} = 1$ and $(-3, -3)$ is in quadrant III, one can choose $\theta = 225°$. Thus, $(r, \theta) = (3\sqrt{2}, 225°)$.

45. Since $r = \sqrt{1^2 + 4^2} = \sqrt{17}$ and $\theta = \cos^{-1}\left(\dfrac{x}{r}\right) = \cos^{-1}\left(\dfrac{1}{\sqrt{17}}\right) \approx 75.96°$ one gets $(r, \theta) = (\sqrt{17}, 75.96°)$.

47. Since $r = \sqrt{(\sqrt{2})^2 + (-2)^2} = \sqrt{6}$ and $\theta = \tan^{-1}\left(\dfrac{y}{x}\right) = \tan^{-1}\left(\dfrac{-2}{\sqrt{2}}\right) \approx -54.7°$, one obtains $(r, \theta) = (\sqrt{6}, -54.7°)$.

49. $(x, y) = (4\cos 26°, 4\sin 26°) \approx (3.60, 1.75)$

51. $(x, y) = (2\cos(\pi/7), 2\sin(\pi/7)) \approx (1.80, 0.87)$

53. $(x, y) = (-2\cos(1.1), -2\sin(1.1)) \approx (-0.91, -1.78)$

55. Since $r = \sqrt{4^2 + 5^2} \approx 6.4$ and $\tan\theta = \dfrac{y}{x} = \dfrac{5}{4}$, we get $\theta = \tan^{-1}\left(\dfrac{5}{4}\right) \approx 51.34°$. Then $(r, \theta) \approx (6.4, 51.34°)$.

57. Note, $r = \sqrt{(-2)^2 + (-7)^2} \approx 7.3$ and $\tan^{-1}\left(\dfrac{-7}{-2}\right) \approx 74.1°$. Since $(-2, -7)$ is a point in the 3rd quadrant, we choose $\theta = 74.1° + 180° = 254.1°$.
Then $(r, \theta) \approx (7.3, 254.1°)$.

59. $r = 2\sin\theta$ is a circle centered at $(x, y) = (0, 1)$. It goes through the following points in polar coordinates: $(0, 0)$, $(1, \pi/6)$, $(2, \pi/2)$, $(1, 5\pi/6)$, $(0, \pi)$.

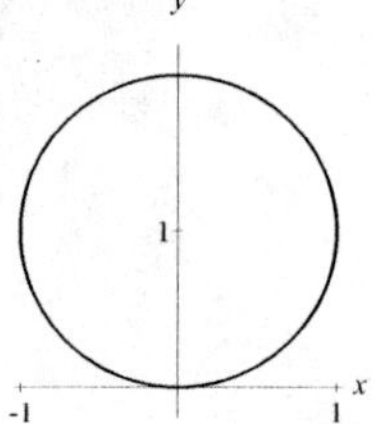

61. $r = 3\cos 2\theta$ is a four-leaf rose that goes through the following points in polar coordinates $(3, 0)$, $(3/2, \pi/6)$, $(0, \pi/4)$, $(-3, \pi/2)$, $(0, 3\pi/4)$, $(3, \pi)$, $(-3, 3\pi/2)$, $(0, 7\pi/4)$.

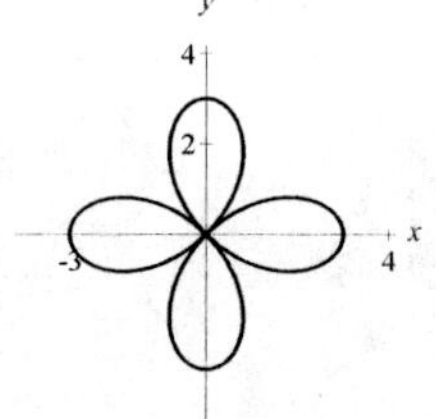

63. $r = 2\theta$ is spiral-shaped and goes through the following points in polar coordinates $(-\pi, -\pi/2)$, $(0, 0)$, $(\pi, \pi/2)$, $(2\pi, \pi)$

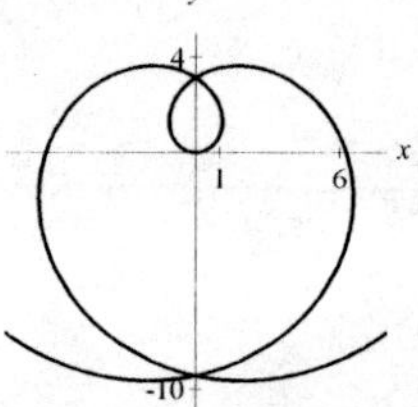

65. $r = 1 + \cos\theta$ goes through the following points in polar coordinates $(2, 0)$, $(1.5, \pi/3)$, $(1, \pi/2)$, $(0.5, 2\pi/3)$, $(0, \pi)$.

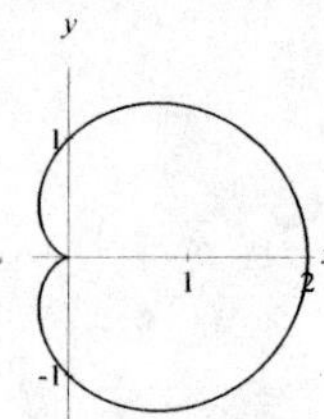

67. $r^2 = 9\cos 2\theta$ goes through the following points in polar coordinates $(0, -\pi/4)$, $(\pm\sqrt{3}/2, -\pi/6)$, $(\pm 3, 0)$, $(\pm\sqrt{3}/2, \pi/6)$, $(0, \pi/4)$.

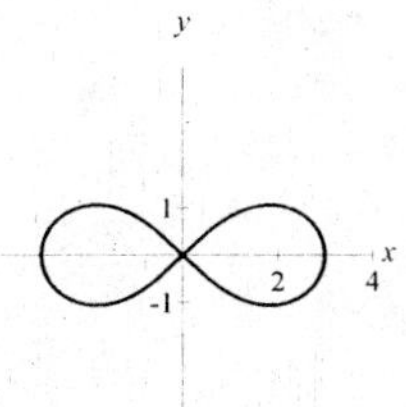

69. $r = 4\cos 2\theta$ is a four-leaf rose that goes through the following points in polar coordinates $(4, 0)$, $(2, \pi/6)$, $(0, \pi/4)$, $(-4, \pi/2)$, $(2, 5\pi/6)$, $(4, \pi)$, $(0, 5\pi/4)$, $(-4, 3\pi/2)$, $(0, 7\pi/4)$.

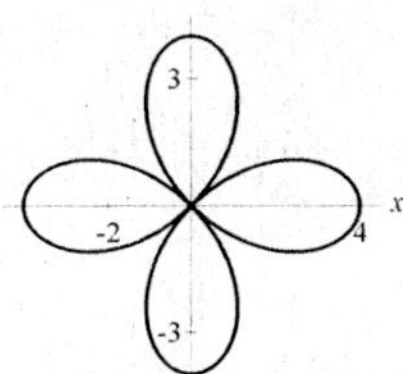

71. $r = 2\sin 3\theta$ is a three-leaf rose that goes through the following points in polar coordinates $(0, 0)$, $(2, \pi/6)$, $(-2, \pi/2)$, $(2, 5\pi/6)$.

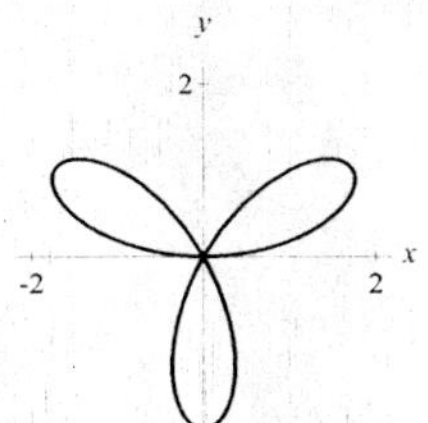

73. $r = 1 + 2\cos\theta$ goes through the following points in polar coordinates $(3, 0)$, $(2, \pi/3)$, $(1, \pi/2)$, $(0, 2\pi/3)$, $(1 - \sqrt{3}, 5\pi/6)$, $(-1, \pi)$.

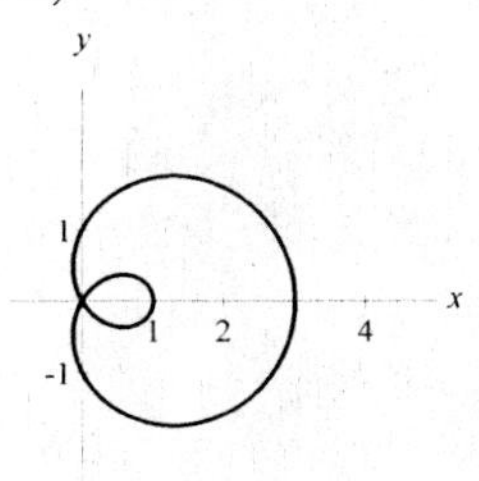

75. $r = 3.5$ is a circle centered at the origin with radius 3.5.

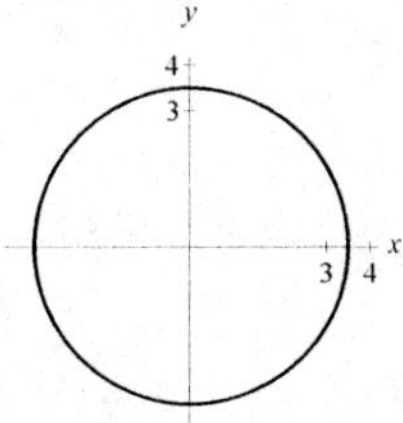

77. $\theta = 30°$ is a line through the origin that makes a $30°$ angle with the positive x-axis.

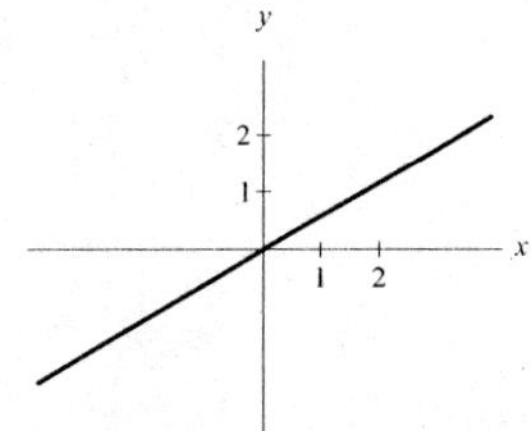

79. Multiply equation by r.

$$\begin{aligned} r^2 &= 4r\cos\theta \\ x^2 + y^2 &= 4x \\ x^2 - 4x + y^2 &= 0 \end{aligned}$$

81. Multiply equation by $\sin\theta$.

$$\begin{aligned} r\sin\theta &= 3 \\ y &= 3 \end{aligned}$$

83. Multiply equation by $\cos\theta$.

$$\begin{aligned} r\cos\theta &= 3 \\ x &= 3 \end{aligned}$$

85. Since $r = 5$, $\sqrt{x^2 + y^2} = 5$ and by squaring one gets $x^2 + y^2 = 25$.

87. Note $\tan\theta = \dfrac{\sin\theta}{\cos\theta} = \dfrac{y/r}{x/r} = \dfrac{y}{x}$.

Since $\tan\pi/4 = 1$, $\dfrac{y}{x} = 1$ or $y = x$.

89. Multiply equation by $1 - \sin\theta$.

$$\begin{aligned} r(1 - \sin\theta) &= 2 \\ r - r\sin\theta &= 2 \\ r - y &= 2 \\ \pm\sqrt{x^2 + y^2} &= y + 2 \\ x^2 + y^2 &= y^2 + 4y + 4 \\ x^2 - 4y &= 4 \end{aligned}$$

91. $r\cos\theta = 4$

93. Note $\tan\theta = y/x$.

$$\begin{aligned} y &= -x \\ \frac{y}{x} &= -1 \\ \tan\theta &= -1 \\ \theta &= -\pi/4 \end{aligned}$$

95. Note $x = r\cos\theta$ and $y = r\sin\theta$.

$$\begin{aligned} (r\cos\theta)^2 &= 4r\sin\theta \\ r^2\cos^2\theta &= 4r\sin\theta \\ r &= \frac{4\sin\theta}{\cos^2\theta} \\ r &= 4\tan\theta\sec\theta \end{aligned}$$

97. $r = 2$

99. Note $x = r\cos\theta$ and $y = r\sin\theta$.

$$\begin{aligned} y &= 2x - 1 \\ r\sin\theta &= 2r\cos\theta - 1 \\ r(\sin\theta - 2\cos\theta) &= -1 \\ r(2\cos\theta - \sin\theta) &= 1 \\ r &= \frac{1}{2\cos\theta - \sin\theta} \end{aligned}$$

101. Note that $y = r\sin\theta$ and $x^2 + y^2 = r^2$.

$$\begin{aligned} x^2 + (y^2 - 2y + 1) &= 1 \\ x^2 + y^2 - 2y &= 0 \\ r^2 - 2r\sin\theta &= 0 \\ r^2 &= 2r\sin\theta \\ r &= 2\sin\theta \end{aligned}$$

103. There are six points of intersection and in polar coordinates these are approximately $(1, 0.17)$, $(1, 2.27)$, $(1, 4.36)$, $(1, 0.87)$, $(1, 2.97)$, $(1, 5.06)$

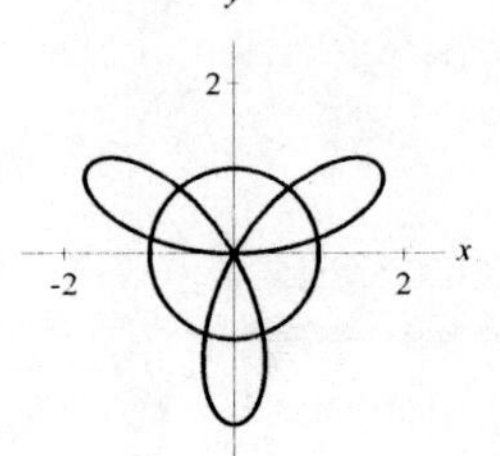

105. There are seven points of intersection and in polar coordinates these are approximately $(0, 0)$, $(0.9, 1.4)$, $(1.2, 1.8)$, $(1.9, 2.8)$, $(1.9, 3.5)$, $(1.2, 4.5)$, $(0.8, 4.9)$

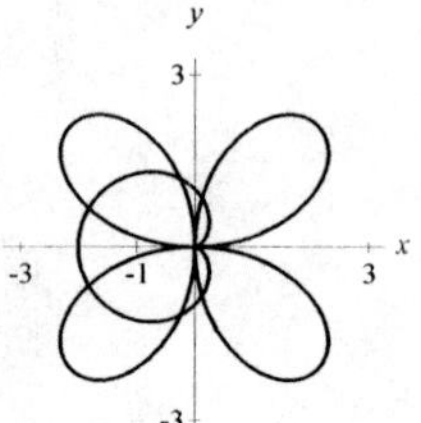

109. Since $1 + i = \sqrt{2}\,(\cos\frac{\pi}{4} + i\sin\frac{\pi}{4})$, we find

$$\begin{aligned} (1+i)^{12} &= (\sqrt{2})^{12}\left(\cos\frac{12\pi}{4} + i\sin\frac{12\pi}{4}\right) \\ &= 64\,(\cos 3\pi + i\sin 3\pi) \\ &= -64 \end{aligned}$$

111. Note, $\arcsin(0.88) \approx 1.08$ and $\pi - \arcsin(0.88) \approx 2.07$. Then

$$x \approx 1.08 + 2k\pi \quad \text{or} \quad x \approx 2.07 + 2k\pi$$

where k is an integer.

113. Use a cofunction identity and the fact that cosine is an even function.

$$\begin{aligned} \sin\left(2x - \frac{\pi}{4}\right) &= \cos\left(\frac{\pi}{2} - \left(2x - \frac{\pi}{4}\right)\right) \\ &= \cos\left(-2x + \frac{3\pi}{4}\right) \\ &= \cos\left(2x - \frac{3\pi}{4}\right) \\ \sin\left(2x - \frac{\pi}{4}\right) &= \cos\left(2\left(x - \frac{3\pi}{8}\right)\right) \end{aligned}$$

115. In the figure below,

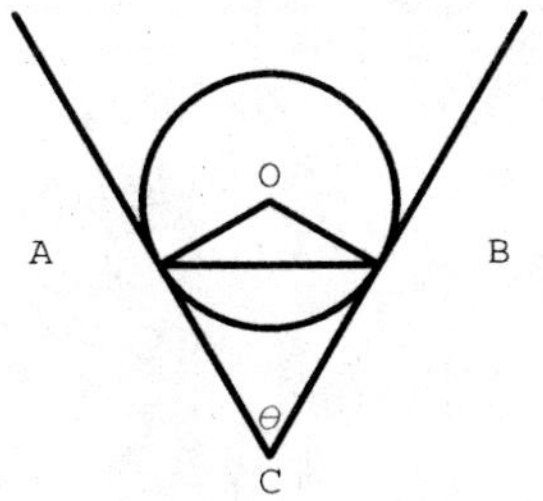

the angle at the center O is

$$\langle AOB = \pi - \theta.$$

Then the area of triangle ΔAOB is

$$A_t = \frac{\sin\theta}{2}$$

and the area of sector $\widetilde{AB}$ is

$$A_s = \frac{\pi - \theta}{2}.$$

a) The area inside the circle and below line AB is

$$\begin{aligned} A_s - A_t &= \frac{\pi}{2} - \frac{\theta}{2} - \frac{\sin\theta}{2} \\ &= \frac{\pi}{2} - \frac{\theta}{2} - \sin(\theta/2)\cos(\theta/2) \end{aligned}$$

b) Applying the Pythagorean Theorem to ΔAOC, we find

$$AC = 1/\tan(\theta/2).$$

Then the area of ΔAOC is

$$\frac{1}{2\tan(\theta/2)}$$

and the area of the quadrilateral $ABCO$ is

$$A_q = \frac{1}{\tan(\theta/2)}.$$

Hence, the area below the circle and inside the trench is

$$A_q - A_s = \frac{1}{\tan(\theta/2)} - \frac{\pi}{2} + \frac{\theta}{2}$$

c) Using a calculator, we find that as θ approaches π, the ratio

$$\frac{A_s - A_t}{A_q - A_s}$$

approaches 2.

For Thought

1. False, t is the parameter.

2. True, graphs of parametric equations are sketched in a rectangular coordinate system in this book.

3. True, since $2x = t$ and $y = 2t + 1 = 2(2x) + 1$ $= 4x + 1$.

4. False, it is a circle of radius 1.

5. True, since if $t = \frac{1}{3}$ then $x = 3\left(\frac{1}{3}\right) + 1 = 2$ and $y = 6\left(\frac{1}{3}\right) - 1 = 1$.

6. False, for if $w^2 - 3 = 1$ then $w = \pm 2$ and this does not satisfy $-2 < w < 2$.

7. True, since x and y take only positive values.

8. True

9. False, since e^t is non-negative while $\ln(t)$ can be negative.

10. True

7.7 Exercises

1. parametric

3. If $t = 0$, then $x = 4(0) + 1 = 1$ and $y = 0 - 2 = -2$. If $t = 1$, then $x = 4(1) + 1 = 5$ and $y = 1 - 2 = -1$.

If $x = 7$, then $7 = 4t + 1$. Solving for t, we get $t = 1.5$. Substitute $t = 1.5$ into $y = t - 2$. Then $y = 1.5 - 2 = -0.5$.

If $y = 1$, then $1 = t - 2$. Solving for t, we get $t = 3$. Consequently $y = 4(3) + 1 = 13$.

We tabulate the results as follows.

t	x	y
0	1	-2
1	5	-1
1.5	7	-0.5
3	13	1

5. If $t = 1$, then $x = 1^2 = 1$ and $y = 3(1) - 1 = 2$.
If $t = 2.5$, then $x = (2.5)^2 = 6.25$ and $y = 3(2.5) - 1 = 6.5$.

If $x = 5$, then $5 = t^2$ and $t = \sqrt{5}$. Consequently, $y = 3\sqrt{5} - 1$.

If $y = 11$, $11 = 3t - 1$. Solving for t, we get $t = 4$. Consequently $x = 4^2 = 16$.

If $x = 25$, then $25 = t^2$ and $t = 5$. Consequently, $y = 3(5) - 1 = 14$.

We tabulate the results as follows.

t	x	y
1	1	2
2.5	6.25	6.5
$\sqrt{5}$	5	$3\sqrt{5} - 1$
4	16	11
5	25	14

7.

Some points are given by

t	x	y
0	-2	3
4	10	7

.

The domain is $[-2, 10]$ and the range is $[3, 7]$.

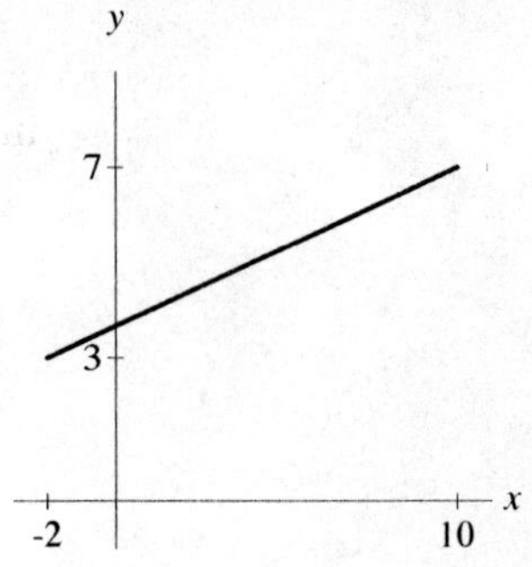

9.

Some points are given by

t	x	y
0	-1	0
2	1	4

.

The domain is $(-\infty, \infty)$ and the range is $[0, \infty)$.

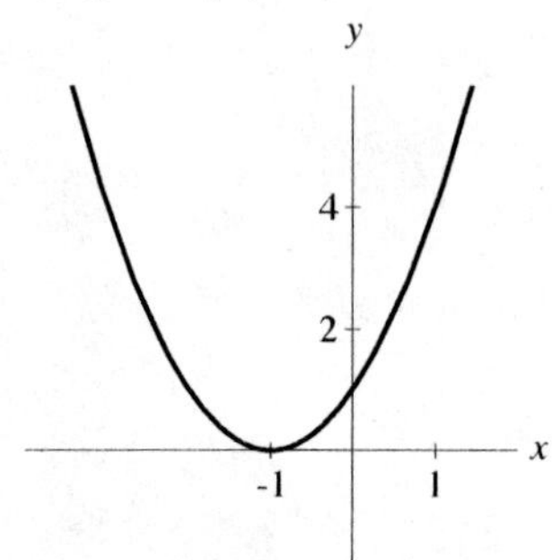

11. A few points are approximated by

w	x	y
0.2	0.4	0.9
0.8	0.9	0.4

.

The domain is $(0, 1)$ and the range is $(0, 1)$.

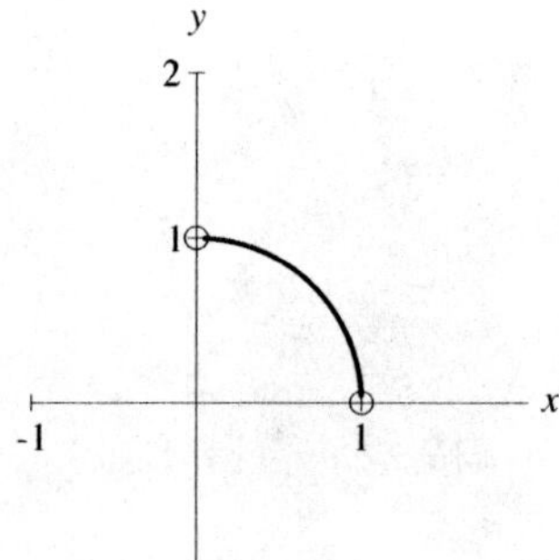

13. A circle of radius 1 and centered at the origin. The domain is $[-1, 1]$ and the range is $[-1, 1]$.

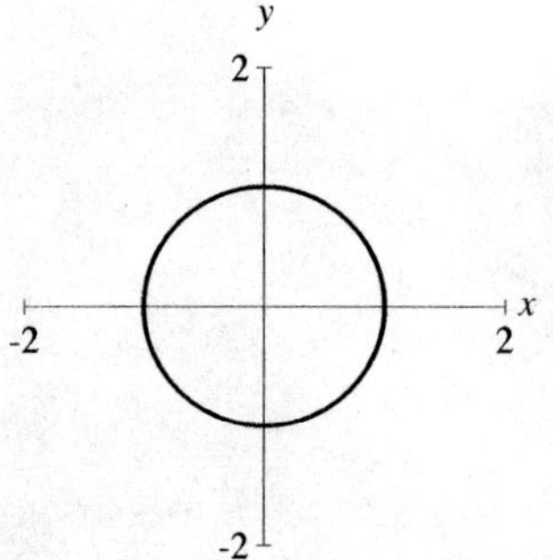

15. Since $t = \dfrac{x+5}{4}$, we obtain

$$y = 3 - 4\left(\frac{x+5}{4}\right) = -x - 2 \text{ or } x + y = -2.$$

The graph is a straight line with domain $(-\infty, \infty)$ and range $(-\infty, \infty)$.

17. Since $x^2 + y^2 = 16\sin^2(3t) + 16\cos^2(3t) = 16$, the graph is a circle with radius 4 and with center at the origin.

Domain $[-4, 4]$ and range $[-4, 4]$

19. Since $t = 4x$, we find $y = e^{4x}$ and the graph is an exponential graph.

Domain $(-\infty, \infty)$ and range $(0, \infty)$

21. $y = 2x + 3$ represents the graph of a straight line

Domain $(-\infty, \infty)$ and range $(-\infty, \infty)$

23. An equation (in terms of x and y) of the line through $(2, 3)$ and $(5, 9)$ is $y = 2x - 1$.

An equation (in terms of t and x) of the line through $(0, 2)$ and $(2, 5)$ is $x = \frac{3}{2}t + 2$.

Parametric equations are $x = \frac{3}{2}t + 2$ and

$y = 2\left(\frac{3}{2}t + 2\right) - 1 = 3t + 3$ where $0 \le t \le 2$.

25. $x = 2\cos t$, $y = 2\sin t$, $\pi < t < \frac{3\pi}{2}$

27. $x = 3$, $y = t$, $-\infty < t < \infty$

29. Since $x = r\cos t$, $y = r\sin t$, and $r = 2\sin t$,

we get $x = 2\sin t\cos t = \sin 2t$ and

$y = 2\sin t\sin t = 2\sin^2 t$ where $0 \le t \le 2\pi$.

Then $x = \sin 2t$ and $y = 2\sin^2 t$.

31. For $-\pi \le t \le \pi$, one obtains the given graph (for a larger range of values for t, more points are filled and the graph would be different)

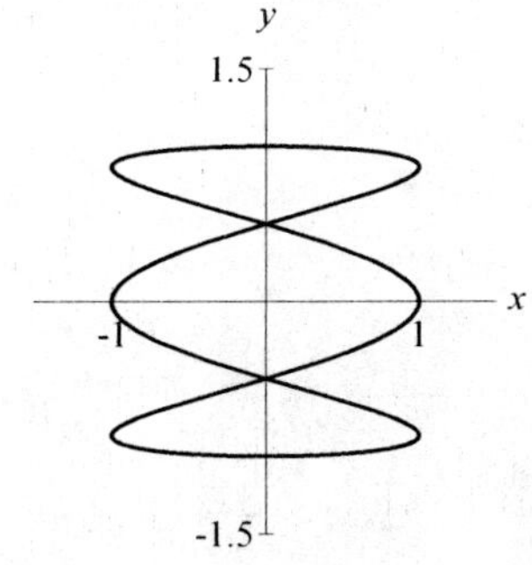

33. For $-15 \le t \le 15$, one finds

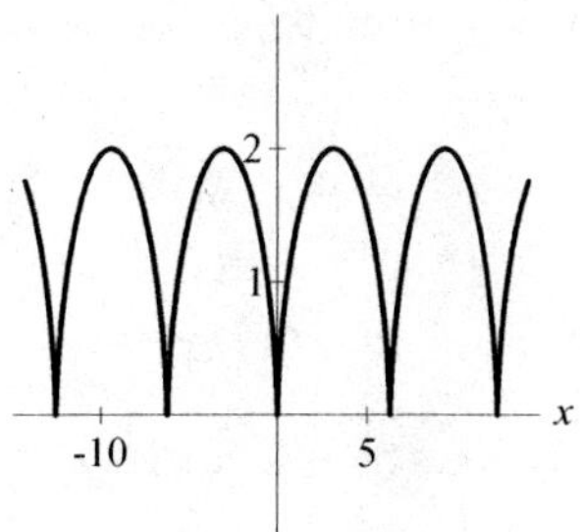

35. For $-10 \le t \le 10$, one obtains

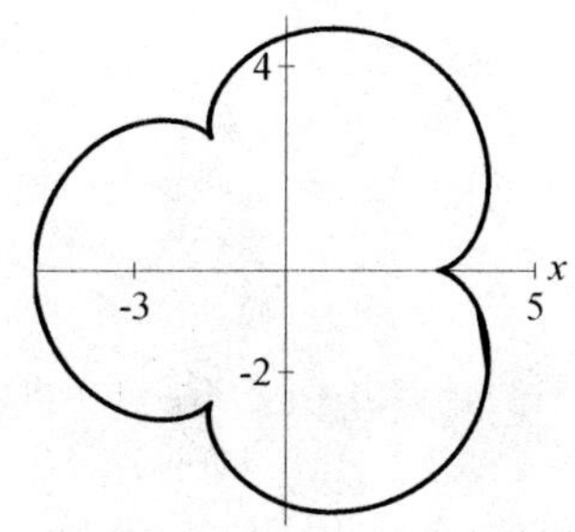

37. A graph of the parametric equations $x = 150\sqrt{3}t$ and $y = -16t^2 + 150t + 5$ (for $0 \le t \le 10$) is given

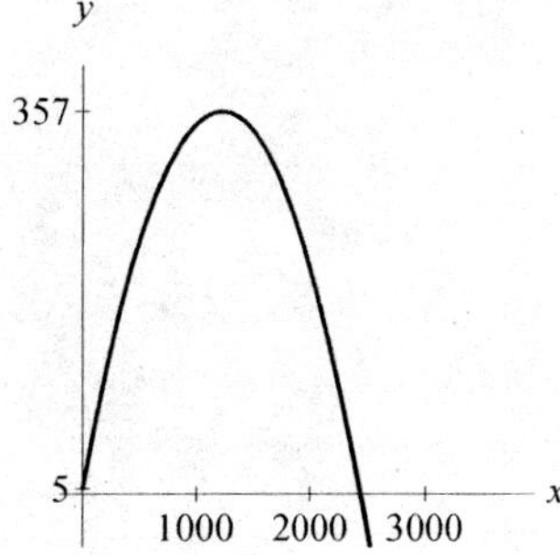

39. Solving $y = -16t^2 + 150t + 5 = 0$, one finds

$$\frac{-150 \pm \sqrt{150^2 - 4(-16)(5)}}{-32} \approx 9.41, -0.03$$

The arrow is in the air for 9.4 seconds.

41. Multiply both sides by r:

$$\begin{aligned} r^2 &= 8r\cos\theta \\ x^2 + y^2 &= 8x \\ x^2 - 8x + y^2 &= 0 \end{aligned}$$

43. The magnitude is

$$\sqrt{3^2 + (-3\sqrt{3})^2} = \sqrt{9+27} = \sqrt{36} = 6.$$

Note, $\arctan \frac{-3\sqrt{3}}{3} = \arctan(-\sqrt{3}) = -\pi/3$. Since $(3, -3\sqrt{3})$ lies in quadrant 4, we may take the argument to be $5\pi/3$. Then

$$3 - 3\sqrt{3} = 6\left(\cos(5\pi/3) + i\sin(5\pi/3)\right).$$

45. Solve by factoring as follows:

$$\begin{aligned}(2\cos x - 1)(\cos x + 1) &= 0\\ \cos x &= -1, \frac{1}{2}\end{aligned}$$

Then $x = \frac{\pi}{3}, \pi, \frac{5\pi}{3}$.

47. The sides of the three squares are $\sqrt{8}$, $\sqrt{13}$, and $\sqrt{17}$. These are also the sides of the triangle. We use Heron's formula to find the area of the triangle. Let

$$s = \frac{\sqrt{8}+\sqrt{13}+\sqrt{17}}{2}.$$

The area of the triangle is

$$\begin{aligned}\text{Area} &= \sqrt{s(s-\sqrt{8})(s-\sqrt{13})(s-\sqrt{17})}\\ &= 5 \text{ acres}\\ &= 5(43,560)\text{ ft}^2\\ \text{Area} &= 217,800\text{ ft}^2.\end{aligned}$$

Review Exercises

1. Draw a triangle with $\gamma = 48°$, $a = 3.4$, $b = 2.6$.

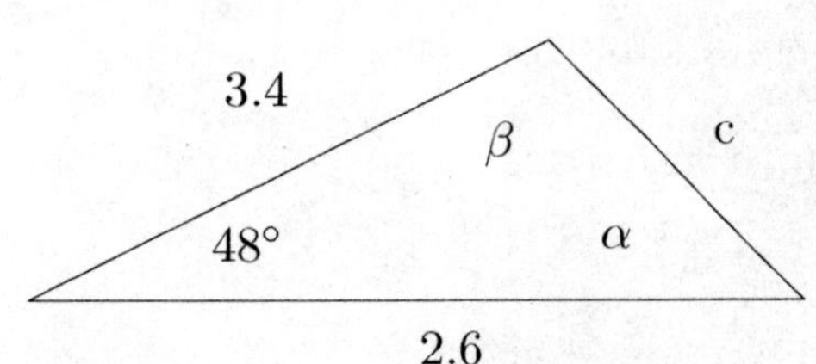

By the cosine law, we obtain $c = \sqrt{2.6^2 + 3.4^2 - 2(2.6)(3.4)\cos 48°} \approx 2.5475 \approx 2.5$. By the sine law, we find

$$\begin{aligned}\frac{2.5475}{\sin 48°} &= \frac{2.6}{\sin\beta}\\ \sin\beta &= \frac{2.6\sin 48°}{2.5475}\\ \sin\beta &\approx 0.75846\\ \beta &\approx \sin^{-1}(0.75846)\\ \beta &\approx 49.3°.\end{aligned}$$

Also, $\alpha = 180° - (49.3° + 48°) = 82.7°$.

3. Draw a triangle with $\alpha = 13°$, $\beta = 64°$, $c = 20$.

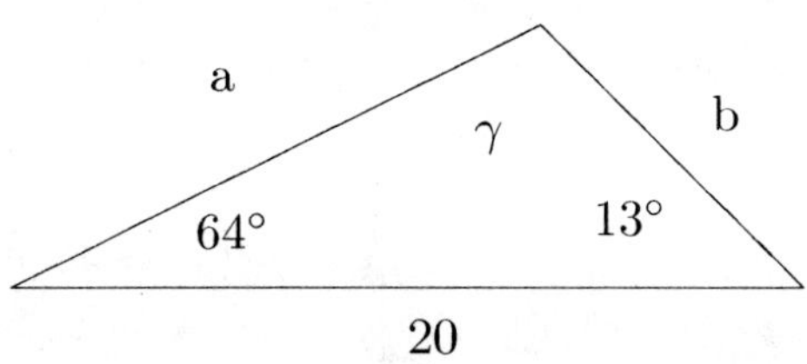

Note $\gamma = 180° - (64° + 13°) = 103°$. By the sine law, we get

$$\frac{20}{\sin 103°} = \frac{a}{\sin 13°}$$

and

$$\frac{20}{\sin 103°} = \frac{b}{\sin 64°}.$$

Then $a = \frac{20}{\sin 103°}\sin 13° \approx 4.6$ and $b = \frac{20}{\sin 103°}\sin 64° \approx 18.4$.

5. Draw a triangle with $a = 3.6$, $b = 10.2$, $c = 5.9$.

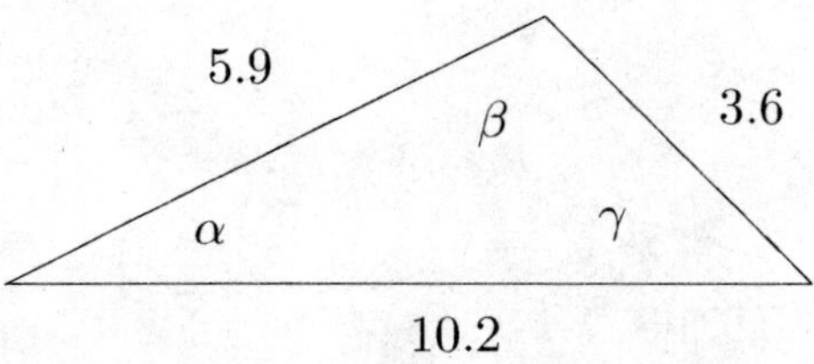

By the cosine law one gets

$$\cos\beta = \frac{5.9^2 + 3.6^2 - 10.2^2}{2(5.9)(3.6)} \approx -1.3.$$

This is a contradiction since the range of cosine is $[-1, 1]$. No triangle exists.

7. Draw a triangle with sides $a = 30.6$, $b = 12.9$, and $c = 24.1$.

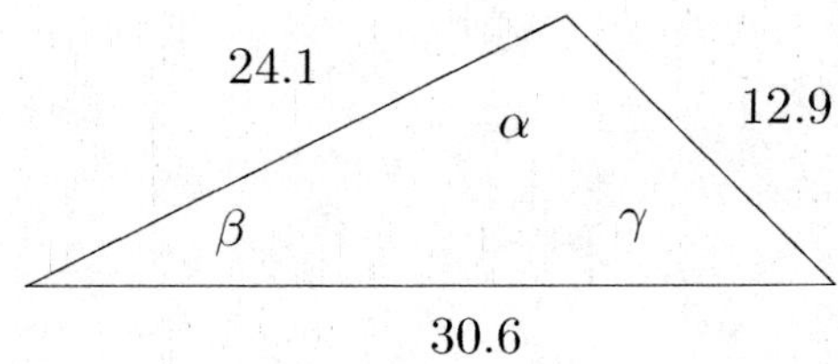

By the cosine law, we get

$\cos\alpha = \dfrac{24.1^2 + 12.9^2 - 30.6^2}{2(24.1)(12.9)} \approx -0.3042.$

So $\alpha = \cos^{-1}(-0.3042) \approx 107.7°$.

Similarly, we find

$\cos\beta = \dfrac{24.1^2 + 30.6^2 - 12.9^2}{2(24.1)(30.6)} \approx 0.9158.$

So $\beta = \cos^{-1}(0.9158) \approx 23.7°$.
Also, $\gamma = 180° - (107.7° + 23.7°) = 48.6°$.

9. Draw angle $\beta = 22°$ and let h be the height.

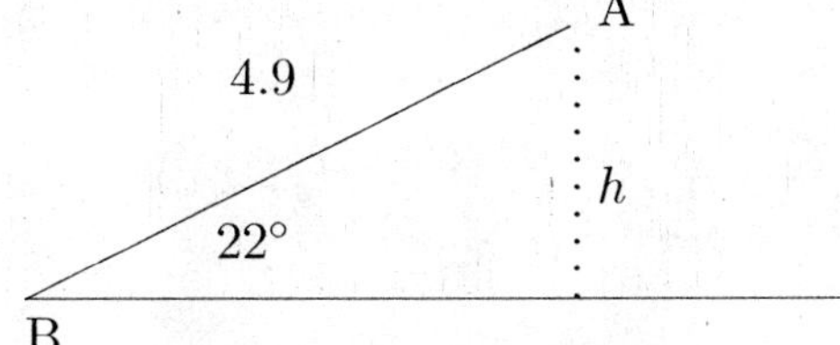

Since $h = 4.9\sin 22° \approx 1.8$ and $1.8 < b < 4.9$, we have two triangles and they are given by

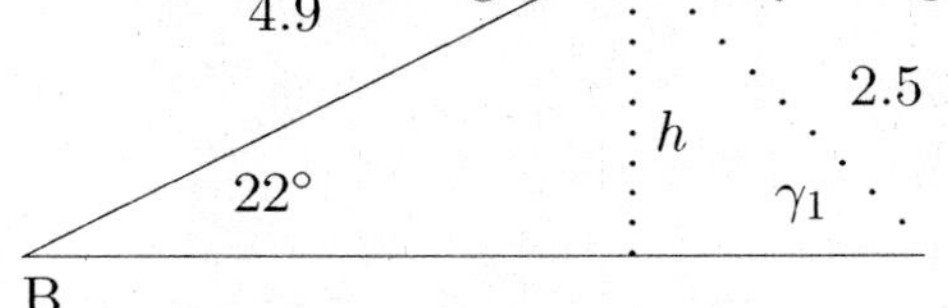

and

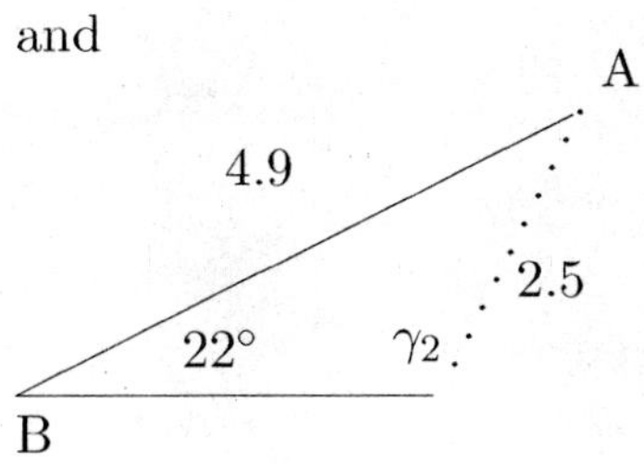

Apply the sine law to case 1.

$$\begin{aligned}\frac{4.9}{\sin\gamma_1} &= \frac{2.5}{\sin 22°}\\ \sin\gamma_1 &= \frac{4.9\sin 22°}{2.5}\\ \sin\gamma_1 &\approx 0.7342\\ \gamma_1 = \sin^{-1}(0.7342) &\approx 47.2°\end{aligned}$$

So $\alpha_1 = 180° - (22° + 47.2°) = 110.8°$.

By the sine law, $a_1 = \dfrac{2.5}{\sin 22°}\sin 110.8° \approx 6.2$.

In case 2, $\gamma_2 = 180° - \gamma_1 = 132.8°$ and $\alpha_2 = 180° - (22° + 132.8°) = 25.2°$.

By the sine law, $a_2 = \dfrac{2.5}{\sin 22°}\sin 25.2° \approx 2.8$.

11. Area is $A = \dfrac{1}{2}(12.2)(24.6)\sin 38° \approx 92.4$ ft^2.

13. Since $S = \dfrac{5.4 + 12.3 + 9.2}{2} = 13.45$, the area is

$\sqrt{13.45(13.45 - 5.4)(13.45 - 12.3)(13.45 - 9.2)}$

≈ 23.0 km^2.

15. $|\,\boldsymbol{v_x}\,| = |6\cos 23.3°| \approx 5.5$,
$|\,\boldsymbol{v_y}\,| = |6\sin 23.3°| \approx 2.4$

17. $|\,\boldsymbol{v_x}\,| = |3.2\cos 231.4°| \approx 2.0$,
$|\,\boldsymbol{v_y}\,| = |3.2\sin 231.4°| \approx 2.5$

19. Magnitude $\sqrt{2^2 + 3^2} = \sqrt{13}$, direction angle $\tan^{-1}(3/2) \approx 56.3°$

21. The magnitude is $\sqrt{(-3.2)^2 + (-5.1)^2} \approx 6.0$. Since $\tan^{-1}(5.1/3.2) \approx 57.9°$, the direction angle is $180° + 57.9° = 237.9°$.

23. $\langle\sqrt{2}\cos 45°, \sqrt{2}\sin 45°\rangle = \langle 1, 1\rangle$

25. $\langle 9.1\cos 109.3°, 9.1\sin 109.3°\rangle \approx \langle -3.0, 8.6\rangle$

27. $\langle -6, 8\rangle$

29. $\langle 2 - 2, -5 - 12\rangle = \langle 0, -17\rangle$

31. $\langle -1, 5\rangle \cdot \langle 4, 2\rangle = -4 + 10 = 6$

33. $-4\,\boldsymbol{i} + 8\,\boldsymbol{j}$

35. $(7.2\cos 30°)\,\boldsymbol{i} + (7.2\sin 30°)\,\boldsymbol{j} \approx 3.6\sqrt{3}\,\boldsymbol{i} + 3.6\,\boldsymbol{j}$

37. $|3-5i| = \sqrt{3^2+(-5)^2} = \sqrt{34}$

39. $|\sqrt{5}+i\sqrt{3}| = \sqrt{(\sqrt{5})^2+(\sqrt{3})^2} = \sqrt{8} = 2\sqrt{2}$

41. Note $\sqrt{(-4.2)^2+(4.2)^2} \approx 5.94$. If the terminal side of α goes through $(-4.2, 4.2)$, then $\tan\alpha = -1$. Choose $\alpha = 135°$. So $-4.2+4.2i = 5.94\left[\cos 135° + i\sin 135°\right]$.

43. Note $\sqrt{(-2.3)^2+(-7.2)^2} \approx 7.6$. If the terminal side of α goes through $(-2.3, -7.2)$ and since $\tan^{-1}(7.2/2.3) \approx 72.3°$, then one can choose $\alpha = 180° + 72.3° = 252.3°$. So $-2.3-7.2i \approx 7.6\left[\cos 252.3° + i\sin 252.3°\right]$.

45. $\sqrt{3}\left(-\frac{\sqrt{3}}{2}+\frac{1}{2}i\right) = -\frac{3}{2}+\frac{\sqrt{3}}{2}i$

47. $6.5[0.8377 + (0.5461)i] \approx 5.4 + 3.5i$

49. Since $z_1 = 2.5\sqrt{2}\left[\cos 45° + i\sin 45°\right]$ and $z_2 = 3\sqrt{2}\left[\cos 225° + i\sin 225°\right]$, we have $z_1 z_2 = 15\left[\cos 270° + i\sin 270°\right] = -15i$ and

$$\begin{aligned}\frac{z_1}{z_2} &= \frac{(5/2)\sqrt{2}}{3\sqrt{2}}\left[\cos(-180°)+i\sin(-180°)\right] \\ &= -\frac{5}{6}.\end{aligned}$$

51. Let α and β be angles whose terminal sides go through $(2,1)$ and $(3,-2)$, respectively. Since $|2+i| = \sqrt{5}$ and $|3-2i| = \sqrt{13}$, we get $\cos\alpha = 2/\sqrt{5}$, $\sin\alpha = 1/\sqrt{5}$, $\cos\beta = 3/\sqrt{13}$, and $\sin\beta = -2/\sqrt{13}$. From the sum and difference identities, we obtain

$$\cos(\alpha+\beta) = \frac{8}{\sqrt{65}},\ \sin(\alpha+\beta) = -\frac{1}{\sqrt{65}},$$

$$\cos(\alpha-\beta) = \frac{4}{\sqrt{65}},\ \sin(\alpha-\beta) = \frac{7}{\sqrt{65}}.$$

Note $z_1 = \sqrt{5}(\cos\alpha + i\sin\alpha)$ and $z_2 = \sqrt{13}(\cos\beta + i\sin\beta)$. Then

$$\begin{aligned}z_1 z_2 &= \sqrt{65}\left(\cos(\alpha+\beta)+i\sin(\alpha+\beta)\right) \\ &= \sqrt{65}\left(\frac{8}{\sqrt{65}}-\frac{1}{\sqrt{65}}i\right) \\ z_1 z_2 &= 8-i\end{aligned}$$

and

$$\begin{aligned}\frac{z_1}{z_2} &= \frac{\sqrt{5}}{\sqrt{13}}\left(\cos(\alpha-\beta)+i\sin(\alpha-\beta)\right) \\ &= \frac{\sqrt{65}}{13}\left(\frac{4}{\sqrt{65}}+\frac{7}{\sqrt{65}}i\right) \\ \frac{z_1}{z_2} &= \frac{4}{13}+\frac{7}{13}i.\end{aligned}$$

53. $2^3\left[\cos 135° + i\sin 135°\right] =$

$$8\left[-\frac{\sqrt{2}}{2}+i\frac{\sqrt{2}}{2}\right] = -4\sqrt{2}+4\sqrt{2}i$$

55. $(4+4i)^3 = (4\sqrt{2})^3\left[\cos 45° + i\sin 45°\right]^3 =$ $128\sqrt{2}\left[\cos 135° + i\sin 135°\right] =$ $128\sqrt{2}\left[-\frac{1}{\sqrt{2}}+i\frac{1}{\sqrt{2}}\right] = -128+128i$

57. Square roots of i are given by

$$\cos\left(\frac{90°+k360°}{2}\right)+i\sin\left(\frac{90°+k360°}{2}\right) =$$

$\cos(45° + k\cdot 180°) + i\sin(45° + k\cdot 180°)$.

When $k = 0, 1$ one gets

$\cos 45° + i\sin 45° = \frac{\sqrt{2}}{2}+\frac{\sqrt{2}}{2}i$ and

$\cos 225° + i\sin 225° = -\frac{\sqrt{2}}{2}-\frac{\sqrt{2}}{2}i$.

59. Since $|\sqrt{3}+i| = 2$, the cube roots are

$$\sqrt[3]{2}\left[\cos\left(\frac{30°+k360°}{3}\right)+i\sin\left(\frac{30°+k360°}{3}\right)\right]$$

When $k = 0, 1, 2$ one gets $\sqrt[3]{2}\left[\cos\alpha + i\sin\alpha\right]$ where $\alpha = 10°, 130°, 250°$.

61. Since $|2+i| = \sqrt{5}$ and $\tan^{-1}(1/2) \approx 26.6°$, the arguments of the cube roots are

$$\frac{26.6°+k360°}{3} \approx 8.9° + k120°$$

When $k = 0, 1, 2$ one gets $\sqrt[6]{5}\left[\cos\alpha + i\sin\alpha\right]$ where $\alpha = 8.9°, 128.9°, 248.9°$.

63. Since $\sqrt[4]{625} = 5$, the fourth roots of $625i$ are

$$5\left[\cos\left(\frac{90°+k360°}{4}\right)+i\sin\left(\frac{90°+k360°}{4}\right)\right]$$

When $k = 0, 1, 2, 3$ the fourth roots are $5\left[\cos\alpha + i\sin\alpha\right]$ where $\alpha = 22.5°, 112.5°, 202.5°, 292.5°$.

65. $(5\cos 60°, 5\sin 60°) = \left(\frac{5}{2}, \frac{5\sqrt{3}}{2}\right)$

67. $(\sqrt{3}\cos 100°, \sqrt{3}\sin 100°) \approx (-0.3, 1.7)$

69. Note $r = \sqrt{(-2)^2 + (-2\sqrt{3})^2} = \sqrt{16} = 4$. Since $\tan\theta = \sqrt{3}$ and the terminal side of θ goes through $(-2, -2\sqrt{3})$, we have

$\theta = 4\pi/3$. Then $(r, \theta) = \left(4, \dfrac{4\pi}{3}\right)$.

71. Note $r = \sqrt{2^2 + (-3)^2} = \sqrt{13}$. Since $\theta = \tan^{-1}(-3/2) \approx -0.98$, we have $(r, \theta) \approx \left(\sqrt{13}, -0.98\right)$.

73. Circle centered at $(r, \theta) = (1, -\pi/2)$

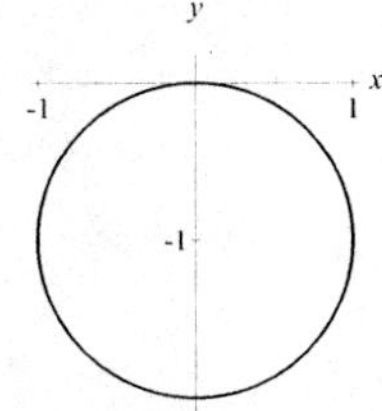

75. four-leaf rose

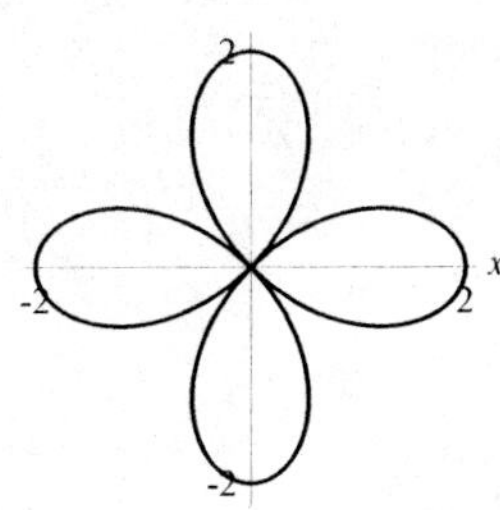

77. Limacon $r = 500 + \cos\theta$

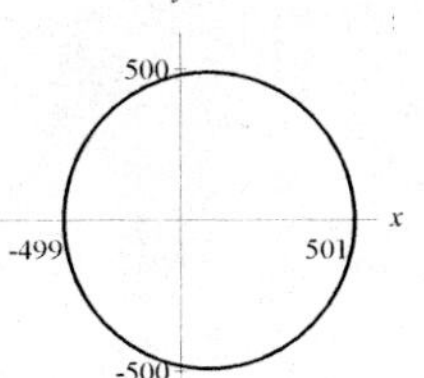

79. Horizontal line $y = 1$

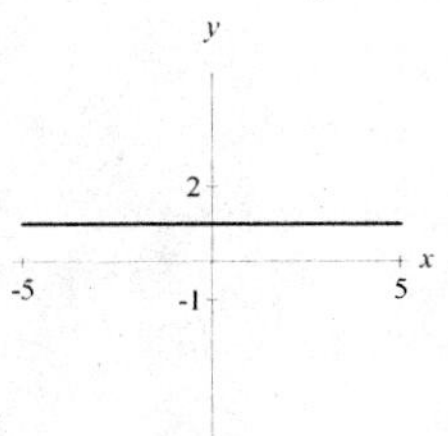

81. Since $r = \dfrac{1}{\sin\theta + \cos\theta}$, we obtain

$$\begin{aligned} r\sin\theta + r\cos\theta &= 1 \\ y + x &= 1. \end{aligned}$$

83. $x^2 + y^2 = 25$

85. Since $y = 3$, we find $r\sin\theta = 3$ and $r = \dfrac{3}{\sin\theta}$.

87. $r = 7$

89. The boundary points are given by

t	x	y
0	0	3
1	3	2

Note, the boundary points do not lie on the graph.

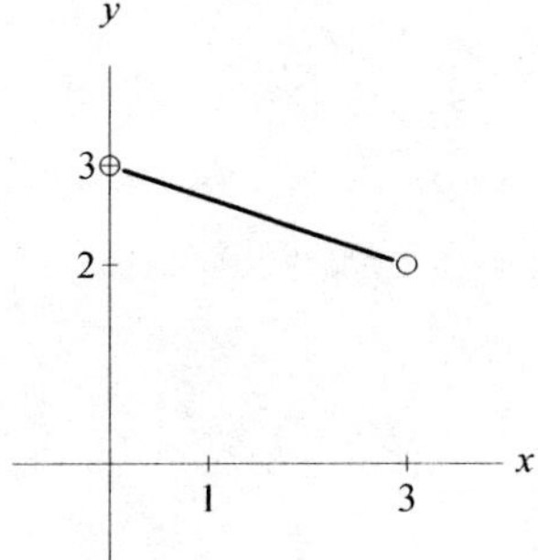

91. The graph is a quarter of a circle.

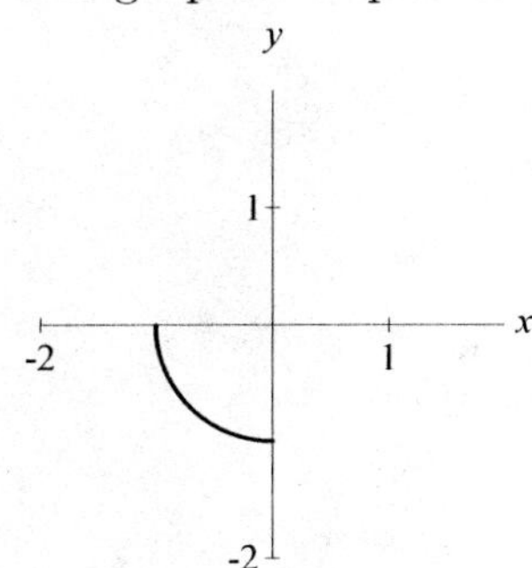

93. Draw two vectors with magnitudes 7 and 12 that act at an angle of 30° with each other.

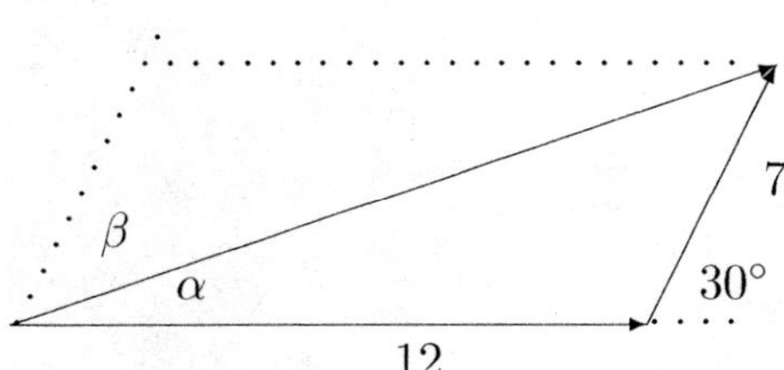

By the cosine law, the magnitude of the resultant force is

$$\sqrt{12^2 + 7^2 - 2(12)(7)\cos 150^\circ} \approx 18.4 \text{ lb}.$$

By the sine law, we find

$$\begin{aligned}\frac{7}{\sin\alpha} &= \frac{18.4}{\sin 150^\circ}\\ \sin\alpha &= \frac{7\sin 150^\circ}{18.4}\\ \sin\alpha &\approx 0.19\\ \alpha\approx\sin^{-1}(0.19) &\approx 11.0^\circ.\end{aligned}$$

The angles between the resultant and each force are 11.0° and $\beta = 180^\circ - 150^\circ - 11^\circ = 19.0^\circ$.

95. Since $\dfrac{482+364+241}{2} = 543.5$, then by using Heron's formula the area of Susan's lot is

$\sqrt{543.5(543.5-482)(543.5-364)(543.5-241)}$ $\approx 42,602\text{ ft}^2$.

Similarly, since $\dfrac{482+369+238}{2} = 544.5$, the area of Seth's lot is

$\sqrt{544.5(544.5-482)(544.5-369)(544.5-238)}$ $\approx 42,785\text{ ft}^2$.

Then Seth got the larger piece.

97. Consider triangle below.

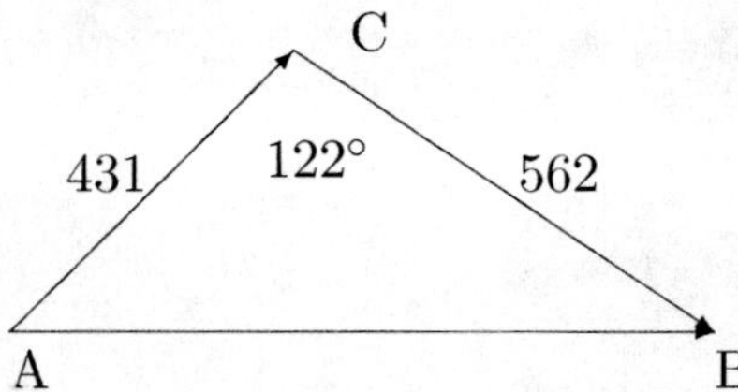

The distance between A and B is

$$\sqrt{431^2+562^2-2(431)(562)\cos 122^\circ} \approx 870.82\text{ ft}.$$

The extra amount spent is

$$(431+562-870.82)(\$21.60) \approx \$2639.$$

99. Let α be the base angle of the larger isosceles triangle. Drop a perpendicular from the top vertex to the base.

The perpendicular bisects the base of unit length into two equal parts.

Using right triangle trigonometry, we find $\cos\alpha = \frac{1}{4}$. The area of the shaded triangle is

$$\begin{aligned}\text{Area} &= \frac{1}{2}ab\sin C\\ &= \frac{1}{2}\sin\alpha\\ &= \frac{1}{2}\sqrt{1-\frac{1}{16}}\\ \text{Area} &= \frac{\sqrt{15}}{8}\end{aligned}$$

101. Let r be the radius of the sun and earth, which we assume are equal. Let a and α be the length of a chord and the corresponding central angle, respectively.

We are given $a = 0.8(2r)$. Then the length of the chord satisfies

$$\begin{aligned}a &= r\sqrt{2-2\cos\alpha}\\ 1.6r &= r\sqrt{2-2\cos\alpha}\\ 1.6^2 &= 2-2\cos\alpha\\ \cos\alpha &= -0.28.\end{aligned}$$

The part of the sun that is blocked is two times the area of a lens-like region. See Exercise 59, Section 7.2. Then

$$(\text{Twice Area of Lens}) = r^2(\alpha - \sin\alpha)$$

The percentage of the sun that is blocked is given by

$$\frac{\text{Twice Area of Lens}}{\pi r^2} = \frac{\alpha-\sin\alpha}{\pi} \approx 28\%.$$

103. Put the moon's center initially at $(0,0)$. Assume the moon and sun are circles of radii 1. Put the sun's center at $(2,0)$. At time $t=1$, put the moon's center at $(4,0)$. Then at time t, the moons center is at $C_t = (4t, 0)$.

At time t, the moon and sun intersects at some point P_t in the first quadrant. Drop a perpendicular from P_t to the x-axis. Label the foot of this perpendicular by F_t which is a point on the x-axis.

Let $\alpha_t = \angle P_t C_t F_t$ be the angle at C_T in the right triangle $\triangle P_t C_t F_t$. Notice, $\overline{C_t F_t} = 1 - 2t$.

Using right triangle trigonometry,

$$\cos \frac{\alpha_t}{2} = 1 - 2t.$$

By a double-angle identity for cosine,

$$\begin{aligned} \cos \alpha_t &= 2\cos^2 \alpha_t - 1 \\ &= 2(1-2t)^2 - 1 \\ &= 1 - 8t + 8t^2. \end{aligned}$$

The area of the sun that is blocked is two times the area of a lens-like region.

$$\begin{aligned} \text{(Twice Area of Lens)} &= r^2(\alpha_t - \sin \alpha_t) \\ &= \cos^{-1} u - \sin(\cos^{-1} u). \end{aligned}$$

where $r = 1$, and $u = 1 - 8t + 8t^2$. Hence, the portion of the sun that is blocked is the above area divided by the area of the sun, i.e.,

$$\frac{\cos^{-1} u - \sin(\cos^{-1} u)}{\pi}.$$

105.

a) First, use four tiles to make a 4-by-4 square. Then construct three more 4-by-4 square squares. Now, you have four 4-by-4 squares. Then put these four squares together to make a 8-by-8 square.

b) By elimination, you will not be able to make a 6-by-6 square. There are only a few possibilities and none of them will make a 6-by-6 square.

Chapter 7 Test

1. Draw a triangle with $\alpha = 30°$, $b = 4$, $a = 2$.

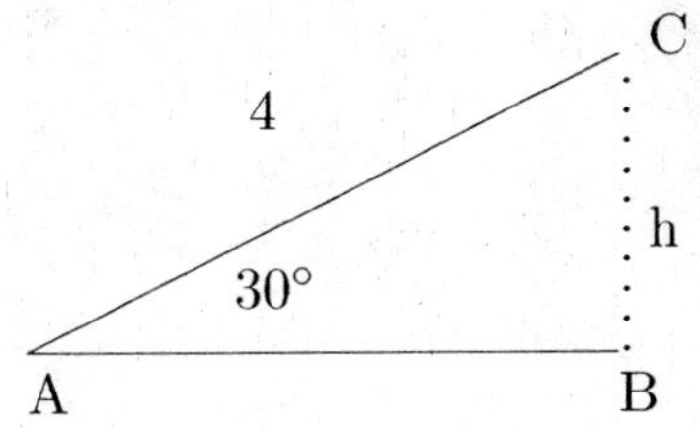

Since $h = 4 \sin 30° = 2$ and $a = 2$, there is only one triangle and $\beta = 90°$. So $\gamma = 90° - 30° = 60°$. Since $c^2 + 2^2 = 4^2$, we get $c = 2\sqrt{3}$.

2. Draw angle $\alpha = 60°$ and let h be the height.

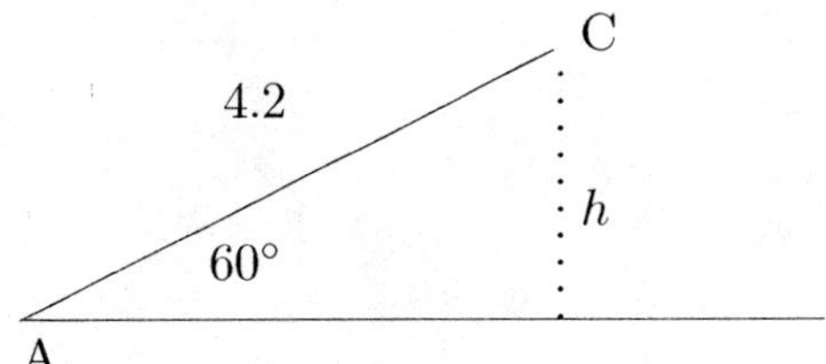

Since $h = 4.2 \sin 60° \approx 3.6$ and $3.6 < a < 4.2$, there are two triangles and they are given by

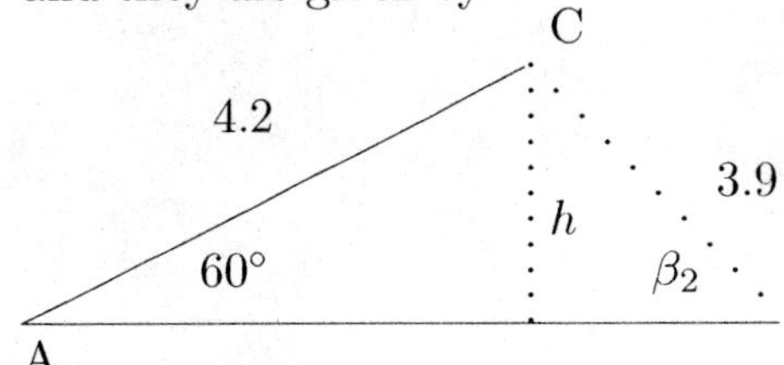

and

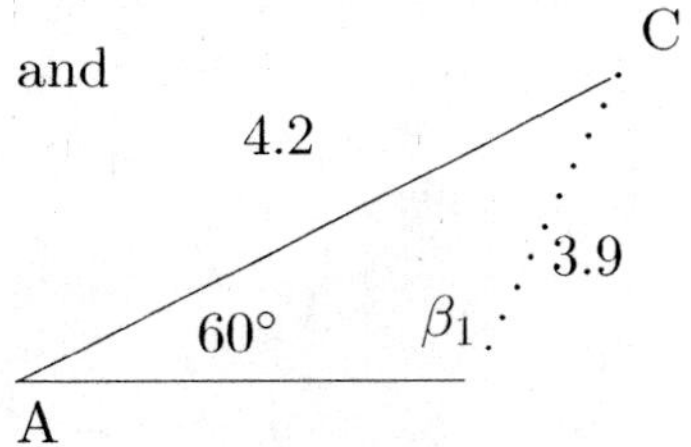

Apply the sine law to the acute triangle.

$$\begin{aligned} \frac{3.9}{\sin 60°} &= \frac{4.2}{\sin \beta_2} \\ \sin \beta_2 &= \frac{4.2 \sin 60°}{3.9} \\ \beta_2 &\approx 68.9° \end{aligned}$$

So $\gamma_2 = 180° - (68.9° + 60°) = 51.1°$.

By the sine law, $c_2 = \dfrac{3.9}{\sin 60°} \sin 51.1° \approx 3.5$.

In the obtuse triangle, $\beta_1 = 180° - \beta_2 = 111.1°$ and $\gamma_1 = 180° - (111.1° + 60°) = 8.9°$.

By the sine law, $c_1 = \dfrac{3.9}{\sin 60°} \sin 8.9° \approx 0.7$.

3. Draw the only triangle with $a = 3.6$, $\alpha = 20.3°$, and $\beta = 14.1°$.

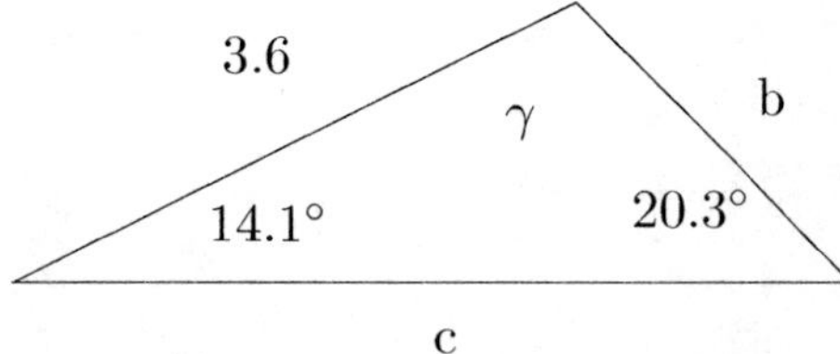

Note $\gamma = 180° - 14.1° - 20.3° = 145.6°$.
By the sine law, we find

$$\frac{b}{\sin 14.1°} = \frac{3.6}{\sin 20.3°} \text{ and } \frac{c}{\sin 145.6°} = \frac{3.6}{\sin 20.3°}.$$

Then $b \approx 2.5$ and $c \approx 5.9$.

4. Draw the only triangle with $a = 2.8$, $b = 3.9$, and $\gamma = 17°$.

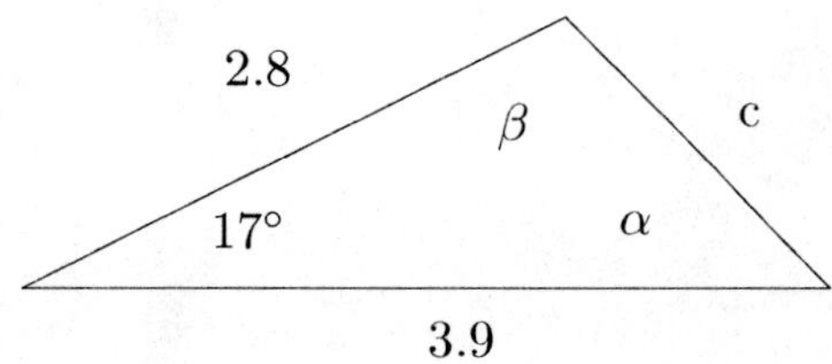

By the cosine law, we get $c = \sqrt{3.9^2 + 2.8^2 - 2(3.9)(2.8)\cos 17°} \approx 1.47 \approx 1.5$. By the sine law,

$$\begin{aligned} \frac{1.47}{\sin 17°} &= \frac{2.8}{\sin \alpha} \\ \sin \alpha &= \frac{2.8 \sin 17°}{1.47} \\ \sin \alpha &\approx 0.5569 \\ \alpha \approx \sin^{-1}(0.5569) &\approx 33.8°. \end{aligned}$$

Also, $\beta = 180° - (33.8° + 17°) = 129.2°$.

5. Draw the only triangle with the given sides $a = 4.1$, $b = 8.6$, and $c = 7.3$.

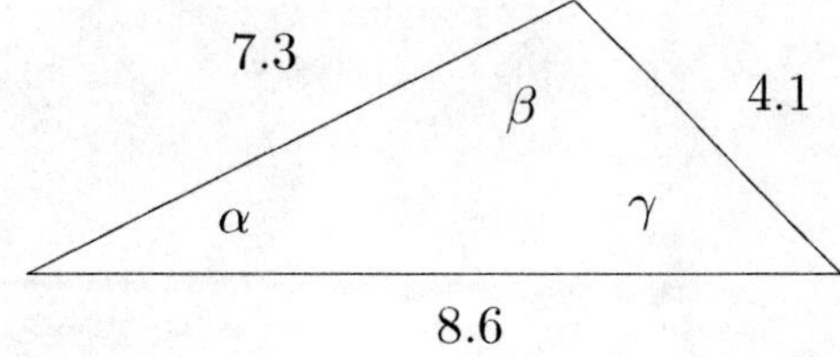

First, find the largest angle β by the cosine law.

$$\begin{aligned} \cos \beta &= \frac{7.3^2 + 4.1^2 - 8.6^2}{2(7.3)(4.1)} \\ \cos \beta &\approx -0.06448 \\ \beta &\approx \cos^{-1}(-0.06448) \\ \beta &\approx 93.7°. \end{aligned}$$

By the sine law,

$$\begin{aligned} \frac{8.6}{\sin 93.7°} &= \frac{7.3}{\sin \gamma} \\ \sin \gamma &= \frac{7.3 \sin 93.7°}{8.6} \\ \sin \gamma &\approx 0.8471 \\ \gamma \approx \sin^{-1}(0.8471) &\approx 57.9°. \end{aligned}$$

Also $\alpha = 180° - (57.9° + 93.7°) = 28.4°$.

6. The magnitude of $\boldsymbol{A} + \boldsymbol{B} = \langle -2, 6 \rangle$ is $\sqrt{(-2)^2 + 6^2} = \sqrt{40} = 2\sqrt{10}$.
Direction angle is $\cos^{-1}(-2/\sqrt{40}) \approx 108.4°$.

7. The magnitude of $\boldsymbol{A} - \boldsymbol{B} = \langle -4, -2 \rangle$ is $\sqrt{(-4)^2 + (-2)^2} = \sqrt{20} = 2\sqrt{5}$.
Since $\tan^{-1}(2/4) \approx 26.6°$, the direction angle is $180° + 26.6° = 206.6°$.

8. Magnitude of $3\boldsymbol{B} = \langle 3, 12 \rangle$ is $\sqrt{3^2 + 12^2} = \sqrt{153} = 3\sqrt{17}$.
Direction angle is $\tan^{-1}(12/3) \approx 76.0°$.

9. Since $|3 + 3i| = 3\sqrt{2}$ and $\tan^{-1}(3/3) = 45°$, we have $3 + 3i = 3\sqrt{2}\,[\cos 45° + i \sin 45°]$.

10. Since $|-1 + i\sqrt{3}| = 2$ and $\cos^{-1}(-1/2) = 120°$, we have $|-1 + i\sqrt{3}| = 2\,[\cos 120° + i \sin 120°]$.

11. Note $|-4 - 2i| = \sqrt{20} = 2\sqrt{5}$.
Since $\tan^{-1}(2/4) = 26.6°$, the direction angle of $-4 - 2i$ is $180° + 26.6° = 206.6°$.
So $-4 - 2i = 2\sqrt{5}\,[\cos 206.6° + i \sin 206.6°]$.

12. $6\,[\cos 45° + i \sin 45°] =$

$$6\left[\frac{\sqrt{2}}{2} + i\frac{\sqrt{2}}{2}\right] = 3\sqrt{2} + 3i\sqrt{2}$$

13. $2^9\,[\cos 90° + i \sin 90°] =$
$512\,[0 + i] = 512i$

14. $\frac{3}{2}\left[\cos 45^\circ + i\sin 45^\circ\right] =$

$\frac{3}{2}\left[\frac{\sqrt{2}}{2} + \frac{\sqrt{2}}{2}i\right] = \frac{3\sqrt{2}}{4} + \frac{3\sqrt{2}}{4}i$

15. $(5\cos 30^\circ, 5\sin 30^\circ) = \left(5\frac{\sqrt{3}}{2}, 5\frac{1}{2}\right) =$

$\left(\frac{5\sqrt{3}}{2}, \frac{5}{2}\right)$

16. $(-3\cos(-\pi/4), -3\sin(-\pi/4)) =$

$\left(-3\frac{\sqrt{2}}{2}, 3\frac{\sqrt{2}}{2}\right) = \left(-\frac{3\sqrt{2}}{2}, \frac{3\sqrt{2}}{2}\right)$

17. $(33\cos 217^\circ, 33\sin 217^\circ) \approx (-26.4, -19.9)$

18. Circle of radius 5/2 with center at $(r, \theta) = (5/2, 0)$

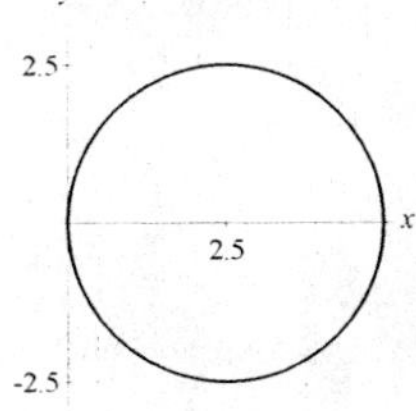

19. Four-leaf rose.

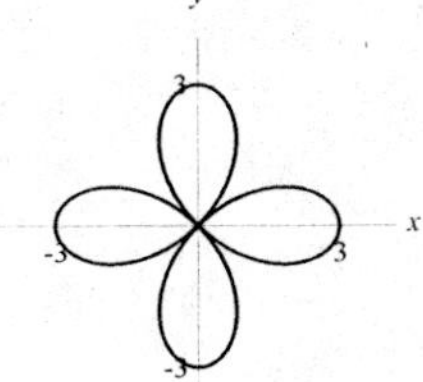

20. We apply Heron's formula.
Since $\frac{4.1+6.8+9.5}{2} = 10.2$, the area is
$\sqrt{10.2(10.2-4.1)(10.2-6.8)(10.2-9.5)}$
$\approx 12.2 \text{ m}^2$.

21. Since $a_1 = 4.6\cos 37.2^\circ \approx 3.66$ and $a_2 = 4.6\sin 37.2^\circ \approx 2.78$, we have
$\boldsymbol{v} \approx 3.66\ \boldsymbol{i} + 2.78\ \boldsymbol{j}$.

22. Fourth roots of -81 are given by

$3\left[\cos\left(\frac{180^\circ + k360^\circ}{4}\right) + i\sin\left(\frac{180^\circ + k360^\circ}{4}\right)\right]$

$= 3\left[\cos(45^\circ + k90^\circ) + i\sin(45^\circ + k90^\circ)\right].$

When $k = 0, 1, 2, 3$ one gets

$$3\left[\cos 45^\circ + i\sin 45^\circ\right] = \frac{3\sqrt{2}}{2} + i\frac{3\sqrt{2}}{2},$$

$$3\left[\cos 135^\circ + i\sin 135^\circ\right] = -\frac{3\sqrt{2}}{2} + i\frac{3\sqrt{2}}{2},$$

$$3\left[\cos 225^\circ + i\sin 225^\circ\right] = -\frac{3\sqrt{2}}{2} - i\frac{3\sqrt{2}}{2},$$

$$3\left[\cos 315^\circ + i\sin 315^\circ\right] = \frac{3\sqrt{2}}{2} - i\frac{3\sqrt{2}}{2}.$$

23. Since $x^2 + y^2 + 5y = 0$, we obtain

$$\begin{aligned} r^2 + 5r\sin\theta &= 0 \\ r + 5\sin\theta &= 0 \\ r &= -5\sin\theta \end{aligned}$$

24. Since $r = 5(2\sin\theta\cos\theta)$, we find

$$\begin{aligned} r^3 &= 10(r\sin\theta)(r\cos\theta) \\ r^3 &= 10yx \\ (x^2+y^2)^{3/2} &= 10xy. \end{aligned}$$

25. The line that passes through through $(-2, -3)$ and $(4, 5)$ is given by

$$y = \frac{4}{3}x - \frac{1}{3}.$$

The linear function $x = f(t)$ that satisfies

i) $x = -2$ when $t = 0$, and

ii) $x = 4$ when $t = 1$

is defined by

$$x = 6t - 2.$$

Then

$$y = \frac{4}{3}(6t - 2) - \frac{1}{3} = 8t - 3.$$

Thus, the parametric equations are

$$x = 6t - 2, \ y = 8t - 3$$

for $0 \le t \le 1$.

26. Draw two vectors with magnitudes 240 and 30.

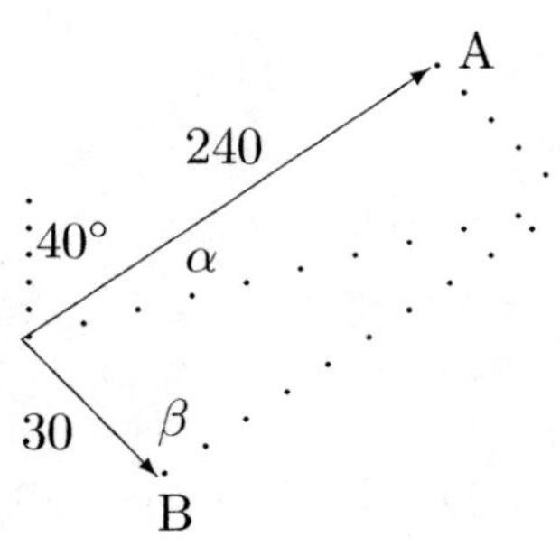

Ground speed of the airplane is the magnitude of the resultant force. From the difference of the two bearings, one sees that 95° is the angle between the two vectors. Since the sum of the angles in a parallelogram is 360°, we get $\beta = \frac{360° - 2 \cdot 95°}{2} = 85°$. By the cosine law, the ground speed is

$$\sqrt{30^2 + 240^2 - 2(30)(240)\cos 85°} \approx 239.3 \text{ mph}.$$

By the sine law, we find

$$\begin{aligned} \frac{30}{\sin\alpha} &= \frac{239.3}{\sin 85°} \\ \sin\alpha &= \frac{30\sin 85°}{239.3} \\ \sin\alpha &\approx 0.12489 \\ \alpha &\approx \sin^{-1}(0.12489) \\ \alpha &\approx 7.2°. \end{aligned}$$

The bearing of the course is $\alpha + 40° = 47.2°$.

Tying It All Together

1. Use the quadratic formula to find the zeros of the second degree factor in

$$x(x^3 - 1) = x(x-1)(x^2 + x + 1).$$

Then

$$\begin{aligned} x &= \frac{-1 \pm \sqrt{1^2 - 4(1)(1)}}{2} \\ x &= \frac{-1 \pm \sqrt{-3}}{2}. \end{aligned}$$

Roots are $x = 0, 1, -\frac{1}{2} \pm \frac{\sqrt{3}}{2}i$.

2. The zeros are $-2, 1, 3$ since

1	1	-2	-5	6
		1	-1	-6
	1	-1	-6	0

and $x^2 - x - 6 = (x-3)(x+2)$

3. Factoring by grouping, we obtain $x^5(x+2) - (x+2) = (x^5 - 1)(x+2) = 0$.
Then $x^5 = 1$ and $x = -2$.
The fifth roots of 1 are given by
$\cos\left(\frac{k360°}{5}\right) + i\sin\left(\frac{k360°}{5}\right) =$
$\cos(k72°) + i\sin(k72°)$.
When $k = 0, 1, 2, 3, 4$ one obtains
$\cos 0° + i\sin 0° = 1$, $\cos 72° + i\sin 72°$, $\cos 144° + i\sin 144°$, $\cos 216° + i\sin 216°$, and $\cos 288° + i\sin 288°$.

4. Factoring by grouping, we get $x^4(x^3 - 1) + 2(x^3 - 1) = (x^4 + 2)(x^3 - 1) = 0$.
So $x^4 = -2$ and $x^3 = 1$. The fourth roots of -2 are given by
$\sqrt[4]{2}\left[\cos\left(\frac{180° + k360°}{4}\right) + i\sin\left(\frac{180° + k360°}{4}\right)\right] =$
$\sqrt[4]{2}\left[\cos(45° + k90°) + i\sin(45° + k90°)\right]$.
When $k = 0, 1, 2, 3$ one obtains
$\sqrt[4]{2}\left[\cos 45° + i\sin 45°\right] = \frac{1}{\sqrt[4]{2}} + i\frac{1}{\sqrt[4]{2}}$,
$\sqrt[4]{2}\left[\cos 135° + i\sin 135°\right] = -\frac{1}{\sqrt[4]{2}} + i\frac{1}{\sqrt[4]{2}}$,
$\sqrt[4]{2}\left[\cos 225° + i\sin 225°\right] = -\frac{1}{\sqrt[4]{2}} - i\frac{1}{\sqrt[4]{2}}$, and
$\sqrt[4]{2}\left[\cos 315° + i\sin 315°\right] = \frac{1}{\sqrt[4]{2}} - i\frac{1}{\sqrt[4]{2}}$.
The cube roots of 1 are given by
$\cos\left(\frac{k360°}{3}\right) + i\sin\left(\frac{k360°}{3}\right)$
$= \cos(k120°) + i\sin(k120°)$
When $k = 0, 1, 2$ one obtains
$\cos 0° + i\sin 0° = 1$,
$\cos 120° + i\sin 120° = -\frac{1}{2} + \frac{\sqrt{3}}{2}i$, and

$$\cos 240° + i \sin 240° = -\frac{1}{2} - \frac{\sqrt{3}}{2}i.$$

The solutions are the fourth roots of -2 and the cube roots of 1.

5. By a double-angle identity, we find

$$2(2\sin x \cos x) - 2\cos x + 2\sin x - 1 = 0$$

$$\begin{aligned} 2\cos x(2\sin x - 1) + (2\sin x - 1) &= 0 \\ (2\cos x + 1)(2\sin x - 1) &= 0 \\ \cos x = -\frac{1}{2} \text{ or } \sin x &= \frac{1}{2}. \end{aligned}$$

The solutions are $x = \frac{2\pi}{3} + 2k\pi$, $\frac{4\pi}{3} + 2k\pi$, $\frac{\pi}{6} + 2k\pi$, and $\frac{5\pi}{6} + 2k\pi$.

6.

$$\begin{aligned} 4x\sin x - 2x + 2\sin x - 1 &= 0 \\ 2x(2\sin x - 1) + (2\sin x - 1) &= 0 \\ (2x + 1)(2\sin x - 1) &= 0 \\ x = -\frac{1}{2} \text{ or } \sin x &= \frac{1}{2} \end{aligned}$$

The solutions are $x = -\frac{1}{2}, \frac{\pi}{6} + 2k\pi, \frac{5\pi}{6} + 2k\pi$.

7. Since $e^{\sin x} = 1$, we get $\sin x = 0$. The solutions are $x = k\pi$ where k is an integer.

8. Since $\sin(e^x) = 1/2$, the solutions are given by

$$\begin{aligned} e^x = \frac{\pi}{6} + 2k\pi \quad &\text{or} \quad e^x = \frac{5\pi}{6} + 2k\pi \\ x = \ln\left(\frac{\pi}{6} + 2k\pi\right) \quad &\text{or} \quad x = \ln\left(\frac{5\pi}{6} + 2k\pi\right) \end{aligned}$$

where k is a nonnegative integer.

9. Using a common base, we obtain

$$\begin{aligned} 2^{2x-3} &= 2^5 \\ 2x - 3 &= 5 \\ 2x &= 8 \\ x &= 4. \end{aligned}$$

10. Express the left-hand side as a single logarithm.

$$\begin{aligned} \log\left(\frac{x-1}{x+2}\right) &= 2 \\ \frac{x-1}{x+2} &= 10^2 \\ x - 1 &= 100x + 200 \\ -201 &= 99x \end{aligned}$$

Checking $x = -201/99$ one gets $\log(-201/99 - 1)$ which is undefined. The solution set is $\emptyset$.

11. $y = \sin x$ has amplitude 1 and period 2π.

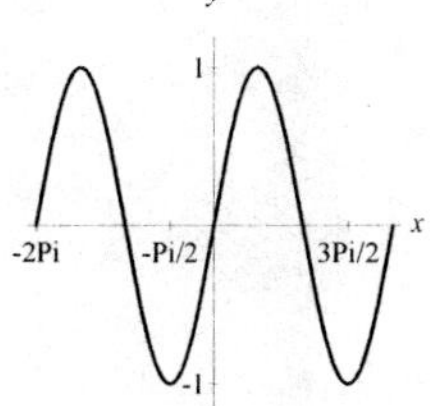

12. $y = e^x$ goes through $(-1, 1/e)$, $(0, 1)$, $(1, e)$.

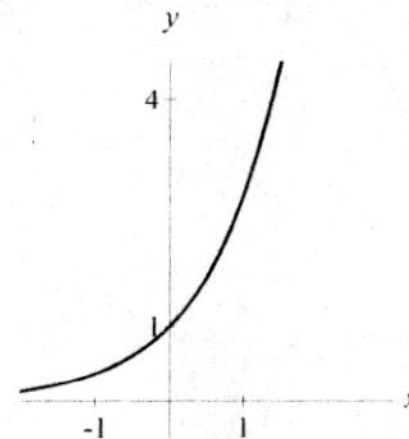

13. $r = \sin\theta$ in polar coordinates is a circle through the origin and $(0, 1)$ with radius $1/2$.

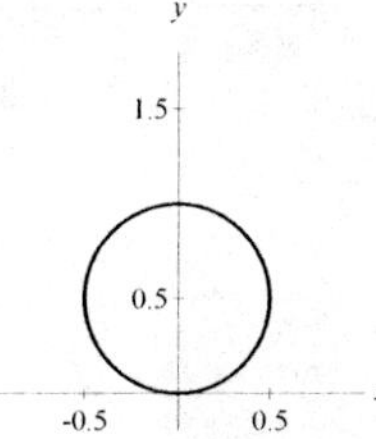

14. $r = \theta$ in polar coordinates goes through $(-\pi, -\pi)$, $(0, 0)$, (π, π).

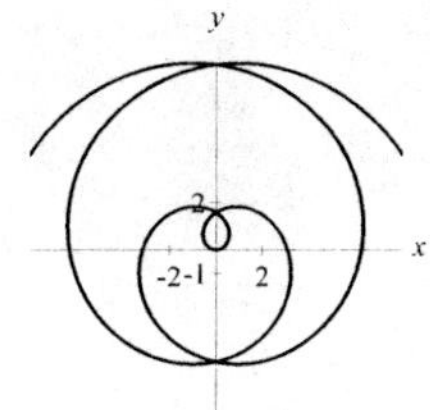

15. $y = \sqrt{\sin x}$ is only defined for values of x for which $\sin x$ is nonnegative.

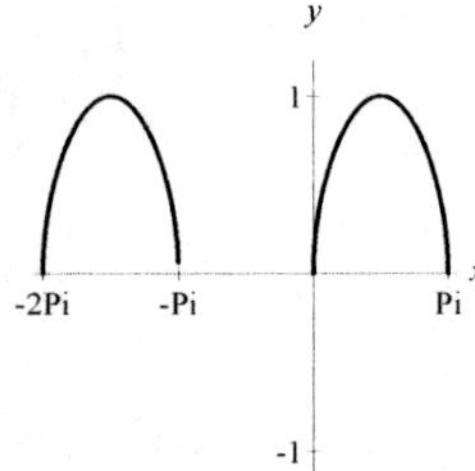

16. $y = \ln(\sin x)$ has vertical asymptotes at $x = k$ where $\sin k = 0$ and is only defined for values of x for which $\sin x > 0$.

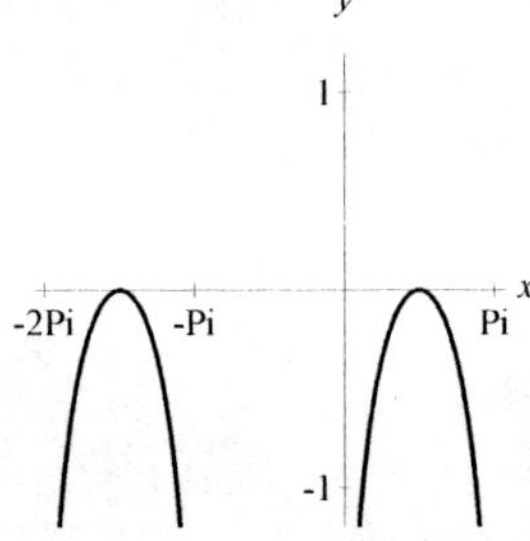

17. $r = \sin(\pi/3) = \sqrt{3}/2$ is a circle with radius $\sqrt{3}/2$ and center at the origin.

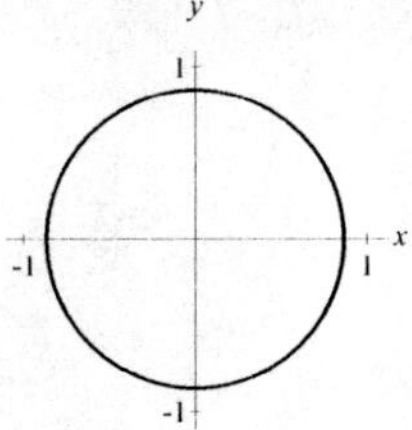

18. $y = x^{1/3}$ goes through $(-8, -2)$, $(0, 0)$, $(8, 2)$.

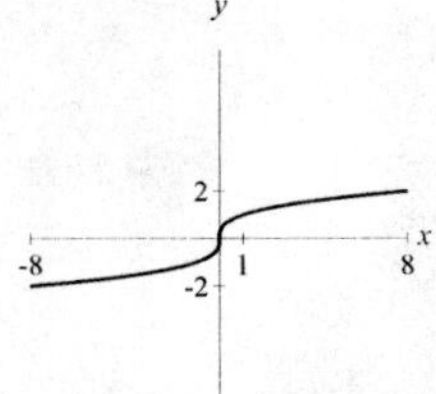

19. $\log(1) = 0$ **20.** $\sin(0) = 0$

21. $\cos(\pi) = -1$ **22.** $\ln(1) = 0$

23. $\sin^{-1}(1/2) = \pi/6$ **24.** $\cos^{-1}(1/2) = \pi/3$

25. $\tan^{-1}(-1) = -\pi/4$ **26.** $\tan^{-1}(1) = \pi/4$

27. $\dfrac{(x-2)+(x+2)}{(x+2)(x-2)} = \dfrac{2x}{x^2-4}$

28. $\dfrac{(1+\sin x)+(1-\sin x)}{1-\sin^2 x} = \dfrac{2}{\cos^2 x} = 2\sec^2 x$

29. $\dfrac{(2-\sqrt{3})+(2+\sqrt{3})}{2^2-3} = \dfrac{4}{1} = 4$

30.

$$\log\left(\frac{1}{(x+2)(x-2)}\right) = \log\left(\frac{1}{x^2-4}\right) =$$
$$-\log(x^2-4)$$

31.

$$\frac{(x-3)(x+3)\cdot 4(x+3)}{2(x-3)\cdot(x+3)^2} = 2$$

32.

$$\frac{\cos^2 x(\cos x - \sin x)(\cos x + \sin x)}{(2\cos^2 x - 2\sin x\cos x)(2\cos^2 x + 2\sin x \cos x)}$$
$$= \frac{\cos^2 x(\cos x - \sin x)(\cos x + \sin x)}{4\cos^2 x(\cos x - \sin x)(\cos x + \sin x)} = \frac{1}{4}$$

33.

$$\frac{\sqrt{16}}{(2-\sqrt{3})(4+2\sqrt{3})} =$$
$$\frac{4}{(2-\sqrt{3})\cdot 2(2+\sqrt{3})} =$$
$$\frac{2}{(2-\sqrt{3})(2+\sqrt{3})} = \frac{2}{4-3} = 2$$

34.

$$\log\left(\frac{(x+1)(x+2)}{(x+3)(x+2)}\right) - \log\left(\frac{x+1}{x+3}\right) =$$
$$\log\left(\frac{x+1}{x+3}\right) - \log\left(\frac{x+1}{x+3}\right) = 0$$

35. Since $x^2 - 4 = (x-2)(x+2)$, we find

$$\frac{\dfrac{1}{x-2} + \dfrac{1}{x+2}}{\dfrac{1}{x^2-4} - \dfrac{1}{x+2}} \cdot \frac{x^2-4}{x^2-4} = \frac{(x+2)+(x-2)}{1-(x-2)}$$

$$\frac{2x}{3-x} = \frac{-2x}{x-3}.$$

36. Express in terms of $\sin x$ and $\cos x$.

$$\begin{aligned} \frac{\dfrac{1}{2\sin x\cos x}-\dfrac{\sin x}{\cos x}}{\dfrac{\cos x}{2\sin x}-\dfrac{\sin x}{2\cos x}}\cdot\frac{2\sin x\cos x}{2\sin x\cos x} &= \\ \frac{1-2\sin^2 x}{\cos^2 x-\sin^2 x} &= \\ \frac{\cos(2x)}{\cos(2x)} &= \\ 1 & \end{aligned}$$

37. frequency

38. vertical asymptote

39. even

40. odd

41. opposite, hypotenuse

42. adjacent, hypotenuse

43. one

44. period

45. Pythagorean

46. even, odd

For Thought

1. True, since if $x = 2$ and $y = 3$ we get $2 + 3 = 5$.

2. False, since $(2, 3)$ does not satisfy $x - y = 1$.

3. False, it is independent since the two lines are perpendicular and intersect at only one point.

4. True. Multiply $x - 2y = 4$ by -3 and add to the second equation.

$$\begin{aligned} -3x + 6y &= -12 \\ 3x - 6y &= 8 \\ \hline 0 &= -4 \end{aligned}$$

Since $0 = -4$ is false, there is no solution.

5. True, adding gives $2x = 6$. **6.** True

7. False, it is dependent because substituting $x = 5 + 3y$ results in an identity.

$$\begin{aligned} 9y - 3(5 + 3y) &= -15 \\ 9y - 15 - 9y &= -15 \\ -15 &= -15 \end{aligned}$$

8. False, it is dependent and the solution set is $\{(5 + 3y, y)|y \text{ is any real number}\}$.

9. False, it is dependent, there are an infinite number of solutions.

10. True, since both lines have slope $1/2$.

8.1 Exercises

1. system

3. inconsistent

5. dependent

7. Note, $x = 1$ and $y = 3$ satisfies both $x + y = 4$ and $x - y = -2$. Yes, $(1, 2)$ is a solution.

9. Note, $x = -1$ and $y = 5$ does not satisfy $x - 2y = -9$. Thus, $(-1, 5)$ is not a solution.

11. $\{(1, 2)\}$

13. No solution since the lines are parallel. The solution set is $\emptyset$.

15. 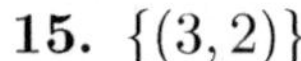 $\{(3, 2)\}$

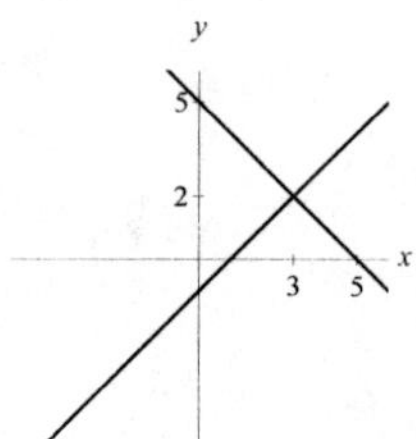

17. $\{(3, 1)\}$

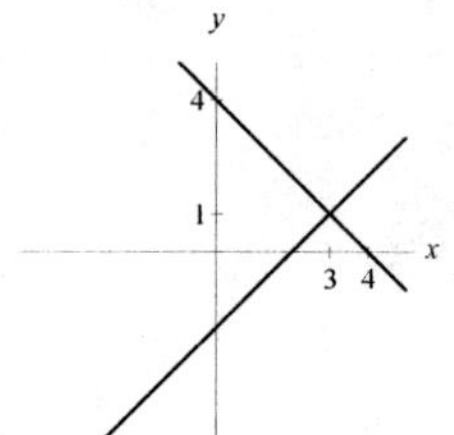

19. No solution since the lines are parallel. The solution set is $\emptyset$.

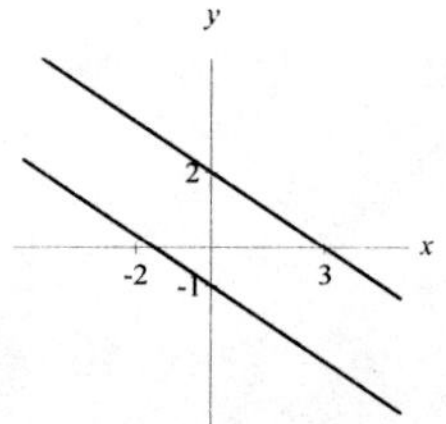

21. Since the lines are identical, the solution set is $\{(x, y) \; |x - 2y = 6\}$.

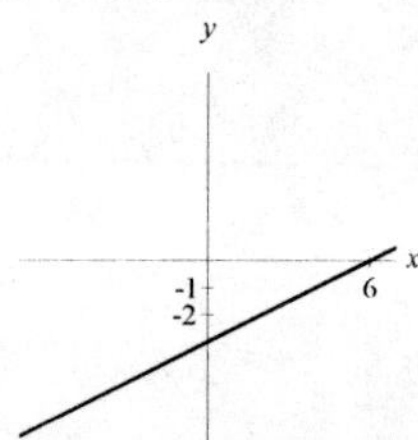

23. Substitute $y = 2x + 1$ into $3x - 4y = 1$.

$$\begin{aligned} 3x - 4(2x + 1) &= 1 \\ 3x - 8x - 4 &= 1 \\ -5x &= 5 \\ x &= -1 \end{aligned}$$

From $y = 2x + 1$, $y = 2(-1) + 1 = -1$. Independent and solution set is $\{(-1, -1)\}$.

25. Substitute $y = 1 - x$ into $2x - 3y = 8$.

$$\begin{aligned} 2x - 3(1 - x) &= 8 \\ 2x - 3 + 3x &= 8 \end{aligned}$$

$$\begin{aligned} 5x &= 11 \\ x &= 11/5 \end{aligned}$$

From $y = 1 - x$, $y = 1 - 11/5 = -6/5$. Independent and solution set is $\{(11/5, -6/5)\}$.

27. Substitute $y = 3x + 5$ into $3(x + 1) = y - 2$.

$$\begin{aligned} 3x + 3 &= (3x + 5) - 2 \\ 3x + 3 &= 3x + 3 \\ 3 &= 3 \end{aligned}$$

Dependent and solution set is $\{(x, y) \mid y = 3x + 5\}$.

29. Multiplying $\frac{1}{2}x + \frac{1}{3}y = 3$ by 6, we get $3x + 2y = 18$. Then substitute $2y = 6 - 3x$ into $3x + 2y = 18$. So

$$\begin{aligned} 3x + (6 - 3x) &= 18 \\ 6 &= 18 \end{aligned}$$

Inconsistent and the solution set is $\emptyset$.

31. Multiplying $0.05x + 0.06y = 10.50$ by 100, we obtain $5x + 6y = 1050$. Then substitute $y = 200 - x$ into $5x + 6y = 1050$.

$$\begin{aligned} 5x + 6(200 - x) &= 1050 \\ 5x + 1200 - 6x &= 1050 \\ -x &= -150 \\ x &= 150 \end{aligned}$$

From $y = 200 - x$, $y = 200 - 150 = 50$. Independent and solution set is $\{(150, 50)\}$.

33. Since $3x + 1 = 3x - 7$ leads to $1 = -7$, the system is inconsistent and the solution set is $\emptyset$.

35. Multiplying the first and second equations by 6 and 4, respectively, we have $3x - 2y = 72$ and $x - 2y = 4$. Substitute $2y = x - 4$ into $3x - 2y = 72$.

$$\begin{aligned} 3x - (x - 4) &= 72 \\ 2x + 4 &= 72 \\ 2x &= 68 \\ x &= 34 \end{aligned}$$

From $2y = x - 4$, $y = \dfrac{34 - 4}{2} = 15$. Independent and solution set is $\{(34, 15)\}$.

37. Adding the two equations, we get $2x = 26$. So $x = 13$ and from $x + y = 20$, we obtain $13 + y = 20$ or $y = 7$. Independent and solution set is $\{(13, 7)\}$.

39. Multiplying $x - y = 5$ by 2 and by adding to the second equation, we obtain

$$\begin{array}{rcr} 2x - 2y &=& 10 \\ 3x + 2y &=& 10 \\ \hline 5x &=& 20 \\ x &=& 4 \end{array}$$

From $x - y = 5$, $4 - y = 5$ or $y = -1$. Independent and solution set is $\{(4, -1)\}$.

41. Adding the two equations leads to $0 = 12$. Inconsistent and the solution set is $\emptyset$.

43. Multiply $2x + 3y = 1$ by -3 and $3x - 5y = -8$ by 2. Then add the equations.

$$\begin{array}{rcr} -6x - 9y &=& -3 \\ 6x - 10y &=& -16 \\ \hline -19y &=& -19 \\ y &=& 1 \end{array}$$

Since $2x + 3y = 1$, $2x + 3 = 1$ or $x = -1$. Independent and solution set is $\{(-1, 1)\}$.

45. Multiply $0.05x + 0.1y = 0.6$ by -100 and $x + 2y = 12$ by 5. Then add the equations.

$$\begin{array}{rcr} -5x - 10y &=& -60 \\ 5x + 10y &=& 60 \\ \hline 0 &=& 0 \end{array}$$

Dependent and the solution set is $\{(x, y) \mid x + 2y = 12\}$.

47. Multiplying $\dfrac{x}{2} + \dfrac{y}{2} = 5$ by 2 and $\dfrac{3x}{2} - \dfrac{2y}{3} = 2$ by 6, we have $x + y = 10$ and $9x - 4y = 12$, respectively. Then multiply $x + y = 10$ by 4 and add to the second equation.

$$\begin{array}{rcr} 4x + 4y &=& 40 \\ 9x - 4y &=& 12 \\ \hline 13x &=& 52 \\ x &=& 4 \end{array}$$

Since $x + y = 10$, $4 + y = 10$ and $y = 6$. Independent and the solution set is $\{(4, 6)\}$.

49. Multiply $3x - 2.5y = -4.2$ by -4 and $0.12x + 0.09y = 0.4932$ by 100. Then add the equations.

$$\begin{aligned} -12x + 10y &= 16.8 \\ 12x + 9y &= 49.32 \\ \hline 19y &= 66.12 \\ y &= 3.48 \end{aligned}$$

From $3x - 2.5y = -4.2$, we find

$$\begin{aligned} 3x - 2.5(3.48) &= -4.2 \\ 3x - 8.7 &= -4.2 \\ 3x &= 4.5 \\ x &= 1.5. \end{aligned}$$

Independent and the solution set is $\{(1.5, 3.48)\}$.

51. Independent

53. Dependent

55. The point of intersection is $(-1000, -497)$.

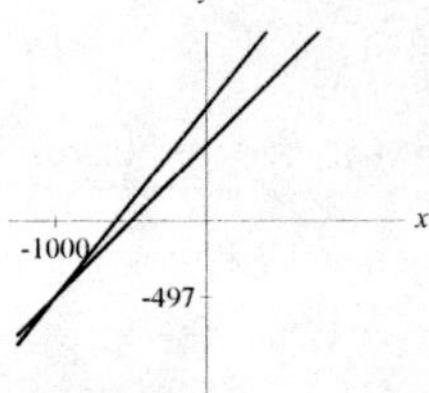

57. The point of intersection is approximately $(6.18, -0.54)$

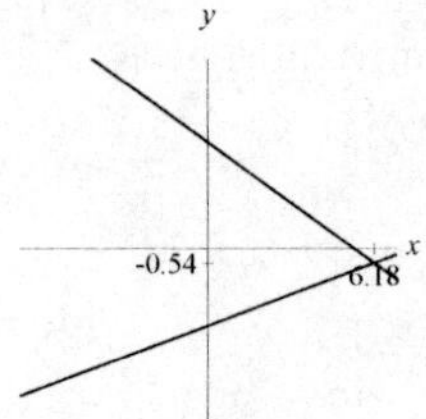

59. Let x and y be Althea's and Vaughn's incomes, respectively. Then $x+y = 82,000$ and $x-y = 16,000$. By adding these two equations, we get $2x = 98,000$ or $x = 49,000$. Thus, Althea's income is \$49,000 and Vaughn's income is \$33,000.

61. Let x and y be the amounts invested at 10% and 8%, respectively.

$$\begin{aligned} x + y &= 25,000 \\ 0.1x + 0.08y &= 2200 \end{aligned}$$

Multiply the second equation by -10 and add to the first.

$$\begin{aligned} x + y &= 25,000 \\ -x - 0.8y &= -22,000 \\ \hline 0.2y &= 3000 \\ y &= 15,000 \end{aligned}$$

Carmen invested \$15,000 at 8% and \$10,000 at 10%.

63. Let x and y be the prices of an adult ticket and child ticket, respectively.

$$\begin{aligned} 2x + 5y &= 33 \\ x + 3y &= 18.50 \end{aligned}$$

Multiply the second equation by -2 and add to the first equation.

$$\begin{aligned} 2x + 5y &= 33 \\ -2x - 6y &= -37 \\ \hline -y &= -4 \\ y &= 4 \end{aligned}$$

A child's ticket costs \$4. Since $x + 3y = 18.50$, $x + 12 = 18.50$ and so an adult's ticket costs $x = \$6.50$.

65. If m and f are the number of male and female memberships, respectively, then a system of equations is

$$\begin{aligned} m + f &= 12 \\ 500m + 500f &= 6,000. \end{aligned}$$

Dividing the second equation by 500,one finds $m + f = 12$; the system of equations is dependent. Therefore, one cannot conclude the number of female memberships and male memberships.

All we know is that m and $12-m$ are the number of male and female memberships where $0 \leq m \leq 12$.

67. Let x and y be the number of cows and ostriches, respectively.

$$\begin{aligned} 2x + 2y &= 84 \\ 4x + 2y &= 122 \end{aligned}$$

Subtract the first equation from the second equation.

$$\begin{aligned} 2x &= 38 \\ x &= 19 \end{aligned}$$

Consequently, we have

$$\begin{aligned} 2x + 2y &= 84 \\ 38 + 2y &= 84 \\ 2y &= 46 \\ y &= 23 \end{aligned}$$

Hence, there are $x = 19$ cows and $y = 23$ ostriches.

69. Let x and y be the number of cows and horses, respectively.

$$\begin{aligned} 4x + 4y &= 96 \\ x + y &= 24 \end{aligned}$$

Divide the first equation by 4.

$$\begin{aligned} x + y &= 24 \\ x + y &= 24 \end{aligned}$$

Since the two equations are identical, the system of equations is dependent and there are infinitely many possible solutions.

71. If c and m are the prices of a coffee and a muffin, respectively, then a system of equations is

$$\begin{aligned} 3c + 7m &= 7.77 \\ 6c + 14m &= 14.80. \end{aligned}$$

But if one multiplies the second equation by two, then one obtains $6c + 14m = 15.54$; this last equation contradicts the second equation in the system. Therefore, the system of equations is inconsistent and has no solution.

73. Let x and y be the number of male and female students, respectively.

$$\begin{aligned} 0.5x + 0.3y &= 230 \\ 0.2x + 0.6y &= 260 \end{aligned}$$

Multiply the first equation by -2 and the second by 5. Then add the resulting equations.

$$\begin{aligned} -x - 0.6y &= -460 \\ x + 3y &= 1300 \\ \hline 2.4y &= 840 \\ y &= 350 \end{aligned}$$

From $0.2x + 0.6y = 260$, $0.2x + 210 = 260$ and so $x = 250$. There are $250 + 350 = 600$ students at CHS.

75. Let x and y be the number of nickels and pennies, respectively. We obtain

$$\begin{aligned} x + y &= 87 \\ 0.05x + 0.01y &= 1.75. \end{aligned}$$

Multiply the second equation by -100 and then add it to the first equation.

$$\begin{aligned} x + y &= 87 \\ -5x - y &= -175 \\ \hline -4x &= -88 \\ x &= 22 \end{aligned}$$

From $x + y = 87$, $22 + y = 87$ and $y = 65$. Isabelle has 22 nickels and 65 pennies.

77. The weights x and y must satisfy

$$\begin{aligned} 5x &= 3y \\ 3(4 + x + y) &= 6y. \end{aligned}$$

The second equation can be written as $12 + 3x = 3y$. Substituting into the first equation one finds

$$\begin{aligned} 12 + 3x &= 5x \\ 12 &= 2x \\ 6 &= x. \end{aligned}$$

Then $x = 6$ oz and $y = 10$ oz since

$$y = \frac{5x}{3} = \frac{5(6)}{3}.$$

79. Let x be the number of months. Plan A costs $150x + 800$ and Plan B costs $200x + 200$. Thus, Plan A is cheaper in the long run. The number of months for which the costs are the same is given by

$$\begin{aligned} 150x + 800 &= 200x + 200 \\ 600 &= 50x \\ x &= 12 \text{ months.} \end{aligned}$$

81. If we set the formulas equal to each other, then we find

$$\begin{aligned}0.08aD &= \frac{D(a+1)}{24}\\ 0.08a &= \frac{(a+1)}{24}\\ 1.92a &= a+1\\ a &= \frac{1}{0.92}\\ a &\approx 1.09.\end{aligned}$$

The dosage is the same if the age is 1.1 yr.

83. Suppose the fronts of the trucks are at the same points. Let t be the number of hours before the trucks pass each other. They will be passing each other when the ends of the trucks are at the same position, i.e., when the total distance driven by the trucks is 100 feet.

$$\begin{aligned}40(5280)t + 50(5280)t &= 100\\ 40t + 50t &= \frac{100}{5280}\\ 90t &= \frac{100}{5280}\\ t &= \frac{100}{(90)5280} \text{ hour}\\ t &= \frac{100(3600)}{(90)5280} \text{ sec}\\ t &= \frac{25}{33} \text{ sec}\\ t &\approx 0.76 \text{ sec}\end{aligned}$$

Hence, the trucks pass each other in 0.76 sec, approximately.

85. Solve for a and b.

$$\begin{aligned}-3a+b &= 9\\ 2a+b &= -1.\end{aligned}$$

Multiply the second equation by -1 and add to the first equation.

$$\begin{aligned}-3a+b &= 9\\ -2a-b &= 1\\ \hline -5a &= 10\\ a &= -2\end{aligned}$$

Substituting $a=-2$ into $2a+b=-1$, one finds $b=3$. An equation of the line is $y=-2x+3$.

87. Solve for a and b.

$$\begin{aligned}-2a+b &= 3\\ 4a+b &= -7\end{aligned}$$

Multiply the first equation by -1 and add to the second equation.

$$\begin{aligned}2a-b &= -3\\ 4a+b &= -7\\ \hline 6a &= -10\\ a &= -\frac{5}{3}\end{aligned}$$

Substituting $a=-\frac{5}{3}$ into $-2a+b=3$, one gets $b=-\frac{1}{3}$. An equation of the line is $y=-\frac{5}{3}x-\frac{1}{3}$.

91. Independent system with solution $(2,-3)$:

$$\begin{aligned}x+y &= -1\\ x-y &= 5\end{aligned}$$

93. **a)** $f(\frac{2}{3}) = \left(8^{1/3}\right)^2 = 2^2 = 4$

b) $g(3) = 4^{2-3} = 4^{-1} = \frac{1}{4}$

c) $(f\circ g)(2) = f(g(2)) = f(1) = 8$

95. Since $2^3=8$ and $2^2=4$, we find

$$\begin{aligned}\left(2^3\right)^{x-3} &= \left(2^2\right)^{x+5}\\ 2^{3x-9} &= 2^{2x+10}\\ 3x-9 &= 2x+10\\ x &= 19\end{aligned}$$

The solution set is $\{19\}$.

97. The zeros of

$$f(x) = (3x-2)(5x-6).$$

are $x = 2/3, 6/5$.

If $x = 0$, then $f(0) > 0$.
If $x = 1$, then $f(1) < 0$.
If $x = 2$, then $f(2) > 0$.

$$\begin{array}{ccccccccc} + & & 0 & & - & & 0 & & + \\ 0 & & \frac{2}{3} & & 1 & & \frac{6}{5} & & 2 \end{array}$$

The solution set of $f(x) \leq 0$ is $\left[\frac{2}{3}, \frac{6}{5}\right]$.

99. Let p and f be the number of students who passed and failed, respectively. Since the mean score for the passing students is 65, the mean score for the failing students is 53, and the mean score for all students is 53, we obtain

$$\frac{65p + 35f}{p + f} = 53.$$

Solving for f, we find

$$f = \frac{2p}{3}.$$

Then the percentage of passing students in the class is

$$\frac{p}{p+f} \cdot 100 = \frac{p}{p + 2p/3} \cdot 100 = \frac{3}{5} \cdot 100 = 60\%.$$

For Thought

1. True

2. False, since $(1, 1, 0)$ does not satisfy $-x - y + z = 4$.

3. True, adding the first two equations gives $0 = 6$ which is false.

4. False, since $(2, 3, -1)$ does not satisfy $x - y - z = 8$.

5. True

6. True, adding the first two equations gives an identity. Also, multiplying the first by -2 and then adding to the third equation gives an identity.

$$\begin{array}{rcr} x - y + z & = & 1 \\ -x + y - z & = & -1 \\ \hline 0 & = & 0 \end{array}$$

$$\begin{array}{rcr} -2x + 2y - 2z & = & -2 \\ 2x - 2y + 2z & = & 2 \\ \hline 0 & = & 0 \end{array}$$

7. True. The calculations in Exercise 6 above show system (c) is dependent.

8. True

9. True, if $x = 1$ then $(x + 2, x, x - 1) = (3, 1, 0)$.

10. False, the value is $0.05x + 0.10y + 0.25z$ dollars.

8.2 Exercises

1. linear

3. dependent

5. Points on the plane are $(5, 0, 0)$, $(0, 5, 0)$, and $(0, 0, 5)$.

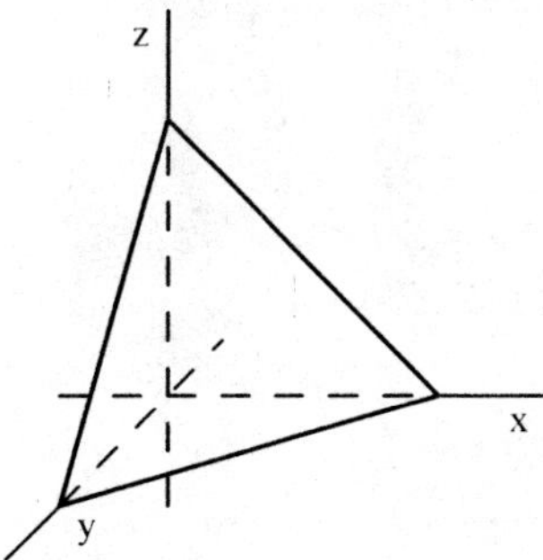

7. Points on the plane are $(3, 0, 0)$, $(0, 3, 0)$, and $(0, 0, -3)$.

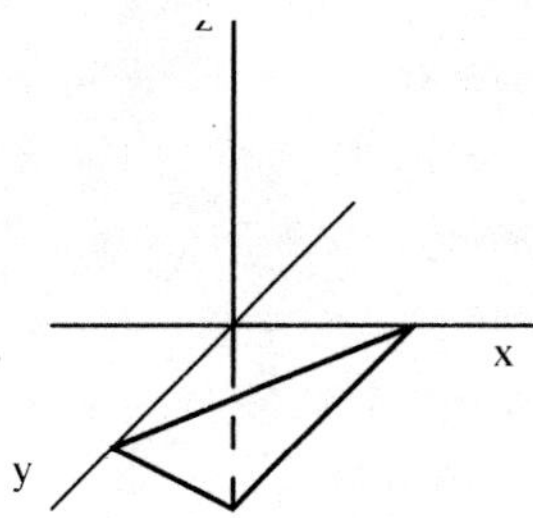

9. Note, $x = 1$, $y = 3$, and $z = 2$ satisfy all the three equations in the system . Yes, $(1, 3, 2)$ is a solution.

11. Note, $x = -1$, $y = 5$, and $z = 2$ does not satisfy $x - y - 2z = -8$. No, $(-1, 5, 2)$ is not a solution.

13. Add the first and second equations and add the first and third equations.

$$\begin{array}{rcr} x + y + z & = & 6 \\ 2x - 2y - z & = & -5 \\ \hline 3x - y & = & 1 \end{array}$$

$$\begin{array}{rcl} x+y+z &=& 6 \\ 3x+y-z &=& 2 \\ \hline 4x+2y &=& 8 \end{array}$$

Divide $4x+2y=8$ by 2 and add to $3x-y=1$.

$$\begin{array}{rcl} 2x+y &=& 4 \\ 3x-y &=& 1 \\ \hline 5x &=& 5 \end{array}$$

So $x=1$. From $3x-y=1$, $3-y=1$ and $y=2$. From $x+y+z=6$, $1+2+z=6$ and $z=3$. The solution set is $\{(1,2,3)\}$.

15. Multiply the first equation by 2 and add to the second. Then multiply the first equation by -3 and add to the third equation.

$$\begin{array}{rcl} 6x+4y+2z &=& 2 \\ x+y-2z &=& -4 \\ \hline 7x+5y &=& -2 \end{array}$$

$$\begin{array}{rcl} -9x-6y-3z &=& -3 \\ 2x-3y+3z &=& 1 \\ \hline -7x-9y &=& -2 \end{array}$$

Add $7x+5y=-2$ to $-7x-9y=-2$.

$$\begin{array}{rcl} 7x+5y &=& -2 \\ -7x-9y &=& -2 \\ \hline -4y &=& -4 \end{array}$$

So $y=1$. From $7x+5y=-2$, $7x+5=-2$ and $x=-1$. From $3x+2y+z=1$, $-3+2+z=1$ and $z=2$.
The solution set is $\{(-1,1,2)\}$.

17. Add first equation to third equation. Multiply second equation by 2 and add to the first.

$$\begin{array}{rcl} 2x+y-2z &=& -15 \\ x+3y+2z &=& -5 \\ \hline 3x+4y &=& -20 \end{array}$$

$$\begin{array}{rcl} 8x-4y+2z &=& 30 \\ 2x+y-2z &=& -15 \\ \hline 10x-3y &=& 15 \end{array}$$

Multiply $3x+4y=-20$ by 3 and $10x-3y=15$ by 4 and then add the equations.

$$\begin{array}{rcl} 9x+12y &=& -60 \\ 40x-12y &=& 60 \\ \hline 49x &=& 0 \end{array}$$

So $x=0$. From $3x+4y=-20$, $4y=-20$ and $y=-5$. From $2x+y-2z=-15$, we get $-5-2z=-15$ and $z=5$.
The solution set is $\{(0,-5,5)\}$.

19. If we substitute $x=1,2,3$ in $(x,x+3,x-5)$, we obtain

$$(1,4,-4),\ (2,5,-3),\ \text{and } (3,6,-2),$$

respectively.

21. If we substitute $y=1,2,3$ in $(2y,y,y-7)$, we obtain

$$(2,1,-6),\ (4,2,-5),\ \text{ and } (6,3,-4),$$

respectively.

23. Let $y=x+3$. Then $x=y-3$, $x-5=y-8$, and

$$(x,x+3,x-5)=(y-3,y,y-8).$$

25. Let $z=x-1$. Then $x=z+1$, $x+1=z+2$, and

$$(x,x+1,x-1)=(z+1,z+2,z).$$

27. Let $y=2x+1$. Then $x=(y-1)/2$, $3x-1=(3y-5)/2$, and

$$(x,2x+1,3x-1)=\left(\frac{y-1}{2},y,\frac{3y-5}{2}\right).$$

29. Adding the two equations, we have $4x-4z=-20$ or $z=x+5$.
From $x+2y-3z=-17$, we obtain

$$\begin{array}{rcl} x+2y-3(x+5) &=& -17 \\ 2y-2x-15 &=& -17 \\ 2y &=& 2x-2 \\ y &=& x-1. \end{array}$$

The solution set is
$\{(x,x-1,x+5)|x \text{ is any real number}\}$.

31. Adding the two equations, we obtain $2x-y=7$ or $y=2x-7$.
From $x+y-z=2$, we obtain

$$\begin{array}{rcl} z &=& x+y-2 \\ z &=& x+(2x-7)-2 \\ z &=& 3x-9. \end{array}$$

The solution set is
$\{(x, 2x-7, 3x-9) | x$ is any real number$\}$.

33. Adding the two equations, we find
$2x + 2z = 12$ or $x = 6 - z$.
From $y + z = 5$, we obtain

$$y = 5 - z.$$

The solution set is
$\{(6-z, 5-z, z) | z$ is any real number$\}$.

35. Adding the first two equations, we have $0 = 0$, which is an identity. Multiply the second equation by 2 and add to the third equation.

$$\begin{array}{rcr} -2x - 4y + 6z & = & -10 \\ 2x + 4y - 6z & = & 10 \\ \hline 0 & = & 0 \end{array}$$

The system is dependent and the solution set is $\{(x, y, z) | x + 2y - 3z = 5\}$.

37. Multiply first equation by -2 and add to the second equation.

$$\begin{array}{rcr} -2x + 4y - 6z & = & -10 \\ 2x - 4y + 6z & = & 3 \\ \hline 0 & = & -7 \end{array}$$

This is false, so the solution set is $\emptyset$.

39. Adding the two equations, we get $3x = 6$ or $x = 2$. Substitute into the two equations. Then

$$\begin{array}{rcl} 2 + y - z & = & 2 \\ y - z & = & 0 \end{array}$$

and

$$\begin{array}{rcl} 4 - y + z & = & 4 \\ -y + z & = & 0. \end{array}$$

In any case $y = z$. The solution set is
$\{(2, y, y) | y$ is any real number$\}$.

41. If the second equation is multiplied by -1 and added to the first equation, the result is the third equation.

$$\begin{array}{rcr} x + y & = & 5 \\ -y + z & = & -2 \\ \hline x + z & = & 3 \end{array}$$

There ae infinitely number many solutions. Since $y = 5 - x$ and $z = 3 - x$, the solution set $\{(x, 5-x, 3-x) | x$ is any real number$\}$.

43. Multiply the first equation by 2 and add to the second equation.

$$\begin{array}{rcr} 2x - 2y + 2z & = & 14 \\ 2y - 3z & = & -13 \\ \hline 2x - z & = & 1 \end{array}$$

Multiply $2x - z = 1$ by -2 and add to the third equation.

$$\begin{array}{rcr} -4x + 2z & = & -2 \\ 3x - 2z & = & -3 \\ \hline -x & = & -5 \end{array}$$

So $x = 5$. From $2x - z = 1$, we get $10 - z = 1$ and $z = 9$. Since $2y - 3z = -13$, $2y - 27 = -13$ and $y = 7$. The solution set is $\{(5, 7, 9)\}$.

45. Multiply first equation by -2 and add to third equation. Then multiply first equation by -4 and add to the second.

$$\begin{array}{rcr} -2x - 2y - 4z & = & -15 \\ 5x + 2y + 5z & = & 21 \\ \hline 3x + z & = & 6 \end{array}$$

$$\begin{array}{rcr} -4x - 4y - 8z & = & -30 \\ 3x + 4y + z & = & 12 \\ \hline -x - 7z & = & -18 \end{array}$$

Multiply $-x - 7z = -18$ by 3 and add to $3x + z = 6$.

$$\begin{array}{rcr} -3x - 21z & = & -54 \\ 3x + z & = & 6 \\ \hline -20z & = & -48 \end{array}$$

So $z = 48/20 = 2.4$. From $3x + z = 6$, we get $3x + 2.4 = 6$ and $x = 1.2$. From $x + y + 2z = 7.5$, $1.2 + y + 4.8 = 7.5$ and $y = 1.5$.
The solution set is $\{(1.2, 1.5, 2.4)\}$.

47. Multiply the first equation by -5 and add to 100 times the second one.

$$\begin{array}{rcr} -5x - 5y - 5z & = & -45,000 \\ 5x + 6y + 9z & = & 71,000 \\ \hline y + 4z & = & 26,000 \end{array}$$

Substitute $z = 3y$ into $y + 4z = 26,000$.

$$\begin{aligned} y + 12y &= 26,000 \\ 13y &= 26,000 \\ y &= 2,000 \end{aligned}$$

From $z = 3y$, we obtain $z = 6000$. From $x + y + z = 9,000$, we have $x + 2000 + 6000 = 9000$ and $x = 1000$. The solution set is $\{(1000, 2000, 6000)\}$.

49. Substitute $x = 2y - 1$ into $z = 2x - 3$ to get $z = 4y - 5$. Then substitute $z = 4y - 5$ into $y = 3z + 2$.

$$\begin{aligned} y &= 3(4y - 5) + 2 \\ y &= 12y - 13 \\ -11y &= -13 \\ y &= 13/11 \end{aligned}$$

From $z = 4y - 5$, we obtain $z = 52/11 - 5 = -3/11$. Since $x = 2y - 1$, $x = 26/11 - 1 = 15/11$. The solution set is $\{(15/11, 13/11, -3/11)\}$.

51. Substitute $(-1, -2)$, $(2, 1)$, $(-2, 1)$ into $y = ax^2 + bx + c$. So

$$\begin{aligned} a - b + c &= -2 \\ 4a + 2b + c &= 1 \\ 4a - 2b + c &= 1. \end{aligned}$$

Multiply first equation by -1 and add to the second and third equations.

$$\begin{array}{rcl} -a + b - c &=& 2 \\ 4a + 2b + c &=& 1 \\ \hline 3a + 3b &=& 3 \end{array}$$

$$\begin{array}{rcl} -a + b - c &=& 2 \\ 4a - 2b + c &=& 1 \\ \hline 3a - b &=& 3 \end{array}$$

Multiply $3a + 3b = 3$ by -1 and add to $3a - b = 3$.

$$\begin{array}{rcl} -3a - 3b &=& -3 \\ 3a - b &=& 3 \\ \hline -4b &=& 0 \end{array}$$

So $b = 0$. From $3a - b = 3$, we get $3a = 3$ and $a = 1$. From $a - b + c = -2$, $1 + c = -2$ and $c = -3$. Since the solution is $(a, b, c) = (1, 0, -3)$, the parabola is $y = x^2 - 3$.

53. Substitute $(0, 0)$, $(1, 3)$, $(2, 2)$ into $y = ax^2 + bx + c$. Then

$$\begin{aligned} c &= 0 \\ a + b + c &= 3 \\ 4a + 2b + c &= 2. \end{aligned}$$

Multiply second equation by -2 and add to third equation.

$$\begin{array}{rcl} -2a - 2b - 2c &=& -6 \\ 4a + 2b + c &=& 2 \\ \hline 2a - c &=& -4 \end{array}$$

Substituting $c = 0$ into $2a - c = -4$, we get $2a = -4$ and $a = -2$. From $a + b + c = 3$, $-2 + b = 3$ and $b = 5$. Since $(a, b, c) = (-2, 5, 0)$, the parabola is $y = -2x^2 + 5x$.

55. Substitute $(0, 4)$, $(-2, 0)$, $(-3, 1)$ into $y = ax^2 + bx + c$. Then

$$\begin{aligned} c &= 4 \\ 4a - 2b + c &= 0 \\ 9a - 3b + c &= 1. \end{aligned}$$

Multiply second and third equations by 3 and -2, respectively, then add the equations.

$$\begin{array}{rcl} 12a - 6b + 3c &=& 0 \\ -18a + 6b - 2c &=& -2 \\ \hline -6a + c &=& -2 \end{array}$$

Substituting $c = 4$ into $-6a + c = -2$, $-6a + 4 = -2$ and $a = 1$. From $4a - 2b + c = 0$, $4 - 2b + 4 = 0$ and $b = 4$. Since $(a, b, c) = (1, 4, 4)$, the parabola is $y = x^2 + 4x + 4$.

57. By substituting the coordinates of the points $(1, 0, 0)$, $(0, 1, 0)$, and $(0, 0, 1)$ into $ax + by + cz = 1$, we get $a = 1$, $b = 1$, and $c = 1$. A linear equation satisfied by the ordered triples is $x + y + z = 1$.

59. By substituting the coordinates of the points $(1, 1, 1)$, $(0, 2, 0)$, and $(1, 0, 0)$ into $ax + by + cz = 1$, we get $a + b + c = 1$, $2b = 1$, and $a = 1$. Then $b = \frac{1}{2}$ and $c = -\frac{1}{2}$. A linear equation satisfied by the ordered triples is $x + \frac{1}{2}y - \frac{1}{2}z = 1$ or $2x + y - z = 2$

61. Let x, y, z be the three numbers listed in increasing order.

$$\begin{aligned} x+y+z &= 40 \\ -x \quad +z &= 12 \\ x+y-z &= 0. \end{aligned}$$

If we add the first two equations and add the last two equations, we obtain (by subtracting the 2nd sum from the 1st sum)

$$\begin{aligned} y+2z &= 52 \\ y \quad &= 12 \\ \hline 2z &= 40 \end{aligned}$$

Then $z = 20$. Since $-x + z = 12$, we find $-x+20 = 12$ or $x = 8$. Thus, the numbers are $8, 12$, and 20.

63. Let x, y, z be the scores in the 1st, 2nd, and 3rd quizzes, respectively.

$$\begin{aligned} x+y+z &= 21 \\ x-y \quad &= -1 \\ y-z &= -4. \end{aligned}$$

If we add the first two equations and add the last two equations, we obtain

$$\begin{aligned} 2x+z &= 20 \\ x-z &= -5 \\ \hline 3x &= 15 \end{aligned}$$

Since $x = 5$, the scores on the quizzes are 5, 6, and 10.

65. Let x, y, and z be the amounts invested in stocks, bonds, and a mutual fund. Then

$$\begin{aligned} x+y+z &= 25,000 \\ 0.08x+0.10y+0.06z &= 1,860 \\ 2y &= z. \end{aligned}$$

Multiply the first equation by -8 and add to 100 times the second.

$$\begin{aligned} -8x-8y-8z &= -200,000 \\ 8x+10y+6z &= 186,000 \\ \hline 2y-2z &= -14,000 \end{aligned}$$

Substitute $z = 2y$ into $2y - 2z = -14,000$.

$$\begin{aligned} 2y-4y &= -14,000 \\ -2y &= -14,000 \\ y &= 7,000 \end{aligned}$$

Since $z = 2y$, we obtain $z = 14,000$.
Since $x + y + z = 25,000$, $x = 4000$.
Marita invested \$4,000 in stocks, \$7000 in bonds, and \$14,000 in a mutual fund.

67. Let x, y, and z be the prices last year of a hamburger, fries, and a Coke, respectively. So

$$\begin{aligned} x+y+z &= 3.80 \\ 1.2x+1.1y+1.25z &= 4.49 \\ 1.25z &= 1.1y-0.07. \end{aligned}$$

Multiply first equation by -12 and add to 10 times the second equation.

$$\begin{aligned} -12x-12y-12z &= -45.6 \\ 12x+11y+12.5z &= 44.9 \\ \hline -y+0.5z &= -0.7 \end{aligned}$$

Substitute $y = 0.5z + 0.7$ into $1.25z = 1.1y - 0.07$.

$$\begin{aligned} 1.25z &= 1.1(0.5z+0.7)-0.07 \\ 0.7z &= 0.7 \\ z &= 1 \end{aligned}$$

Since $y = 0.5z+0.7$, we find $y = 0.5(1)+0.7 = 1.20$. From $x+y+z = 3.80$, we have $x = 1.60$. The prices last year of a hamburger, fries and Coke are \$1.60, \$1.20, and \$1, respectively.

69. Let L_f and L_r be the weights on the left front tire and left rear tire, respectively. Let R_f and R_r be the weights on the right front tire and right rear tire, respectively. Since $1200(0.51) = 612$ and $1200(0.48) = 576$, we obtain

$$\begin{aligned} L_f+L_r &= 612 \\ R_r+L_r &= 576 \\ L_f, L_r, R_f, R_r &\geq 280. \end{aligned}$$

Three possible weight distributions are $(L_r, L_f, R_r, R_f) = (280, 332, 296, 292)$, $(L_r, L_f, R_r, R_f) = (285, 327, 291, 297)$, and $(L_r, L_f, R_r, R_f) = (290, 322, 286, 302)$.

71. Let x, y, and z be the number of pennies, nickels, and dimes, respectively. Then

$$\begin{aligned} x+y+z &= 232 \\ y+z &= x \\ 0.01x+0.05y+0.10z &= 10.36. \end{aligned}$$

Multiply first equation by -1 and add to 10 times the third equation. Also combine first two equations.

$$\begin{aligned} -x-y-z &= -232 \\ 0.1x+0.5y+z &= 103.6 \\ \hline -0.9x-0.5y &= -128.4 \end{aligned}$$

$$\begin{aligned} -x-y-z &= -232 \\ -x+y+z &= 0 \\ \hline -2x &= -232 \end{aligned}$$

Then $x = 116$. Substituting into $-0.9x - 0.5y = -128.4$, we obtain

$$\begin{aligned} -0.9(116)-0.5y &= -128.4 \\ -104.4-0.5y &= -128.4 \\ 24 &= 0.5y \\ 48 &= y. \end{aligned}$$

Since $x+y+z = 232$, $116+48+z = 232$ and $z = 68$. Emma used 116 pennies, 48 nickels, and 68 dimes.

73. Let x, y, and z be the prices of a carton of milk, a cup of coffee, and a doughnut, respectively. So

$$\begin{aligned} 3x+4y+7z &= 5.45 \\ 4x+2y+8z &= 5.30 \\ 2x+5y+6z &= 5.15. \end{aligned}$$

Multiply third equation by -2 and add to second equation. Also, multiply first equation by -4 and add to 3 times the second equation.

$$\begin{aligned} -4x-10y-12z &= -10.30 \\ 4x+2y+8z &= 5.30 \\ \hline -8y-4z &= -5 \end{aligned}$$

$$\begin{aligned} -12x-16y-28z &= -21.80 \\ 12x+6y+24z &= 15.90 \\ \hline -10y-4z &= -5.90 \end{aligned}$$

Multiply $-8y-4z = -5$ by -1 and add to $-10y-4z = -5.90$.

$$\begin{aligned} 8y+4z &= 5 \\ -10y-4z &= -5.90 \\ \hline -2y &= -0.90 \end{aligned}$$

Then $y = 0.45$. Substitute into $-8y-4z = -5$ to get $-3.60-4z = -5$ and $z = 0.35$. Since $3x+4y+7z = 5.45$, we have $3x+1.80+2.45 = 5.45$ and $x = 0.40$.

Alphonse's bill was $5(0.40)+2(0.45)+9(0.35) = \$6.05$. His change is \$3.95.

75. A system of equations is

$$\begin{aligned} 4x &= 6y \\ 2(x+y) &= 15(8) \\ 6(x+y+15+10) &= 10z. \end{aligned}$$

Rewriting the first two equations, one obtains

$$\begin{aligned} 2x-3y &= 0 \\ x+y &= 60. \end{aligned}$$

Solving this smaller system, one finds $x = 36$ lb, $y = 24$ lb. Substituting into the third equation, one finds

$$z = \frac{6(x+y+25)}{10} = \frac{6(60+25)}{10} = 51 \text{ lb}.$$

77.

a) Substitute $(0,0)$, $(10,40)$, $(20,70)$ into $y = ax^2+bx+c$. Consequently, we obtain the following system of equations:

$$\begin{aligned} c &= 0 \\ 100a+10b+c &= 40 \\ 400a+20b+c &= 70. \end{aligned}$$

Multiply the second equation by -2 and add to third equation.

$$\begin{aligned} -200a-20b-2c &= -80 \\ 400a+20b+c &= 70 \\ \hline 200a-c &= -10 \end{aligned}$$

Substitute $c = 0$ into $200a - c = -10$.

Then $a = -\frac{1}{20}$. From $100a+10b+c = 40$, we obtain $-5 + 10b = 40$ and $b = \frac{9}{2}$.
The parabola is $y = -\frac{1}{20}x^2 + \frac{9}{2}x$.

b) Since $-b/(2a) = 45$, the maximum height is

$$-\frac{1}{20}(45)^2 + \frac{9}{2}(45) = 101.25 \text{ m}.$$

c) Since the zeros of $y = -\frac{x}{20}(x - 90)$ are $x = 0, 90$, the missile will strike 90 m from the origin.

81. Multiply the second equation by -2 and add the result to the first equation.

$$\begin{array}{rcl} 2x - 3y & = & 20 \\ -2x - 8y & = & 2 \\ \hline -11y & = & 22 \end{array}$$

Then $y = -2$ and $x = -4y-1 = -4(-2)-1 = 7$. The solution set is $\{(7, -2)\}$.

83. $10,000\left(1 + \frac{0.035}{4}\right)^{24} = \$12,325.52$

85. No, the Vertical Line Test fails for a circle.

87. Let h, f, and c be the costs of a hamburger, an order of french fries, and a coke, respectively.

$$\begin{array}{rcl} 8h + 5f + 2c & = & 14.25 \\ 5h + 3f + c & = & 8.51 \end{array}$$

If you subtract the 2nd equation from the 1st equation, we obtain

$$3h + 2f + c = 5.74.$$

If you multiply the 2nd equation by -2 and add the result to the first equation, we find

$$-2h - f = -2.77.$$

When we add the last two equations, we obtain

$$h + f + c = \$2.97$$

which is the cost of 1 hamburger, 1 order of french fries, and a coke.

For Thought

1. True, since the line $y = x$ passes through the center of the circle.

2. False, when a line is tangent to a circle it intersects the circle at only one point.

3. True, the parabola $y = 2x^2 - 4$ intersects the circle $x^2 + y^2 = 16$ at three points.

4. False, they intersect at $(\pm 1, 0)$.

5. False, they intersect at $(1, 1)$ and $(-1, -1)$.

6. False, since three noncollinear points determine a unique circle through the points.

7. True, since either leg can serve as base and the other leg as altitude.

8. True **9.** True

10. False, two such numbers are $\frac{1}{2}\left(7 \pm \sqrt{45}\right)$.

8.3 Exercises

1. nonlinear

3. Since $x = -1$ and $y = 4$ satisfy both equations in the system , $(-1, 4)$ is a solution.

5. Since $x = 4$ and $y = -5$ does not satisfy $x - y = 1$, $(4, -5)$ is not a solution.

7. Since $y = x$ and $y = x^2$, we obtain

$$\begin{array}{rcl} x & = & x^2 \\ x - x^2 & = & 0 \\ x(x - 1) & = & 0. \end{array}$$

Then $x = 0, 1$ and the solution set is

$$\{(0, 0),\ (1, 1)\}.$$

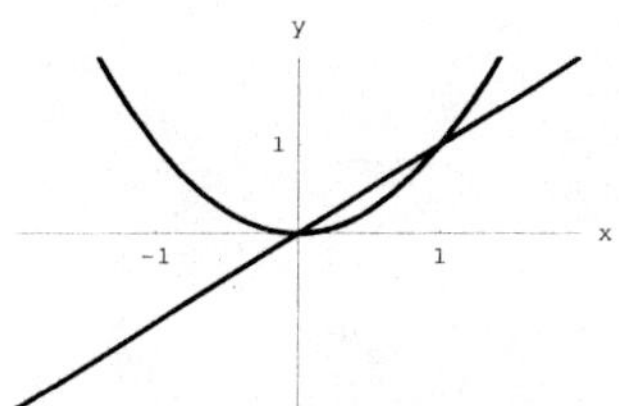

9. Substitute $y = x^2$ into $5x - y = 6$.

$$\begin{aligned} 5x - x^2 &= 6 \\ x^2 - 5x &= -6 \\ x^2 - 5x + 6 &= 0 \\ (x-3)(x-2) &= 0 \end{aligned}$$

If $x = 3, 2$ in $y = x^2$, then $y = 9, 4$.
The solution set is $\{(2,4), (3,9)\}$.

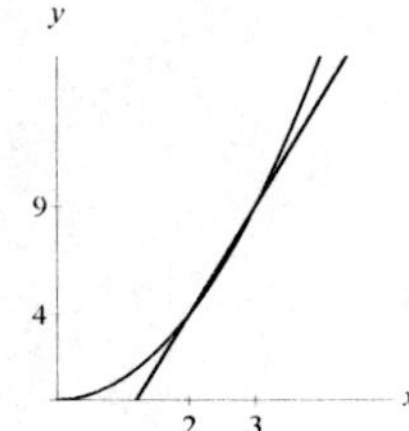

11. Substitute $y = |x|$ into $y = x + 3$.

$$\begin{aligned} |x| &= x + 3 \\ x = x + 3 \quad &\text{or} \quad -x = x + 3 \\ 0 = 3 \quad &\text{or} \quad -2x = 3 \end{aligned}$$

Then $x = -3/2$. Since $y = |x|$, $y = 3/2$.
The solution set is $\{(-3/2, 3/2)\}$.

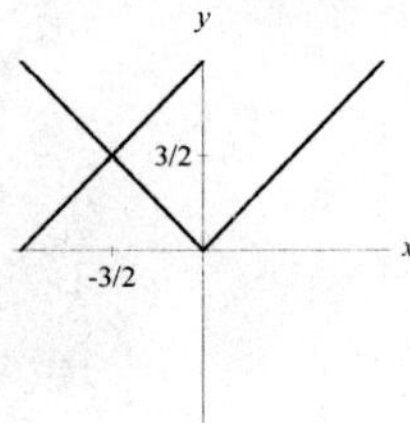

13. Substitute $y = x^2$ into $y = |x|$.

$$\begin{aligned} x^2 &= |x| \\ x^2 = x \quad &\text{or} \quad x^2 = -x \\ x^2 - x = 0 \quad &\text{or} \quad x^2 + x = 0 \\ x(x-1) = 0 \quad &\text{or} \quad x(x+1) = 0 \\ x = 0, 1 \quad &\text{or} \quad x = 0, -1 \end{aligned}$$

Using $x = 0, \pm 1$ in $y = x^2$, one finds $y = 0, 1$. The solution set is $\{(-1,1), (0,0), (1,1)\}$.

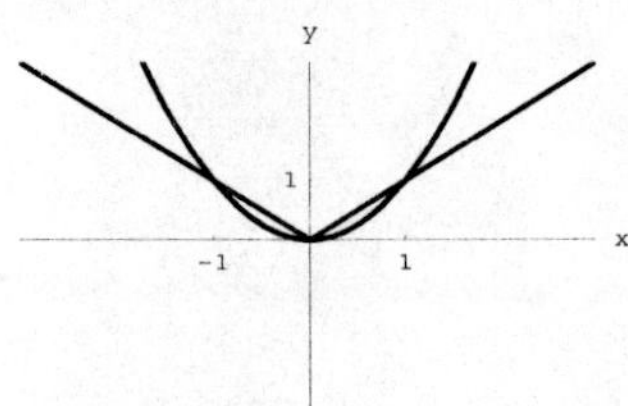

15. Substitute $y = \sqrt{x}$ into $y = 2x$ to obtain

$$\begin{aligned} \sqrt{x} &= 2x \\ x &= 4x^2 \\ x - 4x^2 &= 0 \\ x(1 - 4x) &= 0. \end{aligned}$$

Using $x = 0, 1/4$ in $y = 2x$, $y = 0, 1/2$.

The solution set is

$$\{(0,0), (1/4, 1/2)\}.$$

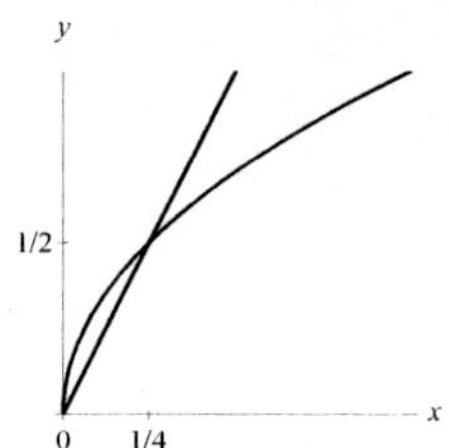

17. Substitute $y = x^3$ into $y = 4x$ to get

$$\begin{aligned} x^3 &= 4x \\ x^3 - 4x &= 0 \\ x(x^2 - 4) &= 0 \\ x &= 0, \pm 2. \end{aligned}$$

Substituting $x = 0, 2, -2$ into $y = 4x$, we get $y = 0, 8, -8$.
The solution set is

$$\{(0,0), (2,8), (-2,-8).$$

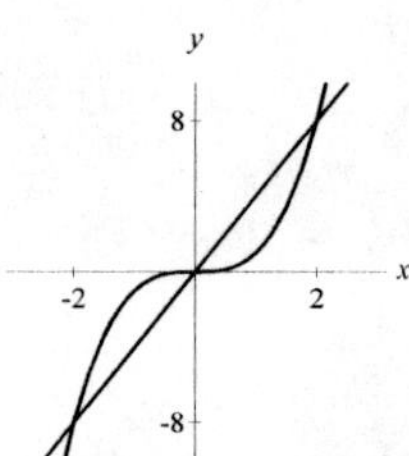

19. Substitute $y = x$ into $y = x^3 - x$.

$$\begin{aligned} x &= x^3 - x \\ 2x - x^3 &= 0 \\ x(2 - x^2) &= 0 \\ x &= 0, \sqrt{2}, -\sqrt{2} \end{aligned}$$

Since $y = x$, the solution set is

$$\{(0,0), (\sqrt{2}, \sqrt{2}), (-\sqrt{2}, -\sqrt{2})\}.$$

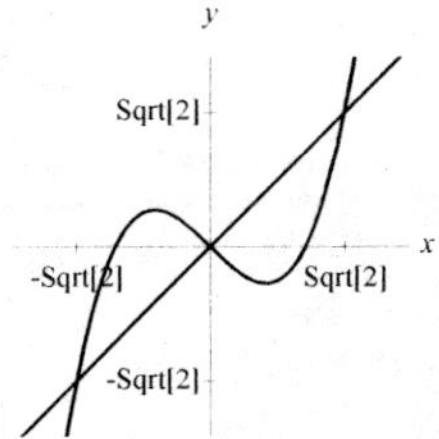

21. Substitute $y = x$ into $x^2 + y^2 = 1$.

$$\begin{aligned} x^2 + x^2 &= 1 \\ 2x^2 &= 1 \\ x^2 &= \frac{1}{2} \\ x &= \pm\frac{\sqrt{2}}{2} \end{aligned}$$

Since $y = x$, the solution set is

$$\left\{\left(\frac{\sqrt{2}}{2}, \frac{\sqrt{2}}{2}\right), \left(-\frac{\sqrt{2}}{2}, -\frac{\sqrt{2}}{2}\right)\right\}.$$

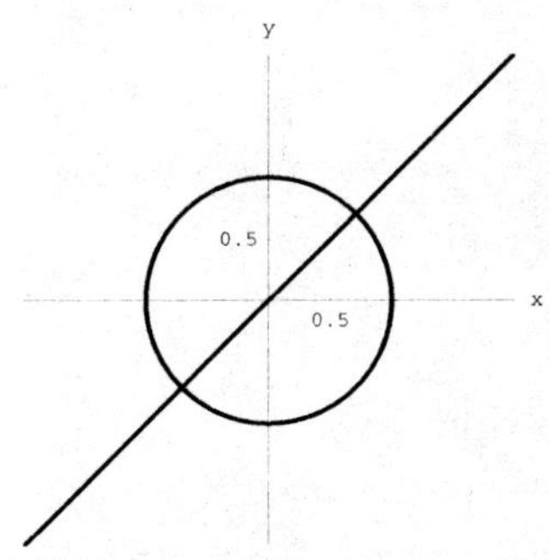

23. Substitute $y = -x - 4$ into $xy = 1$.

$$\begin{aligned} x(-x-4) &= 1 \\ -x^2 - 4x &= 1 \\ x^2 + 4x &= -1 \\ x^2 + 4x + 4 &= 3 \\ (x+2)^2 &= 3 \\ x &= -2 \pm \sqrt{3} \end{aligned}$$

Using $x = -2 + \sqrt{3}, -2 - \sqrt{3}$ in $y = -x - 4$, we find $y = -2 - \sqrt{3}, -2 + \sqrt{3}$.
The solution set is
$\left\{(-2+\sqrt{3}, -2-\sqrt{3}), (-2-\sqrt{3}, -2+\sqrt{3})\right\}$.

25. Substitute $x^2 = 2y^2 - 1$ into $2x^2 - y^2 = 1$.

$$\begin{aligned} 2(2y^2 - 1) - y^2 &= 1 \\ 3y^2 - 2 &= 1 \\ 3y^2 &= 3 \\ y^2 &= 1 \\ y &= \pm 1 \end{aligned}$$

Using $y = 1$ in $x^2 = 2y^2 - 1$, we get $x^2 = 1$ or $x = \pm 1$. Also, if $y = -1$ then $x = \pm 1$.
The solution set is $\{(1, \pm 1), (-1, \pm 1)\}$.

27. Since $2x - y = 1$, we obtain $y = 2x - 1$. Then substitute into $xy - 2x = 2$.

$$\begin{aligned} x(2x-1) - 2x &= 2 \\ 2x^2 - 3x - 2 &= 0 \\ (2x+1)(x-2) &= 0 \\ x &= -\frac{1}{2}, 2 \end{aligned}$$

Since $y = 2x - 1$, we get $y = 2\cdot\left(-\frac{1}{2}\right) - 1 = -2$ and $x = 2(2) - 1 = 3$. The solution set is $\left\{\left(-\frac{1}{2}, -2\right), (2, 3)\right\}$.

29. Multiply the first equation by 2 and add to the second.

$$\begin{aligned} \frac{6}{x} - \frac{2}{y} &= \frac{26}{10} \\ \frac{1}{x} + \frac{2}{y} &= \frac{9}{10} \\ \hline \frac{7}{x} &= \frac{35}{10} \end{aligned}$$

So $70 = 35x$ and $x = 2$.
Substituting into $\frac{3}{x} - \frac{1}{y} = \frac{13}{10}$, we obtain

$$\begin{aligned} \frac{3}{2} - \frac{1}{y} &= \frac{13}{10} \\ -\frac{1}{y} &= \frac{13}{10} - \frac{15}{10} \\ -\frac{1}{y} &= -\frac{2}{10} \\ y &= 5. \end{aligned}$$

The solution set is $\{(2, 5)\}$.

31. Substitute $y = 1 - x$ into $x^2 + xy - y^2 = -5$.

$$x^2 + x(1-x) - (1-x)^2 = -5$$

$$\begin{aligned} x^2+x-x^2-(1-2x+x^2) &= -5 \\ -x^2+3x-1 &= -5 \\ x^2-3x+1 &= 5 \\ x^2-3x-4 &= 0 \\ (x-4)(x+1) &= 0 \\ x &= 4,-1 \end{aligned}$$

Using $x=4,-1$ in $y=1-x$, we get $y=-3,2$. The solution set is $\{(4,-3),(-1,2)\}$.

33. Add the two given equations to obtain $xy=-2$. Substitute $y=-2/x$ into $x^2+2xy-2y^2=-11$.

$$\begin{aligned} x^2+2x\left(-\frac{2}{x}\right)-2\cdot\frac{4}{x^2} &= -11 \\ x^2-4-\frac{8}{x^2} &= -11 \\ x^4+7x^2-8 &= 0 \\ (x^2+8)(x^2-1) &= 0 \\ x &= \pm 1 \end{aligned}$$

Using $x=1,-1$ in $y=-2/x$, $y=-2,2$. The solution set is $\{(1,-2),(-1,2)\}$.

35. Multiply the first equation by 7 and multiply the second equation by -5.

$$\begin{aligned} \frac{28}{x}+\frac{35}{y^2} &= 84 \\ -\frac{15}{x}-\frac{35}{y^2} &= -110 \\ \hline \frac{13}{x} &= -26 \\ x &= -\frac{1}{2} \end{aligned}$$

Then substitute into $\frac{4}{x}+\frac{5}{y^2}=12$.

$$\begin{aligned} -8+\frac{5}{y^2} &= 12 \\ \frac{5}{y^2} &= 20 \\ y^2 &= \frac{1}{4} \\ y &= \pm\frac{1}{2}. \end{aligned}$$

The solution set is $\left\{\left(-\frac{1}{2},\frac{1}{2}\right),\left(-\frac{1}{2},-\frac{1}{2}\right)\right\}$.

37. Since $x=\frac{10^{11}}{y^2}$ and $\frac{x^3}{y}=10^{12}$, we find

$$\begin{aligned} \frac{10^{33}}{y^7} &= 10^{12} \\ 10^{21} &= y^7 \\ y &= 10^3. \end{aligned}$$

Substitute into $x=\frac{10^{11}}{y^2}$. Then

$$x=\frac{10^{11}}{10^6}=10^5.$$

The solution set is $\{(10^5,10^3)\}$.

39. Substitute $y=2^{x+1}$ into $y=4^{-x}$.

$$\begin{aligned} 2^{x+1} &= (2^2)^{-x} \\ 2^{x+1} &= 2^{-2x} \\ x+1 &= -2x \\ 3x &= -1 \end{aligned}$$

Using $x=-1/3$ in $y=2^{x+1}$, we get $y=2^{2/3}$. The solution set is $\left\{\left(-\frac{1}{3},2^{2/3}\right)\right\}$.

41. Substitute $y=\log_2(x)$ into $y=\log_4(x+2)$ and use the base-changing formula.

$$\begin{aligned} \log_2(x) &= \log_4(x+2) \\ \log_2(x) &= \frac{\log_2(x+2)}{\log_2(4)} \\ \log_2(x) &= \frac{\log_2(x+2)}{2} \\ 2\cdot\log_2(x) &= \log_2(x+2) \\ 2\cdot\log_2(x)-\log_2(x+2) &= 0 \\ \log_2\left(\frac{x^2}{x+2}\right) &= 0 \\ \frac{x^2}{x+2} &= 1 \\ x^2 &= x+2 \\ x^2-x-2 &= 0 \\ (x-2)(x+1) &= 0 \\ x &= 2,-1 \end{aligned}$$

But $x = -1$ is an extraneous root since $\log_2(-1)$ is undefined. Using $x = 2$ in $y = \log_2(x)$, we have $y = 1$. The solution set is $\{(2, 1)\}$.

43. Substitute $y = \log_2(x+2)$ into $y = 3 - \log_2(x)$.

$$\begin{aligned} \log_2(x+2) &= 3 - \log_2(x) \\ \log_2(x+2) + \log_2(x) &= 3 \\ \log_2(x^2 + 2x) &= 3 \\ x^2 + 2x &= 2^3 \\ x^2 + 2x - 8 &= 0 \\ (x+4)(x-2) &= 0 \\ x &= -4, 2 \end{aligned}$$

But $x = -4$ is an extraneous root since $\log_2(-4)$ is undefined. Using $x = 2$ in $y = \log_2(x+2)$, we get $y = \log_2(4) = 2$. The solution set is $\{(2, 2)\}$.

45. Substitute $y = 3^x$ into $y = 2^x$.

$$\begin{aligned} 3^x &= 2^x \\ \log(3^x) &= \log(2^x) \\ x \cdot \log(3) &= x \cdot \log(2) \\ x\left[\log(3) - \log(2)\right] &= 0 \\ x &= 0 \end{aligned}$$

Using $x = 0$ in $y = 3^x$, we obtain $y = 3^0 = 1$. The solution set is $\{(0, 1)\}$.

47. Substitute $y = 2^x$ into $x = \log_4(y)$.

$$x = \log_4(2^x) = x \log_4(2) = x \cdot \frac{1}{2}$$

SInce $x = \frac{x}{2}$, we find $x = 0$. Then

$$y = 2^x = 2^0 = 1.$$

The solution set is $\{(0, 1)\}$.

49. If we subtract the second equation from the first, we find

$$\begin{aligned} x + \log_{16}(y+1) &= \frac{1}{2} \\ x + \log_{16}(y) &= \frac{1}{4} \\ \hline \log_{16}\left(\frac{y+1}{y}\right) &= \frac{1}{4} \\ \frac{y+1}{y} &= 2 \\ y &= 1 \end{aligned}$$

Substitute into $x + \log_{16} y = \frac{1}{4}$. Then

$$x + 0 = \frac{1}{4}.$$

The solution set is $\left\{\left(\frac{1}{4}, 1\right)\right\}$.

51. From the graphs, the solution set is $\{(2, 1), (0.3, -1.8)\}$.

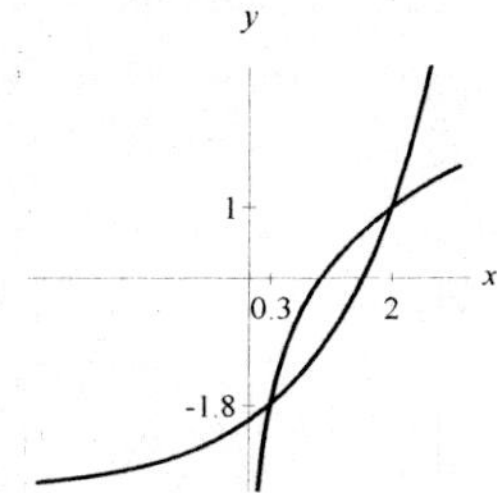

53. From the graphs, the solution set is $\{(1.9, 0.6), (0.1, -2.0)\}$.

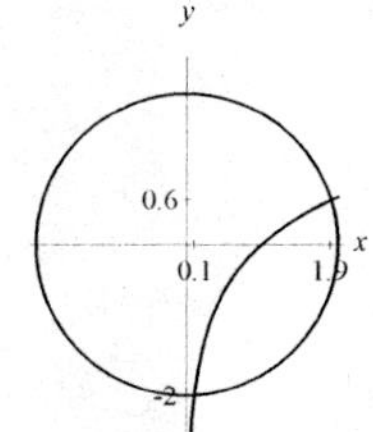

55. From the graphs, the solution set is $\{(-0.8, 0.6), (2, 4), (4, 16)\}$.

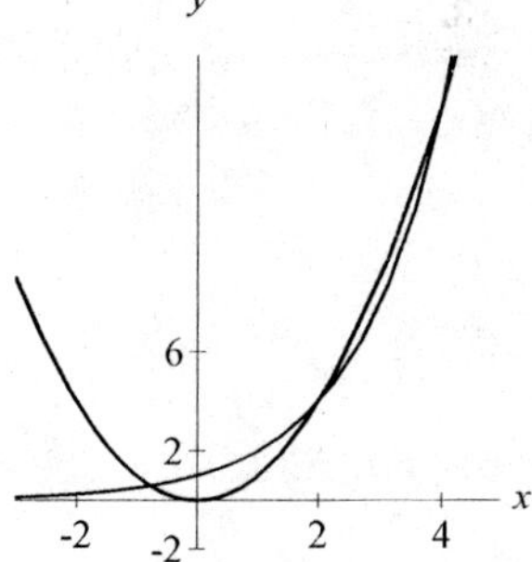

57. From the first equation, we find

$$x + 2y = z + 5.$$

Substitute into the third equation:

$$\begin{aligned} (x+2y)^2 - z^2 &= 15 \\ (z+5)^2 - z^2 &= 15 \end{aligned}$$

Solving for z, we find $z = -1$. Then substitute $z = -1$ into the first two equations, and subtract as follows:

$$\begin{aligned} x + 2y &= 4 \\ 3x + 2y &= 12 \\ \hline -2x &= -8 \\ x &= 4 \end{aligned}$$

Since $x + 2y = 4$ and $x = 4$, we find $y = 0$. The solution set is $\{(4, 0, -1)\}$.

59. If we square the second equation and use the third equation

$$x^2 + y^2 + z^2 = 133$$

we obtain

$$\begin{aligned} (x + y + z)^2 &= 49 \\ 133 + 2(xy + yz + xz) &= 49 \\ 2(xy + yz + xz) &= -84 \end{aligned}$$

Using the first equation $xy = z^2$, we find

$$\begin{aligned} 2(z^2 + yz + xz) &= -84 \\ (z + y + x)z &= -42 \\ 7z &= -42 \\ z &= -6 \end{aligned}$$

for $x + y + z = 7$.

Since $z = -6$, we may rewrite the first two given equations as follows:

$$\begin{aligned} xy &= 36 \\ x + y &= 13 \end{aligned}$$

Since $y = \dfrac{36}{x}$, we substitute and find:

$$\begin{aligned} x + \frac{36}{x} &= 13 \\ x^2 - 13x + 36z &= 0 \\ (x - 4)(x - 9) &= 0 \\ x &= 4, 9 \end{aligned}$$

If $x = 4$, then $y = \dfrac{36}{x} = \dfrac{36}{4} = 9$.

If $x = 9$, then $y = \dfrac{36}{x} = \dfrac{36}{9} = 4$.

The solution set is $\{(4, 9, -6), (9, 4, -6)\}$.

61. Let x and y be the base and height of a 42-inch LCD TV, respectively.

$$\begin{aligned} \frac{y}{x} &= \frac{9}{15} \\ x^2 + y^2 &= 42^2 \end{aligned}$$

Then we find $x \approx 36.0$ in. and $y \approx 21.6$ in.

63. Let x and $6 - x$ be the two numbers.

$$\begin{aligned} x(6 - x) &= -16 \\ x^2 - 6x &= 16 \\ x^2 - 6x + 9 &= 25 \\ (x - 3)^2 &= 25 \\ x - 3 &= \pm 5 \\ x &= 3 \pm 5 \\ x &= -2, 8 \end{aligned}$$

The numbers are -2 and 8.

65. Let x and y be the lengths of the legs of the triangle.

$$\begin{aligned} x^2 + y^2 &= 15^2 \\ \frac{1}{2}xy &= 54 \end{aligned}$$

Substitute $y = \dfrac{108}{x}$ into $x^2 + y^2 = 225$ and use the quadratic formula.

$$\begin{aligned} x^2 + \frac{11664}{x^2} &= 225 \\ x^4 - 225x^2 + 11,664 &= 0 \end{aligned}$$

$$x^2 = \frac{225 \pm \sqrt{225^2 - 4(11,664)}}{2}$$

$$\begin{aligned} x^2 &= \frac{225 \pm 63}{2} \\ x^2 &= 144, 81 \\ x &= 12, 9 \end{aligned}$$

Using $x = 12, 9$ in $y = \dfrac{108}{x}$, we get $y = 9, 12$.

The sides are 9 m and 12 m.

67. If x is the length of the hypotenuse, then $\frac{x}{2}$ and $\frac{x\sqrt{3}}{2}$ are the lengths of the sides opposite the 30° and 60° angles.

$$\begin{aligned} x + \frac{x}{2} + \frac{x\sqrt{3}}{2} &= 12 \\ 2x + x + \sqrt{3}x &= 24 \\ (3+\sqrt{3})x &= 24 \\ x &= \frac{24}{3+\sqrt{3}} \cdot \frac{3-\sqrt{3}}{3-\sqrt{3}} \\ x &= 12 - 4\sqrt{3} \end{aligned}$$

Substituting $x = 12 - 4\sqrt{3}$ in $\frac{x}{2}$ and $\frac{x\sqrt{3}}{2}$, we find $6 - 2\sqrt{3}$ ft and $6\sqrt{3} - 6$ ft are the lengths of the two sides and the hypotenuse is $12 - 4\sqrt{3}$ ft.

69. The values of x and y must satisfy

$$\begin{aligned} 6y &= 6x \\ x(6+y) &= 7(4+12). \end{aligned}$$

Since $x = y$ as seen from the first equation, upon substitution into the second equation one obtains

$$\begin{aligned} 6x + x^2 &= 112 \\ x^2 + 6x - 112 &= 0 \\ (x+14)(x-8) &= 0. \end{aligned}$$

Then $x = 8$ in. and $y = 8$ oz. One must exclude the negative value $x = -14$.

71. Let x and y be the number of minutes it takes for pump A and pump B, respectively, to fill the vat.

$$\begin{aligned} \frac{1}{x} + \frac{1}{y} &= \frac{1}{8} \\ \frac{1}{x} - \frac{1}{y} &= \frac{1}{12} \end{aligned}$$

Adding the two equations, we get

$$\begin{aligned} \frac{2}{x} &= \frac{3}{24} + \frac{2}{24} \\ \frac{2}{x} &= \frac{5}{24} \\ 5x &= 48 \\ x &= 9.6. \end{aligned}$$

Substituting $x = \frac{48}{5}$ into $\frac{1}{x} + \frac{1}{y} = \frac{1}{8}$, we find

$$\begin{aligned} \frac{5}{48} + \frac{1}{y} &= \frac{1}{8} \\ \frac{1}{y} &= \frac{6}{48} - \frac{5}{48} \\ \frac{1}{y} &= \frac{1}{48} \\ y &= 48. \end{aligned}$$

Pump A can fill the vat by itself in $x = 9.6$ min while Pump B will take $y = 48$ min.

73. Let x and y be two numbers satisfying

$$\begin{aligned} x + y &= 6 \\ xy &= 10. \end{aligned}$$

Substituting $y = 6 - x$ into $xy = 10$,

$$\begin{aligned} x(6-x) &= 10 \\ 6x - x^2 &= 10 \\ x^2 - 6x &= -10 \\ x^2 - 6x + 9 &= -1 \\ (x-3)^2 &= -1 \\ x &= 3 \pm i. \end{aligned}$$

If $x = 3 + i$, then $y = 6 - (3+i) = 3 - i$.
If $x = 3 - i$, then $y = 6 - (3-i) = 3 + i$.
The two numbers are $3 + i$ and $3 - i$.

75. Let x and y be the length and width.

$$\begin{aligned} 20xy &= 36,000 \\ 40x + 40y + 2xy &= 7200 \end{aligned}$$

Substitute $x = \frac{1800}{y}$ into $40x + 40y + 2xy = 7200$.

$$\begin{aligned} 40 \cdot \frac{1800}{y} + 40y + 2 \cdot \frac{1800}{y} \cdot y &= 7200 \\ \frac{72,000}{y} + 40y + 3600 &= 7200 \\ \frac{72,000}{y} + 40y &= 3600 \\ 40y^2 - 3600y + 72,000 &= 0 \\ y^2 - 90y + 1800 &= 0 \\ (y-60)(y-30) &= 0 \end{aligned}$$

Using $y = 30$ and $y = 60$ in $x = \dfrac{1800}{w}$, we get $x = 60$ and $x = 30$. Thus, the length is 60 ft and the width is 30 ft.

77. From the graphs, the two models give the same population for $t = 0$ and $t = 9.65$ years. The exponential population model is twice the linear model when $t \approx 29.5$ years.

79. Let x be the time of the sunrise or the number of hours since midnight. Let S and B be Sally's and Bob's speeds. The number of miles Sally and Bob drove are $(16-x)S$ and $(21-x)B$, respectively. Since they met at noon, the distance between Sally's house and Bob's house is $(12-x)S + (12-x)B$; or equivalently $(12-x)(S+B)$. Then

$$\begin{aligned}(16-x)S &= (12-x)(S+B)\\ (16-x)S &= (16-x)(S+B) - 4(S+B)\\ 0 &= (16-x)B - 4(S+B)\\ (16-x)B &= 4(S+B).\end{aligned}$$

Likewise,

$$\begin{aligned}(21-x)B &= (21-x)(S+B) - 9(S+B)\\ (21-x)S &= 9(S+B).\end{aligned}$$

Combining, one obtains
$\dfrac{(16-x)B}{(21-x)S} = \dfrac{4(S+B)}{9(S+B)}$. So, $\dfrac{(16-x)9B}{(21-x)4S} = 1$.
Furthermore, since $(16-x)S = (21-x)B$ one finds $\dfrac{B}{S} = \dfrac{16-x}{21-x}$. Then

$$\begin{aligned}\frac{9(16-x)}{4(21-x)}\cdot\frac{B}{S} &= 1\\ \frac{9}{4}\left(\frac{16-x}{21-x}\right)^2 &= 1\\ \frac{16-x}{21-x} &= \frac{2}{3} \quad \text{since } 16-x>0\\ 48-3x &= 42-2x\\ x &= 6.\end{aligned}$$

The sunrise was at 6:00 A.M.

81. By the time the mouse reaches the northeast corner (50 feet from the southwest corner) of the train, the train would have traveled 20 feet which is one-half the distance the train travels before the mouse returns to the southeast corner. So, when the mouse reaches the northwest corner (40 feet from the southwest corner) the train (in proportion to 20 feet) would have traveled 16 feet.

As the distance traveled by the train ranges from 16 feet to 20 feet, the path the mouse takes will be the hypotenuse of a right triangle whose sides are 10 feet and 4 feet. So, in this range, the mouse travels a distance of $\sqrt{10^2+4^2}$ feet on the ground, or equivalently $2\sqrt{29}$ feet on the ground.

Thus, the total diagonal ground distance (including the diagonal ground distance as the mouse moves from the southeast corner to the southwest corner) traveled by the mouse is twice of $2\sqrt{29}$ feet, or $4\sqrt{29}$ feet. Since the north-south distance traveled by the mouse is 80 feet on the ground, the total distance on the ground traveled by the mouse is $80+4\sqrt{29}$ feet, or about 101.54 ft.

83. Let $C(h,k)$ and r be the center and radius of the fifth circle, respectively. We consider three right triangles each of which has a vertex at C. The first right triangle has another vertex at the origin.

The other two right triangles have sides that are perpendicular to the coordinate axes. Also, the sides of the other two right triangles pass through through the centers of the larger circles that are tangent to the fifth circle.

Applying the Pythagorean Theorem, we obtain a system of equations

$$\begin{aligned}(2-r)^2 &= h^2+k^2\\ (r+2/3)^2 &= (h-4/3)^2+k^2\\ (r+1)^2 &= h^2+(1-k)^2.\end{aligned}$$

The solutions are $r = 1/3$, $h = 4/3$, and $k = 1$.

a) Since the radius of the fifth circle is $r = 1/3$, the diameter is $2/3$ cm.

b) Since $h = 4/3$ and $k = 1$, an equation for the fifth circle is

$$\begin{aligned}(x-4/3)^2+(y-1)^2 &= (1/3)^2\\(x-4/3)^2+(y-1)^2 &= 1/9\end{aligned}$$

85. Let $C(h,k)$ and r be the center and radius of the smallest circle, respectively. Then $k = -r$. We consider two right triangles each of which has a vertex at C.

The right triangles have sides that are perpendicular to the coordinate axes. Also, one side of each right triangle passes through the center of a larger circle.

Applying the Pythagorean Theorem, we list a system of equations

$$\begin{aligned}(r+1)^2 &= h^2+(1-r)^2\\(2-r)^2 &= h^2+r^2.\end{aligned}$$

The solutions are $r = 1/2$, $h = \sqrt{2}$, and $k = -r = -1/2$.

The equation of the smallest circle is

$$(x-\sqrt{2})^2+(y+1/2)^2 = 1/4.$$

87. Let $y = mx$ be an equation of the tangent line. At the point of tangency, the equations of the line and circle are simultaneously satisfied.

$$\begin{aligned}(x-3)^2+(mx)^2 &= 1\\(m^2+1)x^2-6x+8 &= 0.\end{aligned}$$

The discriminant of the left side is

$$b^2-4ac = 36-32(m^2+1)$$

Since the tangent line and the circle has only one point of intersection, $b^2-4ac = 0$. This happens exactly when $m = \pm\sqrt{2}/4$.

Thus, the tangent lines are $y = \pm\sqrt{2}x/4$.

91. If we add the first and third equations, we obtain

$$2x+3y = 51$$

If we add the second equation to two times the third equation, we find

$$5x+5y = 100.$$

If we multiply the above equation by $-2/5$, we obtain

$$-2x-2y = -40.$$

If we add the last equation to $2x+3y = 51$, we obtain $y = 11$. Working backwards, we find $x = 9$ and $z = 13$. The solution set is $\{(9,11,13)\}$.

93. Combine the logarithms as follows:

$$\begin{aligned}\log_3((x+1)(x-5)) &= 3\\\log_3(x^2-4x-5) &= 3\\x^2-4x-5 &= 27\\x^2-4x-32 &= 0\\(x-8)(x+4) &= 0\\x &= 8,-4\end{aligned}$$

Note, -4 is an extraneous root. The solution set is $\{8\}$.

95. Let $p(x) = 2x^3+x^2-41x+20$.

4	2	1	-41	20
		8	36	-20
	2	9	-5	0

Since the quotient factors as

$$2x^2+9x-5 = (2x-1)(x+5)$$

the solution set is $\left\{\left(4,\frac{1}{2},-5\right)\right\}$

97. Since $x+\sqrt{y} = 32$ and $y+\sqrt{x} = 54$, we obtain $\sqrt{y} = 32-x$ and $x^2 = (54-y)^2$. Then

$$y = (32-(54-y))^2.$$

Using a graphing calculator, we find $y = 49$, $y \approx 47.8$, $y \approx 58.9$, or $y \approx 60.3$

Note, $y \approx 58.9$ and $y \approx 60.3$ cannot satisfy $y+\sqrt{x} = 54$ for any real number x.

If $y \approx 47.8$, then $\sqrt{x} \approx 54-47.8 = 6.2$ or $x \approx 38.44$. Note, $x \approx 38.44$ does not satisfy $x+\sqrt{y} = 32$ for any real number y.

However, if $y = 49$ then $\sqrt{x} = 54-49 = 5$ or $x = 25$. Thus, the solution is $(x,y) = (25,49)$.

For Thought

1. True, $\frac{1}{x}+\frac{3}{x+1}=\frac{(x+1)+3x}{x(x+1)}=\frac{4x+1}{x(x+1)}$.

2. True, $x+\frac{3x}{x^2-1}=\frac{x(x^2-1)+3x}{x^2-1}=\frac{x^3+2x}{x^2-1}$.

3. False, by using long division we obtain $\frac{x^2}{x^2-9}=1+\frac{9}{x^2-9}=1+\frac{A}{x-3}+\frac{B}{x+3}$

4. True, since $\frac{1}{2}+\frac{1}{2^3}=\frac{2^2+1}{2^3}=\frac{5}{8}$.

5. False, since $\frac{3x-1}{x^3+x}=\frac{A}{x}+\frac{Bx+C}{x^2+1}$.

6. False, since $\frac{1}{x^2-1}=\frac{1/2}{x-1}-\frac{1/2}{x+1}$.

7. True, by using long division we get

$$\begin{array}{r|l} & x-1 \\ x^2+x-2 & x^3+0x^2+0x+1 \\ & \underline{x^3+x^2-2x} \\ & -x^2+2x+1 \\ & \underline{-x^2-x+2} \\ & 3x-1 \end{array}$$

So $\frac{x^3+1}{x^2+x-2}=x-1+\frac{3x-1}{x^2+x-2}$.

8. False, since $x^3-8=(x-2)(x^2+2x+4)$.

9. True, since $\frac{1}{x-1}+\frac{1}{x^2+x+1}=\frac{(x^2+x+1)+(x-1)}{x^3-1}=\frac{x^2+2x}{x^3-1}$.

10. True, it is already in the form $\frac{Ax+B}{x^2+9}$.

8.4 Exercises

1. partial

3. $\frac{3(x+1)+4(x-2)}{(x-2)(x+1)}=\frac{7x-5}{(x-2)(x+1)}$

5. $\frac{(x^2+2)-3(x-1)}{(x-1)(x^2+2)}=\frac{x^2-3x+5}{(x-1)(x^2+2)}$

7.

$$\frac{(2x+1)(x^2+3)+(x^3+2x+2)}{(x^2+3)^2}=$$

$$\frac{(2x^3+x^2+6x+3)+(x^3+2x+2)}{(x^2+3)^2}=$$

$$\frac{3x^3+x^2+8x+5}{(x^2+3)^2}$$

9.

$$\frac{(x-1)^2+(2x+3)(x-1)+(x^2+1)}{(x-1)^3}=$$

$$\frac{(x^2-2x+1)+(2x^2+x-3)+(x^2+1)}{(x-1)^3}=$$

$$\frac{4x^2-x-1}{(x-1)^3}$$

11. Multiply the equation by $(x-3)(x+3)$.

$$\begin{aligned} 12 &= A(x+3)+B(x-3) \\ 12 &= (A+B)x+(3A-3B) \end{aligned}$$

$A+B=0$ and $3A-3B=12$

Divide $3A-3B=12$ by 3 and add to $A+B=0$.

$$\begin{array}{rcl} A-B &=& 4 \\ A+B &=& 0 \\ \hline 2A &=& 4 \end{array}$$

Using $A=2$ in $A+B=0$, $B=-2$. Then $A=2$ and $B=-2$.

13. $\frac{5x-1}{(x+1)(x-2)}=\frac{A}{x+1}+\frac{B}{x-2}$

$$\begin{aligned} 5x-1 &= A(x-2)+B(x+1) \\ 5x-1 &= (A+B)x+(-2A+B) \end{aligned}$$

$A+B=5$ and $-2A+B=-1$

Multiply $-2A+B=-1$ by -1 and add to $A+B=5$.

$$\begin{array}{rcl} 2A-B &=& 1 \\ A+B &=& 5 \\ \hline 3A &=& 6 \end{array}$$

Using $A=2$ in $A+B=5$, $B=3$.

The answer is $\frac{2}{x+1}+\frac{3}{x-2}$.

15. $\dfrac{2x+5}{(x+4)(x+2)} = \dfrac{A}{x+4} + \dfrac{B}{x+2}$

$$\begin{aligned} 2x+5 &= A(x+2)+B(x+4) \\ 2x+5 &= (A+B)x+(2A+4B) \end{aligned}$$
$$A+B=2 \quad \text{and} \quad 2A+4B=5$$

Multiply $A+B=2$ by -2 and add to $2A+4B=5$.

$$\begin{aligned} -2A-2B &= -4 \\ 2A+4B &= 5 \\ \hline 2B &= 1 \end{aligned}$$

Using $B=1/2$ in $A+B=2$, we find $A=3/2$.

The answer is $\dfrac{3/2}{x+4} + \dfrac{1/2}{x+2}$.

17. $\dfrac{2}{(x-3)(x+3)} = \dfrac{A}{x-3} + \dfrac{B}{x+3}$

$$\begin{aligned} 2 &= A(x+3)+B(x-3) \\ 2 &= (A+B)x+(3A-3B) \end{aligned}$$
$$A+B=0 \quad \text{and} \quad 3A-3B=2$$

Multiply $A+B=0$ by 3 and add to $3A-3B=2$.

$$\begin{aligned} 3A+3B &= 0 \\ 3A-3B &= 2 \\ \hline 6A &= 2 \end{aligned}$$

Using $A=1/3$ in $A+B=0$, we get $B=-1/3$.

The answer is $\dfrac{1/3}{x-3} + \dfrac{-1/3}{x+3}$.

19.

$$\begin{aligned} \frac{1}{x(x-1)} &= \frac{A}{x} + \frac{B}{x-1} \\ 1 &= A(x-1)+Bx \\ 1 &= (A+B)x-A \end{aligned}$$
$$A+B=0 \quad \text{and} \quad -A=1$$

Using $A=-1$ in $A+B=0$, we find $B=1$.

The answer is $\dfrac{-1}{x} + \dfrac{1}{x-1}$.

21. Multiplying the equation by $(x+3)^2(x-2)$, we obtain $x^2+x-31 =$

$$= A(x+3)(x-2)+B(x-2)+C(x+3)^2$$
$$= A(x^2+x-6)+B(x-2)+C(x^2+6x+9)$$
$$= (A+C)x^2+(A+B+6C)x+(-6A-2B+9C).$$

Equate the coefficients and solve the system.

$$\begin{aligned} A+C &= 1 \\ A+B+6C &= 1 \\ -6A-2B+9C &= -31 \end{aligned}$$

Multiply $A+B+6C=1$ by 2 and add to $-6A-2B+9C=-31$.

$$\begin{aligned} 2A+2B+12C &= 2 \\ -6A-2B+9C &= -31 \\ \hline -4A+21C &= -29 \end{aligned}$$

Multiply $A+C=1$ by 4 and add to $-4A+21C=-29$.

$$\begin{aligned} 4A+4C &= 4 \\ -4A+21C &= -29 \\ \hline 25C &= -25 \end{aligned}$$

Using $C=-1$ in $A+C=1$, we obtain $A=2$. From $A+B+6C=1$, $2+B-6=1$ and $B=5$. So $A=2, B=5$, and $C=-1$.

23. $\dfrac{4x-1}{(x-1)^2(x+2)} = \dfrac{A}{x-1} + \dfrac{B}{(x-1)^2} + \dfrac{C}{x+2}$

$$\begin{aligned} 4x-1 &= A(x-1)(x+2)+B(x+2)+C(x-1)^2 \\ 4x-1 &= A(x^2+x-2)+B(x+2)+C(x^2-2x+1 \\ 4x-1 &= (A+C)x^2+(A+B-2C)x+ \\ &\quad (-2A+2B+C) \end{aligned}$$

If we equate the coefficients, we obtain

$$\begin{aligned} A+C &= 0 \\ A+B-2C &= 4 \\ -2A+2B+C &= -1. \end{aligned}$$

Solving the system, we get $A=1$, $B=1$, and $C=-1$. The answer is

$$\frac{1}{x-1} + \frac{1}{(x-1)^2} + \frac{-1}{x+2}.$$

25. $\dfrac{20-4x}{(x-2)^2(x+4)} = \dfrac{A}{x-2} + \dfrac{B}{(x-2)^2} + \dfrac{C}{x+4}$

$$20-4x = A(x-2)(x+4)+B(x+4)+$$

$$\begin{aligned} & C(x-2)^2 \\ 20-4x &= A(x^2+2x-8)+B(x+4)+ \\ & C(x^2-4x+4) \\ 20-4x &= (A+C)x^2+(2A+B-4C)x+ \\ & (-8A+4B+4C) \end{aligned}$$

If we equate the coefficients, we obtain

$$\begin{aligned} A+C &= 0 \\ 2A+B-4C &= -4 \\ -8A+4B+4C &= 20 \end{aligned}$$

Solving the system, we get $A=-1$, $B=2$, and $C=1$. The answer is

$$\frac{-1}{x-2}+\frac{2}{(x-2)^2}+\frac{1}{x+4}.$$

27. Note, $\dfrac{3x^2+3x-2}{(x+1)^2(x-1)}=\dfrac{A}{x+1}+\dfrac{B}{(x+1)^2}+\dfrac{C}{x-1}$

$$\begin{aligned} 3x^2+3x-2 &= A(x+1)(x-1)+B(x-1)+ \\ & C(x+1)^2 \\ 3x^2+3x-2 &= A(x^2-1)+B(x-1)+ \\ & C(x^2+2x+1) \\ 3x^2+3x-2 &= (A+C)x^2+(B+2C)x+ \\ & (-A-B+C) \end{aligned}$$

If we equate the coefficients, we obtain

$$\begin{aligned} A+C &= 3 \\ B+2C &= 3 \\ -A-B+C &= -2. \end{aligned}$$

Solving the system, we get $A=2$, $B=1$, and $C=1$. The answer is $\dfrac{2}{x+1}+\dfrac{1}{(x+1)^2}+\dfrac{1}{x-1}$.

29. Multiplying the equation by $(x+1)(x^2+4)$, we get

$$\begin{aligned} x^2-x-7 &= A(x^2+4)+(Bx+C)(x+1) \\ x^2-x-7 &= (A+B)x^2+(B+C)x+(4A+C). \end{aligned}$$

Equating the coefficients, we have

$$\begin{aligned} A+B &= 1 \\ B+C &= -1 \\ 4A+C &= -7. \end{aligned}$$

Multiply $A+B=1$ by -1 and add to $B+C=-1$.

$$\begin{aligned} -A-B &= -1 \\ B+C &= -1 \\ \hline -A+C &= -2 \end{aligned}$$

Multiply $4A+C=-7$ by -1 and add to $-A+C=-2$.

$$\begin{aligned} -A+C &= -2 \\ -4A-C &= 7 \\ \hline -5A &= 5 \end{aligned}$$

Using $A=-1$ in $A+B=1$, $B=2$.
Using $B=2$ in $B+C=-1$, $C=-3$.
So $A=-1$, $B=2$, and $C=-3$.

31. Note, $\dfrac{5x^2+5x}{(x+2)(x^2+1)}=\dfrac{A}{x+2}+\dfrac{Bx+C}{x^2+1}$.

$$\begin{aligned} 5x^2+5x &= A(x^2+1)+(Bx+C)(x+2) \\ 5x^2+5x &= (A+B)x^2+(2B+C)x+ \\ & (A+2C) \end{aligned}$$

If we equate the coefficients, we obtain

$$\begin{aligned} A+B &= 5 \\ 2B+C &= 5 \\ A+2C &= 0 \end{aligned}$$

Solving the system, we obtain $A=2$, $B=3$, and $C=-1$. The answer is $\dfrac{2}{x+2}+\dfrac{3x-1}{x^2+1}$.

33. Note,

$$\frac{x^2-2}{(x+1)(x^2+x+1)}=\frac{A}{x+1}+\frac{Bx+C}{x^2+x+1}.$$

$$\begin{aligned} x^2-2 &= A(x^2+x+1)+(Bx+C)(x+1) \\ x^2-2 &= (A+B)x^2+(A+B+C)x+ \\ & (A+C) \end{aligned}$$

If we equate the coefficients, we obtain

$$\begin{aligned} A+B &= 1 \\ A+B+C &= 0 \\ A+C &= -2. \end{aligned}$$

Solving the system, we obtain $A=-1$, $B=2$, and $C=-1$. The answer is

$$\frac{-1}{x+1}+\frac{2x-1}{x^2+x+1}.$$

35. $\dfrac{-2x-7}{(x+2)^2} = \dfrac{A}{x+2} + \dfrac{B}{(x+2)^2}$

$$\begin{aligned} -2x-7 &= A(x+2)+B \\ -2x-7 &= Ax+(2A+B) \end{aligned}$$

$$A=-2 \quad \text{and} \quad 2A+B=-7$$

Using $A=-2$ in $2A+B=-7$, we get $B=-3$. The answer is

$$\frac{-3}{(x+2)^2}+\frac{-2}{x+2}.$$

37. Note that $x^3+x^2+x+1 = x^2(x+1)+(x+1) = (x^2+1)(x+1)$. Then we obtain

$$\frac{6x^2-x+1}{(x^2+1)(x+1)} = \frac{A}{x+1}+\frac{Bx+C}{x^2+1}$$

$$\begin{aligned} 6x^2-x+1 &= A(x^2+1)+(Bx+C)(x+1) \\ 6x^2-x+1 &= (A+B)x^2+(B+C)x+(A+C). \end{aligned}$$

Equating the coefficients, we get

$$\begin{aligned} A+B &= 6 \\ B+C &= -1 \\ A+C &= 1. \end{aligned}$$

Multiply $A+B=6$ by -1 and add to $B+C=-1$.

$$\begin{array}{rcl} -A-B &=& -6 \\ B+C &=& -1 \\ \hline -A+C &=& -7 \end{array}$$

Adding $-A+C=-7$ and $A+C=1$, $2C=-6$. Using $C=-3$ in $B+C=-1$ and $A+C=1$, we obtain $B=2$ and $A=4$.

The answer is $\dfrac{4}{x+1}+\dfrac{2x-3}{x^2+1}$.

39. Note,

$$\frac{3x^3-x^2+19x-9}{(x^2+9)^2} = \frac{Ax+B}{x^2+9}+\frac{Cx+D}{(x^2+9)^2}.$$

So $3x^3-x^2+19x-9 = (Ax+B)(x^2+9)+(Cx+D) = Ax^3+Bx^2+(9A+C)x+(9B+D)$.

Then $A=3$ and $B=-1$. Since $9A+C=19$ and $9B+D=-9$, we get $C=-8$ and $D=0$.

The answer is $\dfrac{-8x}{(x^2+9)^2}+\dfrac{3x-1}{x^2+9}$.

41. Observe that

$$\frac{3x^2+17x+14}{(x-2)(x^2+2x+4)} = \frac{A}{x-2}+\frac{Bx+C}{x^2+2x+4}.$$

Then $3x^2+17x+14 = A(x^2+2x+4)+(Bx+C)(x-2) = (A+B)x^2+(2A-2B+C)x+(4A-2C)$

Equating the coefficients, we obtain

$$\begin{aligned} A+B &= 3 \\ 2A-2B+C &= 17 \\ 4A-2C &= 14. \end{aligned}$$

Multiply $A+B=3$ by 2 and add to $2A-2B+C=17$.

$$\begin{array}{rcl} 2A+2B &=& 6 \\ 2A-2B+C &=& 17 \\ \hline 4A+C &=& 23 \end{array}$$

Multiplying $4A-2C=14$ by -1 and adding to $4A+C=23$, $3C=9$. So $C=3$ and from $4A-2C=14$, $A=5$. Using these values in $2A-2B+C=17$, we get $B=-2$. The answer is $\dfrac{5}{x-2}+\dfrac{-2x+3}{x^2+2x+4}$.

43. Divide $2x^3+x^2+3x-2$ by x^2-1 by long division.

$$\begin{array}{r|l} & \quad 2x+1 \\ \hline x^2-1 & 2x^3+x^2+3x-2 \\ & 2x^3+0x^2-2x \\ \hline & \qquad x^2+5x-2 \\ & \qquad x^2+0x-1 \\ \hline & \qquad\quad 5x-1 \end{array}$$

Then $\dfrac{2x^3+x^2+3x-2}{x^2-1} = 2x+1+\dfrac{5x-1}{x^2-1}$.

Decompose $\dfrac{5x-1}{x^2-1} = \dfrac{A}{x-1}+\dfrac{B}{x+1}$.

$$\begin{aligned} 5x-1 &= A(x+1)+B(x-1) \\ 5x-1 &= (A+B)x+(A-B) \end{aligned}$$

So $A+B=5$ and $A-B=-1$.

Adding $A+B=5$ and $A-B=-1$, $2A=4$. Using $A=2$ in $A+B=5$, we find $B=3$.

The answer is $2x+1+\dfrac{2}{x-1}+\dfrac{3}{x+1}$.

45.

Since $\dfrac{3x^3-2x^2+x-2}{(x^2+x+1)^2} = \dfrac{Ax+B}{x^2+x+1}+\dfrac{Cx+D}{(x^2+x+1)^2}$, we get

$3x^3-2x^2+x-2=(Ax+B)(x^2+x+1)+(Cx+D)$
$= Ax^3+(A+B)x^2+(A+B+C)x+(B+D)$.

Equating the coefficients, we find $A=3$. Since $A+B=-2$, we get $B=-5$. From $A+B+C=1$ and $B+D=-2$, we have $C=3$ and $D=3$.

The answer is $\dfrac{3x-5}{x^2+x+1}+\dfrac{3x+3}{(x^2+x+1)^2}$.

47.

Since $\dfrac{3x^3+4x^2-12x+16}{(x-2)(x+2)(x^2+4)} = \dfrac{A}{x-2}+\dfrac{B}{x+2}+\dfrac{Cx+D}{x^2+4}$, we obtain

$3x^3+4x^2-12x+16=$
$=A(x+2)(x^2+4)+B(x-2)(x^2+4)+(Cx+D)(x^2-4)$
$=(A+B+C)x^3+(2A-2B+D)x^2+(4A+4B-4C)x+(8A-8B-4D)$.

Equating the coefficients, we get

$$\begin{aligned} A+B+C &= 3\\ 2A-2B+D &= 4\\ 4A+4B-4C &= -12\\ 8A-8B-4D &= 16. \end{aligned}$$

Multiply first equation by -4 and add to the third. Multiply second equation by -4 and add to the fourth. Also multiply first equation by 2 and add to the second.

$$\begin{array}{rcl} -4A-4B-4C &=& -12\\ 4A+4B-4C &=& -12\\ \hline -8C &=& -24 \end{array}$$

$$\begin{array}{rcl} -8A+8B-4D &=& -16\\ 8A-8B-4D &=& 16\\ \hline -8D &=& 0 \end{array}$$

$$\begin{array}{rcl} 2A+2B+2C &=& 6\\ 2A-2B+D &=& 4\\ \hline 4A+2C+D &=& 10 \end{array}$$

So $C=3$ and $D=0$. From $4A+2C+D=10$, we find $A=1$ and from $2A-2B+D=4$, we get $B=-1$.

The answer is $\dfrac{1}{x-2}+\dfrac{-1}{x+2}+\dfrac{3x}{x^2+4}$.

49. $\dfrac{5x^3+x^2+x-3}{x^3(x-1)}=\dfrac{A}{x}+\dfrac{B}{x^2}+\dfrac{C}{x^3}+\dfrac{D}{x-1}$,

$5x^3+x^2+x-3=$
$=Ax^2(x-1)+Bx(x-1)+C(x-1)+Dx^3$
$=(A+D)x^3+(-A+B)x^2+(-B+C)x-C$

Equating the coefficients, we get $C=3$. From $-B+C=1$, we find $B=2$. From $-A+B=1$, we obtain $A=1$. From $A+D=5$, we have $D=4$.

The answer is $\dfrac{1}{x}+\dfrac{2}{x^2}+\dfrac{3}{x^3}+\dfrac{4}{x-1}$.

51. Note,
$\dfrac{6x^2-28x+33}{(x-2)^2(x-3)}=\dfrac{A}{x-2}+\dfrac{B}{(x-2)^2}+\dfrac{C}{x-3}$.

Then $6x^2-28x+33=$
$=A(x-2)(x-3)+B(x-3)+C(x-2)^2$
$=(A+C)x^2+(-5A+B-4C)x+(6A-3B+4C)$.

Equating the coefficients, we obtain

$$\begin{aligned} A+C &= 6\\ -5A+B-4C &= -28\\ 6A-3B+4C &= 33. \end{aligned}$$

Multiply second equation by 3 and add to the third.

$$\begin{array}{rcl} -15A+3B-12C &=& -84\\ 6A-3B+4C &=& 33\\ \hline -9A-8C &=& -51 \end{array}$$

Multiply $A+C=6$ by 8 and add to $-9A-8C=-51$

$$\begin{array}{rcl} -9A-8C &=& -51\\ 8A+8C &=& 48\\ \hline -A &=& -3 \end{array}$$

Using $A=3$ in $A+C=6$, we find $C=3$. From $6A-3B+4C=33$, we obtain $B=-1$.

The answer is $\dfrac{3}{x-2}+\dfrac{-1}{(x-2)^2}+\dfrac{3}{x-3}$.

53. Use synthetic division to factor the denominator.

-5	1	4	-11	-30
		-5	5	30
	1	-1	-6	0

$$\begin{aligned} x^3+4x^2-11x-30 &= (x+5)(x^2-x-6) \\ &= (x+5)(x+2)(x-3) \end{aligned}$$

Decomposing, $\dfrac{9x^2+21x-24}{(x+5)(x+2)(x-3)} =$

$$= \frac{A}{x+5}+\frac{B}{x+2}+\frac{C}{x-3}.$$

Then $9x^2+21x-24 =$
$= A(x+2)(x-3)+B(x+5)(x-3)+$
$C(x+5)(x+2)$

and substituting $x=-2,3,-5$, we obtain

$$\begin{aligned} -30 &= -15B \\ 2 &= B \end{aligned}$$

$$\begin{aligned} 120 &= 40C \\ 3 &= C \end{aligned}$$

$$\begin{aligned} 96 &= -24A \\ 4 &= A. \end{aligned}$$

The answer is $\dfrac{4}{x+5}+\dfrac{2}{x+2}+\dfrac{3}{x-3}$.

55. Note that $x^3-3x^2+3x-1=(x-1)^3$.

Then $\dfrac{x^2-2}{(x-1)^3}=\dfrac{A}{x-1}+\dfrac{B}{(x-1)^2}+\dfrac{C}{(x-1)^3}$.

$$\begin{aligned} x^2-2 &= A(x-1)^2+B(x-1)+C \\ &= Ax^2+(-2A+B)x+(A-B+C) \end{aligned}$$

Then $A=1$. Since $-2A+B=0$, $B=2$. Since $A-B+C=-2$, we find $1-2+C=-2$ and $C=-1$.

The answer is $\dfrac{1}{x-1}+\dfrac{2}{(x-1)^2}+\dfrac{-1}{(x-1)^3}$.

57.

$$\begin{aligned} \frac{x}{(ax+b)^2} &= \frac{A}{ax+b}+\frac{B}{(ax+b)^2} \\ x &= A(ax+b)+B \\ &= aAx+(bA+B) \end{aligned}$$

So $aA=1$ and $A=1/a$. Since $bA+B=0$, we obtain $\dfrac{b}{a}+B=0$ and $B=-b/a$.

The answer is $\dfrac{-b/a}{(ax+b)^2}+\dfrac{1/a}{ax+b}$.

59. Since $\dfrac{x+c}{x(ax+b)}=\dfrac{A}{x}+\dfrac{B}{ax+b}$, we have

$$\begin{aligned} x+c &= A(ax+b)+Bx \\ &= (aA+B)x+bA. \end{aligned}$$

So $bA=c$ and $A=\dfrac{c}{b}$. Since $aA+B=1$, we have $\dfrac{ac}{b}+B=1$ and $B=1-\dfrac{ac}{b}$.

Answer is $\dfrac{c/b}{x}+\dfrac{1-ac/b}{ax+b}$.

61. Since $\dfrac{1}{x^2(ax+b)}=\dfrac{A}{x}+\dfrac{B}{x^2}+\dfrac{C}{ax+b}$, we get
$1=Ax(ax+b)+B(ax+b)+Cx^2$
$1=(aA+C)x^2+(bA+aB)x+bB$.

So $bB=1$ and $B=\dfrac{1}{b}$. Since $bA+aB=0$, we obtain $bA+\dfrac{a}{b}=0$ and $A=-\dfrac{a}{b^2}$.

Since $aA+C=0$, $-\dfrac{a^2}{b^2}+C=0$ and $C=\dfrac{a^2}{b^2}$.

The answer is

$$\frac{-a/b^2}{x}+\frac{1/b}{x^2}+\frac{a^2/b^2}{ax+b}.$$

63. If we substitute $y=x+2$ into the other equation, we find

$$\begin{aligned} x^2+(x+2)^2 &= 34 \\ 2x^2+4x-30 &= 0 \\ x^2+2x-15 &= 0 \\ (x+5)(x-3) &= 0 \\ x &= -5,3 \end{aligned}$$

If $x=-5$, then $y=x+2=-5+2=-3$. Similarly, if $x=3$ then $y=5$.

The solution set is $\{(-5,-3),(3,5)\}$.

65. We may rewrite $2y = 10x - 6$ as $16 = 10x - 2y$ or $8 = 5x - y$. Then the system of two equations are dependent. Since $y = 5x - 8$, the solution set is

$$\{(x, 5x - 8) | x \text{ is a real number}\}.$$

67. Apply the method of completing the square.

$$y = \frac{1}{2}\left(x^2 + 8x\right) - 9$$

$$y = \frac{1}{2}\left(x^2 + 8x + 16\right) - 9 - 8$$

$$y = \frac{1}{2}(x + 4)^2 - 17$$

69. Open the cylinder to form a 4ft-by-6ft rectangle in such a way that the lizard is on the left side and 1 ft from the base of the rectangle, and the fly is 1 ft from the top of the rectangle and 3 ft from the left side.

If we imagine the rectangle as a piece of paper, then the lizard is on the front page of the paper and the fly in on the back page. Cut the rectangle vertically in the middle into two pieces, then hold the right piece and turn it over. Now, the lizard and fly are on the same side of the pages. Next, move the right piece up by 4 ft.

At this point, we have two 4 ft-by-3 ft rectangles that are joined together at the right top corner of one rectangle (with the lizard) and the left bottom corner of the other rectangle (with the fly). Recall, the lizard and fly are on the same side of the pages.

If the lizard moves 3ft to the right and 4ft up, then the lizard would have reached the fly. However, the shortest path for the lizard is the path along the hypotenuse of a right triangle whose sides are the 3 ft horizontal side and the 4 ft vertical side. The length of the hypotenuse is 5 ft, which is the shortest path from the lizard to the fly.

For Thought

1. False, since $3 > 1 + 2$ is false.

2. False **3.** True

4. False, because $x^2 + y^2 > 5$ is the region outside of a circle of radius $\sqrt{5}$.

5. True, since $(-2, 1)$ satisfies both equations in system (a).

6. True

7. False, $(-2, 0)$ does not satisfy $y < x + 2$.

8. False, $(-1, 2)$ lies on the line $y - 3x = 5$.

9. True

10. True

8.5 Exercises

1. linear inequality

3. c **5.** d

7. $y < 2x$

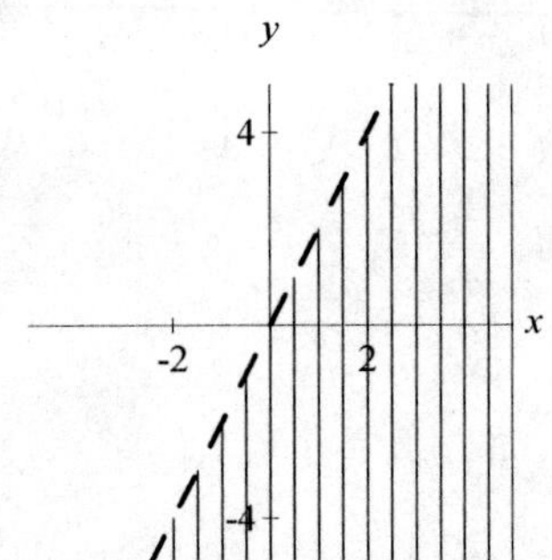

9. $x + y > 3$

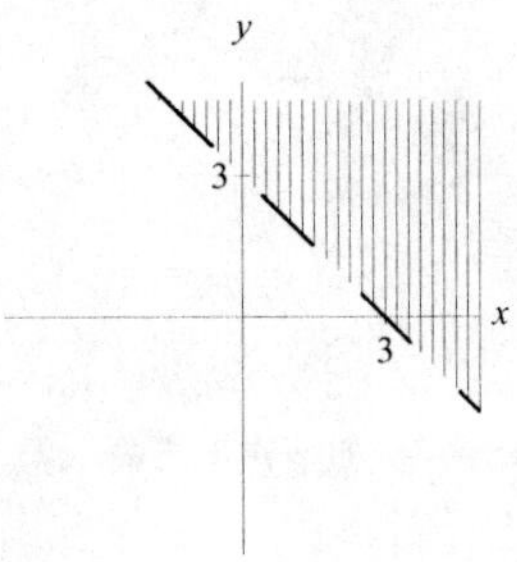

11. $2x - y \leq 4$

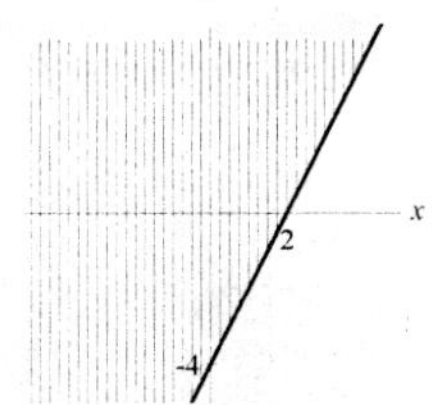

13. $y < -3x - 4$

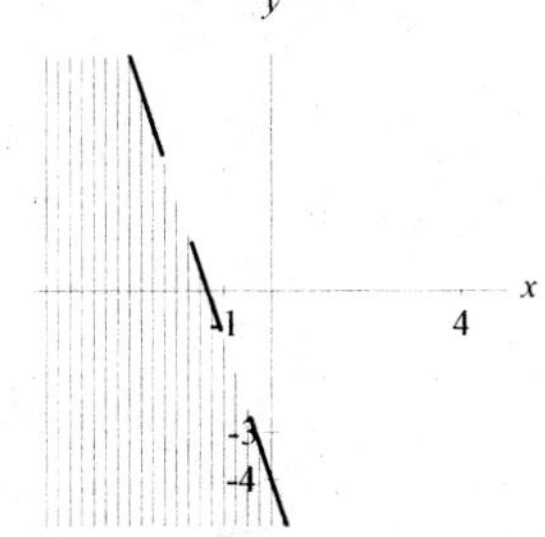

15. $x - 3 \geq 0$

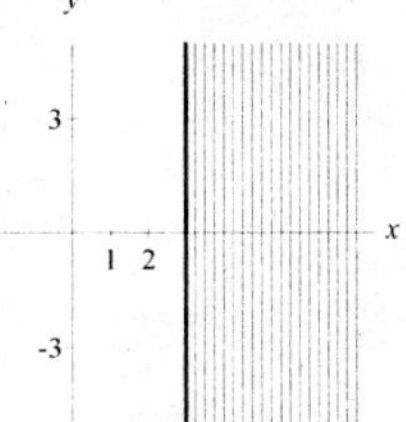

17. $20x - 30y \leq 6000$

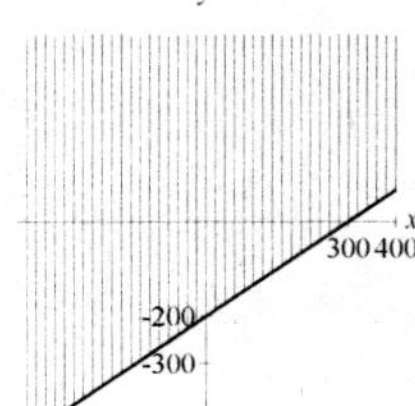

19. $y < 3$

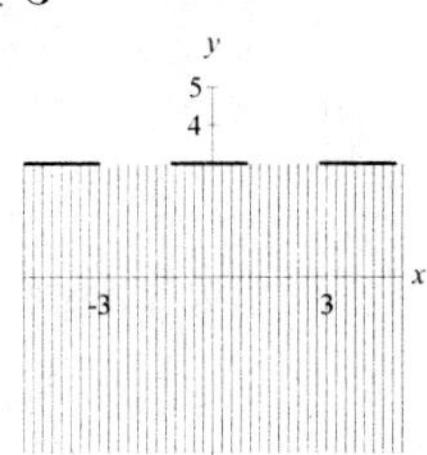

21. $y > -x^2$

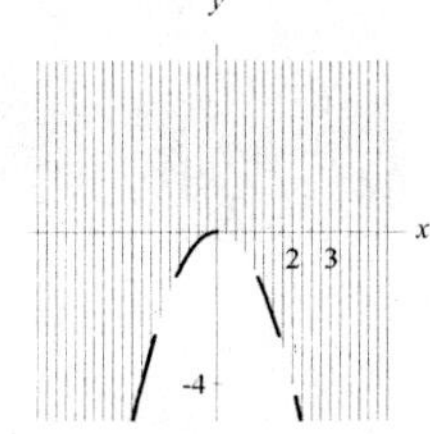

23. $x^2 + y^2 \geq 1$

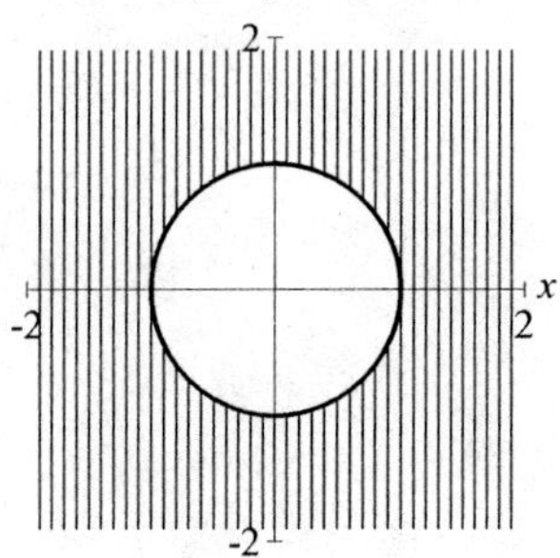

25. $x > |y|$

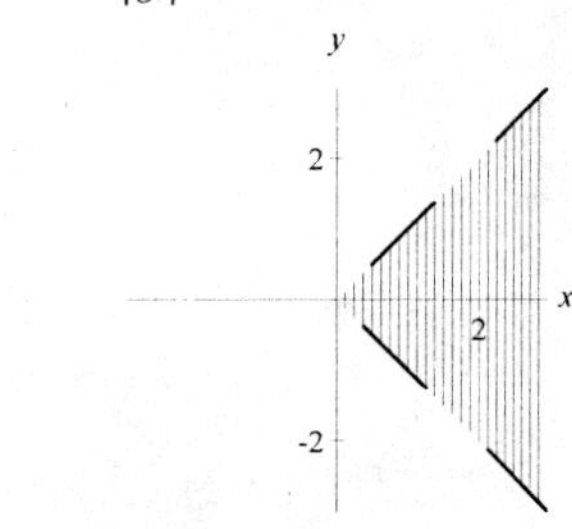

27. $x \geq y^2$

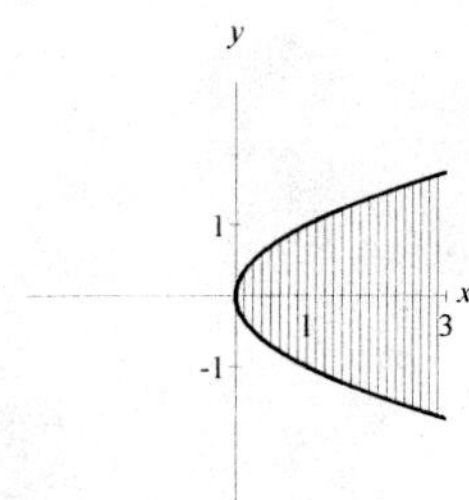

29. $y \geq x^3$

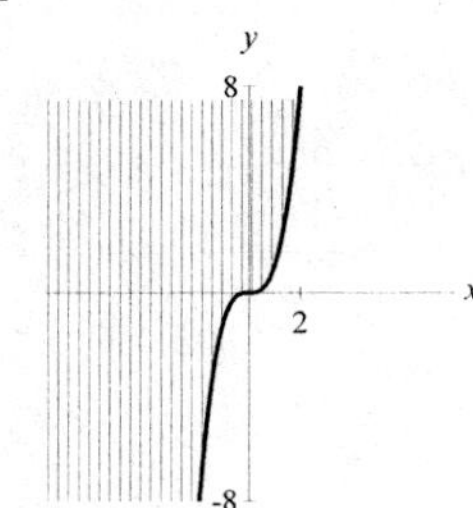

31. $y > 2^x$

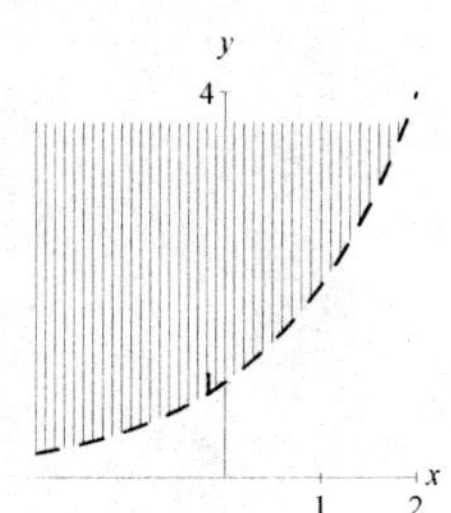

33. $y > x - 4$, $y < -x - 2$

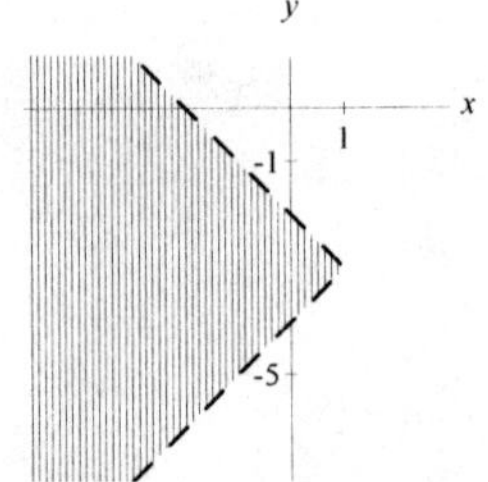

35. $3x - 4y \leq 12$, $x + y \geq -3$

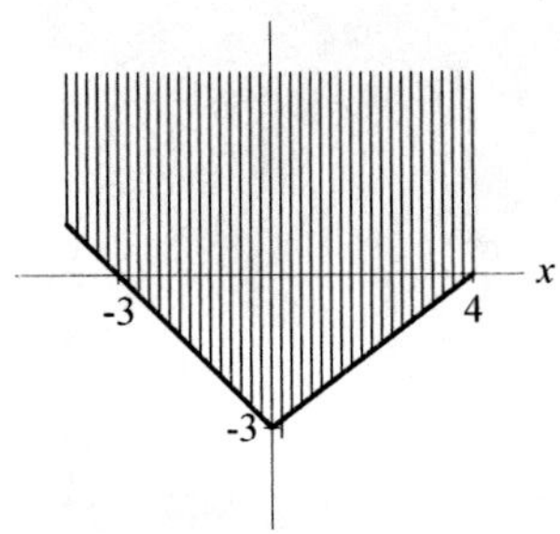

37. $3x - y < 4$, $y < 3x + 5$

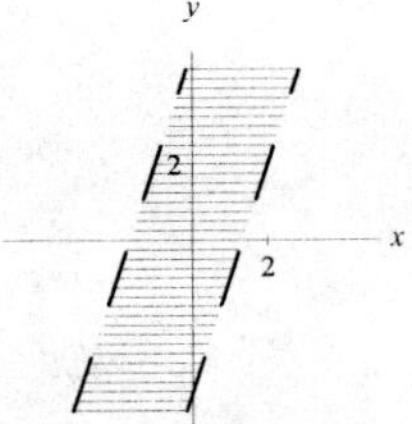

39. No solution since the graphs of $y + x < 0$ and $y > 3 - x$ do not overlap.

41. $x + y < 5$, $y \geq 2$

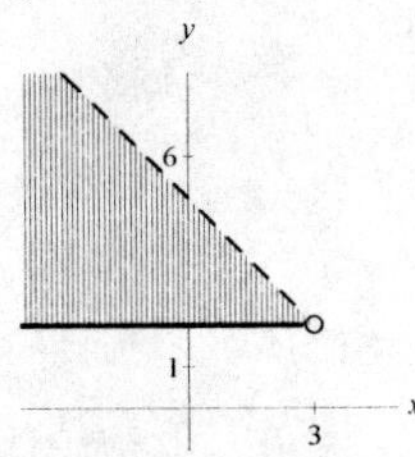

43. $y < x - 3$, $x \leq 4$

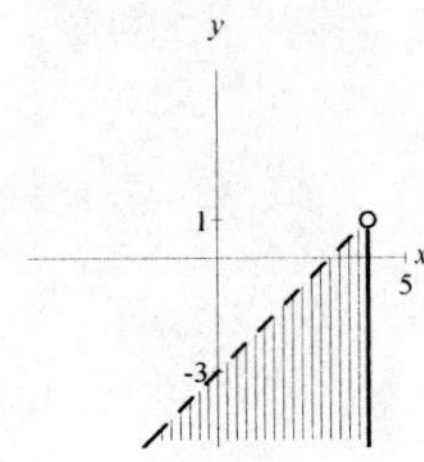

45. $y > x^2 - 3$, $y < x + 1$

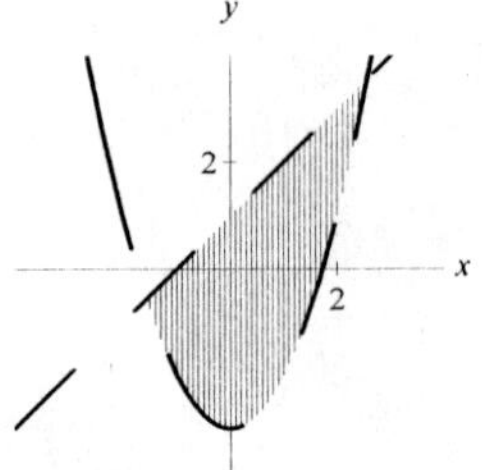

47. $x^2 + y^2 \geq 4$, $x^2 + y^2 \leq 16$

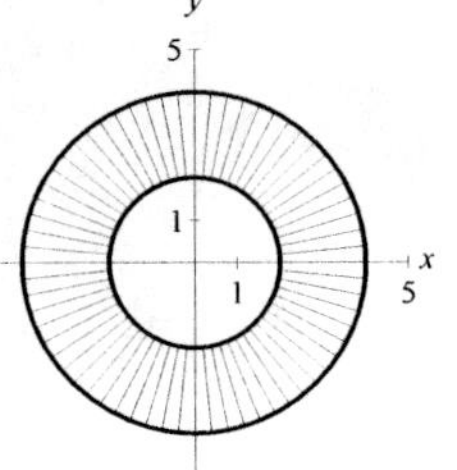

49. $(x - 3)^2 + y^2 \leq 25$, $(x + 3)^2 + y^2 \leq 25$

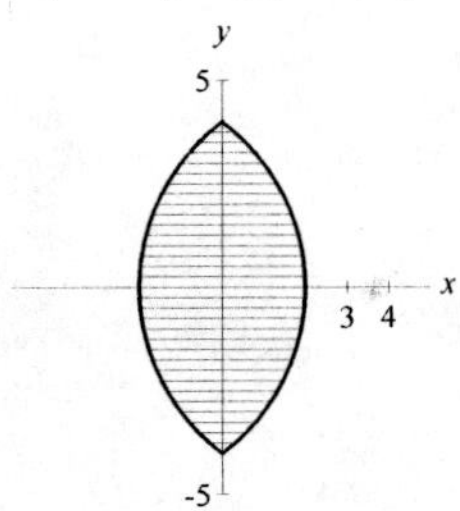

51. $x^2 + y^2 > 4$, $|x| \leq 4$

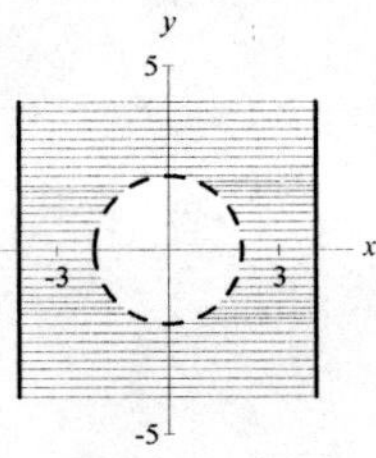

53. $y > |2x| - 4$, $y \leq \sqrt{4 - x^2}$

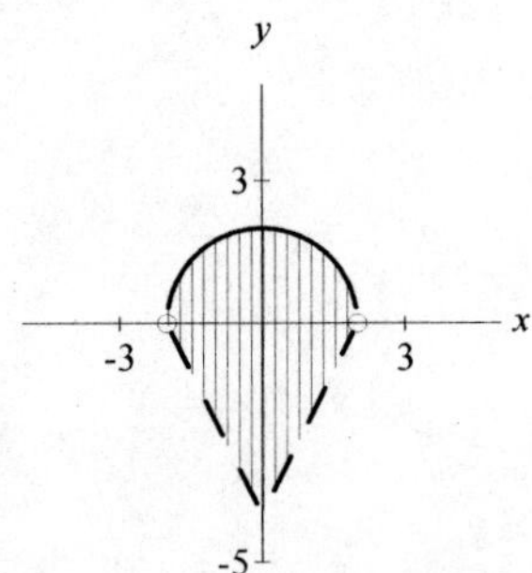

55. $|x-1|<2,\ |y-1|<4$

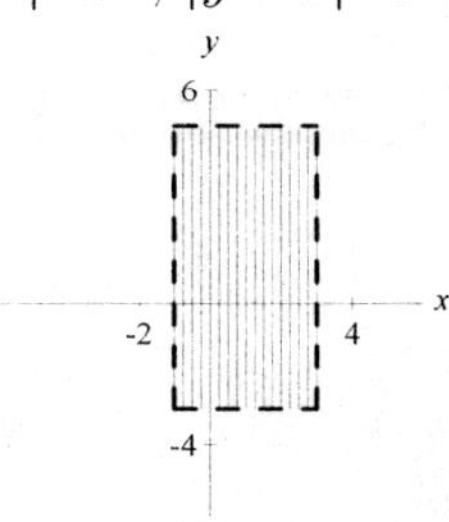

57. $x \geq 0,\ y \geq 0,\ x+y \leq 4$

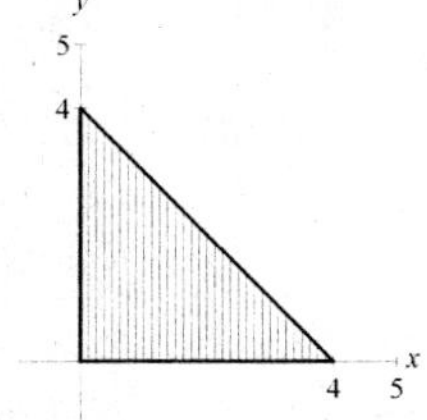

59. $x \geq 0,\ y \geq 0,\ x+y \geq 4,\ y \geq -2x+6$

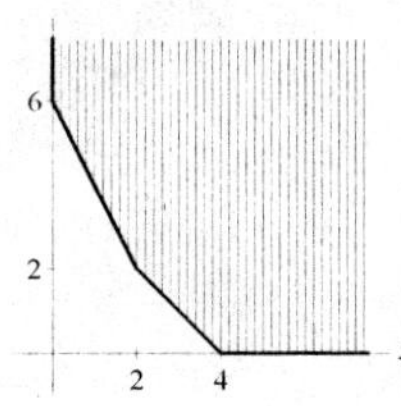

61. $x^2+y^2 \geq 9,\ x^2+y^2 \leq 25,\ y \geq |x|$

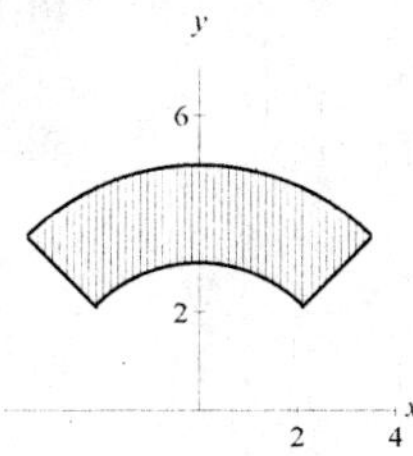

63. $y > (x-1)^3,\ y > 1,\ x+y > -2$

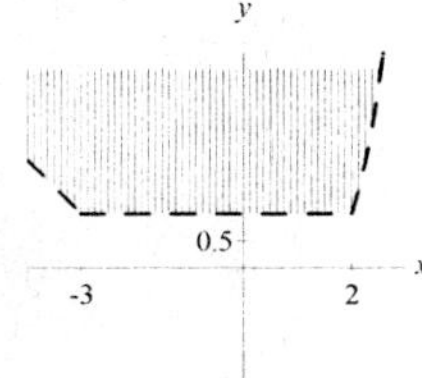

65. $y > 2^x,\ y < 6-x^2,\ x+y > 0$

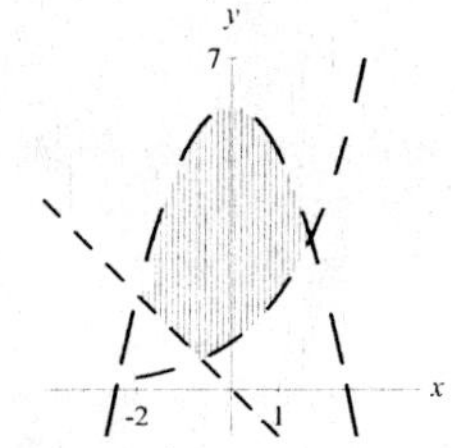

67. $x \geq 0,\ y \geq 0,\ y \leq -\dfrac{2}{3}x+5,\ y \leq -3x+12$

69. $x \geq 0,\ y \geq 0,\ y \geq -\dfrac{1}{2}x+3,\ y \geq -\dfrac{3}{2}x+5$

71. The system is

$$\begin{aligned} |x| &< 2 \\ |y| &< 2. \end{aligned}$$

73. Since a circle of radius 9 with center at the origin is given by $x^2+y^2=81$, the system is

$$\begin{aligned} x^2+y^2 &< 81 \\ x &> 0 \\ y &> 0. \end{aligned}$$

75. $(-1.17,\ 1.84)$ is a solution.

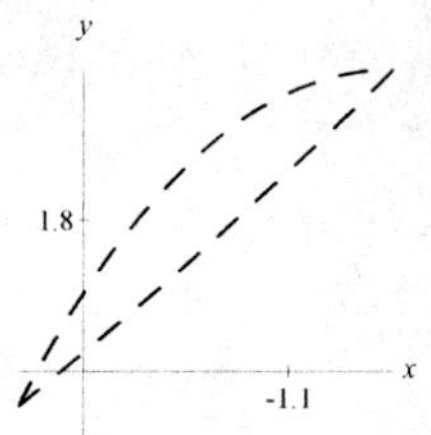

77. $(150,\ 22.4)$ is a solution.

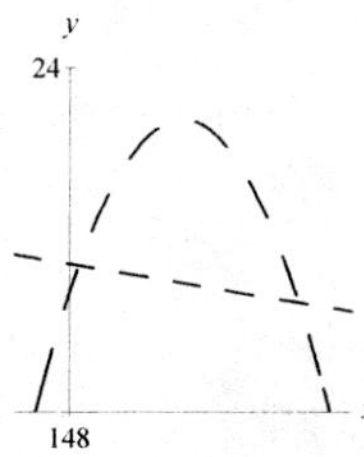

79. Let w and h be the width and height, respectively. Then $50+2w+2h \leq 130$. The system is

$$\begin{aligned} w+h &\leq 40 \\ w,h &\geq 0. \end{aligned}$$

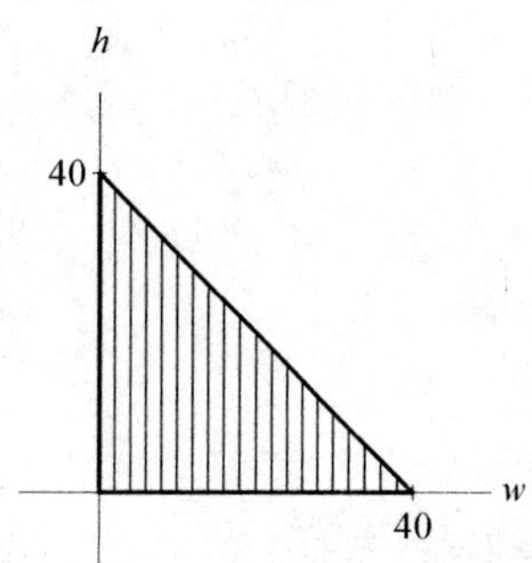

81. Let x and y be the number of mid-size and full-size cars, respectively. Divide $10,000x + 15,000y \leq 1,500,000$ by 1000. The system is

$$\begin{aligned} x + y &\leq 110 \\ x + 1.5y &\leq 150 \\ x &\geq 0 \\ y &\geq 0. \end{aligned}$$

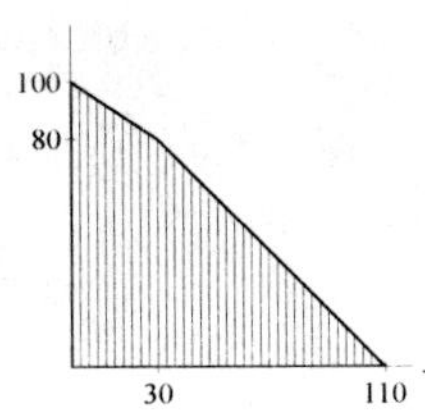

83. Let x and y be the number of \$50 tickets and \$100 tickets, respectively. Simplifying $y \leq 0.2(x + y)$, we get $0.8y \leq 0.2x$. Then $y \leq \frac{1}{4}x$. The system is

$$\begin{aligned} y &\leq \frac{1}{4}x \\ x + y &\leq 500 \\ x \geq 0, y &\geq 0. \end{aligned}$$

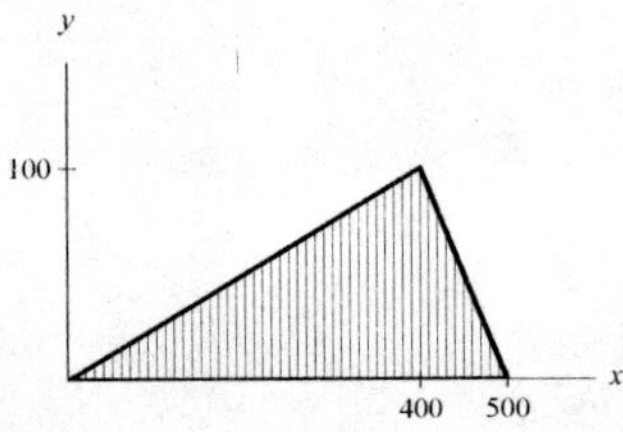

85. From the equation

$$\frac{x^2 + 5x + 3}{x^2(x + 1)} = \frac{A}{x} + \frac{B}{x^2} + \frac{C}{x + 1}$$

we find

$$\begin{aligned} x^2 + 5x + 3 &= Ax(x + 1) + B(x + 1) + Cx^2 \\ x^2 + 5x + 3 &= (A + C)x^2 + (A + B)x + B. \end{aligned}$$

Equating the coefficients, we obtain a system of equations.

$$\begin{aligned} B &= 3 \\ A + B &= 5 \\ A + C &= 1 \end{aligned}$$

The solutions of the above system of equations are $B = 3$, $A = 2$, and $C = -1$.

The partial fraction decomposition is

$$\frac{2}{x} + \frac{3}{x^2} + \frac{-1}{x + 1}.$$

87. The second equation $18y - 10x = 20$ is equivalent to $9y - 5x = 10$ or $5x - 9y = -10$. This contradicts the first equation $5x - 9y = 12$. The solution set is $\emptyset$.

89. $\$20,000e^{0.05(3.5)} = \$23,824.92$

91. A rower can either step over another rower or slide to an adjacent empty seat. The total number of moves is the number of slides plus the number of step overs. Each man in the front must step over a woman or be stepped over by a woman. So the minimum number of step overs is $5 \cdot 5$ or 25. Each rower must move a total of 6 spaces. So the total number of spaces moved is $10 \cdot 6$ or 60. Since each step over is two spaces, the number of slides is $60 - 2 \cdot 25$ or 10.

So the minimum number of moves is 10+25 or 35. Represent the five women in the back and five men in the front as $BBBBB\_\_FFFFF$

B will only move to the right, F will only move to the left, and step overs will only occur between opposite types. The 35 moves can be represented as follows:

$$B, F, F, B, B, B, F, F, F, F, B, B, B, B, B,$$

$$F, F, F, F, F, B, B, B, B, B, F, F, F, F,$$

$$B, B, B, F, F, B$$

For Thought

1. False, $x \geq 0$ include points on the x-axis and the first and fourth quadrants.

2. False, $y \geq 2$ include points on or above the line $y = 2$. **3.** False

4. False, since x-intercept is $(6, 0)$ and y-intercept is $(0, 4)$. **5.** True **6.** True **7.** False

8. True, since $R(1, 3) = 30(1) + 15(3) = 75$.

9. False, since $C(0, 5) = 7(0) + 9(5) + 3 = 48$.

10. True

8.6 Exercises

1. constraints

3. natural

5. Vertices are $(0, 0), (0, 4), (4, 0)$

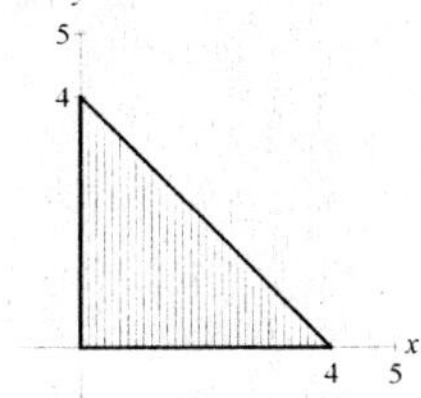

7. Vertices are $(0, 0), (1, 3), (1, 0), (0, 3)$

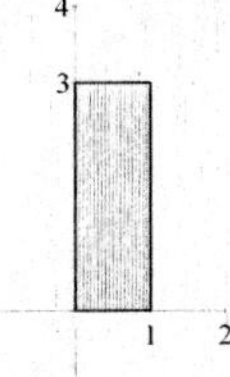

9. Vertices are $(0, 0), (2, 2), (0, 4), (3, 0)$

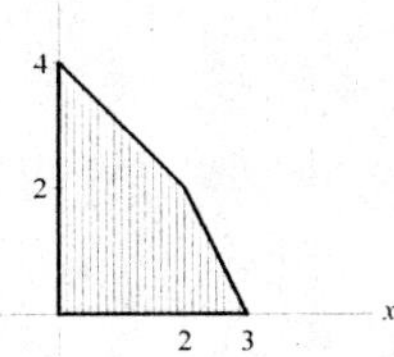

11. Vertices are $(3, 0), (1, 2), (0, 4)$

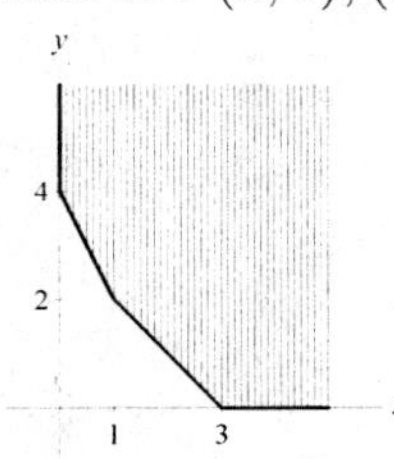

13. Vertices are $(1, 3), (4, 0), (0, 6)$

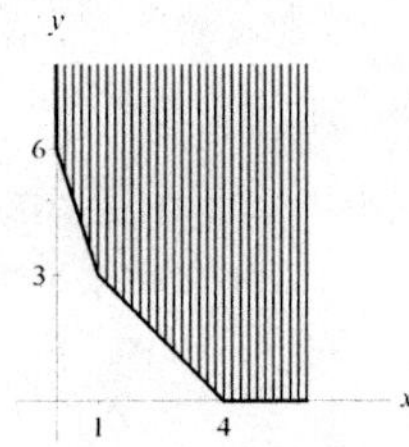

15. Vertices are $(1, 5), (6, 0), (0, 8)$

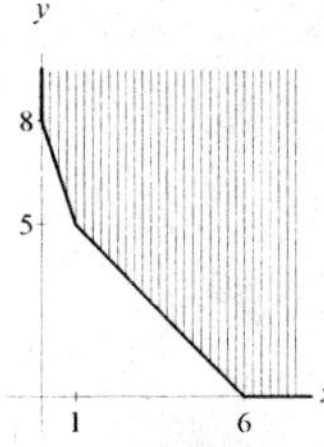

17. The values of $T(x, y) = 2x + 3y$ at the vertices are $T(0, 0) = 0$, $T(0, 4) = 12$, $T(3, 3) = 15$, and $T(5, 0) = 10$. The maximum value is 15.

19. The values of $H(x, y) = 2x + 2y$ at the vertices are $T(0, 6) = 12$, $T(2, 2) = 8$, and $T(5, 0) = 10$. The minimum value is 8.

21. The values of $P(x, y) = 5x + 9y$ at the vertices are $P(0, 0) = 0$, $P(6, 0) = 30$, $P(0, 3) = 27$. Maximum value is 30.

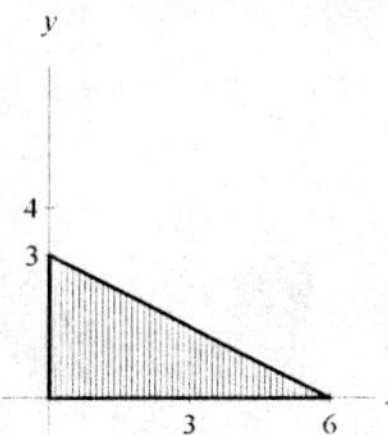

23. The values of $C(x, y) = 3x + 2y$ at the vertices are $C(0, 4) = 8$ and $C(4, 0) = 12$.
The minimum value is 8.

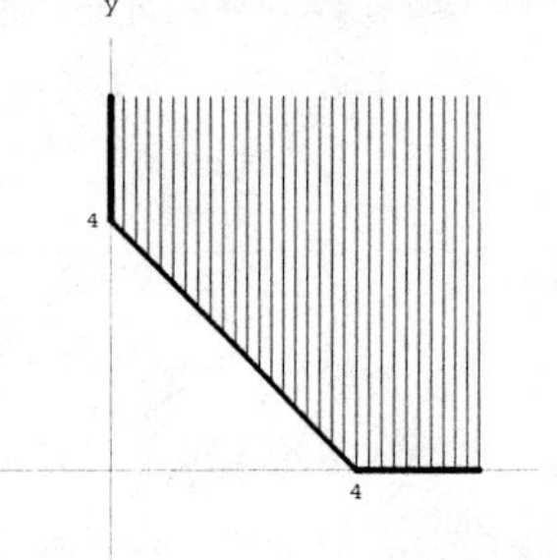

25. The values of $C(x, y) = 10x + 20y$ at the vertices are $C(0, 8) = 160$, $C(5, 3) = 110$, and $C(10, 0) = 100$. Minimum value is 100.

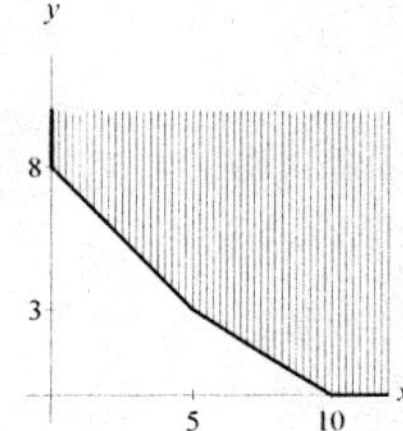

27. Let x and y be the number of bird houses and mailboxes, respectively.

$$\begin{aligned} \text{Maximize } 12x + 20y & \\ \text{subject to } 3x + 4y &\leq 48 \\ x + 2y &\leq 20 \\ x, y &\geq 0 \end{aligned}$$

The values of $R(x, y) = 12x + 20y$ at the vertices are $R(0, 0) = 0$, $R(0, 10) = 200$, $R(8, 6) = 216$, and $R(16, 0) = 192$. To maximize revenue, they must sell 8 bird houses and 6 mailboxes.

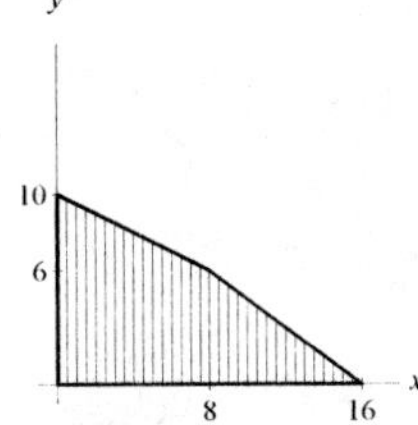

29. Let x and y be the number of bird houses and mailboxes, respectively. The values of $R(x, y) = 18x + 20y$ at the vertices are $R(0, 0) = 0$, $R(0, 10) = 200$, $R(8, 6) = 264$, and $R(16, 0) = 288$. To maximize revenue, they must sell 16 bird houses and 0 mailboxes.

31. Let x and y be the number of small and large truck loads, respectively.

$$\begin{aligned} \text{Minimize } 70x + 60y & \\ \text{subject to } 12x + 20y &\geq 120 \\ x + y &\geq 8 \\ x, y &\geq 0 \end{aligned}$$

The values of $C(x, y) = 70x + 60y$ at the vertices are $C(0, 8) = 480$, $C(10, 0) = 700$, and $C(5, 3) = 530$, To minimize costs, they must make 8 large truck loads and 0 small truck loads.

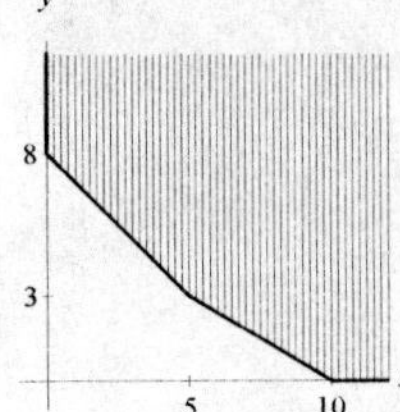

33. Let x and y be the number of small and large truck loads, respectively. The values of $C(x, y) = 70x + 75y$ at the vertices are $C(0, 8) = 600$, $C(10, 0) = 700$, and $C(5, 3) = 575$.

To minimize costs, they must make 5 small truck loads and 3 large truck loads.

35. The points (x, y) satisfying $y > x^2 - 2x$ lie in the region enclosed by the parabola $y = x^2 - 2x$. While the points (x, y) satisfying $y < -1 - x$ lie in the region below the line $y = -1 - x$. Since the line lies to the left of the parabola, the system $y > x^2 - 2x$ and $y < -1 - x$ has no solution. The solution set is $\emptyset$.

37. Multiply by the LCD as follows:

$$\begin{aligned} \frac{1}{x} + \frac{1}{x-1} &= \frac{3}{2} \\ 2(x-1) + 2x &= 3x(x-1) \\ 4x - 2 &= 3x^2 - 3x \\ 0 &= 3x^2 - 7x + 2 \\ 0 &= (3x-1)(x-2) \end{aligned}$$

Then the solution set is $\{1/3, 2\}$.

39. Rewrite the inequality.

$$\begin{aligned} \frac{3x-2}{x-2} - 1 &> 0 \\ \frac{2x}{x-2} &> 0 \end{aligned}$$

Let $f(x) = \dfrac{2x}{x-2} > 0$, and use test points.

If $x = -1$, then $f(-1) > 0$.
If $x = 1$, then $f(1) < 0$.
If $x = 3$, then $f(3) > 0$.

+ 0 − UD +

-1 0 1 2 3

The solution set is $(-\infty, 0) \cup (2, \infty)$.

41. Let x be the distance between B and the point P_1 of tangency along side BC of the circle on the left, and let y be the distance between D and P_1. Then the distance between B and D is $\overline{BD} = x + y$.

Let P_2 be the point of tangency along side AB of the circle on the left. Since $\overline{AB} = 7$, we find $\overline{AP_2} = 7 - x$.

Let Q_1 be the point of tangency along side AC of the circle on the right. Since $\overline{BC} = 10$, we find $\overline{CQ_1} = 10 - 2y - x$. Since $\overline{AC} = 12$, we obtain $\overline{AQ_1} = 2 + x + 2y$.

Since $\overline{AP_2} = \overline{AQ_1}$, we find

$$2 + x + 2y = 7 - x.$$

Then $2x + 2y = 5$ and $x + y = 2.5$. Thus, $\overline{BD} = 2.5$.

Review Exercises

1. The solution set is $\{(3, 5)\}$.

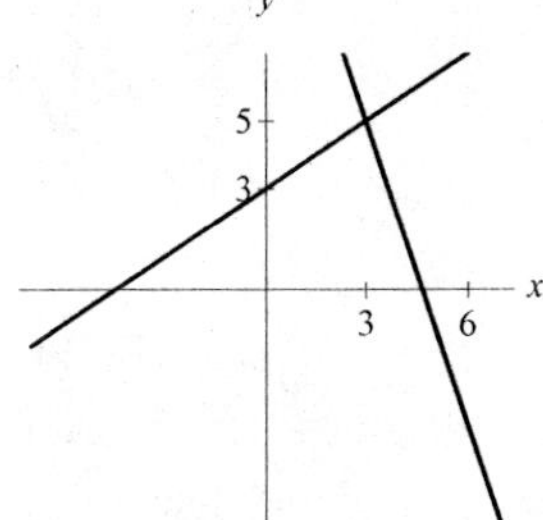

3. The solution set is $\{(-1, 3)\}$

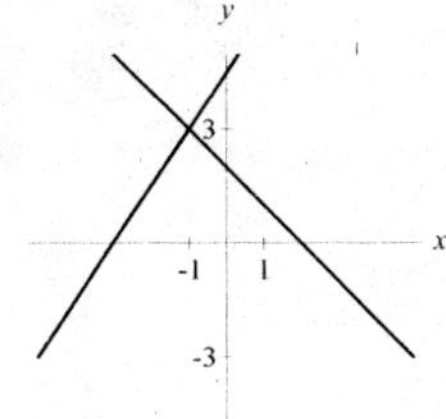

5. Substitute $y = x$ into $3x - 5y = 19$.

$$\begin{aligned} 3x - 5x &= 19 \\ -2x &= 19 \\ x &= -19/2 \end{aligned}$$

Independent and the solution set is $\{(-19/2, -19/2)\}$.

7. Multiply $4x - 3y = 6$ by 2 and $3x + 2y = 9$ by 3, then add the equations.

$$\begin{aligned} 8x - 6y &= 12 \\ 9x + 6y &= 27 \\ \hline 17x &= 39 \end{aligned}$$

Substitute $x = 39/17$ into $4x - 3y = 6$.

$$\begin{aligned} \frac{156}{17} - 3y &= 6 \\ \frac{54}{17} &= 3y \\ \frac{18}{17} &= y \end{aligned}$$

Independent and the solution set is $\{(39/17, 18/17)\}$.

9. Substitute $y = -3x + 1$ into $6x + 2y = 2$.

$$\begin{aligned} 6x + 2(-3x + 1) &= 2 \\ 6x - 6x + 2 &= 2 \\ 2 &= 2 \end{aligned}$$

Dependent and the solution set is $\{(x, y) \mid y = -3x + 1\}$.

11. Multiply $3x - 4y = 12$ by 2 and add to the second equation.

$$\begin{aligned} 6x - 8y &= 24 \\ -6x + 8y &= 9 \\ \hline 0 &= 33 \end{aligned}$$

Inconsistent and the solution set is $\emptyset$.

13. Add the first and second equations. Multiply $2x + y + z = 1$ by -3 and add to the third equation.

$$\begin{aligned} x + y - z &= 8 \\ 2x + y + z &= 1 \\ \hline 3x + 2y &= 9 \end{aligned}$$

$$\begin{aligned} -6x - 3y - 3z &= -3 \\ x + 2y + 3z &= -5 \\ \hline -5x - y &= -8 \end{aligned}$$

Multiply $-5x - y = -8$ by 2 and add to $3x + 2y = 9$.

$$\begin{aligned} -10x - 2y &= -16 \\ 3x + 2y &= 9 \\ \hline -7x &= -7 \end{aligned}$$

Using $x = 1$ in $3x + 2y = 9$, $3 + 2y = 9$ or $y = 3$. From $x + y - z = 8$, $1 + 3 - z = 8$ or $z = -4$. The Solution set is $\{(1, 3, -4)\}$.

15. Multiply first equation by -2 and add to the second equation. Multiply first equation by -2 and add to the third one.

$$\begin{array}{rcl} -2x-2y-2z &=& -2 \\ 2x-y+2z &=& 2 \\ \hline -3y &=& 0 \\ y &=& 0 \end{array}$$

$$\begin{array}{rcl} -2x-2y-2z &=& -2 \\ 2x+2y+2z &=& 2 \\ \hline 0 &=& 0 \end{array}$$

Using $y = 0$ in $x + y + z = 1$, $x + z = 1$ and $z = 1 - x$. The solution set is $\{(x, 0, 1-x) \mid x \text{ is any real number}\}$.

17. Multiply first equation by -1 and add to the third equation.

$$\begin{array}{rcl} -x-y-z &=& -1 \\ x+y+z &=& 4 \\ \hline 0 &=& 3 \end{array}$$

Inconsistent and the solution set is $\emptyset$.

19. Substitute $x = y^2$ into $x^2 + y^2 = 4$ and use the quadratic formula.

$$\begin{aligned} y^4 + y^2 &= 4 \\ y^4 + y^2 - 4 &= 0 \\ y^2 &= \frac{-1+\sqrt{17}}{2} \\ y &= \pm\sqrt{\frac{-1+\sqrt{17}}{2}} \end{aligned}$$

Thus, $x = y^2 = \dfrac{-1+\sqrt{17}}{2}$.

The solution set is

$$\left\{\left(\frac{-1+\sqrt{17}}{2}, \pm\sqrt{\frac{-1+\sqrt{17}}{2}}\right)\right\}.$$

21. Substitute $y = x^2$ into $y = |x|$.

$$\begin{aligned} x^2 &= \sqrt{x^2} \\ x^4 &= x^2 \\ x^2(x^2-1) &= 0 \\ x &= 0, \pm 1 \end{aligned}$$

Using $x = 0, 1, -1$ in $y = x^2$, we get $y = 0, 1, 1$, respectively. The solution set is $\{(0,0), (1,1), (-1,1)\}$.

23. Note, $\dfrac{7x-7}{(x-3)(x+4)} = \dfrac{A}{x-3} + \dfrac{B}{x+4}$.

Then

$$\begin{aligned} 7x - 7 &= A(x+4) + B(x-3) \\ 7x - 7 &= (A+B)x + (4A - 3B). \end{aligned}$$

Equating the coefficients, we obtain

$$\begin{aligned} A + B &= 7 \\ 4A - 3B &= -7. \end{aligned}$$

The solution of this system is $A = 2$, $B = 5$.

The answer is $\dfrac{2}{x-3} + \dfrac{5}{x+4}$.

25. Factoring the denominator, we obtain

$$\begin{aligned} x^3 - 3x^2 + 4x - 12 &= x^2(x-3) + 4(x-3) \\ &= (x^2+4)(x-3), \end{aligned}$$

and so $\dfrac{7x^2 - 7x + 23}{(x-3)(x^2+4)} = \dfrac{A}{x-3} + \dfrac{Bx+C}{x^2+4}$.

Then $7x^2 - 7x + 23 =$
$A(x^2+4) + (Bx+C)(x-3) =$
$= (A+B)x^2 + (-3B+C)x + (4A-3C)$.
Equating the coefficients, we have

$$\begin{aligned} A + B &= 7 \\ -3B + C &= -7 \\ 4A - 3C &= 23. \end{aligned}$$

The solution of this system is $A = 5$, $B = 2$, and $C = -1$. The answer is $\dfrac{5}{x-3} + \dfrac{2x-1}{x^2+4}$.

27. $x^2 + (y-3)^2 < 9$

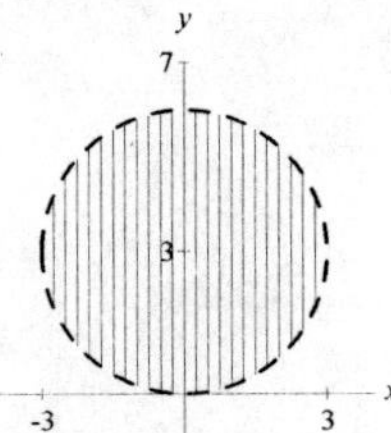

29. $x \le (y-1)^2$

31. $2x - 3y \geq 6,\ x \leq 2$

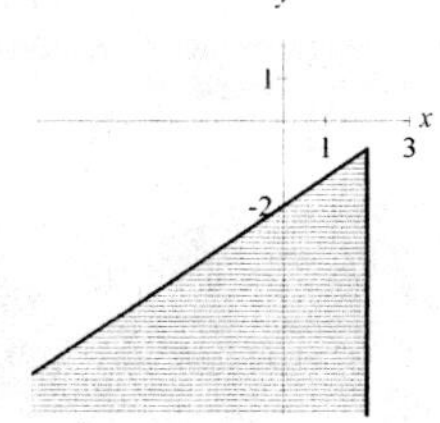

33. $y \geq 2x^2 - 6,\ x^2 + y^2 \leq 9$

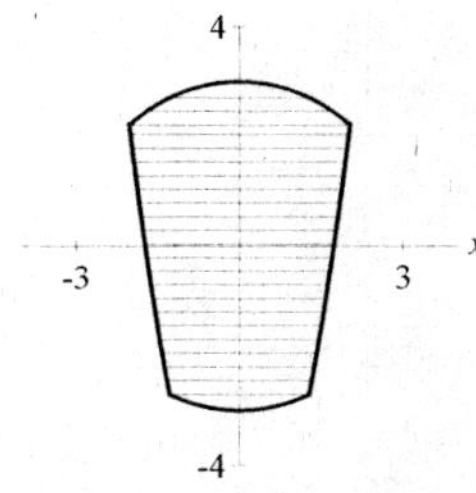

35. $x \geq 0,\ y \geq 1,\ x + 2y \leq 10,\ 3x + 4y \leq 24$

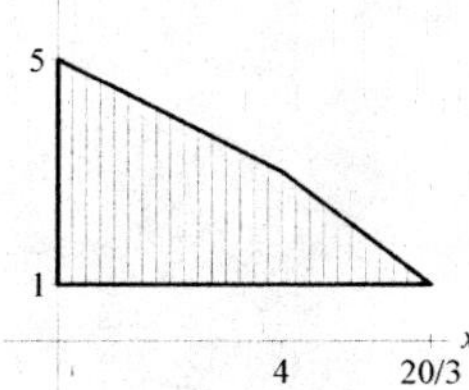

37. $x \geq 0,\ y \geq 0,\ x + 6y \geq 60,\ x + y \geq 35$

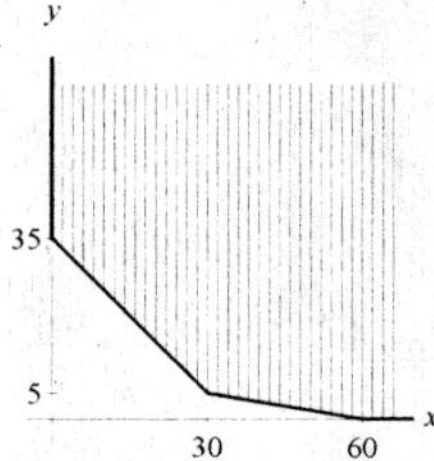

39. Substitute $(-2, 3)$ and $(4, -1)$ into $y = mx + b$.

$$\begin{aligned} -2m + b &= 3 \\ 4m + b &= -1 \end{aligned}$$

Multiply first equation by -1 and add to the second equation.

$$\begin{aligned} 2m - b &= -3 \\ 4m + b &= -1 \\ \hline 6m &= -4 \end{aligned}$$

Using $m = -2/3$ in $-2m + b = 3$, $4/3 + b = 3$ and $b = 5/3$. Equation of line is $y = -\dfrac{2}{3}x + \dfrac{5}{3}$.

41. Substitute $(1, 4)$, $(3, 20)$, and $(-2, 25)$ into $y = ax^2 + bx + c$.

$$\begin{aligned} a + b + c &= 4 \\ 9a + 3b + c &= 20 \\ 4a - 2b + c &= 25 \end{aligned}$$

The solution of the above system is $a = 3$, $b = -4$, $c = 5$. The parabola is given by $y = 3x^2 - 4x + 5$.

43. Let x and y be the number of tacos and burritos, respectively.

$$\begin{aligned} x + 2y &= 181 \\ 2x + 3y &= 300 \end{aligned}$$

Solving the above system, we find $x = 57$ tacos and $y = 62$ burritos.

45. Let x, y and z be the selling price of a daisy, carnation, and a rose, respectively. Then

$$\begin{aligned} 5x + 3y + 2z &= 3.05 \\ 3x + y + 4z &= 2.75 \\ 4x + 2y + z &= 2.10. \end{aligned}$$

Solving the above system, we find $x = 0.30$, $y = 0.25$, and $z = 0.40$. Esther's economy special sells for $x + y + z = \$0.95$.

47. The values of $C(x, y) = 0.42x + 0.84y$ at the vertices are $C(0, 35) = 29.4$, $C(30, 5) = 16.8$, and $C(60, 0) = 25.2$. The minimum value is 16.8.

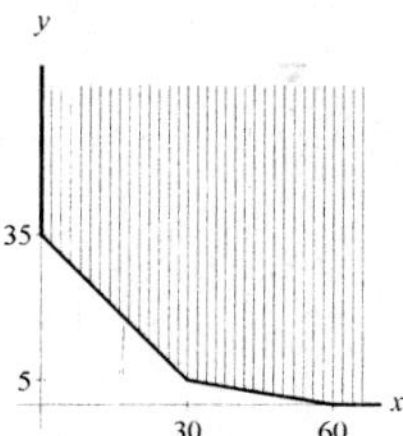

49. Let x and y be the number of barrels of oil obtained through the pipeline and barges, respectively.

$$\begin{aligned} \text{Minimize } 100x + 90y & \\ \text{subject to } x + y &\geq 12{,}000{,}000 \\ x &\leq 12{,}000{,}000 \\ x &\geq 6{,}000{,}000 \\ y &\leq 8{,}000{,}000 \\ x, y &\geq 0 \end{aligned}$$

The values of $C(x, y) = 20x + 18y$ at the vertices are
$C(12 \text{ million}, 0) = 1200$ million,
$C(12 \text{ million}, 8 \text{ million}) = 1920$ million,
$C(6 \text{ million}, 8 \text{ million}) = 1320$ million, and
$C(6 \text{ million}, 6 \text{ million}) = 1140$ million.
To minimize cost, purchase 6 million barrels from each source.

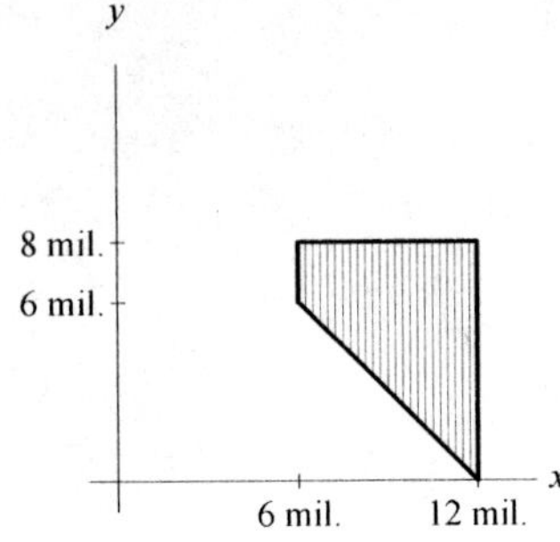

51. Dan can get 30,000 miles by rotating the tires every 6000 miles as follows:

$$S \to LR \to LF \to RF \to RR \to S$$

where S is the spare tire.

Chapter 8 Test

1. Solution set is $\{(-3, 4)\}$.

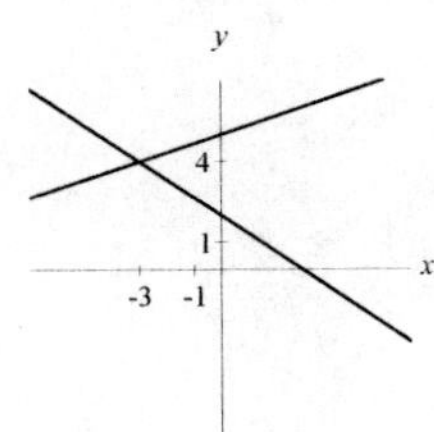

2. Substitute $y = 4 - 2x$ into $3x - 4y = 9$.

$$\begin{aligned} 3x - 4(4 - 2x) &= 9 \\ 3x - 16 + 8x &= 9 \\ 11x &= 25 \\ x &= 25/11 \end{aligned}$$

Using $x = 25/11$ in $y = 4 - 2x$, we have $y = 4 - 50/11 = -6/11$. Solution set is $\{(25/11, -6/11)\}$.

3. Multiply $10x - 3y = 22$ by 2 and $7x + 2y = 40$ by 3. Then add the equations.

$$\begin{aligned} 20x - 6y &= 44 \\ 21x + 6y &= 120 \\ \hline 41x &= 164 \\ x &= 4 \end{aligned}$$

Using $x = 4$ in $10x - 3y = 22$, we get $40 - 3y = 22$ and $y = 6$. The solution set is $\{(4, 6)\}$.

4. Substitute $x = 6 - y$ into $3x + 3y = 4$.

$$\begin{aligned} 3(6 - y) + 3y &= 4 \\ 18 - 3y + 3y &= 4 \\ 18 &= 4 \end{aligned}$$

The system is inconsistent.

5. Substitute $y = \frac{1}{2}x + 3$ into $x - 2y = -6$.

$$\begin{aligned} x - 2\left(\frac{1}{2}x + 3\right) &= -6 \\ x - x - 6 &= -6 \\ -6 &= -6 \end{aligned}$$

The system is dependent.

6. Substitute $y = 2x - 1$ into $y = 3x + 20$.

$$\begin{aligned} 2x - 1 &= 3x + 20 \\ -21 &= x \end{aligned}$$

Using $x = -21$ in $y = 2x - 1$, we get $y = -43$. The system is independent.

7. Substitute $y = -x + 2$ into $y = -x + 5$.

$$\begin{aligned} -x + 2 &= -x + 5 \\ 2 &= 5 \end{aligned}$$

The system is inconsistent.

8. Add the two equations.

$$\begin{aligned} 2x - y + z &= 4 \\ -x + 2y - z &= 6 \\ \hline x + y &= 10 \end{aligned}$$

Using $y = 10 - x$ in $2x - y + z = 4$, we find $2x - (10 - x) + z = 4$ and $z = 14 - 3x$. The solution set is $\{(x, 10 - x, 14 - 3x) \mid x \text{ is any real number}\}$.

9. Add the first two equations. Also, multiply second equation by 3 and add to the third.

$$\begin{aligned} x - 2y - z &= 2 \\ 2x + 3y + z &= -1 \\ \hline 3x + y &= 1 \end{aligned}$$

$$\begin{aligned} 6x + 9y + 3z &= -3 \\ 3x - y - 3z &= -4 \\ \hline 9x + 8y &= -7 \end{aligned}$$

Multiply $3x + y = 1$ by -3 and add to $9x + 8y = -7$.

$$\begin{aligned} -9x - 3y &= -3 \\ 9x + 8y &= -7 \\ \hline 5y &= -10 \end{aligned}$$

Using $y = -2$ in $3x + y = 1$, we get $3x - 2 = 1$ or $x = 1$. From $x - 2y - z = 2$, we have $1 + 4 - z = 2$ or $z = 3$. The solution set is $\{(1, -2, 3)\}$.

10. Add the second and third equations.

$$\begin{aligned} x + y - z &= 4 \\ -x - y + z &= 2 \\ \hline 0 &= 6 \end{aligned}$$

Inconsistent and the solution set is $\emptyset$.

11. Multiply $x^2 + y^2 = 16$ by -1 and add to $x^2 - 4y^2 = 16$.

$$\begin{aligned} -x^2 - y^2 &= -16 \\ x^2 - 4y^2 &= 16 \\ \hline -5y^2 &= 0 \\ y &= 0 \end{aligned}$$

Using $y = 0$ in $x^2 + y^2 = 16$, we find $x^2 = 16$ and $x = \pm 4$. Solution set is $\{(4, 0), (-4, 0)\}$.

12. Substitute $y = x^2 - 5x$ into $x + y = -2$.

$$\begin{aligned} x + (x^2 - 5x) &= -2 \\ x^2 - 4x &= -2 \\ x^2 - 4x + 4 &= -2 + 4 \\ (x - 2)^2 &= 2 \\ x &= 2 \pm \sqrt{2} \end{aligned}$$

Using $x = 2 + \sqrt{2}$ and $x = 2 - \sqrt{2}$ in $y = -2 - x$, we have $y = -4 - \sqrt{2}$ and $y = -4 + \sqrt{2}$, respectively. The solution set is

$$\{(2 + \sqrt{2}, -4 - \sqrt{2}), (2 - \sqrt{2}, -4 + \sqrt{2})\}.$$

13.

$$\begin{aligned} \frac{2x + 10}{(x - 4)(x + 2)} &= \frac{A}{x - 4} + \frac{B}{x + 2} \\ 2x + 10 &= A(x + 2) + B(x - 4) \\ 2x + 10 &= (A + B)x + (2A - 4B) \end{aligned}$$

Equating the coefficients, we obtain

$$\begin{aligned} A + B &= 2 \\ 2A - 4B &= 10. \end{aligned}$$

The solution of this system is $A = 3, B = -1$.

The answer is $\dfrac{3}{x - 4} + \dfrac{-1}{x + 2}$.

14. Note, $\dfrac{4x^2 + x - 2}{x^2(x - 1)} = \dfrac{A}{x} + \dfrac{B}{x^2} + \dfrac{C}{x - 1}$. Then

$$\begin{aligned} 4x^2 + x - 2 &= Ax(x - 1) + B(x - 1) + Cx^2 \\ &= (A + C)x^2 + (-A + B)x - B. \end{aligned}$$

Equating the coefficients, we have

$$\begin{aligned} A + C &= 4 \\ -A + B &= 1 \\ -B &= -2. \end{aligned}$$

The solution of this system is $B = 2, A = 1$, and $C = 3$. The answer is $\dfrac{1}{x} + \dfrac{2}{x^2} + \dfrac{3}{x - 1}$.

15. $2x - y < 8$

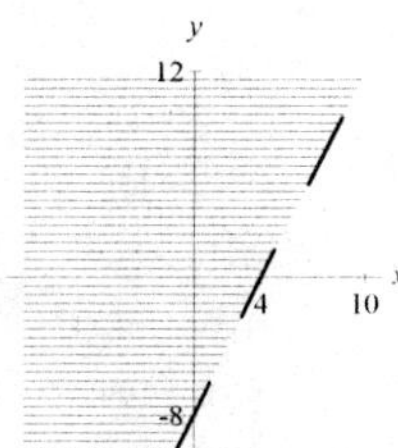

16. $x + y \le 5,\ x - y < 0$

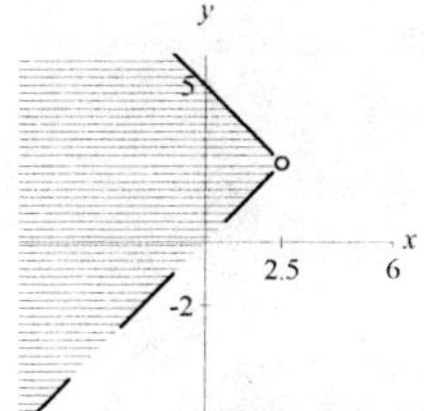

17. $x^2 + y^2 \le 9,\ y \le 1 - x^2$

18. Let x and y be the number of male and female students, respectively. Then

$$\begin{aligned} \frac{1}{3}x + \frac{1}{4}y &= 15 \\ x + y &= 52. \end{aligned}$$

Multiply the first equation by 12 and multiply the second equation by -3. Then add the resulting equations.

$$\begin{aligned} 4x + 3y &= 180 \\ -3x - 3y &= -156 \\ \hline x &= 24 \end{aligned}$$

The solution of this system is $x = 24$ males and $y = 28$ females.

19. Let x and y be the number of TV commercials and newspaper ads, respectively. The linear program is given below.

$$\begin{aligned} \text{Maximize } & 14,000x + 6000y \\ \text{subject to } 9000x + 3000y &\le 99,000 \\ x + y &\le 23 \\ x, y &\ge 0 \end{aligned}$$

The values of $N(x, y) = 14,000x + 6000y$ at the vertices are $N(0, 0) = 0$, $N(0, 23) = 138,000$, $N(5, 18) = 178,000$, and $N(11, 0) = 154,000$. To obtain maximum audience exposure, the hospital must have 5 TV commercials and 18 newspaper ads.

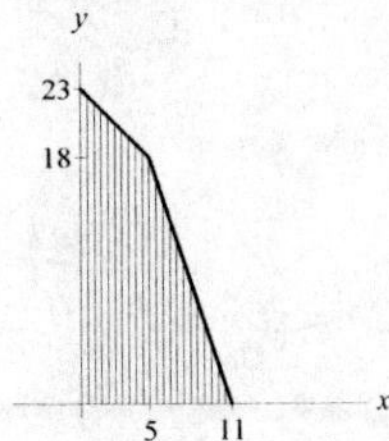

Tying It All Together

1. Multiply the equation by $24(x + 5)$.

$$\begin{aligned} 24(x - 2) &= 11(x + 5) \\ 24x - 48 &= 11x + 55 \\ 13x &= 103 \end{aligned}$$

The solution set is $\{103/13\}$.

2. Multiply the equation by $24x(x + 5)$.

$$\begin{aligned} 24(x + 5) + 24x(x - 2) &= 11x(x + 5) \\ 24x + 120 + 24x^2 - 48x &= 11x^2 + 55x \\ 13x^2 - 79x + 120 &= 0 \\ (13x - 40)(x - 3) &= 0 \end{aligned}$$

The solution set is $\{40/13, 3\}$.

3.

$$\begin{aligned} 5 - 3x - 6 - 2x + 4 &= 7 \\ 3 - 5x &= 7 \\ -5x &= 4 \end{aligned}$$

The solution set is $\{-4/5\}$.

4. We will solve an equivalent statement without absolute values.

$$\begin{aligned} 3 - 2x = 5 \quad &\text{or} \quad 3 - 2x = -5 \\ -2x = 2 \quad &\text{or} \quad -2x = -8 \\ x = -1 \quad &\text{or} \quad x = 4 \end{aligned}$$

The solution set is $\{4, -1\}$.

5. Square both sides of the equation.

$$\begin{aligned} 3 - 2x &= 25 \\ -2x &= 22 \end{aligned}$$

The solution set is $\{-11\}$.

6. Isolate x^2 on one side and take the square roots.

$$\begin{aligned} 3x^2 &= 4 \\ x^2 &= \frac{4}{3} \\ x &= \pm\frac{2}{\sqrt{3}} \\ x &= \pm\frac{2}{\sqrt{3}} \cdot \frac{\sqrt{3}}{\sqrt{3}} \end{aligned}$$

The solution set is $\left\{\pm\frac{2\sqrt{3}}{3}\right\}$.

7. Multiply equation by x^2.

$$\begin{aligned}(x-2)^2 &= x^2\\ x^2-4x+4 &= x^2\\ -4x &= -4\end{aligned}$$

The solution set is $\{1\}$.

8. Since $2^{x-1}=9$, we obtain $x-1=\log_2(9)$ by using the definition of a logarithm. The solution set is $\{1+\log_2(9)\}$.

9. Write left-hand side as a single logarithm.

$$\begin{aligned}\log((x+1)(x+4)) &= 1\\ x^2+5x+4 &= 10^1\\ x^2+5x-6 &= 0\\ (x+6)(x-1) &= 0\\ x &= -6,1\end{aligned}$$

But $\log(x+1)$ is undefined when $x=-6$. The solution set is $\{1\}$.

10. Raise both sides of equation to the power $-3/2$.

$$\begin{aligned}x^{-2/3} &= \frac{1}{4}\\ x &= \pm\left(\frac{1}{4}\right)^{-3/2}\\ x &= \pm(4)^{3/2}\\ x &= \pm(4^{1/2})^3\\ x &= \pm(2)^3\end{aligned}$$

The solution set is $\{\pm 8\}$.

11. Use the quadratic formula.

$$\begin{aligned}x^2-3x-6 &= 0\\ x &= \frac{3\pm\sqrt{33}}{2}\end{aligned}$$

The solution set is $\left\{\frac{3\pm\sqrt{33}}{2}\right\}$.

12. By using the square root property, we obtain

$$\begin{aligned}(x-3)^2 &= \frac{1}{2}\\ x-3 &= \pm\frac{\sqrt{2}}{2}\\ x &= \frac{6}{2}\pm\frac{\sqrt{2}}{2}.\end{aligned}$$

The solution set is $\left\{\frac{6\pm\sqrt{2}}{2}\right\}$.

13. Since $3-2x>0$, we get $3>2x$ and $x<3/2$. The solution set is $(-\infty,3/2)$ and its graph is

3/2

14. Since we must exclude $x=\frac{3}{2}$, the solution set is $(-\infty,1.5)\cup(1.5,\infty)$ and the graph is

1.5

15. The sign graph of $(x-3)(x+3)\geq 0$ is shown below.

```
- - - - - - - - - 0 + + + +
- - - - 0 + + + + + + + + +
-3                3
```

So the solution set is $(-\infty,-3]\cup[3,\infty)$ and the graph is

-3 3

16. Note that $x^2+2x-8\leq 27$ is equivalent to $x^2+2x-35\leq 0$. The sign graph of $(x+7)(x-5)\leq 0$ is

```
- - - - - - - - - 0 + + + +
- - - - 0 + + + + + + + + +
-7                5
```

So the solution set is $[-7,5]$ and the graph is

-7 5

17. $3-2x>y$

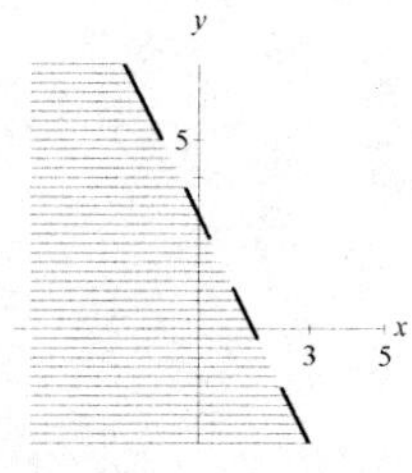

18. $|3-2x|>y$

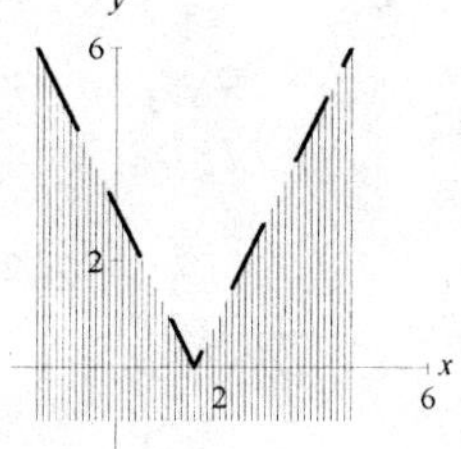

19. $x^2 \geq 9$

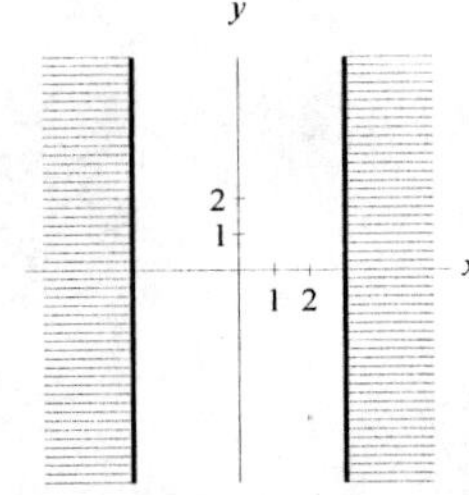

20. $(x-2)(x+4) \leq y$

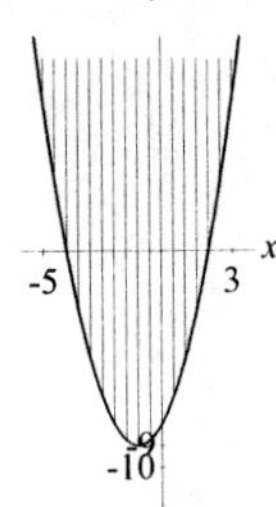

21. consistent

22. inconsistent

23. independent

24. dependent

25. substitution

26. addition

27. inconsistent

28. identity

29. linear, three

30. ordered triple

For Thought

1. False, the augmented matrix is a 2×3 matrix.

2. False, the required matrix is $\left[\begin{array}{cc|c} 1 & -1 & 4 \\ 3 & 1 & 5 \end{array}\right]$.

3. True **4.** True

5. True, since the row operation done is $R_1 + R_2 \to R_2$.

6. True, since the row operation done is $-R_1 + R_2 \to R_2$.

7. False, since it corresponds to

$$\begin{aligned} x &= 2 \\ y &= 7. \end{aligned}$$

8. True, since $0 \cdot x + 0 \cdot y = 7$ has no solution.

9. False, the system is dependent. **10.** True

9.1 Exercises

1. matrix

3. size

5. entry, element

7. diagonal

9. 1×3 **11.** 1×1 **13.** 3×2

15.

$$\left[\begin{array}{cc|c} 1 & -2 & 4 \\ 3 & 2 & -5 \end{array}\right]$$

17.

$$\left[\begin{array}{ccc|c} 1 & -1 & -1 & 4 \\ 1 & 3 & -1 & 1 \\ 0 & 2 & -5 & -6 \end{array}\right]$$

19.

$$\left[\begin{array}{ccc|c} 1 & 3 & -1 & 5 \\ 1 & 0 & 1 & 0 \end{array}\right]$$

21.

$$\begin{aligned} 3x + 4y &= -2 \\ 3x - 5y &= 0 \end{aligned}$$

23.

$$\begin{aligned} 5x &= 6 \\ -4x + 2z &= -1 \\ 4x + 4y &= 7 \end{aligned}$$

25.

$$\begin{aligned} x - y + 2z &= 1 \\ y + 4z &= 3 \end{aligned}$$

27. Interchange R_1 and R_2

$$\left[\begin{array}{cc|c} 1 & 2 & 0 \\ -2 & 4 & 1 \end{array}\right]$$

29. Multiply $\frac{1}{2}$ to R_1

$$\left[\begin{array}{cc|c} 1 & 4 & 1 \\ 0 & 3 & 6 \end{array}\right]$$

31. Multiply 3 to R_1 then add the product to R_2. This is the new R_2.

$$\left[\begin{array}{cc|c} 1 & -2 & 1 \\ -3 & 5 & 0 \end{array}\right] =$$

$$\left[\begin{array}{cc|c} 1 & -2 & 1 \\ 3 \cdot 1 + (-3) & 3 \cdot (-2) + 5 & 3 \cdot 1 + 0 \end{array}\right] =$$

$$\left[\begin{array}{cc|c} 1 & -2 & 1 \\ 0 & -1 & 3 \end{array}\right]$$

33. From the augmented matrix

$$\left[\begin{array}{cc|c} 2 & 4 & 14 \\ 5 & 4 & 5 \end{array}\right]$$

the system is

$$\begin{aligned} 2x + 4y &= 14 \\ 5x + 4y &= 5. \end{aligned}$$

From the final augmented matrix

$$\left[\begin{array}{cc|c} 1 & 0 & -3 \\ 0 & 1 & 5 \end{array}\right]$$

we obtain that the solution set is $\{(-3, 5)\}$.

The row operations are $\frac{1}{2}R_1 \to R_1$, $-5R_1 + R_2 \to R_2$, $-\frac{1}{6}R_2 \to R_2$, and $-2R_2 + R_1 \to R_1$.

35. On $\left[\begin{array}{rr|r} 1 & 1 & 5 \\ -2 & 1 & -1 \end{array}\right]$ use $2R_1 + R_2 \to R_2$ to get

$\left[\begin{array}{rr|r} 1 & 1 & 5 \\ 0 & 3 & 9 \end{array}\right]$, use $\frac{1}{3}R_2 \to R_2$ to get

$\left[\begin{array}{rr|r} 1 & 1 & 5 \\ 0 & 1 & 3 \end{array}\right]$, use $-R_2 + R_1 \to R_1$ to get

$\left[\begin{array}{rr|r} 1 & 0 & 2 \\ 0 & 1 & 3 \end{array}\right]$, solution set is $\{(2,3)\}$,

and the system is independent.

37. On $\left[\begin{array}{rr|r} 2 & 2 & 8 \\ -3 & -1 & -6 \end{array}\right]$ use $\frac{1}{2}R_1 \to R_1$ to get

$\left[\begin{array}{rr|r} 1 & 1 & 4 \\ -3 & -1 & -6 \end{array}\right]$, use $3R_1 + R_2 \to R_2$ to get

$\left[\begin{array}{rr|r} 1 & 1 & 4 \\ 0 & 2 & 6 \end{array}\right]$, use $\frac{1}{2}R_2 \to R_2$ to get

$\left[\begin{array}{rr|r} 1 & 1 & 4 \\ 0 & 1 & 3 \end{array}\right]$, use $-1R_2 + R_1 \to R_1$ to get

$\left[\begin{array}{rr|r} 1 & 0 & 1 \\ 0 & 1 & 3 \end{array}\right]$, solution set is $\{(1,3)\}$,

and the system is independent.

39. On $\left[\begin{array}{rr|r} 2 & -1 & 3 \\ 3 & 2 & 15 \end{array}\right]$ use $-R_1 + R_2 \to R_1$ to get

$\left[\begin{array}{rr|r} 1 & 3 & 12 \\ 3 & 2 & 15 \end{array}\right]$, use $-3R_1 + R_2 \to R_2$ to get

$\left[\begin{array}{rr|r} 1 & 3 & 12 \\ 0 & -7 & -21 \end{array}\right]$, use $-\frac{1}{7}R_2 \to R_2$ to get

$\left[\begin{array}{rr|r} 1 & 3 & 12 \\ 0 & 1 & 3 \end{array}\right]$, use $-3R_2 + R_1 \to R_1$ to get

$\left[\begin{array}{rr|r} 1 & 0 & 3 \\ 0 & 1 & 3 \end{array}\right]$, solution set is $\{(3,3)\}$,

and the system is independent.

41. On $\left[\begin{array}{rr|r} 0.4 & -0.2 & 0 \\ 1 & 1.5 & 2 \end{array}\right]$ use $5R_1 \to R_1$ and $2R_2 \to R_2$ to get

$\left[\begin{array}{rr|r} 2 & -1 & 0 \\ 2 & 3 & 4 \end{array}\right]$, use $R_1 + (-R_2) \to R_2$ to get

$\left[\begin{array}{rr|r} 2 & -1 & 0 \\ 0 & -4 & -4 \end{array}\right]$, use $-\frac{1}{4}R_2 \to R_2$ to get

$\left[\begin{array}{rr|r} 2 & -1 & 0 \\ 0 & 1 & 1 \end{array}\right]$, use $R_1 + R_2 \to R_1$ to get

$\left[\begin{array}{rr|r} 2 & 0 & 1 \\ 0 & 1 & 1 \end{array}\right]$, use $\frac{1}{2}R_1 \to R_1$ to get

$\left[\begin{array}{rr|r} 1 & 0 & 0.5 \\ 0 & 1 & 1 \end{array}\right]$, solution set is $\{(0.5,1)\}$,

and the system is independent.

43. On $\left[\begin{array}{rr|r} 3 & -5 & 7 \\ -3 & 5 & 4 \end{array}\right]$ use $R_1 + R_2 \to R_2$ to get

$\left[\begin{array}{rr|r} 3 & -5 & 7 \\ 0 & 0 & 11 \end{array}\right]$, inconsistent system, and

the solution set is $\emptyset$.

45. On $\left[\begin{array}{rr|r} 0.5 & 1.5 & 2 \\ 3 & 9 & 12 \end{array}\right]$ use $2R_1 \to R_1$ to get

$\left[\begin{array}{rr|r} 1 & 3 & 4 \\ 3 & 9 & 12 \end{array}\right]$, use $-3R_1 + R_2 \to R_2$ to get

$\left[\begin{array}{rr|r} 1 & 3 & 4 \\ 0 & 0 & 0 \end{array}\right]$, dependent system, and

solution set is $\{(u,v) \mid u + 3v = 4\}$.

47. Rewrite system as

$$\begin{aligned} 2x + y &= 4 \\ x - y &= 8. \end{aligned}$$

On $\left[\begin{array}{rr|r} 2 & 1 & 4 \\ 1 & -1 & 8 \end{array}\right]$ use $R_1 + (-2R_2) \to R_2$ to get

$\left[\begin{array}{rr|r} 2 & 1 & 4 \\ 0 & 3 & -12 \end{array}\right]$, use $\frac{1}{3}R_2 \to R_2$ to get

$\left[\begin{array}{rr|r} 2 & 1 & 4 \\ 0 & 1 & -4 \end{array}\right]$, use $-1R_2 + R_1 \to R_1$ to get

$\left[\begin{array}{rr|r} 2 & 0 & 8 \\ 0 & 1 & -4 \end{array}\right]$, use $\frac{1}{2}R_1 \to R_1$ to get

$\left[\begin{array}{rr|r} 1 & 0 & 4 \\ 0 & 1 & -4 \end{array}\right]$, solution set is $\{(4,-4)\}$,

and the system is independent.

49.

On $\left[\begin{array}{ccc|c} 1 & 1 & 1 & 6 \\ 1 & -1 & -1 & 0 \\ 0 & 2 & -1 & 3 \end{array}\right]$ use $-1R_2+R_1 \to R_2$ to get

$\left[\begin{array}{ccc|c} 1 & 1 & 1 & 6 \\ 0 & 2 & 2 & 6 \\ 0 & 2 & -1 & 3 \end{array}\right]$, use $\frac{1}{2}R_2 \to R_2$ to get

$\left[\begin{array}{ccc|c} 1 & 1 & 1 & 6 \\ 0 & 1 & 1 & 3 \\ 0 & 2 & -1 & 3 \end{array}\right]$, use $-1R_2 + R_1 \to R_1$ to get

$\left[\begin{array}{ccc|c} 1 & 0 & 0 & 3 \\ 0 & 1 & 1 & 3 \\ 0 & 2 & -1 & 3 \end{array}\right]$, use $-2R_2 + R_3 \to R_3$ to get

$\left[\begin{array}{ccc|c} 1 & 0 & 0 & 3 \\ 0 & 1 & 1 & 3 \\ 0 & 0 & -3 & -3 \end{array}\right]$, use $-\frac{1}{3}R_3 \to R_3$ to get

$\left[\begin{array}{ccc|c} 1 & 0 & 0 & 3 \\ 0 & 1 & 1 & 3 \\ 0 & 0 & 1 & 1 \end{array}\right]$, use $-1R_3 + R_2 \to R_2$ to get

$\left[\begin{array}{ccc|c} 1 & 0 & 0 & 3 \\ 0 & 1 & 0 & 2 \\ 0 & 0 & 1 & 1 \end{array}\right]$, solution set is $\{(3, 2, 1)\}$,

and the system is independent.

51.

Rewrite system as

$$\begin{aligned} 2x + y - z &= 2 \\ x + 2y - z &= 2 \\ x - y + 2z &= 2. \end{aligned}$$

On $\left[\begin{array}{ccc|c} 2 & 1 & -1 & 2 \\ 1 & 2 & -1 & 2 \\ 1 & -1 & 2 & 2 \end{array}\right]$, use $-1R_2+R_1 \to R_1$

and $-1R_3 + R_2 \to R_3$ to get

$\left[\begin{array}{ccc|c} 1 & -1 & 0 & 0 \\ 1 & 2 & -1 & 2 \\ 0 & 3 & -3 & 0 \end{array}\right]$, use $-1R_2 + R_1 \to R_2$

and $\frac{1}{3}R_3 \to R_3$ to get

$\left[\begin{array}{ccc|c} 1 & -1 & 0 & 0 \\ 0 & -3 & 1 & -2 \\ 0 & 1 & -1 & 0 \end{array}\right]$, use $R_1 + R_3 \to R_1$

and $R_3 + R_2 \to R_2$ to get

$\left[\begin{array}{ccc|c} 1 & 0 & -1 & 0 \\ 0 & -2 & 0 & -2 \\ 0 & 1 & -1 & 0 \end{array}\right]$, use $-\frac{1}{2}R_2 \to R_2$ to get

$\left[\begin{array}{ccc|c} 1 & 0 & -1 & 0 \\ 0 & 1 & 0 & 1 \\ 0 & 1 & -1 & 0 \end{array}\right]$, use $-1R_3 + R_2 \to R_3$ to get

$\left[\begin{array}{ccc|c} 1 & 0 & -1 & 0 \\ 0 & 1 & 0 & 1 \\ 0 & 0 & 1 & 1 \end{array}\right]$, use $R_1 + R_3 \to R_1$ to get

$\left[\begin{array}{ccc|c} 1 & 0 & 0 & 1 \\ 0 & 1 & 0 & 1 \\ 0 & 0 & 1 & 1 \end{array}\right]$, solution set is $\{(1, 1, 1)\}$,

and the system is independent.

53.

Interchange rows of $\left[\begin{array}{ccc|c} 2 & -2 & 1 & -2 \\ 1 & 1 & -3 & 3 \\ 1 & -3 & 1 & -5 \end{array}\right]$ to get

$\left[\begin{array}{ccc|c} 1 & 1 & -3 & 3 \\ 1 & -3 & 1 & -5 \\ 2 & -2 & 1 & -2 \end{array}\right]$, use $-1R_2 + R_1 \to R_2$

and $-1R_3 + R_1 + R_2 \to R_3$ to get

$\left[\begin{array}{ccc|c} 1 & 1 & -3 & 3 \\ 0 & 4 & -4 & 8 \\ 0 & 0 & -3 & 0 \end{array}\right]$, use $\frac{1}{4}R_2 \to R_2$

and $-\frac{1}{3}R_3 \to R_3$ to get

$\left[\begin{array}{ccc|c} 1 & 1 & -3 & 3 \\ 0 & 1 & -1 & 2 \\ 0 & 0 & 1 & 0 \end{array}\right]$, use $R_2 + R_3 \to R_2$ to get

$\left[\begin{array}{ccc|c} 1 & 1 & -3 & 3 \\ 0 & 1 & 0 & 2 \\ 0 & 0 & 1 & 0 \end{array}\right]$, use $-1R_2 + R_1 \to R_1$ to get

$\left[\begin{array}{ccc|c} 1 & 0 & -3 & 1 \\ 0 & 1 & 0 & 2 \\ 0 & 0 & 1 & 0 \end{array}\right]$, use $3R_3 + R_1 \to R_1$ to get

$\left[\begin{array}{ccc|c} 1 & 0 & 0 & 1 \\ 0 & 1 & 0 & 2 \\ 0 & 0 & 1 & 0 \end{array}\right]$, solution set is $\{(1, 2, 0)\}$,

and the system is independent.

55.

Rewrite system as

$$\begin{aligned} x-3y+z &= 0 \\ x-y-3z &= 4 \\ x+y+2z &= -1. \end{aligned}$$

On $\left[\begin{array}{ccc|c} 1 & -3 & 1 & 0 \\ 1 & -1 & -3 & 4 \\ 1 & 1 & 2 & -1 \end{array}\right]$, use

$-1R_2+R_1 \to R_2$ and $-1R_3+R_1 \to R_3$ to get

$\left[\begin{array}{ccc|c} 1 & -3 & 1 & 0 \\ 0 & -2 & 4 & -4 \\ 0 & -4 & -1 & 1 \end{array}\right]$, use $-\frac{1}{2}R_2 \to R_2$

and $-1R_3 \to R_3$ to get

$\left[\begin{array}{ccc|c} 1 & -3 & 1 & 0 \\ 0 & 1 & -2 & 2 \\ 0 & 4 & 1 & -1 \end{array}\right]$, use $-1R_3+R_1 \to R_1$

to get

$\left[\begin{array}{ccc|c} 1 & -7 & 0 & 1 \\ 0 & 1 & -2 & 2 \\ 0 & 4 & 1 & -1 \end{array}\right]$, use $-4R_2+R_3 \to R_3$

to get

$\left[\begin{array}{ccc|c} 1 & -7 & 0 & 1 \\ 0 & 1 & -2 & 2 \\ 0 & 0 & 9 & -9 \end{array}\right]$, use $\frac{1}{9}R_3 \to R_3$ to get

$\left[\begin{array}{ccc|c} 1 & -7 & 0 & 1 \\ 0 & 1 & -2 & 2 \\ 0 & 0 & 1 & -1 \end{array}\right]$, use $2R_3+R_2 \to R_2$

to get

$\left[\begin{array}{ccc|c} 1 & -7 & 0 & 1 \\ 0 & 1 & 0 & 0 \\ 0 & 0 & 1 & -1 \end{array}\right]$, use $R_1+7R_2 \to R_1$ to

get

$\left[\begin{array}{ccc|c} 1 & 0 & 0 & 1 \\ 0 & 1 & 0 & 0 \\ 0 & 0 & 1 & -1 \end{array}\right]$, solution set is $\{(1,0,-1)\}$,

and the system is independent.

57.

On $\left[\begin{array}{ccc|c} 1 & -2 & 3 & 1 \\ 2 & -4 & 6 & 2 \\ -3 & 6 & -9 & -3 \end{array}\right]$, use $\frac{1}{2}R_2 \to R_2$

and $-\frac{1}{3}R_3 \to R_3$ to get

$\left[\begin{array}{ccc|c} 1 & -2 & 3 & 1 \\ 1 & -2 & 3 & 1 \\ 1 & -2 & 3 & 1 \end{array}\right]$, use $-1R_1+R_2 \to R_2$

and $-1R_1+R_3 \to R_3$ to get

$\left[\begin{array}{ccc|c} 1 & -2 & 3 & 1 \\ 0 & 0 & 0 & 0 \\ 0 & 0 & 0 & 0 \end{array}\right]$, dependent system,

and solution set is $\{(x,y,z) \mid x-2y+3z=1\}$.

59.

On $\left[\begin{array}{ccc|c} 1 & -1 & 1 & 2 \\ 2 & 1 & -1 & 1 \\ 2 & -2 & 2 & 5 \end{array}\right]$, use $-\frac{1}{2}R_3+R_1 \to R_3$

to get $\left[\begin{array}{ccc|c} 1 & -1 & 1 & 2 \\ 2 & 1 & -1 & 1 \\ 0 & 0 & 0 & -1/2 \end{array}\right]$, inconsistent,

and the solution set is $\emptyset$.

61.

On $\left[\begin{array}{ccc|c} 1 & 1 & -1 & 3 \\ 3 & 1 & 1 & 7 \\ 1 & -1 & 3 & 1 \end{array}\right]$, use $-1R_3+R_1 \to R_3$

and $3R_1+(-1R_2) \to R_2$ to get

$\left[\begin{array}{ccc|c} 1 & 1 & -1 & 3 \\ 0 & 2 & -4 & 2 \\ 0 & 2 & -4 & 2 \end{array}\right]$, use $-1R_3+R_2 \to R_3$ to

get

$\left[\begin{array}{ccc|c} 1 & 1 & -1 & 3 \\ 0 & 2 & -4 & 2 \\ 0 & 0 & 0 & 0 \end{array}\right]$, use $\frac{1}{2}R_2 \to R_2$ to get

$\left[\begin{array}{ccc|c} 1 & 1 & -1 & 3 \\ 0 & 1 & -2 & 1 \\ 0 & 0 & 0 & 0 \end{array}\right]$, use $-1R_2+R_1 \to R_1$ to

get

$\left[\begin{array}{ccc|c} 1 & 0 & 1 & 2 \\ 0 & 1 & -2 & 1 \\ 0 & 0 & 0 & 0 \end{array}\right]$. Substitute $z=2-x$

into $y=1+2z$. Then $y=1+2(2-x)=5-2x$. The solution set is

$$\{(x,5-2x,2-x) \mid x \text{ is any real number}\}$$

and the system is dependent.

63.

On $\left[\begin{array}{rrr|r} 2 & -1 & 3 & 1 \\ 1 & 1 & -1 & 4 \end{array}\right]$, use $-2R_2+R_1 \to R_2$ to get

$\left[\begin{array}{rrr|r} 2 & -1 & 3 & 1 \\ 0 & -3 & 5 & -7 \end{array}\right]$, use $-3R_1 + R_2 \to R_1$ to get

$\left[\begin{array}{rrr|r} -6 & 0 & -4 & -10 \\ 0 & -3 & 5 & -7 \end{array}\right]$, use $-\frac{1}{6}R_1 \to R_1$ and $-\frac{1}{3}R_2 \to R_2$ to get

$\left[\begin{array}{rrr|r} 1 & 0 & 2/3 & 5/3 \\ 0 & 1 & -5/3 & 7/3 \end{array}\right]$. Note $x = \dfrac{5-2z}{3}$ and $y = \dfrac{7+5z}{3}$. Solving for z, we get $z = \dfrac{5-3x}{2}$.

Then $y = \dfrac{7+5\cdot\frac{5-3x}{2}}{3} = \dfrac{7+5\cdot\frac{5-3x}{2}}{3}\cdot\dfrac{2}{2} = \dfrac{39-15x}{6} = \dfrac{13-5x}{2}$. The solution set is

$\left\{\left(x, \dfrac{13-5x}{2}, \dfrac{5-3x}{2}\right) \mid x \text{ is any real number}\right\}$

and the system is dependent.

65.

On $\left[\begin{array}{rrrr|r} 1 & -1 & 1 & -1 & 2 \\ -1 & 2 & -1 & -1 & -1 \\ 2 & -1 & -1 & 1 & 4 \\ 1 & 3 & -2 & -3 & 6 \end{array}\right]$,

use $2R_2 + R_3 \to R_3$, $R_2 + R_4 \to R_4$, and $R_1 + R_2 \to R_2$ to get

$\left[\begin{array}{rrrr|r} 1 & -1 & 1 & -1 & 2 \\ 0 & 1 & 0 & -2 & 1 \\ 0 & 3 & -3 & -1 & 2 \\ 0 & 5 & -3 & -4 & 5 \end{array}\right]$, use

$-3R_2+R_3 \to R_3$ and $-5R_2+R_4 \to R_4$ to get

$\left[\begin{array}{rrrr|r} 1 & -1 & 1 & -1 & 2 \\ 0 & 1 & 0 & -2 & 1 \\ 0 & 0 & -3 & 5 & -1 \\ 0 & 0 & -3 & 6 & 0 \end{array}\right]$, use

$R_1 + R_2 \to R_1$ and $-\frac{1}{3}R_4 \to R_4$ to get

$\left[\begin{array}{rrrr|r} 1 & 0 & 1 & -3 & 3 \\ 0 & 1 & 0 & -2 & 1 \\ 0 & 0 & -3 & 5 & -1 \\ 0 & 0 & 1 & -2 & 0 \end{array}\right]$,

use $R_3 \leftrightarrow R_4$ to get

$\left[\begin{array}{rrrr|r} 1 & 0 & 1 & -3 & 3 \\ 0 & 1 & 0 & -2 & 1 \\ 0 & 0 & 1 & -2 & 0 \\ 0 & 0 & -3 & 5 & -1 \end{array}\right]$, use

$-3R_3 + (-1R_4) \to R_4$ to get

$\left[\begin{array}{rrrr|r} 1 & 0 & 1 & -3 & 3 \\ 0 & 1 & 0 & -2 & 1 \\ 0 & 0 & 1 & -2 & 0 \\ 0 & 0 & 0 & 1 & 1 \end{array}\right]$, use $-1R_3 + R_1 \to R_1$ to get

$\left[\begin{array}{rrrr|r} 1 & 0 & 0 & -1 & 3 \\ 0 & 1 & 0 & -2 & 1 \\ 0 & 0 & 1 & -2 & 0 \\ 0 & 0 & 0 & 1 & 1 \end{array}\right]$, use $2R_4 + R_3 \to R_3$ to get

$\left[\begin{array}{rrrr|r} 1 & 0 & 0 & -1 & 3 \\ 0 & 1 & 0 & -2 & 1 \\ 0 & 0 & 1 & 0 & 2 \\ 0 & 0 & 0 & 1 & 1 \end{array}\right]$, use $2R_4 + R_2 \to R_2$ to get

$\left[\begin{array}{rrrr|r} 1 & 0 & 0 & -1 & 3 \\ 0 & 1 & 0 & 0 & 3 \\ 0 & 0 & 1 & 0 & 2 \\ 0 & 0 & 0 & 1 & 1 \end{array}\right]$, use $R_4 + R_1 \to R_1$ to get

$\left[\begin{array}{rrrr|r} 1 & 0 & 0 & 0 & 4 \\ 0 & 1 & 0 & 0 & 3 \\ 0 & 0 & 1 & 0 & 2 \\ 0 & 0 & 0 & 1 & 1 \end{array}\right]$, the solution set is

$\{(4, 3, 2, 1)\}$, and the system is independent.

67. Let x and y be the number of hours Mike worked at Burgers and the Soap Opera, respectively. The augmented matrix is

$A = \left[\begin{array}{rr|r} 1 & 1 & 60 \\ 8 & 9 & 502 \end{array}\right]$. On A use $-8R_1 + R_2 \to R_2$ to get

$\left[\begin{array}{rr|r} 1 & 1 & 60 \\ 0 & 1 & 22 \end{array}\right]$, use $-R_2 + R_1 \to R_1$ to get

$\left[\begin{array}{rr|r} 1 & 0 & 38 \\ 0 & 1 & 22 \end{array}\right]$. Mike worked $x = 38$ hours at Burgers and $y = 22$ hours at Soap Opera.

69. Let x, y, and z be the amounts invested in a mutual fund, in treasury bills, and in bonds, respectively. Augmented matrix is

$$A = \left[\begin{array}{ccc|c} 1 & 1 & 1 & 40,000 \\ 0.08 & 0.09 & 0.12 & 3,660 \\ 1 & -1 & -1 & 0 \end{array}\right].$$

On A use $100R_2 \to R_2$ to get

$$\left[\begin{array}{ccc|c} 1 & 1 & 1 & 40,000 \\ 8 & 9 & 12 & 366,000 \\ 1 & -1 & -1 & 0 \end{array}\right], \text{ use}$$

$-8R_1 + R_2 \to R_2$ to get

$$\left[\begin{array}{ccc|c} 1 & 1 & 1 & 40,000 \\ 0 & 1 & 4 & 46,000 \\ 1 & -1 & -1 & 0 \end{array}\right], \text{ use}$$

$-1R_3 + R_1 \to R_3$ and $-1R_2 + R_1 \to R_1$ to get

$$\left[\begin{array}{ccc|c} 1 & 0 & -3 & -6,000 \\ 0 & 1 & 4 & 46,000 \\ 0 & 2 & 2 & 40,000 \end{array}\right], \text{ use } \frac{1}{2}R_3 \to R_3 \text{ to}$$

get

$$\left[\begin{array}{ccc|c} 1 & 0 & -3 & -6,000 \\ 0 & 1 & 4 & 46,000 \\ 0 & 1 & 1 & 20,000 \end{array}\right], \text{ use}$$

$-1R_3 + R_2 \to R_3$ to get

$$\left[\begin{array}{ccc|c} 1 & 0 & -3 & -6,000 \\ 0 & 1 & 4 & 46,000 \\ 0 & 0 & 3 & 26,000 \end{array}\right], \text{ use}$$

$-1R_3 + R_2 \to R_2$ and $\frac{1}{3}R_3 \to R_3$ to get

$$\left[\begin{array}{ccc|c} 1 & 0 & -3 & -6,000 \\ 0 & 1 & 1 & 20,000 \\ 0 & 0 & 1 & 8,666.67 \end{array}\right], \text{ use } 3R_3 + R_1 \to R_1$$

and $-1R_3 + R_2 \to R_2$ to get

$$\left[\begin{array}{ccc|c} 1 & 0 & 0 & 20,000 \\ 0 & 1 & 0 & 11,333.33 \\ 0 & 0 & 1 & 8,666.67 \end{array}\right]. \text{ Investments were}$$

$x = \$20,000$ in a mutual fund, $y = \$11,333.33$ in treasury bills, and $z = \$8,666.67$ in bonds.

71. Since the three given points satisfy the cubic equation, the augmented matrix is

$$A = \left[\begin{array}{ccc|c} -1 & -1 & 1 & 4 \\ 1 & 1 & 1 & 2 \\ 8 & 2 & 1 & 7 \end{array}\right].$$

On A use $R_1 \leftrightarrow R_2$ to get

$$\left[\begin{array}{ccc|c} 1 & 1 & 1 & 2 \\ -1 & -1 & 1 & 4 \\ 8 & 2 & 1 & 7 \end{array}\right], \text{ use}$$

$-\frac{8}{3}R_2 + \left(-\frac{1}{3}R_3\right) \to R_2$ and

$8R_1 + (-1R_3) \to R_3$ to get

$$\left[\begin{array}{ccc|c} 1 & 1 & 1 & 2 \\ 0 & 2 & -3 & -13 \\ 0 & 6 & 7 & 9 \end{array}\right], \text{ use } -3R_2 + R_3 \to R_3$$

to get

$$\left[\begin{array}{ccc|c} 1 & 1 & 1 & 2 \\ 0 & 2 & -3 & -13 \\ 0 & 0 & 16 & 48 \end{array}\right], \text{ use } \frac{1}{2}R_2 \to R_2$$

and $\frac{1}{16}R_3 \to R_3$ to get

$$\left[\begin{array}{ccc|c} 1 & 1 & 1 & 2 \\ 0 & 1 & -3/2 & -13/2 \\ 0 & 0 & 1 & 3 \end{array}\right], \text{ use}$$

$\frac{3}{2}R_3 + R_2 \to R_2$ to get

$$\left[\begin{array}{ccc|c} 1 & 1 & 1 & 2 \\ 0 & 1 & 0 & -2 \\ 0 & 0 & 1 & 3 \end{array}\right], \text{ use } -1R_3 + R_1 \to R_1$$

to get

$$\left[\begin{array}{ccc|c} 1 & 1 & 0 & -1 \\ 0 & 1 & 0 & -2 \\ 0 & 0 & 1 & 3 \end{array}\right], \text{ use } -1R_2 + R_1 \to R_1$$

to get

$$\left[\begin{array}{ccc|c} 1 & 0 & 0 & 1 \\ 0 & 1 & 0 & -2 \\ 0 & 0 & 1 & 3 \end{array}\right].$$

Then $a = 1, b = -2$, and $c = 3$.

73. Since the number of cars entering M.L. King Dr. and Washington St. is 750 and $x + y$ is the number of cars leaving the intersection of M.L. King Dr. and Washington St. then $x + y = 750$.

On the intersection of M.L. King Dr. and JFK Blvd., the number of cars entering this intersection is $450 + x$ and the number of cars leaving is $700 + z$. So $450 + x = 700 + z$.

Simplifying, one gets $y = 750 - x$ and $z = x - 250$; and since y and z are nonnegative, $250 \le x \le 750$. The values of x, y, and z that realize this traffic flow must satisfy

$$\begin{aligned} y &= 750 - x \\ z &= x - 250 \\ 250 \le\ & x \le 750 \end{aligned}$$

If $z = 50$, then $50 = x - 250$ or $x = 300$, and $y = 750 - 300 = 450$.

75. No, since $(-2, 1)$ does not satisfy $y > x + 1$.

77.

On $\left[\begin{array}{ccc|c} 1 & 1 & 0 & 4 \\ 0 & 1 & 1 & -3 \\ 1 & 0 & 1 & -17 \end{array}\right]$, use $-R_1 + R_3 \to R_3$

to get $\left[\begin{array}{ccc|c} 1 & 1 & 0 & 4 \\ 0 & 1 & 1 & -3 \\ 0 & -1 & 1 & -21 \end{array}\right]$.

Use $R_2 + R_3 \to R_3$ to get

$$\left[\begin{array}{ccc|c} 1 & 1 & 0 & 4 \\ 0 & 1 & 1 & -3 \\ 0 & 0 & 2 & -24 \end{array}\right]$$

Use $\frac{1}{2}R_3 \to R_3$ to get

$$\left[\begin{array}{ccc|c} 1 & 1 & 0 & 4 \\ 0 & 1 & 1 & -3 \\ 0 & 0 & 1 & -12 \end{array}\right]$$

Use $-R_3 + R_2 \to R_2$ to get

$$\left[\begin{array}{ccc|c} 1 & 1 & 0 & 4 \\ 0 & 1 & 0 & 9 \\ 0 & 0 & 1 & -12 \end{array}\right].$$

Use $-R_2 + R_1 \to R_1$ to get

$$\left[\begin{array}{ccc|c} 1 & 0 & 0 & -5 \\ 0 & 1 & 0 & 9 \\ 0 & 0 & 1 & -12 \end{array}\right]$$

The solution set is $\{(-5, 9, -12)\}$.

79. Rewrite without absolute values:

$$\begin{aligned} x - 3 &= 2x + 5 \quad \text{or} \quad x - 3 = -2x - 5 \\ -8 &= x \quad \text{or} \quad 3x = -2 \end{aligned}$$

The solution set is $\{-2/3, -8\}$.

81. Let x be the amount of water that is already in, y the amount leaking in per hour, and z the amount pumped out per hour by one pump. Then we obtain a system of equations

$$\begin{aligned} 12(3)(z) &= x + 3y \\ 5(10)(z) &= x + 10y \\ n(2)(z) &= x + 2y. \end{aligned}$$

From the first two equations, we find

$$14z = 7y \text{ or } 2z = y.$$

Using the first equation, we get

$$36z = x + 6z \text{ or } 30z = x.$$

Solving for n, we find

$$n = \frac{x + 2y}{2z} = \frac{x + 4z}{2z} = \frac{30z + 4z}{2z} = 17.$$

Thus, 17 pumps are needed.

For Thought

1. True

2. False, since the orders of matrices A and C are different.

3. False, $A + B = \begin{bmatrix} 2 \\ 6 \end{bmatrix}$.

4. True, $C + D = \begin{bmatrix} 1-3 & 1+5 \\ 3+1 & 3-2 \end{bmatrix} = \begin{bmatrix} -2 & 6 \\ 4 & 1 \end{bmatrix}$.

5. True, $A - B = \begin{bmatrix} 1-1 \\ 3-3 \end{bmatrix} = \begin{bmatrix} 0 \\ 0 \end{bmatrix}$.

6. False, $3B = 3\begin{bmatrix} 1 \\ 3 \end{bmatrix} = \begin{bmatrix} 3 \\ 9 \end{bmatrix}$.

7. False, $-A = -\begin{bmatrix} 1 \\ 3 \end{bmatrix} = \begin{bmatrix} -1 \\ -3 \end{bmatrix}$.

8. False, matrices of different orders cannot be added.

9. False, matrices of different orders cannot be subtracted.

10. False, $C+2D = \begin{bmatrix} 1 & 1 \\ 3 & 3 \end{bmatrix} + \begin{bmatrix} -6 & 10 \\ 2 & -4 \end{bmatrix} =$

$= \begin{bmatrix} -5 & 11 \\ 5 & -1 \end{bmatrix}.$

9.2 Exercises

1. size, equal

3. additive

5. $x = 2,\ y = 5$

7. Since $2x = 6$ and $4y = 16$, $x = 3$ and $y = 4$. Also $3z = z + y$ and so $z = y/2 = 4/2 = 2$. Then $x = 3$, $y = 4$, and $z = 2$.

9. $\begin{bmatrix} 3+2 \\ 5+1 \end{bmatrix} = \begin{bmatrix} 5 \\ 6 \end{bmatrix}$

11. $\begin{bmatrix} -0.5+2 & -0.03+1 \\ 2-0.05 & -0.33+1 \end{bmatrix} = \begin{bmatrix} 1.5 & 0.97 \\ 1.95 & 0.67 \end{bmatrix}$

13. $\begin{bmatrix} 2+1 & -3-1 & 4+1 \\ 4+0 & -6+1 & 8-1 \\ 6+0 & -3+0 & 1+1 \end{bmatrix} = \begin{bmatrix} 3 & -4 & 5 \\ 4 & -5 & 7 \\ 6 & -3 & 2 \end{bmatrix}$

15. $-A = -\begin{bmatrix} 1 & -4 \\ -5 & 6 \end{bmatrix} = \begin{bmatrix} -1 & 4 \\ 5 & -6 \end{bmatrix}$

and $A + (-A) = \begin{bmatrix} 0 & 0 \\ 0 & 0 \end{bmatrix}$

17. $-A = -\begin{bmatrix} 3 & 0 & -1 \\ 8 & -2 & 1 \\ -3 & 6 & 3 \end{bmatrix} =$

$\begin{bmatrix} -3 & 0 & 1 \\ -8 & 2 & -1 \\ 3 & -6 & -3 \end{bmatrix}$ and

$A + (-A) = \begin{bmatrix} 0 & 0 & 0 \\ 0 & 0 & 0 \\ 0 & 0 & 0 \end{bmatrix}$

19. $B - A = \begin{bmatrix} -1+4 & -2-1 \\ 7-3 & 4-0 \end{bmatrix} = \begin{bmatrix} 3 & -3 \\ 4 & 4 \end{bmatrix}$

21. $B - C = \begin{bmatrix} -1+3 & -2+4 \\ 7-2 & 4+5 \end{bmatrix} = \begin{bmatrix} 2 & 2 \\ 5 & 9 \end{bmatrix}$

23. $B-E$ is undefined since B and E have different sizes

25. $3A = 3\begin{bmatrix} -4 & 1 \\ 3 & 0 \end{bmatrix} = \begin{bmatrix} -12 & 3 \\ 9 & 0 \end{bmatrix}$

27. $-1D = -1\begin{bmatrix} -4 \\ 5 \end{bmatrix} = \begin{bmatrix} 4 \\ -5 \end{bmatrix}$

29.

$3A + 3C = \begin{bmatrix} -12 & 3 \\ 9 & 0 \end{bmatrix} + \begin{bmatrix} -9 & -12 \\ 6 & -15 \end{bmatrix} =$

$\begin{bmatrix} -21 & -9 \\ 15 & -15 \end{bmatrix}$

31.

$2A - B = \begin{bmatrix} -8 & 2 \\ 6 & 0 \end{bmatrix} - \begin{bmatrix} -1 & -2 \\ 7 & 4 \end{bmatrix} =$

$\begin{bmatrix} -7 & 4 \\ -1 & -4 \end{bmatrix}$

33.

$2D - 3E = \begin{bmatrix} -8 \\ 10 \end{bmatrix} - \begin{bmatrix} -3 \\ 6 \end{bmatrix} = \begin{bmatrix} -5 \\ 4 \end{bmatrix}$

35. $D+A$ is undefined since D and A have different sizes

37.

$(A+B)+C = \begin{bmatrix} -5 & -1 \\ 10 & 4 \end{bmatrix} + \begin{bmatrix} -3 & -4 \\ 2 & -5 \end{bmatrix} =$

$\begin{bmatrix} -8 & -5 \\ 12 & -1 \end{bmatrix}$

39. $\begin{bmatrix} 0.2+0.2 & 0.1+0.05 \\ 0.4+0.3 & 0.3+0.8 \end{bmatrix} = \begin{bmatrix} 0.4 & 0.15 \\ 0.7 & 1.1 \end{bmatrix}$

41. $\begin{bmatrix} 1/2 & 3/2 \\ 3 & -12 \end{bmatrix}$

43. $\begin{bmatrix} -2 & 4 \\ 6 & 8 \end{bmatrix} - \begin{bmatrix} -12 & 4 \\ 8 & -8 \end{bmatrix} = \begin{bmatrix} 10 & 0 \\ -2 & 16 \end{bmatrix}$

45. Undefined since we cannot add matrices with different sizes

47. $\begin{bmatrix} -1 & 13 \\ -9 & 3 \\ 6 & -2 \end{bmatrix}$

49. $\begin{bmatrix} 3\sqrt{2} & 2 & 3\sqrt{3} \end{bmatrix}$

51.
$$\begin{bmatrix} 2a \\ 2b \end{bmatrix} + \begin{bmatrix} 6a \\ 12b \end{bmatrix} + \begin{bmatrix} 5a \\ -15b \end{bmatrix} = \begin{bmatrix} 13a \\ -b \end{bmatrix}$$

53.
$$\begin{bmatrix} -0.4x - 0.6x & 0.4y - 0.9y \\ 0.8x - 1.5x & 3.2y + 0.3y \end{bmatrix} =$$
$$\begin{bmatrix} -x & -0.5y \\ -0.7x & 3.5y \end{bmatrix}$$

55.
$$\begin{bmatrix} 2x & 2y & 2z \\ -2x & 4y & 6z \\ 2x & -2y & -6z \end{bmatrix} - \begin{bmatrix} -x & 0 & 3z \\ 4x & y & -z \\ 2x & 5y & z \end{bmatrix} =$$
$$\begin{bmatrix} 3x & 2y & -z \\ -6x & 3y & 7z \\ 0 & -7y & -7z \end{bmatrix}$$

57. Equate the corresponding entries.

$$\begin{aligned} x + y &= 5 \\ x - y &= 1 \end{aligned}$$

Adding the two equations, we get $2x = 6$. Substitute $x = 3$ into $x+y = 5$. Then $3+y = 5$ and $y = 2$. Solution set is $\{(3, 2)\}$.

59. Equate the corresponding entries.

$$\begin{aligned} 2x + 3y &= 7 \\ x - 4y &= -13 \end{aligned}$$

Multiply second equation by -2 and add to the first one.

$$\begin{aligned} 2x + 3y &= 7 \\ -2x + 8y &= 26 \\ \hline 11y &= 33 \\ y &= 3 \end{aligned}$$

Substitute $y = 3$ into $x - 4y = -13$. Then $x - 12 = -13$ and $x = -1$. The solution set is $\{(-1, 3)\}$.

61. Equate the corresponding entries.

$$\begin{aligned} x + y + z &= 8 \\ x - y - z &= -7 \\ x - y + z &= 2 \end{aligned}$$

Adding the first and second equations, $2x = 1$ and $x = 0.5$. Multiply second equation by -1 and add to the third.

$$\begin{aligned} -x + y + z &= 7 \\ x - y + z &= 2 \\ \hline 2z &= 9 \\ z &= 4.5 \end{aligned}$$

Substitute $x = 0.5$ and $z = 4.5$ into $x + y + z = 8$. Then $y + 5 = 8$ and $y = 3$. The solution set is $\{(0.5, 3, 4.5)\}$.

63. The matrices for January,February and March are, respectively,

$$J = \begin{bmatrix} 120 \\ 30 \\ 40 \end{bmatrix}, \; F = \begin{bmatrix} 130 \\ 70 \\ 50 \end{bmatrix}, \text{ and}$$

$$M = \begin{bmatrix} 140 \\ 60 \\ 45 \end{bmatrix}. \text{ The sum}$$

$$J + F + M = \begin{bmatrix} \$390 \\ \$160 \\ \$135 \end{bmatrix} \text{ represents the}$$

total expenses on food, clothing and utilities for the three months.

65. The supply matrix for the first week is

$$S = \begin{bmatrix} 40 & 80 \\ 30 & 90 \\ 80 & 200 \end{bmatrix}. \text{ Next week's supply matrix}$$

$$\text{is } S + 0.5S = \begin{bmatrix} 40 & 80 \\ 30 & 90 \\ 80 & 200 \end{bmatrix} + \begin{bmatrix} 20 & 40 \\ 15 & 45 \\ 40 & 100 \end{bmatrix}$$

$$= \begin{bmatrix} 60 & 120 \\ 45 & 135 \\ 120 & 300 \end{bmatrix}.$$

67. Yes, yes

69. Yes, yes

71. Yes, yes

73. $\begin{bmatrix} 0 & 0 \\ 0 & 0 \end{bmatrix}$

75. On $\left[\begin{array}{cc|c} 2 & 3 & 4 \\ 1 & -4 & -31 \end{array}\right]$, apply

$-2R_2 + R_1 \to R_2$ to get $\left[\begin{array}{cc|c} 2 & 3 & 4 \\ 0 & 11 & 66 \end{array}\right]$.

Use $\frac{1}{11}R_2 \to R_2$ to get $\left[\begin{array}{cc|c} 2 & 3 & 4 \\ 0 & 1 & 6 \end{array}\right]$.

Use $-3R_2 + R_1 \to R_2$ to get

$$\left[\begin{array}{cc|c} 2 & 0 & -14 \\ 0 & 1 & 6 \end{array}\right].$$

Use $\frac{1}{2}R_1 \to R_1$ to get $\left[\begin{array}{cc|c} 1 & 0 & -7 \\ 0 & 1 & 6 \end{array}\right]$.

The solution set is $\{(-7, 6)\}$, and independent.

77. **a)** $bx(ax + y) + z(ax + y) = (ax + y)(bx + z)$
b) $x(6x^2 - 23xy + 20y^2) = x(3x - 4y)(2x - 5y)$

79. $\dfrac{\ln x}{\ln 7}$

81. Let the ordered triple (x, y, z) represent the amount of antifreeze in the 8-quart radiator, the 5-quart container, and the 3-quart container, respectively.

To leave four quarts of antifreeze in the radiator, perform the sequence of operations defined by the ordered triples:

$$(8, 0, 0), (5, 0, 3), (5, 3, 0), (2, 3, 3),$$
$$(2, 5, 1), (7, 0, 1), (7, 1, 0), (4, 1, 3).$$

Afterwards, fill the radiator with water.

For Thought

1. True **2.** True **3.** False, they cannot be multiplied since the number of columns of A is not the same as the number of rows of C.

4. False, the order of CA is 2×1. **5.** True

6. True, $BC = [7 \cdot 2 + 9 \cdot 4 \quad 7 \cdot 3 + 9 \cdot 5] = [14 + 36 \quad 21 + 45] = [50 \quad 66]$.

7. True, $AB = \begin{bmatrix} 1 \\ 6 \end{bmatrix} [7 \quad 9] = \begin{bmatrix} 1 \cdot 7 & 1 \cdot 9 \\ 6 \cdot 7 & 6 \cdot 9 \end{bmatrix} = \begin{bmatrix} 7 & 9 \\ 42 & 54 \end{bmatrix}$.

8. True, $\begin{bmatrix} 2 & 3 \\ 4 & 5 \end{bmatrix} \begin{bmatrix} 2 & -1 \\ 0 & 3 \end{bmatrix} = \begin{bmatrix} 4+0 & -2+9 \\ 8+0 & -4+15 \end{bmatrix} = \begin{bmatrix} 4 & 7 \\ 8 & 11 \end{bmatrix}$.

9. True, $BA = [7 \quad 9] \begin{bmatrix} 1 \\ 6 \end{bmatrix} = [7 + 54] = [61]$.

10. False, since $EC = \begin{bmatrix} 2 & -1 \\ 0 & 3 \end{bmatrix} \begin{bmatrix} 2 & 3 \\ 4 & 5 \end{bmatrix} = \begin{bmatrix} 4-4 & 6-5 \\ 0+12 & 0+15 \end{bmatrix} = \begin{bmatrix} 0 & 1 \\ 12 & 15 \end{bmatrix}$ and from Exercise 8 one sees $EC \neq CE$.

9.3 Exercises

1. product

3. 3×5 **5.** 1×1 **7.** 5×5

9. 3×3

11. undefined

13. $[-3(4) + 2(1)] = [-10]$

15. $\begin{bmatrix} 1(1) + 3(3) \\ 2(1) + (-4)(3) \end{bmatrix} = \begin{bmatrix} 10 \\ -10 \end{bmatrix}$

17. $\begin{bmatrix} 5(1) + 1(3) & 5(2) + 1(1) \\ 2(1) + 1(3) & 2(2) + 1(1) \end{bmatrix} = \begin{bmatrix} 8 & 11 \\ 5 & 5 \end{bmatrix}$

19. $\begin{bmatrix} 3(5) & 3(6) \\ 1(5) & 1(6) \end{bmatrix} = \begin{bmatrix} 15 & 18 \\ 5 & 6 \end{bmatrix}$

21. $AB = \begin{bmatrix} 1(1) + 3(-1) & 1(0) + 3(1) & 1(1) + 3(0) \\ 2(1) + 4(-1) & 2(0) + 4(1) & 2(1) + 4(0) \\ 5(1) + 6(-1) & 5(0) + 6(1) & 5(1) + 6(0) \end{bmatrix} = \begin{bmatrix} -2 & 3 & 1 \\ -2 & 4 & 2 \\ -1 & 6 & 5 \end{bmatrix}$ and

$BA = \begin{bmatrix} 1(1) + 0(2) + 1(5) & 1(3) + 0(4) + 1(6) \\ -1(1) + 1(2) + 0(5) & -1(3) + 1(4) + 0(6) \end{bmatrix} = \begin{bmatrix} 6 & 9 \\ 1 & 1 \end{bmatrix}$

23. $AB =$

$$\begin{bmatrix} 1(1)+2(0)+3(0) & 1(1)+2(1)+3(0) & 1(1)+2(1)+3(1) \\ 2(1)+1(0)+3(0) & 2(1)+1(1)+3(0) & 2(1)+1(1)+3(1) \\ 3(1)+2(0)+1(0) & 3(1)+2(1)+1(0) & 3(1)+2(1)+1(1) \end{bmatrix}$$

$$= \begin{bmatrix} 1 & 3 & 6 \\ 2 & 3 & 6 \\ 3 & 5 & 6 \end{bmatrix} \text{ and}$$

$BA =$

$$\begin{bmatrix} 1(1)+1(2)+1(3) & 1(2)+1(1)+1(2) & 1(3)+1(3)+1(1) \\ 0(1)+1(2)+1(3) & 0(2)+1(1)+1(2) & 0(3)+1(3)+1(1) \\ 0(1)+0(2)+1(3) & 0(2)+0(1)+1(2) & 0(3)+0(3)+1(1) \end{bmatrix}$$

$$= \begin{bmatrix} 6 & 5 & 7 \\ 5 & 3 & 4 \\ 3 & 2 & 1 \end{bmatrix}$$

25.

$$AB = \begin{bmatrix} 2\cdot 2 & 2\cdot 3 & 2\cdot 4 \\ -3\cdot 2 & -3\cdot 3 & -3\cdot 4 \\ 1\cdot 2 & 1\cdot 3 & 1\cdot 4 \end{bmatrix} =$$

$$\begin{bmatrix} 4 & 6 & 8 \\ -6 & -9 & -12 \\ 2 & 3 & 4 \end{bmatrix}$$

27. $\begin{bmatrix} 2+0+0 & 2+3+0 & 2+3+4 \end{bmatrix} =$

$\begin{bmatrix} 2 & 5 & 9 \end{bmatrix}$

29. undefined

31.

$$EC = \begin{bmatrix} 2+4+1 & 3+5+0 \\ 0+4+1 & 0+5+0 \\ 0+0+1 & 0+0+0 \end{bmatrix} = \begin{bmatrix} 7 & 8 \\ 5 & 5 \\ 1 & 0 \end{bmatrix}$$

33.

$$DC = \begin{bmatrix} 4-4+1 & 6-5+0 \\ 0+12+2 & 0+15+0 \end{bmatrix} =$$

$$\begin{bmatrix} 1 & 1 \\ 14 & 15 \end{bmatrix}$$

35. Undefined

37.

$$EA = \begin{bmatrix} 2-3+1 \\ 0-3+1 \\ 0+0+1 \end{bmatrix} = \begin{bmatrix} 0 \\ -2 \\ 1 \end{bmatrix}$$

39. Undefined

41. We will use the answer in Exercise 25.

$AB + 2E =$

$$\begin{bmatrix} 4 & 6 & 8 \\ -6 & -9 & -12 \\ 2 & 3 & 4 \end{bmatrix} + \begin{bmatrix} 2 & 2 & 2 \\ 0 & 2 & 2 \\ 0 & 0 & 2 \end{bmatrix}$$

$$= \begin{bmatrix} 6 & 8 & 10 \\ -6 & -7 & -10 \\ 2 & 3 & 6 \end{bmatrix}$$

43. $A^2 = \begin{bmatrix} 1 & 0 \\ 1 & 1 \end{bmatrix}\begin{bmatrix} 1 & 0 \\ 1 & 1 \end{bmatrix} =$

$$\begin{bmatrix} 1(1)+0(1) & 1(0)+0(1) \\ 1(1)+1(1) & 1(0)+1(1) \end{bmatrix} = \begin{bmatrix} 1 & 0 \\ 2 & 1 \end{bmatrix}$$

45. We will use the answer in Exercise 44.

$$A^4 = A^3 A = \begin{bmatrix} 1 & 0 \\ 3 & 1 \end{bmatrix}\begin{bmatrix} 1 & 0 \\ 1 & 1 \end{bmatrix} =$$

$$\begin{bmatrix} 1(1)+0(1) & 1(0)+0(1) \\ 3(1)+1(1) & 3(0)+1(1) \end{bmatrix} = \begin{bmatrix} 1 & 0 \\ 4 & 1 \end{bmatrix}$$

47.

$$\begin{bmatrix} 2\cdot 1+0\cdot 0 & 2\cdot 1+0\cdot 1 \\ 3\cdot 1+1\cdot 0 & 3\cdot 1+1\cdot 1 \end{bmatrix} = \begin{bmatrix} 2 & 2 \\ 3 & 4 \end{bmatrix}$$

49.

$$\begin{bmatrix} 7\cdot 3+4\cdot(-5) & 7\cdot(-4)+4\cdot 7 \\ 5\cdot 3+3\cdot(-5) & 5\cdot(-4)+3\cdot 7 \end{bmatrix} =$$

$$\begin{bmatrix} 1 & 0 \\ 0 & 1 \end{bmatrix}$$

51.

$$\begin{bmatrix} -0.5\cdot 1+4\cdot 0 & -0.5\cdot 0+4\cdot 1 \\ 9\cdot 1+0.7\cdot 0 & 9\cdot 0+0.7\cdot 1 \end{bmatrix} =$$

$$\begin{bmatrix} -0.5 & 4 \\ 9 & 0.7 \end{bmatrix}$$

53. $[-2a+6a \quad -6b+3b] = [4a \quad -3b]$

55.

$$\begin{bmatrix} -2a+0\cdot 1 & 5a+0\cdot 4 & 3a+0\cdot 6 \\ 0\cdot(-2)+b & 0\cdot 5+4b & 0\cdot 3+6b \end{bmatrix} =$$

$$\begin{bmatrix} -2a & 5a & 3a \\ b & 4b & 6b \end{bmatrix}$$

57. $[1{\cdot}1{+}2{\cdot}0{+}3{\cdot}1 \quad 1{\cdot}0{+}2{\cdot}1{+}3{\cdot}0 \quad 1{\cdot}1{+}2{\cdot}1{+}3{\cdot}1]$
$= [4 \quad 2 \quad 6]$

59. $[-1 \cdot (-5) + 0 \cdot 1 + 3 \cdot 4] = [17]$

61.
$$\begin{bmatrix} x^2 & xy \\ xy & y^2 \end{bmatrix}$$

63.
$$\begin{bmatrix} -1 \cdot \sqrt{2} + 2 \cdot 0 + 3\sqrt{2} \\ 3 \cdot \sqrt{2} + 4 \cdot 0 + 4\sqrt{2} \end{bmatrix} = \begin{bmatrix} 2\sqrt{2} \\ 7\sqrt{2} \end{bmatrix}$$

65.
$$\begin{bmatrix} (1/2) \cdot (-8) + (1/3) \cdot (-5) & 6 + (1/3) \cdot 15 \\ (1/4) \cdot (-8) + (1/5) \cdot (-5) & 3 + (1/5) \cdot 15 \end{bmatrix}$$
$$= \begin{bmatrix} -4 - (5/3) & 6+5 \\ -2-1 & 3+3 \end{bmatrix} = \begin{bmatrix} -17/3 & 11 \\ -3 & 6 \end{bmatrix}$$

67. Undefined

69.
$$\begin{bmatrix} 9+0-7 & 8+0-8 & 10+0-4 \\ 0+3+0 & 0+5+0 & 0+2+0 \\ 9+3+7 & 8+5+8 & 10+2+4 \end{bmatrix} =$$
$$\begin{bmatrix} 2 & 0 & 6 \\ 3 & 5 & 2 \\ 19 & 21 & 16 \end{bmatrix}$$

71.
$$\begin{bmatrix} 1-0.6-0.6 & 1.5+1.2-1 \\ 0.8+0.4+1.8 & 1.2-0.8+3 \\ 0.4+0.6-2.4 & 0.6-1.2-4 \end{bmatrix} =$$
$$\begin{bmatrix} -0.2 & 1.7 \\ 3 & 3.4 \\ -1.4 & -4.6 \end{bmatrix}$$

73. A system of equations is
$$\begin{aligned} 2x - 3y &= 0 \\ x + 2y &= 7. \end{aligned}$$
Multiply second equation by -2 and add to the first one.
$$\begin{aligned} 2x - 3y &= 0 \\ -2x - 4y &= -14 \\ \hline -7y &= -14 \end{aligned}$$
Substitute $y = 2$ into $x + 2y = 7$. Then $x + 4 = 7$ and $x = 3$. Solution set is $\{(3, 2)\}$.

75. A system of equations is
$$\begin{aligned} 2x + 3y &= 5 \\ 4x + 6y &= 9 \end{aligned}$$
Multiply first equation by -2 and add to the second one.
$$\begin{aligned} -4x - 6y &= -10 \\ 4x + 6y &= 9 \\ \hline 0 &= -1 \end{aligned}$$
Inconsistent and the solution set is $\emptyset$.

77. System of equations is
$$\begin{aligned} x + y + z &= 4 \\ y + z &= 5 \\ z &= 6. \end{aligned}$$
Substitute $z = 6$ into $y + z = 5$ to get $y + 6 = 5$ and $y = -1$. From $x + y + z = 4$, we have $x - 1 + 6 = 4$ and $x = -1$. Solution set is $\{(-1, -1, 6)\}$.

79.
$$\begin{bmatrix} 2 & 3 \\ 4 & -1 \end{bmatrix} \begin{bmatrix} x \\ y \end{bmatrix} = \begin{bmatrix} 9 \\ 6 \end{bmatrix}$$

81.
$$\begin{bmatrix} 1 & 2 & -1 \\ 3 & -1 & 3 \\ 2 & 1 & -4 \end{bmatrix} \begin{bmatrix} x \\ y \\ z \end{bmatrix} = \begin{bmatrix} 3 \\ 1 \\ 0 \end{bmatrix}$$

83.
$$A = \begin{bmatrix} \$24,000 & \$40,000 \\ \$38,000 & \$70,000 \end{bmatrix}, \ Q = \begin{bmatrix} 4 \\ 7 \end{bmatrix}, \text{ and}$$
matrix product $AQ = \begin{bmatrix} \$376,000 \\ \$642,000 \end{bmatrix}$
represents the costs for labor and material for building 4 economy houses and 7 deluxe models.

85. False

87. True

89. True

91. $A + 3B = \begin{bmatrix} 1 & 2 \\ -3 & 5 \end{bmatrix} + \begin{bmatrix} -3 & 9 \\ 6 & 12 \end{bmatrix} =$
$$\begin{bmatrix} -2 & 11 \\ 3 & 17 \end{bmatrix}$$

93. Independent since the lines have different slopes, namely, -9 and 500.

95. Let $f(x) = \dfrac{x-9}{x+99} \le 0$.
If $x = 10$, then $f(10) > 0$.
If $x = 0$, then $f(0) < 0$.
If $x = -100$, then $f(100) > 0$.

+	U	−	0	+
0	−99	10	9	100

The solution set is $(-99, 9]$.

97. There were 15 soccer teams in the north section, and 9 teams in the south section. In the south section, it is possible that Springville won three games and tied five games for a total of 5.5 points.

For Thought

1. True, $AB = \begin{bmatrix} 10-9 & -6+6 \\ 15-15 & -9+10 \end{bmatrix} =$
$$= \begin{bmatrix} 1 & 0 \\ 0 & 1 \end{bmatrix} = \begin{bmatrix} 10-9 & 15-15 \\ -6+6 & -9+10 \end{bmatrix} = BA.$$

2. True, $AB = BA = I$ by Exercise 1.

3. True, A is the inverse of B by Exercise 1.

4. False, AC is undefined, although CA is defined.

5. True

6. False, a non-square matrix has no inverse.

7. False, the coefficient matrix is $\begin{bmatrix} 2 & 3 \\ 3 & 1 \end{bmatrix}$.

8. True, since $A^{-1} = B$ then $A^{-1}D =$
$$= \begin{bmatrix} 5 & -3 \\ -3 & 2 \end{bmatrix}\begin{bmatrix} 11 \\ 19 \end{bmatrix} = \begin{bmatrix} -2 \\ 5 \end{bmatrix}.$$

9. False, $(-2, 5)$ does not satisfy $3x + y = 19$.

10. False, the solution is $\begin{bmatrix} 2 & 3 \\ 3 & 1 \end{bmatrix}^{-1}\begin{bmatrix} 3 \\ -7 \end{bmatrix}$.

9.4 Exercises

1. identity

3.
$$\begin{aligned} AI &= \begin{bmatrix} 1 & 3 \\ 4 & 6 \end{bmatrix}\begin{bmatrix} 1 & 0 \\ 0 & 1 \end{bmatrix} \\ &= \begin{bmatrix} 1(1)+3(0) & 1(0)+3(1) \\ 4(1)+6(0) & 4(0)+6(1) \end{bmatrix} \\ &= \begin{bmatrix} 1 & 3 \\ 4 & 6 \end{bmatrix} \\ &= A \end{aligned}$$

and

$$\begin{aligned} IA &= \begin{bmatrix} 1 & 0 \\ 0 & 1 \end{bmatrix}\begin{bmatrix} 1 & 3 \\ 4 & 6 \end{bmatrix} \\ &= \begin{bmatrix} 1(1)+0(4) & 1(3)+0(6) \\ 0(1)+1(4) & 0(3)+1(6) \end{bmatrix} \\ &= \begin{bmatrix} 1 & 3 \\ 4 & 6 \end{bmatrix} \\ &= A \end{aligned}$$

5. $AI = \begin{bmatrix} 3 & 2 & 1 \\ 5 & 6 & 2 \\ 7 & 8 & 3 \end{bmatrix}\begin{bmatrix} 1 & 0 & 0 \\ 0 & 1 & 0 \\ 0 & 0 & 1 \end{bmatrix} =$
$$\begin{bmatrix} 3(1)+2(0)+1(0) & 3(0)+2(1)+1(0) & 3(0)+2(0)+1(1) \\ 5(1)+6(0)+2(0) & 5(0)+6(1)+2(0) & 5(0)+6(0)+2(1) \\ 7(1)+8(0)+3(0) & 7(0)+8(1)+3(0) & 7(0)+8(0)+3(1) \end{bmatrix}$$
$$= \begin{bmatrix} 3 & 2 & 1 \\ 5 & 6 & 2 \\ 7 & 8 & 3 \end{bmatrix} = A$$

and

$$IA = \begin{bmatrix} 1 & 0 & 0 \\ 0 & 1 & 0 \\ 0 & 0 & 1 \end{bmatrix}\begin{bmatrix} 3 & 2 & 1 \\ 5 & 6 & 2 \\ 7 & 8 & 3 \end{bmatrix} =$$
$$\begin{bmatrix} 1(3)+0(5)+0(7) & 1(2)+0(6)+0(8) & 1(1)+0(2)+0(3) \\ 0(3)+1(5)+0(7) & 0(2)+1(6)+0(8) & 0(1)+1(2)+0(3) \\ 0(3)+0(5)+1(7) & 0(2)+0(6)+1(8) & 0(1)+0(2)+1(3) \end{bmatrix}$$
$$= \begin{bmatrix} 3 & 2 & 1 \\ 5 & 6 & 2 \\ 7 & 8 & 3 \end{bmatrix} = A$$

7.
$$I\begin{bmatrix} -3 & 5 \\ 12 & 6 \end{bmatrix} = \begin{bmatrix} -3 & 5 \\ 12 & 6 \end{bmatrix}$$

9.
$$\begin{bmatrix} -8+9 & -6+6 \\ 12-12 & 9-8 \end{bmatrix} = \begin{bmatrix} 1 & 0 \\ 0 & 1 \end{bmatrix}$$

11.
$$\begin{bmatrix} 5-4 & -4+4 \\ 5-5 & -4+5 \end{bmatrix} = \begin{bmatrix} 1 & 0 \\ 0 & 1 \end{bmatrix}$$

13.
$$\begin{bmatrix} 3 & 5 & 1 \\ 4 & 5 & 7 \\ 4 & 9 & 2 \end{bmatrix} I = \begin{bmatrix} 3 & 5 & 1 \\ 4 & 5 & 7 \\ 4 & 9 & 2 \end{bmatrix}$$

15.
$$\begin{bmatrix} 0+0+1 & 1+0-1 & -3+0+3 \\ 0+0+0 & 1+0+0 & -3+3+0 \\ 0+0+0 & 0+0+0 & 0+1+0 \end{bmatrix} = \begin{bmatrix} 1 & 0 & 0 \\ 0 & 1 & 0 \\ 0 & 0 & 1 \end{bmatrix}$$

17.
$$\begin{bmatrix} 0.5+0.5+0 & -0.5+0.5+0 & 0.5-0.5+0 \\ 0+0.5-0.5 & 0+0.5+0.5 & 0-0.5+0.5 \\ 0.5+0-0.5 & -0.5+0+0.5 & 0.5+0+0.5 \end{bmatrix} = \begin{bmatrix} 1 & 0 & 0 \\ 0 & 1 & 0 \\ 0 & 0 & 1 \end{bmatrix}$$

19.
Yes, since $\begin{bmatrix} 3 & 1 \\ 11 & 4 \end{bmatrix} \begin{bmatrix} 4 & -1 \\ -11 & 3 \end{bmatrix} =$

$\begin{bmatrix} 12-11 & -3+3 \\ 44-44 & -11+12 \end{bmatrix} = \begin{bmatrix} 1 & 0 \\ 0 & 1 \end{bmatrix}$ and

similarly $\begin{bmatrix} 4 & -1 \\ -11 & 3 \end{bmatrix} \begin{bmatrix} 3 & 1 \\ 11 & 4 \end{bmatrix} = I.$

21.
No, since $\begin{bmatrix} 1/2 & -1 \\ 3 & -12 \end{bmatrix} \begin{bmatrix} 4 & 2 \\ 1 & 1 \end{bmatrix} =$

$\begin{bmatrix} 2-1 & 1-1 \\ 12-12 & 6-12 \end{bmatrix} = \begin{bmatrix} 1 & 0 \\ 0 & -6 \end{bmatrix} \neq I.$

23. No, since only square matrices may have inverses.

25.
On $\left[\begin{array}{cc|cc} 1 & 4 & 1 & 0 \\ 0 & 2 & 0 & 1 \end{array}\right]$, use $-2R_2 + R_1 \to R_1$ to get

$\left[\begin{array}{cc|cc} 1 & 0 & 1 & -2 \\ 0 & 2 & 0 & 1 \end{array}\right]$, use $\frac{1}{2}R_2 \to R_2$ to get

$\left[\begin{array}{cc|cc} 1 & 0 & 1 & -2 \\ 0 & 1 & 0 & 1/2 \end{array}\right]$. Then $A^{-1} = \begin{bmatrix} 1 & -2 \\ 0 & 1/2 \end{bmatrix}$.

27.
On $\left[\begin{array}{cc|cc} 1 & 6 & 1 & 0 \\ 1 & 9 & 0 & 1 \end{array}\right]$, use $-1R_1 + R_2 \to R_2$ to get

$\left[\begin{array}{cc|cc} 1 & 6 & 1 & 0 \\ 0 & 3 & -1 & 1 \end{array}\right]$, use $-2R_2 + R_1 \to R_1$ to get

$\left[\begin{array}{cc|cc} 1 & 0 & 3 & -2 \\ 0 & 3 & -1 & 1 \end{array}\right]$, use $\frac{1}{3}R_2 \to R_2$ to get

$\left[\begin{array}{cc|cc} 1 & 0 & 3 & -2 \\ 0 & 1 & -1/3 & 1/3 \end{array}\right]$.

Thus, $A^{-1} = \begin{bmatrix} 3 & -2 \\ -1/3 & 1/3 \end{bmatrix}$.

29.
On $\left[\begin{array}{cc|cc} -2 & -3 & 1 & 0 \\ 3 & 4 & 0 & 1 \end{array}\right]$, use $R_2 + R_1 \to R_1$ to get

$\left[\begin{array}{cc|cc} 1 & 1 & 1 & 1 \\ 3 & 4 & 0 & 1 \end{array}\right]$, use $-3R_1 + R_2 \to R_2$ to get

$\left[\begin{array}{cc|cc} 1 & 1 & 1 & 1 \\ 0 & 1 & -3 & -2 \end{array}\right]$, use $-1R_2 + R_1 \to R_1$ to get

$\left[\begin{array}{cc|cc} 1 & 0 & 4 & 3 \\ 0 & 1 & -3 & -2 \end{array}\right]$. So $A^{-1} = \begin{bmatrix} 4 & 3 \\ -3 & -2 \end{bmatrix}$.

31.
On $\left[\begin{array}{cc|cc} 1 & -5 & 1 & 0 \\ -1 & 3 & 0 & 1 \end{array}\right]$, use $R_1 + R_2 \to R_2$ to get

$\left[\begin{array}{cc|cc} 1 & -5 & 1 & 0 \\ 0 & -2 & 1 & 1 \end{array}\right]$, use $-\frac{1}{2}R_2 \to R_2$ to get

$\left[\begin{array}{cc|cc} 1 & -5 & 1 & 0 \\ 0 & 1 & -1/2 & -1/2 \end{array}\right]$, use $5R_2 + R_1 \to R_1$ to get

$\left[\begin{array}{cc|cc} 1 & 0 & -3/2 & -5/2 \\ 0 & 1 & -1/2 & -1/2 \end{array}\right]$.

Then $A^{-1} = \begin{bmatrix} -3/2 & -5/2 \\ -1/2 & -1/2 \end{bmatrix}$.

33.
On $\left[\begin{array}{cc|cc} -1 & 5 & 1 & 0 \\ 2 & -10 & 0 & 1 \end{array}\right]$, use $R_1 + R_2 \to R_1$ and $\frac{1}{2}R_2 \to R_2$ to get

$\left[\begin{array}{cc|cc} 1 & -5 & 1 & 1 \\ 1 & -5 & 0 & 1/2 \end{array}\right]$, use $-1R_2 + R_1 \to R_1$ to get

$\left[\begin{array}{cc|cc} 0 & 0 & 1 & 1/2 \\ 1 & -5 & 0 & 1/2 \end{array}\right]$. So A has no inverse.

35.
On $\left[\begin{array}{ccc|ccc} 1 & 1 & 0 & 1 & 0 & 0 \\ 0 & -1 & -1 & 0 & 1 & 0 \\ 1 & 0 & -1 & 0 & 0 & 1 \end{array}\right]$, use

$-1R_3 + R_1 \to R_3$ and $R_2 + R_1 \to R_1$ to get

$\left[\begin{array}{ccc|ccc} 1 & 0 & -1 & 1 & 1 & 0 \\ 0 & -1 & -1 & 0 & 1 & 0 \\ 0 & 1 & 1 & 1 & 0 & -1 \end{array}\right]$, use

$R_2 + R_3 \to R_2$ to get

$\left[\begin{array}{ccc|ccc} 1 & 0 & -1 & 1 & 1 & 0 \\ 0 & 0 & 0 & 1 & 1 & -1 \\ 0 & 1 & 1 & 1 & 0 & 1 \end{array}\right]$. So A^{-1} does not exist.

37.
On $\left[\begin{array}{ccc|ccc} 1 & 1 & 1 & 1 & 0 & 0 \\ 1 & -1 & -1 & 0 & 1 & 0 \\ 1 & -1 & 1 & 0 & 0 & 1 \end{array}\right]$, use

$R_2 + (-1R_3) \to R_3$ and $R_1 + R_2 \to R_2$ to get

$\left[\begin{array}{ccc|ccc} 1 & 1 & 1 & 1 & 0 & 0 \\ 2 & 0 & 0 & 1 & 1 & 0 \\ 0 & 0 & -2 & 0 & 1 & -1 \end{array}\right]$, use

$-2R_1 + R_2 \to R_2$ and $-\frac{1}{2}R_3 \to R_3$ to get

$\left[\begin{array}{ccc|ccc} 1 & 1 & 1 & 1 & 0 & 0 \\ 0 & -2 & -2 & -1 & 1 & 0 \\ 0 & 0 & 1 & 0 & -1/2 & 1/2 \end{array}\right]$, use

$-\frac{1}{2}R_2 \to R_2$ to get

$\left[\begin{array}{ccc|ccc} 1 & 1 & 1 & 1 & 0 & 0 \\ 0 & 1 & 1 & 1/2 & -1/2 & 0 \\ 0 & 0 & 1 & 0 & -1/2 & 1/2 \end{array}\right]$, use

$-1R_2 + R_1 \to R_1$ to get

$\left[\begin{array}{ccc|ccc} 1 & 0 & 0 & 1/2 & 1/2 & 0 \\ 0 & 1 & 1 & 1/2 & -1/2 & 0 \\ 0 & 0 & 1 & 0 & -1/2 & 1/2 \end{array}\right]$, use

$-1R_3 + R_2 \to R_2$ to get

$\left[\begin{array}{ccc|ccc} 1 & 0 & 0 & 1/2 & 1/2 & 0 \\ 0 & 1 & 0 & 1/2 & 0 & -1/2 \\ 0 & 0 & 1 & 0 & -1/2 & 1/2 \end{array}\right]$.

So $A^{-1} = \left[\begin{array}{ccc} 1/2 & 1/2 & 0 \\ 1/2 & 0 & -1/2 \\ 0 & -1/2 & 1/2 \end{array}\right]$.

39.
On $\left[\begin{array}{ccc|ccc} 0 & 2 & 0 & 1 & 0 & 0 \\ 3 & 3 & 2 & 0 & 1 & 0 \\ 2 & 5 & 1 & 0 & 0 & 1 \end{array}\right]$, use

$-1R_3 + R_2 \to R_3$ to get

$\left[\begin{array}{ccc|ccc} 0 & 2 & 0 & 1 & 0 & 0 \\ 3 & 3 & 2 & 0 & 1 & 0 \\ 1 & -2 & 1 & 0 & 1 & -1 \end{array}\right]$, use

$R_3 \leftrightarrow R_1$ to get

$\left[\begin{array}{ccc|ccc} 1 & -2 & 1 & 0 & 1 & -1 \\ 3 & 3 & 2 & 0 & 1 & 0 \\ 0 & 2 & 0 & 1 & 0 & 0 \end{array}\right]$, use

$R_3 + R_1 \to R_1$ and $-1R_3 + R_2 \to R_2$ to get

$\left[\begin{array}{ccc|ccc} 1 & 0 & 1 & 1 & 1 & -1 \\ 3 & 1 & 2 & -1 & 1 & 0 \\ 0 & 2 & 0 & 1 & 0 & 0 \end{array}\right]$, use

$-3R_1 + R_2 \to R_2$ to get

$\left[\begin{array}{ccc|ccc} 1 & 0 & 1 & 1 & 1 & -1 \\ 0 & 1 & -1 & -4 & -2 & 3 \\ 0 & 2 & 0 & 1 & 0 & 0 \end{array}\right]$, use

$-2R_2 + R_3 \to R_3$ to get

$\left[\begin{array}{ccc|ccc} 1 & 0 & 1 & 1 & 1 & -1 \\ 0 & 1 & -1 & -4 & -2 & 3 \\ 0 & 0 & 2 & 9 & 4 & -6 \end{array}\right]$, use

$\frac{1}{2}R_3 \to R_3$ to get

$\left[\begin{array}{ccc|ccc} 1 & 0 & 1 & 1 & 1 & -1 \\ 0 & 1 & -1 & -4 & -2 & 3 \\ 0 & 0 & 1 & 9/2 & 2 & -3 \end{array}\right]$, use

$R_2 + R_3 \to R_2$ to get

$\left[\begin{array}{ccc|ccc} 1 & 0 & 1 & 1 & 1 & -1 \\ 0 & 1 & 0 & 1/2 & 0 & 0 \\ 0 & 0 & 1 & 9/2 & 2 & -3 \end{array}\right]$, use

$-1R_3 + R_1 \to R_1$ to get

$$\left[\begin{array}{ccc|ccc} 1 & 0 & 0 & -7/2 & -1 & 2 \\ 0 & 1 & 0 & 1/2 & 0 & 0 \\ 0 & 0 & 1 & 9/2 & 2 & -3 \end{array}\right].$$

So $A^{-1} = \begin{bmatrix} -7/2 & -1 & 2 \\ 1/2 & 0 & 0 \\ 9/2 & 2 & -3 \end{bmatrix}$.

41.

On $\left[\begin{array}{ccc|ccc} 1 & 0 & 1 & 1 & 0 & 0 \\ 0 & 2 & 2 & 0 & 1 & 0 \\ 2 & 1 & 0 & 0 & 0 & 1 \end{array}\right]$, use

$2R_1 + (-1R_3) \to R_3$ and $\frac{1}{2}R_2 \to R_2$ to get

$$\left[\begin{array}{ccc|ccc} 1 & 0 & 1 & 1 & 0 & 0 \\ 0 & 1 & 1 & 0 & 1/2 & 0 \\ 0 & -1 & 2 & 2 & 0 & -1 \end{array}\right], \text{ use}$$

$R_2 + R_3 \to R_3$ to get

$$\left[\begin{array}{ccc|ccc} 1 & 0 & 1 & 1 & 0 & 0 \\ 0 & 1 & 1 & 0 & 1/2 & 0 \\ 0 & 0 & 3 & 2 & 1/2 & -1 \end{array}\right], \text{ use}$$

$\frac{1}{3}R_3 \to R_3$ to get

$$\left[\begin{array}{ccc|ccc} 1 & 0 & 1 & 1 & 0 & 0 \\ 0 & 1 & 1 & 0 & 1/2 & 0 \\ 0 & 0 & 1 & 2/3 & 1/6 & -1/3 \end{array}\right], \text{ use}$$

$-1R_3 + R_2 \to R_2$ and $-1R_3 + R_1 \to R_1$ to get

$$\left[\begin{array}{ccc|ccc} 1 & 0 & 0 & 1/3 & -1/6 & 1/3 \\ 0 & 1 & 0 & -2/3 & 1/3 & 1/3 \\ 0 & 0 & 1 & 2/3 & 1/6 & -1/3 \end{array}\right].$$

Then $A^{-1} = \begin{bmatrix} 1/3 & -1/6 & 1/3 \\ -2/3 & 1/3 & 1/3 \\ 2/3 & 1/6 & -1/3 \end{bmatrix}$.

43.

On $\left[\begin{array}{ccc|ccc} 0 & 4 & 2 & 1 & 0 & 0 \\ 0 & 3 & 2 & 0 & 1 & 0 \\ 1 & -1 & 1 & 0 & 0 & 1 \end{array}\right]$, use

$R_1 \leftrightarrow R_3$ to get

$$\left[\begin{array}{ccc|ccc} 1 & -1 & 1 & 0 & 0 & 1 \\ 0 & 3 & 2 & 0 & 1 & 0 \\ 0 & 4 & 2 & 1 & 0 & 0 \end{array}\right], \text{ use}$$

$-1R_2 + R_3 \to R_2$ to get

$$\left[\begin{array}{ccc|ccc} 1 & -1 & 1 & 0 & 0 & 1 \\ 0 & 1 & 0 & 1 & -1 & 0 \\ 0 & 4 & 2 & 1 & 0 & 0 \end{array}\right], \text{ use}$$

$R_1 + R_2 \to R_1$ to get

$$\left[\begin{array}{ccc|ccc} 1 & 0 & 1 & 1 & -1 & 1 \\ 0 & 1 & 0 & 1 & -1 & 0 \\ 0 & 4 & 2 & 1 & 0 & 0 \end{array}\right], \text{ use}$$

$-4R_2 + R_3 \to R_3$ to get

$$\left[\begin{array}{ccc|ccc} 1 & 0 & 1 & 1 & -1 & 1 \\ 0 & 1 & 0 & 1 & -1 & 0 \\ 0 & 0 & 2 & -3 & 4 & 0 \end{array}\right], \text{ use}$$

$\frac{1}{2}R_3 \to R_3$ to get

$$\left[\begin{array}{ccc|ccc} 1 & 0 & 1 & 1 & -1 & 1 \\ 0 & 1 & 0 & 1 & -1 & 0 \\ 0 & 0 & 1 & -3/2 & 2 & 0 \end{array}\right], \text{ use}$$

$-1R_3 + R_1 \to R_1$ to get

$$\left[\begin{array}{ccc|ccc} 1 & 0 & 0 & 5/2 & -3 & 1 \\ 0 & 1 & 0 & 1 & -1 & 0 \\ 0 & 0 & 1 & -3/2 & 2 & 0 \end{array}\right].$$

Then $A^{-1} = \begin{bmatrix} 5/2 & -3 & 1 \\ 1 & -1 & 0 \\ -3/2 & 2 & 0 \end{bmatrix}$.

45.

On $\left[\begin{array}{cccc|cccc} 1 & 2 & 3 & 4 & 1 & 0 & 0 & 0 \\ 0 & 1 & 2 & 3 & 0 & 1 & 0 & 0 \\ 0 & 0 & 1 & 2 & 0 & 0 & 1 & 0 \\ 0 & 0 & 0 & 1 & 0 & 0 & 0 & 1 \end{array}\right]$, use

$-2R_2 + R_1 \to R_1$ to get

$$\left[\begin{array}{cccc|cccc} 1 & 0 & -1 & -2 & 1 & -2 & 0 & 0 \\ 0 & 1 & 2 & 3 & 0 & 1 & 0 & 0 \\ 0 & 0 & 1 & 2 & 0 & 0 & 1 & 0 \\ 0 & 0 & 0 & 1 & 0 & 0 & 0 & 1 \end{array}\right], \text{ use}$$

$-2R_4 + R_3 \to R_3$ to get

$$\left[\begin{array}{cccc|cccc} 1 & 0 & -1 & -2 & 1 & -2 & 0 & 0 \\ 0 & 1 & 2 & 3 & 0 & 1 & 0 & 0 \\ 0 & 0 & 1 & 0 & 0 & 0 & 1 & -2 \\ 0 & 0 & 0 & 1 & 0 & 0 & 0 & 1 \end{array}\right], \text{ use}$$

$-2R_3 + R_2 \to R_2$ and $-2R_4 + R_3 \to R_3$ to get

$$\left[\begin{array}{cccc|cccc} 1 & 0 & -1 & -2 & 1 & -2 & 0 & 0 \\ 0 & 1 & 0 & 3 & 0 & 1 & -2 & 4 \\ 0 & 0 & 1 & -2 & 0 & 0 & 1 & -4 \\ 0 & 0 & 0 & 1 & 0 & 0 & 0 & 1 \end{array}\right], \text{ use}$$

$R_1 + R_3 \to R_1$ to get

$$\left[\begin{array}{cccc|cccc} 1 & 0 & 0 & -4 & 1 & -2 & 1 & -4 \\ 0 & 1 & 0 & 3 & 0 & 1 & -2 & 4 \\ 0 & 0 & 1 & -2 & 0 & 0 & 1 & -4 \\ 0 & 0 & 0 & 1 & 0 & 0 & 0 & 1 \end{array}\right], \text{ use}$$

$2R_4 + R_3 \to R_3$, $-3R_4 + R_2 \to R_2$ and $4R_4 + R_1 \to R_1$ to get

$$\left[\begin{array}{cccc|cccc} 1 & 0 & 0 & 0 & 1 & -2 & 1 & 0 \\ 0 & 1 & 0 & 0 & 0 & 1 & -2 & 1 \\ 0 & 0 & 1 & 0 & 0 & 0 & 1 & -2 \\ 0 & 0 & 0 & 1 & 0 & 0 & 0 & 1 \end{array}\right].$$

Then $A^{-1} = \begin{bmatrix} 1 & -2 & 1 & 0 \\ 0 & 1 & -2 & 1 \\ 0 & 0 & 1 & -2 \\ 0 & 0 & 0 & 1 \end{bmatrix}$.

47.

Since the coefficient matrix is $A = \begin{bmatrix} 1 & 6 \\ 1 & 9 \end{bmatrix}$,

$$A^{-1}\begin{bmatrix} -3 \\ -6 \end{bmatrix} = \begin{bmatrix} 3 & -2 \\ -1/3 & 1/3 \end{bmatrix}\begin{bmatrix} -3 \\ -6 \end{bmatrix} =$$

$= \begin{bmatrix} 3 \\ -1 \end{bmatrix}$. The solution set is $\{(3, -1)\}$.

49.

Since the coefficient matrix is $A = \begin{bmatrix} 1 & 6 \\ 1 & 9 \end{bmatrix}$,

$$A^{-1}\begin{bmatrix} 4 \\ 5 \end{bmatrix} = \begin{bmatrix} 3 & -2 \\ -1/3 & 1/3 \end{bmatrix}\begin{bmatrix} 4 \\ 5 \end{bmatrix} =$$

$= \begin{bmatrix} 2 \\ 1/3 \end{bmatrix}$. The solution set is $\{(2, 1/3)\}$.

51.

Since the coefficient matrix is

$A = \begin{bmatrix} -2 & -3 \\ 3 & 4 \end{bmatrix}$, we get

$$A^{-1}\begin{bmatrix} 1 \\ -1 \end{bmatrix} = \begin{bmatrix} 4 & 3 \\ -3 & -2 \end{bmatrix}\begin{bmatrix} 1 \\ -1 \end{bmatrix} =$$

$= \begin{bmatrix} 1 \\ -1 \end{bmatrix}$. The solution set is $\{(1, -1)\}$.

53.

Since the coefficient matrix is

$A = \begin{bmatrix} 1 & -5 \\ -1 & 3 \end{bmatrix}$, we obtain

$$A^{-1}\begin{bmatrix} -5 \\ 1 \end{bmatrix} = \begin{bmatrix} -3/2 & -5/2 \\ -1/2 & -1/2 \end{bmatrix}\begin{bmatrix} -5 \\ 1 \end{bmatrix} =$$

$= \begin{bmatrix} 5 \\ 2 \end{bmatrix}$. The solution set is $\{(5, 2)\}$.

55.

Since coefficient matrix is

$A = \begin{bmatrix} 1 & 1 & 1 \\ 1 & -1 & -1 \\ 1 & -1 & 1 \end{bmatrix}$, we find $A^{-1}\begin{bmatrix} 3 \\ -1 \\ 5 \end{bmatrix} =$

$$= \begin{bmatrix} 1/2 & 1/2 & 0 \\ 1/2 & 0 & -1/2 \\ 0 & -1/2 & 1/2 \end{bmatrix}\begin{bmatrix} 3 \\ -1 \\ 5 \end{bmatrix} = \begin{bmatrix} 1 \\ -1 \\ 3 \end{bmatrix}.$$

The solution set is $\{(1, -1, 3)\}$.

57.

Since the coefficient matrix is

$A = \begin{bmatrix} 0 & 2 & 0 \\ 3 & 3 & 2 \\ 2 & 5 & 1 \end{bmatrix}$, we get $A^{-1}\begin{bmatrix} 6 \\ 16 \\ 19 \end{bmatrix} =$

$$= \begin{bmatrix} -7/2 & -1 & 2 \\ 1/2 & 0 & 0 \\ 9/2 & 2 & -3 \end{bmatrix}\begin{bmatrix} 6 \\ 16 \\ 19 \end{bmatrix} = \begin{bmatrix} 1 \\ 3 \\ 2 \end{bmatrix}.$$

The solution set is $\{(1, 3, 2)\}$.

59. The coefficient matrix of

$$\begin{aligned} 0.3x + 0.1y &= 3 \\ 2x + 4y &= 7 \end{aligned}$$

is $A = \begin{bmatrix} 0.3 & 0.1 \\ 2 & 4 \end{bmatrix}$. Note $A^{-1}\begin{bmatrix} 3 \\ 7 \end{bmatrix} =$

$$\begin{bmatrix} 4 & -0.1 \\ -2 & 0.3 \end{bmatrix}\begin{bmatrix} 3 \\ 7 \end{bmatrix} = \begin{bmatrix} 11.3 \\ -3.9 \end{bmatrix}.$$

The solution set is $\{(11.3, -3.9)\}$.

61. Use the Gauss-Jordan method. On the augmented matrix $\left[\begin{array}{ccc|c} 1 & -1 & 1 & 5 \\ 2 & -1 & 3 & 1 \\ 0 & 1 & 1 & -9 \end{array}\right]$,

use $-1R_1 + R_2 \to R_1$ to get

$$\left[\begin{array}{ccc|c} 1 & 0 & 2 & -4 \\ 2 & -1 & 3 & 1 \\ 0 & 1 & 1 & -9 \end{array}\right], \text{ use}$$

$-2R_1 + R_2 \to R_2$ to get

$$\left[\begin{array}{rrr|r} 1 & 0 & 2 & -4 \\ 0 & -1 & -1 & 9 \\ 0 & 1 & 1 & -9 \end{array}\right], \text{ use}$$

$R_2 + R_3 \to R_3$ and $-1R_2 \to R_2$ to get

$$\left[\begin{array}{rrr|r} 1 & 0 & 2 & -4 \\ 0 & 1 & 1 & -9 \\ 0 & 0 & 0 & 0 \end{array}\right]. \text{ Since } y + z = -9$$

and $x + 2z = -4$, the solution set is

$\{(-2z - 4, -z - 9, z) \mid z \text{ is any real number }\}$.

63.

Note coefficient matrix is $A = \begin{bmatrix} 1 & 1 & 1 \\ 2 & 4 & 1 \\ 1 & 3 & 6 \end{bmatrix}$

and $A^{-1} \begin{bmatrix} 1 \\ 2 \\ 3 \end{bmatrix} =$

$$\begin{bmatrix} 7/4 & -1/4 & -1/4 \\ -11/12 & 5/12 & 1/12 \\ 1/6 & -1/6 & 1/6 \end{bmatrix} \begin{bmatrix} 1 \\ 2 \\ 3 \end{bmatrix} = \begin{bmatrix} 1/2 \\ 1/6 \\ 1/3 \end{bmatrix}.$$

The solution set is $\{(1/2, 1/6, 1/3)\}$.

65.

Note $A^{-1} = \begin{bmatrix} -55/6 & 5/2 & 5/3 \\ 35/12 & -5/4 & 5/6 \\ 40/3 & 0 & -10/3 \end{bmatrix}$

and $A^{-1} \begin{bmatrix} 27 \\ 9 \\ 16 \end{bmatrix} = \begin{bmatrix} -165 \\ 97.5 \\ 240 \end{bmatrix}$.

The solution set is

$$\{(-165, 97.5, 240)\}.$$

67.

We get $A^{-1} = \begin{bmatrix} 68/133 & -36/133 & 8/19 \\ 10/19 & -12/19 & 6/19 \\ 127/133 & -122/133 & 6/19 \end{bmatrix}$

and $A^{-1} \begin{bmatrix} 16 \\ 24 \\ -8 \end{bmatrix} \approx \begin{bmatrix} -1.6842 \\ -9.2632 \\ -9.2632 \end{bmatrix}$.

The solution set is approximately

$$\{(-1.6842, -9.2632, -9.2632)\}.$$

69.

Since $AA^{-1} = \begin{bmatrix} a & 7 \\ 3 & b \end{bmatrix} \begin{bmatrix} -b & 7 \\ 3 & -a \end{bmatrix} =$

$= \begin{bmatrix} 21 - ab & 0 \\ 0 & 21 - ab \end{bmatrix}$, $21 - ab = 1$ and

$ab = 20$. List of permissible pairs (a, b):

a	1	2	4	5	10	20
b	20	10	5	4	2	1

The matrices are $\begin{bmatrix} 1 & 7 \\ 3 & 20 \end{bmatrix}, \begin{bmatrix} 2 & 7 \\ 3 & 10 \end{bmatrix}$,

$\begin{bmatrix} 4 & 7 \\ 3 & 5 \end{bmatrix}, \begin{bmatrix} 5 & 7 \\ 3 & 4 \end{bmatrix}$,

$\begin{bmatrix} 10 & 7 \\ 3 & 2 \end{bmatrix}$, and $\begin{bmatrix} 20 & 7 \\ 3 & 1 \end{bmatrix}$.

71. Let x and y be the costs of a dozen eggs and a magazine before taxes. Then

$$\begin{aligned} 0.08x + 0.05y &= 0.59 \\ x + y &= 8.79 - 0.59. \end{aligned}$$

If $A = \begin{bmatrix} .08 & .05 \\ 1 & 1 \end{bmatrix}$, then $A^{-1} \begin{bmatrix} .59 \\ 8.20 \end{bmatrix} =$

$= \begin{bmatrix} 100/3 & -5/3 \\ -100/3 & 8/3 \end{bmatrix} \begin{bmatrix} .59 \\ 8.20 \end{bmatrix} = \begin{bmatrix} 6 \\ 2.20 \end{bmatrix}$.

The eggs cost \$2.20 a dozen and the magazine costs \$6.

73. Let x and y be the costs of one load of plywood and a load insulation, respectively. Then

$$\begin{aligned} 4x + 6y &= 2500 \\ 3x + 5y &= 1950. \end{aligned}$$

If $A = \begin{bmatrix} 4 & 6 \\ 3 & 5 \end{bmatrix}$ then $A^{-1} \begin{bmatrix} 2500 \\ 1950 \end{bmatrix} =$

$= \begin{bmatrix} 5/2 & -3 \\ -3/2 & 2 \end{bmatrix} \begin{bmatrix} 2500 \\ 1950 \end{bmatrix} = \begin{bmatrix} 400 \\ 150 \end{bmatrix}$.

One load of plywood costs \$400 and a load of insulation costs \$150 .

75.

We find $A^{-1} = \begin{bmatrix} 2 & -1 \\ -5 & 3 \end{bmatrix}$. To decode the message, we find

$$\begin{bmatrix} 2 & -1 \\ -5 & 3 \end{bmatrix}\begin{bmatrix} 36 \\ 65 \end{bmatrix} = \begin{bmatrix} 7 \\ 15 \end{bmatrix} = \begin{bmatrix} g \\ o \end{bmatrix},$$

$$\begin{bmatrix} 2 & -1 \\ -5 & 3 \end{bmatrix}\begin{bmatrix} 49 \\ 83 \end{bmatrix} = \begin{bmatrix} 15 \\ 4 \end{bmatrix} = \begin{bmatrix} o \\ d \end{bmatrix},$$

$$\begin{bmatrix} 2 & -1 \\ -5 & 3 \end{bmatrix}\begin{bmatrix} 12 \\ 24 \end{bmatrix} = \begin{bmatrix} 0 \\ 12 \end{bmatrix} = \begin{bmatrix} space \\ l \end{bmatrix},$$

$$\begin{bmatrix} 2 & -1 \\ -5 & 3 \end{bmatrix}\begin{bmatrix} 66 \\ 111 \end{bmatrix} = \begin{bmatrix} 21 \\ 3 \end{bmatrix} = \begin{bmatrix} u \\ c \end{bmatrix},$$

$$\begin{bmatrix} 2 & -1 \\ -5 & 3 \end{bmatrix}\begin{bmatrix} 33 \\ 55 \end{bmatrix} = \begin{bmatrix} 11 \\ 0 \end{bmatrix} = \begin{bmatrix} k \\ space \end{bmatrix}.$$

The message is 'Good luck'.

77. Let x and y be the amounts invested in the Asset Manager Fund and Magellan Fund, respectively. Since $0.86(60,000) = 51,600$, we get

$$\begin{aligned} x + y &= 60,000 \\ 0.76x + 0.90y &= 51,600. \end{aligned}$$

If $A = \begin{bmatrix} 1 & 1 \\ 0.76 & 0.90 \end{bmatrix}$, then

$A^{-1} = \begin{bmatrix} 45/7 & -50/7 \\ -38/7 & 50/7 \end{bmatrix}$ and

$A^{-1}\begin{bmatrix} 60,000 \\ 51,600 \end{bmatrix} = \begin{bmatrix} 17,142.86 \\ 42,857.14 \end{bmatrix}$. In the Asset Manager Fund the amount invested was \$17,142.86; in the Magellan Fund the amount invested was \$42,857.14.

79. Let x, y, z be the number of texts, number of minutes of talk time, and the monthly bill.

$$\begin{aligned} 32 + 0.04x + 0.10y &= z \\ 68 + 0.01x + 0.05y &= z \\ 26 + 0.10x + 0.09y &= z. \end{aligned}$$

From the first two equations, and from the last two equations, we obtain, respectively,

$$\begin{aligned} 36 - 0.03x - 0.05y &= 0 \\ 42 - 0.09x - 0.04y &= 0 \end{aligned}$$

The solutions to the above system are $x = 200$ texts, $y = 600$ minutes of talk time.

Substituting into the very first equation, we find $z = \$100$ is the monthly bill.

81. Let $x, y,$ and z be the prices of an animal totem, a trade-bead necklace, and a tribal mask, respectively. Then we obtain the system

$$\begin{aligned} 24x + 33y + 12z &= 202.23 \\ 19x + 40y + 22z &= 209.38 \\ 30x + 9y + 19z &= 167.66. \end{aligned}$$

The inverse of the coefficient matrix A is

$$A^{-1} = \frac{1}{11,007}\begin{bmatrix} 562 & -519 & 246 \\ 299 & 96 & -300 \\ -1029 & 774 & 333 \end{bmatrix}.$$

Since $A^{-1}\begin{bmatrix} 202.23 \\ 209.38 \\ 167.66 \end{bmatrix} \approx \begin{bmatrix} \$4.20 \\ \$2.75 \\ \$0.89 \end{bmatrix}$, we find animal totem costs \$4.20, a necklace costs \$2.75, and a tribal mask costs \$0.89.

83.

On $\left[\begin{array}{ccc|c} 1 & 1 & 2 & 9 \\ 1 & 2 & 3 & 12 \end{array}\right]$, apply $-R_1 + R_2 \to R_2$ to get

$$\left[\begin{array}{ccc|c} 1 & 1 & 2 & 9 \\ 0 & 1 & 1 & 3 \end{array}\right]$$

Use $-R_2 + R_1 \to R_1$ to get

$$\left[\begin{array}{ccc|c} 1 & 0 & 1 & 6 \\ 0 & 1 & 1 & 3 \end{array}\right]$$

Then the solution set is

$$\{(6 - z, 3 - z, z) : z \text{ is real}\}.$$

85.

$$\begin{bmatrix} 0 & 1 & 0 \\ 1 & 0 & 1 \end{bmatrix}\begin{bmatrix} 1 & 2 \\ 3 & 4 \\ 5 & 6 \end{bmatrix} =$$

$$\begin{bmatrix} 1(3) & 1(4) \\ 1(1) + 1(5) & 1(2) + 1(6) \end{bmatrix} =$$

$$\begin{bmatrix} 3 & 4 \\ 6 & 8 \end{bmatrix}$$

87. Use the method of completing the square.

$$\begin{aligned} x^2 - 4x &= -1 \\ x^2 - 4x + 4 &= 3 \\ (x-2)^2 &= 3 \\ x - 2 &= \pm\sqrt{3} \end{aligned}$$

The solution set is $\left\{2 \pm \sqrt{3}\right\}$.

89.

a) Substitute $x = 0, 1, 2, ..., 10$ into $x(10 - x)$. We find that the maximum product is 25. This maximum is achieved when $x = 5$. Thus, the two numbers are 5 and 5.

b) We enumerate all the possible whole numbers satisfying

$$1 \le x \le y \le z \le w \le 10$$

and $x + y + z + w = 10$.
We find that the maximum product for $xyzw$ is 36, and this happens when $x = y = 2$ and $z = w = 3$.

c) The maximum product is 729, and the numbers are six 3's.

For Thought

1. False, $|A| = 12 - (-5) = 17$.

2. True, for $|A| \neq 0$.

3. True, $|B| = 4 \cdot 5 - (-2)(-10) = 0$.

4. False, since $|B| = 0$.

5. True, since the determinant of the coefficient matrix A is nonzero.

6. True, in general $|LM| = |L||M|$ for any square matrices L and M of the same size.

7. False, the system is not linear. **8.** True

9. False, $\begin{vmatrix} 2 & 0.1 \\ 100 & 5 \end{vmatrix} = 2 \cdot 5 - 100(0.1) = 0$.

10. False, because a 2×2 matrix is not equal to the number 27.

9.5 Exercises

1. determinant

3.
$$\begin{vmatrix} 1 & 3 \\ 0 & 2 \end{vmatrix} = 1 \cdot 2 - 0 \cdot 3 = 2$$

5.
$$\begin{vmatrix} 3 & 4 \\ 2 & 9 \end{vmatrix} = 3(9) - 2(4) = 19$$

7.
$$\begin{vmatrix} -0.3 & -0.5 \\ -0.7 & 0.2 \end{vmatrix} = (-0.3)(0.2) - (-0.7)(-0.5)$$
$= -0.41$

9.
$$\begin{vmatrix} 1/8 & -3/8 \\ 2 & -1/4 \end{vmatrix} = (1/8)(-1/4) - (2)(-3/8)$$
$= -1/32 + 3/4 = 23/32$

11.
$$\begin{vmatrix} 0.02 & 0.4 \\ 1 & 20 \end{vmatrix} = (0.02)(20) - (0.4)(1) = 0$$

13.
$$\begin{vmatrix} 3 & -5 \\ -9 & 15 \end{vmatrix} = (3)(15) - (-9)(-5) = 0$$

15.

Since $\begin{vmatrix} a & 2 \\ 3 & 4 \end{vmatrix} = 4a - 6 = 10$, we find

$4a = 16$ or $a = 4$.

17.

Since $\begin{vmatrix} a & 8 \\ 2 & a \end{vmatrix} = a^2 - 16 = 0$, we find

$a^2 = 16$ or $a = \pm 4$.

19.

Note $D = \begin{vmatrix} 2 & 1 \\ 1 & 2 \end{vmatrix} = 3$, $D_x = \begin{vmatrix} 5 & 1 \\ 7 & 2 \end{vmatrix} = 3$,

and $D_y = \begin{vmatrix} 2 & 5 \\ 1 & 7 \end{vmatrix} = 9$.

Then $x = \dfrac{D_x}{D} = \dfrac{3}{3} = 1$ and $y = \dfrac{D_y}{D} = \dfrac{9}{3} = 3$.
Solution set is $\{(1, 3)\}$.

21.

Note $D = \begin{vmatrix} 1 & -2 \\ 1 & 2 \end{vmatrix} = 4$, $D_x = \begin{vmatrix} 7 & -2 \\ -5 & 2 \end{vmatrix} = 4$, and $D_y = \begin{vmatrix} 1 & 7 \\ 1 & -5 \end{vmatrix} = -12$.

So $x = \dfrac{D_x}{D} = \dfrac{4}{4} = 1$ and $y = \dfrac{D_y}{D} = \dfrac{-12}{4} = -3$. The solution set is $\{(1, -3)\}$.

23.

Note $D = \begin{vmatrix} 2 & -1 \\ 1 & 3 \end{vmatrix} = 7$, $D_x = \begin{vmatrix} -11 & -1 \\ 12 & 3 \end{vmatrix} = -21$, and $D_y = \begin{vmatrix} 2 & -11 \\ 1 & 12 \end{vmatrix} = 35$. Then

$x = \dfrac{D_x}{D} = -\dfrac{21}{7} = -3$ and $y = \dfrac{D_y}{D} = \dfrac{35}{7} = 5$.

Solution set is $\{(-3, 5)\}$.

25. Rewrite system as

$$\begin{aligned} x - y &= 6 \\ x + y &= 5. \end{aligned}$$

Note $D = \begin{vmatrix} 1 & -1 \\ 1 & 1 \end{vmatrix} = 2$, $D_x = \begin{vmatrix} 6 & -1 \\ 5 & 1 \end{vmatrix} = 11$, and $D_y = \begin{vmatrix} 1 & 6 \\ 1 & 5 \end{vmatrix} = -1$.

So $x = \dfrac{D_x}{D} = \dfrac{11}{2}$ and $y = \dfrac{D_y}{D} = -\dfrac{1}{2}$.

Solution set is $\{(11/2, -1/2)\}$.

27.

Note $D = \begin{vmatrix} 1/2 & -1/3 \\ 1/4 & 1/2 \end{vmatrix} = 1/3$,

$D_x = \begin{vmatrix} 4 & -1/3 \\ 6 & 1/2 \end{vmatrix} = 4$, and

$D_y = \begin{vmatrix} 1/2 & 4 \\ 1/4 & 6 \end{vmatrix} = 2$. So $x = \dfrac{D_x}{D} = \dfrac{4}{1/3} = 12$ and $y = \dfrac{D_y}{D} = \dfrac{2}{1/3} = 6$.

Solution set is $\{(12, 6)\}$.

29.

Note $D = \begin{vmatrix} 0.2 & 0.12 \\ 1 & 1 \end{vmatrix} = 0.08$,

$D_x = \begin{vmatrix} 148 & 0.12 \\ 900 & 1 \end{vmatrix} = 40$, and

$D_y = \begin{vmatrix} 0.2 & 148 \\ 1 & 900 \end{vmatrix} = 32$. Then

$x = \dfrac{D_x}{D} = \dfrac{40}{0.08} = 500$ and

$y = \dfrac{D_y}{D} = \dfrac{32}{0.08} = 400$.

Solution set is $\{(500, 400)\}$.

31. Cramer's rule does not apply since

$D = \begin{vmatrix} 3 & 1 \\ -6 & -2 \end{vmatrix} = 0$. Dividing the second equation by -2, one gets the first equation. Solution set is $\{(x, y) \mid 3x + y = 6\}$.

33. Cramer's Rule does not apply since the determinant D is zero.

Adding the two equations, one gets $0 = 19$. Inconsistent and the solution set is $\emptyset$.

35. We use Cramer's Rule on the system

$$\begin{aligned} x - y &= 3 \\ 3x - y &= -9. \end{aligned}$$

Note, $D = \begin{vmatrix} 1 & -1 \\ 3 & -1 \end{vmatrix} = 2$,

$D_x = \begin{vmatrix} 3 & -1 \\ -9 & -1 \end{vmatrix} = -12$, and

$D_y = \begin{vmatrix} 1 & 3 \\ 3 & -9 \end{vmatrix} = -18$.

So $x = \dfrac{D_x}{D} = \dfrac{-12}{2} = -6$ and

$y = \dfrac{D_y}{D} = \dfrac{-18}{2} = -9$.

Solution set is $\{(-6, -9)\}$.

37. Note, $D = \begin{vmatrix} \sqrt{2} & \sqrt{3} \\ 3\sqrt{2} & -2\sqrt{3} \end{vmatrix} = -5\sqrt{6}$,

$D_x = \begin{vmatrix} 4 & \sqrt{3} \\ -3 & -2\sqrt{3} \end{vmatrix} = -5\sqrt{3}$, and

$D_y = \begin{vmatrix} \sqrt{2} & 4 \\ 3\sqrt{2} & -3 \end{vmatrix} = -15\sqrt{2}$.

So $x = \dfrac{D_x}{D} = \dfrac{-5\sqrt{3}}{-5\sqrt{6}} = \dfrac{1}{\sqrt{2}} = \dfrac{\sqrt{2}}{2}$

and $y = \dfrac{D_y}{D} = \dfrac{-15\sqrt{2}}{-5\sqrt{6}} = \dfrac{3}{\sqrt{3}} = \sqrt{3}$.

The solution set is $\{(\sqrt{2}/2, \sqrt{3})\}$.

39. Multiply second equation by -1 and add to the first one.

$$\begin{array}{rcl} x^2 + y^2 & = & 25 \\ -x^2 + y & = & -5 \\ \hline y^2 + y & = & 20 \\ y^2 + y - 20 & = & 0 \\ (y+5)(y-4) & = & 0 \end{array}$$

If $y = -5$, then $x^2 = 0$ and $x = 0$.
If $y = 4$, then $x^2 = 9$ and $x = \pm 3$.
The solution set is $\{(0, -5), (\pm 3, 4)\}$.

41. Solving for x in the second equation, we find $x = 4 + 2y$. Substituting into the first equation, we obtain

$$\begin{array}{rcl} 4 + 2y - 2y & = & y^2 \\ 4 & = & y^2 \\ \pm 2 & = & y. \end{array}$$

Using $y = 2$ in $x = 4 + 2y$, we get $x = 8$.
Similarly, if $y = -2$ then $x = 0$.
Solution set is $\{(8, 2), (0, -2)\}$.

43.

Invertible, since $\begin{vmatrix} 4 & 0.5 \\ 2 & 3 \end{vmatrix} = 12 - 1 = 11 \neq 0$

45.

Not invertible, for $\begin{vmatrix} 3 & -4 \\ 9 & -12 \end{vmatrix} = -36 + 36 = 0$

47. Note, $D = \begin{vmatrix} 3.47 & 23.09 \\ 12.48 & 3.98 \end{vmatrix} = -274.3526$,

$D_x = \begin{vmatrix} 5978.95 & 23.09 \\ 2765.34 & 3.98 \end{vmatrix} = -40{,}055.4796$,

and

$D_y = \begin{vmatrix} 3.47 & 5978.95 \\ 12.48 & 2765.34 \end{vmatrix} = -65{,}021.5662$.

Then $x = \dfrac{D_x}{D} = 146$ and $y = \dfrac{D_y}{D} = 237$.

The solution set is $\{(146, 237)\}$.

49. Let x and y be the number of boys and girls, respectively. Then

$$\begin{array}{rcl} 0.44x + 0.35y & = & 231 \\ x + y & = & 615. \end{array}$$

Note, $D = \begin{vmatrix} 0.44 & 0.35 \\ 1 & 1 \end{vmatrix} = 0.09$,

$D_x = \begin{vmatrix} 231 & 0.35 \\ 615 & 1 \end{vmatrix} = 15.75$, and

$D_y = \begin{vmatrix} 0.44 & 231 \\ 1 & 615 \end{vmatrix} = 39.6$.

There were $x = \dfrac{D_x}{D} = \dfrac{15.75}{0.09} = 175$ boys

and $y = \dfrac{D_y}{D} = \dfrac{39.6}{0.09} = 440$ girls.

51. Let x and y be the measurements of the two acute angles. Then we obtain

$$\begin{array}{rcl} x + y & = & 90 \\ x - 2y & = & 1. \end{array}$$

Note, $D = \begin{vmatrix} 1 & 1 \\ 1 & -2 \end{vmatrix} = -3$,

$D_x = \begin{vmatrix} 90 & 1 \\ 1 & -2 \end{vmatrix} = -181$, and

$D_y = \begin{vmatrix} 1 & 90 \\ 1 & 1 \end{vmatrix} = -89$.

The acute angles are $x = \dfrac{D_x}{D} = \dfrac{181}{3}$ degrees

and $y = \dfrac{D_y}{D} = \dfrac{89}{3}$ degrees.

53. Let x and y be the salaries of the president and vice-president, respectively. Then we have

$$\begin{aligned} x + y &= 400,000 \\ x - y &= 100,000. \end{aligned}$$

Note, $D = \begin{vmatrix} 1 & 1 \\ 1 & -1 \end{vmatrix} = -2$,

$D_x = \begin{vmatrix} 400,000 & 1 \\ 100,000 & -1 \end{vmatrix} = -500,000$, and

$D_y = \begin{vmatrix} 1 & 400,000 \\ 1 & 100,000 \end{vmatrix} = -300,000$.

Thus, $x = \dfrac{D_x}{D} = \dfrac{-500,000}{-2} = 250,000$ and

$y = \dfrac{D_y}{D} = \dfrac{-300,000}{-2} = 150,000.$

The president's salary is \$250,000 and the vice-president's salary is \$150,000.

55. Yes, $|MN| = |M||N|$ since $|M| = 2$, $|N| = 3$, and $|MN| = \begin{vmatrix} 8 & 31 \\ 14 & 55 \end{vmatrix} = 6.$

57.

$$\left| \begin{bmatrix} a & b \\ c & d \end{bmatrix} \begin{bmatrix} e & f \\ g & h \end{bmatrix} \right| =$$

$$\begin{vmatrix} ae + bg & af + bh \\ ce + dg & cf + dh \end{vmatrix} =$$

$(ae + bg)(cf + dh) - (ce + dg)(af + bh) =$
$aecf + bgcf + aedh + bgdh - ceaf - dgaf -$
$cebh - dgbh = bgcf + aedh - dgaf - cebh =$
$ad(eh-gf)-bc(eh-gf) = (ad-bc)(eh-gf) =$

$$= \begin{vmatrix} a & b \\ c & d \end{vmatrix} \begin{vmatrix} e & f \\ g & h \end{vmatrix}$$

59.

No, since $|-2M| = \begin{vmatrix} -6 & -4 \\ -10 & -8 \end{vmatrix} = 8$

and $-2|M| = -4$.

61.

On $\left[\begin{array}{cc|cc} 1 & 1 & 1 & 0 \\ 1 & 3 & 0 & 1 \end{array}\right]$, use $-R_1 + R_2 \to R_2$

to get

$\left[\begin{array}{cc|cc} 1 & 1 & 1 & 0 \\ 0 & 2 & -1 & 1 \end{array}\right]$, use $\frac{1}{2}R_2 \to R_2$

to get

$\left[\begin{array}{cc|cc} 1 & 1 & 1 & 0 \\ 0 & 1 & -\frac{1}{2} & \frac{1}{2} \end{array}\right]$, use $-R_2 + R_1 \to R_1$

to get

$\left[\begin{array}{cc|cc} 1 & 0 & \frac{3}{2} & -\frac{1}{2} \\ 0 & 1 & -\frac{1}{2} & \frac{1}{2} \end{array}\right]$.

Then $A^{-1} = \begin{bmatrix} 3/2 & -1/2 \\ -1/2 & 1/2 \end{bmatrix}$.

63. If we add the first two equations, the sum is $2x + 3y = 15$. Note, the third equation is similar to the sum but the right side is different, i.e., $2x + 36 = 10$.

Thus, there are no solutions. The solution set is the empty set $\emptyset$.

65. Since $0.91x = 72,800$, we find

$$x = \frac{72,800}{0.91} = 80,000.$$

The solution set is $\{80,000\}$.

67. Take out one more of type A. Cut them in half one at a time and take half of each pill. Take the other halves the next day.

For Thought

1. False, the sign array of A is used in evaluating $|A|$.

2. False, the last term should be $1 \cdot \begin{vmatrix} 3 & 4 \\ 0 & 0 \end{vmatrix}$.

3. True

4. True, $|A|$ was expanded about the third row.

5. False, $|A|$ can be expanded only about a row or column. **6.** False, a minor is a 2×2 matrix only if it comes from a 3×3 matrix.

7. False, $x = \dfrac{D_x}{D}$. **8.** True

9. False, it can happen that $D = 0$ and there are infinitely many solutions.

10. False, Cramer's Rule applies only to a system of linear equations.

9.6 Exercises

1. minor

3.
$$\begin{vmatrix} 5 & -6 \\ 9 & -8 \end{vmatrix} = -40 + 54 = 14$$

5.
$$\begin{vmatrix} 4 & 5 \\ 7 & 9 \end{vmatrix} = 36 - 35 = 1$$

7.
$$\begin{vmatrix} 2 & 1 \\ 7 & -8 \end{vmatrix} = -16 - 7 = -23$$

9.
$$\begin{vmatrix} 2 & 1 \\ 4 & -6 \end{vmatrix} = -12 - 4 = -16$$

11.
$$1\begin{vmatrix} 1 & -2 \\ -1 & 5 \end{vmatrix} - (-3)\begin{vmatrix} -4 & 0 \\ -1 & 5 \end{vmatrix} + 3\begin{vmatrix} -4 & 0 \\ 1 & -2 \end{vmatrix}$$
$$= 1(3) - (-3)(-20) + 3(8) = -33$$

13.
$$3\begin{vmatrix} 4 & -1 \\ 1 & -2 \end{vmatrix} - 0\begin{vmatrix} -1 & 2 \\ 1 & -2 \end{vmatrix} + 5\begin{vmatrix} -1 & 2 \\ 4 & -1 \end{vmatrix}$$
$$= 3(-7) - 0 + 5(-7) = -56$$

15.
$$-2\begin{vmatrix} 0 & -1 \\ 2 & -7 \end{vmatrix} - (-3)\begin{vmatrix} 5 & 1 \\ 2 & -7 \end{vmatrix} + 0\begin{vmatrix} 5 & 1 \\ 0 & -1 \end{vmatrix}$$
$$= -2(2) - (-3)(-37) + 0 = -115$$

17.
$$0.1\begin{vmatrix} 20 & 6 \\ 90 & 8 \end{vmatrix} - 0.4\begin{vmatrix} 30 & 1 \\ 90 & 8 \end{vmatrix} + 0.7\begin{vmatrix} 30 & 1 \\ 20 & 6 \end{vmatrix}$$
$$= 0.1(-380) - 0.4(150) + 0.7(160) = 14$$

19. Expanding about the second row,
$$D = -(-2)\begin{vmatrix} 3 & 5 \\ 3 & -4 \end{vmatrix} = -(-2)(-27) = -54.$$

21. Expanding about the first row,
$$D = 1\begin{vmatrix} 2 & 2 \\ 4 & 4 \end{vmatrix} - 1\begin{vmatrix} 2 & 2 \\ 4 & 4 \end{vmatrix} + 1\begin{vmatrix} 2 & 2 \\ 4 & 4 \end{vmatrix}$$
$$= 1(0) - 1(0) + 1(0) = 0.$$

23. Expanding about the first row,
$$D = -(-1)\begin{vmatrix} 3 & 6 \\ -2 & -5 \end{vmatrix} = -(-1)(-3) = -3.$$

25. Expanding about the second column,
$$D = -9\begin{vmatrix} 2 & 1 \\ 4 & 6 \end{vmatrix} = -9(8) = -72.$$

27. Expanding about the first row,
$$3\begin{vmatrix} -3 & 2 & 0 \\ 3 & 1 & 2 \\ -4 & 1 & 3 \end{vmatrix} + 0 + 1\begin{vmatrix} 2 & -3 & 0 \\ -2 & 3 & 2 \\ 2 & -4 & 3 \end{vmatrix} -$$
$$5\begin{vmatrix} 2 & -3 & 2 \\ -2 & 3 & 1 \\ 2 & -4 & 1 \end{vmatrix} =$$
$$= 3(-37) + 1(4) - 5(6) = -137.$$

29. Expand the determinant about the second row.
$$-(1)\begin{vmatrix} -3 & 4 & 6 \\ 3 & 1 & -3 \\ 0 & 2 & 1 \end{vmatrix} + (-5)\begin{vmatrix} 2 & 4 & 6 \\ 1 & 1 & -3 \\ -2 & 2 & 1 \end{vmatrix} =$$
$$= -(1)(3) + (-5)(58) = -293.$$

31.

One finds $D = \begin{vmatrix} 1 & 1 & 1 \\ 1 & -1 & 1 \\ 2 & 1 & 1 \end{vmatrix} = 2,$

$D_x = \begin{vmatrix} 6 & 1 & 1 \\ 2 & -1 & 1 \\ 7 & 1 & 1 \end{vmatrix} = 2,\ D_y = \begin{vmatrix} 1 & 6 & 1 \\ 1 & 2 & 1 \\ 2 & 7 & 1 \end{vmatrix} =$

4, and $D_z = \begin{vmatrix} 1 & 1 & 6 \\ 1 & -1 & 2 \\ 2 & 1 & 7 \end{vmatrix} = 6.$

Since $x = \dfrac{D_x}{D} = 2/2 = 1,\ y = \dfrac{D_y}{D} = 4/2 = 2,$

and $z = \dfrac{D_z}{D} = 6/2 = 3,$

the solution set is $\{(1, 2, 3)\}$.

33.

One finds $D = \begin{vmatrix} 1 & 2 & 0 \\ 1 & -3 & 1 \\ 2 & -1 & 0 \end{vmatrix} = 5,$

$D_x = \begin{vmatrix} 8 & 2 & 0 \\ -2 & -3 & 1 \\ 1 & -1 & 0 \end{vmatrix} = 10,$

$$D_y = \begin{vmatrix} 1 & 8 & 0 \\ 1 & -2 & 1 \\ 2 & 1 & 0 \end{vmatrix} = 15,$$

and $D_z = \begin{vmatrix} 1 & 2 & 8 \\ 1 & -3 & -2 \\ 2 & -1 & 1 \end{vmatrix} = 25.$

Since $x = \dfrac{D_x}{D} = 10/5 = 2,$

$y = \dfrac{D_y}{D} = 15/5 = 3,$

and $z = \dfrac{D_z}{D} = 25/5 = 5,$

the solution set is $\{(2, 3, 5)\}$.

35. One finds $D = \begin{vmatrix} 2 & -3 & 1 \\ 1 & 4 & -1 \\ 3 & -1 & 2 \end{vmatrix} = 16,$

$$D_x = \begin{vmatrix} 1 & -3 & 1 \\ 0 & 4 & -1 \\ 0 & -1 & 2 \end{vmatrix} = 7,$$

$$D_y = \begin{vmatrix} 2 & 1 & 1 \\ 1 & 0 & -1 \\ 3 & 0 & 2 \end{vmatrix} = -5,$$

and $D_z = \begin{vmatrix} 2 & -3 & 1 \\ 1 & 4 & 0 \\ 3 & -1 & 0 \end{vmatrix} = -13.$

Then $x = \dfrac{D_x}{D} = 7/16,$

$y = \dfrac{D_y}{D} = -5/16,$ and $z = \dfrac{D_z}{D} = -13/16.$

The solution set is $\{(7/16, -5/16, -13/16)\}$.

37. One finds $D = \begin{vmatrix} 1 & 1 & 1 \\ 2 & -1 & 3 \\ 3 & 1 & -1 \end{vmatrix} = 14,$

$$D_x = \begin{vmatrix} 2 & 1 & 1 \\ 0 & -1 & 3 \\ 0 & 1 & -1 \end{vmatrix} = -4,$$

$$D_y = \begin{vmatrix} 1 & 2 & 1 \\ 2 & 0 & 3 \\ 3 & 0 & -1 \end{vmatrix} = 22,$$

and $D_z = \begin{vmatrix} 1 & 1 & 2 \\ 2 & -1 & 0 \\ 3 & 1 & 0 \end{vmatrix} = 10.$

Then $x = \dfrac{D_x}{D} = -4/14 = -2/7,$

$y = \dfrac{D_y}{D} = 22/14 = 11/7,$ and

$z = \dfrac{D_z}{D} = 10/14 = 5/7.$

The solution set is $\{(-2/7, 11/7, 5/7)\}$.

39. This system is dependent since the sum of the first two equations is the third equation. Adding the first and third equations, we get $3x - 3z = 4$ or $z = x - 4/3$. Substituting into the first equation, we find

$$\begin{aligned} x + y - 2\left(x - \frac{4}{3}\right) &= 1 \\ y &= 1 - \frac{8}{3} + x \\ y &= x - \frac{5}{3}. \end{aligned}$$

Solution set is

$$\left\{\left(x, x - \frac{5}{3}, x - \frac{4}{3}\right) \mid x \text{ is any real number}\right\}.$$

41. Multiply the first equation by -2 and add to the third equation.

$$\begin{aligned} -2x + 2y - 2z &= -10 \\ 2x - 2y + 2z &= 16 \\ \hline 0 &= 6 \end{aligned}$$

Inconsistent and the solution set is $\emptyset$.

43. Let $x, y,$ and z be the ages of Jackie, Rochelle, and Alisha, respectively. Since $\dfrac{x+y}{2} = 33,$ $\dfrac{y+z}{2} = 25,$ and $\dfrac{x+z}{2} = 19,$ we obtain

$$\begin{aligned} x + y &= 66 \\ y + z &= 50 \\ x + z &= 38. \end{aligned}$$

One finds $D = \begin{vmatrix} 1 & 1 & 0 \\ 0 & 1 & 1 \\ 1 & 0 & 1 \end{vmatrix} = 2,$

$D_x = \begin{vmatrix} 66 & 1 & 0 \\ 50 & 1 & 1 \\ 38 & 0 & 1 \end{vmatrix} = 54,$ $D_y = \begin{vmatrix} 1 & 66 & 0 \\ 0 & 50 & 1 \\ 1 & 38 & 1 \end{vmatrix} =$

78, and $D_z = \begin{vmatrix} 1 & 1 & 66 \\ 0 & 1 & 50 \\ 1 & 0 & 38 \end{vmatrix} = 22.$

So Jackie is $x = \dfrac{D_x}{D} = 54/2 = 27$ years old,

Rochelle is $y = \dfrac{D_y}{D} = 78/2 = 39$ years old,

and Alisha is $z = \dfrac{D_z}{D} = 22/2 = 11$ years old.

45. Let x, y, and z be the scores in the first test, second test, and final exam, respectively. Then

$$\begin{aligned} x + y + z &= 180 \\ 0.2x + 0.2y + 0.6z &= 76 \\ 0.1x + 0.2y + 0.7z &= 83. \end{aligned}$$

One finds $D = \begin{vmatrix} 1 & 1 & 1 \\ 0.2 & 0.2 & 0.6 \\ 0.1 & 0.2 & 0.7 \end{vmatrix} = -0.04,$

$$D_x = \begin{vmatrix} 180 & 1 & 1 \\ 76 & 0.2 & 0.6 \\ 83 & 0.2 & 0.7 \end{vmatrix} = -1.2,$$

$$D_y = \begin{vmatrix} 1 & 180 & 1 \\ 0.2 & 76 & 0.6 \\ 0.1 & 83 & 0.7 \end{vmatrix} = -2,$$

and $D_z = \begin{vmatrix} 1 & 1 & 180 \\ 0.2 & 0.2 & 76 \\ 0.1 & 0.2 & 83 \end{vmatrix} = -4.$

Test scores are $x = \dfrac{D_x}{D} = 1.2/(0.04) = 30,$

$y = \dfrac{D_y}{D} = 2/(0.04) = 50$, and

the final exam is $z = \dfrac{D_z}{D} = 4/(0.04) = 100.$

47. One finds $D = \begin{vmatrix} 0.2 & -0.3 & 1.2 \\ 0.25 & 0.35 & -0.9 \\ 2.4 & -1 & 1.25 \end{vmatrix} = -\dfrac{527}{800},$

$$D_x = \begin{vmatrix} 13.11 & -0.3 & 1.2 \\ -1.575 & 0.35 & -0.9 \\ 42.02 & -1 & 1.25 \end{vmatrix} = -\frac{11,067}{1000},$$

$$D_y = \begin{vmatrix} 0.2 & 13.11 & 1.2 \\ 0.25 & -1.575 & -0.9 \\ 2.4 & 42.02 & 1.25 \end{vmatrix} = -\frac{64,821}{8000},$$

and $D_z = \begin{vmatrix} 0.2 & -0.3 & 13.11 \\ 0.25 & 0.35 & -1.575 \\ 2.4 & -1 & 42.02 \end{vmatrix} = -\dfrac{3689}{500}.$

Note, $\dfrac{D_x}{D} = 16.8$, $\dfrac{D_y}{D} = 12.3$, and $\dfrac{D_z}{D} = 11.2$. The solution set is $\{(16.8, 12.3, 11.2)\}$.

49. Let x, y, and z be the prices per gallon of regular, plus, and supreme gasoline, respectively. One finds

$$D = \begin{vmatrix} 1270 & 980 & 890 \\ 1450 & 1280 & 1050 \\ 1340 & 1190 & 1060 \end{vmatrix} =$$

18,038,000,

$$D_x = \begin{vmatrix} 12,204.86 & 980 & 890 \\ 14,698.22 & 1280 & 1050 \\ 13,969.41 & 1190 & 1060 \end{vmatrix} =$$

68,526,362,

$$D_y = \begin{vmatrix} 1270 & 12,204.86 & 890 \\ 1450 & 14,698.22 & 1050 \\ 1340 & 13,969.41 & 1060 \end{vmatrix} =$$

70,330,162,

$$D_z = \begin{vmatrix} 1270 & 980 & 12,204.86 \\ 1450 & 1280 & 14,698.22 \\ 1340 & 1190 & 13,969.41 \end{vmatrix} =$$

72,133,962.

Regular gas costs $\dfrac{D_x}{D} \approx \$3.799$,

plus costs $\dfrac{D_y}{D} \approx \$3.899$, and

supreme costs $\dfrac{D_z}{D} \approx \$3.999$ per gallon.

51.

One finds $D = \begin{vmatrix} 1 & 1 & 1 & 1 \\ 2 & -1 & 1 & 3 \\ 1 & 2 & -1 & 2 \\ 1 & -1 & -1 & 4 \end{vmatrix} = 11,$

$$D_w = \begin{vmatrix} 4 & 1 & 1 & 1 \\ 13 & -1 & 1 & 3 \\ -2 & 2 & -1 & 2 \\ 8 & -1 & -1 & 4 \end{vmatrix} = 11,$$

$$D_x = \begin{vmatrix} 1 & 4 & 1 & 1 \\ 2 & 13 & 1 & 3 \\ 1 & -2 & -1 & 2 \\ 1 & 8 & -1 & 4 \end{vmatrix} = -22,$$

$$D_y = \begin{vmatrix} 1 & 1 & 4 & 1 \\ 2 & -1 & 13 & 3 \\ 1 & 2 & -2 & 2 \\ 1 & -1 & 8 & 4 \end{vmatrix} = 33,$$

and $D_z = \begin{vmatrix} 1 & 1 & 1 & 4 \\ 2 & -1 & 1 & 13 \\ 1 & 2 & -1 & -2 \\ 1 & -1 & -1 & 8 \end{vmatrix} = 22.$

Note, $\frac{D_w}{D} = 1$, $\frac{D_x}{D} = -2$, $\frac{D_y}{D} = 3$, and $\frac{D_z}{D} = 2$. The solution set is $\{(1, -2, 3, 2)\}$.

53.

Note, $\begin{vmatrix} x & y & 1 \\ 3 & -5 & 1 \\ -2 & 6 & 1 \end{vmatrix} = 8 - 11x - 5y = 0.$

Since both points $(3, -5)$ and $(-2, 6)$ satisfies $8 - 11x - 5y = 0$, this is an equation of the line through the two points.

55. We will show it for one particular case and the other cases can proved similarly.

Let us suppose $A = \begin{bmatrix} a & b & c \\ 0 & 0 & 0 \\ g & h & i \end{bmatrix}$.

Then $|A| = a \cdot 0 \cdot i + b \cdot 0 \cdot g + c \cdot 0 \cdot h - g \cdot 0 \cdot c - h \cdot 0 \cdot a - i \cdot 0 \cdot b = 0$

57. We will prove it for one particular case and the other cases can be shown similarly.

Let us suppose $A = \begin{bmatrix} a & b & c \\ d & e & f \\ g & h & i \end{bmatrix}$ and

$B = \begin{bmatrix} a & b & c \\ d & e & f \\ kg & kh & ki \end{bmatrix}$. Then $|B| =$

$aeki + bfkg + cdkh - kgec - khfa - kidb = k(aei + bfg + cdh - gec - hfa - idb) = k|A|$.

59.

$$\begin{vmatrix} 3 & -2 \\ -1 & 4 \end{vmatrix} = 3(4) - (-1)(-2) = 10$$

61. For each value of z, the system below

$$\begin{aligned} 5x - 9y &= 44 - 11z \\ 13x - 12y &= -9 - 31z \end{aligned}$$

has unique solution since the coefficient matrix $\begin{bmatrix} 5 & -9 \\ 13 & -12 \end{bmatrix}$ has a nonzero determinant. Then the given system of three equations is not independent since the solution is not unique. Hence, in addition since the system is not inconsistent, it follows by elimination that the system is dependent.

63. If $(x - 4)^2 = x^2$, then $x - 4 = \pm x$. Since $x - 4 = x$ is inconsistent, we have $x - 4 = -x$ or $2x = 4$. Then $x = 2$. Thus, $y = x^2 = 2^2 = 4$. The solution set is $\{(2, 4)\}$.

65. It makes no sense to do this computation. No one is holding \$29. Rather, three times \$9 (i.e., total paid by the three boys) is the same as \$25 (rent) plus \$2 (bellboy).

Review Exercises

1.

$$\begin{bmatrix} 2+3 & -3+7 \\ -2+1 & 4+2 \end{bmatrix} = \begin{bmatrix} 5 & 4 \\ -1 & 6 \end{bmatrix}$$

3.

$$\begin{bmatrix} 4 & -6 \\ -4 & 8 \end{bmatrix} - \begin{bmatrix} 3 & 7 \\ 1 & 2 \end{bmatrix} = \begin{bmatrix} 1 & -13 \\ -5 & 6 \end{bmatrix}$$

5.

$$AB = \begin{bmatrix} 6-3 & 14-6 \\ -6+4 & -14+8 \end{bmatrix} = \begin{bmatrix} 3 & 8 \\ -2 & -6 \end{bmatrix}$$

7. $D + E$ is undefined

9.

$$AC = \begin{bmatrix} -2-9 \\ 2+12 \end{bmatrix} = \begin{bmatrix} -11 \\ 14 \end{bmatrix}$$

11.

$$EF = \begin{bmatrix} 3 & 2 & -1 \\ -12 & -8 & 4 \\ 9 & 6 & -3 \end{bmatrix}$$

13. $FG = [-3+2+2 \quad 2-3 \quad -1] = [1 \quad -1 \quad -1]$

15. GF is undefined

17.

On $\left[\begin{array}{rr|rr} 2 & -3 & 1 & 0 \\ -2 & 4 & 0 & 1 \end{array}\right]$, use $R_1 + R_2 \to R_2$ to get

$\left[\begin{array}{rr|rr} 2 & -3 & 1 & 0 \\ 0 & 1 & 1 & 1 \end{array}\right]$, use $3R_2 + R_1 \to R_1$ to get

$\left[\begin{array}{rr|rr} 2 & 0 & 4 & 3 \\ 0 & 1 & 1 & 1 \end{array}\right]$, use $\frac{1}{2}R_1 \to R_1$ to get

$\left[\begin{array}{rr|rr} 1 & 0 & 2 & 1.5 \\ 0 & 1 & 1 & 1 \end{array}\right]$. So $A^{-1} = \left[\begin{array}{rr} 2 & 1.5 \\ 1 & 1 \end{array}\right]$.

19.

On $\left[\begin{array}{rrr|rrr} -1 & 0 & 0 & 1 & 0 & 0 \\ 1 & 1 & 0 & 0 & 1 & 0 \\ -2 & 3 & 1 & 0 & 0 & 1 \end{array}\right]$,

use $2R_2 + R_3 \to R_3$, $R_1 + R_2 \to R_2$, and $-1R_1 \to R_1$ to get

$\left[\begin{array}{rrr|rrr} 1 & 0 & 0 & -1 & 0 & 0 \\ 0 & 1 & 0 & 1 & 1 & 0 \\ 0 & 5 & 1 & 0 & 2 & 1 \end{array}\right]$,

use $-5R_2 + R_3 \to R_3$ to get

$\left[\begin{array}{rrr|rrr} 1 & 0 & 0 & -1 & 0 & 0 \\ 0 & 1 & 0 & 1 & 1 & 0 \\ 0 & 0 & 1 & -5 & -3 & 1 \end{array}\right]$.

So $G^{-1} = \left[\begin{array}{rrr} -1 & 0 & 0 \\ 1 & 1 & 0 \\ -5 & -3 & 1 \end{array}\right]$.

21.

From Exercise 5, $(AB)^{-1} = \left[\begin{array}{rr} 3 & 8 \\ -2 & -6 \end{array}\right]^{-1}$.

On $\left[\begin{array}{rr|rr} 3 & 8 & 1 & 0 \\ -2 & -6 & 0 & 1 \end{array}\right]$, use $-\frac{1}{2}R_2 \to R_2$ to get

$\left[\begin{array}{rr|rr} 3 & 8 & 1 & 0 \\ 1 & 3 & 0 & -0.5 \end{array}\right]$, use $-3R_2 + R_1 \to R_1$ to get

$\left[\begin{array}{rr|rr} 0 & -1 & 1 & 1.5 \\ 1 & 3 & 0 & -0.5 \end{array}\right]$, use $3R_1 + R_2 \to R_2$ to get

$\left[\begin{array}{rr|rr} 0 & -1 & 1 & 1.5 \\ 1 & 0 & 3 & 4 \end{array}\right]$, use $R_1 \leftrightarrow R_2$ to get

$\left[\begin{array}{rr|rr} 1 & 0 & 3 & 4 \\ 0 & -1 & 1 & 1.5 \end{array}\right]$, use $-1R_2 \to R_2$ to get

$\left[\begin{array}{rr|rr} 1 & 0 & 3 & 4 \\ 0 & 1 & -1 & -1.5 \end{array}\right]$.

So $(AB)^{-1} = \left[\begin{array}{rr} 3 & 4 \\ -1 & -1.5 \end{array}\right]$.

23.

$AA^{-1} = I = \left[\begin{array}{rr} 1 & 0 \\ 0 & 1 \end{array}\right]$

25. $|A| = 8 - 6 = 2$

27. Expanding about the third column,

$|G| = 1 \cdot \left|\begin{array}{rr} -1 & 0 \\ 1 & 1 \end{array}\right| = 1(-1) = -1.$

29. The solution set is $\{(10/3, 17/3)\}$.

First solution is by Gaussian elimination.

On $\left[\begin{array}{rr|r} 1 & 1 & 9 \\ 2 & -1 & 1 \end{array}\right]$, use $-2R_1 + R_2 \to R_2$ to get

$\left[\begin{array}{rr|r} 1 & 1 & 9 \\ 0 & -3 & -17 \end{array}\right]$, use $-\frac{1}{3}R_2 \to R_2$ to get

$\left[\begin{array}{rr|r} 1 & 1 & 9 \\ 0 & 1 & 17/3 \end{array}\right]$, use $-1R_2 + R_1 \to R_1$ to get

$\left[\begin{array}{rr|r} 1 & 0 & 10/3 \\ 0 & 1 & 17/3 \end{array}\right]$.

Secondly, by matrix inversion note $A^{-1} = \left[\begin{array}{rr} 1/3 & 1/3 \\ 2/3 & -1/3 \end{array}\right]$ and $A^{-1}\left[\begin{array}{r} 9 \\ 1 \end{array}\right] = \left[\begin{array}{r} 10/3 \\ 17/3 \end{array}\right]$.

Thirdly, by Cramer's Rule note

$D = \left|\begin{array}{rr} 1 & 1 \\ 2 & -1 \end{array}\right| = -3$, $D_x = \left|\begin{array}{rr} 9 & 1 \\ 1 & -1 \end{array}\right| = -10$,

and $D_y = \left|\begin{array}{rr} 1 & 9 \\ 2 & 1 \end{array}\right| = -17$.

So $\frac{D_x}{D} = 10/3$, $\frac{D_y}{D} = 17/3$.

31. The solution set is $\{(-2, 3)\}$.

First solution is by Gaussian elimination. On $\left[\begin{array}{rr|r} 2 & 1 & -1 \\ 3 & 2 & 0 \end{array}\right]$, use $-\frac{3}{2}R_1 + R_2 \to R_2$ to get

$\left[\begin{array}{cc|c} 2 & 1 & -1 \\ 0 & 1/2 & 3/2 \end{array}\right]$, use $-2R_2 + R_1 \to R_1$

and $2R_2 \to R_2$ to get

$\left[\begin{array}{cc|c} 2 & 0 & -4 \\ 0 & 1 & 3 \end{array}\right]$, use $\frac{1}{2}R_2 \to R_2$ to get

$\left[\begin{array}{cc|c} 1 & 0 & -2 \\ 0 & 1 & 3 \end{array}\right]$.

Secondly, by matrix inversion note $A^{-1} =$

$\begin{bmatrix} 2 & -1 \\ -3 & 2 \end{bmatrix}$ and $A^{-1}\begin{bmatrix} -1 \\ 0 \end{bmatrix} = \begin{bmatrix} -2 \\ 3 \end{bmatrix}$.

Thirdly, by Cramer's Rule note

$D = \begin{vmatrix} 2 & 1 \\ 3 & 2 \end{vmatrix} = 1$, $D_x = \begin{vmatrix} -1 & 1 \\ 0 & 2 \end{vmatrix} = -2$,

and $D_y = \begin{vmatrix} 2 & -1 \\ 3 & 0 \end{vmatrix} = 3$.

So $\frac{D_x}{D} = -2$, $\frac{D_y}{D} = 3$.

33. Solution set is $\{(x, y)|x - 5y = 9\}$.

Solution is by Gaussian elimination.

On $\left[\begin{array}{cc|c} 1 & -5 & 9 \\ -2 & 10 & -18 \end{array}\right]$, use $2R_1 + R_2 \to R_2$

to get $\left[\begin{array}{cc|c} 1 & -5 & 9 \\ 0 & 0 & 0 \end{array}\right]$. Dependent system.

This cannot be solved by matrix inversion since $A^{-1} = \begin{bmatrix} 1 & -5 \\ -2 & 10 \end{bmatrix}^{-1}$ does not exist.

Nor can it be solved by Cramer's rule

since $|D| = \begin{vmatrix} 1 & -5 \\ -2 & 10 \end{vmatrix} = 0$.

35. The solution set is $\emptyset$.

First, apply the Gaussian elimination method.

On $\left[\begin{array}{cc|c} 0.05 & 0.1 & 1 \\ 10 & 20 & 20 \end{array}\right]$, use

$-200R_1 + R_2 \to R_2$ to get

$\left[\begin{array}{cc|c} 0.05 & 0.1 & 1 \\ 0 & 0 & -180 \end{array}\right]$, which is inconsistent.

This cannot be solved by matrix inversion since

$A^{-1} = \begin{bmatrix} 0.05 & 0.1 \\ 10 & 20 \end{bmatrix}^{-1}$ does not exist.

Nor can it be solved by Cramer's rule

since $|D| = \begin{vmatrix} 0.05 & 0.1 \\ 10 & 20 \end{vmatrix} = 0$.

37. Solution set is $\{(1, 2, 3)\}$.

First solution is by Gaussian elimination.

On $\left[\begin{array}{ccc|c} 1 & 1 & -2 & -3 \\ -1 & 2 & -1 & 0 \\ -1 & -1 & 3 & 6 \end{array}\right]$, use

$R_1 + R_2 \to R_2$ and $R_1 + R_3 \to R_3$to get

$\left[\begin{array}{ccc|c} 1 & 1 & -2 & -3 \\ 0 & 3 & -3 & -3 \\ 0 & 0 & 1 & 3 \end{array}\right]$, use

$\frac{1}{3}R_2 \to R_2$ to get

$\left[\begin{array}{ccc|c} 1 & 1 & -2 & -3 \\ 0 & 1 & -1 & -1 \\ 0 & 0 & 1 & 3 \end{array}\right]$, use

$R_3 + R_2 \to R_2$ and $2R_3 + R_1 \to R_1$ to get

$\left[\begin{array}{ccc|c} 1 & 1 & 0 & 3 \\ 0 & 1 & 0 & 2 \\ 0 & 0 & 1 & 3 \end{array}\right]$, use $-1R_2 + R_1 \to R_1$

to get $\left[\begin{array}{ccc|c} 1 & 0 & 0 & 1 \\ 0 & 1 & 0 & 2 \\ 0 & 0 & 1 & 3 \end{array}\right]$.

Secondly, by matrix inversion note $A^{-1} =$

$\begin{bmatrix} 5/3 & -1/3 & 1 \\ 4/3 & 1/3 & 1 \\ 1 & 0 & 1 \end{bmatrix}$ and $A^{-1}\begin{bmatrix} -3 \\ 0 \\ 6 \end{bmatrix} = \begin{bmatrix} 1 \\ 2 \\ 3 \end{bmatrix}$.

Thirdly, by Cramer's Rule note

$D = \begin{vmatrix} 1 & 1 & -2 \\ -1 & 2 & -1 \\ -1 & -1 & 3 \end{vmatrix} = 3$,

$D_x = \begin{vmatrix} -3 & 1 & -2 \\ 0 & 2 & -1 \\ 6 & -1 & 3 \end{vmatrix} = 3$,

$D_y = \begin{vmatrix} 1 & -3 & -2 \\ -1 & 0 & -1 \\ -1 & 6 & 3 \end{vmatrix} = 6$, and $D_z = \begin{vmatrix} 1 & 1 & -3 \\ -1 & 2 & 0 \\ -1 & -1 & 6 \end{vmatrix} = 9.$

Then $\dfrac{D_x}{D} = 1$, $\dfrac{D_y}{D} = 2$, and $\dfrac{D_z}{D} = 3$.

39. The solution set is $\{(-3, 4, 1)\}$.

First solution is by Gaussian elimination.

On $\left[\begin{array}{ccc|c} 0 & 1 & -3 & 1 \\ 1 & 2 & 0 & 5 \\ 1 & 0 & 4 & 1 \end{array}\right]$, use

$-1R_3 + R_2 \to R_2$ and $R_1 \leftrightarrow R_3$ to get

$\left[\begin{array}{ccc|c} 1 & 0 & 4 & 1 \\ 0 & 2 & -4 & 4 \\ 0 & 1 & -3 & 1 \end{array}\right]$, use $\dfrac{1}{2}R_2 \to R_2$ to get

$\left[\begin{array}{ccc|c} 1 & 0 & 4 & 1 \\ 0 & 1 & -2 & 2 \\ 0 & 1 & -3 & 1 \end{array}\right]$, use $-1R_2 + R_3 \to R_3$

to get

$\left[\begin{array}{ccc|c} 1 & 0 & 4 & 1 \\ 0 & 1 & -2 & 2 \\ 0 & 0 & -1 & -1 \end{array}\right]$, use

$-2R_3 + R_2 \to R_2$, $4R_3 + R_1 \to R_1$, and

$-1R_3 \to R_3$ to get

$\left[\begin{array}{ccc|c} 1 & 0 & 0 & -3 \\ 0 & 1 & 0 & 4 \\ 0 & 0 & 1 & 1 \end{array}\right]$.

Secondly, by matrix inversion

$$A^{-1} = \begin{bmatrix} 4 & -2 & 3 \\ -2 & 3/2 & -3/2 \\ -1 & 1/2 & -1/2 \end{bmatrix}$$

and $A^{-1}\begin{bmatrix} 1 \\ 5 \\ 1 \end{bmatrix} = \begin{bmatrix} -3 \\ 4 \\ 1 \end{bmatrix}$.

Thirdly, by Cramer's Rule note

$$D = \begin{vmatrix} 0 & 1 & -3 \\ 1 & 2 & 0 \\ 1 & 0 & 4 \end{vmatrix} = 2,$$

$$D_x = \begin{vmatrix} 1 & 1 & -3 \\ 5 & 2 & 0 \\ 1 & 0 & 4 \end{vmatrix} = -6,$$

$$D_y = \begin{vmatrix} 0 & 1 & -3 \\ 1 & 5 & 0 \\ 1 & 1 & 4 \end{vmatrix} = 8,$$

and $D_z = \begin{vmatrix} 0 & 1 & 1 \\ 1 & 2 & 5 \\ 1 & 0 & 1 \end{vmatrix} = 2.$

Then $\dfrac{D_x}{D} = -3$, $\dfrac{D_y}{D} = 4$, and $\dfrac{D_z}{D} = 1$.

41. The solution set is

$$\left\{\left(\frac{3y+3}{2}, y, \frac{1-y}{2}\right) \mid y \text{ is any real number}\right\}.$$

First, apply the Gaussian elimination method.

On $\left[\begin{array}{ccc|c} 1 & -1 & 1 & 2 \\ 1 & -2 & -1 & 1 \\ 2 & -3 & 0 & 3 \end{array}\right]$, use

$-2R_1 + R_3 \to R_3$ and $-1R_1 + R_2 \to R_2$ to get

$\left[\begin{array}{ccc|c} 1 & -1 & 1 & 2 \\ 0 & -1 & -2 & -1 \\ 0 & -1 & -2 & -1 \end{array}\right]$, use

$-1R_2 + R_3 \to R_3$, $-1R_2 + R_1 \to R_1$,

and $-1R_2 \to R_2$ to get

$\left[\begin{array}{ccc|c} 1 & 0 & 3 & 3 \\ 0 & 1 & 2 & 1 \\ 0 & 0 & 0 & 0 \end{array}\right]$. Since $z = \dfrac{1-y}{2}$, we get

$x = 3 - 3z = 3 - 3\left(\dfrac{1-y}{2}\right) = \dfrac{3y+3}{2}$.

This cannot be solved by matrix inversion

since $A^{-1} = \begin{bmatrix} 1 & -1 & 1 \\ 1 & -2 & -1 \\ 2 & -3 & 0 \end{bmatrix}^{-1}$ does not exist.

Nor can it be solved by Cramer's rule

since $|D| = \begin{vmatrix} 1 & -1 & 1 \\ 1 & -2 & -1 \\ 2 & -3 & 0 \end{vmatrix} = 0.$

43. Solution set is $\emptyset$ as seen by an application of the Gaussian elimination method.

On $\left[\begin{array}{ccc|c} 1 & -3 & -1 & 2 \\ 1 & -3 & -1 & 1 \\ 1 & -3 & -1 & 0 \end{array}\right]$, use

$-1R_1 + R_2 \to R_2$ and $-1R_1 + R_3 \to R_3$ to get

$\left[\begin{array}{ccc|c} 1 & -3 & -1 & 2 \\ 0 & 0 & 0 & -1 \\ 0 & 0 & 0 & -2 \end{array}\right]$. Inconsistent.

This cannot be solved by matrix inversion

since $A^{-1} = \begin{bmatrix} 1 & -3 & -1 \\ 1 & -3 & -1 \\ 1 & -3 & -1 \end{bmatrix}^{-1}$ does not exist.

Nor can it be solved by Cramer's rule

since $|D| = \begin{vmatrix} 1 & -3 & -1 \\ 1 & -3 & -1 \\ 1 & -3 & -1 \end{vmatrix} = 0.$

45. Using $x = 9$ in $x + y = -3$, we find $9 + y = -3$ and $y = -12$. The solution set is $\{(9, -12)\}$.

47.

Note $\begin{bmatrix} x \\ y \end{bmatrix} = \begin{bmatrix} 1 & 1 \\ 2 & 1 \end{bmatrix}^{-1} \begin{bmatrix} 6 \\ 8 \end{bmatrix} =$

$\begin{bmatrix} -1 & 1 \\ 2 & -1 \end{bmatrix} \begin{bmatrix} 6 \\ 8 \end{bmatrix} = \begin{bmatrix} 2 \\ 4 \end{bmatrix}.$

The solution set is $\{(2, 4)\}$.

49. System of equations can be written as

$$\begin{aligned} x + y &= -3 \\ -x &= 0. \end{aligned}$$

Using $x = 0$ in $x + y = -3$, we get $y = -3$. Solution set is $\{(0, -3)\}$.

51. System of equations can be written as

$$\begin{bmatrix} 1 & 1 & 0 \\ 0 & 1 & 1 \\ 1 & 0 & 1 \end{bmatrix} \begin{bmatrix} x \\ y \\ z \end{bmatrix} = \begin{bmatrix} 1 \\ 1 \\ 1 \end{bmatrix}.$$

By using matrix inversion, we obtain

$\begin{bmatrix} x \\ y \\ z \end{bmatrix} = \begin{bmatrix} 1/2 & -1/2 & 1/2 \\ 1/2 & 1/2 & -1/2 \\ -1/2 & 1/2 & 1/2 \end{bmatrix} \begin{bmatrix} 1 \\ 1 \\ 1 \end{bmatrix} =$

$\begin{bmatrix} 1/2 \\ 1/2 \\ 1/2 \end{bmatrix}$. The solution set is $\{(1/2, 1/2, 1/2)\}$.

53. By using the inverse of the coefficient matrix, we get

$\begin{bmatrix} x \\ y \\ z \end{bmatrix} = \begin{bmatrix} 3/5 & -3/5 & 2/5 \\ 2/5 & 3/5 & -2/5 \\ -1/5 & 1/5 & 1/5 \end{bmatrix} \begin{bmatrix} -1 \\ 7 \\ 17 \end{bmatrix} =$

$\begin{bmatrix} 2 \\ -3 \\ 5 \end{bmatrix}$. The solution set is $\{(2, -3, 5)\}$.

55. Let x and y be the number of gallons of pollutant A and pollutant B, respectively. Then

$$\begin{aligned} 10x + 6y &= 4060 \\ \frac{3}{4} &= \frac{x}{y}. \end{aligned}$$

Substitute $x = \frac{3}{4}y$ in $10x + 6y = 4060$. Solving for x, one finds the quantities discharged are $x \approx 225.56$ gallons of pollutant A and $y \approx 300.74$ gallons of pollutant B.

57. Let $x, y,$ and z be the expenses including tax for water, gas, and electricity, respectively.

$$\begin{aligned} x + y + z &= 189.83 \\ \frac{x}{1.04} + \frac{y}{1.05} + \frac{z}{1.06} &= 180 \\ z &= 2y \end{aligned}$$

Solving the system, one finds the expenses including taxes are $x = \$22.88$ for water, $y = \$55.65$ for gas, and $z = \$111.30$ for electricity.

59. Let $1 \le x \le 199$ be the number of dogs remaining. The only number less than 200 that is divisible by four of the five denominators below

$$3, 4, 5, 7, \text{ and } 9$$

is $x = 180$. Then there are $180/9 = 20$ dachshunds remaining, $3(20) = 60$ original beagles, $180/5 = 36$ beagles remaining, and $60 - 36 = 24$ beagles escaped.

Chapter 9 Test

1.

On $\left[\begin{array}{cc|c} 2 & -3 & 1 \\ 1 & 9 & 4 \end{array}\right]$, use $-2R_2 + R_1 \to R_2$

to get

$\left[\begin{array}{cc|c} 2 & -3 & 1 \\ 0 & -21 & -7 \end{array}\right]$, use $\frac{1}{2}R_1 \to R_1$ and

$-\frac{1}{21}R_2 \to R_2$ to get

$\left[\begin{array}{rr|r} 1 & -3/2 & 1/2 \\ 0 & 1 & 1/3 \end{array}\right]$, use $\frac{3}{2}R_2 + R_1 \to R_1$

to get

$\left[\begin{array}{rr|r} 1 & 0 & 1 \\ 0 & 1 & 1/3 \end{array}\right]$. Solution set is $\{(1, 1/3)\}$.

2.

On $\left[\begin{array}{rrr|r} 2 & -1 & 1 & 5 \\ 1 & -2 & -1 & -2 \\ 3 & -1 & -1 & 6 \end{array}\right]$, use

$-\frac{1}{2}R_1 + R_2 \to R_2$ and

$-\frac{3}{2}R_1 + R_3 \to R_3$ to get

$\left[\begin{array}{rrr|r} 2 & -1 & 1 & 5 \\ 0 & -3/2 & -3/2 & -9/2 \\ 0 & 1/2 & -5/2 & -3/2 \end{array}\right]$, use

$3R_3 + R_2 \to R_3$ and $-\frac{2}{3}R_2 \to R_2$ to get

$\left[\begin{array}{rrr|r} 2 & -1 & 1 & 5 \\ 0 & 1 & 1 & 3 \\ 0 & 0 & -9 & -9 \end{array}\right]$,

use $-\frac{1}{9}R_3 \to R_3$ and $R_2 + R_1 \to R_1$ to get

$\left[\begin{array}{rrr|r} 2 & 0 & 2 & 8 \\ 0 & 1 & 1 & 3 \\ 0 & 0 & 1 & 1 \end{array}\right]$, use

$-1R_3 + R_2 \to R_2$ and $\frac{1}{2}R_1 \to R_1$ to get

$\left[\begin{array}{rrr|r} 1 & 0 & 1 & 4 \\ 0 & 1 & 0 & 2 \\ 0 & 0 & 1 & 1 \end{array}\right]$, use $-1R_3 + R_1 \to R_1$

to get $\left[\begin{array}{rrr|r} 1 & 0 & 0 & 3 \\ 0 & 1 & 0 & 2 \\ 0 & 0 & 1 & 1 \end{array}\right]$.

The solution set is $\{(3, 2, 1)\}$.

3.

On $\left[\begin{array}{rrr|r} 1 & -1 & -1 & 1 \\ 2 & 1 & -1 & 0 \\ 5 & -2 & -4 & 3 \end{array}\right]$, use

$-2R_1 + R_2 \to R_2$ and

$-5R_1 + R_3 \to R_3$ to get

$\left[\begin{array}{rrr|r} 1 & -1 & -1 & 1 \\ 0 & 3 & 1 & -2 \\ 0 & 3 & 1 & -2 \end{array}\right]$, use

$-1R_3 + R_2 \to R_3$ and $\frac{1}{3}R_2 \to R_2$ to get

$\left[\begin{array}{rrr|r} 1 & -1 & -1 & 1 \\ 0 & 1 & 1/3 & -2/3 \\ 0 & 0 & 0 & 0 \end{array}\right]$, use

$R_2 + R_1 \to R_1$ to get

$\left[\begin{array}{rrr|r} 1 & 0 & -2/3 & 1/3 \\ 0 & 1 & 1/3 & -2/3 \\ 0 & 0 & 0 & 0 \end{array}\right]$.

Since $x = \frac{2}{3}z + \frac{1}{3}$, we find $3x = 2z + 1$ and $z = \frac{3x-1}{2}$. Since $y = -\frac{1}{3}z - \frac{2}{3}$, we get

$$\begin{aligned} y &= -\frac{1}{3}\left(\frac{3x-1}{2}\right) - \frac{2}{3} \\ y &= \frac{-3x-3}{6} \\ y &= \frac{-x-1}{2} \end{aligned}$$

The solution set is

$\left\{\left(x, \frac{-x-1}{2}, \frac{3x-1}{2}\right) \mid x \text{ is any real number}\right\}$.

4.

$A + B = \begin{bmatrix} 3 & -4 \\ -6 & 10 \end{bmatrix}$

5.

$2A - B = \begin{bmatrix} 2 & -2 \\ -4 & 8 \end{bmatrix} - \begin{bmatrix} 2 & -3 \\ -4 & 6 \end{bmatrix} =$

$\begin{bmatrix} 0 & 1 \\ 0 & 2 \end{bmatrix}$

6.

$AB = \begin{bmatrix} 2+4 & -3-6 \\ -4-16 & 6+24 \end{bmatrix} = \begin{bmatrix} 6 & -9 \\ -20 & 30 \end{bmatrix}$

7.

$AC = \begin{bmatrix} -2-1 \\ 4+4 \end{bmatrix} = \begin{bmatrix} -3 \\ 8 \end{bmatrix}$

8. CB is undefined

9. $FG = [-2 \quad 3-2 \quad 1+1] = [-2 \quad 1 \quad 2]$

10.

$EF = \begin{bmatrix} 2(1) & 2(0) & 2(-1) \\ 3(1) & 3(0) & 3(-1) \\ -1(1) & -1(0) & -1(-1) \end{bmatrix}$

$$= \begin{bmatrix} 2 & 0 & -2 \\ 3 & 0 & -3 \\ -1 & 0 & 1 \end{bmatrix}$$

11.

On $\left[\begin{array}{rr|rr} 1 & -1 & 1 & 0 \\ -2 & 4 & 0 & 1 \end{array}\right]$, use

$2R_1 + R_2 \to R_2$ to get

$\left[\begin{array}{rr|rr} 1 & -1 & 1 & 0 \\ 0 & 2 & 2 & 1 \end{array}\right]$, use

$\frac{1}{2}R_2 + R_1 \to R_1$ and $\frac{1}{2}R_2 \to R_2$

to get $\left[\begin{array}{rr|rr} 1 & 0 & 2 & 1/2 \\ 0 & 1 & 1 & 1/2 \end{array}\right]$.

Then $A^{-1} = \begin{bmatrix} 2 & 1/2 \\ 1 & 1/2 \end{bmatrix}$.

12.

On $\left[\begin{array}{rrr|rrr} -2 & 3 & 1 & 1 & 0 & 0 \\ -3 & 1 & 3 & 0 & 1 & 0 \\ 0 & 2 & -1 & 0 & 0 & 1 \end{array}\right]$,

use $-\frac{3}{2}R_1 + R_2 \to R_2$ and $-\frac{1}{2}R_1 \to R_1$ to get

$\left[\begin{array}{rrr|rrr} 1 & -3/2 & -1/2 & -1/2 & 0 & 0 \\ 0 & -7/2 & 3/2 & -3/2 & 1 & 0 \\ 0 & 2 & -1 & 0 & 0 & 1 \end{array}\right]$, use

$\frac{4}{7}R_2 + R_3 \to R_3$, $-\frac{3}{7}R_2 + R_1 \to R_1$,

and $-\frac{2}{7}R_2 \to R_2$ to get

$\left[\begin{array}{rrr|rrr} 1 & 0 & -8/7 & 1/7 & -3/7 & 0 \\ 0 & 1 & -3/7 & 3/7 & -2/7 & 0 \\ 0 & 0 & -1/7 & -6/7 & 4/7 & 1 \end{array}\right]$, use

$-3R_3 + R_2 \to R_2$, $-8R_3 + R_1 \to R_1$,

and $-7R_3 \to R_3$ to get

$\left[\begin{array}{rrr|rrr} 1 & 0 & 0 & 7 & -5 & -8 \\ 0 & 1 & 0 & 3 & -2 & -3 \\ 0 & 0 & 1 & 6 & -4 & -7 \end{array}\right]$.

Then $G^{-1} = \begin{bmatrix} 7 & -5 & -8 \\ 3 & -2 & -3 \\ 6 & -4 & -7 \end{bmatrix}$.

13. $|A| = 4 - 2 = 2$

14. $|B| = 12 - 12 = 0$

15. Expanding about the third row, we find
$|G| = -2\begin{vmatrix} -2 & 1 \\ -3 & 3 \end{vmatrix} + (-1)\begin{vmatrix} -2 & 3 \\ -3 & 1 \end{vmatrix} =$

$-2(-3) + (-1)(7) = -1$

16.

One finds $D = \begin{vmatrix} 1 & -1 \\ -2 & 4 \end{vmatrix} = 2$,

$D_x = \begin{vmatrix} 2 & -1 \\ 2 & 4 \end{vmatrix} = 10$, $D_y = \begin{vmatrix} 1 & 2 \\ -2 & 2 \end{vmatrix} = 6$.

Then $\frac{D_x}{D} = 5$ and $\frac{D_y}{D} = 3$.

The solution set is $\{(5, 3)\}$.

17. Cramer's Rule is not applicable since
$D = \begin{vmatrix} 2 & -3 \\ -4 & 6 \end{vmatrix} = 0$. Rather, multiply first

equation by 2 and add to the second one.

$$\begin{array}{rcr} 4x - 6y & = & 12 \\ -4x + 6y & = & 1 \\ \hline 0 & = & 13 \end{array}$$

Inconsistent and the solution set is $\emptyset$.

18.

One finds $D = \begin{vmatrix} -2 & 3 & 1 \\ -3 & 1 & 3 \\ 0 & 2 & -1 \end{vmatrix} = -1$,

$D_x = \begin{vmatrix} -2 & 3 & 1 \\ -4 & 1 & 3 \\ 0 & 2 & -1 \end{vmatrix} = -6$,

$D_y = \begin{vmatrix} -2 & -2 & 1 \\ -3 & -4 & 3 \\ 0 & 0 & -1 \end{vmatrix} = -2$, and

$D_z = \begin{vmatrix} -2 & 3 & -2 \\ -3 & 1 & -4 \\ 0 & 2 & 0 \end{vmatrix} = -4$.

Then $\frac{D_x}{D} = 6$, $\frac{D_y}{D} = 2$, and $\frac{D_z}{D} = 4$.

The solution set is $\{(6, 2, 4)\}$.

19. The inverse of coefficient matrix is given
by $A^{-1} = \begin{bmatrix} 2 & 1/2 \\ 1 & 1/2 \end{bmatrix}$. Since

$A^{-1}\begin{bmatrix} 1 \\ -8 \end{bmatrix} = \begin{bmatrix} -2 \\ -3 \end{bmatrix}$, the solution set is $\{(-2,-3)\}$.

20. The inverse of coefficient matrix was found in Exercise 12: $A^{-1} = \begin{bmatrix} 7 & -5 & -8 \\ 3 & -2 & -3 \\ 6 & -4 & -7 \end{bmatrix}$.

Since $A^{-1}\begin{bmatrix} 1 \\ 0 \\ -1 \end{bmatrix} = \begin{bmatrix} 15 \\ 6 \\ 13 \end{bmatrix}$, the solution set is $\{(15,6,13)\}$.

21. Corresponding system of equations is

$$\begin{aligned} x - y &= 12 \\ 35y - 10x &= 730. \end{aligned}$$

Solving this system, one finds $x = 46$ copies were bought and $y = 34$ copies were sold.

22. Substitute $(0,3)$, $(1,-1/2)$, and $(4,3)$ into $y = ax^2 + b\sqrt{x} + c$. Then we obtain

$$\begin{aligned} c &= 3 \\ a + b + c &= -\frac{1}{2} \\ 16a + 2b + c &= 3. \end{aligned}$$

Solving this system, one finds $a = 0.5$, $b = -4$, $c = 3$, and the graph is given by

$$y = 0.5x^2 - 4\sqrt{x} + 3.$$

Tying It All Together

1. Simplify the left-hand side.

$$\begin{aligned} 2x + 6 - 5x &= 7 \\ -3x &= 1 \end{aligned}$$

Solution set is $\{-1/3\}$.

2. Multiply each side of the equation by 30.

$$\begin{aligned} \frac{1}{2}x - \frac{1}{6} &= \frac{4}{5} \\ 15x - 5 &= 24 \\ 15x &= 29 \end{aligned}$$

Solution set is $\{29/15\}$.

3. Multiply $\frac{1}{2}$ to $(2x-2)$.

$$\begin{aligned} (x-1)(6x-8) &= 4 \\ 6x^2 - 14x + 8 &= 4 \\ 6x^2 - 14x + 4 &= 0 \\ 2(3x^2 - 7x + 2) &= 0 \\ 2(3x-1)(x-2) &= 0 \end{aligned}$$

Solution set is $\{1/3, 2\}$.

4. Simplify left-hand side.

$$\begin{aligned} 1 - (4x - 2) &= 9 \\ -4x + 3 &= 9 \\ -4x &= 6 \end{aligned}$$

Solution set is $\{-3/2\}$.

5. The solution set is $\{(4,-2)\}$ as can be seen from the point of intersection.

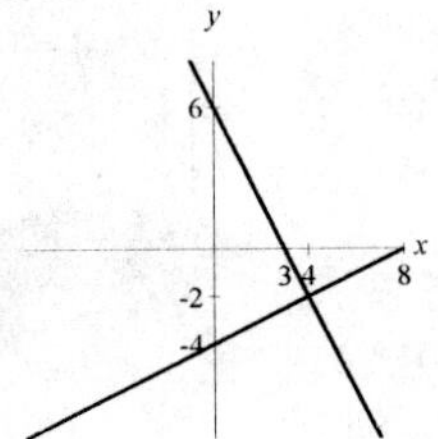

6. Using $y = 1 - 2x$ in $2x + 6y = 2$, we get

$$\begin{aligned} 2x + 6(1-2x) &= 2 \\ 2x + 6 - 12x &= 2 \\ -10x &= -4. \end{aligned}$$

Using $x = \frac{2}{5}$ in $y = 1 - 2x$, we have $y = 1 - \frac{4}{5} = \frac{1}{5}$. Solution set is $\left\{\left(\frac{2}{5}, \frac{1}{5}\right)\right\}$.

7. Multiply second equation by 6 and add to the first one.

$$\begin{aligned} 2x - 0.06y &= 20 \\ 18x + 0.06y &= 120 \\ \hline 20x &= 140 \end{aligned}$$

Using $x = 7$ in $2x - 0.06y = 20$, we find

$$\begin{aligned} 14 - 0.06y &= 20 \\ -0.06y &= 6 \\ y &= -100. \end{aligned}$$

The solution set is $\{(7,-100)\}$.

8.

On $\left[\begin{array}{rr|r} 2 & -1 & -1 \\ 1 & 3 & -11 \end{array}\right]$, use $-2R_2 + R_1 \to R_2$

to get

$\left[\begin{array}{rr|r} 2 & -1 & -1 \\ 0 & -7 & 21 \end{array}\right]$, use $-\frac{1}{7}R_2 \to R_2$

and $\frac{1}{2}R_1 \to R_1$ to get

$\left[\begin{array}{rr|r} 1 & -1/2 & -1/2 \\ 0 & 1 & -3 \end{array}\right]$, use $\frac{1}{2}R_2 + R_1 \to R_1$

to get $\left[\begin{array}{rr|r} 1 & 0 & -2 \\ 0 & 1 & -3 \end{array}\right]$.

The solution set is $\{(-2, -3)\}$.

9. Inverse of the coefficient matrix A is

$$A^{-1} = \left[\begin{array}{rr} -1/2 & -5/2 \\ -1/2 & -3/2 \end{array}\right].$$

Since $A^{-1}\left[\begin{array}{r} -7 \\ 1 \end{array}\right] = \left[\begin{array}{r} 1 \\ 2 \end{array}\right]$,

the solution set is $\{(1, 2)\}$.

10.

One finds $D = \left|\begin{array}{rr} 4 & -3 \\ 3 & -5 \end{array}\right| = -11$,

$D_x = \left|\begin{array}{rr} 5 & -3 \\ 1 & -5 \end{array}\right| = -22$, and

$D_y = \left|\begin{array}{rr} 4 & 5 \\ 3 & 1 \end{array}\right| = -11$.

Then $\frac{D_x}{D} = 2$ and $\frac{D_y}{D} = 1$.

The solution set is $\{(2, 1)\}$.

11. Since $6 = 3x - 5y$ and $1 = 3x - 5y$, we get $6 = 1$. A contradiction. The solution set is $\emptyset$.

12. Multiply first equation by 2 and add the result to second equation.

$$\begin{array}{rcr} 2x - 4y & = & 6 \\ -2x + 4y & = & -6 \\ \hline 0 & = & 0 \end{array}$$

A dependent system and the solution set is

$$\{(x, y) \mid x - 2y = 3\}.$$

13. Substitute $x = y - 1$ in $x^2 + y^2 = 25$. Then

$$\begin{aligned} (y^2 - 2y + 1) + y^2 &= 25 \\ 2y^2 - 2y - 24 &= 0 \\ 2(y^2 - y - 12) &= 0 \\ 2(y - 4)(y + 3) &= 0. \end{aligned}$$

If $y = 4$, then $x = y - 1 = 4 - 1 = 3$.

If $y = -3$, then $x = (-3) - 1 = -4$.

The solution set is $\{(3, 4), (-4, -3)\}$.

14. Substituting $y = x^2 - 1$ in $x + y = 1$, we get

$$\begin{aligned} x + (x^2 - 1) &= 1 \\ x^2 + x - 2 &= 0 \\ (x + 2)(x - 1) &= 0. \end{aligned}$$

If $x = -2$, then $y = x^2 - 1 = (-2)^2 - 1 = 3$.

If $x = 1$, then $y = 1^2 - 1 = 0$.

Solution set is $\{(-2, 3), (1, 0)\}$.

15. Adding the given equations, we obtain

$$\begin{array}{rcr} x^2 - y^2 & = & 1 \\ x^2 + y^2 & = & 3 \\ \hline 2x^2 & = & 4 \\ x^2 & = & 2 \\ x & = & \pm\sqrt{2}. \end{array}$$

Substitute $x = \pm\sqrt{2}$ into $x^2 + y^2 = 3$. Then $2 + y^2 = 3$ and so $y = \pm 1$. The solution set is $\{(\sqrt{2}, \pm 1), (-\sqrt{2}, \pm 1)\}$.

16. Multiply the first equation by -2 and add it to the second equation.

$$\begin{array}{rcr} -2x^2 + 2y^2 & = & -2 \\ 2x^2 + 3y^2 & = & 2 \\ \hline 5y^2 & = & 0 \\ y & = & 0 \end{array}$$

Substitute $y = 0$ into $x^2 - y^2 = 1$. Then $x^2 = 1$ and so $x = \pm 1$. The solution set is $\{(\pm 1, 0)\}$.

17. plane

18. nonlinear, linear

19. partial fraction decomposition

20. linear

21. constraints

22. objective

23. matrix

24. row

25. column

26. augmented

27. equivalent

28. Gaussian elimination

For Thought

1. False, vertex is $(0, -1/2)$. **2.** True

3. True, since $p = 3/2$ and the focus is $(4, 5)$, vertex is $(4 - 3/2, 5) = (5/2, 5)$.

4. False, focus is at $(0, 1/4)$ since parabola opens upward. **5.** True, $p = 1/4$.

6. False. Since $p = 4$ and the vertex is $(2, -1)$, equation of parabola is $y = \frac{1}{16}(x-2)^2 - 1$ and x-intercepts are $(6, 0)$, $(-2, 0)$.

7. False, if $x = 0$ then $y = 0$ and y-intercept is $(0, 0)$. **8.** True

9. False. Since $p = 1/4$ and the vertex is $(5, 4)$, the focus is $(5, 4 + 1/4) = (-5, 17/4)$.

10. False, it opens to the left.

10.1 Exercises

1. conic section

3. axis of symmetry

5. Note $p = 1$.
Vertex $(0, 0)$, focus $(0, 1)$, and directrix $y = -1$

7. Note $p = -1/2$. Vertex $(1, 2)$, focus $(1, 3/2)$, and directrix $y = 5/2$

9. Note $p = 3/4$. Vertex $(3, 1)$, focus $(15/4, 1)$, and directrix $x = 9/4$

11. Since the vertex is equidistant from $y = -2$ and $(0, 2)$, the vertex is $(0, 0)$ and $p = 2$. Then $a = \frac{1}{4p} = \frac{1}{8}$ and an equation is $y = \frac{1}{8}x^2$.

13. Since the vertex is equidistant from $y = 3$ and $(0, -3)$, the vertex is $(0, 0)$ and $p = -3$. Then $a = \frac{1}{4p} = -\frac{1}{12}$ and an equation is $y = -\frac{1}{12}x^2$.

15. One finds $p = \frac{3}{2}$. So $a = \frac{1}{4p} = \frac{1}{6}$ and vertex is $\left(3, 5 - \frac{3}{2}\right) = \left(3, \frac{7}{2}\right)$. Parabola is given by $y = \frac{1}{6}(x-3)^2 + \frac{7}{2}$.

17. One finds $p = -\frac{5}{2}$. So $a = \frac{1}{4p} = -\frac{1}{10}$ and vertex is $\left(1, -3 + \frac{5}{2}\right) = \left(1, -\frac{1}{2}\right)$. Parabola is given by $y = -\frac{1}{10}(x-1)^2 - \frac{1}{2}$.

19. One finds $p = 0.2$. So $a = \frac{1}{4p} = 1.25$ and vertex is $(-2, 1.2 - 0.2) = (-2, 1)$. Parabola is given by $y = 1.25(x+2)^2 + 1$.

21. Since $p = 1$, we get $a = \frac{1}{4p} = \frac{1}{4}$.
An equation is $y = \frac{1}{4}x^2$.

23. Since $p = -\frac{1}{4}$, we get $a = \frac{1}{4p} = -1$.
An equation is $y = -x^2$

25. Note, vertex is $(1, 0)$ and since $a = \frac{1}{4p} = 1$, we find $p = \frac{1}{4}$. The focus is $(1, 0+p) = \left(1, \frac{1}{4}\right)$ and the directrix is $y = 0 - p$ or $y = -\frac{1}{4}$.

27. Note, vertex is $(3, 0)$ and since $a = \frac{1}{4p} = \frac{1}{4}$, we find $p = 1$. The focus is $(3, 0 + p) = (3, 1)$ and the directrix is $y = 0 - p$ or $y = -1$.

29. Note, vertex is $(3, 4)$ and since $a = \frac{1}{4p} = -2$, we find $p = -\frac{1}{8}$. The focus is $(3, 4+p) = \left(3, \frac{31}{8}\right)$ and the directrix is $y = 4 - p$ or $y = \frac{33}{8}$.

31. Completing the square, we obtain

$$\begin{aligned} y &= (x^2 - 8x + 16) - 16 + 3 \\ y &= (x-4)^2 - 13. \end{aligned}$$

Since $\frac{1}{4p} = 1$, $p = 0.25$.

Since vertex is $(4, -13)$, focus is $(4, -13 + 0.25) = (4, -51/4)$, and directrix is $y = -13 - p = -53/4$.

33. Completing the square, we obtain

$$\begin{aligned} y &= 2(x^2 + 6x + 9) + 5 - 18 \\ y &= 2(x + 3)^2 - 13. \end{aligned}$$

Since $\dfrac{1}{4p} = 2$, $p = 1/8$.

Since vertex is $(-3, -13)$, the focus is $(-3, -13 + 1/8) = (-3, -103/8)$, and directrix is $y = -13 - 1/8 = -105/8$.

35. Completing the square, we get

$$\begin{aligned} y &= -2(x^2 - 3x + 9/4) + 1 + 9/2 \\ y &= -2(x - 3/2)^2 + 11/2. \end{aligned}$$

Since $\dfrac{1}{4p} = -2$, $p = -1/8$.

Since vertex is $(3/2, 11/2)$, the focus is $(3/2, 11/2 - 1/8) = (3/2, 43/8)$, and directrix is $y = 11/2 + 1/8 = 45/8$.

37. Completing the square,

$$\begin{aligned} y &= 5(x^2 + 6x + 9) - 45 \\ y &= 5(x + 3)^2 - 45. \end{aligned}$$

Since $\dfrac{1}{4p} = 5$, we have $p = 0.05$.

Since vertex is $(-3, -45)$, the focus is $(-3, -45 + 0.05) = (-3, -44.95)$, and directrix is $y = -45 - .05 = -45.05$.

39. Completing the square, we get

$$\begin{aligned} y &= \frac{1}{8}(x^2 - 4x + 4) + \frac{9}{2} - \frac{1}{2} \\ y &= \frac{1}{8}(x - 2)^2 + 4. \end{aligned}$$

Since $\dfrac{1}{4p} = 1/8$, we find $p = 2$.

Since vertex is $(2, 4)$, the focus is $(2, 4 + 2) = (2, 6)$, and directrix is $y = 4 - 2$ or $y = 2$.

41. Since $a = 1$ and $b = -4$, we find $x = \dfrac{-b}{2a} = \dfrac{4}{2} = 2$. Since $\dfrac{1}{4p} = a = 1$, we obtain $p = 1/4$.

Substituting $x = 2$, we get $y = 2^2 - 4(2) + 3 = -1$. Thus, the vertex is $(2, -1)$, the focus is $(2, -1 + p) = (2, -3/4)$, the directrix is $y = -1 - p = -5/4$, and the parabola opens upward since $a > 0$.

43. Note, $a = -1$, $b = 2$. So $x = \dfrac{-b}{2a} = \dfrac{-2}{-2} = 1$ and since $\dfrac{1}{4p} = a = -1$, $p = -1/4$.

Substituting $x = 1$, $y = -(1)^2 + 2(1) - 5 = -4$.

The vertex is $(1, -4)$ and focus is $(1, -4 + p) = (1, -17/4)$, directrix is $y = -4 - p = -15/4$, and parabola opens down since $a < 0$.

45. Note, $a = 3$, $b = -6$. So $x = \dfrac{-b}{2a} = \dfrac{6}{6} = 1$ and since $\dfrac{1}{4p} = a = 3$, $p = 1/12$. Substituting $x = 1$, $y = 3(1)^2 - 6(1) + 1 = -2$.

The vertex is $(1, -2)$ and focus is $(1, -2 + p) = (1, -23/12)$, directrix is $y = -2 - p = -25/12$, and parabola opens up since $a > 0$.

47. Note, $a = -1/2$, $b = -3$. So $x = \dfrac{-b}{2a} = \dfrac{3}{-1} = -3$ and since $\dfrac{1}{4p} = a = -1/2$, $p = -1/2$. Substituting $x = -3$, we have $y = -\dfrac{1}{2}(-3)^2 - 3(-3) + 2 = \dfrac{13}{2}$. The vertex is $\left(-3, \dfrac{13}{2}\right)$ and focus is $\left(-3, \dfrac{13}{2} + p\right) = (-3, 6)$, directrix is $y = \dfrac{13}{2} - p = 7$, and parabola opens down since $a < 0$.

49. Note, $y = \dfrac{1}{4}x^2 + 5$ is of the form $y = a(x - h)^2 + k$. So $h = 0$, $k = 5$, and $\dfrac{1}{4p} = a = \dfrac{1}{4}$ from which we have $p = 1$. The vertex is $(h,k) = (0,5)$, focus is $(0, 5+p) = (0,6)$, directrix is $y = 5 - p = 4$ and parabola opens up since $a > 0$.

51. From the given focus and directrix, one finds $p = 1/4$. So $a = \dfrac{1}{4p} = 1$, vertex is $(h,k) = (1/2, -2-p) = (1/2, -9/4)$, axis of symmetry is $x = 1/2$, and parabola is given by $y = a(x-h)^2 + k = \left(x - \dfrac{1}{2}\right)^2 - \dfrac{9}{4}$.

If $y = 0$ then $x - \dfrac{1}{2} = \pm\dfrac{3}{2}$ or $x = 2, -1$.

The x-intercepts are $(2,0), (-1,0)$.

If $x = 0$ then $y = \left(0 - \dfrac{1}{2}\right)^2 - \dfrac{9}{4} = -2$ and y-intercept is $(0,-2)$.

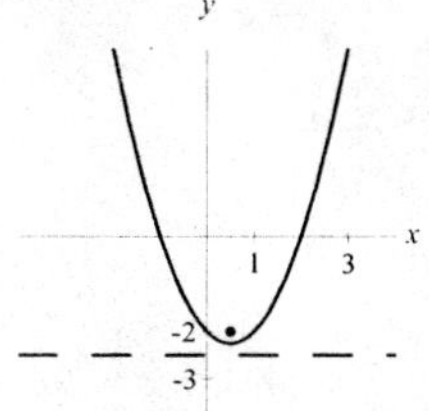

53. From the given focus and directrix one finds $p = -1/4$. So $a = \dfrac{1}{4p} = -1$, vertex is $(h,k) = (-1/2, 6-p) = (-1/2, 25/4)$, axis of symmetry is $x = -1/2$, and parabola is given by

$$y = a(x-h)^2 + k = -\left(x + \frac{1}{2}\right)^2 + \frac{25}{4}.$$

If $y = 0$ then $x + \dfrac{1}{2} = \pm\dfrac{5}{2}$ or $x = -3, 2$.

The x-intercepts are $(-3,0), (2,0)$.

If $x = 0$ then $y = -\left(0 + \dfrac{1}{2}\right)^2 + \dfrac{25}{4} = 6$ and y-intercept is $(0,6)$.

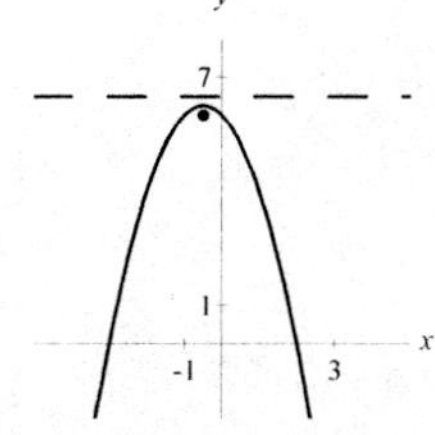

55. Since $\dfrac{1}{2}(x+2)^2 + 2$ is of the form $a(x-h)^2 + k$, vertex is $(h,k) = (-2,2)$ and axis of symmetry is $x = -2$. If $y = 0$ then $0 = \dfrac{1}{2}(x+2)^2 + 2$; this has no solution since left-hand side is always positive. No x-intercept. If $x = 0$ then $y = \dfrac{1}{2}(0+2)^2 + 2 = 4$. y-intercept is $(0,4)$.

Since $\dfrac{1}{4p} = a = \dfrac{1}{2}$, $p = \dfrac{1}{2}$, focus is $(h, k+p) = (-2, 5/2)$, and directrix is $y = k - p = 3/2$.

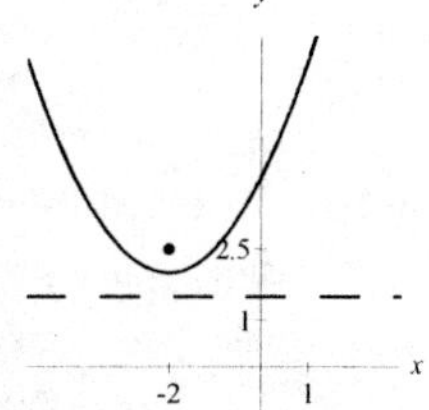

57. Since $-\dfrac{1}{4}(x+4)^2 + 2$ is of the form $a(x-h)^2 + k$, vertex is $(h,k) = (-4,2)$ and axis of symmetry is $x = -4$. If $y = 0$ then

$$\begin{aligned} \frac{1}{4}(x+4)^2 &= 2 \\ x + 4 &= \pm\sqrt{8}. \end{aligned}$$

x-intercepts are $(-4 \pm 2\sqrt{2}, 0)$. If $x = 0$, then $y = -\dfrac{1}{4}(0+4)^2 + 2 = -2$. The y-intercept is $(0,-2)$. Since $\dfrac{1}{4p} = a = -\dfrac{1}{4}$, $p = -1$, focus is $(h, k+p) = (-4, 1)$, and directrix is $y = k - p = 3$.

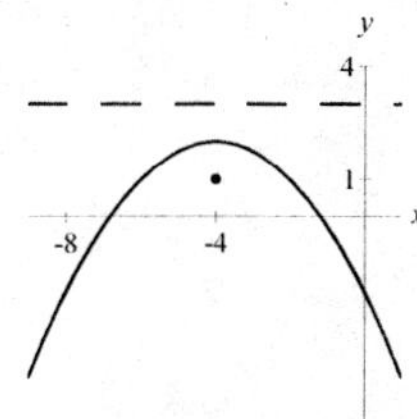

59. Since $\dfrac{1}{2}x^2 - 2$ is of the form $a(x-h)^2 + k$, vertex is $(h,k) = (0,-2)$ and axis of symmetry is $x = 0$. If $y = 0$, then

$$\begin{aligned} \frac{1}{2}x^2 &= 2 \\ x^2 &= 4. \end{aligned}$$

x-intercepts are $(\pm 2, 0)$. If $x = 0$ then $y = \frac{1}{2}(0)^2 - 2 = -2$. The y-intercept is $(0, -2)$. Since $\frac{1}{4p} = a = \frac{1}{2}$, $p = 1/2$, focus is $(h, k + p) = (0, -3/2)$, and directrix is $y = k - p = -5/2$.

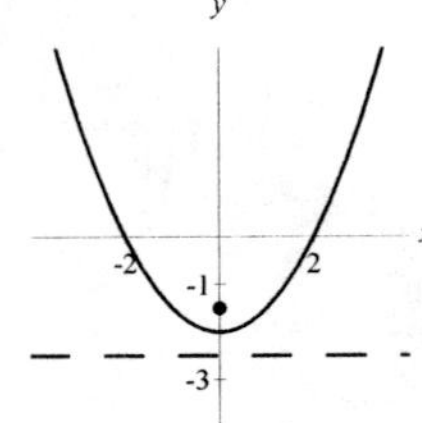

61. Since $y = (x - 2)^2$ is of the form $a(x - h)^2 + k$, vertex is $(h, k) = (2, 0)$, and axis of symmetry is $x = 2$. If $y = 0$ then $(x - 2)^2 = 0$ and x-intercept is $(2, 0)$. If $x = 0$ then $y = (0 - 2)^2 = 4$ and y-intercept is $(0, 4)$. Since $\frac{1}{4p} = a = 1$, $p = 1/4$, focus is $(h, k + p) = (2, 1/4)$, and directrix is $y = k - p = -1/4$.

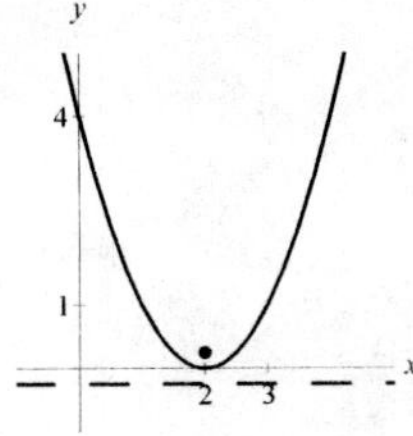

63. By completing the square, we obtain

$$y = \frac{1}{3}(x - 3/2)^2 - 3/4.$$

Vertex is $(h, k) = (3/2, -3/4)$ and axis of symmetry is $x = 3/2$. If $y = 0$, then

$$\begin{aligned} \frac{1}{3}\left(x - \frac{3}{2}\right)^2 &= \frac{3}{4} \\ \left(x - \frac{3}{2}\right)^2 &= \frac{9}{4} \\ x &= \frac{3}{2} \pm \frac{3}{2}. \end{aligned}$$

x-intercepts are $(3, 0), (0, 0)$. If $x = 0$, then $y = \frac{1}{3}\left(0 - \frac{3}{2}\right)^2 - \frac{3}{4} = 0$ and y-intercept is $(0, 0)$. Since $\frac{1}{4p} = a = 1/3$, $p = 3/4$, focus is $(h, k + p) = (3/2, 0)$, and directrix is $y = k - p = -3/2$ or $y = -3/2$.

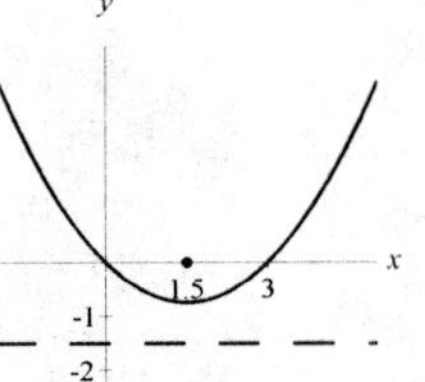

65. Since $x = -y^2$ is of the form $x = a(y - h)^2 + k$, vertex is $(k, h) = (0, 0)$ and axis of symmetry is $y = 0$. If $y = 0$ then $x = -0^2 = 0$ and x-intercept is $(0, 0)$. If $x = 0$ then $0 = -y^2$ and y-intercept is $(0, 0)$. Since $\frac{1}{4p} = a = -1$, $p = -1/4$, focus is $(k + p, h) = (-1/4, 0)$, and directrix is $x = k - p = 1/4$.

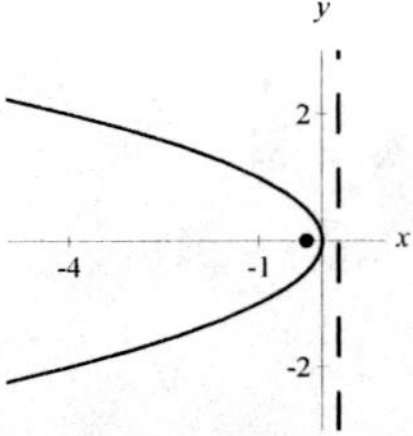

67. Since $x = -\frac{1}{4}y^2 + 1$ is of the form

$$x = a(y - h)^2 + k,$$

vertex is $(k, h) = (1, 0)$ and axis of symmetry is $y = 0$. If $y = 0$ then $x = 1$ and x-intercept is $(1, 0)$. If $x = 0$ then $\frac{1}{4}y^2 = 1$, $y^2 = 4$, and y-intercepts are $(0, \pm 2)$. Since $\frac{1}{4p} = a = -\frac{1}{4}$, $p = -1$, focus is $(k + p, h) = (0, 0)$, and directrix is $x = k - p = 2$ or $x = 2$.

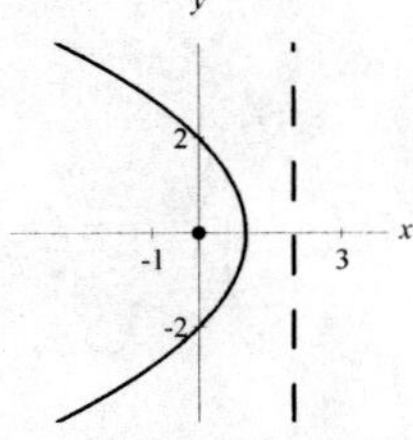

69. By completing the square, we obtain

$$x = (y + 1/2)^2 - 25/4.$$

Vertex is $(k, h) = (-25/4, -1/2)$ and axis of symmetry is $y = -1/2$. If $y = 0$, then $x =$

$1/4 - 25/4 = -6$. The x-intercept is $(-6, 0)$. If $x = 0$, then

$$\begin{aligned} \left(y + \frac{1}{2}\right)^2 &= \frac{25}{4} \\ y &= -\frac{1}{2} \pm \frac{5}{2} \\ y &= 2, -3. \end{aligned}$$

y-intercepts are $(0, 2), (0, -3)$. Since $\frac{1}{4p} = a = 1$, $p = 1/4$, focus is $(k + p, h) = (-6, -1/2)$, and directrix is $x = k - p = -13/2$.

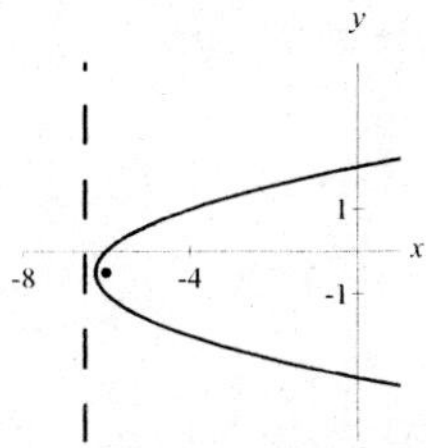

71. By completing the square, we get

$$x = -\frac{1}{2}(y + 1)^2 - 7/2.$$

Vertex is $(k, h) = (-7/2, -1)$ and axis of symmetry is $y = -1$. If $y = 0$, then $x = -1/2 - 7/2 = -4$. The x-intercept is $(-4, 0)$. If $x = 0$, then

$$0 = -\frac{1}{2}(y + 1)^2 - 7/2 < 0$$

which is inconsistent and so there is no y-intercept. Since $\frac{1}{4p} = a = -1/2$, $p = -1/2$, focus is $(k + p, h) = (-4, -1)$, and directrix is $x = k - p = -3$.

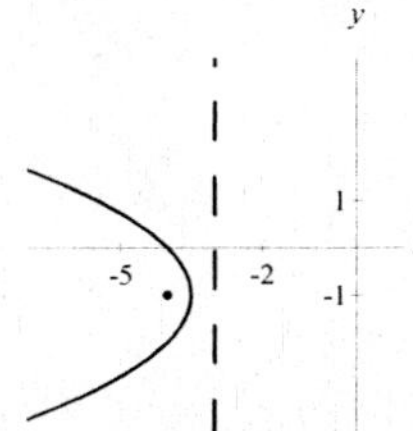

73. Since $x = 2(y - 1)^2 + 3$ is of the form $a(y-h)^2+k$, we find that the vertex is $(k, h) = (3, 1)$ and axis of symmetry is $y = 1$. If $y = 0$ then $x = 2(-1)^2 + 3 = 5$ and x-intercept is $(5, 0)$. If $x = 0$, we obtain $2(y - 1)^2 + 3 = 0$ which is inconsistent since the left-hand side is always positive. No y-intercept.

Since $\frac{1}{4p} = a = 2$, $p = 1/8$, focus is $(k+p, h) = (25/8, 1)$, and directrix is $x = k - p = 23/8$.

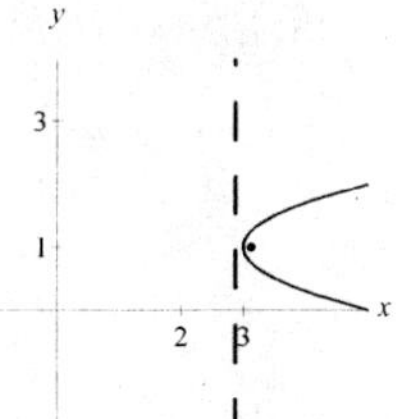

75. Since $x = -\frac{1}{2}(y + 2)^2 + 1$ is of the form $a(y - h)^2 + k$, vertex is $(k, h) = (1, -2)$ and axis of symmetry is $y = -2$. If $y = 0$, then $x = -\frac{1}{2}(2)^2 + 1 = -1$ and x-intercept is $(-1, 0)$. If $x = 0$, then

$$\begin{aligned} \frac{1}{2}(y + 2)^2 &= 1 \\ (y + 2)^2 &= 2 \\ y &= -2 \pm \sqrt{2}. \end{aligned}$$

The y-intercepts are $(0, -2 \pm \sqrt{2})$. Since $\frac{1}{4p} = a = -\frac{1}{2}$, we find $p = -\frac{1}{2}$, focus is $(k + p, h) = (1/2, -2)$, and directrix is $x = k - p = 3/2$.

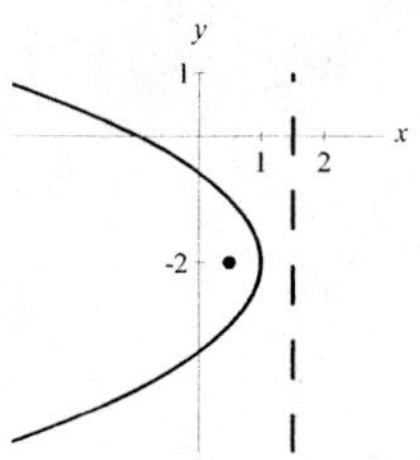

77. Since focus is 1 unit above the vertex $(1, 4)$, we obtain $p = 1$. Then $a = \frac{1}{4p} = \frac{1}{4}$ and parabola is given by $y = \frac{1}{4}(x - 1)^2 + 4$.

79. Since vertex $(0, 0)$ is 2 units to the right of the directrix, we find $p = 2$ and parabola opens to the right. Then $a = \frac{1}{4p} = \frac{1}{8}$ and the parabola is given by $x = \frac{1}{8}y^2$.

81. Since the parabola opens up, $p = 55(12)$ inches, and the vertex is $(0,0)$. Note, $\frac{1}{4p} = \frac{1}{2640}$. Thus, the parabola is given by $y = \frac{1}{2640}x^2$. Thickness at the outside edge is $23 + \frac{1}{2640}(100)^2 \approx 26.8$ in.

83. $y = x^2$ has vertex $(0,0)$ and opens up. The second parabola can be written as

$$y = 2(x-1)^2 + 3$$

and its vertex is $(1,3)$. In the given viewing window these graphs look alike.

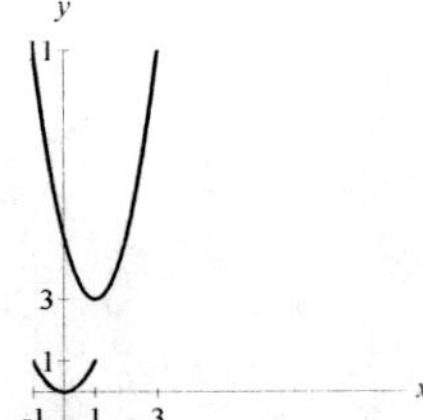

85. Two functions are $f_1(x) = \sqrt{-x}$ and $f_2(x) = -\sqrt{-x}$ where $x \le 0$.

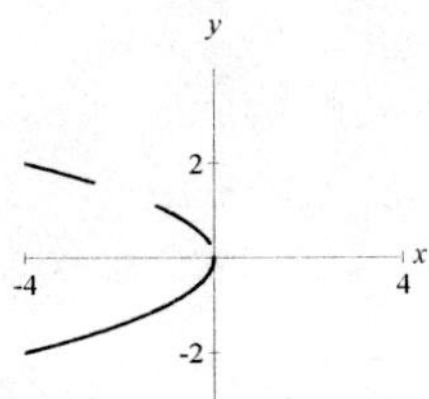

89. $\frac{\sqrt{3}}{\sqrt{6}-\sqrt{3}} \cdot \frac{\sqrt{6}+\sqrt{3}}{\sqrt{6}+\sqrt{3}} = \frac{\sqrt{18}+3}{3} = \frac{3\sqrt{2}+3}{3} = \sqrt{2}+1$

91.

$$\begin{aligned} \frac{9x-15}{(x-3)(x+3)} &= \frac{A}{x-3} + \frac{B}{x+3} \\ 9x - 15 &= A(x+3) + B(x-3) \end{aligned}$$

If $x = 3$ in the above equation, we find

$$\begin{aligned} 12 &= 6A \\ 2 &= A \end{aligned}$$

Likewise, if $x = -3$ then

$$\begin{aligned} -42 &= -6B \\ 7 &= B \end{aligned}$$

The partial fraction decomposition is

$$\frac{2}{x-3} + \frac{7}{x+3}.$$

93. Use the method of completing the square.

$$\begin{aligned} f(x) &= 3\left(x^2 - \frac{4}{3}x\right) + 7 \\ f(x) &= 3\left(x^2 - \frac{4}{3}x + \frac{4}{9}\right) + 7 - \frac{4}{3} \\ f(x) &= 3\left(x - \frac{2}{3}\right)^2 + \frac{17}{3} \end{aligned}$$

95. Let r be the radius of the pipe that is placed on top of two pipes with radii 2 ft and 3 ft. Draw a triangle with vertices at the center of the three circles (cross section of the pipes). Let θ be the angle of the triangle at the center of the circle with radius 2 ft. Using the Law of Cosines, we obtain

$$(r+3)^2 = 5^2 + (r+2)^2 - 2(5)(r+2)\cos\theta.$$

Solving for r, we find

$$r = \frac{10(1-\cos\theta)}{5\cos\theta + 1}.$$

Note, the angle between the side of the triangle joining the centers of the circle with radii 2 ft and 3 ft and the horizontal is $\cos^{-1}(\sqrt{24}/5)$. Since the top circle is tangent to the circle with radius 3 feet, θ must lie in the interval

$$\left[0, \frac{\pi}{2} - \cos^{-1}(\sqrt{24}/5)\right].$$

The maximum of r occurs at the right end-point of the above closed interval. That is, the maximum radius is

$$r = \frac{10(1-\cos\theta)}{5\cos\theta + 1} = \frac{10(1-1/5)}{5(1/5)+1} = 4\text{ft}.$$

For Thought

1. False, y-intercepts are $(\pm 3, 0)$.

2. True, since it can be written as $\dfrac{x^2}{1/2} + y^2 = 1$.

3. True, length of the major axis is $2a = 2(5) = 10$.

4. True, if $y = 0$ then $x^2 = \dfrac{1}{0.5} = 2$ and $x = \pm\sqrt{2}$.

5. True, if $x = 0$ then $y^2 = 3$ and $y = \pm\sqrt{3}$.

6. False, the center is not a point on the circle.

7. True **8.** False, $(3, -1)$ satisfies equation.

9. False. No point satisfies the equation since the left-hand side is always positive.

10. False, since the circle can be written as $(x-2)^2 + (y+1/2)^2 = 53/4$ and so the radius is $\sqrt{53}/2$.

10.2 Exercises

1. ellipse

3. circle

5. Foci $(\pm\sqrt{5}, 0)$, vertices $(\pm 3, 0)$, center $(0, 0)$

7. Foci $(2, 1 \pm \sqrt{5})$, vertices $(2, 4)$ and $(2, -2)$, center $(2, 1)$

9. Since $c = 2$ and $b = 3$, we get $a^2 = b^2 + c^2 = 9 + 4 = 13$ and $a = \sqrt{13}$.

Ellipse is given by $\dfrac{x^2}{13} + \dfrac{y^2}{9} = 1$.

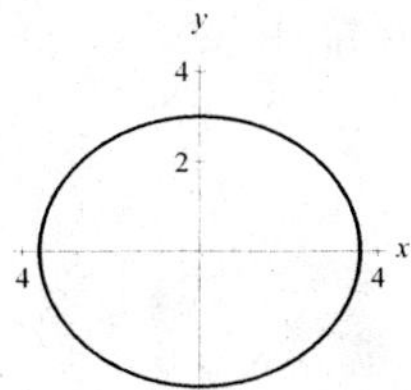

11. Since $c = 4$ and $a = 5$, we find $b^2 = a^2 - c^2 = 25 - 16 = 9$ and $b = 3$.

Ellipse is given by $\dfrac{x^2}{25} + \dfrac{y^2}{9} = 1$.

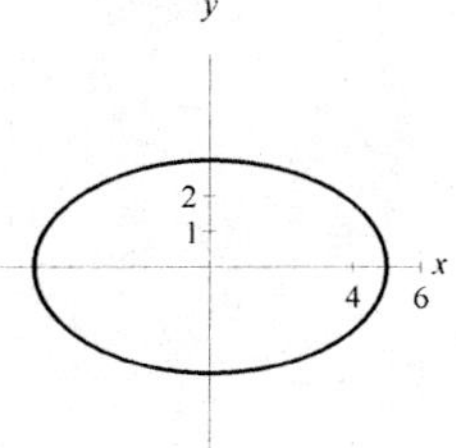

13. Since $c = 2$ and $b = 2$, we obtain $a^2 = b^2 + c^2 = 4 + 4 = 8$ and $a = \sqrt{8}$.

Ellipse is given by $\dfrac{x^2}{4} + \dfrac{y^2}{8} = 1$.

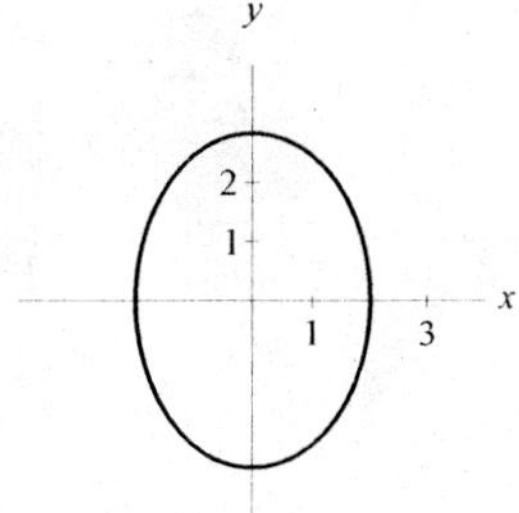

15. Since $c = 4$ and $a = 7$, we obtain $b^2 = a^2 - c^2 = 49 - 16 = 33$ and $b = \sqrt{33}$.

Ellipse is given by $\dfrac{x^2}{33} + \dfrac{y^2}{49} = 1$.

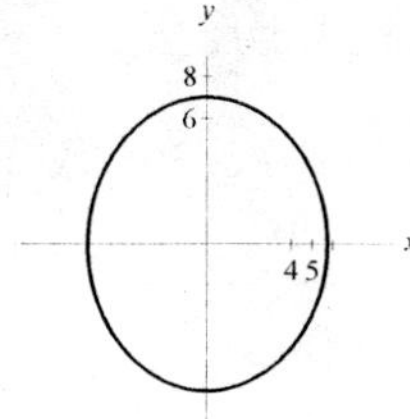

17. Since $c = \sqrt{a^2 - b^2} = \sqrt{16 - 4} = 2\sqrt{3}$, the foci are $(\pm 2\sqrt{3}, 0)$

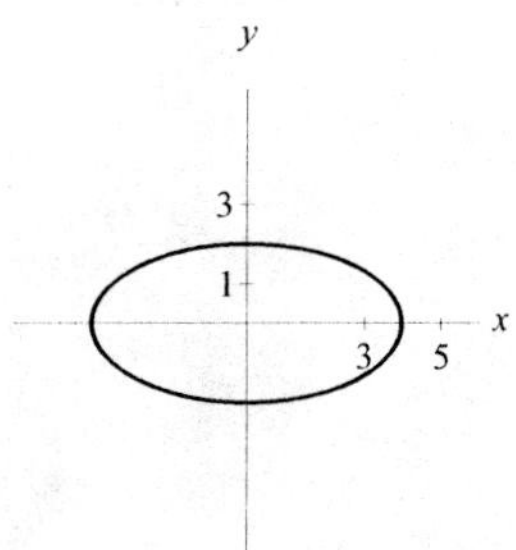

19. Since $c = \sqrt{a^2 - b^2} = \sqrt{36 - 9} = \sqrt{27}$, the foci are $(0, \pm 3\sqrt{3})$

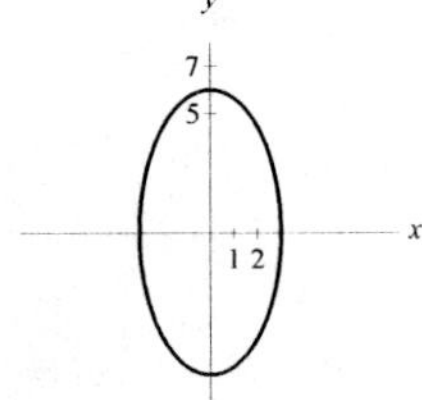

21. Since $c = \sqrt{a^2 - b^2} = \sqrt{25 - 1} = \sqrt{24}$, the foci are $(\pm 2\sqrt{6}, 0)$

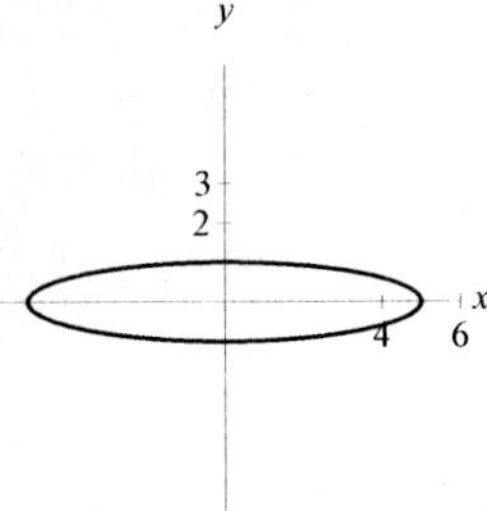

23. Since $c = \sqrt{a^2 - b^2} = \sqrt{25 - 9} = \sqrt{16}$, the foci are $(0, \pm 4)$

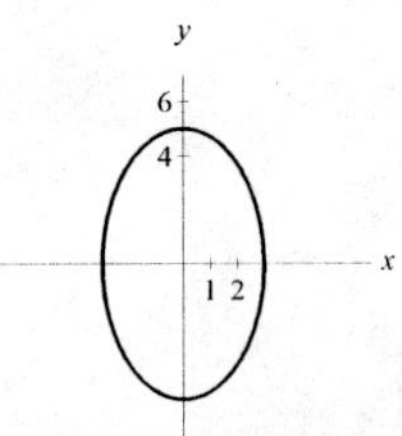

25. From $x^2 + \dfrac{y^2}{9} = 1$, one finds $c = \sqrt{a^2 - b^2}$ $= \sqrt{9 - 1} = \sqrt{8}$ and foci are $(0, \pm 2\sqrt{2})$.

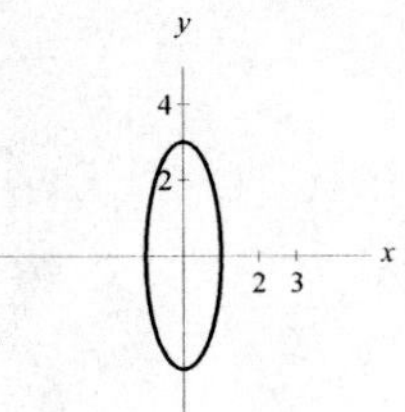

27. From $\dfrac{x^2}{9} + \dfrac{y^2}{4} = 1$, one finds $c = \sqrt{a^2 - b^2}$ $= \sqrt{9 - 4} = \sqrt{5}$ and foci are $(\pm\sqrt{5}, 0)$.

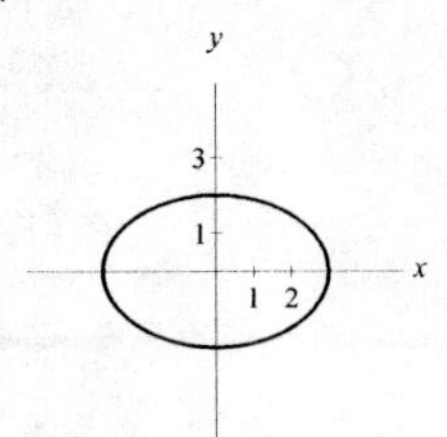

29. Since $c = \sqrt{a^2 - b^2} = \sqrt{16 - 9} = \sqrt{7}$, the foci are $(1 \pm c, -3) = (1 \pm \sqrt{7}, -3)$.

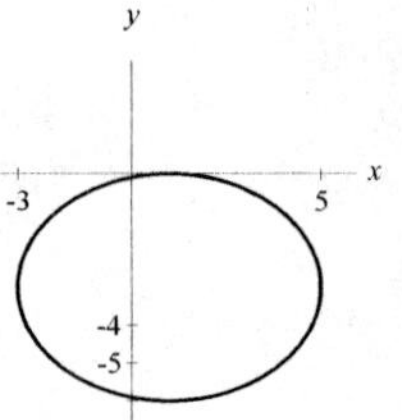

31. Since $c = \sqrt{a^2 - b^2} = \sqrt{25 - 9} = \sqrt{16} = 4$, the foci are $(3, -2 \pm c)$, or $(3, 2)$ and $(3, -6)$.

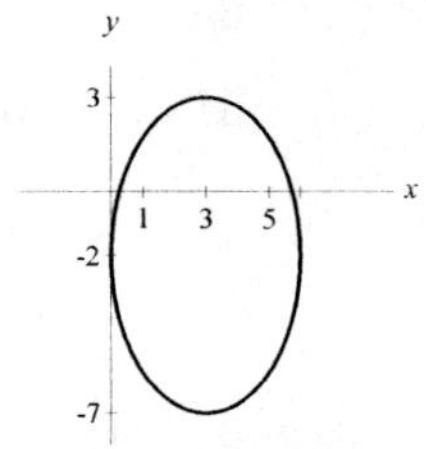

33. Since $\dfrac{(x+4)^2}{36} + (y+3)^2 = 1$, we get

$$c = \sqrt{a^2 - b^2} = \sqrt{36 - 1} = \sqrt{35}$$

and the foci are

$$(-4 \pm \sqrt{35}, -3).$$

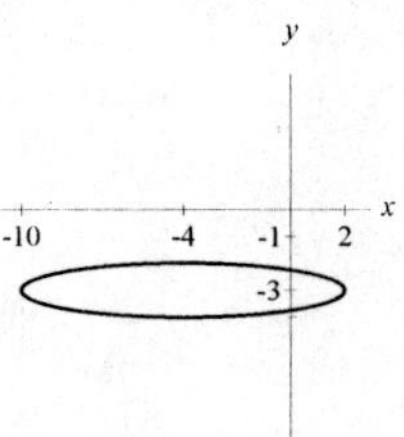

35. If one applies the method of completing the square, one obtains

$$\begin{aligned} 9(x^2 - 2x + 1) + 4(y^2 + 4y + 4) &= 11 + 9 + 16 \\ 9(x-1)^2 + 4(y+2)^2 &= 36 \\ \frac{(x-1)^2}{4} + \frac{(y+2)^2}{9} &= 1. \end{aligned}$$

From $a^2 = b^2 + c^2$ with $a = 3$ and $b = 2$, one finds $c = \sqrt{5}$. The foci are $(1, -2 \pm \sqrt{5})$ and a sketch of the ellipse is given.

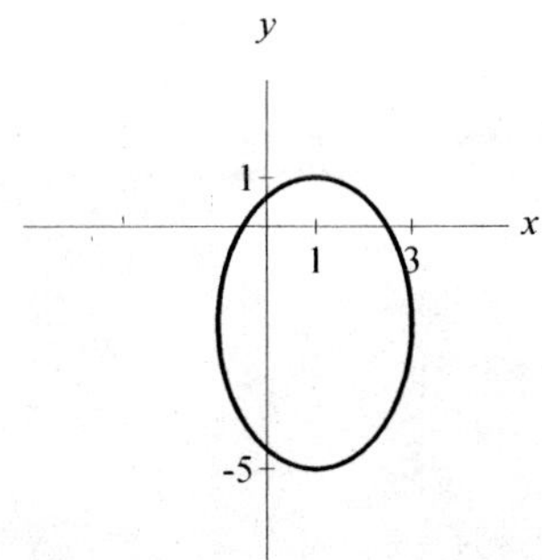

37. Applying the method of completing the square, we find

$$\begin{aligned} 9(x^2-6x+9)+4(y^2+4y+4) &= -61+81+16 \\ 9(x-3)^2+4(y+2)^2 &= 36 \\ \frac{(x-3)^2}{4}+\frac{(y+2)^2}{9} &= 1. \end{aligned}$$

From $a^2 = b^2 + c^2$ with $a = 3$ and $b = 2$, we find $c = \sqrt{5}$. The foci are $(3, -2 \pm \sqrt{5})$ and a sketch of the ellipse is given.

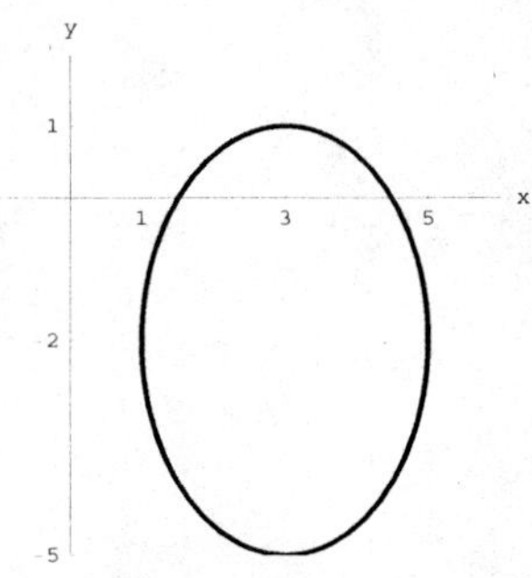

39. Since $\dfrac{x^2}{16}+\dfrac{y^2}{4}=1$, we get $c=\sqrt{a^2-b^2}$ $=\sqrt{16-4}=\sqrt{12}$, and the foci are $(\pm 2\sqrt{3}, 0)$.

41. Since $\dfrac{(x+1)^2}{4}+\dfrac{(y+2)^2}{16}=1$, we get $c=\sqrt{a^2-b^2}=\sqrt{16-4}=\sqrt{12}$, and the foci are $(-1, -2 \pm 2\sqrt{3})$.

43. $x^2 + y^2 = 4$

45. Since $r=\sqrt{(4-0)^2+(5-0)^2}=\sqrt{41}$, the circle is given by $x^2+y^2=41$.

47. Since $r=\sqrt{(4-2)^2+(1+3)^2}=\sqrt{20}$, the circle is given by $(x-2)^2+(y+3)^2=20$.

49. Since center is $\left(\dfrac{3-1}{2}, \dfrac{4+2}{2}\right)=(1,3)$ and $r=\sqrt{(3-1)^2+(4-3)^2}=\sqrt{5}$, the circle is given by $(x-1)^2+(y-3)^2=5$.

51. center $(0, 0)$, radius 10

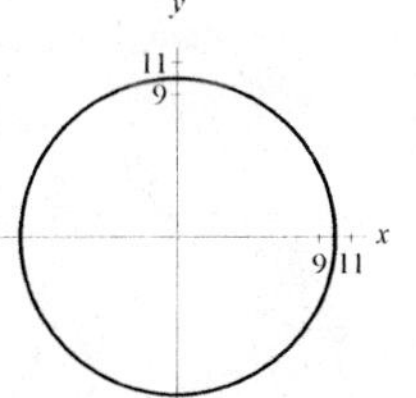

53. center $(1, 2)$, radius 2

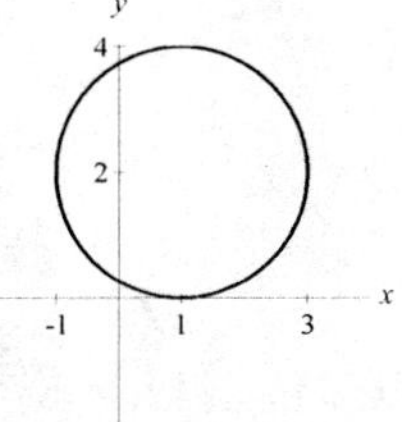

55. center $(-2, -2)$, radius $\sqrt{8}$ or $2\sqrt{2}$

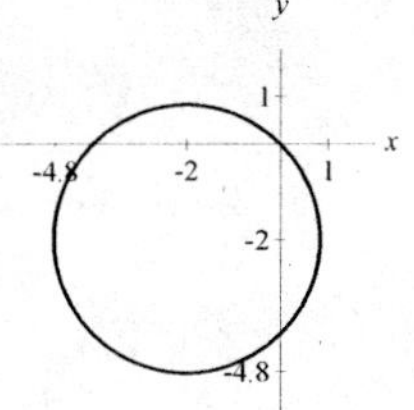

57. Completing the square, we obtain

$$\begin{aligned} x^2+(y^2+2y+1) &= 8+1 \\ x^2+(y+1)^2 &= 9. \end{aligned}$$

The center is $(0, -1)$ with radius 3.

59. Completing the square, we find

$$\begin{aligned} x^2+8x+16+y^2-10y+25 &= 16+25 \\ (x+4)^2+(y-5)^2 &= 41. \end{aligned}$$

The center is $(-4, 5)$ with radius $\sqrt{41}$.

61. Completing the square, we find

$$\begin{aligned} (x^2+4x+4)+y^2 &= 5+4 \\ (x+2)^2+y^2 &= 9. \end{aligned}$$

The center is $(-2, 0)$ with radius 3.

63. Completing the square, we have

$$\begin{aligned} x^2-x+\frac{1}{4}+y^2+y+\frac{1}{4} &= \frac{1}{2}+\frac{1}{4}+\frac{1}{4} \\ (x-1/2)^2+(y+1/2)^2 &= 1. \end{aligned}$$

The center is $(0.5, -0.5)$ with radius 1.

65. Completing the square, we get

$$\begin{aligned} x^2+\frac{2}{3}x+\frac{1}{9}+y^2+\frac{1}{3}y+\frac{1}{36} &= \frac{1}{9}+\frac{1}{9}+\frac{1}{36} \\ (x+1/3)^2+(y+1/6)^2 &= 1/4. \end{aligned}$$

The center is $(-1/3, -1/6)$ with radius $1/2$.

67. Divide equation by 2 and complete the square.

$$\begin{aligned} x^2+2x+y^2 &= 1/2 \\ (x^2+2x+1)+y^2 &= 1/2+1 \\ (x+1)^2+y^2 &= 3/2 \end{aligned}$$

The center is $(-1, 0)$ with radius $\sqrt{\dfrac{3}{2}}$ or $\dfrac{\sqrt{6}}{2}$.

69. Completing the square, we get

$$\begin{aligned} y^2-y+x^2 &= 0 \\ y^2-y+\frac{1}{4}+x^2 &= \frac{1}{4} \\ (y-1/2)^2+x^2 &= \frac{1}{4}, \end{aligned}$$

which is a circle.

71. Divide equation by 4.

$$\begin{aligned} x^2+3y^2 &= 1 \\ x^2+\frac{y^2}{1/3} &= 1 \end{aligned}$$

This is an ellipse.

73. Solve for y and complete the square.

$$\begin{aligned} y &= -2x^2-4x+4 \\ y &= -2(x^2+2x)+4 \\ y &= -2(x^2+2x+1)+4+2 \\ y &= -2(x+1)^2+6 \end{aligned}$$

We find a parabola.

75. Note, $(y-2)^2 = (2-y)^2$. Then we solve for x.

$$\begin{aligned} 2-x &= (y-2)^2 \\ x &= -(y-2)^2+2 \end{aligned}$$

This is a parabola.

77. Simplify and note $(x-4)^2 = (4-x)^2$.

$$\begin{aligned} 2(x-4)^2 &= 4-y^2 \\ 2(x-4)^2+y^2 &= 4 \\ \frac{(x-4)^2}{2}+\frac{y^2}{4} &= 1 \end{aligned}$$

We find an ellipse.

79. Divide given equation by 9 to get the circle given by $x^2+y^2 = \dfrac{1}{9}$.

81. From the foci, we obtain $c = 2$ and $a^2 = b^2+c^2 = b^2+4$. Equation of the ellipse is of the form $\dfrac{x^2}{b^2+4}+\dfrac{y^2}{b^2} = 1$. Substitute $x = 2$ and $y = 3$.

$$\begin{aligned} \frac{4}{b^2+4}+\frac{9}{b^2} &= 1 \\ 4b^2+9(b^2+4) &= b^2(b^2+4) \\ 0 &= b^4-9b^2-36 \\ 0 &= (b^2-12)(b^2+3) \end{aligned}$$

So $b^2 = 12$ and $a^2 = 12+4 = 16$.

The ellipse is given by $\dfrac{x^2}{16}+\dfrac{y^2}{12} = 1$.

83. If c is the distance between the center and focus $(0, 0)$ then the other focus is $(2c, 0)$. Since the distance between the x-intercepts is $6+2c$, which is also the length of the major axis, then $6+2c = 2a$. Since $2a$ is the sum of the distances of $(0, 5)$ from the foci,

$$\begin{aligned} 5+\sqrt{25+4c^2} &= 2a \\ 5+\sqrt{25+4c^2} &= 6+2c \\ \sqrt{25+4c^2} &= 1+2c \\ 25+4c^2 &= 1+4c+4c^2 \\ 24 &= 4c \\ 6 &= c. \end{aligned}$$

The other focus is $(2c, 0) = (12, 0)$.

85. Since the sun is a focus of the elliptical orbit, the length of the major axis is $2a = 521$ (the sum of the shortest distance, $P = 1$ AU, and longest distance, $A = 520$ AU, between the orbit and the sun, respectively). In addition, $c = 259.5$ AU (which is the distance from the center to a focus). The eccentricity is given by $e = \frac{c}{a} = \frac{259.5}{260.5} \approx 0.996$. The orbit's equation is

$$\frac{x^2}{260.5^2} + \frac{y^2}{520} = 1.$$

87. If $2a$ is the sum of the distances from Halley's comet to the two foci and c is the distance from the sun to the center of the ellipse then, $c = a - 8 \times 10^7$.

Since $0.97 = c/a$, we get

$$0.97 = \frac{a - 8 \times 10^7}{a}$$

and the solution of this equation is $a \approx 2.667 \times 10^9$. So $c = a(0.97) \approx 2.587 \times 10^9$. The maximum distance from the sun is $c + a \approx 5.25 \times 10^9$ km.

89. Solving for y, one finds

$$\begin{aligned} y^2 &= 6360^2 - x^2 \\ y &= \pm\sqrt{6360^2 - x^2}. \end{aligned}$$

A sketch of the circle is given.

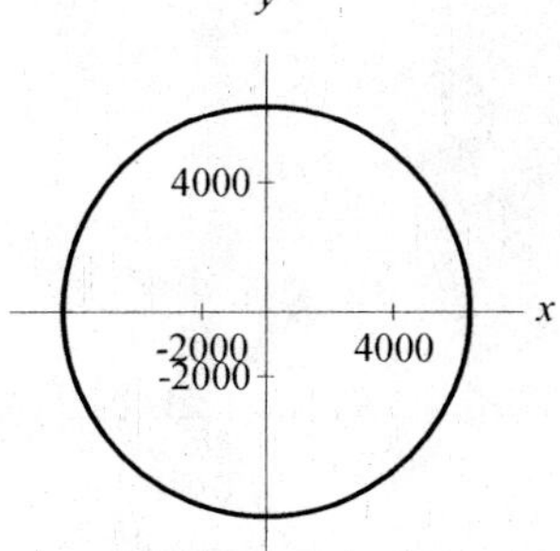

91.

a) As derived below, an equation of the tangent line is given by

$$\begin{aligned} \frac{x_1 x}{a^2} + \frac{y_1 y}{b^2} &= 1 \\ \frac{-4x}{25} + \frac{\frac{9}{5}y}{9} &= 1 \\ \frac{1}{5}y &= 1 + \frac{4x}{25} \\ y &= \frac{4x}{5} + 5. \end{aligned}$$

b) The tangent line and the ellipse intersect at the point $\left(-4, \frac{9}{5}\right)$ as shown.

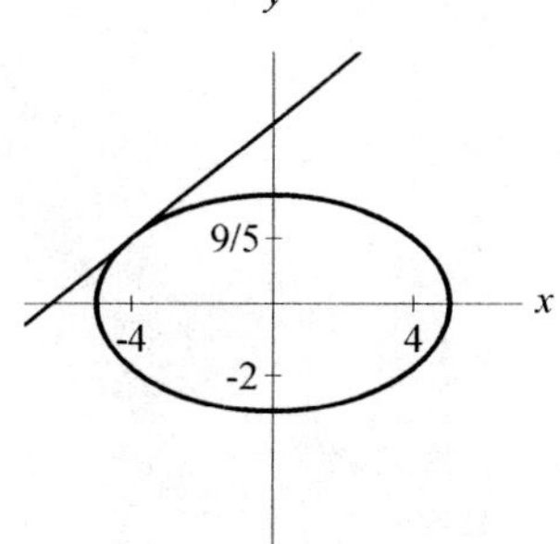

93. A parabolic reflector is preferable since otherwise one would have to place the moving quarterback on one focus of the ellipse.

95. If (x, y) is any point on the ellipse and the sum of the distances from the two foci $(0, \pm c)$ is $2a$, then

$$\sqrt{x^2 + (y - c)^2} + \sqrt{x^2 + (y + c)^2} = 2a.$$

Simplify the inside of the radical and transpose a term to the right-hand side.

$$\sqrt{x^2 + y^2 - 2yc + c^2} = 2a - \sqrt{x^2 + y^2 + 2cy + c^2}$$

Squaring both sides, we find $x^2 + y^2 - 2yc + c^2 = 4a^2 - 4a\sqrt{x^2 + y^2 + 2cy + c^2} + x^2 + y^2 + 2cy + c^2$. Cancel like terms and simplify.

$$\begin{aligned} -4yc - 4a^2 &= -4a\sqrt{x^2 + y^2 + 2cy + c^2} \\ yc + a^2 &= a\sqrt{x^2 + y^2 + 2cy + c^2} \\ y^2c^2 + 2yca^2 + a^4 &= a^2(x^2 + y^2 + 2cy + c^2) \\ y^2c^2 + a^4 &= a^2(x^2 + y^2 + c^2) \\ a^4 - a^2c^2 &= a^2x^2 + y^2(a^2 - c^2) \end{aligned}$$

Let $b^2 = a^2 - c^2$. So $a^2b^2 = a^2x^2 + b^2y^2$ and $\frac{x^2}{b^2} + \frac{y^2}{a^2} = 1$ is an ellipse with foci $(0, \pm c)$ and one finds that the x-intercepts are $(\pm b, 0)$.

97. Factoring, we find

$$\begin{aligned} y &= (2x - 1)^2 \\ y &= 4\left(x - \frac{1}{2}\right)^2. \end{aligned}$$

If $\frac{1}{4p} = 4$, then $p = \frac{1}{16}$.

Since the vertex is $(\frac{1}{2}, 0)$, the focus is $(\frac{1}{2}, 0 + p) = (\frac{1}{2}, \frac{1}{16})$, and directrix is $y = 0 - p = -\frac{1}{16}$.

99. The distance is

$$\begin{aligned} \sqrt{\left(\frac{1}{2}-\frac{1}{4}\right)^2 + \left(-\frac{1}{2}-1\right)^2} &= \\ \sqrt{\left(\frac{1}{4}\right)^2 + \left(-\frac{3}{2}\right)^2} &= \\ \sqrt{\frac{1}{16}+\frac{9}{4}} &= \\ \sqrt{\frac{37}{16}} &= \\ \frac{\sqrt{37}}{4}. \end{aligned}$$

The midpoint is

$$\left(\frac{\frac{1}{2}+\frac{1}{4}}{2}, \frac{-\frac{1}{2}+1}{2}\right) = \left(\frac{3}{8}, \frac{1}{4}\right).$$

101. Since $f(-x) = -f(x)$, the graph is symmetric about the origin.

103. Let (h, k) be the center of the smallest circle. Since the smallest circle is tangent to the y-axis, its radius is $r = h$

Draw a right triangle such that its hypotenuse passes through (h, k) and the center $(1/2, 0)$ of the circle with radius $1/2$. By the Pythagorean Theorem and since the circles are tangent to each other, we have

$$\begin{aligned} (h + 1/2)^2 &= (1/2 - h)^2 + k^2 \\ 2h &= k^2 \end{aligned}$$

after squaring and canceling.

Likewise, draw a right triangle such that its hypotenuse passes through (h, k) and the center $(1/4, \sqrt{2}/2)$ of the circle with radius $1/4$. Then

$$\begin{aligned} (1/4 - h)^2 + (\sqrt{2}/2 - k)^2 &= (h + 1/4)^2 \\ \frac{1}{2} - \sqrt{2}k + k^2 &= h. \end{aligned}$$

after squaring and canceling.

The solutions of the above system of two quadratic equations are $h = 3/2 - \sqrt{2}$ and $k = \sqrt{2} - 1$.

Thus, the center of the smallest circle is $(h, k) = (3/2 - \sqrt{2}, \sqrt{2} - 1)$, and the radius is $r = h$ or $r = 3/2 - \sqrt{2}$.

For Thought

1. False, it is a parabola.

2. False, there is no y-intercept.

3. True **4.** True

5. False, $y = \frac{b}{a}x$ is an asymptote.

6. True **7.** True, for $c = \sqrt{16 + 9} = 5$.

8. True, since $c = \sqrt{3 + 5} = \sqrt{8}$.

9. False, for $y = \frac{2}{3}x$ is an asymptote.

10. False, it is a circle centered at $(0, 0)$.

10.3 Exercises

1. hyperbola

3. center

5. Vertices $(\pm 1, 0)$, foci $(\pm\sqrt{2}, 0)$, asymptotes $y = \pm x$

7. Vertices $(1, \pm 3)$, foci $(1, \pm\sqrt{10})$

Since the slope of the asymptotes are ± 3, the equations of the asymptotes are of the form $y = \pm 3x + b$. If we substitute $(1, 0)$ into $y = \pm 3x + b$, then $0 = \pm 3(1) + b$ or $b = \mp 3$. Thus, the asymptotes are $y = 3x - 3$ and $y = -3x + 3$.

9. Note, $c = \sqrt{a^2 + b^2} = \sqrt{2^2 + 3^2} = \sqrt{13}$.

Foci $(\pm\sqrt{13}, 0)$, asymptotes $y = \pm\frac{b}{a}x = \pm\frac{3}{2}x$

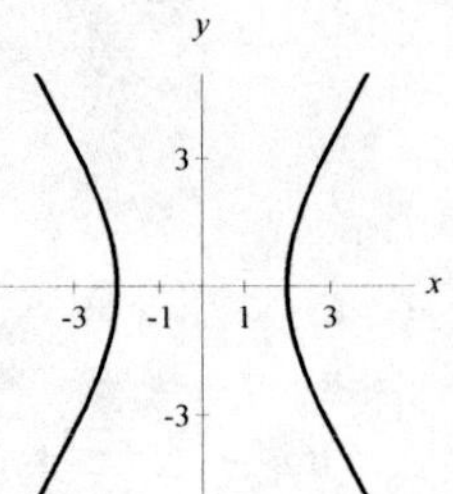

11. Note, $c = \sqrt{a^2+b^2} = \sqrt{2^2+5^2} = \sqrt{29}$.

Foci $(0, \pm\sqrt{29})$, asymptotes $y = \pm\frac{a}{b}x = \pm\frac{2}{5}x$

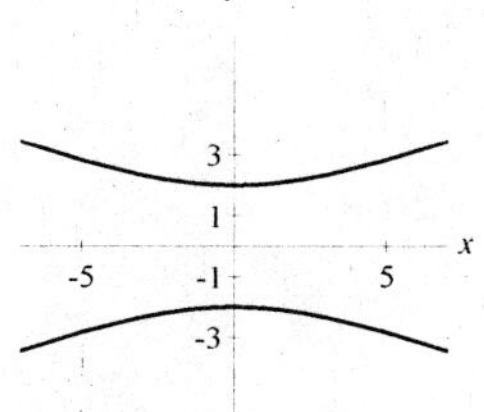

13. Note, $c = \sqrt{a^2+b^2} = \sqrt{2^2+1^2} = \sqrt{5}$.

Foci $(\pm\sqrt{5}, 0)$, asymptotes $y = \pm\frac{b}{a}x = \pm\frac{1}{2}x$

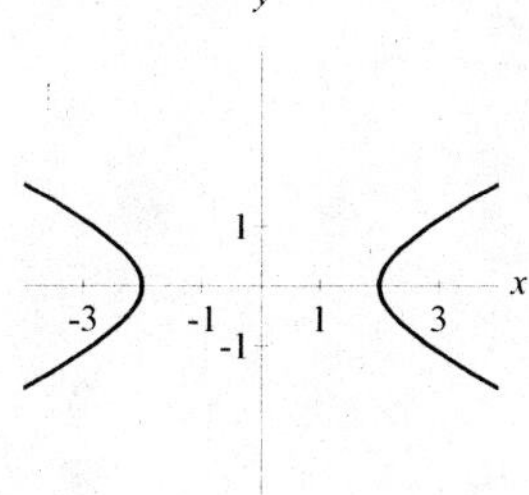

15. Note, $c = \sqrt{a^2+b^2} = \sqrt{1^2+3^2} = \sqrt{10}$.

Foci $(\pm\sqrt{10}, 0)$, asymptotes $y = \pm\frac{b}{a}x = \pm 3x$

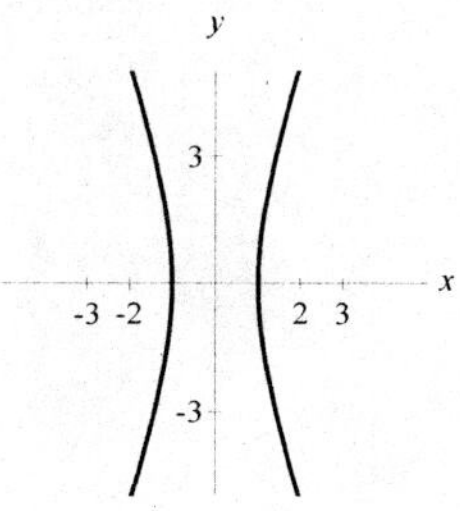

17. Dividing by 144, we get $\frac{x^2}{9} - \frac{y^2}{16} = 1$.

Note, $c = \sqrt{a^2+b^2} = \sqrt{3^2+4^2} = 5$.

Foci $(\pm 5, 0)$, asymptotes $y = \pm\frac{b}{a}x = \pm\frac{4}{3}x$

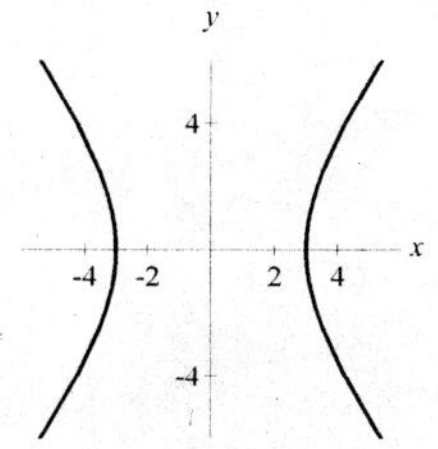

19. Note, $c = \sqrt{a^2+b^2} = \sqrt{1^2+1^2} = \sqrt{2}$.

Foci $(\pm\sqrt{2}, 0)$, asymptotes $y = \pm\frac{b}{a}x = \pm x$

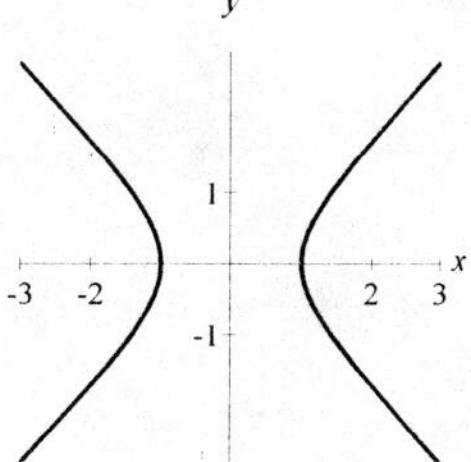

21. Note, $c = \sqrt{a^2+b^2} = \sqrt{2^2+3^2} = \sqrt{13}$. Since the center is $(-1, 2)$, we find that the foci are $(-1 \pm \sqrt{13}, 2)$. Solving for y in

$$y - 2 = \pm\frac{3}{2}(x+1)$$

we obtain that the asymptotes are $y = \frac{3}{2}x + \frac{7}{2}$ and $y = -\frac{3}{2}x + \frac{1}{2}$.

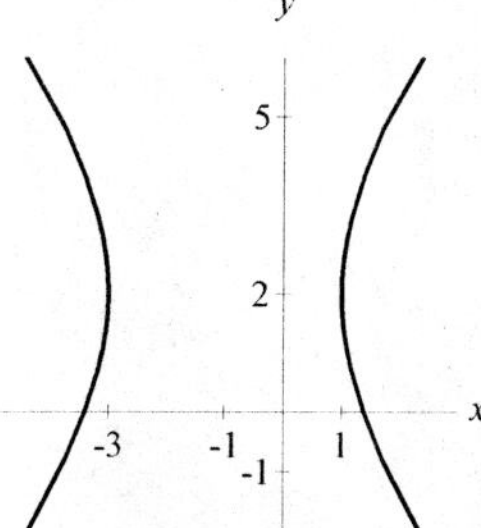

23. Note, $c = \sqrt{a^2+b^2} = \sqrt{2^2+1^2} = \sqrt{5}$. Since the center is $(-2, 1)$, the foci are $(-2, 1 \pm \sqrt{5})$. Solving for y in $y - 1 = \pm 2(x+2)$, we obtain that the asymptotes are $y = 2x + 5$ and $y = -2x - 3$.

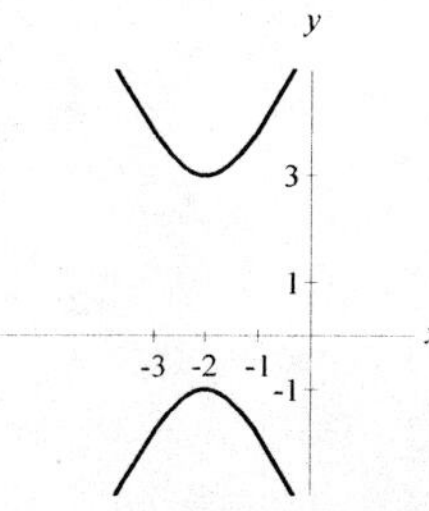

25. Note, $c = \sqrt{a^2+b^2} = \sqrt{4^2+3^2} = 5$. Since the center is $(-2, 3)$, the foci are $(3, 3)$ and $(-7, 3)$. Solving for y in $y - 3 = \pm\frac{3}{4}(x+2)$, we find that the asymptotes are $y = \frac{3}{4}x + \frac{9}{2}$ and $y = -\frac{3}{4}x + \frac{3}{2}$.

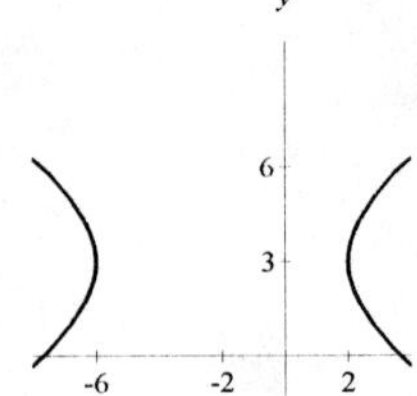

27. Note, $c = \sqrt{a^2 + b^2} = \sqrt{1^2 + 1^2} = \sqrt{2}$. Since the center is $(3, 3)$, the foci are $(3, 3 \pm \sqrt{2})$. Solving for y in $y-3 = \pm(x-3)$, we obtain that the asymptotes are $y = x$ and $y = -x + 6$.

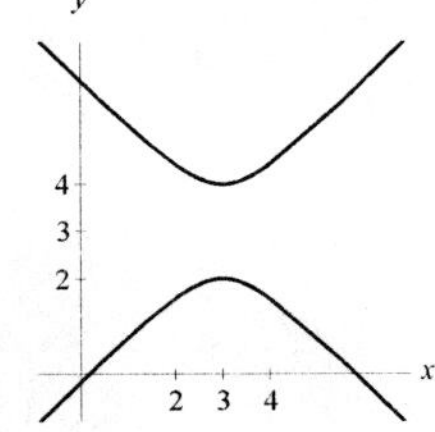

29. Since the x-intercepts are $(\pm 6, 0)$, the hyperbola is given by $\frac{x^2}{6^2} - \frac{y^2}{b^2} = 1$.

From the asymptotes one gets $\frac{1}{2} = \frac{b}{6}$ and $b = 3$. An equation is $\frac{x^2}{36} - \frac{y^2}{9} = 1$.

31. Since the x-intercepts are $(\pm 3, 0)$, the hyperbola is given by $\frac{x^2}{3^2} - \frac{y^2}{b^2} = 1$.

From the foci, $c = 5$ and $b^2 = c^2 - a^2$ $= 5^2 - 3^2 = 16$. An equation is $\frac{x^2}{9} - \frac{y^2}{16} = 1$.

33. By using the vertices of the fundamental rectangle and since it opens sideways, one gets $a = 3$, $b = 5$, and the center is at the origin. An equation is $\frac{x^2}{9} - \frac{y^2}{25} = 1$.

35. $\frac{x^2}{9} - \frac{y^2}{16} = 1$

37. $\frac{y^2}{9} - \frac{(x-1)^2}{9} = 1$

39. Completing the square,

$$\begin{aligned} y^2 - (x^2 - 2x) &= 2 \\ y^2 - (x^2 - 2x + 1) &= 2 - 1 \\ y^2 - (x-1)^2 &= 1 \end{aligned}$$

we obtain a hyperbola.

41. $y = x^2 + 2x$ is a parabola.

43. Simplifying,

$$\begin{aligned} 25x^2 + 25y^2 &= 2500 \\ x^2 + y^2 &= 100 \end{aligned}$$

we obtain a circle.

45. Simplifying,

$$\begin{aligned} 25x &= -100y^2 + 2500 \\ x &= -4y^2 + 100 \end{aligned}$$

we find a parabola.

47. Completing the square,

$$\begin{aligned} 2(x^2 - 2x) + 2(y^2 - 4y) &= -9 \\ 2(x^2 - 2x + 1) + 2(y^2 - 4y + 4) &= -9 + 2 + 8 \\ 2(x-1)^2 + 2(y-2)^2 &= 1 \\ (x-1)^2 + (y-2)^2 &= \frac{1}{2} \end{aligned}$$

we find a circle.

49. Completing the square,

$$\begin{aligned} 2(x^2 + 2x) + y^2 + 6y &= -7 \\ 2(x^2 + 2x + 1) + y^2 + 6y + 9 &= -7 + 2 + 9 \\ 2(x+1)^2 + (y+3)^2 &= 4 \\ \frac{(x+1)^2}{2} + \frac{(y+3)^2}{4} &= 1 \end{aligned}$$

we get an ellipse.

51. Completing the square, we find

$$\begin{aligned} 25(x^2 - 6x + 9) - 4(y^2 + 2y + 1) &= -121 + 225 - 4 \\ 25(x-3)^2 - 4(y+1)^2 &= 100 \\ \frac{(x-3)^2}{4} - \frac{(y+1)^2}{25} &= 1. \end{aligned}$$

We have a hyperbola.

53. From the center $(0,0)$ and vertex $(0,8)$ one gets $a = 8$. From the foci $(0, \pm 10)$, $c = 10$. So $b^2 = c^2 - a^2 = 10^2 - 8^2 = 36$.

Hyperbola is given by $\dfrac{y^2}{64} - \dfrac{x^2}{36} = 1$.

55. Multiply $16y^2 - x^2 = 16$ by 9 and add to $9x^2 - 4y^2 = 36$.

$$\begin{aligned} -9x^2 + 144y^2 &= 144 \\ 9x^2 - 4y^2 &= 36 \\ \hline 140y^2 &= 180 \\ y^2 &= \frac{9}{7} \\ y &= \frac{3\sqrt{7}}{7} \end{aligned}$$

Using $y^2 = \dfrac{9}{7}$ in $x^2 = 16(y^2 - 1)$, we get

$x^2 = 16\left(\dfrac{2}{7}\right)$ or $x = \dfrac{4\sqrt{14}}{7}$.

The exact location is $\left(\dfrac{4\sqrt{14}}{7}, \dfrac{3\sqrt{7}}{7}\right)$.

57. Since $c^2 = a^2 + b^2 = 1^2 + 1^2 = 2$, the foci of $x^2 - y^2 = 1$ are $A(\sqrt{2}, 0)$ and $B(-\sqrt{2}, 0)$. Note, $y^2 = x^2 - 1$. Suppose (x, y) is a point on the hyperbola whose distance from B is twice the distance between (x, y) and A. Then

$$\begin{aligned} 2\sqrt{(x - \sqrt{2})^2 + y^2} &= \sqrt{(x + \sqrt{2})^2 + y^2} \\ 4((x - \sqrt{2})^2 + y^2) &= (x + \sqrt{2})^2 + y^2 \end{aligned}$$

$$\begin{aligned} 4(x - \sqrt{2})^2 - (x + \sqrt{2})^2 + 3y^2 &= 0 \\ 3x^2 - 10\sqrt{2}x + 6 + 3y^2 &= 0 \\ 3x^2 - 10\sqrt{2}x + 6 + 3(x^2 - 1) &= 0 \\ 6x^2 - 10\sqrt{2}x + 3 &= 0. \end{aligned}$$

Solving for x, one finds $x = \dfrac{3\sqrt{2}}{2}$ and $x = \dfrac{\sqrt{2}}{6}$; the second value must be excluded since it is out of the domain. Substituting $x = \dfrac{3\sqrt{2}}{2}$ into $y^2 = x^2 - 1$, one obtains $y = \pm\dfrac{\sqrt{14}}{2}$.

By the symmetry of the hyperbola, there are four points that are twice as far from one focus as they are from the other focus. Namely, these points are

$$\left(\pm\frac{3\sqrt{2}}{2}, \pm\frac{\sqrt{14}}{2}\right).$$

59. Note, the asymptotes are $y = \pm x$. The difference is

$$50 - \sqrt{50^2 - 1} \approx 0.01$$

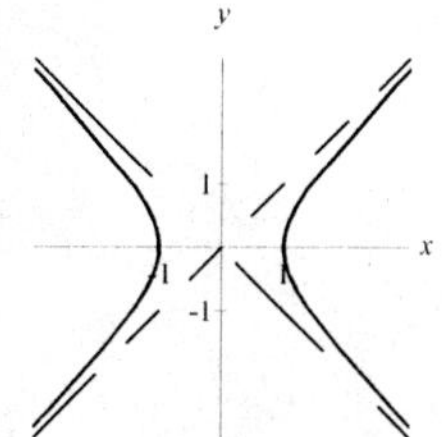

63. $\begin{bmatrix} 1 & 2 \\ -3 & 5 \end{bmatrix} \begin{bmatrix} -1 & 3 \\ 2 & 4 \end{bmatrix} =$

$\begin{bmatrix} 1(-1) + 2(2) & 1(3) + 2(4) \\ -3(-1) + 5(2) & -3(3) + 5(4) \end{bmatrix} =$

$\begin{bmatrix} 3 & 11 \\ 13 & 11 \end{bmatrix}$

65. Since $a = 6$ and $c = 5$, we find $b^2 = a^2 - c^2 = 36 - 25 = 11$. Then the ellipse is

$$\begin{aligned} \frac{x^2}{a^2} + \frac{y^2}{b^2} &= 1 \\ \frac{x^2}{36} + \frac{y^2}{11} &= 1. \end{aligned}$$

67. Let $f(x) = x(x - 1) > 0$.

We apply the test-point method.

Note, $f(-1) > 0, f(1/2) < 0, f(2) > 0$.

$$\begin{array}{ccccc} + & 0 & - & 0 & + \\ \hline -1 & 0 & \frac{1}{2} & 1 & 2 \end{array}$$

The solution set is $(-\infty, 0) \cup (1, \infty)$.

69. On the xyz-space, the points $F(x, 0, 0)$, $T(0, \frac{8x}{x-8}, 8)$, and $C(8, 8, \frac{8(x-8)}{x})$ are collinear. The points F and T are the endpoints of an iron pipe where F lies on the floor of one of the corridors, and T lies on the ceiling of the other corridor.

Using the Pythagorean Theorem, the length of the iron pipe (distance from F to T) is

$$d = \sqrt{x^2 + \left(\frac{8x}{x-8}\right)^2 + 8^2}.$$

Using a calculator, the minimum value of d is 24, and this occurs when $x = 16$. Hence, the longest length of an iron pipe that can be transported is 24 feet.

Review Exercises

1. If $y = 0$, then by factoring we get $0 = (x+6)(x-2)$ and the x-intercepts are $(2,0)$ and $(-6,0)$.

If $x = 0$, then $y = -12$ and the y-intercept is $(0,-12)$.

By completing the square, we obtain $y = (x+2)^2 - 16$, vertex $(h,k) = (-2,-16)$, and the axis of symmetry is $x = -2$.

Since $p = \dfrac{1}{4a} = \dfrac{1}{4}$, the focus is $(h, k+p) = (-2, -63/4)$ and directrix is $y = k - p$ or $y = -65/4$.

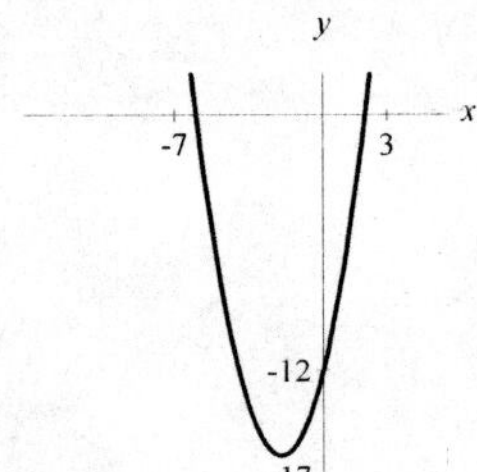

3. If $y = 0$, then by factoring we find $0 = x(6-2x)$ and the x-intercepts are $(0,0)$ and $(3,0)$.

If $x = 0$, then $y = 0$ and the y-intercept is $(0,0)$.

By completing the square, one gets $y = -2(x-3/2)^2 + 9/2$, with vertex $(h,k) = (3/2, 9/2)$, and axis of symmetry is $x = 3/2$.

Since $p = \dfrac{1}{4a} = -\dfrac{1}{8}$, the focus is $(h, k+p) = (3/2, 35/8)$ and the directrix is $y = k - p$ or $y = 37/8$.

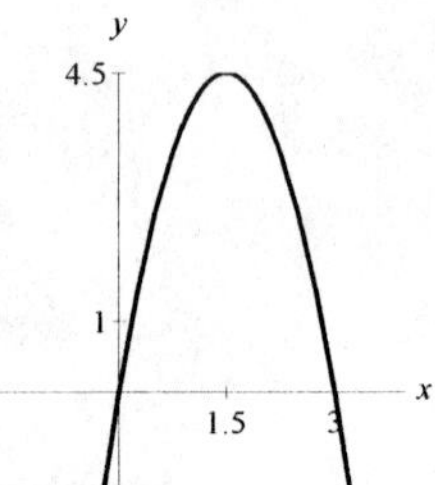

5. By completing the square, we have $x = (y+2)^2 - 10$. If $x = 0$, then $y + 2 = \pm\sqrt{10}$ and the y-intercepts are $(0, -2 \pm \sqrt{10})$.

If $y = 0$, then $x = -6$ and the x-intercept is $(-6,0)$. Since $x = (y+2)^2 - 10$ is of the form $x = a(y-h)^2 + k$, the vertex is $(k,h) = (-10,-2)$, and the axis of symmetry is $y = -2$.

Since $p = \dfrac{1}{4a} = \dfrac{1}{4}$, the focus is

$$(k+p, h) = (-39/4, -2)$$

and directrix is $x = k - p$ or $x = -41/4$.

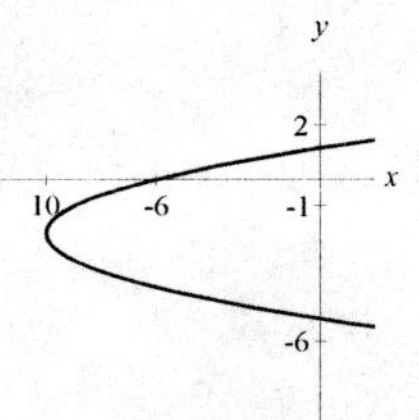

7. Since $c = \sqrt{a^2 - b^2} = \sqrt{36 - 16} = 2\sqrt{5}$, the foci are $(0, \pm 2\sqrt{5})$

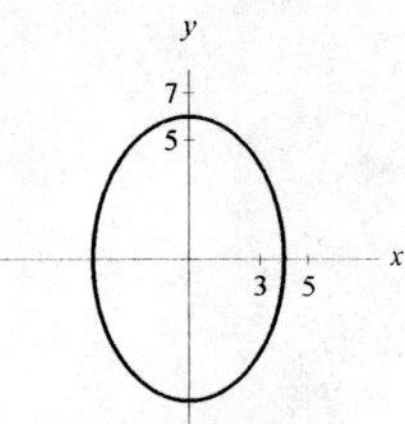

9. Since $c = \sqrt{a^2 - b^2} = \sqrt{24 - 8} = 4$, the foci are $(1, 1 \pm 4)$, or $(1,5)$ and $(1,-3)$.

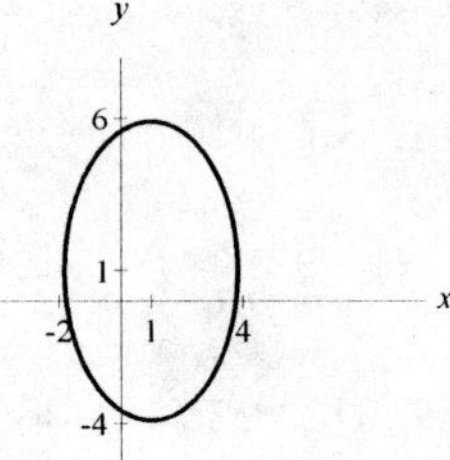

11. Rewrite equation as $\dfrac{(x-1)^2}{8} + \dfrac{(y+3)^2}{10} = 1$.

Since $c = \sqrt{a^2 - b^2} = \sqrt{10-8} = \sqrt{2}$, the foci are $(1, -3 \pm \sqrt{2})$.

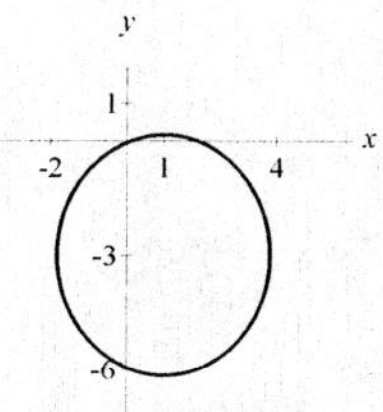

13. center $(0,0)$, radius 9

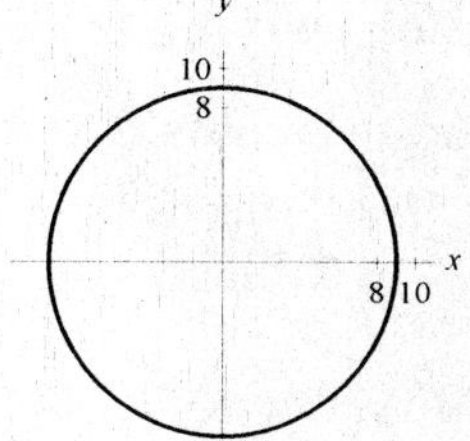

15. center $(-1,0)$, radius 2

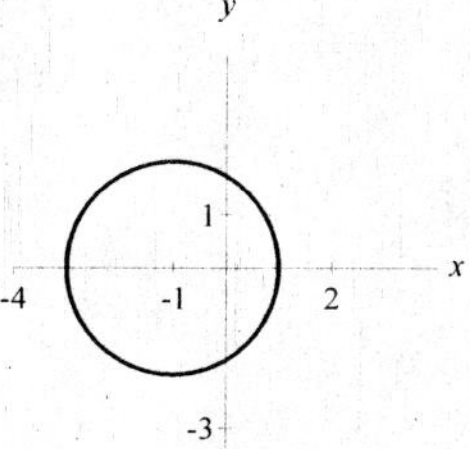

17. Completing the square, we have

$$\begin{aligned} x^2 + 5x + \frac{25}{4} + y^2 &= -\frac{1}{4} + \frac{25}{4} \\ \left(x + \frac{5}{2}\right)^2 + y^2 &= 6, \end{aligned}$$

and so center is $(-5/2, 0)$ and radius is $\sqrt{6}$.

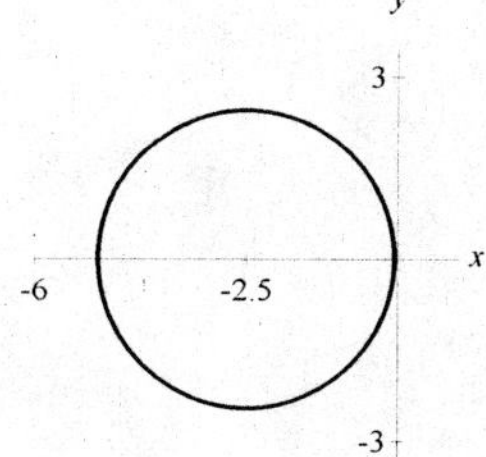

19. $x^2 + (y+4)^2 = 9$

21. $(x+2)^2 + (y+7)^2 = 6$

23. Since $c = \sqrt{a^2 + b^2} = \sqrt{8^2 + 6^2} = 10$,

foci $(\pm 10, 0)$, asymptotes $y = \pm\dfrac{b}{a} = \pm\dfrac{3}{4}x$

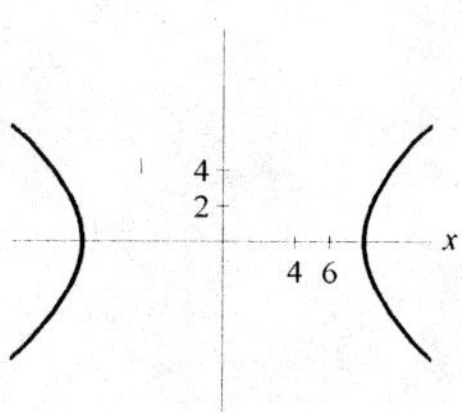

25. Since $c = \sqrt{a^2 + b^2} = \sqrt{8^2 + 4^2} = 4\sqrt{5}$,

the foci are $(4, 2 \pm 4\sqrt{5})$. Solving

for y in $y - 2 = \pm\dfrac{8}{4}(x-4)$,

asymptotes are $y = 2x - 6$ and $y = -2x + 10$.

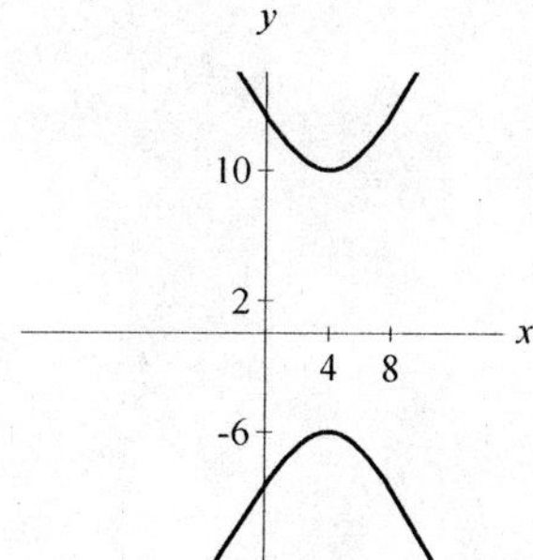

27. Completing the square, we have

$$\begin{aligned} x^2 - 4x - 4(y^2 - 8y) &= 64 \\ x^2 - 4x + 4 - 4(y^2 - 8y + 16) &= 64 + 4 - 64 \\ (x-2)^2 - 4(y-4)^2 &= 4 \\ \frac{(x-2)^2}{4} - (y-4)^2 &= 1, \end{aligned}$$

and so $c = \sqrt{a^2 + b^2} = \sqrt{2^2 + 1^2} = \sqrt{5}$, and the foci are $(2 \pm \sqrt{5}, 4)$. Solving for y

in $y - 4 = \pm\dfrac{1}{2}(x-2)$, we get that the

asymptotes are $y = \dfrac{1}{2}x + 3$ and $y = -\dfrac{1}{2}x + 5$.

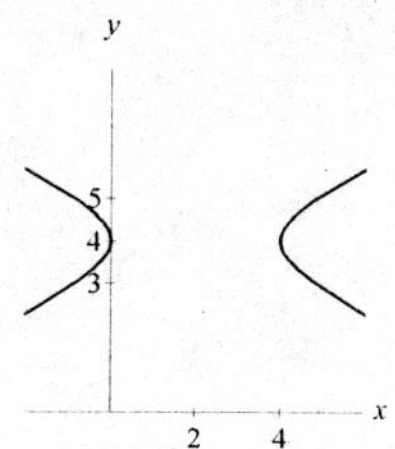

29. Hyperbola

31. Ellipse

33. Parabola

35. Hyperbola

37. $x^2 + y^2 = 4$ is a circle

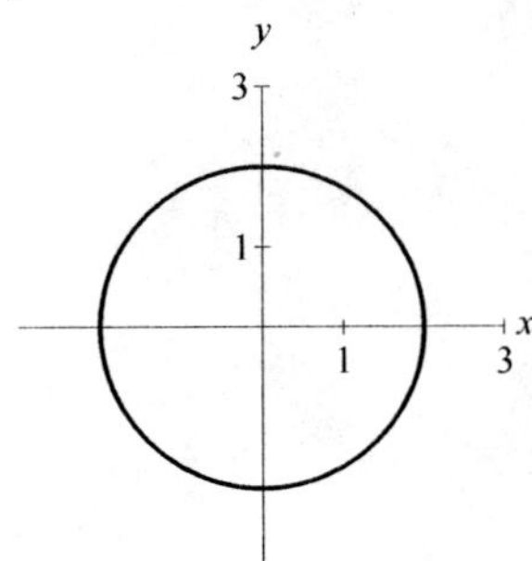

39. Since $4y = x^2 - 4$, $y = \dfrac{1}{4}x^2 - 1$.

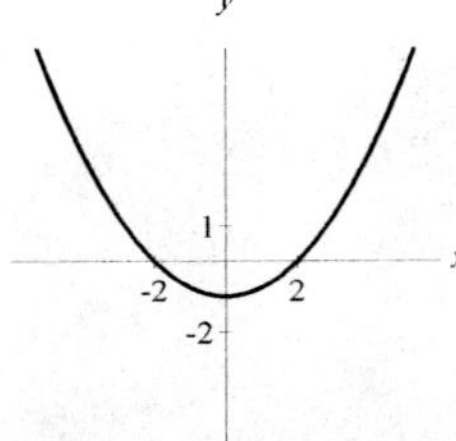

41. Since $x^2 + 4y^2 = 4$, $\dfrac{x^2}{4} + y^2 = 1$.

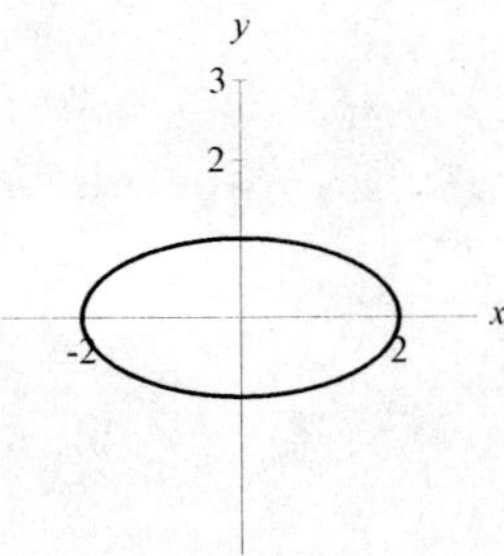

43. Since $x^2 - 4x + 4 + 4y^2 = 4$, we find

$(x-2)^2 + 4y^2 = 4$ and $\dfrac{(x-2)^2}{4} + y^2 = 1$.

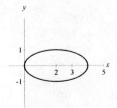

45. Since the vertex is midway between the focus $(1, 3)$ and directrix $x = \dfrac{1}{2}$, the vertex is $\left(\dfrac{3}{4}, 3\right)$ and $p = \dfrac{1}{4}$. Since $a = \dfrac{1}{4p} = 1$, parabola is given by

$x = (y-3)^2 + \dfrac{3}{4}$.

47. From the foci and vertices, we get $c = 4$ and $a = 6$, respectively. Since $b^2 = a^2 - c^2 = 36 - 16 = 20$, the ellipse is given by $\dfrac{x^2}{36} + \dfrac{y^2}{20} = 1$.

49. Radius is $\sqrt{(-1-1)^2 + (-1-3)^2} = \sqrt{20}$. Equation is $(x-1)^2 + (y-3)^2 = 20$.

51. From the foci and x-intercepts one gets $c = 3$ and $a = 2$, respectively. Since $b^2 = c^2 - a^2 = 9 - 4 = 5$, the hyperbola is given by $\dfrac{x^2}{4} - \dfrac{y^2}{5} = 1$.

53. Since the center of the circle is $(-2, 3)$ and the raidus is 3, we have $(x+2)^2 + (y-3)^2 = 9$.

55. Note, the center of the ellipse is $(-2, 1)$. Using the lengths of the major axis, we get $a = 3$ and $b = 1$. Thus, an equation is

$$\frac{(x+2)^2}{9} + (y-1)^2 = 1.$$

57. Note, the center of the hyperbola is $(2, 1)$. By using the fundamental rectangle, we find $a = 3$ and $b = 2$. Thus, an equation is

$$\frac{(y-1)^2}{9} - \frac{(x-2)^2}{4} = 1.$$

59. The equation is of the form $\dfrac{x^2}{100^2} - \dfrac{y^2}{b^2} = 1$. Since the graph passes through $(120, 24\sqrt{11})$, we get

$$\begin{aligned} \frac{120^2}{100^2} - \frac{(24\sqrt{11})^2}{b^2} &= 1 \\ 1.44 - \frac{6336}{b^2} &= 1 \\ b^2 &= \frac{6336}{0.44} \\ b^2 &= 120^2. \end{aligned}$$

Equation is $\dfrac{x^2}{100^2} - \dfrac{y^2}{120^2} = 1$.

61. Note $c = 30$ and $a = 34$. Then an equation we can use is of the form $\dfrac{x^2}{34^2} + \dfrac{y^2}{b^2} = 1$.
Since $b^2 = a^2 - c^2 = 34^2 - 30^2 = 16^2$, the equation is

$$\frac{x^2}{34^2} + \frac{y^2}{16^2} = 1.$$

To find h, let $x = 32$.

$$\begin{aligned} \frac{32^2}{34^2} + \frac{y^2}{16^2} &= 1 \\ y^2 &= \left(1 - \frac{32^2}{34^2}\right) 16^2 \\ y &\approx 5.407. \end{aligned}$$

Thus, $h = 2y \approx 10.81$ feet.

63. Since 311 is an odd number and the difference between two even numbers is even, we conclude $y = 0$. Then $2^x - 1 = 311$ or $2^x = 312$. The latter equation has no integer solutions. Thus, there are no integers x, y such that $2^x - 2^y = 311$.

Chapter 10 Test

1. Circle $x^2 + y^2 = 8$

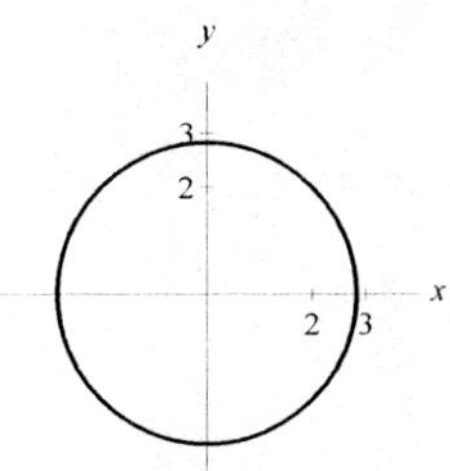

2. Ellipse $\dfrac{x^2}{9} + \dfrac{y^2}{100} = 1$

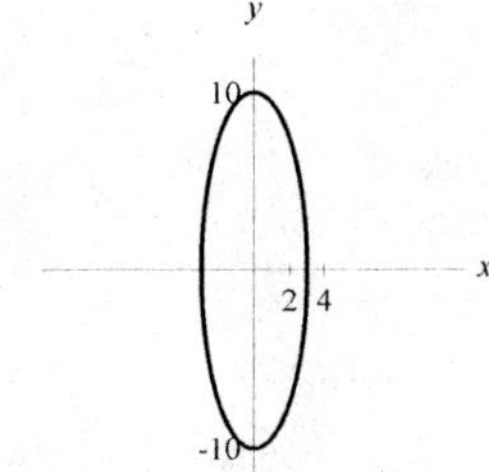

3. Parabola $y = x^2 + 6x + 8$

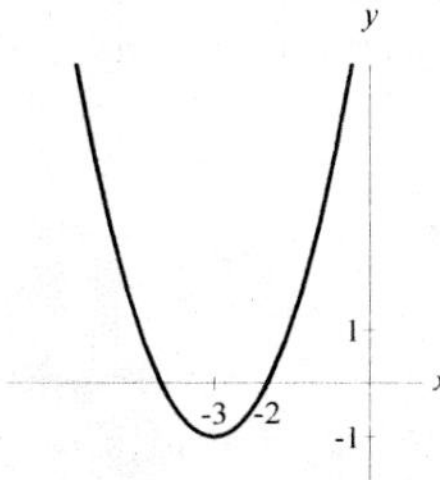

4. Hyperbola $\dfrac{y^2}{25} - \dfrac{x^2}{9} = 1$

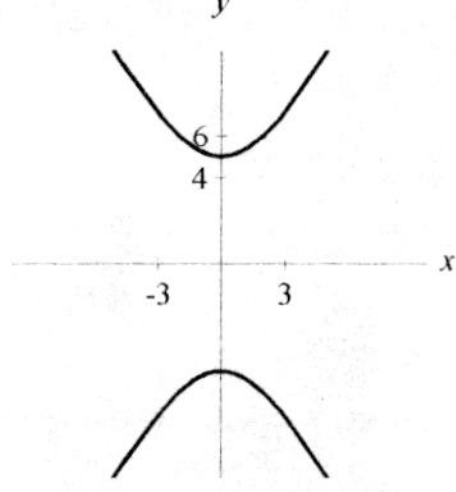

5. Circle $(x + 3)^2 + (y - 1)^2 = 10$

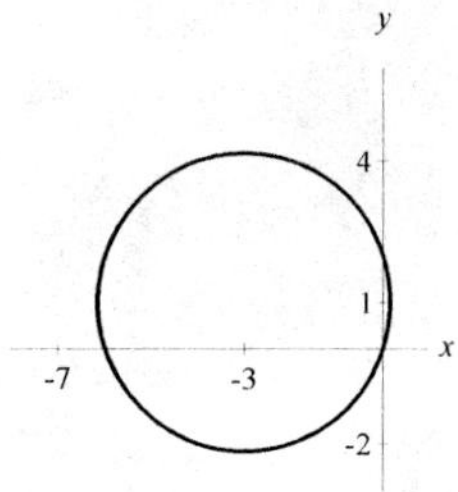

6. Hyperbola $\dfrac{(x - 2)^2}{9} - \dfrac{(y + 3)^2}{4} = 1$

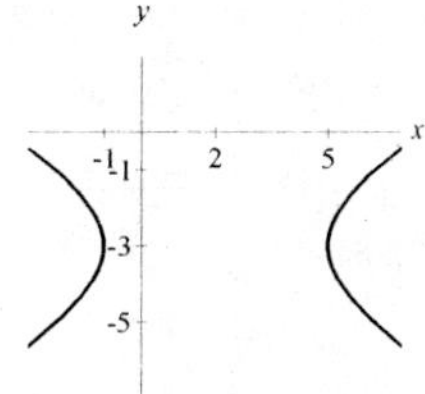

7. By completing the square, the equation can be written as $(x - 4)^2 - 16 = y^2$. This is a hyperbola.

8. By completing the square, we can rewrite the equation as $(x - 4)^2 - 16 = y$. This is a parabola.

9. By completing the square, we can rewrite the equation as $-(x - 4)^2 + 16 = y^2$. This is a circle.

10. By completing the square, the equation can be written as $-(x-4)^2+16=8y^2$. This is an ellipse.

11. Since $(2\sqrt{3})^2=12$, the circle is given by $(x+3)^2+(y-4)^2=12$.

12. Midway between the focus $(2,0)$ and directrix $x=-2$ is the vertex $(0,0)$. Since $p=2$,

$a=\dfrac{1}{4p}=\dfrac{1}{8}$. Equation is $x=\dfrac{1}{8}y^2$.

13. Equation is of the form $\dfrac{x^2}{2^2}+\dfrac{y^2}{a^2}=1$.

Since $a^2=b^2+c^2=2^2+\sqrt{6}^2=10$,

equation is $\dfrac{x^2}{4}+\dfrac{y^2}{10}=1$.

14. From the foci and vertices one gets $c=8$ and $a=6$. Since $b^2=c^2-a^2=8^2-6^2=28$,

equation is $\dfrac{x^2}{36}-\dfrac{y^2}{28}=1$.

15. Complete the square to get $y=(x-2)^2-4$.

So vertex is $(h,k)=(2,-4)$, axis of

symmetry $x=2$, $a=1$, and $p=\dfrac{1}{4a}=\dfrac{1}{4}$.

Focus is $(h,k+p)=(2,-15/4)$ and directrix is $y=k-p=-17/4$.

16. Since $\dfrac{x^2}{16}+\dfrac{y^2}{4}=1$, we get
$c=\sqrt{a^2-b^2}=\sqrt{4^2-2^2}=2\sqrt{3}$.
Foci are $(\pm 2\sqrt{3},0)$, length of major axis is $2a=8$, length of minor axis is $2b=4$.

17. Since $y^2-\dfrac{x^2}{16}=1$, we obtain
$c=\sqrt{a^2+b^2}=\sqrt{1^2+4^2}=\sqrt{17}$.
Foci are $(0,\pm\sqrt{17})$, vertices $(0,\pm 1)$,

asymptotes $y=\pm\dfrac{1}{4}x$, length of transverse

axis is $2a=2$, length of conjugate axis is $2b=8$.

18. Completing the square, we find

$$\begin{aligned} x^2+x+\frac{1}{4}+y^2-3y+\frac{9}{4} &= -\frac{1}{4}+\frac{1}{4}+\frac{9}{4} \\ \left(x+\frac{1}{2}\right)^2+\left(y-\frac{3}{2}\right)^2 &= \frac{9}{4}. \end{aligned}$$

The center is $\left(-\dfrac{1}{2},\dfrac{3}{2}\right)$ and the radius is $\dfrac{3}{2}$.

19. Since

$$\frac{x^2}{225}+\frac{y^2}{81}=1$$

we find $a=15$ and $b=9$. Note,

$$c=\sqrt{a^2-c^2}=\sqrt{15^2-9^2}=12.$$

Since the foci are $(\pm 12,0)$, the distance from the point of generation of the waves to the kidney stones is $2c=24$ cm.

Tying It All Together

1. Parabola $y=6x-x^2$

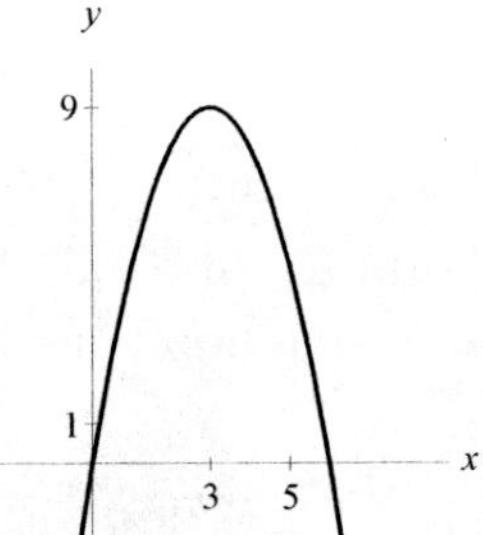

2. Line $y=6x$

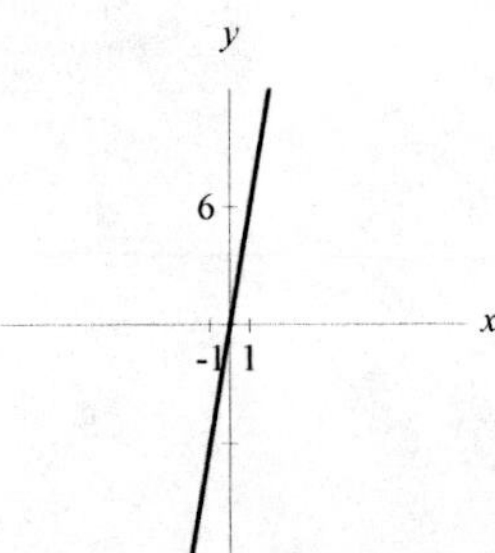

3. Parabola $y=6-x^2$

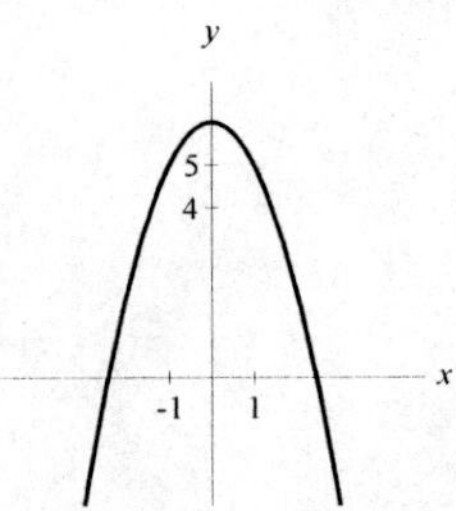

4. Circle $x^2 + y^2 = 6$

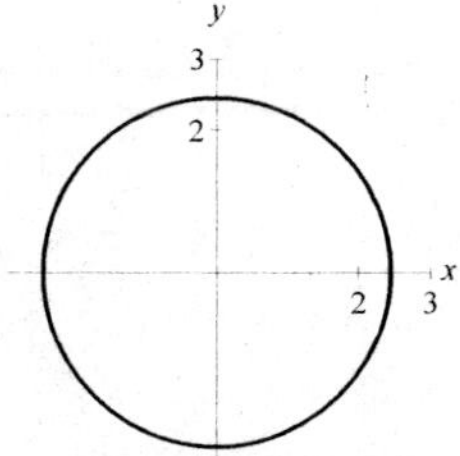

5. Line $y = x + 6$

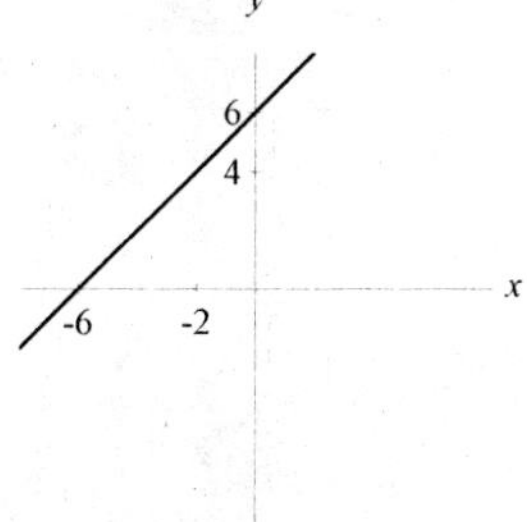

6. Parabola $x = 6 - y^2$

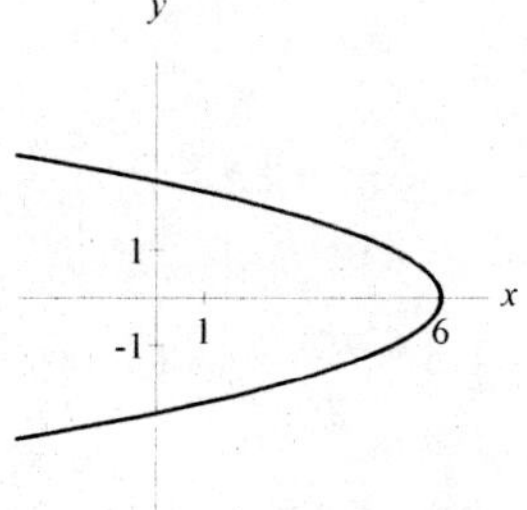

7. Parabola $y = (6 - x)^2 = (x - 6)^2$

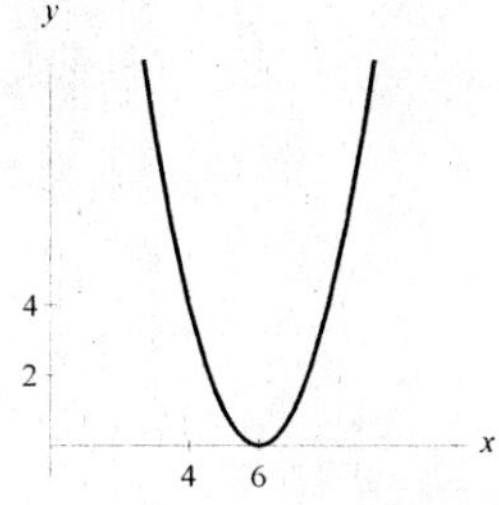

8. $y = |x + 6|$ goes through $(-6, 0)$, $(-5, 1)$

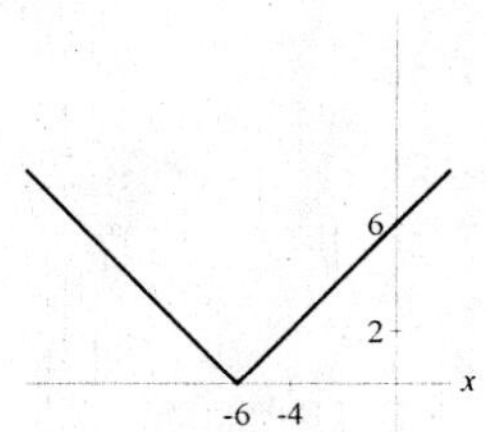

9. $y = 6^x$ goes through $(0, 1)$, $(1, 6)$

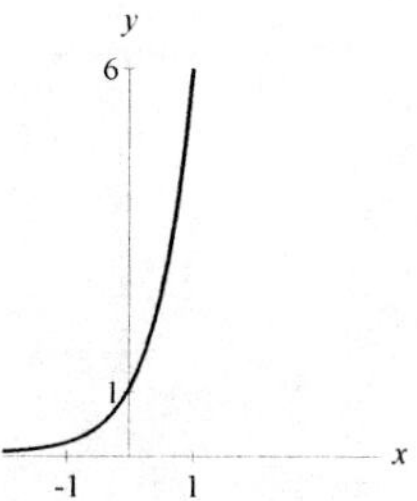

10. $y = \log_6(x)$ goes through $(6, 1)$, $(1/6, -1)$

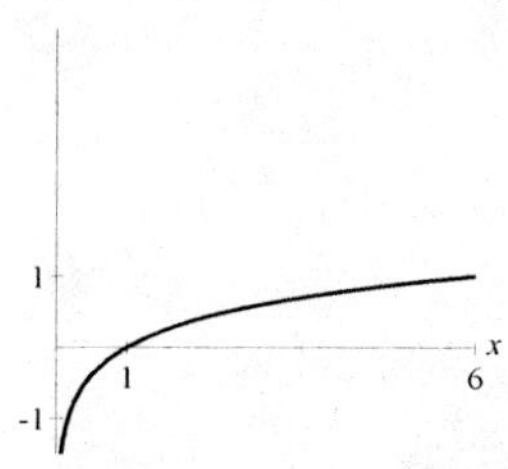

11. $y = \dfrac{1}{x^2 - 6}$ has asymptotes $x = \pm\sqrt{6}$

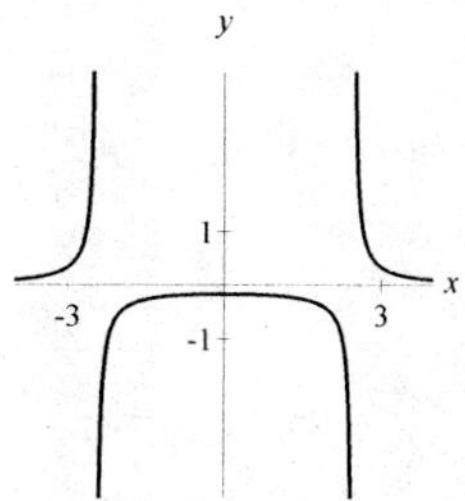

12. Ellipse $\dfrac{x^2}{9} + \dfrac{y^2}{4} = 1$

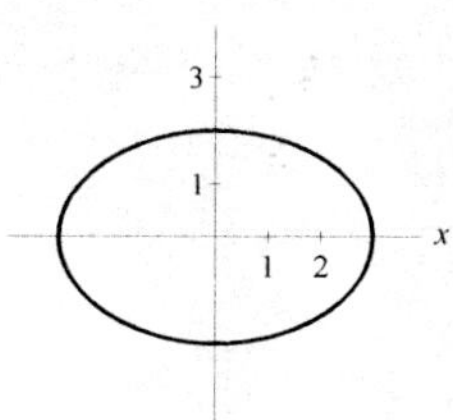

13. Hyperbola $\dfrac{x^2}{9} - \dfrac{y^2}{4} = 1$

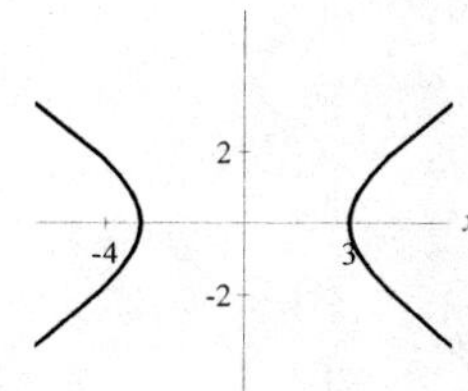

14. $y = x(6 - x^2)$ goes through $(0, 0)$, $(\pm\sqrt{6}, 0)$

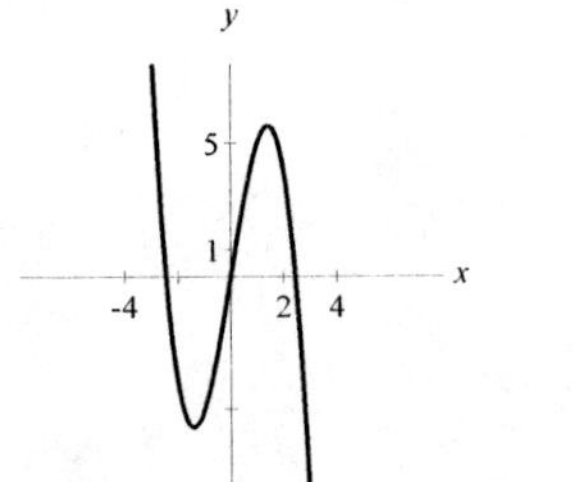

15. Line $2x + 3y = 6$

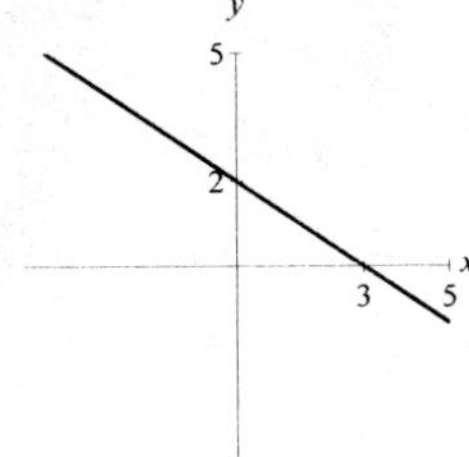

16. By completing the square, the equation may be written as $\dfrac{(x-3)^2}{3} - \dfrac{y^2}{3} = 1$. This is a hyperbola.

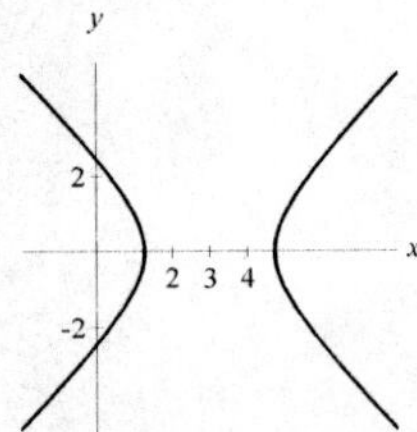

17. Solving for x, we get

$$\begin{aligned} 3x - 9 + 5 &= -9 \\ 3x &= -5. \end{aligned}$$

The solution set is $\{-5/3\}$.

18. Solving for x, we find

$$\begin{aligned} 5x - 8x + 12 &= 17 \\ -3x &= 5. \end{aligned}$$

The solution set is $\{-5/3\}$.

19. Solving for x, we obtain

$$\begin{aligned} 2x - \frac{2}{3} - \frac{3}{2} + 3x &= \frac{3}{2} \\ 5x - \frac{13}{6} &= \frac{3}{2} \\ 5x &= \frac{11}{3}. \end{aligned}$$

The solution set is $\{11/15\}$.

20. Solving for x, we have

$$\begin{aligned} -4x + 2 &= 3x + 2 \\ -7x &= 0. \end{aligned}$$

The solution set is $\{0\}$.

21. Solving for x, we find

$$\begin{aligned} \frac{1}{2}x - \frac{1}{4}x &= \frac{3}{2} + \frac{1}{3} \\ \frac{1}{4}x &= \frac{11}{6} \\ x &= \frac{44}{6}. \end{aligned}$$

Solution set is $\{22/3\}$.

22. Multiplying both sides by 100, we obtain

$$\begin{aligned} 5(x - 20) + 2(x + 10) &= 270 \\ 5x - 100 + 2x + 20 &= 270 \\ 7x &= 350. \end{aligned}$$

The solution set is $\{50\}$.

23. By using the quadratic formula to solve $2x^2 + 31x - 51 = 0$, one obtains

$$\begin{aligned} x &= \frac{-31 \pm \sqrt{31^2 - 4(2)(-51)}}{4} \\ &= \frac{-31 \pm \sqrt{1369}}{4} \\ &= \frac{-31 \pm 37}{4} \\ &= \frac{3}{2}, -17. \end{aligned}$$

The solution set is $\left\{\frac{3}{2}, -17\right\}$.

24. Since $x(2x + 31) = 0$, the solution set is $\left\{0, -\frac{31}{2}\right\}$.

25. By using the quadratic formula to solve $x^2 - 34x + 286 = 0$, one obtains

$$\begin{aligned} x &= \frac{34 \pm \sqrt{34^2 - 4(1)(286)}}{2} \\ x &= \frac{34 \pm \sqrt{12}}{2} \\ x &= \frac{34 \pm 2\sqrt{3}}{2} \\ x &= 17 \pm \sqrt{3}. \end{aligned}$$

The solution set is $\left\{17 \pm \sqrt{3}\right\}$.

26. In solving $x^2 - 34x + 290 = 0$, one can use the quadratic formula. Then

$$\begin{aligned} x &= \frac{34 \pm \sqrt{34^2 - 4(1)(290)}}{2} \\ x &= \frac{34 \pm \sqrt{-4}}{2} \\ x &= \frac{34 \pm 2i}{2} \\ x &= 17 \pm i. \end{aligned}$$

The solution set is $\{17 \pm i\}$.

27. identity

28. scalar

29. $m \times p$

30. identity

31. determinant

32. determinant

33. Cramer's rule

34. parabola

35. axis of symmetry

36. vertex

37. ellipse

38. circle

39. hyperbola

For Thought

1. True

2. False, the domain of a finite sequence is $\{1, 2, ..., n\}$ for some positive integer n.

3. True

4. False, n is the independent variable.

5. False, the first four terms are $1, -8, 27, -64$.

6. False, $a_5 = -3 + 4 \cdot 6 = 21$.

7. False, the common difference is $d = 4 - 7 = -3$.

8. False, there is no common difference.

9. False, since if $14 = 4 + 2d$ then $d = 5$ and $a_4 = 4 + 3 \cdot 5 = 19$. **10.** True

11.1 Exercises

1. finite sequence

3. terms

5. arithmetic

7. $a_1 = 1^2 = 1$, $a_2 = 2^2 = 4$, $a_3 = 3^2 = 9$, $a_4 = 4^2 = 16$, $a_5 = 5^2 = 25$, $a_6 = 6^2 = 36$, $a_7 = 7^2 = 49$.

First seven terms are $1, 4, 9, 16, 25, 36, 49$.

9. $b_1 = \dfrac{(-1)^2}{2} = \dfrac{1}{2}$, $b_2 = \dfrac{(-1)^3}{3} = -\dfrac{1}{3}$,

$b_3 = \dfrac{(-1)^4}{4} = \dfrac{1}{4}$, $b_4 = \dfrac{(-1)^5}{5} = -\dfrac{1}{5}$,

$b_5 = \dfrac{(-1)^6}{6} = \dfrac{1}{6}$, $b_6 = \dfrac{(-1)^7}{7} = -\dfrac{1}{7}$,

$b_7 = \dfrac{(-1)^8}{8} = \dfrac{1}{8}$, $b_8 = \dfrac{(-1)^9}{9} = -\dfrac{1}{9}$.

First eight terms are

$\dfrac{1}{2}, -\dfrac{1}{3}, \dfrac{1}{4}, -\dfrac{1}{5}, \dfrac{1}{6}, -\dfrac{1}{7}, \dfrac{1}{8}, -\dfrac{1}{9}$.

11. $c_1 = (-2)^0 = 1$, $c_2 = (-2)^1 = -2$, $c_3 = (-2)^2 = 4$, $c_4 = (-2)^3 = -8$, $c_5 = (-2)^4 = 16$, $c_6 = (-2)^5 = -32$.

First six terms are $1, -2, 4, -8, 16, -32$

13. $a_1 = 2^1 = 2$, $a_2 = 2^0 = 1$,

$a_3 = 2^{-1} = \dfrac{1}{2}$, $a_4 = 2^{-2} = \dfrac{1}{4}$, $a_5 = 2^{-3} = \dfrac{1}{8}$.

First five terms are $2, 1, \dfrac{1}{2}, \dfrac{1}{4}, \dfrac{1}{8}$.

15. $a_1 = -6 + 0 = -6$, $a_2 = -6 - 4 = -10$,

$a_3 = -6 - 8 = -14$, $a_4 = -6 - 12 = -18$,

$a_5 = -6 - 16 = -22$. First five

terms are $-6, -10, -14, -18, -22$.

17. $b_1 = 5 + 0 = 5$, $b_2 = 5 + 0.5 = 5.5$,

$b_3 = 5 + 1 = 6$, $b_4 = 5 + 1.5 = 6.5$,

$b_5 = 5 + 2 = 7$, $b_6 = 5 + 2.5 = 7.5$,

$b_7 = 5 + 3 = 8$. First seven

terms are $5, 5.5, 6, 6.5, 7, 7.5, 8$.

19. $a_1 = -0.1 + 9 = 8.9$,

$a_2 = -0.2 + 9 = 8.8$, $a_3 = -0.3 + 9 = 8.7$,

$a_4 = -0.4 + 9 = 8.6$, $a_{10} = -1 + 9 = 8$.

First four terms are $8.9, 8.8, 8.7, 8.6$ and

and 10th term is 8.

21. $a_1 = 8 + 0 = 8$, $a_2 = 8 - 3 = 5$,

$a_3 = 8 - 6 = 2$, $a_4 = 8 - 9 = -1$, and

$a_{10} = 8 - 27 = -19$. First four terms are $8, 5, 2, -1$ and 10th term is -19.

23. One gets $a_1 = \dfrac{4}{2+1} = \dfrac{4}{3}$,

$a_2 = \dfrac{4}{4+1} = \dfrac{4}{5}$, $a_3 = \dfrac{4}{6+1} = \dfrac{4}{7}$,

$a_4 = \dfrac{4}{8+1} = \dfrac{4}{9}$, $a_{10} = \dfrac{4}{20+1} = \dfrac{4}{21}$

The first four terms are

$\dfrac{4}{3}, \dfrac{4}{5}, \dfrac{4}{7}, \dfrac{4}{9}$ and 10th term is $\dfrac{4}{21}$.

25. One gets $a_1 = \dfrac{(-1)^1}{2 \cdot 3} = -\dfrac{1}{6}$,

$a_2 = \dfrac{(-1)^2}{3 \cdot 4} = \dfrac{1}{12}$, $a_3 = \dfrac{(-1)^3}{4 \cdot 5} = -\dfrac{1}{20}$,

$a_4 = \dfrac{(-1)^4}{5 \cdot 6} = \dfrac{1}{30}$, $a_{10} = \dfrac{(-1)^{10}}{11 \cdot 12} = \dfrac{1}{132}$.

The first four terms are

$-\dfrac{1}{6}, \dfrac{1}{12}, -\dfrac{1}{20}, \dfrac{1}{30}$ and 10th term is $\dfrac{1}{132}$.

27. $a_1 = 2! = 2$, $a_2 = 4! = 24$, $a_3 = 6! = 720$,

$a_4 = 8! = 40,320$, and $a_5 = 10! = 3,628,800$

29. $a_1 = \frac{2^1}{1!} = 2$, $a_2 = \frac{2^2}{2!} = 2$, $a_3 = \frac{2^3}{3!} = \frac{4}{3}$,

$a_4 = \frac{2^4}{4!} = \frac{2}{3}$, and $a_5 = \frac{2^5}{5!} = \frac{4}{15}$

31. $b_1 = \frac{1!}{0!} = 1$, $b_2 = \frac{2!}{1!} = 2$, $b_3 = \frac{3!}{2!} = 3$,

$b_4 = \frac{4!}{3!} = 4$, and $b_5 = \frac{5!}{4!} = \frac{5 \cdot 4!}{4!} = 5$

33. $c_1 = \frac{(-1)^1}{2!} = -\frac{1}{2}$, $c_2 = \frac{(-1)^2}{3!} = \frac{1}{6}$,

$c_3 = \frac{(-1)^3}{4!} = -\frac{1}{24}$, $c_4 = \frac{(-1)^4}{5!} = \frac{1}{120}$,

and $c_5 = \frac{(-1)^5}{6!} = -\frac{1}{720}$.

35. Note $t_n = \frac{1}{(n+2)!e^n}$. Then we obtain

$t_1 = \frac{1}{3!e} = \frac{1}{6e}$, $t_2 = \frac{1}{4!e^2} = \frac{1}{24e^2}$,

$t_3 = \frac{1}{5!e^3} = \frac{1}{120e^3}$, $t_4 = \frac{1}{6!e^4} = \frac{1}{720e^4}$,

and $t_5 = \frac{1}{7!e^5} = \frac{1}{5040e^5}$.

37. $a_n = 2n$

39. $a_n = 2n + 7$

41. $a_n = (-1)^{n+1}$

43. $a_n = n^3$

45. $a_n = e^n$

47. $a_n = \frac{1}{2^{n-1}}$

49. $a_2 = 3a_1 + 2 = 3(-4) + 2 = -10$,
$a_3 = 3a_2 + 2 = 3(-10) + 2 = -28$,
$a_4 = 3a_3 + 2 = 3(-28) + 2 = -82$,
$a_5 = 3a_4 + 2 = 3(-82) + 2 = -244$,
$a_6 = 3a_5 + 2 = 3(-244) + 2 = -730$,
$a_7 = 3a_6 + 2 = 3(-730) + 2 = -2188$,
$a_8 = 3a_7 + 2 = 3(-2188) + 2 = -6562$.
First four terms $-4, -10, -28, -82$ and 8th term is -6562.

51. One finds $a_2 = a_1^2 - 3 = 2^2 - 3 = 1$,
$a_3 = a_2^2 - 3 = 1^2 - 3 = -2$,
$a_4 = a_3^2 - 3 = (-2)^2 - 3 = 1$.
By a repeating pattern, $a_8 = 1$. First four terms are $2, 1, -2, 1$ and 8th term is 1.

53. $a_2 = a_1 + 7 = (-15) + 7 = -8$,
$a_3 = a_2 + 7 = (-8) + 7 = -1$,
$a_4 = a_3 + 7 = (-1) + 7 = 6$.
There is a common difference of 7.
So $a_8 = 34$. First four terms are $-15, -8, -1, 6$ and 8th term is 34.

55. First four terms are $6, 3, 0, -3$ and 10th term is $a_{10} = 6 + 9(-3) = -21$.

57. First four terms are $1, 0.9, 0.8, 0.7$ and 10th term is $c_{10} = 1 + 9(-0.1) = 0.1$.

59. One finds $w_1 = -\frac{1}{3}(1) + 5 = \frac{14}{3}$,

$w_2 = -\frac{1}{3}(2) + 5 = \frac{13}{3}$,

$w_3 = -\frac{1}{3}(3) + 5 = \frac{12}{3}$,

$w_4 = -\frac{1}{3}(4) + 5 = \frac{11}{3}$, and

$w_{10} = -\frac{1}{3}(10) + 5 = \frac{5}{3}$.

First four terms are $\frac{14}{3}, \frac{13}{3}, \frac{12}{3}, \frac{11}{3}$ and 10th term is $\frac{5}{3}$.

61. Yes, $d = 1$ is the common difference.

63. No, there is no common difference.

65. No, there is no common difference.

67. Yes, $d = \frac{\pi}{4}$ is the common difference.

69. Since $d = 5$ and $a_1 = 1$, $a_n = a_1 + (n-1)d$ $= 1 + (n-1)5$. Then $a_n = 5n - 4$.

71. Since $d = 2$ and $a_1 = 0$, $a_n = a_1 + (n-1)d$ $= 0 + (n-1)2$. So $a_n = 2n - 2$.

73. Since $d = -4$ and $a_1 = 5$, $a_n = a_1 + (n-1)d$ $= 5 + (n-1)(-4)$. So $a_n = -4n + 9$.

75. Since $d = 0.1$ and $a_1 = 1$, $a_n = a_1 + (n-1)d$ $= 1 + (n-1)(0.1)$. So $a_n = 0.1n + 0.9$.

77. Since $d = \frac{\pi}{6}$ and $a_1 = \frac{\pi}{6}$, $a_n = a_1 + (n-1)d$ $= \frac{\pi}{6} + (n-1)\frac{\pi}{6}$. So $a_n = \frac{\pi}{6}n$.

79. Since $d = 15$ and $a_1 = 20$, we have

$a_n = a_1 + (n-1)d = 20 + (n-1)15$.

Then $a_n = 15n + 5$.

81. Since $a_n = a_1 + (n-1)d$, we get $a_8 = (-3) + (7)5 = 32$.

83. Since $a_n = a_1 + (n-1)d$, we find

$$\begin{aligned} a_1 + 2d &= 6 \\ a_1 + 6d &= 18 \end{aligned}$$

Subtracting the first equation from the second, we get $4d = 12$ and $d = 3$. From the first equation, $a_1 + 6 = 6$ and $a_1 = 0$. Then $a_{10} = a_1 + 9d = 0 + 9 \cdot 3 = 27$. So $a_{10} = 27$.

85. Since $a_n = a_1 + (n-1)d$, we get $a_{21} = 12 + 20d = 96$. So $20d = 84$ and the common difference is $d = 4.2$.

87. Since $a_n = a_1 + (n-1)d$, we find

$$\begin{aligned} a_1 + 2d &= 10 \\ a_1 + 6d &= 20. \end{aligned}$$

Subtracting the first equation from the second, we obtain $4d = 10$ and $d = 2.5$. From the first equation, we get $a_1 + 5 = 10$ and $a_1 = 5$. Then $a_n = 5 + (n-1)(2.5) =$ $2.5n + 2.5$. So $a_n = 2.5n + 2.5$.

89. $a_n = a_{n-1} + 9$, $a_1 = 3$

91. $a_n = 3a_{n-1}$, $a_1 = \frac{1}{3}$

93. $a_n = \sqrt{a_{n-1}}$, $a_1 = 16$

95. A formula for the MSRP is

$$a_n = 43,440(1.06)^n$$

where n is the number of years since 2008. For the years 2009-2013, the MSRP are $a_1 = \$46,046$, $a_2 = \$48,809$, $a_3 = \$51,738$, $a_4 = \$54,842$, and $a_5 = \$58,133$.

97. The number of pages read on the nth day of November is $a_n = 5 + 3(n-1)$ or $a_n = 3n + 2$. On November 30, they will read $a_{30} = 92$ pages.

99. In year n, the annual cost is

$$a_n = 70,810 + 4000(n - 2008).$$

In 2017, the cost is projected to be $a_{2017} =$ $70,810 + 4000(2017 - 2008) = \$106,810$.

101. Since there are four corners, $C_n = 4$.
If the corner tiles are in place, the number of edge tiles needed for each side is $2(n-1)$. Since there are four sides, $E_n = 8(n-1)$.

If the corner tiles and edge tiles are in place, the interior tiles will occupy an $(n-1)$-by-$(n-1)$ square.

Note, the area of a tile is $\frac{1}{4}$ ft^2. Then the number of interior tiles is $I_n = \dfrac{(n-1)^2}{\frac{1}{4}}$ or equivalently $I_n = 4n^2 - 8n + 4$ for $n = 1, 2, 3, ...$

105. Since $a = 5$ and $b = 6$, we find $c^2 = a^2 + b^2 =$ $25 + 36 = 61$. Then the foci are $(\pm\sqrt{61}, 0)$, asymptotes are $y = \pm\frac{b}{a}x$ or $y = \pm\frac{6}{5}x$, and the intercepts are $(\pm a, 0) = (\pm 5, 0)$.

107. Use the method of completing the square:

$$\begin{aligned} y &= -2(x^2 - 2x) \\ &= -2(x^2 - 2x + 1) + 2 \\ &= -2(x-1)^2 + 2 \end{aligned}$$

Since $\frac{1}{4p} = -2$, we solve $p = -\frac{1}{8}$. Since the vertex is $(h,k) = (1,2)$, the focus is $(h, k+p) =$ $\left(1, \frac{15}{8}\right)$, the directrix is $y = k - p$ or $y = \frac{17}{8}$, and the axis of symmetry is $x = h$ or $x = 1$.

If $y = 0$, then

$$\begin{aligned} 0 &= -2(x-1)^2 + 2 \\ (x-1)^2 &= 1 \\ x &= 0, 2. \end{aligned}$$

Thus, the x-intercepts are $(0,0)$ and $(2,0)$.

109.

In general, the inverse of a 2-by-2 matrix is

$$\begin{bmatrix} a & b \\ c & d \end{bmatrix}^{-1} = \frac{1}{ad-bc}\begin{bmatrix} d & -b \\ -c & a \end{bmatrix}$$

if $ad - bc \neq 0$. In particular,

$$\begin{bmatrix} 6 & 7 \\ 5 & 6 \end{bmatrix}^{-1} = \frac{1}{36-35}\begin{bmatrix} 6 & -7 \\ -5 & 6 \end{bmatrix}$$

$$= \begin{bmatrix} 6 & -7 \\ -5 & 6 \end{bmatrix}$$

111. Notice $n!$ is a multiple of 100 for $n \geq 10$. Then the units and tens digit in $0! + 1! + ... + 10000!$ are the same as that for

$$0! + 1! + ... + 9! = 409,114.$$

Thus, the units digit is 4, and the tens digit 1.

For Thought

1. True, $\sum_{i=1}^{3}(-2)^i = (-2)^1 + (-2)^2 + (-2)^3 = -6$

2. True, $\sum_{i=1}^{6}(0 \cdot i + 5) = (0 \cdot 1 + 5) + (0 \cdot 2 + 5) +$ $(0 \cdot 3+5)+(0 \cdot 4+5)+(0 \cdot 5+5)+(0 \cdot 6+5) = 30.$

3. True, since $\sum_{i=1}^{k}(5i) = (5 \cdot 1) + ... + (5 \cdot k) =$ $5(1 + 2 + ... + k) = 5\sum_{i=1}^{k}(i).$

4. True, $\sum_{i=1}^{k}(i^2 + 1) = (1^2 + 1) + ... + (k^2 + 1) =$ $(1^2 + 2^2 + ... + k^2) + \underbrace{(1 + ... + 1)}_{k \text{ times}} = \sum_{i=1}^{k} i^2 + k.$

5. False, there are ten terms.

6. True, if $j = i - 1$ and $i = 2$ then $j = 1$.

7. True

8. True, $1 + 2 + ... + n = \frac{n}{2}(1 + n)$ represents an arithmetic series.

9. False, for if $a_1 = 8$, $a_n = 68$, $d = 2$ then $68 = 8 + (n - 1)2$ and $n = 31$. The sum of the arithmetic series is $S_{31} = \frac{31}{2}(8 + 68)$.

10. False, $\sum_{i=1}^{10} i^2$ is not an arithmetic series.

11.2 Exercises

1. series, sequence

3. mean

5. $1^2 + 2^2 + 3^2 + 4^2 + 5^2 = 55$

7. $2^0 + 2^1 + 2^2 + 2^3 = 15$

9. $1 + 1 + 2 + 6 + 24 = 34$

11. $\underbrace{3 + ... + 3}_{10 \text{ times}} = 30$

13. $(2 + 5) + (4 + 5) + (6 + 5) + (8 + 5) + (10 + 5)$ $+(12 + 5) + (14 + 5) = 91$

15. $[(-1)^7 + (-1)^8] + ... + [(-1)^{43} + (-1)^{44}] =$ $0 + ... + 0 = 0$

17. $0 + 0 + 0 + 3(2)(1) + 4(3)(2) + 5(4)(3) = 90$

19. $\sum_{i=1}^{6} i$

21. $\sum_{i=1}^{5}(-1)^i(2i - 1)$

23. $\sum_{i=1}^{5} i^2$

25. $\sum_{i=1}^{\infty}\left(-\frac{1}{2}\right)^{i-1}$

27. $\sum_{i=1}^{\infty} \ln(x_i)$

29. $\sum_{i=1}^{11} ar^{i-1}$

31. Let $i = j - 1$. Since $1 \leq i = j - 1 \leq 3$, we find $2 \leq j \leq 4$. Then we find

$$\sum_{i=1}^{3} i^2 = \sum_{i=2}^{4}(j - 1).$$

The equation is true.

33. We find $\sum_{x=1}^{5}(2x-1)=25$ and $\sum_{y=3}^{7}(2y-1)=45$. The equation is false.

35. Let $x+5=j+6$. Since $1\le x=j+1\le 5$, we find $0\le j\le 4$. Then we obtain

$$\sum_{i=1}^{5}(x+5)=\sum_{j=0}^{4}(j+6).$$

The equation is true.

37. Let $j=i-1$. New series is $\sum_{j=0}^{31}(-1)^{j+1}$

39. Let $j=i-3$. New series is $\sum_{j=1}^{10}(2j+7)$

41. Let $j=x-2$. New series is $\sum_{j=0}^{8}\dfrac{10!}{(j+2)!(8-j)!}$

43. Let $j=n+3$. New series is $\sum_{j=5}^{\infty}\dfrac{5^{j-3}\cdot e^{-5}}{(j-3)!}$

45. $0.5+0.5r+0.5r^2+0.5r^3+0.5r^4+0.5r^5$

47. $a^4+a^3b+a^2b^2+ab^3+b^4$

49. $a^2+2ab+b^2$

51. $\dfrac{6+23+45}{3}=\dfrac{74}{3}$

53. $\dfrac{-6+0+3+4+3+92}{6}=16$

55. $\dfrac{\sqrt{2}+\pi+33.6-19.4+52}{5}\approx 14.151$

57. Since $a_1=-3$, $a_{12}=6(12)-9=63$, and $n=12$, the sum is $S_{12}=\dfrac{12}{2}(-3+63)=360$.

59. Since $a_3=0.7$, $a_{15}=-0.5$, and there are $n=13$ terms in the series, the sum is $S_{13}=\dfrac{13}{2}(0.7-0.5)=1.3$.

61. Note $a_1=1$, $a_n=47$, $n=47$. Then the sum is $S_{47}=\dfrac{47}{2}(1+47)=1128$.

63. Since $95=5+(n-1)5$, we get $n=19$. Then the sum is $S_{19}=\dfrac{19}{2}(5+95)=950$.

65. Since $-238=10+(n-1)(-1)$, we get $n=249$. The sum is $S_{249}=\dfrac{249}{2}(10-238)=-28,386$.

67. Since $-16=8+(n-1)(-3)$, we get $n=9$. Thus, the sum is $S_9=\dfrac{9}{2}(8+(-16))=-36$.

69. Since $55=3+(n-1)4$, we obtain $n=14$. The sum is $S_{14}=\dfrac{14}{2}(3+55)=406$.

71. Since $5=\dfrac{1}{2}+(n-1)\dfrac{1}{4}$, we get $n=19$. Thus, the sum is $S_{19}=\dfrac{19}{2}(1/2+5)=52.25$.

73. $1+2+3+\cdots+n=\dfrac{n(n+1)}{2}$

75. Note, the mean of an arithmetic sequence with n terms is $\bar{a}=\dfrac{a_1+a_n}{2}$. Thus, $\bar{a}=\dfrac{2+16}{2}=9$.

77. Note, the mean of an arithmetic sequence with n terms is $\bar{a}=\dfrac{a_1+a_n}{2}$. Then $\bar{a}=\dfrac{2+142}{2}=72$.

79. Since $a_1=30,000$, $d=1,000$, and $a_{30}=30,000+29(1,000)=59,000$, we find his total salary for 30 years of work is

$$S_{30}=\frac{30}{2}(30,000+59,000)=\$1,335,000.$$

The mean annual salary for 30 years is $\dfrac{S_{30}}{30}=\$44,500$.

81. In the nth level, the number of cans is $a_n=12(10-n)$. In the mountain of cans there are $\sum_{i=1}^{9}(120-12i)=\sum_{j=1}^{9}(120-12(10-j))=\sum_{j=1}^{9}12j=\dfrac{9}{2}(12+108)=540$ cans.

83. At 5% compounded annually, the future value of \$1000 after i years is $\$1000(1.05)^i$. Then the amount in the account on January 1, 2000 is $\sum_{i=1}^{10}1000(1.05)^i$.

85. With a calculator, one finds $200+200(.63)+200(.63)^2+200(.63)^3\approx 455$ mg.

87. Since $a_9 = 101$, $a_{60} = 356$, and there are $n = 52$ terms from a_9 to a_{60}, we get that the mean of the 9th through the 60th terms is $\frac{1}{52}\sum_{i=9}^{60} a_i = \frac{1}{52}\frac{52}{2}(101 + 356) = 228.5$.

89. One finds $a_2 = a_1^2 - 3 = (-2)^2 - 3 = 1$ and $a_3 = a_2^2 - 3 = 1^2 - 3 = -2$. By a repeating pattern, one derives $a_7 = a_9 = -2$ and $a_8 = a_{10} = 1$. The mean from the 7th term through the 10th terms is

$$\frac{-2+1-2+1}{4} = -\frac{1}{2}.$$

93. $a_n = (-2)^{n-1}$

95. **a)** $f(0.01) = \log(0.01) = -2$

b) The exponential form of $\log a = 3$ is $a = 10^3$. Then $a = 1000$.

97. Take the ln of both sides.

$$\begin{aligned} \ln\left(8^{x^2-1}\right) &= \ln 30 \\ x^2 - 1 &= \frac{\ln 30}{\ln 8} \\ x &= \pm\sqrt{1 + \frac{\ln 30}{\ln 8}} \end{aligned}$$

The solution set is approximately $\{\pm 1.623\}$.

99. $\sum_{i=10^7}^{10^8-1} i + \sum_{i=10^9}^{10^{10}-1} i =$

$$\left(\sum_{i=1}^{10^8-1} i - \sum_{i=1}^{10^7-1} i\right) + \left(\sum_{i=1}^{10^{10}-1} i - \sum_{i=1}^{10^9-1} i\right) =$$

$$\left(\frac{(10^8-1)10^8}{2} - \frac{(10^7-1)10^7}{2}\right) +$$

$$\left(\frac{(10^{10}-1)10^{10}}{2} - \frac{(10^9-1)10^9}{2}\right) =$$

49,504,949,995,455,000,000

For Thought

1. False, the ratios $\frac{6}{2}$ and $\frac{24}{6}$ are not equal

2. True, $a_n = 3 \cdot 2^3 \cdot 2^{-n} = 24\left(\frac{1}{2}\right)^n$.

3. True, $a_1 = 5(0.3)^1 = 1.5$.

4. False, since $a_n = \left(\frac{1}{5}\right)^n$, the common ratio is $\frac{1}{5}$.

5. True

6. False, note that the ratio 2 does not satisfy $|r| < 1$ and we cannot use the formula $S = \frac{a}{1-r}$.

7. False, $\sum_{i=1}^{9} 3(0.6)^i = \frac{1.8(1-0.6^9)}{1-0.6}$.

8. True, $\sum_{i=0}^{4} 2(10)^i = \frac{2(1-10^5)}{1-10} = 22,222$.

9. True, $\sum_{i=1}^{\infty} 3(0.1)^i = \frac{0.3}{1-0.1} = \frac{1}{3}$.

10. True, $\sum_{i=1}^{\infty}\left(\frac{1}{2}\right)^i = \frac{1/2}{1-1/2} = 1$.

11.3 Exercises

1. geometric

3. $a_1 = 3 \cdot 2^0 = 3$, $a_2 = 3 \cdot 2^1 = 6$, $a_3 = 3 \cdot 2^2 = 12$, $a_4 = 3 \cdot 2^3 = 24$

First four terms are $3, 6, 12, 24$, and common ratio is 2.

5. $a_1 = 800 \cdot \left(\frac{1}{2}\right)^1 = 400$, $a_2 = 800 \cdot \left(\frac{1}{2}\right)^2 = 200$,

$a_3 = 800 \cdot \left(\frac{1}{2}\right)^3 = 100$, $a_4 = 800 \cdot \left(\frac{1}{2}\right)^4 = 50$

First four terms are $400, 200, 100, 50$, and common ratio is $\frac{1}{2}$.

7. $a_1 = \left(-\frac{2}{3}\right)^0 = 1$, $a_2 = \left(-\frac{2}{3}\right)^1 = -\frac{2}{3}$,

$a_3 = \left(-\frac{2}{3}\right)^2 = \frac{4}{9}$, $a_4 = \left(-\frac{2}{3}\right)^3 = -\frac{8}{27}$

First four terms are $1, -\frac{2}{3}, \frac{4}{9}, -\frac{8}{27}$, and common ratio is $-\frac{2}{3}$.

9. $\frac{1}{2}$ **11.** 10

13. -2 **15.** -1

17. Since $r = \frac{1/3}{1/6} = 2$, we get $a_n = \frac{1}{6}2^{n-1}$.

19. Since $r = \frac{0.09}{0.9} = 0.1$, we find $a_n = (0.9)(0.1)^{n-1}$.

21. Since $r = \frac{-12}{4} = -3$, we have $a_n = 4{\cdot}(-3)^{n-1}$.

23. Arithmetic, since $4 - 2 = 6 - 4 = ... = 2$.

25. Neither, since there is no common difference or ratio.

27. Geometric, since $\frac{-4}{2} = \frac{8}{-4} = ... = -2$.

29. Arithmetic, since $\frac{1}{3} - \frac{1}{6} = \frac{1}{2} - \frac{1}{3} = ... = \frac{1}{6}$.

31. Geometric, since $\frac{1/3}{1/6} = \frac{2/3}{1/3} = ... = 2$.

33. Neither, since there is no common difference or ratio.

35. $a_1 = 2 \cdot 1 = 2$, $a_2 = 2 \cdot 2 = 4$, $a_3 = 2 \cdot 3 = 6$, $a_4 = 2 \cdot 4 = 8$. First four terms are $2, 4, 6, 8$. Arithmetic sequence.

37. One finds $a_1 = 1^2 = 1$, $a_2 = 2^2 = 4$, $a_3 = 3^2 = 9$, $a_4 = 4^2 = 16$. First four terms are $1, 4, 9, 16$. Neither geometric nor arithmetic.

39. $a_1 = \left(\frac{1}{2}\right)^1 = \frac{1}{2}$, $a_2 = \left(\frac{1}{2}\right)^2 = \frac{1}{4}$, $a_3 = \left(\frac{1}{2}\right)^3 = \frac{1}{8}$, $a_4 = \left(\frac{1}{2}\right)^4 = \frac{1}{16}$. First four terms are $\frac{1}{2}, \frac{1}{4}, \frac{1}{8}, \frac{1}{16}$. Geometric sequence.

41. $b_1 = 2^3 = 8$, $b_2 = 2^5 = 32$, $b_3 = 2^7 = 128$, $b_4 = 2^9 = 512$. First four terms are $8, 32, 128, 512$. Geometric sequence.

43. $c_2 = -3c_1 = -3 \cdot 3 = -9$, $c_3 = -3c_2 = -3 \cdot (-9) = 27$, $c_4 = -3c_3 = -3 \cdot 27 = -81$. First four terms are $3, -9, 27, -81$. Geometric sequence.

45. Since $a_n = a_1 r^{n-1}$, we obtain

$$\begin{aligned} \frac{3}{1024} &= 3\left(\frac{1}{2}\right)^{n-1} \\ \frac{1}{1024} &= \left(\frac{1}{2}\right)^{n-1} \\ \left(\frac{1}{2}\right)^{10} &= \left(\frac{1}{2}\right)^{n-1} \end{aligned}$$

So $n - 1 = 10$ and the number of terms is $n = 11$.

47. Since $a_n = a_1 r^{n-1}$, we have $\frac{1}{81} = a_1\left(\frac{1}{3}\right)^5$. Solving, one finds the first term is $a_1 = 3$.

49. Since $a_n = a_1 r^{n-1}$, we obtain $6 = \frac{2}{3}r^2$. So $9 = r^2$ and the common ratio is $r = \pm 3$.

51. Since $a_6 = a_3 r^3$, we obtain $96 = -12r^3$ and $r = -2$. Since $a_3 = a_1 r^2$, we get $-12 = a_1(-2)^2$. Then $a_1 = -3$ and $a_n = (-3)(-2)^{n-1}$.

53. Since $a = 1$, $r = 2$, and $n = 5$, we find that the sum is $S = \frac{1(1 - 2^5)}{1 - 2} = 31$.

55. Since $a = 2$, $r = 2$, and $n = 6$, we obtain that the sum is $S = \frac{2(1 - 2^6)}{1 - 2} = 126$.

57. Since $a = 9$, $r = 1/3$, and $n = 6$, the sum is $S = \frac{9(1 - (1/3)^6)}{1 - 1/3} = 364/27$.

59. Since $a = 6$, $r = \frac{1}{3}$, and $n = 5$, we find that the sum is $S = \frac{6\left(1 - \left(\frac{1}{3}\right)^5\right)}{1 - \frac{1}{3}} = \frac{242}{27}$.

61. $\sum_{i=1}^{8} 1.5(-2)^{i-1} = \frac{1.5(1-(-2)^8)}{1-(-2)} = -127.5$

63. $\frac{2(1-1.05^{12})}{1-1.05} \approx 31.8343$

65. $\frac{200(1-1.01^8)}{1-1.01} \approx 1657.1341$

67. $\sum_{n=1}^{5} 3\left(-\frac{1}{3}\right)^{n-1}$

69. $\sum_{n=1}^{\infty} 0.6(0.1)^{n-1}$

71. $\sum_{n=1}^{\infty} (-4.5)\left(-\frac{1}{3}\right)^{n-1}$

73. $S = \frac{a_1}{1-r} = \frac{3}{1-(-1/3)} = \frac{3}{4/3} = \frac{9}{4}$

75. $S = \frac{a_1}{1-r} = \frac{0.9}{1-0.1} = \frac{0.9}{0.9} = 1$

77. $S = \frac{a_1}{1-r} = \frac{-9.9}{1-(-1/3)} = \frac{-9.9}{4/3} =$
$-\frac{99}{10} \cdot \frac{3}{4} = -\frac{297}{40}$

79. $S = \frac{a_1}{1-r} = \frac{0.34}{1-0.01} = \frac{0.34}{0.99} = \frac{34}{99}$

81. No sum, since $|r| = |-1.06| > 1$.

83. $S = \frac{a_1}{1-r} = \frac{0.6}{1-0.1} = \frac{0.6}{0.9} = \frac{2}{3}$

85. $S = \frac{a_1}{1-r} = \frac{34}{1-(-0.7)} = \frac{34}{1.7} = 20$

87. $\sum_{i=2}^{\infty} 4(0.1)^i = \frac{0.04}{1-0.1} = \frac{0.04}{0.9} = \frac{4}{90} = \frac{2}{45}$

89. $8.2 + 0.05454... = 8.2 + 0.1(0.5454...) =$
$8.2 + 0.1 \sum_{i=0}^{\infty} 0.54(0.01)^i =$
$8.2 + (0.1)\frac{0.54}{1-0.01} = 8.2 + \frac{0.054}{0.99} =$
$8.2 + \frac{54}{990} = \frac{8118+54}{990} = \frac{8172}{990} = \frac{454}{55}$

91. Using a formula for S_n, the sum is

$$\sum_{n=1}^{25} 100(0.69)^{n-1} = \frac{100\,(1-(0.69)^{25})}{1-0.69} \approx 322.55 \text{ mg.}$$

93. The total number of subscriptions in June is
$\sum_{i=1}^{30} 2^{i-1} = \frac{1-2^{30}}{1-2} = 1,073,741,823.$

95. At the end of the nth quarter the amount is $a_n = 4000(1.02)^n$. At the end of the 37th quarter, it is $a_{37} = 4000(1.02)^{37} = \$8,322.74$.

97. At the end of the 12th month, the amount in the account is $\sum_{i=1}^{12} 200(1.01)^i =$
$\frac{200(1.01)(1-1.01^{12})}{1-1.01} = \$2561.87.$

99. The value of the annuity immediately after the last payment is $\sum_{i=0}^{359} 100\left(1+\frac{0.09}{12}\right)^i =$
$\frac{100(1-1.0075^{360})}{1-1.0075} = \$183,074.35.$

101. Assume the ball has a small radius. Approximately, the distance it travels before it comes to a rest is
$9 + \frac{2}{3}\cdot 9 + \frac{2}{3}\cdot 9 + \left(\frac{2}{3}\right)^2 \cdot 9 + \left(\frac{2}{3}\right)^2 \cdot 9 + ... =$
$9 + 18\left[\frac{2}{3} + \left(\frac{2}{3}\right)^2 + \left(\frac{2}{3}\right)^3 + ...\right] =$
$9 + 18\frac{2/3}{1-2/3} = 45$ feet.

103. The first amount spent in Hammond is \$2 million, then 75% of \$2 million is the next amount spent, and so on. So the total economic impact is
$\sum_{i=0}^{\infty} 2(10^6)(0.75)^i = \frac{2,000,000}{1-0.75} =$
$\$8,000,000.$

105. Yes, since if $a = d = 0$ and r is any positive number, then the arithmetic sequence is also a geometric sequence. This sequence is $0, 0, ...$

107. If $0 < r < 1$, then $r^x \to 0$ as $x \to \infty$.
If $r > 1$, then $|r^x| \to \infty$ as $x \to \infty$.

The graphs of $y = 2^x$ and $y = \left(\frac{1}{2}\right)^x$ are shown below.

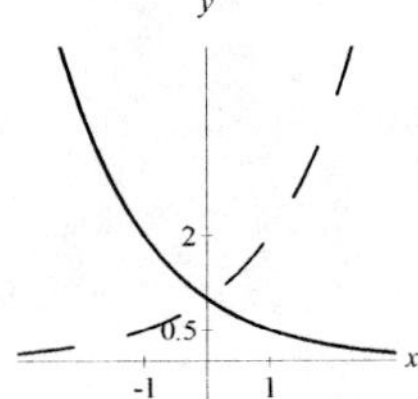

109. A formula is $a_n = 3n$. The sum of the first $n = 100$ terms in an arithmetic series is

$$\frac{n}{2}(a_1 + a_{100}) = \frac{100}{2}(3 + 300) = 15,150.$$

111. Let $a_n = a_1 + d(n-1)$. Then $a_4 = a_1 + 3d = 5$ and $a_8 = a_1 + 7d = 11$. The solutions of the system of equations are $a_1 = \frac{1}{2}$ and $d = \frac{3}{2}$.

Then $a_{20} = \frac{1}{2} + \frac{3}{2}(20 - 1) = 29$.

113. The domain is $(-\infty, \infty)$ since the domain of the absolute value is the set of all real numbers.

Since the range of the negative absolute value function is $(-\infty, 0]$, the range of $f(x) = -30|x - 50| + 200$ is $(-\infty, 200]$.

115. First, the ant crawls a distance of $12r$ ft to the 13-ft hypotenuse where $r = 5/13$.

Then the ant crawls a distance of $12r^2$ ft to the hypotenuse of the smaller right triangle.

Continuing in this manner, the total distance that the ant crawls is

$$12r + 12r^2 + 12r^3 + \ldots = \frac{12r}{1-r} = \frac{12(5/13)}{1 - 5/13} = 7.5\text{ft}.$$

For Thought

1. False, there are $90 \cdot 26$ different codes.

2. False, there are 24^3 different possible fraternity names.

3. False, there are $3 \cdot 5 \cdot 3 \cdot 1 = 45$ different outfits.

4. True, $5! = 120$.

5. True, since

$$P(10, 2) = \frac{10!}{8!} = 10 \cdot 9 = 90.$$

6. False, the number of ways is 4^{20}.

7. True

8. True, since $\dfrac{1000(999)998!}{998!} = 999,000$.

9. False, since

$$P(10, 1) = \frac{10!}{9!} = 10$$

and

$$P(10, 9) = \frac{10!}{1!} = 10!.$$

10. False, $P(29, 1) = \dfrac{29!}{28!} = \dfrac{29 \cdot 28!}{28!} = 29$.

11.4 Exercises

1. fundamental counting principle

3. The number of possible outcomes is $2 \cdot 2$ or 4.

5. The number of possible outcomes is $2 \cdot 2 \cdot 2$ or 8.

7. Number of routes is $2 \cdot 4 = 8$.

9. There are $3! = 6$ different schedules.
These schedules are given by:

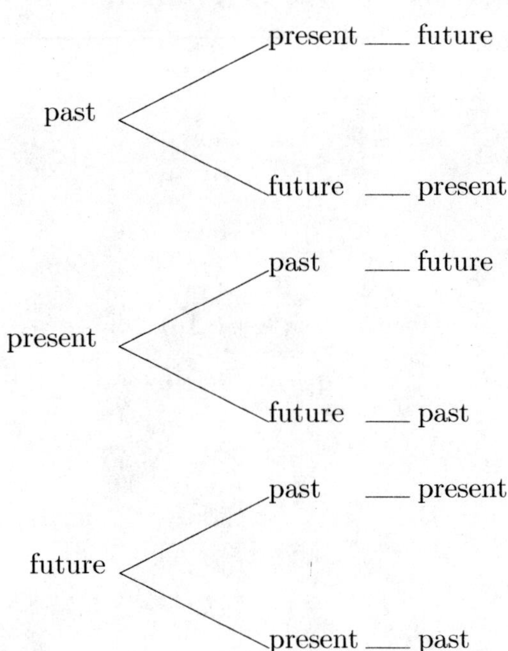

11. There are $4^5 = 1024$ different possible hands.

13. There are 8 optional items since $2^8 = 256$.

15. $\frac{7(6)(5)4!}{4!} = 7(6)(5) = 210$

17. $\frac{5!}{0!} = \frac{120}{1} = 120$

19. $\frac{78(77!)}{77!} = 78$

21. $\frac{4!}{2!} = 12$

23. $\frac{5!}{2!} = 60$

25. $\frac{7!}{4!} = 210$

27. $\frac{9!}{4!} = 15,120$

29. $\frac{5!}{0!} = 120$

31. $\frac{11!}{8!} = 990$

33. $\frac{99!}{99!} = 1$

35. $\frac{105!}{103!} = \frac{105(104)(103!)}{103!} = 105(104) = 10,920$

37. Hercules can perform 12 tasks in $12! = 479,001,600$ ways.

39. The number of ways a first, second, and a third restaurant can be chosen is

$P(15,3) = \frac{15!}{12!} = 2730$ ways.

41. Since there are 8 half-hour shows from 6:00 p.m. to 10:00 p.m., the number of possible different schedules is $P(26,8) \approx 6.3 \times 10^{10}$.

43. Since each question has 4 possible answers, the number of ways to answer 6 questions is 4^6 or 4096.

45. There are $4 \cdot 3 \cdot 6 = 72$ possible prizes.

47. There are $P(26,3) = 15,600$ possible passwords.

49. There are $3 \cdot 10^4 = 30,000$ possible phone numbers.

51. Superman has $4! = 24$ ways to arrange these rescues.

53. Since the true-false questions can be answered in 2^5 ways and the multiple choice questions in 4^6 ways, the test can be answered in $2^5 \cdot 4^6 = 131,072$ ways.

55. On her list will be $P(7,3) = 210$ possible words.

57. The number of subsets is 2^n.

59. A finite geometric series with $n = 10$ terms has the sum

$$\frac{a(1-r^n)}{1-r} = \frac{1(1-2^{11})}{1-2} = 2047.$$

61. If $a_n = ar^{n-1}$, then $ar^2 = \frac{5}{4}$ and $ar^4 = \frac{5}{16}$. Solving for a and r, we find $a = 5$ and $r = \pm\frac{1}{2}$. Thus, $a_{10} = 5\left(\pm\frac{1}{2}\right)^9 = \pm\frac{5}{512}$.

63. The two equations are multiples of each other. The solution set is $\{(x,y) : y = 5x - 12\}$.

65. The number of distinct routes from the corner of First Ave and First Street are

$$\frac{16!}{8!8!} = 12,870.$$

Since Ms. Peabody takes two distinct routes each day, it will take her

$$\frac{12,870}{2(365)} = 17 \text{ years, } 230 \text{ days}$$

to walk all possible routes.

For Thought

1. True, the number of ways is $C(5,3) = C(5,2) = 10$.

2. False, the number of ways is 3^5.

3. True, $P(5,2) = 20$.

4. False, it contains $n+1$ terms. **5.** True

6. True, $C(n, r)$ is the number of subsets of size r, and a set with n elements has 2^n subsets.

7. False, $P(8, 3) = 336$ and $C(8, 3) = 56$.

8. True **9.** False, since $P(8, 3) = 336$ and $P(8, 5) = 6720$. **10.** True

11.5 Exercises

1. combination

3. binomial expansion

5. 10 **7.** 7

9. $\frac{5!}{4!1!} = 5$

11. $\frac{8!}{4!4!} = 70$

13. $\frac{7!}{3!4!} = 35$

15. $\frac{10!}{10!0!} = 1$

17. $\frac{12!}{0!12!} = 1$

19. There are $C(4, 2) = 6$ possible selections. They are Alice and Brenda, Alice and Carol, Alice and Dolores, Brenda and Carol, Brenda and Dolores, Carol and Dolores.

21. There are $C(5, 2) = 10$ possible selections.

23. $C(5, 3) = 10$ ways for an in-depth interview

25. $C(49, 6) = 13,983,816$ ways to choose six numbers

27. $C(52, 5) = 2,598,960$ possible poker hands

29. These assignments can be done in $\frac{10!}{5!3!2!} = 2520$ ways.

31. Since A occurs 4 times, the number of permutations is $\frac{7!}{4!} = 210$.

33. There are 3, 6, 10, and $C(n, 2)$ distinct chords, respectively.

35. There are $\frac{10!}{3!2!5!} = 2520$ possible assignments.

37. There are $3^3 = 27$ ways to watch.

39. $C(8, 3) = 56$ ways to make the selection

41. Since 3 men can be selected in $C(9, 3)$ ways and 2 women in $C(6, 2)$ ways, the team can be selected in $C(9, 3) \cdot C(6, 2) = 1260$ ways.

43. $12! = 479,001,600$ ways to return the papers

45. $6 \cdot 6 = 36$ possible outcomes

47. Since there are 4! ways to arrange the bands in a line and 3! ways to line up the floats, the parade can be lead in $4!3! = 144$ ways.

49. Use the numbers 1,2,1 from Pascal's triangle. Then $(x+y)^2 = 1x^2+2xy+1y^2 = x^2+2xy+y^2$.

51. Use the numbers 1,2,1 from Pascal's triangle. Then $(2a + (-3))^2 =$

$1(2a)^2 + 2(2a)(-3) + 1(-3)^2 = 4a^2 - 12a + 9$.

53. Use the numbers 1,3,3,1 from Pascal's triangle. Then $(a - 2)^3 =$

$1(a^3) + 3(a^2)(-2) + 3(a)(-2)^2 + 1(-2)^3 =$

$a^3 - 6a^2 + 12a - 8$.

55. Use the numbers 1,3,3,1 from Pascal's triangle. Thus, $(2a + b^2)^3 =$

$1(2a)^3 + 3(2a)^2(b^2) + 3(2a)(b^2)^2 + 1(b^2)^3 =$

$8a^3 + 12a^2b^2 + 6ab^4 + b^6$.

57. Use the numbers 1,4,6,4,1 from Pascal's triangle. Then we get $(x - 2y)^4 =$

$1(x)^4+4(x)^3(-2y)+6(x)^2(-2y)^2+4(x)(-2y)^3$
$+1(-2y)^4 =$

$x^4 - 8x^3y + 24x^2y^2 - 32xy^3 + 16y^4$.

59. Use the numbers 1,4,6,4,1 from Pascal's triangle. Then we obtain $(x^2 + 1)^4 =$

$1(x^2)^4 + 4(x^2)^3(1) + 6(x^2)^2(1)^2 + 4(x^2)(1)^3$
$+1(1)^4 = x^8 + 4x^6 + 6x^4 + 4x^2 + 1$.

61. Use the numbers 1,5,10,10,5,1 from Pascal's triangle. Then we obtain $(a - 3)^5 =$

$1(a)^5 - 5a^4(3) + 10a^3(3)^2 - 10a^2(3)^3+$
$5a(3)^4 - 1(3)^4 =$
$a^5 - 15a^4 + 90a^3 - 270a^2 + 405a - 243$.

63. Use the numbers 1,6,15,20,15,6,1 from Pascal's triangle. Then we obtain $(x+2a)^6 =$

$1(x)^6 + 6x^5(2a) + 15x^4(2a)^2 + 20x^3(2a)^3 +$
$15x^2(2a)^4 + 6x(2a)^5 + 1(2a)^6 =$
$x^6 + 12ax^5 + 60a^2x^4 + 160a^3x^3 +$
$240a^4x^2 + 192a^5x + 64a^6$.

65. Use the Binomial Theorem. So $(x+y)^9 =$

$$\binom{9}{0}x^9 + \binom{9}{1}x^8y + \binom{9}{2}x^7y^2 + ... =$$

$x^9 + 9x^8y + 36x^7y^2 + ...$

67. Use the Binomial Theorem. Then $(2x-y)^{12} =$

$$\binom{12}{0}(2x)^{12} + \binom{12}{1}(2x)^{11}(-y) +$$
$$\binom{12}{2}(2x)^{10}(-y)^2 + ... =$$

$4096x^{12} - 24,576x^{11}y + 67,584x^{10}y^2 + ...$

69. Use the Binomial Theorem. So $(2s-0.5t)^8 =$

$$\binom{8}{0}(2s)^8 + \binom{8}{1}(2s)^7(-0.5t) +$$
$$\binom{8}{2}(2s)^6(-0.5t)^2 + ... =$$

$256s^8 - 512s^7t + 448s^6t^2 + ...$

71. Use the Binomial Theorem.

Thus, $(m^2 - 2w^3)^9 =$

$$\binom{9}{0}(m^2)^9 + \binom{9}{1}(m^2)^8(-2w^3) +$$
$$\binom{9}{2}(m^2)^7(-2w^3)^2 + ... =$$

$m^{18} - 18m^{16}w^3 + 144m^{14}w^6 + ...$

73. $\binom{8}{5} = 56$

75. $\binom{13}{5}(-2)^5 = -41,184$

77. By looking at a particular term, one finds

$$(a+[b+c])^{12} = ... + \binom{12}{10}a^2[b+c]^{10} + ... =$$
$$... + \binom{12}{10}a^2\left[... + \binom{10}{6}b^4c^6 + ...\right] + ...$$

So the coefficient of $a^2b^4c^6$ is

$$\binom{12}{10}\binom{10}{6} = \frac{12!}{2!4!6!} = 13,860.$$

79. The coefficient of a^3b^7 in $(a+b+2c)^{10}$ is the number of rearrangements of $aaabbbbbbb$, which is $\dfrac{10!}{3!7!} = 120$.

85. $4^8 = 65,536$ ways

87. $10(9)(8) = 720$

89. $a_2 = 5 - 10 = -5$, $a_3 = 5 - (-5) = 10$, $a_4 = 5 - 10 = -5$, and the 5th term is $a_5 = 5 - (-5) = 10$.

91. Since $1 + 2 + 3 + ... + 1413 = 998,991$, the 998,991th term is 1413. Thus, after 1009 additional terms, the next term is the one millionth term which is 1414.

For Thought

1. False, $P(E) = \dfrac{n(E)}{n(S)}$.

2. False, there are $2^4 = 16$ outcomes.

3. True, since possible outcomes are (H,H), (H,T), (T,H), and (T,T) then $P(\text{at least one tail}) = \dfrac{3}{4} = 0.75$.

4. True, since $\dfrac{5}{6}$ is the probability of not getting a four when a die is tossed then $P(\text{at least one four}) = 1 - P(\text{no four}) =$ $1 - \dfrac{5}{6}\cdot\dfrac{5}{6} = \dfrac{11}{36}$.

5. False, since $\frac{1}{2}$ is the probability of getting a tail when a coin is tossed,

$P(\text{at least one head}) = 1 - P(\text{four tails})$

$= 1 - \left(\frac{1}{2}\right)^4 = \frac{15}{16}.$

6. False, the complement is getting either no head or exactly one head.

7. True, the complement of getting exactly 3 tails in a toss of three coins is getting at least one head.

8. False, the odds in favor of snow is

$\frac{P(\text{snow})}{P(\text{no snow})} = \frac{7/10}{3/10} = \frac{7}{3}$ i.e. 7 to 3.

9. False, $P(E) = 3/7$.

10. False, it is equivalent to 1 to 4.

11.6 Exercises

1. experiment

3. Equally likely

5. complementary

7. $\{(H, H), (H, T), (T, H), (T, T)\}$

9. $\{(H, 1), (H, 2), (H, 3), (H, 4), (H, 5), (H, 6), (T, 1), (T, 2), (T, 3), (T, 4), (T, 5), (T, 6)\}$

11. Note, (H, T) and (T, H) are the only outcomes with exactly one head. Since the sample space has 4 elements, the probability of getting exactly one head is $\frac{n\{(H, T), (T, H)\}}{4} = \frac{1}{2}$.

13. Note, $(6, 6)$ is the only outcome with exactly two sixes. Since the sample space has 36 elements, the probability of getting exactly two sixes is $\frac{n\{(6, 6)\}}{36} = \frac{1}{36}$.

15. Note, $(H, 5)$ is the only outcome with heads and a five. Since the sample space has 12 elements, the probability of getting heads and a five is $\frac{n\{(H, 5)\}}{12} = \frac{1}{12}$,

17. Note, (T, T, T) is the only outcome with all tails. Since the sample space has 8 elements, the probability of getting all tails is

$\frac{n\{(T, T, T)\}}{8} = \frac{1}{8}.$

19. (a) $\frac{n(\{3, 4, 5, 6\})}{6} = \frac{4}{6} = \frac{2}{3}$,

(b) $\frac{n(\{1, 2, 3, 4, 5, 6\})}{6} = \frac{6}{6} = 1$,

(c) $\frac{n(\{1, 2, 3, 5, 6\})}{6} = \frac{5}{6}$,

(d) 0, (e) $\frac{n(\{1\})}{6} = \frac{1}{6}$

21. (a) $\frac{n\{(T, T)\}}{4} = 1/4$,

(b) $1 - P(2\ tails) = 1 - 1/4 = 3/4$,

(c) $\frac{n\{(H, H)\}}{4} = 1/4$,

(d) $\frac{n\{(T, T), (H, T), (T, H)\}}{4} = 3/4$

23. Note there are 36 possible outcomes.

(a) $\frac{n\{(3, 3)\}}{36} = 1/36$,

(b) Since there are 11 possible outcomes with at least one 3, $P(at\ least\ one\ 3) = 11/36$,

(c) Since there are 5 possible outcomes with a sum of 6, $P(sum\ is\ 6) = 5/36$,

(d) $1 - P(sum\ is\ 2) = 1 - 1/36 = 35/36$,

(e) $P(sum\ is\ 2) = 1/36$

25. (a) 1/3, (b) 2/3, (c) 0

27. (a) 3/13, (b) 9/13, (c) 9/13,

(d) $P(marble\ is\ yellow) = 4/13$

29. There are 72 possible outcomes.

(a) $\frac{n\{(1, 9)\}}{72} = 1/72$,

(b) $\frac{n\{(1, 3), (3, 1)\}}{72} = 2/72 = 1/36$,

(c) $\frac{n\{(1, 4), (4, 1), (2, 3), (3, 2)\}}{72} = \frac{4}{72} = \frac{1}{18}$

31. (a) $\dfrac{1}{C(52,5)} = \dfrac{1}{2,598,960}$

(b) By using the counting principle, the probability of one 3, one 4, one 5, one 6, and one 7 is $\dfrac{4^5}{C(52,5)} = \dfrac{1024}{2,598,960}$.

33. $P(A \cup B) = P(A) + P(B) - P(A \cap B) = 0.8 + 0.6 - 0.5 = 0.9$

35. Since $P(E \cup F) = P(E) + P(F) - P(E \cap F)$, we get $0.8 = 0.2 + 0.7 - P(E \cap F)$. Then $P(E \cap F) = 0.2 + 0.7 - 0.8 = 0.1$

37. $P(A \cup B) = P(A) + P(B) - P(A \cap B) = 0.2 + 0.3 - 0 = 0.5$

39. Yes, since an outcome cannot have a sum that is both 4 and 5.

41. No, since the sum in the outcome $(1,1)$ is 2 (which is less than 5) and is even.

43. Yes, since there is no outcome with a two and with a sum greater than nine. Note, in $(2,6)$, the highest possible sum is 8.

45. $P(high{-}risk\ or\ a\ woman) = P(high{-}risk) + P(woman) - P(high - risk\ and\ a\ woman) =$ $38\% + 64\% - 24\% = 78\%$.

47. One finds the probabilities $P(3\ boys) = P(3\ girls) = 1/8$. By the addition rule for mutually exclusive events, we get $P(3\ boys\ or\ 3\ girls) = 1/8 + 1/8 = 1/4$.

49. One obtains the probabilities $P(heart) = 13/52$, $P(king) = 4/52$, $P(heart\ and\ king) = 1/52$. By the addition rule, we get $P(heart\ or\ king) = 13/52 + 4/52 - 1/52 = 4/13$.

51. (a) 34%, (b) $22\% + 18\% = 40\%$, (c) $34\% + 22\% + 18\% = 74\%$

53. $1 - P(surviving) = 1 - 0.001 = 0.999$

55. Note there are 36 possible outcomes.

(a) $\dfrac{n\{(4,4)\}}{36} = \dfrac{1}{36}$,

(b) $1 - P(pair\ of\ 4's) = 1 - 1/36 = \dfrac{35}{36}$,

(c) $\dfrac{35}{36}$, since this is the same event as (b)

57. Since $\dfrac{4/5}{1/5} = 4$, the odds in favor of rain is 4 to 1.

59. Odds in favor of the eye of the hurricane coming ashore is $\dfrac{0.8}{0.2} = 4$, i.e., 4 to 1

61. (a) since $\dfrac{1/4}{3/4} = \dfrac{1}{3}$, the odds in favor of stock market going up is 1 to 3

(b) odds against market going up is 3 to 1

63. If p is the probability of rain today, then

$$\begin{aligned} \frac{p}{1-p} &= \frac{4}{1} \\ p &= 4 - 4p \\ p &= \frac{4}{5}. \end{aligned}$$

The probability of rain today is $\frac{4}{5}$ or 80%.

65. (a) 1 to 9 (b) 9/10

67. Since $P(2\ heads) = \dfrac{C(4,2)}{2^4} = 3/8$ and $\dfrac{3/8}{5/8} = 3/5$, the odds in favor of getting 2 heads in four tosses is 3 to 5

69. Since $\dfrac{P(sum\ of\ 7)}{P(sum\ that\ is\ not\ 7)} = \dfrac{1/6}{5/6} = 1/5$, odds in favor of getting a sum of 7 is 1 to 5.

71. 1 to 1,999,999

73. $\dfrac{1}{1+31} = \dfrac{1}{32}$

75. A and B are mutually exclusive if $P(A \cap B) = 0$. They are complementary events if $P(A) + P(B) = 1$

81. $C(5,3) = 10$

83. $2^{12} = 4096$

85. Geometric with common ratio $\dfrac{1}{2}$

87. We consider a 60-by-60 rectangular region in the xy-plane given by

$$R = \{(x, y) : 0 \le x, y \le 60\}.$$

Let $S = \{(x, y) \in R : |x - y| \le 10\}$. Then the probability that neither friend must wait more than 10 minutes for the other is

$$\frac{\text{Area of } S}{\text{Area of } R} = \frac{1100}{3600} = \frac{11}{36}.$$

For Thought

1. True, for when $n = 1$ we get $4 \cdot 1 - 2 = 2 \cdot 1^2$ which is a true statement.

2. False, when $n = 3$ one gets $27 < 12 + 15$, which is inconsistent.

3. False, for $\frac{100 - 1}{100 + 1} \approx 0.98$.

4. False, for mathematical induction uses integers.

5. True **6.** False, rather $\sum_{i=1}^{k+1} \frac{1}{i(i+1)} = \frac{k+1}{k+2}$.

7. False **8.** False, since $0 > 0$ is inconsistent.

9. True

To see this, let S_n be the inequality $n^2 - n > 0$. Note S_2 is true since $2^2 - 2 > 0$ holds. Suppose S_k is true. We rewrite S_{k+1} as follows:

$$\begin{aligned}(k+1)^2 - (k+1) &= (k^2 + 2k + 1) - k - 1 \\ &= (k^2 - k) + 2k \\ &> 0 + 2k \quad \text{since } S_k \text{ is true} \\ &= 2k \\ &> 0.\end{aligned}$$

Then S_{k+1} is true if S_k is true. Thus, S_n is true for integers $n > 1$.

10. False, when $n = 1$ one gets $1^2 - 1 > 0$, which is inconsistent.

11.7 Exercises

1. If $n = 1$, then $3 - 1 = 2$ and $\frac{3 \cdot 1^2 + 1}{2} = \frac{4}{2}$.

If $n = 2$, then $(3 - 1) + (6 - 1) = 7$ and $\frac{3 \cdot 2^2 + 2}{2} = \frac{14}{2} = 7$.

If $n = 3$, then $(3 - 1) + (6 - 1) + (9 - 1) = 15$ and $\frac{3 \cdot 3^2 + 3}{2} = \frac{30}{2} = 15$.

It is true for $n = 1, 2$, and 3.

3. If $n = 1$, then $\frac{1}{1(1+1)} = \frac{1}{2}$ and $\frac{1}{1+1} = \frac{1}{2}$.

If $n = 2$, then $\frac{1}{2} + \frac{1}{6} = \frac{4}{6}$ and $\frac{2}{2+1} = \frac{2}{3}$.

If $n = 3$, then $\frac{1}{2} + \frac{1}{6} + \frac{1}{12} = \frac{9}{12}$ and $\frac{3}{3+1} = \frac{3}{4}$.

It is true for $n = 1, 2$, and 3.

5. If $n = 1$, one has $1 = 4 - 3$.

If $n = 2$, then $1 + 4 = 5$ and $4(2) - 3 = 5$.

If $n = 3$, then $1 + 4 + 9 = 14$ and $4(3) - 3 = 9$.

It is true for $n = 1, 2$ and false for $n = 3$.

7. If $n = 1$, one has $1 < 1$.

If $n = 2$, one has $4 < 8$. If $n = 3$, $9 < 27$.

It is true for $n = 2, 3$ and false for $n = 1$.

9. $S_1 : 1 = \frac{1(1+1)}{2}$ or $1 = \frac{2}{2}$ is a true statement.

$S_2 : 1 + 2 = \frac{2(2+1)}{2}$ or $3 = \frac{6}{2}$ is a true statement.

$S_3 : 1 + 2 + 3 = \frac{3(3+1)}{2}$ or $6 = \frac{12}{2}$ is a true statement.

$S_4 : 1 + 2 + 3 + 4 = \frac{4(4+1)}{2}$ or $10 = \frac{20}{2}$ is a true statement.

11. $S_1: 1^3 = \frac{1^2(1+1)^2}{4}$ or $1 = \frac{4}{4}$ is a true statement.

$S_2: 1^3 + 2^3 = \frac{2^2(2+1)^2}{4}$ or $9 = \frac{36}{4}$ is a true statement.

$S_3: 1^3 + 2^3 + 3^3 = \frac{3^2(3+1)^2}{4}$ or $36 = \frac{144}{4}$ is a true statement.

$S_4: 1^3 + 2^3 + 3^3 + 4^3 = \frac{4^2(4+1)^2}{4}$ or $100 = \frac{400}{4}$ is a true statement.

13. $S_1: 7^1 - 1$ divisible by 6 is true since the quotient $\frac{7^1 - 1}{6} = \frac{6}{6} = 1$ is a whole number.

$S_2: 7^2 - 1$ divisible by 6 is true since the quotient $\frac{7^2 - 1}{6} = \frac{48}{6} = 8$ is a whole number.

$S_3: 7^3 - 1$ divisible by 6 is true since the quotient $\frac{7^3 - 1}{6} = \frac{342}{6} = 57$ is a whole number.

$S_4: 7^4 - 1$ divisible by 6 is true since the quotient $\frac{7^4 - 1}{6} = \frac{2400}{6} = 400$ is a whole number.

15. $S_1: 2(1) = 1(1+1)$

$S_k: \sum_{i=1}^{k} 2i = k(k+1)$

$S_{k+1}: \sum_{i=1}^{k+1} 2i = (k+1)(k+2)$

17. $S_1: 2 = 2 \cdot 1^2$
$S_k: 2 + 6 + ... + (4k-2) = 2k^2$

$S_{k+1}:$

$$\begin{aligned} 2 + 6 + ... + (4(k+1) - 2) &= \\ 2 + 6 + ... + (4k+2) &= 2(k+1)^2 \end{aligned}$$

19. $S_1: 2 = 2^2 - 2$

$S_k: \sum_{i=1}^{k} 2^i = 2^{k+1} - 2$

$S_{k+1}: \sum_{i=1}^{k+1} 2^i = 2^{k+2} - 2$

21. $S_1: (ab)^1 = a^1 b^1$
$S_k: (ab)^k = a^k b^k$
$S_{k+1}: (ab)^{k+1} = a^{k+1} b^{k+1}$

23. $S_1:$ If $0 < a < 1$ then $0 < a^1 < 1$
$S_k:$ If $0 < a < 1$ then $0 < a^k < 1$
$S_{k+1}:$ If $0 < a < 1$ then $0 < a^{k+1} < 1$

25. Let $T_n: 1 + 2 + ... + n = \frac{n(n+1)}{2}$.

Step 1: If $n = 1$ then $T_1: 1 = \frac{1(2)}{2}$. So T_1 is true.

Step 2: Assume $T_k: 1 + 2 + ... + k = \frac{k(k+1)}{2}$ is true. Add $(k+1)$ to both sides. Then we obtain

$$\begin{aligned} 1 + 2 + ... + k + (k+1) &= \frac{k(k+1)}{2} + (k+1) \\ &= \frac{k(k+1) + 2(k+1)}{2} \\ &= \frac{(k+1)(k+2)}{2}. \end{aligned}$$

Then the truth of T_k implies the truth of T_{k+1}. T_n is true for every positive integer n.

27. Let $T_n: 3 + 7 + ... + (4n-1) = n(2n+1)$.

Step 1: If $n = 1$ then $T_1: 3 = 1(2+1)$. Thus, T_1 is true.

Step 2: Assume T_k is true i.e. $3 + 7 + ... + (4k-1) = k(2k+1)$. Note $4(k+1) - 1 = 4k + 3$. Adding $4k+3$ to both sides, we get

$3 + ... + (4k-1) + (4k+3) =$

$$\begin{aligned} &= k(2k+1) + (4k+3) \\ &= 2k^2 + k + 4k + 3 \\ &= 2k^2 + 5k + 3 \\ &= (k+1)(2k+3) \\ &= (k+1)(2(k+1)+1). \end{aligned}$$

The truth of T_k implies the truth of T_{k+1}. T_n is true for every positive integer n.

29. Let $T_n : \sum_{i=1}^{n} 2^i = 2^{n+1} - 2$.

Step 1: If $n = 1$ then $T_1 : 2 = 2^2 - 2$.

So T_1 is true.

Step 2: Assume $T_k : \sum_{i=1}^{k} 2^i = 2^{k+1} - 2$ is true.

Then add 2^{k+1} to both sides.

$$\begin{aligned}\sum_{i=1}^{k} 2^i + 2^{k+1} &= 2^{k+1} - 2 + 2^{k+1} \\ &= 2 \cdot 2^{k+1} - 2 \\ &= 2^{k+2} - 2 \\ &= 2^{(k+1)+1} - 2\end{aligned}$$

Thus, the truth of T_k implies the truth of T_{k+1}. T_n is true for every positive integer n.

31. Let $T_n : \sum_{i=1}^{n} (3i - 1) = \frac{3n^2 + n}{2}$.

Step 1: If $n = 1$ then $T_1 : 3 - 1 = \frac{3+1}{2}$.

So T_1 is true.

Step 2: Assume $T_k : \sum_{i=1}^{k} (3i - 1) = \frac{3k^2 + k}{2}$ is true. Note $3(k+1) - 1 = 3k + 2$. Then add $3k + 2$ to both sides.

$$\begin{aligned}\sum_{i=1}^{k}(3i-1) + 3k + 2 &= \frac{3k^2 + k}{2} + (3k+2) \\ &= \frac{3k^2 + k + 2(3k+2)}{2} \\ &= \frac{3k^2 + 7k + 4}{2} \\ &= \frac{3(k^2 + 2k + 1) + (k+1)}{2} \\ &= \frac{3(k+1)^2 + (k+1)}{2}\end{aligned}$$

Thus, the truth of T_k implies the truth of T_{k+1}. T_n is true for every positive integer n.

33. Let $T_n : 1^2 + 2^2 + \ldots + n^2 = \frac{n(n+1)(2n+1)}{6}$.

Step 1: If $n = 1$ then $T_1 : 1 = \frac{1(2)(3)}{6}$.

So T_1 is true.

Step 2: Assume T_k is true i.e. we assume $1^2 + 2^2 + \ldots + k^2 = \frac{k(k+1)(2k+1)}{6}$

Add $(k+1)^2$ to both sides.

$1^2 + 2^2 + \ldots + k^2 + (k+1)^2$

$$\begin{aligned}&= \frac{k(k+1)(2k+1)}{6} + (k+1)^2 \\ &= \frac{k(k+1)(2k+1) + 6(k+1)^2}{6} \\ &= \frac{(k+1)[k(2k+1) + 6(k+1)]}{6} \\ &= \frac{(k+1)[2k^2 + 7k + 6]}{6} \\ &= \frac{(k+1)[k+2][2k+3]}{6} \\ &= \frac{(k+1)[(k+1)+1][2(k+1)+1]}{6}\end{aligned}$$

Thus, the truth of T_k implies the truth of T_{k+1}. T_n is true for every positive integer n.

35. Let $T_n : 1 \cdot 3 + 2 \cdot 4 + \ldots + n \cdot (n+2) = \frac{n}{6}(n+1)(2n+7)$.

Step 1: If $n = 1$ then $T_1 : 1 \cdot 3 = \frac{1}{6}(2)(2+7)$.

So T_1 is true.

Step 2: Assume T_k is true i.e. we assume

$1 \cdot 3 + 2 \cdot 4 + \ldots + k \cdot (k+2) = \frac{k}{6}(k+1)(2k+7)$.

Add $(k+1)(k+3)$ to both sides. Then

$1 \cdot 3 + 2 \cdot 4 + \ldots + k \cdot (k+2) + (k+1)(k+3)$

$$\begin{aligned}&= \frac{k}{6}(k+1)(2k+7) + (k+1)(k+3) \\ &= \frac{k(k+1)(2k+7) + 6(k+1)(k+3)}{6} \\ &= \frac{(k+1)(2k^2 + 13k + 18)}{6} \\ &= \frac{(k+1)(2k+9)(k+2)}{6} \\ &= \frac{(k+1)}{6}[k+2][2k+9] \\ &= \frac{(k+1)}{6}[(k+1)+1][2(k+1)+7]\end{aligned}$$

So the truth of T_k implies the truth of T_{k+1}. Then T_n is true for every positive integer n.

37. Let $T_n : \frac{1}{1\cdot 3}+\frac{1}{3\cdot 5}+...+\frac{1}{(2n-1)(2n+1)} = \frac{n}{2n+1}$.

Step 1: If $n=1$ then $T_1 : \frac{1}{1\cdot 3} = \frac{1}{2+1}$.

So T_1 is true.

Step 2: Assume T_k is true i.e. we assume

$\frac{1}{1\cdot 3}+...+\frac{1}{(2k-1)(2k+1)} = \frac{k}{2k+1}$.

Note $\frac{1}{[2(k+1)-1][2(k+1)+1]} = \frac{1}{(2k+1)(2k+3)}$. Then add $\frac{1}{(2k+1)(2k+3)}$ to both sides.

Then $\frac{1}{1\cdot 3}+...$

$$..+\frac{1}{(2k-1)(2k+1)}+\frac{1}{(2k+1)(2k+3)} =$$

$$\begin{aligned}
&= \frac{k}{2k+1}+\frac{1}{(2k+1)(2k+3)}\\
&= \frac{k(2k+3)+1}{(2k+1)(2k+3)}\\
&= \frac{2k^2+3k+1}{(2k+1)(2k+3)}\\
&= \frac{(2k+1)(k+1)}{(2k+1)(2k+3)}\\
&= \frac{k+1}{2(k+1)+1}.
\end{aligned}$$

Thus, the truth of T_k implies the truth of T_{k+1}. T_n is true for every positive integer n.

39. Let T_n : If $0<a<1$ then $0<a^n<1$.

Step 1: Clearly, T_1 is true for $n=1$.

Step 2: Assume $0<a<1$ implies $0<a^k<1$, i.e., assume T_k is true. Multiply $0<a^k<1$ by a to get $0<a^{k+1}<a$. Since $a<1$ and by transitivity, we get $0<a^{k+1}<1$.

Then the truth of T_k implies the truth of T_{k+1}. T_n is true for every positive integer n.

41. Let $T_n : n<2^n$.

Step 1: If $n=1$, then $1<2^1$. So T_1 is true.

Step 2: Assume $k<2^k$, i.e., assume T_k is true. Add 1 to both sides.

$$\begin{aligned}
k+1 &< 2^k+1\\
&< 2^k+2^k\\
&= 2\cdot 2^k\\
&= 2^{k+1}
\end{aligned}$$

Then $k+1<2^{k+1}$.
So, the truth of T_k implies the truth of T_{k+1}.
T_n is true for every positive integer n.

43. Let $T_n : 5^n-1$ is divisible by 4.

Step 1: If $n=1$, then 5^1-1 is divisible by 4. So T_1 is true.

Step 2: Assume 5^k-1 is divisible by 4, i.e., assume T_k is true.
Observe that $5^{k+1}-1=5(5^k-1)+4$. Since sums of multiples of 4 are again multiples of 4, we get that $5^{k+1}-1$ is a multiple of 4.

Then the truth of T_k implies the truth of T_{k+1}. T_n is true for every positive integer n.

45. Let $T_n : (ab)^n = a^nb^n$.

Step 1: If $n=1$ then $(ab)^1=a^1b^1$. So T_1 is true.

Step 2: Assume $(ab)^k=a^kb^k$, i.e., assume T_k is true. Then

$$\begin{aligned}
(ab)^{k+1} &= (ab)^k(ab)\\
&= a^kb^k(ab) \text{ since } T_k \text{ is true}\\
&= a^{k+1}b^{k+1}.
\end{aligned}$$

Thus, the truth of T_k implies the truth of T_{k+1}. T_n is true for every positive integer n.

47. Let $T_n : \sum_{i=0}^{n} x^i = \frac{x^{n+1}-1}{x-1}$, $x\neq 1$.

Step 1: If $n=1$, then $T_1 : 1+x = \frac{x^2-1}{x-1}$.

T_1 is true since $(x^2-1)=(x-1)(x+1)$.

Step 2: Assume $T_k : \sum_{i=0}^{k} x^i = \dfrac{x^{k+1}-1}{x-1}$ is true.
Add x^{k+1} to both sides.

$$\begin{aligned}\sum_{i=0}^{k} x^i + x^{k+1} &= \frac{x^{k+1}-1}{x-1} + x^{k+1} \\ &= \frac{x^{k+1}-1+(x-1)x^{k+1}}{x-1} \\ &= \frac{x^{k+1}-1+x^{k+2}-x^{k+1}}{x-1} \\ &= \frac{x^{k+2}-1}{x-1}\end{aligned}$$

Then the truth of T_k implies the truth of T_{k+1}. T_n is true for every positive integer n.

51. The probability of getting the first answer correct is $\frac{1}{2}$. Since the ten questions are independent, the probability of getting all ten correct answers is $\left(\frac{1}{2}\right)^{10} = \frac{1}{1024}$.

53. The probability is $1 - \dfrac{1}{5,000,000} = \dfrac{4,999,999}{5,000,000}$

55. Neither, since there is no common difference or ratio.

57. Since $(5!)^3 = 1,728,000$, the units digit and tens digits of

$$(1!)^3 + (2!)^3 + ... + (101!)^3$$

are the same as the units digit and tens digits, respectively, of

$$(1!)^3 + (2!)^3 + (3!)^3 + (4!)^3 = 14,049.$$

The units digit is 9, and the tens digit is 4.

Review Exercises

1. One gets $a_1 = 2^0 = 1$, $a_2 = 2^1 = 2$, $a_3 = 2^2 = 4$, $a_4 = 2^3 = 8$, $a_5 = 2^4 = 16$

First five terms are $1, 2, 4, 8, 16$.

3. $a_1 = \dfrac{(-1)^1}{1!} = -1$, $a_2 = \dfrac{(-1)^2}{2!} = \dfrac{1}{2}$, $a_3 = \dfrac{(-1)^3}{3!} = \dfrac{-1}{6}$, $a_4 = \dfrac{(-1)^4}{4!} = \dfrac{1}{24}$.

First four terms are $-1, \dfrac{1}{2}, -\dfrac{1}{6}, \dfrac{1}{24}$.

5. $a_1 = 3(0.5)^0 = 3$, $a_2 = 3(0.5)^1 = 1.5$, and $a_3 = 3(0.5)^2 = 0.75$.

First three terms are $3, 1.5, 0.75$.

7. $c_1 = -3 + 6 = 3$, $c_2 = -6 + 6 = 0$, and $c_3 = -9 + 6 = -3$.

First three terms are $3, 0, -3$.

9. Since $S = \dfrac{a_1(1-r^n)}{1-r}$ is the sum of a geometric series, $S = \dfrac{0.5(1-0.5^4)}{1-0.5} = 0.9375$.

11. Since $S = \dfrac{n}{2}(a_1 + a_n)$ is the sum of an arithmetic series, $S = \dfrac{50}{2}(11 + 207) = 5450$.

13. Since $S = \dfrac{a_1}{1-r}$ is the sum of a geometric series, $S = \dfrac{0.3}{1-0.1} = \dfrac{0.3}{0.9} = \dfrac{1}{3}$.

15. Since $S = \dfrac{a_1(1-r^n)}{1-r}$ is the sum of a geometric series, $S = \dfrac{1000(1-1.05^{20})}{1-1.05} \approx 33,065.9541$

17. $a_n = \dfrac{(-1)^n}{n+2}$

19. $a_n = 6\left(\dfrac{1}{6}\right)^{n-1}$

21. $\sum_{i=1}^{\infty} \dfrac{(-1)^{i+1}}{i+1}$

23. $\sum_{i=1}^{14} 2i$

25. Note, $a_n = a_1 r^{n-1}$. Since $256 = 4r^6$, we get $64 = r^6$ and the common ratio is $r = \pm 2$.

27. $100\left(1 + \dfrac{0.09}{4}\right)^{40} = \$243.52$

29. $\sum_{i=1}^{10} 1000(1.06)^i = \dfrac{1000(1.06)(1-1.06^{10})}{1-1.06}$
$= \$13,971.64$

31. If the pattern continues, the total number of Tummy Masters that could be sold is $\sum_{i=0}^{\infty} 100,000(0.9)^i = \dfrac{100,000}{1-0.9} = 1$ million.

33. $a^4 + \binom{4}{1} a^3(2b) +$

$\binom{4}{2} a^2(2b)^2 + \binom{4}{3} a(2b)^3 + (2b)^4 =$

$a^4 + 8a^3b + 24a^2b^2 + 32ab^3 + 16b^4$

35. $(2a)^5 + \binom{5}{1}(2a)^4(-b) + \binom{5}{2}(2a)^3(-b)^2 +$

$\binom{5}{3}(2a)^2(-b)^3 + \binom{5}{4}(2a)(-b)^4 + (-b)^5 =$

$32a^5 - 80a^4b + 80a^3b^2 - 40a^2b^3 + 10ab^4 - b^5$

37. $a^{10} + \binom{10}{1} a^9b + \binom{10}{2} a^8b^2 + ... =$

$a^{10} + 10a^9b + 45a^8b^2 + ...$

39. $(2x)^8 + \binom{8}{1}(2x)^7\left(\frac{y}{2}\right) + \binom{8}{2}(2x)^6\left(\frac{y}{2}\right)^2 +$ ...

$= 256x^8 + 512x^7y + 448x^6y^2 + ...$

41. $\binom{13}{9} = 715$

43. $\frac{11!}{2!3!6!} 2^3 = 36,960$

45. 24 terms

47. $5^9 = 1,953,125$ ways to mark the answers

49. $P(7,3) = 210$ possible three-letter 'words'

51. $3 \cdot 5 \cdot 4 = 60$ ways to place advertisements

53. $C(8,5) = 56$ possible vacations where order is not taken into account

55. (a) $C(8,4) = 70$ possible councils,
(b) $C(5,4) = 5$ possible councils from Democrats,
(c) $C(5,2) \cdot C(3,2) = 30$ possible councils with two Democrats and two Republicans

57. For KANSAS, $\frac{6!}{2!2!} = 180$ arrangements;

for TEXAS, $5! = 120$ arrangements.

59. $2^7 = 128$ different families

61. Since there are 2^{10} ways to answer the test, the probabilities are

$P(\textit{all 10 correct}) = \frac{1}{1024}$ and

$P(\textit{all 10 wrong}) = \frac{1}{1024}$.

63. (a) 5/13, (b) 8/13, (c) 0, (d) 1

65. Assuming the probabilities of a boy or girl being born are equal, $P(\textit{3 boys}) = \frac{1}{8}$. Then the odds in favor of 3 boys is $\frac{1/8}{7/8}$, i.e., 1 to 7.

67. Odds in favor of catching a perch is $\frac{0.9}{0.1}$, i.e., 9 to 1.

69. By the Addition Rule, $P(\textit{Math or English}) = 0.7 + 0.6 - 0.4 = 0.9$, i.e., 90%.

71. 40, 320

73. 20

75. 1680

77. $\frac{8!}{2!6!} = 28$

79. $\frac{8!}{4!} = 1680$

81. $\frac{8!}{7!1!} = 8$

83. Let $T_n : 3 + 6 + 9 + ... + 3n = \frac{3}{2}(n^2 + n)$.

Step 1: If $n = 1$, then $T_1 : 3 = \frac{3}{2}(1 + 1)$.

So T_1 is true.

Step 2: Assume $T_k : 3 + 6 + ... + 3k = \frac{3}{2}(k^2 + k)$ is true. Add $3(k + 1)$ to both sides.

$3 + 6 + ... + 3k + 3(k + 1) =$

$$\begin{aligned} &= \frac{3}{2}(k^2 + k) + 3(k + 1) \\ &= \frac{3(k^2 + k) + 6(k + 1)}{2} \\ &= \frac{3k^2 + 9k + 6}{2} \end{aligned}$$

$$= \frac{3}{2}(k^2 + 3k + 2)$$
$$= \frac{3}{2}((k+1)^2 + (k+1))$$

Then the truth of T_k implies the truth of T_{k+1}. T_n is true for every positive integer n.

85. The 1st generation ancestor of a male bee is described by the ordered pair (0 male, 1 female).

The 2nd generation ancestors of a male bee is described by the ordered pair (1 male, 1 female).

The 3rd generation ancestors of a male bee is described by the ordered pair (1 male, 2 females).

The 4th generation ancestors of a male bee is described by the ordered pair (2 males, 3 females).

A formula from the nth generation to the $(n+1)$st generation ancestors is given by the rule

(x males, y females) to (y males, $x+y$ females).

Using the above rule, we find that the 10th generation ancestors of a male bee is described by the ordered pair (34 males, 55 females).

Thus, the number of ancestors going back 10 generations is

$$1+2+3+5+8+13+21+34+55+89 = 231.$$

Chapter 11 Test

1. $a_1 = 2.3$, $a_2 = 2.3 + 0.5 = 2.8$, $a_3 = 2.3 + 1 = 3.3$, and $a_4 = 2.3 + 1.5 = 3.8$. First four terms are 2.3, 2.8, 3.3, 3.8.

2. One finds $c_1 = 20$, $c_2 = \frac{1}{2} \cdot 20 = 10$,

$c_3 = \frac{1}{2} \cdot 10 = 5$, and $c_4 = \frac{1}{2} \cdot 5 = 2.5$.

First four terms are 20, 10, 5, 2.5.

3. $a_n = (-1)^{n-1}(n-1)^2$

4. $a_n = 7 + 3(n-1) = 3n + 4$

5. $a_n = \frac{1}{3}\left(-\frac{1}{2}\right)^{n-1}$

6. Since $S = \frac{n}{2}(a_1 + a_n)$, the sum is $\frac{54}{2}(-5 + 154) = 4023$.

7. Since $S = \frac{a_1(1-r^n)}{1-r}$, the sum is

$$\frac{300(1.05)(1 - 1.05^{23})}{1 - 1.05} \approx 13,050.5997.$$

8. Since $S = \frac{a_1}{1-r}$, the sum is $\frac{0.98}{1 - 0.98} = 49$.

9. Since $a_n = a_1 + d(n-1)$, $9 = -3 + 8d$.

Solving for d, one gets $d = 1.5$.

So $a_n = -3 + 1.5(n-1)$ or

$$a_n = 1.5n - 4.5.$$

10. $\frac{9!}{4!2!2!} = 3780$ possible nine-letter 'words'

11. Since $S = \frac{n}{2}(a_1 + a_n)$ is the sum of an arithmetic series, the mean daily sales is

$$\frac{1}{30}\sum_{i=0}^{29} 300 + 10i = \frac{1}{30}\frac{30}{2}(300 + 590) = \$445$$

12. There are $10^4 = 10,000$ possible secret numbers. The probability of guessing the correct number in 3 tries is the probability of getting it in the 1st try plus the probability of getting it in the 2nd try plus the probability of getting it in the 3rd try. Assuming the person remembers what he tries and tries a different number after a failure, this probability is

$$\frac{1}{10,000} + \frac{9,999}{10,000} \cdot \frac{1}{9,999} + \frac{9,999}{10,000} \cdot \frac{9,998}{9,999} \cdot \frac{1}{9,998} = \frac{3}{10,000}.$$

13. At the end of the 300th month the value of the annuity is $\sum_{i=1}^{300} 700\left(1 + \frac{0.06}{12}\right)^i =$

$$\sum_{i=1}^{300} 700(1.005)^i = 700(1.005)\left(\frac{1 - 1.005^{300}}{1 - 1.005}\right)$$

$= \$487,521.25.$

The cost of the house at the end of 25 years is

$$120,000(1.06)^{25} = \$515,024.49.$$

No, they will not have enough money to buy the house.

14. $a^5 + \binom{5}{1}a^4(-2x) + \binom{5}{2}a^3(-2x)^2 +$

$\binom{5}{3}a^2(-2x)^3 + \binom{5}{4}a(-2x)^4 + (-2x)^5 =$

$a^5 - 10a^4x + 40a^3x^2 - 80a^2x^3 + 80ax^4 - 32x^5$

15. $x^{24} + \binom{24}{1}x^{23}(y^2) + \binom{24}{2}x^{22}(y^2)^2 + ... =$

$x^{24} + 24x^{23}y^2 + 276x^{22}y^4 + ...$

16. $\sum_{i=0}^{30} \binom{30}{i} m^{30-i}y^i$

17. Note, $P(sum\ is\ 7) = \frac{6}{36} = \frac{1}{6}$. The odds in favor of 7 is $\frac{6/36}{30/36} = \frac{1}{5}$, i.e., 1 to 5.

18. $C(12,3) = 220$ selections

19. $C(12,2) \cdot C(10,2) = 66 \cdot 45 = 2970$ outcomes

20. $\frac{1}{P(8,3)} = \frac{1}{336}$ is the probability of getting the horses and the order of finish correctly.

21. Let $T_n : \sum_{i=1}^{n} \left(\frac{1}{2}\right)^i = 1 - 2^{-n}$.

Step 1: If $n = 1$, then $T_1 : \frac{1}{2} = 1 - 2^{-1}$.

So T_1 is true.

Step 2: Assume $T_k : \sum_{i=1}^{k} \left(\frac{1}{2}\right)^i = 1 - 2^{-k}$ is true.

Add $\left(\frac{1}{2}\right)^{k+1}$ to both sides.

$$\begin{aligned} \sum_{i=1}^{k} \left(\frac{1}{2}\right)^i + \left(\frac{1}{2}\right)^{k+1} &= 1 - 2^{-k} + \left(\frac{1}{2}\right)^{k+1} \\ &= 1 - 2 \cdot 2^{-(k+1)} + 2^{-(k+1)} \\ &= 1 - 2^{-(k+1)} \end{aligned}$$

Then the truth of T_k implies the truth of T_{k+1}. Thus, T_n is true for every positive integer n.

Since $1 - 2^{-n} < 1$ for all positive integers n, then by transitivity we obtain $\sum_{i=1}^{n} \left(\frac{1}{2}\right)^i < 1$ for all positive integers n.